MW01128991

OPTIMAL CONTROL AND ESTIMATION

Robert F. Stengel

Department of Mechanical and Aerospace Engineering
Princeton University, Princeton, New Jersey

DOVER PUBLICATIONS, INC.
New York

To Pegi, Brooke, and Christopher

Bibliographical Note

This Dover edition, first published in 1994, is an unabridged, corrected republication of the work first published by John Wiley & Sons ("A Wiley–Interscience Publication"), New York, 1986 under the title *Stochastic Optimal Control: Theory and Application*. For this edition the author has corrected a number of errors and added a new Preface and an autobiographical statement.

Library of Congress Cataloging-in-Publication Data

Stengel, Robert F.
[Stochastic optimal control]
Optimal control and estimation / Robert F. Stengel.
p. cm.
Originally published: Stochastic optimal control. New York ; Wiley, c1986. With new pref.
Includes bibliographical references and index.
ISBN 0-486-68200-5
1. Control theory. 2. Mathematical optimatization. 3. Stochastic processes. I. Title.
QA402.3.S76 1994
629.8′312—dc20 94-20406
CIP

Manufactured in the United States by Courier Corporation
68200508
www.doverpublications.com

PREFACE TO THE DOVER EDITION

I am particularly pleased that the Dover edition of my book (formerly entitled *Stochastic Optimal Control*) is being reprinted under the title *Optimal Control and Estimation*. The new title more clearly reflects the structure of the text, which treats control and estimation separately before combining them to form stochastic controllers, and its simplicity reinforces the message that this book is about practical control-system design and state estimation.

Since the original publication of the book in 1986, the field has changed in two important ways. Computer programs that facilitate the design and analysis of multivariable control and estimation logic have become available. The techniques described in the book are easily implemented using computer-aided-design tools.

Additionally, new techniques for designing *robust* linear control systems (i.e., control systems that provide satisfactory stability and performance in the presence of plant uncertainty) have emerged. The book contains considerable material on frequency-domain modeling of linear multivariable systems, modal properties and stability margins of optimally regulated systems, tracking a non-zero set point, choice of cost-function weights (**Q**, **M**, **R**) and control structures (including model following, frequency-shaping, and integral compensation), digital (sampled-data) equivalents of continuous-time controllers, loop transfer recovery, and singular-value analysis. It also introduces probabilistic evaluation of control-system robustness, which provides an attractive approach for relating the likelihood of satisfactory closed-loop performance to uncertainty in plant parameters.

Optimal Control and Estimation bridges the gap between classical methods of control analysis and modern methods of control-system synthesis. The methods it describes are especially valuable in robotics, vehicle control, process control, econometrics, and intelligent-control-system development.

ROBERT F. STENGEL

Princeton, New Jersey
October 1993

PREFACE TO THE FIRST EDITION

Stochastic Optimal Control has an intimidating sound about it, yet the term describes a body of techniques that may well be the most broadly useful in all of mathematics. The notion itself is simple enough: determine the best strategy for controlling a dynamic system in the presence of uncertainty. This objective is common to many disciplines, including engineering, science, economics, sociology, and politics. Where there is such commonality of purpose, it is virtually certain that the most fundamental principles are straightforward. The complexity arises when the principles are reduced to practice, for then the realities of implementing logic, storing information, and transferring data have to be addressed. The symbology must be learned, and care must be exercised in stating each problem and its solution. The payoff for understanding stochastic optimal control can be a deep appreciation for the unifying principles of system dynamics and control.

This book is intended to introduce stochastic optimal control concepts for application to real problems, with sufficient theoretical background to justify their use. As such, the book spans the middle ground between pure mathematics and application. Recognizing the difficulty in serving both the beginner and the advanced reader, I have organized the book so that the flow of information can be maintained for readers in both categories. One objective is to give the student with little background in control theory the tools to design practical control systems and the confidence to tackle the more advanced literature in areas of particular interest. A second is to provide a ready reference for those who have attained working knowledge of the subject.

Courses on optimal control and estimation normally are offered at the college senior or graduate level, although study could begin earlier. It is important to understand something about the types of systems that may be subjected to control. Gathering such information takes time, as well as a degree of experience and maturity. Nevertheless, the mathematics needed to get started is fairly basic, at the level of second-year calculus. This book provides sufficient introductory material to make it possible for college juniors or practicing engineers who may have been away from formal classes recently to study optimal control and estimation without supplemental materials. It also includes bibliographic references to related publications to aid more detailed study.

The first chapter presents a qualitative introduction to optimal control; at its end, the reader has two choices. Chapter 2 provides a review of the mathematics of control and estimation, arranged so that each section corresponds to a similarly numbered chapter. (This results in a minor problem of causality and self-reference, as Section 2.1 provides background for Chapter 1, and Section 2.2 provides background for Chapter 2!) The advanced reader might skip Chapter 2 entirely, proceeding directly to Chapter 3. However, as the need arises, the appropriate section of Chapter 2 is available for reference. Chapter 3 relies only on mathematics appearing through Section 2.3, and so on.

The remaining chapters treat four topics. Chapter 3 addresses optimal control of systems that may be nonlinear and time-varying, but whose inputs and parameters are known without error. It illustrates how open-loop control policies generalize to closed-loop control laws when system dynamics are linear and the cost function is quadratic. Chapter 4 presents methods for estimating the dynamic states of a system that is driven by uncertain forces and is observed with random measurement error. The ultimate objective is to use such estimates to help control a system (via feedback); hence the development stops short of after-the-fact optimal estimation (optimal smoothing), which has little bearing on optimal control. For the most part, it is assumed that random variables and processes have Gaussian probability distributions, which are totally described by their means and covariances (or spectral densities). It also avoids the more complex theoretical issues of optimal estimation for highly nonlinear systems, which only rarely can be put into practice with current computational tools.

Chapter 5 discusses the general problem of stochastic optimal control, where optimal control depends on optimal estimation of feedback information. Conditions under which control and estimation policies can be derived separately are reviewed. In the linear-quadratic-Gaussian (LQG) case, the separation is seen to be certainty-equivalent: the optimal control policy is the same as it would be in the deterministic case, except that the state estimate rather than the state itself is fed back. Chapter 6 focuses on linear, time-invariant systems, for which multivariable controllers can be based on linear-quadratic control laws with linear-Gaussian estimators. The remarkable breadth and power of stochastic optimal control becomes apparent, as interpreted by notions drawn from classical control applied to multi-input/multi-output systems. Although many interesting developments in control system *analysis* have been made recently, optimality remains the most important unifying criterion for control system *synthesis*. The final chapter presents a brief comment on the value of stochastic optimal control.

There is enough material in the book to support a two-term course sequence, and then some. At an advanced level, Chapters 1, 3, 4, 5, and 6 could be presented in order. At an introductory level, it would be reasonable to intersperse sections or parts of sections from Chapter 2; a sample one-term sequence would be Chapter 1, Section 2.1, Sections 2.2 and 2.3, Chapter 3, Section 2.4, and selected parts of Chapter 4. The second term then could be based on the remainder of Chapter 4, Chapter 5, and Chapter 6, with reference to Sections 2.5 and 2.6 as needed.

The book's examples and problems are directed at confidence building. I would prefer that most readers be able to do most problems rather than reserving that accomplishment for only a few. One outcome is that many problems are meant to be solved with the aid of a personal computer. The programming is left to the reader, who should find that even a passing acquaintance with an introductory programming language like BASIC is sufficient for the tasks at hand. The principal challenge is to understand the theory and algorithms of the book well enough to reduce them to practice. Capabilities gained should be readily transferable to more complex problems requiring more sophisticated computation. This is fitting, as implementation of stochastic optimal control in most situations is entirely dependent on the use of computers.

Writing this book has been a profound learning experience for me, and I hope that I have been able to convey a sense of my own excitement in completing this work. I am deeply indebted to more individuals than I can acknowledge here, most notably the many researchers whose work has been my guide. To the extent that this book interprets the ideas of others, I ask their forbearance if I have failed to capture the scope of their intent. Co-workers and students provided an unceasing stream of insights, including all-important "midcourse guidance" on many occasions. I also am grateful to the supervisors, technical monitors, and reviewers who have helped my career evolve; their support and confidence was crucial.

Several individuals have provided invaluable assistance in the preparation of the book. When I was formulating ideas for the book, discussions with Theodore Edelbaum and Jason Speyer, then at the Charles Stark Draper Laboratory; Michael Athans of M.I.T.; and Charles Price, Bard Crawford, John Broussard, Paul Berry, Fred Marcus, and Michael Safonov, then at The Analytic Sciences Corporation were most helpful. Others have provided suggestions and criticisms on the text itself, including Professor Bradley Dickinson and students Mark Psiaki, Prakash Shrivastava, Stephen Lane, David Handelman, and Chien Huang of Princeton University; Muthathamby Sri Jayantha of IBM; and Robert Otto of the Monsanto Corporation. While any faults in the book are wholly my own, I wish to share any compliments that the book may garner with these people.

And finally, warmest thanks go to my chief cheerleaders and cheer providers, Pegi, Brooke, and Christopher, whose continued understanding, encouragement, and sense of humor have been truly optimum.

ROBERT F. STENGEL

Princeton, New Jersey
June 1986

CONTENTS

1 INTRODUCTION

Designing control logic that commands a dynamic system to a desired output or that augments the system's stability is a common objective in many technical fields, ranging from decision making for economic and social systems through trajectory control for robots and vehicles to "knowledge-based engineering" in the application of "artificial intelligence" concepts. If the control objective can be expressed as a quantitative criterion, then optimization of this criterion establishes a feasible design structure for the control logic. This book is about the theory and application of optimal control; particular attention is given to providing a rigorous yet straightforward foundation for mathematical optimization and to illustrating how practical control systems can be designed with this approach.

1.1 FRAMEWORK FOR OPTIMAL CONTROL

Optimal control theory describes the application of forces to a system for the purpose of maximizing some measure of performance or minimizing a cost function. If the information which the control system must use is uncertain or if the dynamic system is forced by random disturbances, it may not be possible to optimize this criterion with certainty; the best one can hope to do is to maximize or minimize the expected value of the criterion, given assumptions about the statistics of the uncertain factors. This leads to the concept of "stochastic" optimal control, or optimal control which recognizes the random behavior of the system and attempts to optimize response or stability on the average rather than with assured precision.

A stochastic control system really performs two functions; it not only *controls* the dynamic system, it *estimates* the system's dynamic state in order to provide the best "feedback" information for closed-loop control. All control systems are intended to optimize some criteria, whether or not the criteria are stated explicitly, and there is some degree of uncertainty in any control system implementation. Consequently, stochastic optimal control theory has broad application to practical control system synthesis, as demonstrated in the remainder of the book.

By way of introduction, consider a few examples of dynamic systems that might be controlled in some optimal fashion (Fig. 1.1-1). The figures are conceptual, and the implied control may not be accomplished easily.

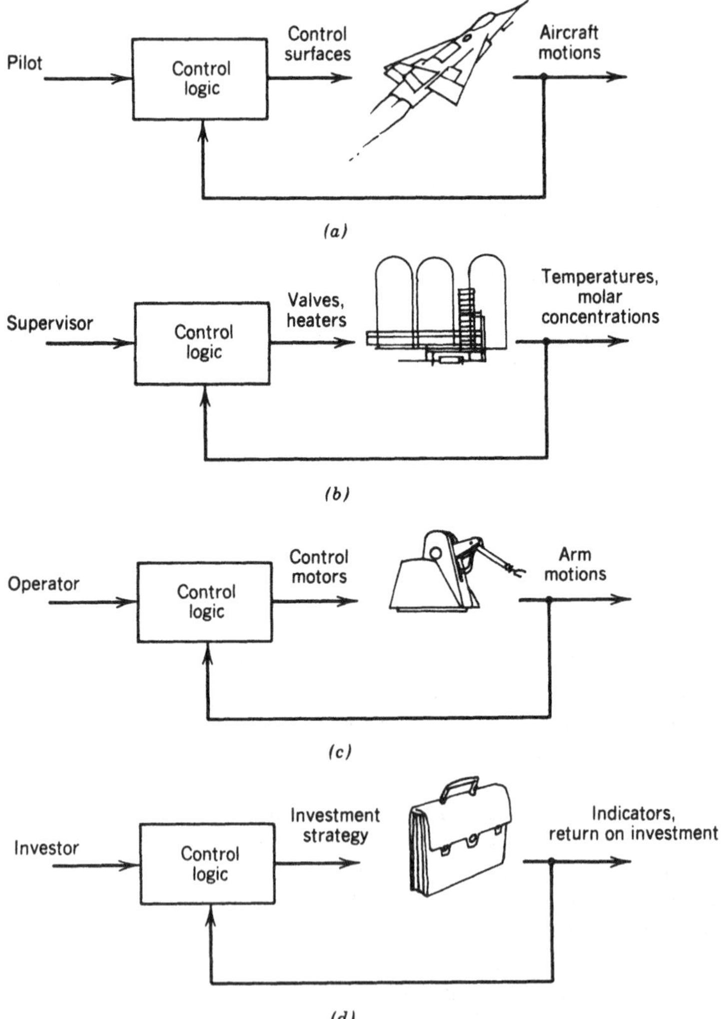

FIGURE 1.1-1 Dynamic processes that could be subject to control. (*a*) Aircraft control; (*b*) chemical process control; (*c*) robotics; (*d*) economic control.

Nevertheless, it should be apparent that seemingly diverse systems are subject to control and that feasible solutions to control problems can be identified using essentially similar methodologies.

The notion of *optimizing the control policy* can be pursued in each case, although the "figures of merit" (or performance criteria, cost functions, penalty functions, etc.) that reflect successful control may be markedly different. (We will generally refer to all of these as "cost functions" whether or not monetary cost is involved. In this broad sense, *cost* implies penalty or value, and it is to be minimized or maximized, as appropriate.) One control objective for aircraft would be to minimize fuel use without any penalties for use of control, and the fuel used defines a *cost function*. Another might be to obtain relative insensitivity to turbulence with limited motion of the control surfaces; hence, the cost function would contain a weighted sum of motion and control magnitudes. The two control objectives lead to different control policies, yet the control structures can be developed within a common framework: *the optimization of a cost function*. Similar reasoning can be applied to the other examples, in which the "cost" of control is weighed against the "cost" of deviations in dynamic variables to formulate an "optimal" control policy.

Let's consider some cost functions that we might wish to maximize or minimize. These fall into two categories: discrete objectives that are optimized "once and for all" and continuing (regulatory) objectives that should be maintained at optimal values during the passage of time (including an infinite time interval). In the first category, we find the following examples:

- Minimize the time required to transfer from one dynamic state to another.
- Minimize the energy or fuel used in transferring from one state to another.
- Minimize the error in arriving at the final state.
- Maximize the return on investment at the end of a fixed period of time.
- Minimize the risk in pursuing an investment policy, as measured at the end of the investment period.

In the second category, we could have the following:

- Regulate a system to remain near a fixed nominal condition in spite of disturbances.
- Follow a nominal path with minimum error, even though system parameters have uncertain values.
- Respond to arbitrary changes in the desired nominal condition with well-behaved transients in state and control variables.

- Minimize inventories required to produce and deliver a product in a changing marketplace.

There is an important third category which uses optimization as an intermediate step, in that the cost function itself is only indirectly related to the design objectives. For example, in the design of a practical servo-mechanism, the numerical tradeoff between state and control perturbations may be less significant than the natural frequency and damping, stability margins, and so on. Weighting factors in the cost function can be used as *design parameters* to achieve the desired result. The rationale for pursuing this approach is that the *feasible designs* produced by optimization generally use available control resources in an efficient manner, requiring little intuition or "prior art" to start the design process. Furthermore, certain guarantees regarding the stability of the controlled system can be made. These characteristics are essential, particularly in the design of controls for systems with many dynamic states and control variables or those which are not inherently stable.

The motions of dynamic systems evolve as time passes; hence the criteria to be optimized are expressed naturally in terms of final values and time-integrals of the motion and control variables (i.e., in the *time domain*). Nevertheless, stability, response to sinusoidal inputs of varying frequencies, and sensitivity of response to system parameter variations are of great practical concern, so equivalent *frequency domain* criteria are useful. The power of time domain techniques for control synthesis is related to their generality; optimal controls for nonlinear, time-varying systems are specified (if not calculated) readily, and the simplifications afforded by linearity and time-invariance are accommodated in the same theoretical framework. When both linearity and time-invariance can be assumed, frequency domain concepts aid interpretation and, in some cases, control synthesis. This book presents stochastic optimal control from a time domain viewpoint, providing frequency domain perspectives where helpful and appropriate.

There is a natural relationship between stochastic optimal control and many mathematical concepts often collected under the rubric of "artificial intelligence" (AI). AI algorithms are applied to interpretation of data, system monitoring and identification, prediction, planning, and design. They may use "production systems" to model dynamic interactions predicated on large data bases, as well as optimization techniques for decision making under uncertainty. Stochastic optimal control theory provides tools for solving many of these AI problems.

What follows cannot include all methodologies for all optimization problems; the book focuses on just those aspects that have demonstrated practical applications or show great promise to do so. Interpretations of the theoretical underpinnings of optimization techniques, as well as detailed examples of their use, are presented.

1.2 MODELING DYNAMIC SYSTEMS

Stochastic optimal control can be applied to a dynamic system that consists of a physical, chemical, economic, or other dynamic process and its observation. In order to design the control logic, a *mathematical model* of the dynamic system must be defined. It should be understood that the model is, at best, a surrogate for the actual system, whose precision is subject to the assumptions and approximations made by the control designer.

Elements of the mathematical model can be arranged in distinct families of variables that identify their respective roles in the system. Each family consists of a set of scalar quantities and is represented by a single symbol. Once the ordering of quantities is established (and fixed), the symbol can be interpreted as a *column vector* (see Section 2.1) that is synonymous with the entire set of quantities.

Symbol (or Vector)	Family of Variables	Dimension
$\mathbf{p}$	Parameters	l
$\mathbf{u}$	Controllable inputs	m
$\mathbf{w}$	Uncontrolled inputs (Disturbances)	s
$\mathbf{x}$	Dynamic states	n
$\mathbf{y}$	Process outputs	r
$\mathbf{n}$	Measurement errors	r
$\mathbf{z}$	Observations	r

The relationships between these variables are illustrated by Fig. 1.2-1. These seven categories of variables serve distinct purposes in the dynamic system. The vector of *parameters* $\mathbf{p}$ scales the process's response to inputs and to its own motions, either in the dynamic process, the observation, or

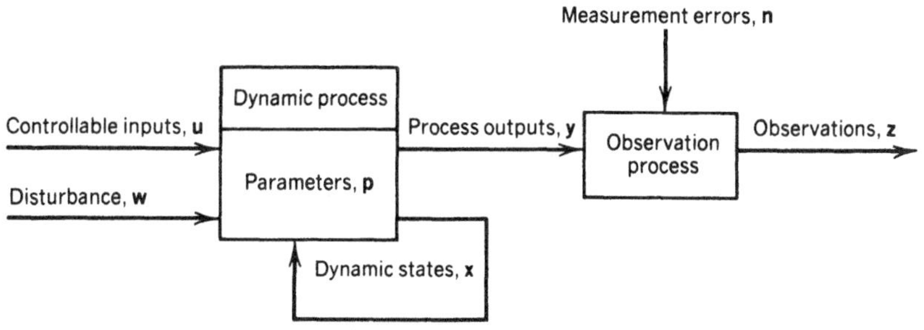

FIGURE 1.2-1 Elements of the dynamic system.

both. The forces on the process (the *inputs*) that can be controlled are contained in **u**, while those that are beyond control are contained in **w**. The *state* vector **x** represents the dynamic condition of the process, including its fundamental response to inputs and initial conditions. The process structure is such that **x** "feeds back" into the system through natural dynamic paths; this feedback can modify process response in several ways (e.g., by shifting the steady-state characteristics, causing oscillations, stabilizing or destabilizing the process).

For examples of specific definitions of parameter, control, disturbance, and state vectors, consider a chemical reactor (Fig. 1.1-1*b*) designed to produce ethylene (W-1). The parameters of the process, **p**, include the physical characteristics of the reactor as well as product market values and separation costs. The controls, **u**, include hydrocarbon feed rate, steam/hydrocarbon ratio in the feed stream, reactor skin temperature, and reactor outlet pressure. Disturbances, **w**, might include variations in the chemical composition of the feedstock and local (uncontrollable) temperature and pressure variations along the walls or within the reactor. The state components, **x**, include the carbon deposition rates at chosen points along the reactor tube and a measure of the total net earnings accumulated from the start of the production interval.

The output vector, **y**, can contain none or all of the preceding families, selected components of each vector, or transformations of these vectors, depending on the process and its instrumentation. In general, the output cannot be measured exactly, so the observation, **z**, is some combination of the output, **y**, and measurement error, **n**. Any dynamic effects associated with control actuation or measurement sensing can be included in the dynamic process model. For the ethylene reactor considered, the output vector, **y**, contains linear and quadratic functions of the controls and the carbon deposition rates, each of which may be measured imprecisely in forming the observation vector, **z**.

If the dynamic process is subjected to control inputs without regard to measured (or estimated) values of the state, the control is called *open-loop control*, as there is no information path (or control loop) from **x** to **u**. With the addition of feedback information, *closed-loop control* results. Open-loop optimal control can be applied only if the dynamic system is characterized by adequate "controllability," and closed-loop control (including stochastic control) requires adequate "observability" as well (Fig. 1.2-2).

A system is *controllable* at time t_0 if there is a *control history*, $\mathbf{u}(t)$, in a finite time interval $[t_0, t_f]$, which transfers each element of an arbitrary initial state, $\mathbf{x}(t_0)$, to zero at t_f. If there is an independent natural path between every component of **x** and at least one element of **u**, then the process is likely to be completely controllable. (There must be adequate control "power" as well.) The path can be either direct or indirect, as long as it can be distinguished from possibly redundant paths in the system. As an example, if the valves and heating controls included in the chemical

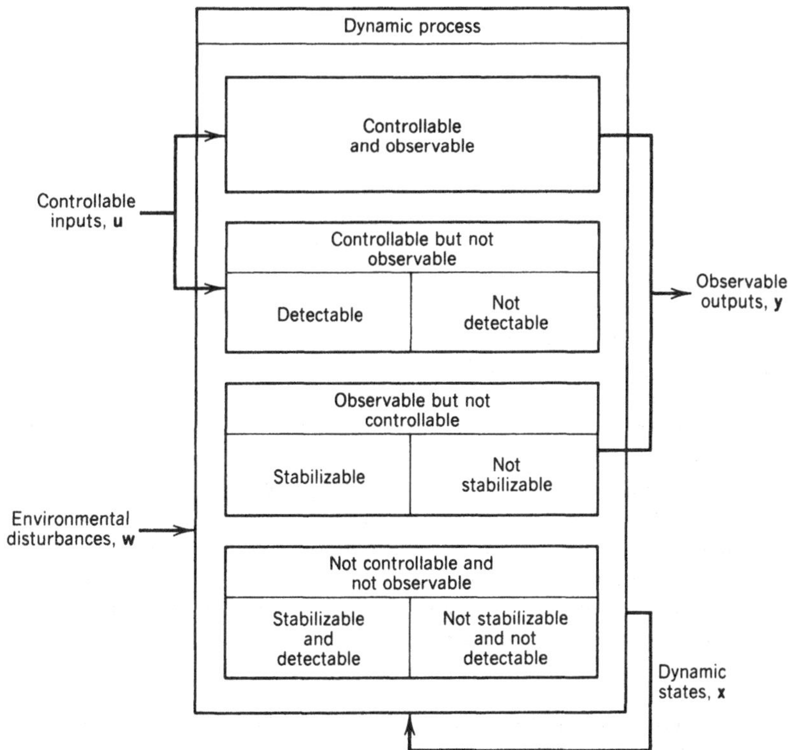

FIGURE 1.2-2 Controllability and observability of the dynamic process.

process of Fig. 1.1-1*b* can affect every dynamic variable in the reactor in a unique fashion, the process is controllable. If a primary valve jams, complete controllability may be lost.

A system is *observable* at t_0 if the output history, $\mathbf{y}(t)$, $t_0 \leq t \leq t_f \leq \infty$ is adequate to reconstruct $\mathbf{x}(t_0)$. If there is an independent path between every component of $\mathbf{x}$ and at least one element of $\mathbf{y}$, then the process is likely to be completely observable. Again, the path may be direct or indirect, but it must be nonidentical to all other natural paths in the dynamic system. If the angle of the elbow of the manipulator arm shown in Fig. 1.1-1*c* is measured, the position of a tool at the end of the arm can be determined; if the angle sensor fails, observability may become incomplete.

These definitions say nothing about the potential quality of control or observation; they merely indicate whether the natural structures for control and observation exist. Mathematical conditions that must be satisfied for complete controllability and observability can be found in Chapter 2. In most cases, the dynamic process is only partially controllable and observable, either as a consequence of the dynamic process itself or because limited resources prevent implementation of complete control and observation. Such a process can be separated into four parts: those subprocesses that are both completely controllable and observable, those that are one but not the other, and those that are neither.

From a practical point of view, complete controllability and observability are not required if the uncontrolled or unobserved states are well-behaved. For example, adequate rigid-body control of a vehicle need not depend on controlling or observing well-damped, high-frequency elastic modes. If an uncontrolled subprocess in an otherwise controllable process is *stable* (i.e., bounded inputs or initial conditions produce bounded response), the process is said to be *stabilizable*, and the subprocess's effects cannot cause the process to diverge. If an unobserved subprocess is stable, the otherwise observable process is *detectable*, as bounds on the observed states can be estimated. Partial controllability and observability may be acceptable, but unstable subprocesses that are neither controllable nor observable must be avoided because closed-loop (presumably stabilizing) control can be applied only to those elements that possess both characteristics. More will be said about the methods for determining stability in Chapter 2.

Dynamic processes normally are continuous functions of time, so it is appropriate to model the evolution of their motions by differential equations. These processes are called *continuous-time systems*. The solution variables for these equations are contained in the state vector, $\mathbf{x}(t)$. The fundamental characteristic of continuous-time dynamic systems is that the rate of change of $\mathbf{x}(t)$ with respect to time, $d\mathbf{x}(t)/dt$, is a function of the system's state, input, and parameter vectors. Although some processes are best described by *partial* differential equations (containing derivatives with respect to more than one independent variable), we will restrict our view to *ordinary* differential equations (containing derivatives with respect to a single independent variable, e.g., time), noting that numerical approximations to the former can be achieved with the latter. Figure 1.1-1*a–c* provides examples of continuous processes.

In an increasing number of applications using digital computers, control settings are calculated and measurements are made at discrete (often periodic) instants of time. Difference equations that are equivalent to the original differential equations can be found, and continuous-time stochastic optimal control solutions are paralleled by discrete-time results. Processes that are rigorously modeled by difference equations are called *discrete-time systems*. Econometric models (e.g., Fig. 1.1-1*d*), sociological models, and biological models that use discrete measurements made at periodic intervals frequently originate as discrete-time models. Consequently, a thorough exposition of discrete-time methods can serve a variety of applications.

For either continuous-time or discrete-time models, the dynamic equations can be classified as in Fig. 1.2-3. If the dynamic coefficients, or parameters, $\mathbf{p}$, are changing rapidly in time, in comparison with the time scale of motions, the dynamic model must be *time-varying*. If the

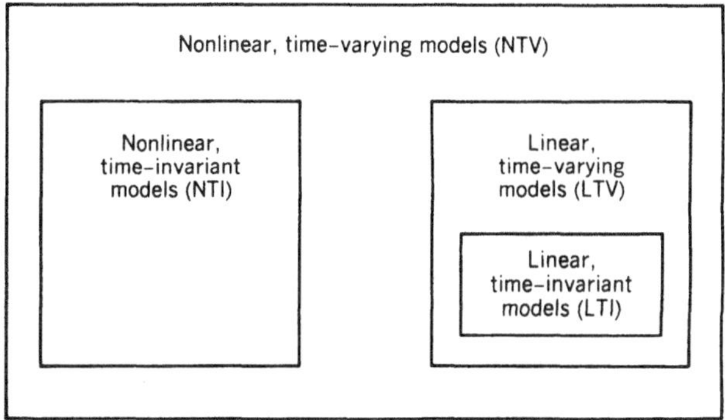

FIGURE 1.2-3 Classification of dynamic equations.

coefficients are relatively constant, a *time-invariant* model will suffice. If motions evidence the superposition characteristic (i.e., doubling the input or initial conditions doubles the corresponding response), then *linear* models can be used; if not, the model must be *nonlinear.*

Consider the dynamic response of a weathervane to changes in wind direction and velocity. The wind-induced torque on the weathervane is proportional to the square of the wind velocity and to the sine of the angle θ between the wind direction and the indicator's angle. For small θ, $\sin(\theta) \simeq \theta$ and $\sin(2\theta) \simeq 2\theta$, so the torque is approximated by a linear function of the angle. If the wind speed is constant but its direction changes slightly, perturbations in the weathervane's orientation can be described by a linear, time-invariant differential equation. If the wind speed is changing as well, the linear differential equation must contain time-varying coefficients. If the wind undergoes a large, rapid change in wind direction, θ becomes large, $\sin(\theta) \neq \theta$, and the weathervane response can be portrayed accurately only by a nonlinear differential equation.

1.3 OPTIMAL CONTROL OBJECTIVES

It will be helpful to preface the mathematical developments of the book with a few pictures that illustrate some ground rules and goals for optimization. While these figures may appear intuitively obvious, they serve as reminders that straightforward objectives are central to optimization. If the notation is unfamiliar, it may be helpful to skim Section 2.1 as you go.

In the simplest case, a scalar cost function J_1 of a scalar variable u is given. We would like to find that value of u which causes J_1 to be a minimum, that is, the point u^* shown in Fig. 1.3-1a. As an alternative, we may wish to find the value u^* which maximizes J_2, as in Fig. 1.3-1b. The

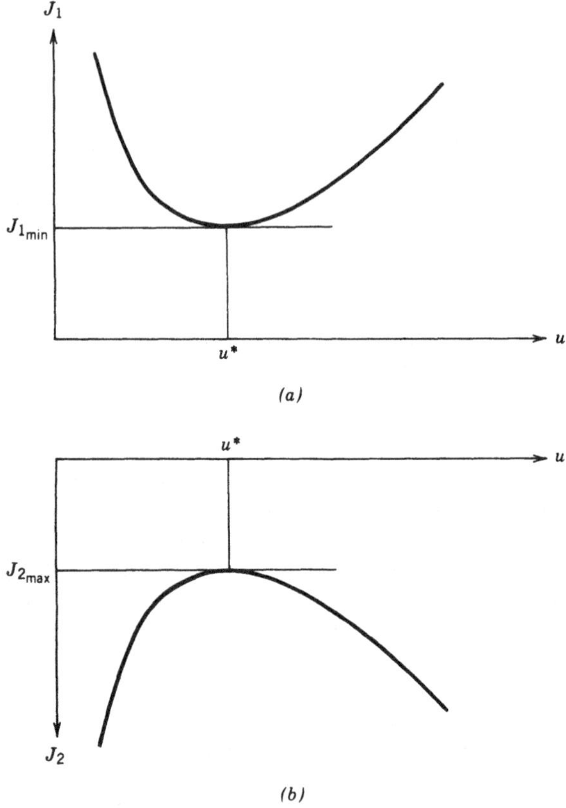

FIGURE 1.3-1 Functions subject to optimization. (*a*) Scalar function with a minimum; (*b*) scalar function with a maximum.

figures have been purposely drawn so $J_2 = -J_1$; hence the u^* that minimizes J_1 also maximizes J_2. It always is possible to change the search for a maximum to the search for a minimum (or vice versa) by changing the sign of the cost function. The value of u^* which minimizes (maximizes) the function J_1 (J_2) is defined by the same characteristic in both cases. What is it? What characteristic of the functions distinguishes the maximum from the minimum?

The cost function can establish a trade-off between two or more variables. Suppose J is the sum of two functions, $J_1(x)$ and $J_2(u)$, as shown in Fig. 1.3-2*a*, where the state, x, is a function of the control u [i.e., $x = x(u)$]. Then u^* establishes some "best" combination of J_1 and J_2 (and, therefore, of x and u). The optimum point depends on the relative importance of J_1 and J_2. If we make the cost of control only half as important as before (Fig. 1.3-2*b*), the optimum point shifts in the direction of more control to achieve

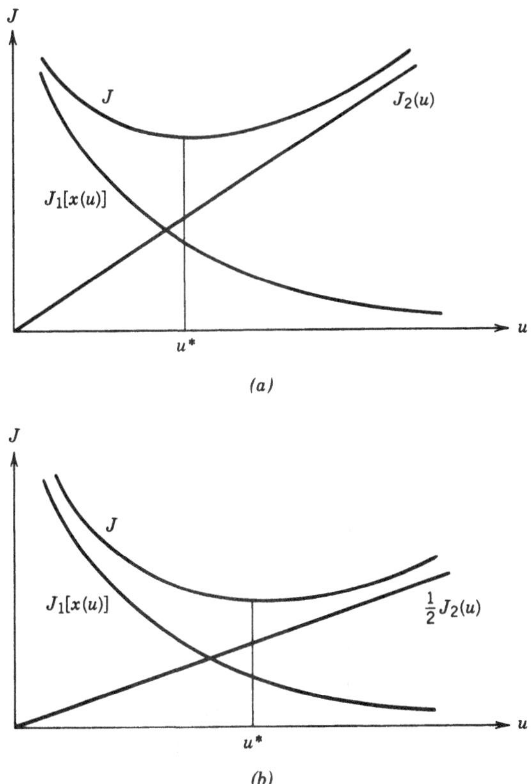

FIGURE 1.3-2 Definition of a cost function that combines the effects of two variables. (*a*) Cost function that trades off effects of one variable against another; (*b*) effect of reduced control cost on the optimum.

lower state-related cost. The effect on u^* would be the same as that of increasing the state-related cost, although the numerical value of J would be different.

If J is a scalar function of not one but two independent controls, u_1 and u_2, its value is described by a surface rather than a curve. A minimum value of J is found at the bottom of a bowl-shaped contour, as in Fig. 1.3-3*a*; the minimum's location is defined by concurrently optimal values of the controls, u_1^* and u_2^*. If one control has the proper value but the other does not, J does not take its minimum value.

Contours of equal cost can be found by making horizontal slices through J's surface. These can be plotted in the control "space" (Fig. 1.3-3*b*), and the optimum control point is defined by the middle of the "thumbprint." Note that if we decided to find the optimum control point by some iterative process, beginning with nonoptimum values of u_1 and u_2, we could reach

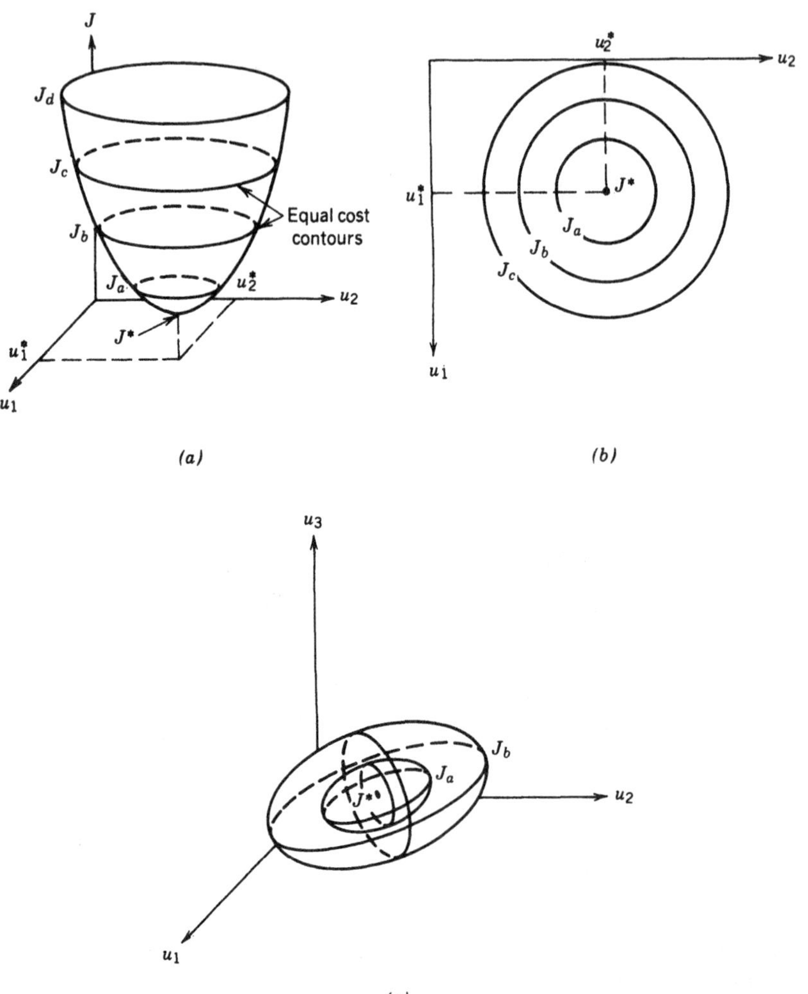

FIGURE 1.3-3 Cost functions with more than one control variable. (*a*) Cost function with two control variables; (*b*) corresponding contours of equal cost with two control variables; (*c*) contours of equal cost with three control variables.

the optimum by always traveling in a direction that is perpendicular to the local constant-cost contour. (One additional piece of information is required. What is it?)

Should there be three or more control variables, the cost surface becomes a *hypersurface*, (i.e., it no longer can be plotted in three dimensions). However, equal-cost contours for the three-control case can be visualized as surfaces that close around the point (u_1^*, u_2^*, u_3^*) at which J is maximum or minimum (Fig. 1.3-3*c*). Efficient paths to the optimum point would be perpendicular to those surfaces. With four or more controls, the equal-cost contours themselves become hypersurfaces.

It may well be that the path to the optimum is constrained to lie along or to one side of a given curve. The former case establishes an *equality constraint*, $f(u_1, u_2) = 0$, an equation relating u_1 and u_2 that must be satisfied at the same time that J is minimized. As shown in Fig. 1.3-4*a*, the smallest allowable value of J is the point at which the constraining curve is tangent to the constant-J contour of lowest cost. It is not an absolute minimum, but it is the best that can be done under the circumstances. An *inequality constraint* [e.g., $f(u_1, u_2) \geq 0$ or $f(u_1, u_2) \leq 0$] separates the control space into acceptable and unacceptable regions. If the acceptable region contains the absolute minimum (Fig. 1.3-4*b*), that is, of course, the solution. If it does not (Fig. 1.3-4*c*), then the equality-constraint minimum is the answer.

The concepts illustrated in the last several figures strictly apply to the optimization of static (unchanging) systems; however, we can extend these ideas to the optimal control of dynamic systems (at least figuratively) by making a few changes. Perhaps the most significant of these relates to the equality constraint. Using the vector notation described in Section 2.1, the previous examples of static optimization considered an equality constraint on the controls that took the form

$$f(\mathbf{u}) = 0 \tag{1.3-1}$$

Had there been more than one constraint, a vector of equalities could have been written as

$$\mathbf{f}(\mathbf{u}) = \mathbf{0} \tag{1.3-2}$$

Optimization of a dynamic system governed by an ordinary differential equation for the state vector, $\mathbf{x}(t)$,

$$\frac{d\mathbf{x}(t)}{dt} \triangleq \dot{\mathbf{x}}(t) = \mathbf{f}[\mathbf{x}(t), \mathbf{u}(t), \mathbf{w}(t), \mathbf{p}(t), t] \tag{1.3-3}$$

is constrained to satisfy this equation during the solution interval, $t_0 \leq t \leq t_f$. The constraint could be rewritten as

$$\mathbf{f}[\mathbf{x}(t), \mathbf{u}(t), \mathbf{p}(t), \mathbf{w}(t), t] - \dot{\mathbf{x}}(t) = \mathbf{0} \tag{1.3-4}$$

In effect, the equality constraint of Fig. 1.3-4*a* can be multidimensional and continually changing in time. Furthermore, it is a function of $\mathbf{p}(t)$, $\mathbf{w}(t)$, $\mathbf{x}(t)$, t, and $\dot{\mathbf{x}}(t)$, as well as the control, $\mathbf{u}(t)$. Consequently,

- The overall cost function for a dynamic system can only be calculated

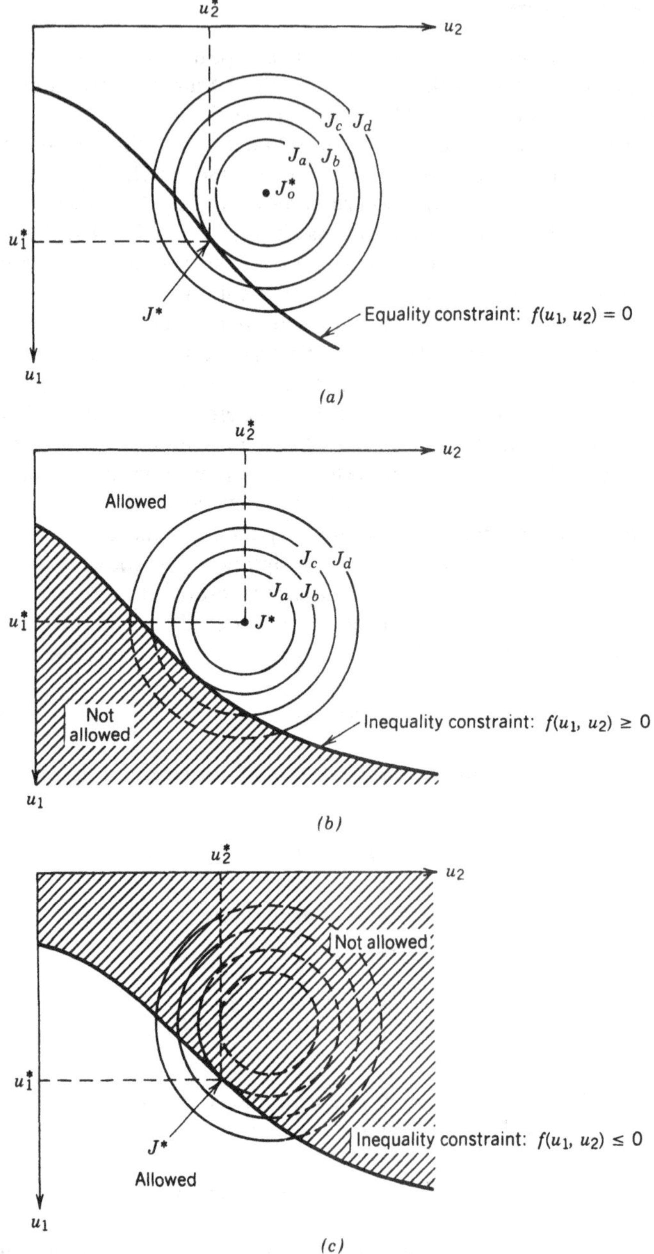

FIGURE 1.3-4 Effects of equality and inequality constraints on control. (*a*) Optimal control subject to an equality constraint; (*b*) optimal control subject to an inequality constraint: absolute optimum in allowable control region; (*c*) optimal control subject to an inequality constraint: absolute optimum in disallowed control region.

at the end of the solution interval, and it may be necessary to sum or integrate instantaneous components over that time interval.

- The resulting optimal control history [$\mathbf{u}(t)$ for $t_0 \le t \le t_f$] is likely to change from one instant to the next.
- The instantaneous optimal control setting may depend on future values of the parameters, disturbances, and state, as well as on the current state and the cost function.

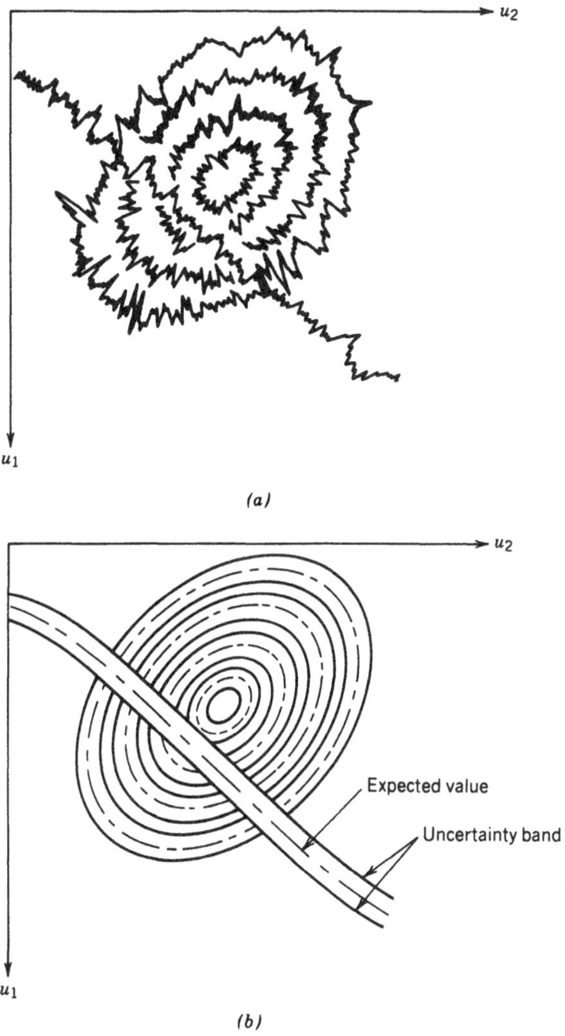

FIGURE 1.3-5 Qualitative effects of uncertainty on cost function and constraint. (*a*) Noisy representation of cost function and constraint; (*b*) statistical representation of cost function and constraint.

It will be seen that the dependence of the instantaneous optimal control on past history is represented entirely through its dependence on the current state.

Our ability to construct a plot such as Fig. 1.3-4*a* from real data may be limited by errors in measurement, uncertainty in the structure of mathematical models, and imprecision in the definition of parameters. Consequently, the observed costs and constraints might look like Fig. 1.3-5*a* or be biased from their actual positions. In such circumstances, it may be desirable to forego deterministic optimization and use a statistical approach, which takes into account our best knowledge of expected values and likely deviations in describing the system (Fig. 1.3-5*b*). As inferred from the figure, having obtained estimates of the expected values of the state and the resulting control, it is possible to use these *stochastic* variables in the optimization. Observations of the system can be expressed as a vector function, $\mathbf{z}(t)$, of the system output and error,

$$\mathbf{z}(t) = \mathbf{j}[\mathbf{y}(t), \mathbf{n}(t), t] \tag{1.3-5}$$

where the output vector $\mathbf{y}(t)$ is a function of the state, control, disturbance, and parameter vectors of the system:

$$\mathbf{y}(t) = \mathbf{h}[\mathbf{x}(t), \mathbf{u}(t), \mathbf{w}(t), \mathbf{p}(t), t] \tag{1.3-6}$$

The objective is to obtain an estimate of $\mathbf{x}(t)$ from $\mathbf{z}(t)$, one that minimizes the degrading effects of $\mathbf{n}(t)$ and of uncertainties in $\mathbf{p}(t)$. This optimal estimate of $\mathbf{x}(t)$ can then be incorporated in the formulation of an optimal control strategy.

In the most general stochastic optimal problem, estimation and control are coupled (i.e., one depends on the other); however, for a wide class of problems, the estimation and control phases can be treated separately. The process of obtaining optimal estimates parallels the computation of optimal controls; the processes are said to be mathematical "duals" in that similar equations and computational sequences are used in both cases.

1.4 OVERVIEW OF THE BOOK

The first five chapters of the book deal with the general problem of controlling and observing nonlinear, time-varying systems, and the sixth chapter concentrates attention on linear, time-invariant systems with constant-coefficient estimation and control. Chapter 2 presents a general mathematical background, with each section supporting the numerically equivalent chapter (Section 2.3 for Chapter 3, etc.). For the most part, it is not necessary to complete all of Chapter 2 before proceeding to the other chapters. The theory of deterministic nonlinear optimal control then is given in Chapter 3. Neighboring-optimal trajectories are introduced as prototypes for feedback control, and the possible need for estimation is noted. Optimal filtering and prediction for discrete-time systems are presented in Chapter 4, with continuous-time estimation developed as the limiting case for a vanishingly small time interval. Stochastic optimal control is formulated as the combined processes of estimation and control (Chapter 5). Next, the special case of controlling and observing linear, time-invariant physical systems with constant-coefficient controllers and estimators is addressed. The design of linear multivariable controllers using optimal control theory is addressed in Chapter 6. A final comment is offered in Chapter 7.

Biographical notes throughout the book are drawn principally from (C-1), (G-1), (G-2), (H-1), (I-1), and (S-1) (see References).

PROBLEMS

1. Define the vectors of a dynamic process model [$\mathbf{x}, \mathbf{u}, \mathbf{w}, \mathbf{p}, \mathbf{y}, \mathbf{n}, \mathbf{z}$] and cost function [$J$] for optimization in your own field of interest. Can the variables of this model change continuously or do they change only at discrete instants of time? If the process is continuous, how fast would discrete measurements have to occur to capture the important details of time variables? Is the system linear? Nonlinear? Time-varying?
2. Consider the group of five pendulums shown in the diagram. Each pendulum oscillates with the same natural frequency. Its motion is represented by two state variable components (angle and angular rate); hence the total state vector has dimension 10. Each pendulum could be affected by a control force, but it is desired to provide a control vector of minimum dimension. What is the smallest dimension control vector that provides complete controllability? Would the answer be different if the pendulums had different natural frequencies?

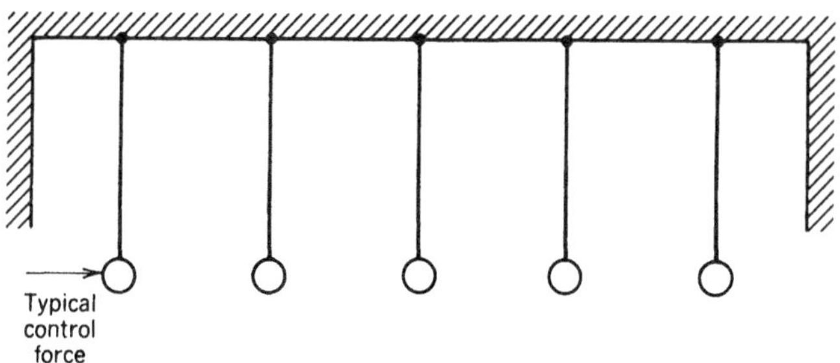

3. The pendulums of Problem 2 can be observed individually by angle and/or rate measurements. What is the smallest dimension observation vector that provides complete observability? Would the answer be different with different natural frequencies?
4. Assume that the pendulum oscillations are stable. Would the system be stabilizable and detectable with no control inputs and observations? Why?
5. Formulate three mathematical functions that have well-defined minima. Define ranges of variables within which the minima are certain to exist. What would the allowable minimum values of the function be if the functions' variables were constrained to lie outside these regions?

REFERENCES

C-1 Cajori, F., *A History of Mathematics*, Chelsea, New York, 1980.

G-1 Gillespie, C. C., ed., *Dictionary of Scientific Biography*, Charles Scribner's Sons, New York, 1970.

G-2 Goldstine, H. H., *A History of the Calculus of Variations from the 17th through the 19th Century*, Springer-Verlag, New York, 1980.

H-1 Harris, W. H. and Levy, J. S., *The New Columbia Encyclopedia*, Columbia University Press, New York, 1975.

I-1 Ireland, N. O., *Index to Scientists of the World from Ancient to Modern Times*, F. W. Faxon Co., Boston, 1962.

S-1 Simmons, G. F., *Differential Equations, with Applications and Historical Notes*, McGraw-Hill, New York, 1972.

S-2 Stefik, M., et al., The Organization of Expert Systems, *Artificial Intelligence*, **18**(2), 135–174, March 1982.

S-3 Stengel, R. F., An Introduction to Stochastic Optimal Control Theory, *Theory and Applications of Optimal Control in Aerospace Systems*, **AGARD-AG-251**, 3-1 to 3-33, July 1981.

W-1 Wilson, J. T. and Bélanger, P. R., Application of Dynamic Optimization Techniques to Reactor Control in the Manufacture of Ethylene, *IEEE Transactions on Automatic Control*, **AC-17**(6), 756–762, December 1972.

2 THE MATHEMATICS OF CONTROL AND ESTIMATION

Optimization is accomplished by using equations that model a dynamic system and represent its control strategy. The language of mathematics is intended to capture broad principles in concise form, and while this may lead to a certain elegance and efficiency in expression, the notation and conventions can appear, at first, a bit baffling. Optimization is a codification of common sense, and mathematics provides the tool for stating and using that code.

This chapter presents a review of the mathematics of dynamic systems, particularly as it relates to optimal estimation and control of these systems. It cannot treat every topic in depth, and the advanced student may progress through this review quickly. The chapter provides a background for understanding the remainder of the book, and it suggests sources for further reading in the mathematical literature. It is a parallel supplement rather than an antecedent to the other chapters.

The mathematical concepts of special concern include multivariable calculus, linear algebra, numerical methods, probability and statistics, and transform analysis. These topics normally are covered in second- and third-year engineering mathematics courses; hence a background in more advanced mathematics is not essential. Nevertheless, these subjects may have been studied from differing viewpoints or with dissimilar notation, so particular attention is devoted to establishing a unified framework for application to control problems.

2.1 SCALARS, VECTORS, AND MATRICES

Scalars

A *real, scalar constant* is a single quantity that does not vary during a specific problem's solution interval. The constant may be represented by a symbol (e.g., an English or Greek letter) that is synonymous with a specific *real number* (as distinct from *complex numbers*, which are discussed in Section 2.2), measured in appropriate units. For example, the dimensionless, *absolute scalar constants*, e and π, represent 2.71828 . . . and 3.14159 . . . , respectively, without ambiguity. A *generic symbol* can be assigned a fixed value that pertains to a specific problem but may change from one case to the next; the mass of an object may be signified by m in two cases, representing 1 kg in the first case and 10 kg in the second. In either case, dimensional values must be consistent with usage in the remainder of the problem; alternate specifications, such as $m = 2.205$ lb-mass or 22.05 lb-mass, may be equivalent but not appropriate if *Système International* (SI) units are used in related computations.

If the single quantity does vary in space or time or as the result of changes in other quantities, it is a *scalar variable*. The quantity (or quantities) on which the variable depends can be stated explicitly in brackets. The mass of a thrusting rocket vehicle decreases with time t and can be denoted by $m(t)$. The magnitude of gravitational acceleration g depends on the distance r from the earth's center, and it can be expressed as $g(r)$. Variables can be functions of other variables; the radius increases as the rocket leaves earth, so, $r = r(t)$ and the gravitational acceleration is $g[r(t)]$. In this book, scalar constants and variables normally are symbolized by lowercase letters. Exceptions are made to agree with widely used symbols (e.g., the cost function J). Explicit expression of a variable's dependence on other variables is made only when necessary for clarity.

In some cases, it is most appropriate to consider the variable to be a function of a *discrete sampling index*, which usually is represented by a subscript. If the varying mass is measured at 11 times with a sampling interval of 1 s, the sequence of measurements could be represented as m_k, $k = 0$ to 10. On the other hand, the continuous description of mass in the same interval is $m(t)$, $0 \leq t \leq 10$.

Vectors

A *set* of scalars may have some common trait, and if these scalars always are cited in fixed relationship to each other, the *ordered set* can be treated as a single, multidimensional quantity. Such a quantity is called a *vector*, and it may contain as many or few components (or elements) as are deemed necessary to characterize the set. A scalar, like the volume of water in a glass, can be considered a vector with one component. The location of a swarm of bees can be considered a vector with thousands of components, three for each bee. Both vector definitions are valid, yet each is appropriate under different circumstances. Consequently, a vector can be represented by a single symbol (usually a boldface, lowercase letter here), but this representation must be accompanied by one more piece of information: the *dimension* of the vector.

When we speak of an ordered set, we mean that the components are written one after the other in fixed order. The location of a single bee is a three-component vector, $\mathbf{r}$, with Cartesian coordinates (x, y, z) which could represent easterly, northerly, and vertical displacement from the beehive. If we agree to write the displacements in fixed order, then the bee's location can be expressed as either

$$\mathbf{r} = (x, y, z) \tag{2.1-1}$$

or

$$\mathbf{r} = (r_1, r_2, r_3) \tag{2.1-2}$$

The subscripted scalar variables are equivalent to the initial ordered set, and the dimension n of this vector quite obviously is three. Once a range of values is determined for each component, a vector of dimension n can be considered to establish a *space* of dimension n. Such a space is difficult to visualize for n greater than 3, and its coordinates may not have identical units; nevertheless, the geometrical interpretation is often useful.

An alternate convention for expressing a vector is to drop the commas and arrange the vector's components in a column:

$$\mathbf{r} = \begin{bmatrix} x \\ y \\ z \end{bmatrix} = \begin{bmatrix} r_1 \\ r_2 \\ r_3 \end{bmatrix} \tag{2.1-3}$$

This is called a *column vector*, and, unless otherwise noted, it is the implicit definition of a vector in this book. We also could choose to write the components in a row (without the commas), forming a *row vector*. The *transpose* of a column vector is a row vector (and vice versa), and the result is written with the superscript T;

$$\mathbf{r}^T = [x \quad y \quad z] = [r_1 \quad r_2 \quad r_3] \tag{2.1-4}$$

The definitions of vector constants and variables correspond to those of scalars. *Vector constants* have constant elements; examples include the *unit vector*, in which every element is one, and the *null vector*, in which every element is zero. *Vector variables* change as other quantities change. The position of the rocket vehicle leaving Earth, taking the Earth's center as the origin, is

$$\mathbf{r}(t) = \begin{bmatrix} x(t) \\ y(t) \\ z(t) \end{bmatrix} = \begin{bmatrix} r_1(t) \\ r_2(t) \\ r_3(t) \end{bmatrix} \tag{2.1-5}$$

and the corresponding gravitational acceleration has direction as well as magnitude:

$$\mathbf{g}[\mathbf{r}(t)] = \begin{bmatrix} g_x[x(t), y(t), z(t)] \\ g_y[x(t), y(t), z(t)] \\ g_z[x(t), y(t), z(t)] \end{bmatrix} = \begin{bmatrix} g_1[\mathbf{r}(t)] \\ g_2[\mathbf{r}(t)] \\ g_3[\mathbf{r}(t)] \end{bmatrix} \tag{2.1-6}$$

In many cases, a vector variable is a *linear function* of another vector [i.e., changes in the first vector are directly (linearly) proportional to changes in the second]. The proportionality between the two is expressed by a *matrix* of coefficients, $\mathbf{A}$ and is

$$\mathbf{y}(\mathbf{x}) = \mathbf{A}\mathbf{x} \tag{2.1-7}$$

which can be written term by term as

$$\begin{bmatrix} y_1 \\ y_2 \\ \cdot \\ \cdot \\ \cdot \\ y_r \end{bmatrix} = \begin{bmatrix} a_{11} & a_{12} & \cdots & a_{1n} \\ a_{21} & a_{22} & \cdots & a_{2n} \\ \cdot & \cdot & & \cdot \\ \cdot & \cdot & & \cdot \\ \cdot & \cdot & & \cdot \\ a_{r1} & a_{r2} & \cdots & a_{rn} \end{bmatrix} \begin{bmatrix} x_1 \\ x_2 \\ \cdot \\ \cdot \\ \cdot \\ x_n \end{bmatrix} \tag{2.1-8}$$

Each component of $\mathbf{y}$ may be a function of all components of $\mathbf{x}$; for example,

$$\begin{aligned} y_1 &= a_{11}x_1 + a_{12}x_2 + \cdots + a_{1n}x_n \\ y_2 &= a_{21}x_1 + a_{22}x_2 + \cdots + a_{2n}x_n \\ \vdots &= \qquad \vdots \\ y_r &= a_{r1}x_1 + a_{r2}x_2 + \cdots + a_{rn}x_n \end{aligned} \tag{2.1-9}$$

making the point that the rows of $\mathbf{A}$ each multiply the column vector $\mathbf{x}$ to produce the column vector $\mathbf{y}$. Hence the number of rows of $\mathbf{A}$ must equal the dimension of $\mathbf{y}$, while the number of columns must equal the dimension of $\mathbf{x}$. $\mathbf{A}$ is said to have dimension $(r \times n)$, and it is *conformable* with $\mathbf{x}$. A simple example follows:

Example 2.1-1 Mixing Apples and Oranges (and Pears)

Four shoppers purchase apples at \$.30 each, oranges at \$.20 each, and pears at \$.25 each. The first shopper buys three apples, two oranges, and four pears; the second buys no apples, three oranges, and one pear; and the third and fourth each buy two apples, four oranges, and two pears. How much did each shopper spend?

Of course, vectors and matrices are not required to solve this problem, but our objective is to provide a tangible example of matrix-vector multiplication. The fruit prices form the vector **x**, the shoppers' expenditures form the vector **y**, and the quantities purchased by each shopper are the coefficients of **A**:

$$\begin{bmatrix} \$2.30 \\ \$.85 \\ \$1.90 \\ \$1.90 \end{bmatrix} = \begin{bmatrix} 3 & 2 & 4 \\ 0 & 3 & 1 \\ 2 & 4 & 2 \\ 2 & 4 & 2 \end{bmatrix} \begin{bmatrix} \$.30 \\ \$.20 \\ \$.25 \end{bmatrix}$$

The scalar equations in Eq. 2.1-9 could represent the product of a row vector and a matrix (rather than a matrix and a column vector), in which case the product would be expressed as a row vector. Using the same convention for multiplication (the rows of the first set multiply the columns of the second),

$$[y_1 \quad y_2 \ldots y_r] = [x_1 \quad x_2 \ldots x_n] \begin{bmatrix} a_{11} & a_{21} \ldots a_{r1} \\ a_{12} & a_{22} \ldots a_{r2} \\ a_{1n} & a_{2n} \ldots a_{rn} \end{bmatrix} \tag{2.1-10}$$

or

$$\mathbf{y}^T = \mathbf{x}^T \mathbf{A}^T \tag{2.1-11}$$

where the rows and columns of **A** have been interchanged to form $\mathbf{A}^T$, the transpose of **A**. The dimension of $\mathbf{A}^T$ is $(n \times r)$. Note that the ordering of **A** and **x** is reversed in expressing the transpose of a product as the product of transposes.

Matrices

Matrices are *two-dimensional arrays of scalars*. They are signified by boldface capital letters here. With r rows and n columns, the dimension of **A** is $(r \times n)$. If the dimensions, r and n, both equal one, **A** reduces to a scalar. If only n is one, **A** reduces to a column vector; if only r is one, **A** reduces to a row vector. If r and n are equal, **A** is *square*. If all the elements of **A** are zero, it is a *null matrix*, regardless of dimension.

Square matrices can be classified by the characteristics of their elements. The principal diagonal of a square matrix **A** contains the elements a_{11}, a_{22}, and so on. If all other elements are zero, $\mathbf{A}$ is a *diagonal matrix*:

$$\mathbf{A} = \begin{bmatrix} a_{11} & 0 & \dots & 0 \\ 0 & a_{22} & \dots & 0 \\ \vdots & \vdots & & \vdots \\ 0 & 0 & \dots & a_{nn} \end{bmatrix} \tag{2.1-12}$$

If all the diagonal elements are one, $\mathbf{A}$ is an *identity matrix*, $\mathbf{I}$, of dimension n:

$$\mathbf{I} = \begin{bmatrix} 1 & 0 & \dots & 0 \\ 0 & 1 & \dots & 0 \\ 0 & 0 & \dots & 1 \end{bmatrix} \tag{2.1-13}$$

In the first instance, Eq. 2.1-7 yields,

$$\begin{bmatrix} y_1 \\ y_2 \\ \vdots \\ y_n \end{bmatrix} = \begin{bmatrix} a_{11}x_1 \\ a_{22}x_2 \\ \vdots \\ a_{nn}x_n \end{bmatrix} \tag{2.1-14}$$

while in the second

$$\mathbf{y} = \mathbf{Ix} \tag{2.1-15}$$

or

$$\begin{bmatrix} y_1 \\ y_2 \\ \vdots \\ y_n \end{bmatrix} = \begin{bmatrix} x_1 \\ x_2 \\ \vdots \\ x_n \end{bmatrix} \tag{2.1-16}$$

The off-diagonal elements of a *symmetric matrix* are "mirror images" of each other (i.e., $a_{ij} = a_{ji}$),

$$\mathbf{A} = \begin{bmatrix} a_{11} & a_{12} \dots a_{1n} \\ a_{12} & a_{22} \dots a_{2n} \\ \vdots & \vdots \quad\quad \vdots \\ a_{1n} & a_{2n} \dots a_{nn} \end{bmatrix} \tag{2.1-17}$$

and $\mathbf{A} = \mathbf{A}^T$. If $\mathbf{A} = -\mathbf{A}^T$, or $a_{ij} = -a_{ji}$, then $a_{ii} = 0$, and the matrix is *skew symmetric*:

$$\mathbf{A} = \begin{bmatrix} 0 & a_{12} & \dots & a_{1n} \\ -a_{12} & 0 & \dots & a_{2n} \\ \vdots & \vdots & & \vdots \\ -a_{1n} & -a_{2n} & \dots & 0 \end{bmatrix} \tag{2.1-18}$$

Whereas a diagonal matrix **A** establishes a one-to-one correspondence between elements of **y** and **x**—a parallelism of sorts—a skew-symmetric matrix produces an orthogonal relationship, as demonstrated by a later example.

Matrices of identical dimension can be summed or differenced term by term. If the shoppers in Example 2.1-1 went back for more apples, oranges, and pears a week later, and if the prices had not changed in the interim, their total expenditures could be expressed by the equation

$$\mathbf{y} = (\mathbf{A}_1 + \mathbf{A}_2)\mathbf{x} \tag{2.1-19}$$

where $\mathbf{A}_1$ and $\mathbf{A}_2$ are the "quantity matrices" for each week. Similarly, if they bought identical quantities three weeks in succession, their expenditures could be written

$$\mathbf{y} = k\mathbf{A}\mathbf{x} = \mathbf{B}\mathbf{x} \tag{2.1-20}$$

where $k = 3$ and $b_{ij} = ka_{ij}$.

The *product of two matrices* is another matrix whose dimension is determined by the number of rows of the first matrix and columns of the second matrix:

$$\mathbf{C} = \mathbf{AB} \tag{2.1-21}$$

A must have as many columns as **B** has rows for the product to exist (i.e., for the two matrices to be conformable). As might be expected by analogy to the vector-matrix products,

$$\mathbf{C}^T = \mathbf{B}^T\mathbf{A}^T \tag{2.1-22}$$

and Eqs. 2.1-7 and 2.1-11 are seen to be special cases of Eqs. 2.1-21 and 2.1-22. Matrix multiplication is not *commutative*, (i.e., $\mathbf{AB} \neq \mathbf{BA}$ except in special cases), so the reverse ordering in the product of transposes is significant.

Inner and Outer Products

Because vectors can be considered matrices with either one row or column, products of vectors must satisfy the conformability requirements of Eq.

2.1-21. The equation allows two kinds of vector products to be formed; both require that the two vectors which form the product have an equal number of components. The *inner product* results when a row vector (or column vector transpose) "pre"-multiplies a column vector, and the result is a *scalar*:

$$\underset{(1\times 1)}{z} = \underset{(1\times n)(n\times 1)}{\mathbf{x}^T\mathbf{y}} \tag{2.1-23}$$

or

$$z = x_1y_1 + x_2y_2 + \cdots + x_ny_n \tag{2.1-24}$$

The *outer product* results when a column vector premultiplies a row vector (or column vector transpose); for vectors of dimension n, the product is an $(n \times n)$ matrix:

$$\underset{(n\times n)}{\mathbf{Z}} = \underset{(n\times 1)(1\times n)}{\mathbf{x}\mathbf{y}^T} \tag{2.1-25}$$

or

$$\begin{bmatrix} z_{11} & z_{12} & \cdots & z_{1n} \\ z_{21} & z_{22} & \cdots & z_{2n} \\ \vdots & \vdots & & \vdots \\ z_{n1} & z_{n2} & \cdots & z_{nn} \end{bmatrix} = \begin{bmatrix} x_1y_1 & x_1y_2 & \cdots & x_1y_n \\ x_2y_1 & x_2y_2 & \cdots & x_2y_n \\ \vdots & \vdots & & \vdots \\ x_ny_1 & x_ny_2 & \cdots & x_ny_n \end{bmatrix} \tag{2.1-26}$$

The inner product provides a means of measuring just how parallel two vectors are, while the outer product plays an important role in determining how closely correlated all the components of one vector are with the other vector. Note that the sum of the diagonal elements of the outer product equals the inner product. The sum of a square matrix's diagonal elements is called the *trace* of the matrix, and it is denoted by $\text{Tr}(\cdot)$; hence, in the current example,

$$z = \text{Tr}(\mathbf{Z}) \tag{2.1-27}$$

The trace of a matrix is a scalar, and for any square product **ABC**, where the individual matrices need not be square,

$$\text{Tr}(\mathbf{ABC}) = \text{Tr}(\mathbf{CAB}) = \text{Tr}(\mathbf{BCA}) \tag{2.1-28}$$

EXAMPLE 2.1-2 LENGTH OF A POSITION VECTOR

A bee is located at the position $\mathbf{r} = (x \quad y \quad z)^T$. What is the magnitude of its distance $|\mathbf{r}|$ from the hive?

$$|\mathbf{r}| = (\mathbf{r}^T\mathbf{r})^{1/2} = \left\{ [x \quad y \quad z] \begin{bmatrix} x \\ y \\ z \end{bmatrix} \right\}^{1/2}$$

$$= (x^2 + y^2 + z^2)^{1/2}$$

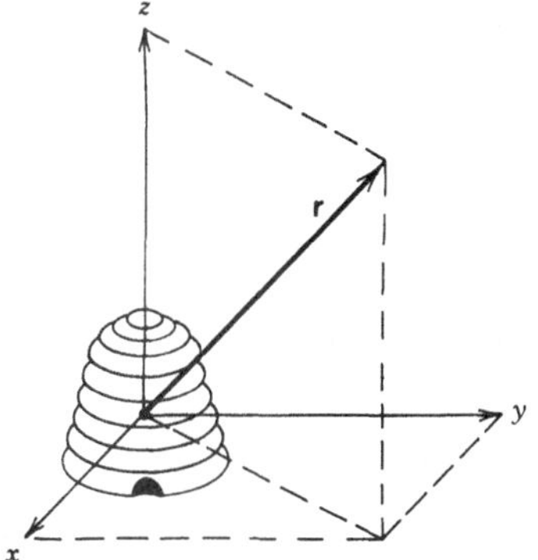

EXAMPLE 2.1-3 ANGLE BETWEEN TWO VECTORS

What is the angle between $\mathbf{r}_1$ and $\mathbf{r}_2$ in the diagram?

From trigonometry, the cosine of θ could be determined from the projection of $\mathbf{r}_2$ on $\mathbf{r}_1$, divided by the magnitude of $\mathbf{r}_2$, or the projection of $\mathbf{r}_1$ on $\mathbf{r}_2$, divided by the magnitude of $\mathbf{r}_1$. Using the former definition, the

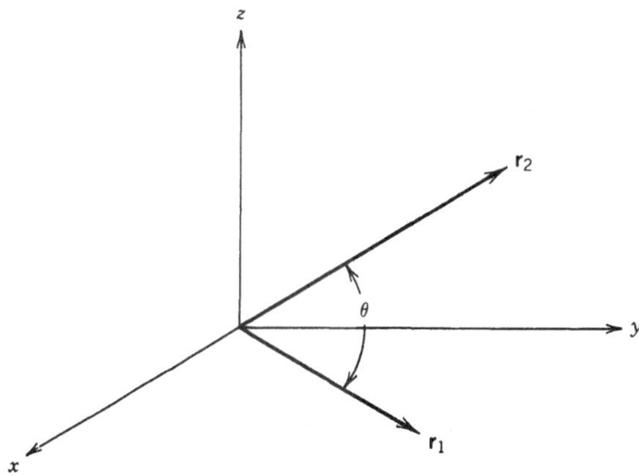

projection of $\mathbf{r}_2$ on $\mathbf{r}_1$ is

$$\frac{\mathbf{r}_2^T\mathbf{r}_1}{|\mathbf{r}_1|} = \frac{\mathbf{r}_2^T\mathbf{r}_1}{(\mathbf{r}_1^T\mathbf{r}_1)^{1/2}}$$

and the angle between the two vectors is

$$\theta = \cos^{-1}\left[\frac{\mathbf{r}_2^T\mathbf{r}_1}{(\mathbf{r}_1^T\mathbf{r}_1)^{1/2}(\mathbf{r}_2^T\mathbf{r}_2)^{1/2}}\right]$$

Vector Lengths, Norms, and Weighted Norms

Example 2.1-2 showed that the length of a three-dimensional vector is readily expressed as the square root of the sum of the squares or *root-mean-squared value*, that is, as the square root of the inner product. This concept of length can be extended to vectors of arbitrary dimension, consisting of whatever components belong in the ordered set. The scalar measure of length (or metric) is called a *Euclidean* or *quadratic norm*.* For the n-dimensional vector $\mathbf{x}$ it is

$$\|\mathbf{x}\| = (\mathbf{x}^T\mathbf{x})^{1/2} = (x_1^2 + x_2^2 + \cdots + x_n^2)^{1/2} \tag{2.1-29}$$

There are no strict limitations on how we define $\mathbf{x}$, and it often occurs that $\mathbf{x}$ really does contain an "apples and oranges" mix of scalar variables. For example, a state vector for a chemical process might contain variables whose natural units are molar concentration, degrees Celsius, cubic meters, and Newtons per square meter. A Euclidean norm could be formed easily enough, but what does it mean? In such instance, it might be better to compute the norm of a normalized vector $\mathbf{y}$ whose components are the original components of $\mathbf{x}$, each divided by its maximum expected value. We could write $\mathbf{y}$ as

$$\mathbf{y} = \mathbf{D}\mathbf{x} \tag{2.1-30}$$

where $\mathbf{D}$ is a diagonal matrix of the inverses of the maximum expected values,

$$\mathbf{D} = \begin{bmatrix} (1/x_{1_{\max}}) & 0 & \cdots & 0 \\ 0 & (1/x_{2_{\max}}) & \cdots & 0 \\ \vdots & \vdots & & \vdots \\ 0 & 0 & \cdots & (1/x_{n_{\max}}) \end{bmatrix} \tag{2.1-31}$$

*Euclid (ca. 300 B.C.) was the Greek mathematician who first documented the principles of geometry.

Then the Euclidean norm of $\mathbf{y}$ can be written as the weighted Euclidean norm of $\mathbf{x}$,

$$\begin{aligned}\|\mathbf{y}\| &= (\mathbf{y}^T\mathbf{y})^{1/2} = \|\mathbf{Dx}\| \\ &= (\mathbf{x}^T\mathbf{D}^T\mathbf{Dx})^{1/2} = (\mathbf{x}^T\mathbf{Qx})^{1/2} \qquad (2.1\text{-}32)\end{aligned}$$

where $\mathbf{Q}$ is defined by $\mathbf{D}^T\mathbf{D}$. In the present case, $\mathbf{Q}$ is a diagonal matrix with elements $(1/x_{i\,\max})^2$. Had $\mathbf{D}$ established a more general relationship between $\mathbf{x}$ and $\mathbf{y}$ (i.e., if $\mathbf{D}$ were not diagonal), $\mathbf{Q}$ also would not be diagonal, but it would be symmetric.

Recall that even though the weighted norm contains a vector-matrix product, it is a scalar variable. The dimension of $\mathbf{x}^T\mathbf{Qx}$ is

$$(1 \times n)(n \times n)(n \times 1) = 1$$

This product is called a *quadratic form* because it contains squared ("quadratic") terms. If the quadratic form is greater than zero for any $\mathbf{x}$ that is not a null vector, it is said to be *positive definite*. If it is less than zero in the same circumstances, it is *negative definite*. If selected nonzero values of $\mathbf{x}$ cause $\mathbf{x}^T\mathbf{Qx}$ to be zero, it is positive or negative semidefinite. "Definiteness" pertains to the (scalar) product, $\mathbf{x}^T\mathbf{Qx}$, but it is governed by characteristics of the *defining matrix* $\mathbf{Q}$, as discussed in Section 2.2. It is useful to note that, from Eq. 2.1-28,

$$\mathbf{x}^T\mathbf{Qx} = \mathrm{Tr}(\mathbf{x}^T\mathbf{Qx}) = \mathrm{Tr}(\mathbf{xx}^T\mathbf{Q}) = \mathrm{Tr}(\mathbf{Qxx}^T) \qquad (2.1\text{-}33)$$

Stationary, Minimum, and Maximum Points of a Scalar Variable (Ordinary Maxima and Minima)

The scalar function, $J(u)$, of a scalar variable u that is sketched in Fig. 2.1-1 contains two minima, an inflection point, and a maximum between its endpoints, that is, at *interior points*. Each of these points is stationary and is characterized by zero slope: $dJ(u)/du = 0$ at the point (Fig. 2.1-2*a*). These points can be classified according to the local curvature, or, equivalently, the second derivative, $d^2J(u)/du^2$, of the function. The curvature is positive at a minimum, zero at the inflection point, and negative at the maximum (Fig. 2.1-2*b*).

Without going into the equations describing $J(u)$ and its derivatives, it is possible to learn several things from these two figures. The first is that the locations of stationary points (including maxima and minima) are defined by points at which the first derivative goes to zero—$dJ(u)/du = 0$ is a necessary condition for stationarity or optimality. Taken alone, stationarity is not sufficient to define optimality. Higher derivatives considered separately, such as $d^2J(u)/du^2$, do not define stationary points; however, the sign of the

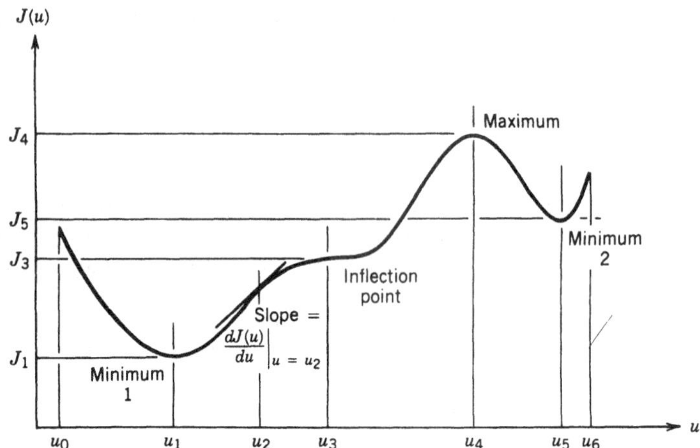

FIGURE 2.1-1 Scalar function of a scalar variable, showing locally minimum, stationary, and maximum values.

second derivative at a stationary point [previously defined by $dJ(u)/du = 0$] is adequate to classify the nature of the stationary point. Considered together, the first and second derivatives establish necessary and sufficient conditions for optimality at interior points.*

If $J(u)$ is defined in a fixed interval, say (u_0, u_6), there is an additional condition for maxima or minima at the endpoints. Suppose $dJ/du < 0$ at u_0; then u_0 is a local maximum even though J is not stationary. J would become larger if u extended below u_0—but it does not, so $J(u_0)$ is a maximum. Similarly, $dJ/du > 0$ at u_6 would indicate a local maximum, and the two endpoint criteria are reversed for local minima.

At best, these conditions identify locally maximum or minimum points. Having found a minimum such as that at $u = u_1$ is no guarantee that other minima like the one at $u = u_5$ do not exist or that J_1 is the smallest value of $J(u)$. The *global minimum* (i.e., the smallest of the small throughout the range of u) is found by simple comparison of the function values at multiple local minima. In the present example, $J_1 = J(u_1)$ is smaller than $J_5 = J(u_5)$ and is, therefore, the global minimum. Note that any number of inflection points (including none) could fall between local minima, but they must be separated by one (and only one) local maximum in the scalar case.

At least three problematical cases exist, as shown in Fig. 2.1-3. In the first case, the function is discontinuous at the minimum (Fig. 2.1-3a); in the

*If $d^2J/du^2 = 0$ at a stationary point, it is possible that the sign of even higher-order derivatives of u (e.g., d^4J/du^4, could provide sufficient information for optimality). As an example, consider the minimum of $J = u^4$. The minimum occurs at $u = 0$, the first three derivatives are zero, and the fourth derivative is positive.

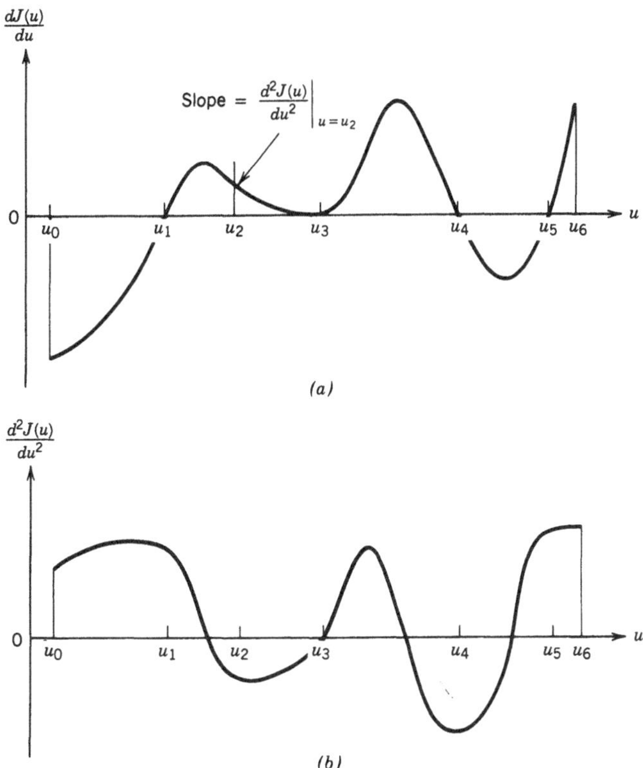

FIGURE 2.1-2 Slopes and curvatures of $J(u)$, $dJ(u)/du$, and $d^2J(u)/du^2$. (*a*) First derivative of $J(u)$, $dJ(u)/du$; (*b*) second derivative of $J(u)$, $d^2J(u)/du^2$.

second, the slope is discontinuous (Fig. 2.1-3*b*); in the third, the slope remains zero over a range of u values (Fig. 2.1-3*c*). To apply the necessary and sufficient conditions noted earlier, it has been assumed that the first two derivatives are continuous; if they are not, it may be necessary to use numerical methods to find and classify stationary points. In the last case, any choice of u in the interval $u_1 \leq u \leq u_2$ is minimizing, and the second derivative is zero, preventing classification of the stationary point. In such an instance, either the slopes and function values immediately outside the interval (u_1, u_2) must be tested or higher derivatives of J with respect to u (if they exist) may infer the stationarity class. Unless otherwise indicated, normally we will assume that all necessary derivatives of functions exist and are continuous.

With this proviso, $J(u)$ can be expanded in a *Taylor series** about a

*Brook Taylor (1685–1731) was an English mathematician whose work led to mathematical descriptions of physical observations, such as the mechanics of vibrating strings.

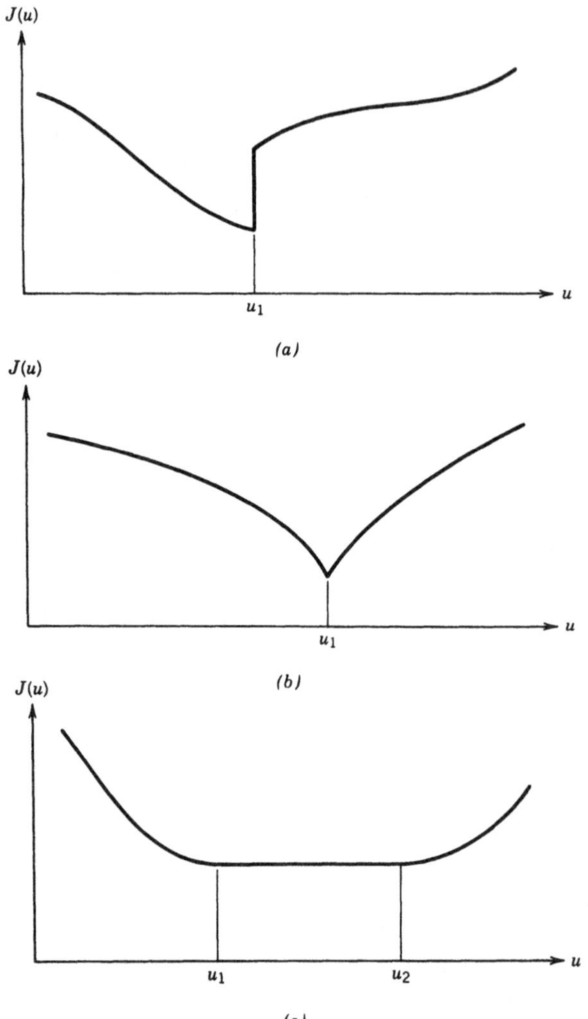

FIGURE 2.1-3 Problematical cases for defining a minimum. (*a*) Discontinuous function and slope; (*b*) discontinuous slope; (*c*) stationary range of values.

nominal point, $J_0 = J(u_0)$:

$$J(u) = J(u_0) + \left.\frac{dJ(u)}{du}\right|_{u=u_0}(u - u_0) + \\ + \frac{1}{2}\left.\frac{d^2J(u)}{du^2}\right|_{u=u_0}(u - u_0)^2 + \text{higher-order terms} \qquad (2.1\text{-}34)$$

Defining perturbations in J and u as

$$\Delta J = J(u) - J(u_0) \tag{2.1-35}$$

$$\Delta u = u - u_0 \tag{2.1-36}$$

and neglecting higher-order terms, Eq. 2.1-34 leads to

$$\Delta J = \left.\frac{dJ(u)}{du}\right|_{u=u_0} \Delta u + \frac{1}{2}\left.\frac{d^2J(u)}{du^2}\right|_{u=u_0} \Delta u^2 \tag{2.1-37}$$

If u_0 is a stationary point, $dJ(u)/du = 0$, and the local variation in J is entirely due to second (and higher) derivatives of the function:

$$\Delta J = \frac{1}{2}\left.\frac{d^2J(u)}{du^2}\right|_{u=u_0} \Delta u^2 \tag{2.1-38}$$

Hence the variation of J in the neighborhood of the stationary point is well approximated as a quadratic form of the scalar variable Δu. This observation will prove useful in later consideration of more complex problems.

To this point, it has been assumed that the scalar variable J is a function of a scalar variable u, but we can extend the discussion to *scalar functions of vector variables* using the material presented earlier. Given the function, $J(\mathbf{u})$, where J is a scalar and $\mathbf{u}$ is a vector of dimension m, we could expand $J(\mathbf{u})$ about the nominal point $J_0 = J(\mathbf{u}_0)$, noting that (a) partial derivatives of scalars with respect to vectors must be defined and (b) all products in the Taylor-series expansion must result in scalars. Retaining terms up to second order,

$$J(\mathbf{u}) = J(\mathbf{u}_0) + \left.\frac{\partial J(\mathbf{u})}{\partial \mathbf{u}}\right|_{\mathbf{u}=\mathbf{u}_0} (\mathbf{u}-\mathbf{u}_0) + \frac{1}{2}(\mathbf{u}-\mathbf{u}_0)^T \left.\frac{\partial^2 J(\mathbf{u})}{\partial \mathbf{u}^2}\right|_{\mathbf{u}=\mathbf{u}_0} (\mathbf{u}-\mathbf{u}_0) \tag{2.1-39}$$

The first derivative of J with respect to $\mathbf{u}$ is defined as a row vector of partial derivatives,

$$\left.\frac{\partial J(\mathbf{u})}{\partial \mathbf{u}}\right|_{\mathbf{u}=\mathbf{u}_0} \triangleq J_{\mathbf{u}} = \left[\frac{\partial J(\mathbf{u})}{\partial u_1} \frac{\partial J(\mathbf{u})}{\partial u_2} \cdots \frac{\partial J(\mathbf{u})}{\partial u_m}\right]_{\mathbf{u}=\mathbf{u}_0} \tag{2.1-40}$$

so that its product with the column vector $(\mathbf{u}-\mathbf{u}_0)$ is a scalar:

$$J_{\mathbf{u}}(\mathbf{u}-\mathbf{u}_0) = \left.\frac{\partial J(\mathbf{u})}{\partial u_1}\right|_{\mathbf{u}=\mathbf{u}_0} (u_1 - u_{1_0}) + \left.\frac{\partial J(\mathbf{u})}{\partial u_2}\right|_{\mathbf{u}=\mathbf{u}_0} (u_2 - u_{2_0}) + \cdots + \left.\frac{\partial J(\mathbf{u})}{\partial u_m}\right|_{\mathbf{u}=\mathbf{u}_0} (u_m - u_{m_0}) \tag{2.1-41}$$

The row vector of partial derivatives is called the *gradient* of J with respect to $\mathbf{u}$. The gradient sometimes is defined as a column vector; this definition necessitates the use of the gradient's transpose to achieve conformability in Eq. 2.1-41.

The second derivative of the scalar J with respect to $\mathbf{u}$ (or the first derivative of the row vector $\partial J(\mathbf{u})/\partial \mathbf{u}$ with respect to $\mathbf{u}$) is the square *Hessian matrix*,* $\partial^2 J(\mathbf{u})/\partial \mathbf{u}^2$,

$$\left.\frac{\partial^2 J(\mathbf{u})}{\partial \mathbf{u}^2}\right|_{\mathbf{u}=\mathbf{u}_0} \triangleq J_{\mathbf{uu}} = \left[\begin{matrix} \dfrac{\partial^2 J(\mathbf{u})}{\partial u_1^2} & \dfrac{\partial^2 J(\mathbf{u})}{\partial u_1 \partial u_2} & \cdots & \dfrac{\partial^2 J(\mathbf{u})}{\partial u_1 \partial u_m} \\ \dfrac{\partial^2 J(\mathbf{u})}{\partial u_2 \partial u_1} & \dfrac{\partial^2 J(\mathbf{u})}{\partial u_2^2} & \cdots & \dfrac{\partial^2 J(\mathbf{u})}{\partial u_2 \partial u_m} \\ \dfrac{\partial^2 J(\mathbf{u})}{\partial u_m \partial u_1} & \dfrac{\partial^2 J(\mathbf{u})}{\partial u_m \partial u_2} & \cdots & \dfrac{\partial^2 J(\mathbf{u})}{\partial u_m^2} \end{matrix}\right]_{\mathbf{u}=\mathbf{u}_0} \tag{2.1-42}$$

and it is a symmetric matrix as long as the order of differentiation is immaterial. To achieve the needed scalar result in the Taylor-series expansion, the Hessian matrix is premultiplied by $(\mathbf{u}-\mathbf{u}_0)^T$ and postmultiplied by $(\mathbf{u}-\mathbf{u}_0)$; hence the product is a quadratic form, as defined earlier.

J is stationary at the point $\mathbf{u}=\mathbf{u}_0$ if simultaneous variations in all the components of $\mathbf{u}$ have negligible effect on the value of J. The gradient of J (each and every component) must be zero at a stationary point, $\mathbf{u}_0$:

$$\left.\frac{\partial J(\mathbf{u})}{\partial \mathbf{u}}\right|_{\mathbf{u}=\mathbf{u}_0} = J_{\mathbf{u}} = \mathbf{0} \tag{2.1-43}$$

The Hessian matrix is seen to be a defining matrix, in the sense of Eq. 2.1-33. The (scalar) product of the quadratic form, $(\mathbf{u}-\mathbf{u}_0)^T J_{\mathbf{uu}}(\mathbf{u}-\mathbf{u}_0)$ is:

Positive definite at a minimum.

Negative definite at a maximum.

Neither positive nor negative definite at an inflection point.

The multidimensional distribution of J with varying $\mathbf{u}$ leads to certain suboptimal conditions of importance, as the following example shows:

EXAMPLE 2.1-4 HEIGHT OF A MOUNTAIN, DEPTH OF A VALLEY

Suppose J represents altitude above sea level, and the two-component vector $\mathbf{u}$ represents easterly and northerly position; then $J(\mathbf{u})$ represents the

*Ludwig Hesse (1811–1874), who studied under Jacobi, developed the Hessian matrix while working on the inflection points of cubic curves.

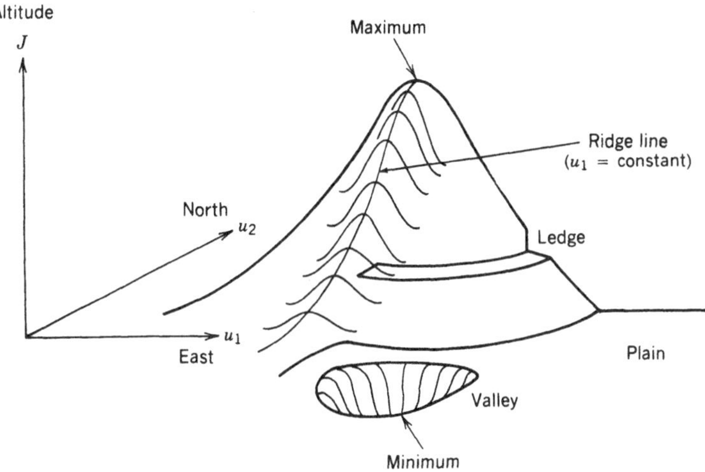

topography of some given area. If the region contains one mountain and one bowl-shaped valley, then J is maximized at the mountain's peak and minimized at the valley's bottom. Assuming a continuous topography that is twice differentiable, the Hessian matrix is negative definite at the former and positive definite at the latter. If there are level plains or ledges in the region, the Hessian is zero in these areas. The gradient is zero at all of these locations. If there is a ridge running due north to the mountain peak, the top of the ridge (or "ridge line") varies in height as u_2 varies, but it represents a maximum in a vertical plane perpendicular to the ridge line; (i.e., in a variable (u) space of reduced dimension). Consequently, $J_{\mathbf{u}}$ is not zero, and $J_{\mathbf{uu}}$ is not negative definite along the ridge line; however, $\partial J(\mathbf{u})/\partial u_1$ is zero and $\partial^2 J(\mathbf{u})/\partial u_1^2$ is negative definite for the appropriate value of u_1. Similar relationships hold along the floor of a meandering valley or river.

Actually defining the value of $\mathbf{u}$ that causes J to be minimized can be difficult if the dimension of $\mathbf{u}$ is high or if the shape of J is intricate. In such cases, an iterative numerical technique such as the *Newton–Raphson* algorithm* must be used. If Eq. 2.1-39 fits J exactly in the vicinity of the minimum, where $\mathbf{u} = \mathbf{u}^*$, then the following should hold:

$$\left.\frac{\partial J}{\partial \mathbf{u}}\right|_{\mathbf{u}=\mathbf{u}^*} = \mathbf{0} = \left.\frac{\partial J}{\partial \mathbf{u}}\right|_{\mathbf{u}=\mathbf{u}_0} + (\mathbf{u}^* - \mathbf{u}_0)^T \left.\frac{\partial^2 J}{\partial \mathbf{u}^2}\right|_{\mathbf{u}=\mathbf{u}_0} \tag{2.1-44}$$

*Isaac Newton (1642–1727) shares the invention of calculus with Leibniz and is most widely known for his laws of gravitation. He and Joseph Raphson (1648–1715) are responsible for the modified algorithm that bears their names.

leading to

$$\mathbf{u}^* = \mathbf{u}_0 - \left(\frac{\partial^2 J}{\partial \mathbf{u}^2}\right)^{-1}_{\mathbf{u}=\mathbf{u}_0} \left(\frac{\partial J}{\partial \mathbf{u}}\right)^T_{\mathbf{u}=\mathbf{u}_0} \tag{2.1-45}$$

When J is not precisely quadratic, this equation will not yield $\mathbf{u}^*$ on the first try; however, an iteration based on this form should converge to the minimizing solution:

$$\mathbf{u}_k = \mathbf{u}_{k-1} - \left(\frac{\partial^2 J}{\partial \mathbf{u}^2}\right)^{-1}_{k-1} \left(\frac{\partial J}{\partial \mathbf{u}}\right)^T_{k-1} \tag{2.1-46}$$

where k is an iteration index.

Constrained Minima and Lagrange Multipliers

$J(\mathbf{u})$ can represent a scalar cost function of the $(m \times 1)$ control vector $\mathbf{u}$; hence the problem of finding the value of $\mathbf{u}$ that minimizes J (an ordinary minimum), provides an introduction to the concept of optimal control. m components of $\mathbf{u}$ must be defined to specify the point of minimum J. If a *scalar equality constraint*,

$$f(\mathbf{u}) = 0 \tag{2.1-47}$$

exists, then there are, at most, $(m-1)$ independent components of $\mathbf{u}$. If J is minimized subject to the constraint, the order of the optimization problem has been reduced by one. Equation 2.1-47 effectively specifies one component of $\mathbf{u}$ in terms of the remaining $(m-1)$ terms, which are varied until the minimum cost, $J^*(\mathbf{u}^*)$, is achieved. The solution is found either by direct substitution of Eq. 2.1-47 in the expression for $J(\mathbf{u})$ or by using Lagrange multipliers, as described below.

A *vector equality constraint* can be defined as

$$\mathbf{f}(\mathbf{u}') = \mathbf{0} \tag{2.1-48}$$

where $\mathbf{f}(\mathbf{u}')$ is itself an n-dimensional vector and the dimension of $\mathbf{u}'$ is $(m+n)$. n components of $\mathbf{u}'$ are specified as functions of the remaining m terms, which are varied until $J(\mathbf{u}')$ is a minimum. By partitioning $\mathbf{u}'$, we could say that $J(\mathbf{u}') = J(\mathbf{u}_a, \mathbf{u}_b)$ is to be minimized subject to $\mathbf{f}(\mathbf{u}_a, \mathbf{u}_b) = \mathbf{0}$, where

$$\mathbf{u}' = \begin{bmatrix} \mathbf{u}_a \\ \mathbf{u}_b \end{bmatrix} \tag{2.1-49}$$

$\mathbf{u}_a$ denotes the n components specified by Eq. 2.1-48, and $\mathbf{u}_b$ represents the m terms that are varied to achieve a minimum. Alternatively, we could make the following changes in notation:

$$\mathbf{x} = \mathbf{u}_a, \tag{2.1-50}$$

$$\mathbf{u} = \mathbf{u}_b \tag{2.1-51}$$

Then $J(\mathbf{x}, \mathbf{u})$ is to be minimized subject to $\mathbf{f}(\mathbf{x}, \mathbf{u}) = \mathbf{0}$. Note that this form of the problem sets the stage for minimizing a cost function that depends on a *state vector* $\mathbf{x}$ and a *control vector* $\mathbf{u}$. In this case, the n-dimensional *state equation*,

$$\mathbf{f}(\mathbf{x}, \mathbf{u}) = \mathbf{0} \tag{2.1-52}$$

simply identifies a static relationship between the n state components and the m control components.

Example 2.1-5 Descent into the Valley

Suppose that the bowl-shaped valley in the previous example had a contour (within its rim) defined by

$$J = u_1^2 - 2u_1u_2 + 3u_2^2 - 40 \text{ (meters)}$$

Where is the bottom of the valley, and what is its depth?

The minimum is defined by

$$\frac{\partial J}{\partial u_1} = 0 = 2u_1 - 2u_2$$

$$\frac{\partial J}{\partial u_2} = 0 = -2u_1 + 6u_2$$

Simultaneous solution of these equations yields $u_1^* = u_2^* = 0$ meters; the greatest depth, J^*, is -40 m.

If there is a straight path through the valley defined by the equation

$$u_2 = u_1 + 2$$

what is the lowest point on the path?

The path equation can be expressed in the form of Eq. 2.1-52 as

$$u_2 - u_1 - 2 = 0$$

Substituting for u_2, J describes the vertical contour of the path as a function of the single coordinate, u_1:

$$J = u_1^2 - 2u_1(u_1+2) + 3(u_1+2)^2 - 40$$
$$= 2u_1^2 + 8u_1 - 28$$

Then the minimum is specified by a single equation:

$$\frac{\partial J}{\partial u_1} = 4u_1 + 8 = 0$$

The values of u and J at the constrained minimum are

$$u_1^* = -\ 2\text{ m}$$
$$u_2^* = \ \ 0\text{ m}$$
$$J^* = -36\text{ m}$$

Direct substitution of the equality constraint to reduce the order of the minimization is rarely as easy as the example might imply. An alternative approach is to form an *augmented cost function*, J_A,

$$J_A = J + \boldsymbol{\lambda}^T \mathbf{f}(\mathbf{x}, \mathbf{u}) \tag{2.1-53}$$

which is to be minimized with respect to $\mathbf{u}$. The equality constraint is *adjoined* (or added) to the original cost function using the vector of *Lagrange multipliers*,* $\boldsymbol{\lambda}$, which, of necessity, has dimension n. Hence the product $\boldsymbol{\lambda}^T\mathbf{f}(\mathbf{x}, \mathbf{u})$ is a scalar, and, from Eq. 2.1-52, it always equals zero for values of $\mathbf{x}$ and $\mathbf{u}$ that satisfy the constraint and for finite values of $\boldsymbol{\lambda}$. J is minimized when J_A is minimized, but the n components of $\boldsymbol{\lambda}$ must be found to effect the latter minimization. A total of $(m+2n)$ equations must be solved: n equations to define the intermediate variable $\boldsymbol{\lambda}$, n equations to find the optimum values of the state, $\mathbf{x}$, and m equations to find the optimizing control $\mathbf{u}$.

In the vicinity of the optimum, J_A^*, perturbations in $\mathbf{x}$ and $\mathbf{u}$ lead to perturbations in the cost. To first order,

$$\Delta J_A = \left.\frac{\partial J_A}{\partial \mathbf{x}}\right|_{\mathbf{x},\,\mathbf{u}=\mathbf{x}^*,\,\mathbf{u}^*} \Delta\mathbf{x} + \left.\frac{\partial J_A}{\partial \mathbf{u}}\right|_{\mathbf{x},\,\mathbf{u}=\mathbf{x}^*,\,\mathbf{u}^*} \Delta\mathbf{u} \tag{2.1-54}$$

*Also called the adjoint vector or costate vector. Comte Joseph Lagrange (1736–1813) derived methods for solving variational problems in classical mechanics.

From Eq. 2.1-53,

$$\frac{\partial J_A}{\partial \mathbf{x}} = \frac{\partial J}{\partial \mathbf{x}} + \boldsymbol{\lambda}^T \frac{\partial \mathbf{f}}{\partial \mathbf{x}} \tag{2.1-55}$$

$$\frac{\partial J_A}{\partial \mathbf{u}} = \frac{\partial J}{\partial \mathbf{u}} + \boldsymbol{\lambda}^T \frac{\partial \mathbf{f}}{\partial \mathbf{u}} \tag{2.1-56}$$

The partial derivative of the n-vector $\mathbf{f}$ with respect to the m-vector $\mathbf{u}$ is the $(n \times m)$ *Jacobian matrix*,* whose scalar elements are $\partial f_i/\partial u_j$, $i = 1$ to n, $j = 1$ to m. Similarly, $\partial \mathbf{f}/\partial \mathbf{x}$ is an $(n \times n)$ Jacobian matrix with elements $\partial f_i/\partial x_j$, $i = 1$ to n, $j = 1$ to n. $\partial J_A/\partial \mathbf{x}$ is identically zero if $\boldsymbol{\lambda}$ is chosen as

$$\begin{aligned} \boldsymbol{\lambda}^* &= -\left[\left(\frac{\partial J}{\partial \mathbf{x}}\right)\left(\frac{\partial \mathbf{f}}{\partial \mathbf{x}}\right)^{-1}\right]^T \\ &= -\left[\left(\frac{\partial \mathbf{f}}{\partial \mathbf{x}}\right)^{-1}\right]^T \left(\frac{\partial J}{\partial \mathbf{x}}\right)^T \end{aligned} \tag{2.1-57}$$

The partial derivatives are evaluated at $\mathbf{x} = \mathbf{x}^*$ and $\mathbf{u} = \mathbf{u}^*$, providing the n equations to find $\boldsymbol{\lambda}$. Then

$$\Delta J_A = \left(\frac{\partial J}{\partial \mathbf{u}} + \boldsymbol{\lambda}^T \frac{\partial \mathbf{f}}{\partial \mathbf{u}}\right) \Delta \mathbf{u} = 0 \tag{2.1-58}$$

at the optimum. To remain zero for arbitrary variations in the control,

$$\frac{\partial J}{\partial \mathbf{u}} + \boldsymbol{\lambda}^{*T} \frac{\partial \mathbf{f}}{\partial \mathbf{u}} = \mathbf{0} \tag{2.1-59a}$$

or

$$\frac{\partial J}{\partial \mathbf{u}} - \frac{\partial J}{\partial \mathbf{x}} \left(\frac{\partial \mathbf{f}}{\partial \mathbf{x}}\right)^{-1} \left(\frac{\partial \mathbf{f}}{\partial \mathbf{u}}\right) = \mathbf{0} \tag{2.1-59b}$$

Equation 2.1-59 provides m equations for the control. The remaining n state equations are provided by the equality constraint, Eq. 2.1-52. In practice, the state solution may come from the control equations, and vice versa, as the next example shows. If J and $\mathbf{f}$ are particularly complicated functions of $\mathbf{x}$ and $\mathbf{u}$, it may be necessary to solve Eq. 2.1-59 iteratively, as by the Newton–Raphson algorithm described earlier.

*Carl G. J. Jacobi (1804–1851), a contemporary of Bessel, Gauss, Fourier, and Poisson, linked many mathematical disciplines in his work.

Example 2.1-6 Descent into the Valley, Part 2

Defining $u = u_1$ and $x = u_2$, the valley contour of Example 2.1-5 is

$$J = u^2 - 2xu + 3x^2 - 40$$

and the equality constraint for the path is

$$x - u - 2 = 0$$

Then

$$\frac{\partial J}{\partial x} = -2u + 6x, \qquad \frac{\partial f}{\partial x} = 1$$

$$\frac{\partial J}{\partial u} = 2u - 2x, \qquad \frac{\partial f}{\partial u} = -1$$

From Eq. 2.1-57,

$$\lambda^* = 2u - 6x$$

From Eq. 2.1-59b,

$$(2u - 2x) + (2u - 6x)(-1) = 0$$

or

$$x^* = 0$$

From Eq. 2.1-52,

$$u^* = -2$$

The greatest depth on the path, $J = -36$ meters, is found by substitution in the original equation.

As in the unconstrained case, a positive definite Hessian matrix representing the second-order effects of control variation on the cost is sufficient (along with Eq. 2.1-59) to define a minimum. The second variation of the augmented cost can be written as

$$\Delta^2 J_A = \frac{1}{2}[\Delta\mathbf{x}^T \Delta\mathbf{u}^T]\begin{bmatrix} J_{A_{\mathbf{xx}}} & J_{A_{\mathbf{xu}}} \\ J_{A_{\mathbf{ux}}} & J_{A_{\mathbf{uu}}} \end{bmatrix}\begin{bmatrix} \Delta\mathbf{x} \\ \Delta\mathbf{u} \end{bmatrix} \geq 0 \tag{2.1-60}$$

at a minimum. For sufficiency with $\Delta^2 J_A = 0$, a higher-order even variation must be positive definite. Because $\mathbf{f}(\mathbf{x}, \mathbf{u}) = \mathbf{0}$,

$$\Delta\mathbf{f}(\mathbf{x}, \mathbf{u}) = \mathbf{0} = \mathbf{f}_\mathbf{x}\Delta\mathbf{x} + \mathbf{f}_\mathbf{u}\Delta_\mathbf{u} + \text{higher-order terms} \tag{2.1-61}$$

For small perturbations,

$$\Delta\mathbf{x} = -\mathbf{f}_\mathbf{x}^{-1}\mathbf{f}_\mathbf{u}\Delta\mathbf{u} \tag{2.1-62}$$

so Eq. 2.1-60 becomes

$$\begin{aligned}
\Delta^2 J_A &= \frac{1}{2}\Delta\mathbf{u}^T[-(\mathbf{f}_\mathbf{x}^{-1}\mathbf{f}_\mathbf{u})^T \quad \mathbf{I}_m]\begin{bmatrix} J_{A_{\mathbf{xx}}} & J_{A_{\mathbf{xu}}} \\ J_{A_{\mathbf{ux}}} & J_{A_{\mathbf{uu}}} \end{bmatrix}\begin{bmatrix} -\mathbf{f}_\mathbf{x}^{-1}\mathbf{f}_\mathbf{u} \\ \mathbf{I}_m \end{bmatrix}\Delta\mathbf{u} \\
&= \frac{1}{2}\Delta\mathbf{u}^T[(\mathbf{f}_\mathbf{x}^{-1}\mathbf{f}_\mathbf{u})^T J_{A_{\mathbf{xx}}}(\mathbf{f}_\mathbf{x}^{-1}\mathbf{f}_\mathbf{u}) - (\mathbf{f}_\mathbf{x}^{-1}\mathbf{f}_\mathbf{u})^T J_{A_{\mathbf{xu}}} \\
&\quad - J_{A_{\mathbf{ux}}}(\mathbf{f}_\mathbf{x}^{-1}\mathbf{f}_\mathbf{u}) + J_{A_{\mathbf{uu}}}]\Delta\mathbf{u} \\
&= \frac{1}{2}\Delta\mathbf{u}^T J'_{A_{\mathbf{uu}}}\Delta\mathbf{u} \quad (>0 \text{ for } \Delta\mathbf{u} \neq \mathbf{0})
\end{aligned} \tag{2.1-63}$$

The distinction between $J'_{A_{\mathbf{uu}}}$ and $J_{A_{\mathbf{uu}}}$ is that the former reflects the $\mathbf{x}, \mathbf{u}$ coupling that is required to maintain $\mathbf{f}(\mathbf{x}, \mathbf{u}) = \mathbf{0}$, while the latter does not. Equations (2.1-52), (2.1-57), (2.1-59), and (2.1-63) provide *necessary and sufficient conditions* for the existence of the constrained, local, static minimum in J.

2.2 MATRIX PROPERTIES AND OPERATIONS

Inverse Vector Relationship

In Section 2.1, matrices were introduced by noting that if one vector variable $\mathbf{y}$ is a linear function of another $\mathbf{x}$, then the two are related by a matrix of coefficients $\mathbf{A}$:

$$\mathbf{y} = \mathbf{A}\mathbf{x} \tag{2.2-1}$$

It was assumed that $\mathbf{A}$ and $\mathbf{x}$ were known and that $\mathbf{y}$ was to be found. Suppose, however, that $\mathbf{A}$ and $\mathbf{y}$ are known, and we want to compute $\mathbf{x}$. A relationship of the form

$$\mathbf{x} = \mathbf{A}^{-1}\mathbf{y} \tag{2.2-2}$$

is required, where $\mathbf{A}^{-1}$ is the *inverse* of $\mathbf{A}$. If $\mathbf{x}$ and $\mathbf{y}$ were scalars, the inverse operation would simply represent division. Given

$$y = ax \tag{2.2-3}$$

the inverse relationship is

$$x = \frac{1}{a} y = a^{-1} y \tag{2.2-4}$$

Loosely speaking, inversion is the matrix equivalent of scalar division, although the process of computing the inverse usually is considerably more complex than simple division.

The diagonal matrix (Eq. 2.1-12) presents an exception. Because components of $\mathbf{y}$ are individually related to components of $\mathbf{x}$ by Eq. 2.2-1, that is,

$$y_i = a_{ii} x_i, \qquad i = 1 \text{ to } n \tag{2.2-5}$$

the inverse is

$$x_i = (a_{ii})^{-1} y_i, \qquad i = 1 \text{ to } n \tag{2.2-6}$$

leading to

$$\mathbf{A}^{-1} = \begin{bmatrix} 1/a_{11} & 0 & \cdots & 0 \\ 0 & 1/a_{22} & \cdots & 0 \\ \vdots & \vdots & & \vdots \\ 0 & 0 & \cdots & 1/a_{nn} \end{bmatrix} \tag{2.2-7}$$

For the nondiagonal two-component case,

$$\begin{bmatrix} y_1 \\ y_2 \end{bmatrix} = \begin{bmatrix} a_{11} & a_{12} \\ a_{21} & a_{22} \end{bmatrix} \begin{bmatrix} x_1 \\ x_2 \end{bmatrix} \tag{2.2-8}$$

a process of elimination would yield

$$\begin{bmatrix} x_1 \\ \\ x_2 \end{bmatrix} = \begin{bmatrix} \dfrac{1}{(a_{11} - a_{12}a_{21}/a_{22})} & \dfrac{1}{(a_{21} - a_{11}a_{22}/a_{12})} \\ \\ \dfrac{1}{(a_{12} - a_{11}a_{22}/a_{21})} & \dfrac{1}{(a_{22} - a_{12}a_{21}/a_{11})} \end{bmatrix} \begin{bmatrix} y_1 \\ \\ y_2 \end{bmatrix} \tag{2.2-9}$$

which also could be written

$$\begin{bmatrix} x_1 \\ x_2 \end{bmatrix} = \frac{1}{(a_{11}a_{22} - a_{12}a_{21})} \begin{bmatrix} a_{22} & -a_{12} \\ -a_{21} & a_{11} \end{bmatrix} \begin{bmatrix} y_1 \\ y_2 \end{bmatrix} \tag{2.2-10}$$

This equation indicates that $\mathbf{A}^{-1}$ can be written as a matrix function of **A** divided by a scalar function of **A**; more specifically,

$$\mathbf{A}^{-1} = \frac{\text{Adj}(\mathbf{A})}{|\mathbf{A}|} \tag{2.2-11}$$

where Adj(**A**) is the *adjoint matrix* corresponding to **A** and $|\mathbf{A}|$ is the matrix's *determinant*. For the inverse to exist, the matrix **A** must be square; therefore, **y** and **x** must have the same dimension for Eq. 2.2-2 to be valid.

Matrix Determinant

Comparing Eqs. 2.2-8 and 2.2-10, it can be seen that the determinant of a (2×2) matrix is the product of elements on the principal diagonal (upper left to lower right) less the product of elements on the secondary diagonal (lower left to upper right). A similar diagonal lattice of multiplications is used to evaluate the determinant of a (3×3) matrix:

$$\begin{vmatrix} a_{11} & a_{12} & a_{13} \\ a_{21} & a_{22} & a_{23} \\ a_{31} & a_{32} & a_{33} \end{vmatrix} = a_{11}a_{22}a_{33} + a_{12}a_{23}a_{31} + a_{13}a_{21}a_{32} - a_{11}a_{23}a_{32} - a_{12}a_{21}a_{33} - a_{13}a_{22}a_{31} \tag{2.2-12}$$

This result is a scalar quantity.

The determinants of larger matrices can be expressed in terms of lower-dimensional determinants. Sequences of mathematical operations, or *algorithms*, for computing the determinant sum the products of permutations of the elements, as defined below. Two algorithms are of particular interest: Laplace expansion and pivotal condensation. In the first approach, the $(n \times n)$ determinant is expressed as a sum of the determinants of $(n-1) \times (n-1)$ submatrices, which are, in turn, expressed as sums of determinants of $(n-2) \times (n-2)$ submatrices, and so on, until only (3×3) or (2×2) matrices are left. In the second approach, the $(n \times n)$ matrix is replaced by an $(n-1) \times (n-1)$ matrix, each of whose elements is a determinant of a (2×2) submatrix. The algorithm continues until only a single (2×2) matrix remains.

Laplace expansion hinges on the definition of matrix cofactors, which themselves are based on matrix minors. Suppose a_{ij} is an element of the $(n \times n)$ matrix $\mathbf{A}$. Form the $(n-1) \times (n-1)$ matrix $\mathbf{B}(i, j)$ by removing the ith row and j^{th} column of $\mathbf{A}$. The determinant of $\mathbf{B}(i, j)$ is the ij^{th} *minor* of $\mathbf{A}$. Multiply $|\mathbf{B}(i, j)|$ by $(-1)^{i+j}$ to form the corresponding "signed minor"; this also is called the ij^{th} *cofactor* of $\mathbf{A}$, or C_{ij}. Then C_{ij} is the $(n-1) \times (n-1)$ determinant,

$$C_{ij} = (-1)^{i+j} |\mathbf{B}(i, j)| \tag{2.2-13}$$

The determinant of $\mathbf{A}$ is expressed in terms of the original elements and their cofactors, expanding along either a row or a column of the original matrix:

$$|\mathbf{A}| = \sum_{j=1}^{n} a_{kj} C_{kj}, \qquad k = \text{any row of } \mathbf{A} \tag{2.2-14}$$

or

$$|\mathbf{A}| = \sum_{i=1}^{n} a_{il} C_{il}, \qquad l = \text{any column of } \mathbf{A} \tag{2.2-15}$$

Of course, C_{kj} or C_{il} must be found before either equation can be applied, so the same process must be repeated until the cofactors of (2×2) or (3×3) submatrices are defined.

Pivotal condensation generally requires fewer multiplications and is more easily implemented. A nonzero element, a_{ij}, of $\mathbf{A}$ is chosen as a "pivotal element" [e.g., see (B-6)]. The (2×2) determinants of a_{ij} and all other elements in the ith row and j^{th} column are computed, yielding a matrix with $(n-1) \times (n-1)$ elements. The matrix is multiplied by the scalar, $(1/a_{ij})^{n-2}$ to form a second $(n-1) \times (n-1)$ matrix. A new pivotal element is found, and the process is repeated until the determinant is condensed to (2×2) or (3×3). Using Eq. 2.2-12 as an illustration and assuming that a_{11} is not zero,

$$|\mathbf{A}| = \frac{1}{a_{11}} \begin{vmatrix} \begin{vmatrix} a_{11} a_{12} \\ a_{21} a_{22} \end{vmatrix} & \begin{vmatrix} a_{11} a_{13} \\ a_{21} a_{23} \end{vmatrix} \\ \begin{vmatrix} a_{11} a_{12} \\ a_{31} a_{32} \end{vmatrix} & \begin{vmatrix} a_{11} a_{13} \\ a_{31} a_{33} \end{vmatrix} \end{vmatrix}$$

$$= \frac{1}{a_{11}} \begin{vmatrix} (a_{11} a_{22} - a_{12} a_{21}) & (a_{11} a_{23} - a_{13} a_{21}) \\ (a_{11} a_{32} - a_{12} a_{31}) & (a_{11} a_{33} - a_{13} a_{31}) \end{vmatrix} \tag{2.2-16}$$

which is equivalent to Eq. 2.2-12.

A matrix is said to be *singular* if its determinant is zero. From the Laplace expansion algorithm, Eq. 2.2-14 or 2.2-15, it is clear that a zero row or column guarantees singularity. If two (or more) rows or columns are equal or linearly dependent (i.e., if one is a multiple of the other), the matrix is singular. Singularity precludes the conventional computation of the inverse (Eq. 2.2-11), which requires division by the determinant.

Before proceeding to the adjoint matrix, an additional application of the determinant is treated: the identification of the "definiteness" of the symmetric defining matrix $\mathbf{Q}$ in the quadratic form $\mathbf{x}^T\mathbf{Q}\mathbf{x}$. The *principal minors* of $\mathbf{Q}$ are the determinants of all the square matrices of $\mathbf{Q}$ (with dimensions of 1 to n) whose diagonal elements lie on the diagonal of $\mathbf{Q}$. If all the principal minors are positive, $\mathbf{Q}$ is positive definite; if all are either positive or zero, $\mathbf{Q}$ is positive semidefinite. If all the *leading* principal minors (i.e., those containing the element q_{11}) are positive, $\mathbf{Q}$ is positive definite. This simplified procedure is not adequate for determining semidefiniteness.

The matrix $\mathbf{Q}$ is negative definite if the leading principal minors alternate sign as the minor dimension increases. Denoting a leading principal minor of dimension n as Δ_n, the sequence is $\Delta_1 < 0$, $\Delta_2 > 0$, $\Delta_3 < 0$, and so on.

Adjoint Matrix

To complete the definition of the matrix inverse, the adjoint matrix must be found, and it is the transpose of the matrix of cofactors of the original matrix:

$$\text{Adj}(\mathbf{A}) = \mathbf{C}^T \tag{2.2-17}$$

Consequently, it is an $(n \times n)$ matrix of $(n-1)\times(n-1)$ determinants. The (3×3) matrix of Eq. 2.2-12 provides a satisfactory example:

$$\text{Adj}(\mathbf{A}) = \begin{bmatrix} C_{11} C_{21} C_{31} \\ C_{12} C_{22} C_{32} \\ C_{13} C_{23} C_{33} \end{bmatrix} = \begin{bmatrix} |\mathbf{B}(1,1)| & -|\mathbf{B}(2,1)| & |\mathbf{B}(3,1)| \\ -|\mathbf{B}(1,2)| & |\mathbf{B}(2,2)| & -|\mathbf{B}(3,2)| \\ |\mathbf{B}(1,3)| & -|\mathbf{B}(2,3)| & |\mathbf{B}(3,3)| \end{bmatrix}$$

$$= \begin{bmatrix} \begin{vmatrix} a_{22} a_{23} \\ a_{32} a_{33} \end{vmatrix} & -\begin{vmatrix} a_{12} a_{13} \\ a_{32} a_{33} \end{vmatrix} & \begin{vmatrix} a_{12} a_{13} \\ a_{22} a_{23} \end{vmatrix} \\ -\begin{vmatrix} a_{21} a_{23} \\ a_{31} a_{33} \end{vmatrix} & \begin{vmatrix} a_{11} a_{13} \\ a_{31} a_{33} \end{vmatrix} & -\begin{vmatrix} a_{11} a_{13} \\ a_{21} a_{23} \end{vmatrix} \\ \begin{vmatrix} a_{21} a_{22} \\ a_{31} a_{32} \end{vmatrix} & -\begin{vmatrix} a_{11} a_{12} \\ a_{31} a_{32} \end{vmatrix} & \begin{vmatrix} a_{11} a_{12} \\ a_{21} a_{22} \end{vmatrix} \end{bmatrix} \tag{2.2-18}$$

Matrix Inverse

In order for the inverse of $\mathbf{A}$ (Eq. 2.2-11) to exist, the matrix must be square and its determinant must not be zero (i.e., it has to be *nonsingular*).

In such a case, the product of a matrix and its inverse form the identity matrix

$$\mathbf{A}\mathbf{A}^{-1} = \mathbf{A}^{-1}\mathbf{A} = \mathbf{I} \tag{2.2-19}$$

since, from Eqs. 2.1-7 and 2.2-1, the desired relationship is

$$\mathbf{y} = \mathbf{A}\mathbf{x} = \mathbf{A}(\mathbf{A}^{-1}\mathbf{y}) = \mathbf{y} \tag{2.2-20}$$

or

$$\mathbf{x} = \mathbf{A}^{-1}\mathbf{y} = \mathbf{A}^{-1}(\mathbf{A}\mathbf{x}) = \mathbf{x} \tag{2.2-21}$$

Equation 2.2-19 can be verified readily for the the two- and three-dimensional cases. It should be apparent that if $\mathbf{A}$ is symmetric, $\mathbf{A}^{-1}$ also is symmetric.

EXAMPLE 2.2-1 WHAT WAS THAT PRICE AGAIN?

The first three shoppers in Example 2.1-1 remember their total bills and how many apples, oranges, and pears they each bought, but they have forgotten the individual prices. Can they reconstruct the prices from these data? Of course! Equation 2.2-2 yields

$$\begin{bmatrix} x_1 \\ x_2 \\ x_3 \end{bmatrix} = \begin{bmatrix} 3 & 2 & 4 \\ 0 & 3 & 1 \\ 2 & 4 & 2 \end{bmatrix}^{-1} \begin{bmatrix} \$2.30 \\ \$.85 \\ \$1.90 \end{bmatrix}$$

$$= \frac{1}{(-14)} \begin{bmatrix} 2 & 12 & -10 \\ 2 & -2 & -3 \\ -6 & -8 & 9 \end{bmatrix} \begin{bmatrix} \$2.30 \\ \$.85 \\ \$1.90 \end{bmatrix} = \begin{bmatrix} \$.30 \\ \$.20 \\ \$.25 \end{bmatrix}$$

Suppose not the first three but the last three try the same trick. Can they reconstruct the prices? Of course? Of course not! The third and fourth shoppers bought the same items, so their data are redundant, and there is insufficient information to deduce the original prices. Mathematically, this leads to a *singular* quantity matrix (i.e., the determinant is zero):

$$\begin{vmatrix} 0 & 3 & 1 \\ 2 & 4 & 2 \\ 2 & 4 & 2 \end{vmatrix} = 0$$

Therefore, the matrix inverse does not exist, and the original prices are indeterminate from the data.

The *inverse of the product of matrices* often is required, and it can be obtained from a product of the inverses. As in the case of the transpose, the order of multiplication is reversed. This is easily seen by considering three vectors of the same dimension, $\mathbf{x}$, $\mathbf{y}$, $\mathbf{z}$, related as follows:

$$\mathbf{y} = \mathbf{A}\mathbf{x} \tag{2.2-22}$$

$$\mathbf{z} = \mathbf{B}\mathbf{y} = \mathbf{B}\mathbf{A}\mathbf{x} \tag{2.2-23}$$

The inverse relationships are

$$\mathbf{y} = \mathbf{B}^{-1}\mathbf{z} \tag{2.2-24}$$

$$\mathbf{x} = \mathbf{A}^{-1}\mathbf{y} = \mathbf{A}^{-1}\mathbf{B}^{-1}\mathbf{z} = (\mathbf{B}\mathbf{A})^{-1}\mathbf{z} \tag{2.2-25}$$

It was mentioned previously that computing the inverse of a diagonal matrix is particularly easy, and there are three commonly occurring additional cases in which inversion is simplified. A diagonal matrix has nonzero elements along the principal diagonal and zeros elsewhere; an $(n \times n)$ *block-diagonal matrix* can be partitioned into square blocks (or submatrices) of arbitrary dimension (less than n) arrayed along the diagonal with zeros in the remaining elements:

$$\mathbf{A} = \begin{bmatrix} \mathbf{A}_1 & \mathbf{0} & \dots & \mathbf{0} \\ \mathbf{0} & \mathbf{A}_2 & \dots & \mathbf{0} \\ \mathbf{0} & \mathbf{0} & \dots & \mathbf{0} \\ \mathbf{0} & \mathbf{0} & \dots & \mathbf{A}_m \end{bmatrix} \tag{2.2-26}$$

The m submatrices are distinct from each other, so the inverse of $\mathbf{A}$ can be constructed from the inverses of the smaller matrices:

$$\mathbf{A}^{-1} = \begin{bmatrix} \mathbf{A}_1^{-1} & \mathbf{0} & \dots & \mathbf{0} \\ \mathbf{0} & \mathbf{A}_2^{-1} & \dots & \mathbf{0} \\ \mathbf{0} & \mathbf{0} & \dots & \mathbf{0} \\ \mathbf{0} & \mathbf{0} & \dots & \mathbf{A}_m^{-1} \end{bmatrix} \tag{2.2-27}$$

$\mathbf{A}$ is an *upper(lower)-block-triangular matrix* if the blocks below (above) the diagonal blocks are zero but those above (below) are not. In this case, Adj($\mathbf{A}$) must be computed as before (Eq. 2.2-17), but

$$|\mathbf{A}| = |\mathbf{A}_1||\mathbf{A}_2| \dots |\mathbf{A}_m| \tag{2.2-28}$$

affording simplification in the computation. If $\mathbf{A}$ represents an *orthonormal transformation*, such as a rotation from one Cartesian coordinate system to another, then,

$$\mathbf{A}^{-1} = \mathbf{A}^T \tag{2.2-29}$$

as demonstrated by the next example.

Example 2.2-2 Linear Velocity of an Aircraft in Two Coordinate Systems

The orientation of an aircraft is expressed by three *Euler angles* representing yaw (ψ), pitch (θ), and roll (ϕ). Its velocity relative to the earth's surface $\mathbf{v}$, can be resolved into orthogonal reference axes that are fixed either to the Earth's surface, $\mathbf{v}_E$, or to the aircraft's body axes, $\mathbf{v}_B$;

$$\mathbf{v}_E = \begin{bmatrix} v_x \\ v_y \\ v_z \end{bmatrix}_E ; \qquad \mathbf{v}_B = \begin{bmatrix} v_x \\ v_y \\ v_z \end{bmatrix}_B$$

The two representations of the velocity vector are related by the equation

$$\mathbf{v}_B = \mathbf{H}_{BE}(\phi, \theta, \psi)\mathbf{v}_E$$

where $\mathbf{H}_{BE}$ represents the rotation *from* Earth-relative axes *to* body axes in three successive rotations (starting from the right),

$$\mathbf{H}_{BE} = \mathbf{H}_3(\phi)\mathbf{H}_2(\theta)\mathbf{H}_1(\psi)$$

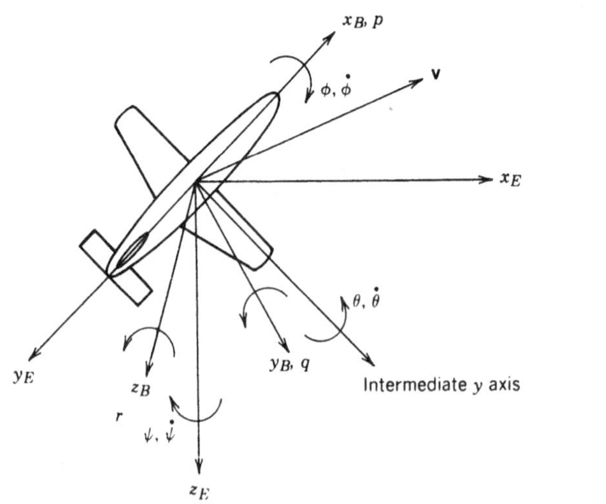

where

$$\mathbf{H}_1(\psi) = \begin{bmatrix} \cos\psi & \sin\psi & 0 \\ -\sin\psi & \cos\psi & 0 \\ 0 & 0 & 1 \end{bmatrix} \qquad \text{(rotation about Earth-relative } z \text{ axis)}$$

$$\mathbf{H}_2(\theta) = \begin{bmatrix} \cos\theta & 0 & -\sin\theta \\ 0 & 1 & 0 \\ \sin\theta & 0 & \cos\theta \end{bmatrix} \qquad \text{(rotation about intermediate } y \text{ axis)}$$

$$\mathbf{H}_3(\phi) = \begin{bmatrix} 1 & 0 & 0 \\ 0 & \cos\phi & \sin\phi \\ 0 & -\sin\phi & \cos\phi \end{bmatrix} \qquad \text{(rotation about final } x \text{ axis)}$$

Given $\mathbf{v}_B$ and $\mathbf{H}_{BE}$, what is $\mathbf{v}_E$?

$$\begin{aligned} \mathbf{v}_E &= (\mathbf{H}_{BE})^{-1}\mathbf{v}_B = \mathbf{H}_{EB}\mathbf{v}_B \\ &= \mathbf{H}_1^{-1}\mathbf{H}_2^{-1}\mathbf{H}_3^{-1}\mathbf{v}_B \end{aligned}$$

A few minutes with pencil and paper will show that

$$\mathbf{H}_1^T\mathbf{H}_1 = \mathbf{H}_2^T\mathbf{H}_2 = \mathbf{H}_3^T\mathbf{H}_3 = \mathbf{I}$$

hence, from Eq. 2.2-19,

$$\mathbf{H}_1^T = \mathbf{H}_1^{-1}, \qquad \mathbf{H}_2^T = \mathbf{H}_2^{-1}, \qquad \mathbf{H}_3^T = \mathbf{H}_3^{-1}$$

Each rotation produces an orthonormal transformation, so the net transformation is orthonormal as well:

$$\mathbf{H}_{EB} = (\mathbf{H}_{BE})^{-1} = (\mathbf{H}_{BE})^T$$

since

$$\mathbf{H}_1^{-1}\mathbf{H}_2^{-1}\mathbf{H}_3^{-1} = \mathbf{H}_1^T\mathbf{H}_2^T\mathbf{H}_3^T$$

As an exercise, derive the nonorthonormal transformation, $\mathbf{L}_{BE}$, from the Euler angle rate vector $(\dot{\phi} \quad \dot{\theta} \quad \dot{\psi})^T$ to the body-axis angular velocity $(p \quad q \quad r)^T$ and its inverse, $\mathbf{L}_{EB}$. Is $\mathbf{L}_{BE}$ nonsingular under all conditions?

Generalized Inverses

In discussing the matrix inverse, it has been required that the dimensions of $\mathbf{y}$ and $\mathbf{x}$ (r and n) in Eq. 2.2-1 be equal; therefore, both $\mathbf{A}$ and $\mathbf{A}^{-1}$ are square matrices. How can an inverse relationship between $\mathbf{y}$ and $\mathbf{x}$ be defined if r and n are not equal? The answer is found in the *generalized inverse* or *pseudoinverse matrix*, denoted by $\mathbf{A}^{PI}$. The pseudoinverse matrix takes one of two forms, depending upon the relative dimensions of $\mathbf{y}$ and $\mathbf{x}$. If $r>n$, Eq.. 2.2-1 represents more scalar equations (for the components of $\mathbf{y}$) than there are unknowns (components of $\mathbf{x}$), so the inverse solution may be *overdetermined*. If $r<n$, the opposite is true, and the solution for $\mathbf{x}$ is *underdetermined*.

Before reviewing the two pseudoinverse expressions, it is useful to introduce the concept of matrix rank. The *rank* of a matrix is defined as the dimension of the largest nonzero determinant of any submatrix of the matrix. This is equivalent to the maximum number of linearly independent rows *or* columns.* The rank of a nonsingular $(n \times n)$ matrix is n because the determinant of the full matrix is not equal to zero. If the rank of an $(n\times n)$ matrix is less than n, the matrix is singular. The rank of an $(r\times n)$ matrix can be no greater than the smaller dimension, which limits the size of the largest square submatrix. When two matrices are multiplied, the rank of the product can be no greater than the smaller rank of the original matrices.

In the following paragraphs, we shall see that products of the form $\mathbf{A}^T\mathbf{A}$ and $\mathbf{A}\mathbf{A}^T$ are involved in pseudoinverse definitions. From Eq. 2.2-1, $\mathbf{A}$ is an $(r\times n)$ matrix; hence, $\mathbf{A}^T\mathbf{A}$ has dimension $(n\times n)$, while $\mathbf{A}\mathbf{A}^T$ has dimension $(r\times r)$. Nevertheless, both products (of the same matrix, $\mathbf{A}$) have rank no greater than *r or n, whichever is less*. If $r>n$, $\mathbf{A}^T\mathbf{A}$ may be invertible, but $\mathbf{A}\mathbf{A}^T$ definitely is singular. Conversely, if $r<n$, $\mathbf{A}\mathbf{A}^T$ may be nonsingular, but the determinant of $\mathbf{A}^T\mathbf{A}$ is zero.

The *left pseudoinverse* is appropriate for the solution of the overdetermined case. Multiplying both sides of Eq. 2.2-1 by $\mathbf{A}^T$,

$$\mathbf{A}^T\mathbf{A}\mathbf{x} = \mathbf{A}^T\mathbf{y} \tag{2.2-30}$$

Premultiplying both sides by $(\mathbf{A}^T\mathbf{A})^{-1}$,

$$\begin{aligned}\mathbf{x} &= (\mathbf{A}^T\mathbf{A})^{-1}\mathbf{A}^T\mathbf{y}\\ &= \mathbf{A}^{PI}\mathbf{y} = \mathbf{A}^{L}\mathbf{y}\end{aligned} \tag{2.2-31}$$

Development of the *right pseudoinverse* for the underdetermined case begins by noting that

$$(\mathbf{A}\mathbf{A}^T)(\mathbf{A}\mathbf{A}^T)^{-1} = \mathbf{I} \tag{2.2-32}$$

*A row (or column) is linearly dependent on other rows (or columns) if it can be derived as a linear combination of the others. For example, if the vector $\mathbf{y}=a\mathbf{x}_a+b\mathbf{x}_b+c\mathbf{x}_c$, it is linearly dependent on $\mathbf{x}_a$, $\mathbf{x}_b$, and $\mathbf{x}_c$. A zero row or column is linearly dependent on any other row or column of equal dimension, as it can be derived through multiplication by zero.

Then Eq. 2.2-1 could be written as

$$\mathbf{Ax} = \mathbf{AA}^T(\mathbf{AA}^T)^{-1}\mathbf{y} \tag{2.2-33}$$

or

$$\begin{aligned}\mathbf{x} &= \mathbf{A}^T(\mathbf{AA}^T)^{-1}\mathbf{y} \\ &= \mathbf{A}^{PI}\mathbf{y} = \mathbf{A}^R\mathbf{y}\end{aligned} \tag{2.2-34}$$

Equations 2.2-31 and 2.2-34 both reduce to the conventional inverse equation when $r=n$.

The overdetermined case contains redundant information; if $\mathbf{y}$ is a vector of consistent, error-free measurements, Eq. 2.2-31 provides the exact value of $\mathbf{x}$ with more computations than necessary. If the measurements are not consistent or contain errors, Eq. 2.2-31 performs an *averaging process* that minimizes the *mean-square error* in the *estimate* of $\mathbf{x}$, as detailed in later sections. The underdetermined case does not contain enough information to provide a unique solution for $\mathbf{x}$, but it does provide a feasible solution that minimizes the quadratic norm of $\mathbf{x}$. These two cases are illustrated in the following examples.

Example 2.2-3 A Resistor Network

Two voltage sources are connected to three ground points through six resistors, as shown in the diagram. The currents at the ground points are measured, and the values of the resistances are known. What are the voltages?

From Ohm's law,

$$\begin{aligned} i_1 &= v_1/r_{11} + v_2/r_{12} \\ i_2 &= v_1/r_{21} + v_2/r_{22} \\ i_3 &= v_1/r_{31} + v_2/r_{32} \end{aligned}$$

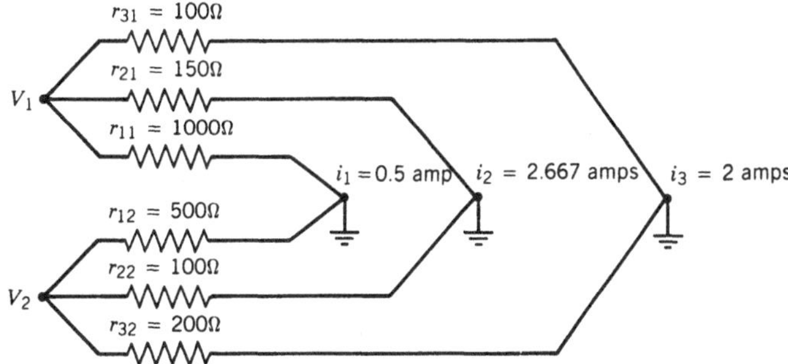

or

$$\mathbf{y} = \mathbf{A}\mathbf{x}$$

where

$$\mathbf{y} = \begin{bmatrix} 0.5 \\ 2.667 \\ 2. \end{bmatrix}$$

$$\mathbf{x} = \begin{bmatrix} v_1 \\ v_2 \end{bmatrix}$$

$$\mathbf{A} = \begin{bmatrix} \frac{1}{1000} & \frac{1}{500} \\ \frac{1}{150} & \frac{1}{100} \\ \frac{1}{100} & \frac{1}{200} \end{bmatrix}$$

The left pseudoinverse matrix is

$$\mathbf{A}^L = \left\{ \begin{bmatrix} \frac{1}{1000} & \frac{1}{150} & \frac{1}{100} \\ \frac{1}{500} & \frac{1}{100} & \frac{1}{200} \end{bmatrix} \begin{bmatrix} \frac{1}{1000} & \frac{1}{500} \\ \frac{1}{150} & \frac{1}{100} \\ \frac{1}{100} & \frac{1}{200} \end{bmatrix} \right\}^{-1} \begin{bmatrix} \frac{1}{1000} & \frac{1}{150} & \frac{1}{100} \\ \frac{1}{500} & \frac{1}{100} & \frac{1}{200} \end{bmatrix}$$

$$= \begin{bmatrix} -23.07 & -69.5 & 148.59 \\ 36.71 & 141.42 & -97.85 \end{bmatrix}$$

Then the voltages are

$$\mathbf{x} = \begin{bmatrix} v_1 \\ v_2 \end{bmatrix} = \mathbf{A}^L \mathbf{y} = \begin{bmatrix} 100\text{ V} \\ 200\text{ V} \end{bmatrix}$$

We could have gotten the same answer using just the first two current measurements, in which case,

$$\begin{bmatrix} v_1 \\ v_2 \end{bmatrix} = \begin{bmatrix} \frac{1}{1000} & \frac{1}{500} \\ \frac{1}{150} & \frac{1}{100} \end{bmatrix}^{-1} \begin{bmatrix} 0.5 \\ 2.667 \end{bmatrix}$$

$$= \begin{bmatrix} -3000. & 600. \\ 2000. & -300. \end{bmatrix} \begin{bmatrix} 0.5 \\ 2.667 \end{bmatrix} = \begin{bmatrix} 100\text{ V} \\ 200\text{ V} \end{bmatrix}$$

The second solution is considerably less tolerant of measurement errors than the first. As an example, assume that there is a 10 percent error in measuring the second current and compute the two solutions.

EXAMPLE 2.2-4 MINIMUM NORM FLOW THROUGH REDUNDANT CONTROL VALVES

Suppose that the volumetric flow of a nonreacting fluid into a tank is controlled by three valves whose fully open flow rates are 10, 20, and 30 m³/s, respectively. How much fluid should flow through each valve to provide a net flow rate of 25 m³/s with a minimum root-mean-square flow rate through all valves?

The governing equation is

$$\mathbf{y} = \mathbf{A}\mathbf{x}$$

or

$$25 = [1 \quad 1 \quad 1]\begin{bmatrix} x_1 \\ x_2 \\ x_3 \end{bmatrix}$$

where $\mathbf{y}$ is the desired flow rate, $\mathbf{A}$ accounts for each of the three valves, and $\mathbf{x}$ represents the flow rate through each valve. There are an infinity of possible solutions, but only one minimizes the norm of the flow rates. Using the right pseudoinverse,

$$\begin{bmatrix} x_1 \\ x_2 \\ x_3 \end{bmatrix} = \begin{bmatrix} 1 \\ 1 \\ 1 \end{bmatrix}\left\{[1 \quad 1 \quad 1]\begin{bmatrix} 1 \\ 1 \\ 1 \end{bmatrix}\right\}^{-1}(25)$$

$$= \begin{bmatrix} \frac{1}{3} \\ \frac{1}{3} \\ \frac{1}{3} \end{bmatrix}(25) = \begin{bmatrix} 8.33 \\ 8.33 \\ 8.33 \end{bmatrix}$$

The corresponding percentage valve openings are 83.33, 41.67, and 27.78%, respectively.

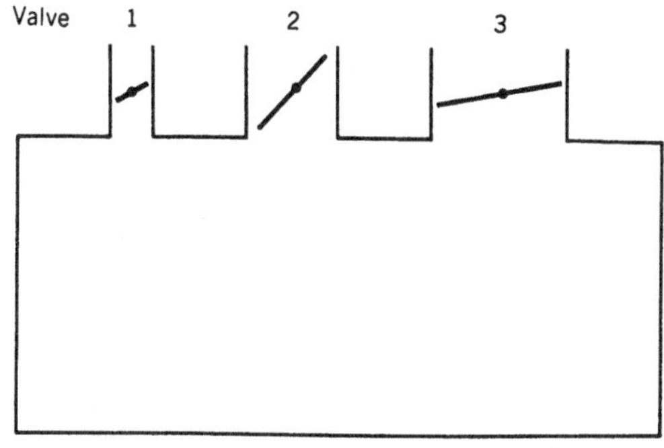

While each valve is providing an equal share of the flow, the unequal valve openings probably are undesirable: it may be better to use the largest valve to establish the bulk of the flow and the smallest to make vernier corrections to flow rate. Put another way, the minimum-norm solution for one set of units (dimensional flow rate) may not be as good as the minimum-norm solution for another set of units (percentage flow rate). It would be easy enough to minimize the latter by redefining **A** such that **x** represented percentage openings; an alternative is to define a *weighted pseudoinverse matrix* that accounts for the relative value of or penalty for using each valve in this underdetermined case. This is done readily using transformations, as discussed below.

Before leaving matrix inversion, it should be noted that using Eqs. 2.2-2 and 2.2-11 in the square, nonsingular case or Eqs. 2.2-31 or 2.2-34 in the nonsquare case is not necessarily the most efficient or accurate way of determining **x** from **y**. *Gaussian elimination** replaces Eq. 2.2-2 with a sequential reduction of equations to upper-triangular form. *Singular-value decomposition*, which is discussed below, provides an alternate means of computing $\mathbf{A}^{PI}$ (even when both $\mathbf{A}^T\mathbf{A}$ and $\mathbf{A}\mathbf{A}^T$ are singular) that is more robust and precise than Eqs. 2.2-31 and 2.2-34.

Transformations

In the most general sense, a *transformation* is a functional relationship between one variable and another, sometimes referred to as a rule for "mapping" every element of one set of numbers (the "domain") to another set (the "codomain"). The mapping rule can include integration, differentiation, and other operations, but our usage of the term normally will be somewhat more limited, reflecting an instantaneous ("memoryless") algebraic or transcendental relationship between the variables. Hence $y = f(x)$ and $\mathbf{y} = \mathbf{A}\mathbf{x}$ are examples of transformations that have been used up to this point. The latter is a linear transformation because there is a direct proportionality between changes in **x** and **y**. The transformation is "one-to-one" if different values of **x** give different values of **y** and it is "onto" if every **y** corresponds to at least one **x**. Example 2.2-2 provides an illustration of an orthogonal transformation, which is both "one-to-one" and "onto."

A particularly simple transformation is used when *rescaling* a vector. For example, the original units of an n-vector, $\mathbf{x}_1$, may be in feet, degrees, and pounds, while the desired n-vector, $\mathbf{x}_2$, has units of meters, radians, and kilograms. The vector is rescaled with the diagonal matrix **D**, whose

*Carl Friedrich Gauss (1777–1855), one of the great mathematicians of all time, directed the astronomical observatory at Göttingen and contributed widely in the areas of number theory, differential geometry, and topology.

nonzero elements are the necessary *conversion factors*:

$$\mathbf{x}_2 = \mathbf{D}\mathbf{x}_1 \tag{2.2-35}$$

By definition, $\mathbf{D}^{-1}$ exists, so $\mathbf{D}$ is "one-to-one" and "onto," allowing $\mathbf{x}_1$ to be uniquely determined from $\mathbf{x}_2$ and vice versa.

If we choose to rescale $\mathbf{x}_1$ so that the elements of $\mathbf{x}_2$, x_{2_i}, range from -1 to $+1$ as the original x_{1_i} range between $\pm x_{1_{i\,\max}}$, then $\mathbf{D}$ is said to *normalize* $\mathbf{x}_1$. Expressing the r-vector $\mathbf{y}$ as a function of $\mathbf{x}_2$ (through Eq. 2.2-1),

$$\begin{aligned}\mathbf{y} &= \mathbf{A}_1\mathbf{D}^{-1}\mathbf{x}_2 \\ &= \mathbf{A}_2\mathbf{x}_2\end{aligned} \tag{2.2-36}$$

Suppose that $r < n$, and the inverse relationship between $\mathbf{x}_2$ and $\mathbf{y}$ is desired. The right pseudoinverse solution is, from Eq. 2.2-34,

$$\begin{aligned}\mathbf{x}_2 &= \mathbf{A}_2^R\mathbf{y} \\ &= \mathbf{A}_2^T(\mathbf{A}_2\mathbf{A}_2^T)^{-1}\mathbf{y}\end{aligned} \tag{2.2-37}$$

and it is the minimum-norm solution for $\mathbf{x}_2$. The corresponding $\mathbf{x}_1$ solution can be found by substitution. From Eq. 2.2-35 to 2.2-37, noting that $\mathbf{D} = \mathbf{D}^T$ and $\mathbf{D}^{-1} = (\mathbf{D}^{-1})^T \triangleq \mathbf{D}^{-T}$,

$$\begin{aligned}\mathbf{x}_1 &= \mathbf{D}^{-1}\mathbf{x}_2 \\ &= \mathbf{D}^{-1}\mathbf{A}_2^R\mathbf{y} \\ &= \mathbf{D}^{-1}[\mathbf{D}^{-1}\mathbf{A}_1^T(\mathbf{A}_1\mathbf{D}^{-1}\mathbf{D}^{-1}\mathbf{A}_1^T)^{-1}]\mathbf{y} \\ &= \mathbf{R}^{-1}\mathbf{A}_1^T(\mathbf{A}_1\mathbf{R}^{-1}\mathbf{A}_1^T)^{-1}\mathbf{y} \\ &= \mathbf{A}^{WR}\mathbf{y}\end{aligned} \tag{2.2-38}$$

where the diagonal matrix $\mathbf{R}$ is the "square" of $\mathbf{D}$:

$$\mathbf{R} = \mathbf{D}\mathbf{D}^T \tag{2.2-39}$$

Equation 2.2-38 is recognized as the *weighted right pseudoinverse solution* for $\mathbf{x}_1$. $\mathbf{R}$ establishes quadratic weighting on the solution, since from Eq. 2.2-39 and the definition of $\mathbf{D}$,

$$\mathbf{R} = \begin{bmatrix} 1/x_{1_{\max}}^2 & 0 & \cdots & 0 \\ 0 & 1/x_{2_{\max}}^2 & \cdots & 0 \\ \vdots & \vdots & & \vdots \\ 0 & 0 & \cdots & 1/x_{n_{\max}}^2 \end{bmatrix} \tag{2.2-40}$$

EXAMPLE 2.2-5 MINIMUM-WEIGHTED-NORM FLOW THROUGH REDUNDANT CONTROL VALVES

Returning to Example 2.2-4, let us determine a normalized solution for the three flow rates with

$$\mathbf{A} = [1 \quad 1 \quad 1]; \qquad \mathbf{y} = 25$$

$$\mathbf{D} = \begin{bmatrix} \frac{1}{10} & 0 & 0 \\ 0 & \frac{1}{20} & 0 \\ 0 & 0 & \frac{1}{30} \end{bmatrix}$$

$$\mathbf{R}^{-1} = \begin{bmatrix} 10^2 & 0 & 0 \\ 0 & 20^2 & 0 \\ 0 & 0 & 30^2 \end{bmatrix}$$

The weighted right pseudoinverse matrix

$$\mathbf{A}^{WR} = \begin{bmatrix} 10^2 & 0 & 0 \\ 0 & 20^2 & 0 \\ 0 & 0 & 30^2 \end{bmatrix} \begin{bmatrix} 1 \\ 1 \\ 1 \end{bmatrix} \left\{ [1 \quad 1 \quad 1] \begin{bmatrix} 10^2 & 0 & 0 \\ 0 & 20^2 & 0 \\ 0 & 0 & 30^2 \end{bmatrix} \begin{bmatrix} 1 \\ 1 \\ 1 \end{bmatrix} \right\}^{-1}$$

$$= \begin{bmatrix} 0.0714 \\ 0.2857 \\ 0.6429 \end{bmatrix}$$

and the three flow rates are

$$\begin{bmatrix} x_1 \\ x_2 \\ x_3 \end{bmatrix} = \begin{bmatrix} 1.786 \\ 7.143 \\ 16.071 \end{bmatrix} \text{m}^3/\text{s}$$

These correspond to valve openings of 17.86, 35.71, and 53.56%, respectively. As an exercise, verify that the same result would have been computed with the unweighted pseudoinverse and $\mathbf{A} = (\frac{1}{10} \quad \frac{1}{20} \quad \frac{1}{30})$.

A third feasible solution would be to open each valve by the same percentage—in this case, $(\frac{25}{60}) \times 100 = 41.67$ percent. What choice of $\mathbf{R}$ would provide this result?

The final transformation to be considered is that associated with a *change of basis* (i.e., a redefinition of the coordinates used to describe a vector equation). In the next section, we will deal with ordinary differential equations of the form,

$$\frac{d\mathbf{x}_1}{dt} \triangleq \dot{\mathbf{x}}_1 = \mathbf{F}_1\mathbf{x}_1 + \text{forcing terms} \tag{2.2-41}$$

where $\mathbf{x}_1$ and $\dot{\mathbf{x}}_1$ are measured in the same coordinate frame. If there is a fixed, nonsingular coordinate transformation,

$$\mathbf{x}_2 = \mathbf{T}\mathbf{x}_1 \tag{2.2-42}$$

it may be desirable to express Eq. 2.2-41 in terms of the transformed vector, $\mathbf{x}_2$. Because $\mathbf{T}$ is unchanging,

$$\dot{\mathbf{x}}_2 = \mathbf{T}\dot{\mathbf{x}}_1 \tag{2.2-43}$$

and Eq. 2.2-41 can be written as

$$\begin{aligned}\dot{\mathbf{x}}_2 &= \mathbf{T}(\mathbf{F}_1\mathbf{x}_1 + \cdots) \\ &= \mathbf{T}\mathbf{F}_1\mathbf{T}^{-1}\mathbf{x}_2 + \mathbf{T}(\ldots)_1 \\ &= \mathbf{F}_2\mathbf{x}_2 + (\ldots)_2\end{aligned} \tag{2.2-44}$$

The relationship

$$\mathbf{F}_2 = \mathbf{T}\mathbf{F}_1\mathbf{T}^{-1} \tag{2.2-45}$$

is called a *similarity transformation* because it preserves the dynamic characteristics established by Eq. 2.2.41 while expressing the result in the alternate coordinate frame. The similarity transformation has application in changing the units, reordering the state vector, reorienting the coordinates, and "diagonalizing" the elements of linear differential equations.

EXAMPLE 2.2-6 COORDINATE REORDERING AND RESCALING

Small perturbations from horizontal equilibrium of the mechanical system shown in the sketch can be described by a second-order linear differential equation. A rotational spring and a linear damper govern the dynamics of the mass according to the equation

$$\ddot{z} = (1/m)(-d\dot{z} - k\theta)$$

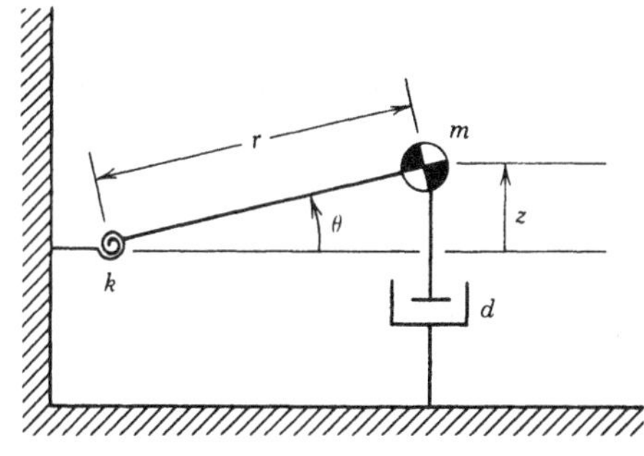

where m is specified in slugs, d in lb/(ft/s), k in lb/deg, z in ft, and θ in deg. Defining the state vector, $\mathbf{x}_a$,

$$\mathbf{x}_a = \begin{bmatrix} x_1 \\ x_2 \end{bmatrix} = \begin{bmatrix} \theta & \text{(deg)} \\ \dot{z} & \text{(ft/s)} \end{bmatrix}$$

the second-order equation can be expressed as two first-order equations,

$$\dot{x}_1 = \dot{\theta} = (180/\pi r)\dot{z} = (180/\pi r)x_2$$
$$\dot{x}_2 = (1/m)(-d\dot{z} - k\theta) = (-d/m)x_2 - (k/m)x_1$$

or

$$\begin{bmatrix} \dot{x}_1 \\ \dot{x}_2 \end{bmatrix} = \begin{bmatrix} 0 & (180/\pi r) \\ (-k/m) & (-d/m) \end{bmatrix} \begin{bmatrix} x_1 \\ x_2 \end{bmatrix}$$

which is

$$\dot{\mathbf{x}}_a = \mathbf{F}_a \mathbf{x}_a$$

We wish to redefine the state vector as

$$\mathbf{x}_b = \begin{bmatrix} \dot{z} \text{ (m/s)} \\ z \text{ (m)} \end{bmatrix}$$

and find the corresponding differential equation. Use a similarity transformation to accomplish this result.

The transformation substitutes z for θ, changes from feet to meters, and reverses the ordering of the state vector:

$$\mathbf{x}_b = \mathbf{T}\mathbf{x}_a$$

where

$$\mathbf{T} = \begin{bmatrix} 0 & 0.3048 \\ 0.3048(\pi r/180) & 0 \end{bmatrix}$$

and

$$\mathbf{T}^{-1} = \begin{bmatrix} 0 & 180/(0.3048\pi r) \\ 1/(0.3048) & 0 \end{bmatrix}$$

Then

$$\dot{\mathbf{x}}_b = \mathbf{F}_b \mathbf{x}_b$$

and

$$\mathbf{F}_b = \mathbf{T}\mathbf{F}_a\mathbf{T}^{-1}$$
$$= \begin{bmatrix} (-d/m) & (-k/m)(180/\pi r) \\ 1 & 0 \end{bmatrix}$$

Differentiation and Integration

As is the case for vectors, matrices are differentiated and integrated term by term. For example, the time derivative and integral of $\mathbf{A}(t)$ are

$$\frac{d\mathbf{A}(t)}{dt} = \begin{bmatrix} da_{11}/dt & da_{12}/dt & \dots & da_{1n}/dt \\ da_{21}/dt & da_{22}/dt & \dots & da_{2n}/dt \\ \vdots & \vdots & & \vdots \\ da_{r1}/dt & da_{r2}/dt & \dots & da_{rn}/dt \end{bmatrix} \tag{2.2-46}$$

$$\int_{t_0}^{t_f} \mathbf{A}(t)\,dt = \begin{bmatrix} \int_{t_0}^{t_f} a_{11}dt & \int_{t_0}^{t_f} a_{12}dt & \dots & \int_{t_0}^{t_f} a_{1n}dt \\ \int_{t_0}^{t_f} a_{21}dt & \int_{t_0}^{t_f} a_{22}dt & \dots & \int_{t_0}^{t_f} a_{2n}dt \\ \vdots & \vdots & & \vdots \\ \int_{t_0}^{t_f} a_{r1}dt & \int_{t_0}^{t_f} a_{r2}dt & \dots & \int_{t_0}^{t_f} a_{rn}dt \end{bmatrix} \tag{2.2-47}$$

Two matrix derivatives are of particular interest. Given a square, nonsingular matrix, $\mathbf{A}(t)$, the product $\mathbf{A}\mathbf{A}^{-1} = \mathbf{I}$ is invariant; hence its derivative must be zero:

$$\frac{d}{dt}[\mathbf{A}\mathbf{A}^{-1}] = \dot{\mathbf{A}}\mathbf{A}^{-1} + \mathbf{A}\dot{\mathbf{A}}^{-1} = \mathbf{0} \tag{2.2-48}$$

Then the *derivative of the inverse* can be expressed as

$$\dot{\mathbf{A}}^{-1} = -\mathbf{A}^{-1}\dot{\mathbf{A}}\mathbf{A}^{-1} \tag{2.2-49}$$

If $\mathbf{H}_{BE}(t)$ represents an *orthonormal transformation* (as in Example 2.2-2), then the magnitude of the matrix's determinant is fixed ($|\mathbf{H}_{BE}| = 1$), so its derivative can represent, at most, a rotation. The time derivative of the matrix can be expressed as a *cross product of the angular rate vector*, $\boldsymbol{\omega} = (p \quad q \quad r)^T$, and $\mathbf{H}_{BE}$,

$$\dot{\mathbf{H}}_{BE}(t) = -\boldsymbol{\omega}(t) \times \mathbf{H}_{BE}(t) \tag{2.2-50}$$

which, in turn, can be represented using the *cross-product equivalent matrix*, $\boldsymbol{\Omega}(t)$:

$$\dot{\mathbf{H}}_{BE}(t) = -\boldsymbol{\Omega}(t)\mathbf{H}_{BE}(t)$$

$$= \begin{bmatrix} 0 & r(t) & -q(t) \\ -r(t) & 0 & p(t) \\ q(t) & -p(t) & 0 \end{bmatrix} \mathbf{H}_{BE}(t) \tag{2.2-51}$$

Note that $\boldsymbol{\Omega}$ is skew symmetric.

The *time derivative of a quadratic form*, $\mathbf{x}^T\mathbf{Q}\mathbf{x}$, with symmetric constant weighting matrix, $\mathbf{Q}$, is the gradient with respect to $\mathbf{x}(t)$ times $\dot{\mathbf{x}}(t)$:

$$\frac{d}{dt}(\mathbf{x}^T\mathbf{Q}\mathbf{x}) = \dot{\mathbf{x}}^T\mathbf{Q}\mathbf{x} + \mathbf{x}^T\mathbf{Q}\dot{\mathbf{x}}$$

$$= 2\mathbf{x}^T\mathbf{Q}\dot{\mathbf{x}} \tag{2.2-52}$$

The result is a scalar, and the gradient is seen to be the row vector $2\mathbf{x}^T\mathbf{Q}$.

Analogous results for the derivative of a matrix trace with respect to a matrix are helpful. Recall that the trace of a matrix is a scalar function; its derivative with respect to an $(n \times m)$ component matrix $\mathbf{A}$ has dimension $(n \times m)$. Given the $(m \times m)$ symmetric matrix, $\mathbf{B}$,

$$\frac{d}{d\mathbf{A}}[\mathrm{Tr}(\mathbf{A}\mathbf{B}\mathbf{A}^T)] = 2\mathbf{A}\mathbf{B} \tag{2.2-53}$$

If $\mathbf{B}$ is an $(n \times m)$ matrix, the product $\mathbf{A}\mathbf{B}^T$ is square, and

$$\frac{d}{d\mathbf{A}}[\mathrm{Tr}(\mathbf{A}\mathbf{B}^T)] = \mathbf{B} \tag{2.2-54}$$

Some Matrix Identities

The matrix equations of control and estimation frequently involve matrix inverses, so it is helpful to establish certain identities for later use. These can be developed by finding alternate ways of expressing the submatrices of a partitioned inverse of the matrix $\mathbf{A}$. If $\mathbf{B} = \mathbf{A}^{-1}$, then

$$\mathbf{A}\mathbf{B} = \mathbf{I} \tag{2.2-55}$$

and

$$\mathbf{B}\mathbf{A} = \mathbf{I} \tag{2.2-56}$$

Let $\mathbf{A}$ be an $(m+n)\times(m+n)$ matrix, partitioned as follows,

$$\mathbf{A}=\begin{bmatrix}\mathbf{A}_1 & \mathbf{A}_2\\ \mathbf{A}_3 & \mathbf{A}_4\end{bmatrix} \tag{2.2-57}$$

where $\mathbf{A}_1$ is a symmetric $(m\times m)$ submatrix, $\mathbf{A}_4$ is a symmetric $(n\times n)$ submatrix, $\mathbf{A}_2$ has dimension $(m\times n)$, and $\mathbf{A}_3$ has dimension $(n\times m)$. The larger matrix, $\mathbf{A}$, need not have any meaning of its own; it is simply a formalism for combining two symmetric matrices ($\mathbf{A}_1$ and $\mathbf{A}_4$) of differing dimension with conformable matrices ($\mathbf{A}_2$ and $\mathbf{A}_3$). $\mathbf{B}$ is similarly partitioned as

$$\mathbf{B}=\begin{bmatrix}\mathbf{B}_1 & \mathbf{B}_2\\ \mathbf{B}_3 & \mathbf{B}_4\end{bmatrix} \tag{2.2-58}$$

with $(m\times m)\,\mathbf{B}_1$, $(n\times n)\,\mathbf{B}_4$, $(m\times n)\,\mathbf{B}_2$, and $(n\times m)\,\mathbf{B}_3$. Then Eqs. 2.2-55 and 2.2-56 are

$$\begin{bmatrix}\mathbf{A}_1 & \mathbf{A}_2\\ \mathbf{A}_3 & \mathbf{A}_4\end{bmatrix}\begin{bmatrix}\mathbf{B}_1 & \mathbf{B}_2\\ \mathbf{B}_3 & \mathbf{B}_4\end{bmatrix}=\begin{bmatrix}(\mathbf{A}_1\mathbf{B}_1+\mathbf{A}_2\mathbf{B}_3) & (\mathbf{A}_1\mathbf{B}_2+\mathbf{A}_2\mathbf{B}_4)\\ (\mathbf{A}_3\mathbf{B}_1+\mathbf{A}_4\mathbf{B}_3) & (\mathbf{A}_3\mathbf{B}_2+\mathbf{A}_4\mathbf{B}_4)\end{bmatrix}=\begin{bmatrix}\mathbf{I}_m & \mathbf{0}\\ \mathbf{0} & \mathbf{I}_n\end{bmatrix} \tag{2.2-59}$$

and

$$\begin{bmatrix}\mathbf{B}_1 & \mathbf{B}_2\\ \mathbf{B}_3 & \mathbf{B}_4\end{bmatrix}\begin{bmatrix}\mathbf{A}_1 & \mathbf{A}_2\\ \mathbf{A}_3 & \mathbf{A}_4\end{bmatrix}=\begin{bmatrix}(\mathbf{B}_1\mathbf{A}_1+\mathbf{B}_2\mathbf{A}_3) & (\mathbf{B}_1\mathbf{A}_2+\mathbf{B}_2\mathbf{A}_4)\\ (\mathbf{B}_3\mathbf{A}_1+\mathbf{B}_4\mathbf{A}_3) & (\mathbf{B}_3\mathbf{A}_2+\mathbf{B}_4\mathbf{A}_4)\end{bmatrix}=\begin{bmatrix}\mathbf{I}_m & \mathbf{0}\\ \mathbf{0} & \mathbf{I}_n\end{bmatrix} \tag{2.2-60}$$

(The subscript on the identity matrix indicates its row and column dimensions.) The four partitioned equations of Eq. 2.2-59 can be solved to find the submatrices of $\mathbf{B}$ by elimination. For example, the left column of the equation yields the following sequence of equations, assuming all indicated inverses exist:

$$\mathbf{B}_3=-\mathbf{A}_4^{-1}\mathbf{A}_3\mathbf{B}_1 \tag{2.2-61a}$$

$$\mathbf{A}_1\mathbf{B}_1-\mathbf{A}_2\mathbf{A}_4^{-1}\mathbf{A}_3\mathbf{B}_1=\mathbf{I}_m \tag{2.2-62a}$$

$$(\mathbf{A}_1-\mathbf{A}_2\mathbf{A}_4^{-1}\mathbf{A}_3)\mathbf{B}_1=\mathbf{I}_m \tag{2.2-62b}$$

$$\mathbf{B}_1=(\mathbf{A}_1-\mathbf{A}_2\mathbf{A}_4^{-1}\mathbf{A}_3)^{-1} \tag{2.2-62c}$$

$$\mathbf{B}_3=-\mathbf{A}_4^{-1}\mathbf{A}_3(\mathbf{A}_1-\mathbf{A}_2\mathbf{A}_4^{-1}\mathbf{A}_3)^{-1} \tag{2.2-61b}$$

The right column provides solutions for $\mathbf{B}_2$ and $\mathbf{B}_4$, leading to

$$\mathbf{B}=\begin{bmatrix}(\mathbf{A}_1-\mathbf{A}_2\mathbf{A}_4^{-1}\mathbf{A}_3)^{-1} & -\mathbf{A}_1^{-1}\mathbf{A}_2(\mathbf{A}_4-\mathbf{A}_3\mathbf{A}_1^{-1}\mathbf{A}_2)^{-1}\\ -\mathbf{A}_4^{-1}\mathbf{A}_3(\mathbf{A}_1-\mathbf{A}_2\mathbf{A}_4^{-1}\mathbf{A}_3)^{-1} & (\mathbf{A}_4-\mathbf{A}_3\mathbf{A}_1^{-1}\mathbf{A}_2)^{-1}\end{bmatrix} \tag{2.2-63}$$

If $\mathbf{A}_3 = \mathbf{A}_2^T$, the first matrix identity appears in the upper right and lower left elements of the matrix $\mathbf{B}$:

$$\begin{aligned}\mathbf{A}_1^{-1}\mathbf{A}_2(\mathbf{A}_4 - \mathbf{A}_3\mathbf{A}_1^{-1}\mathbf{A}_2)^{-1} &= [\mathbf{A}_4^{-1}\mathbf{A}_3(\mathbf{A}_1 - \mathbf{A}_2\mathbf{A}_4^{-1}\mathbf{A}_3)^{-1}]^T \\ &= (\mathbf{A}_1 - \mathbf{A}_2\mathbf{A}_4^{-1}\mathbf{A}_2^T)^{-1}\mathbf{A}_2\mathbf{A}_4^{-1} \qquad (2.2\text{-}64)\end{aligned}$$

Alternate definitions of $\mathbf{B}$'s submatrices can be derived in a similar manner from Eq. 2.2-60:

$$\mathbf{B} = \begin{bmatrix} (\mathbf{A}_1 - \mathbf{A}_2\mathbf{A}_4^{-1}\mathbf{A}_3)^{-1} & -(\mathbf{A}_1 - \mathbf{A}_2\mathbf{A}_4^{-1}\mathbf{A}_3)^{-1}\mathbf{A}_2\mathbf{A}_4^{-1} \\ -(\mathbf{A}_4 - \mathbf{A}_3\mathbf{A}_1^{-1}\mathbf{A}_2)^{-1}\mathbf{A}_3\mathbf{A}_1^{-1} & (\mathbf{A}_4 - \mathbf{A}_3\mathbf{A}_1^{-1}\mathbf{A}_2)^{-1} \end{bmatrix} \qquad (2.2\text{-}65)$$

One particularly useful relationship known as the *matrix inversion lemma* can be derived by manipulating the submatrices of $\mathbf{B}$. From the lower right solution of Eq. 2.2-60,

$$\mathbf{B}_4 = (\mathbf{I}_n - \mathbf{B}_3\mathbf{A}_2)\,\mathbf{A}_4^{-1} \qquad (2.2\text{-}66)$$

Substituting for $\mathbf{B}_3$ and $\mathbf{B}_4$ using Eq. 2.2-63,

$$(\mathbf{A}_4 - \mathbf{A}_3\mathbf{A}_1^{-1}\mathbf{A}_2)^{-1} = \mathbf{A}_4^{-1} - \mathbf{A}_4^{-1}\mathbf{A}_3(\mathbf{A}_2\mathbf{A}_4^{-1}\mathbf{A}_3 - \mathbf{A}_1)^{-1}\mathbf{A}_2\mathbf{A}_4^{-1} \qquad (2.2\text{-}67a)$$

For symmetric $\mathbf{A}$, this can be written as

$$(\mathbf{A}_4 - \mathbf{A}_2^T\mathbf{A}_1^{-1}\mathbf{A}_2)^{-1} = \mathbf{A}_4^{-1} - \mathbf{A}_4^{-1}\mathbf{A}_2^T(\mathbf{A}_2\mathbf{A}_4^{-1}\mathbf{A}_2^T - \mathbf{A}_1)^{-1}\mathbf{A}_2\mathbf{A}_4^{-1} \qquad (2.2\text{-}67b)$$

Eigenvalues and Eigenvectors

The $(n \times n)$ matrix $\mathbf{A}$ could be used to transform one n-vector ($\mathbf{x}$) to another ($\mathbf{y}$):

$$\mathbf{y} = \mathbf{A}\mathbf{x} \qquad (2.2\text{-}68)$$

If we can find a scalar variable, λ_i, and a particular value of $\mathbf{x}$ ($= \mathbf{e}_i$) such that identical transformations are given by

$$\mathbf{y}_{e_i} = \mathbf{A}\mathbf{e}_i = \lambda_i\mathbf{e}_i \qquad (2.2\text{-}69)$$

then λ_i is called an eigenvalue of $\mathbf{A}$, and $\mathbf{e}_i$ is the corresponding eigenvector.* Every $(n \times n)$ matrix has n eigenvalues and up to n eigenvectors

*These also are called a *characteristic value* and a *characteristic vector* in some writings.

that are jointly specified (from Eq. 2.2-69) by

$$(\lambda_i \mathbf{I}_n - \mathbf{A})\mathbf{e}_i = \mathbf{0}, \qquad i = 1, n \tag{2.2-70}$$

If this equation is satisfied, then so is

$$(\lambda_i \mathbf{I}_n - \mathbf{A})\alpha\mathbf{e}_i = \mathbf{0} \tag{2.2-71}$$

where α is any scalar; hence Eq. 2.2–70 specifies eigenvectors within an arbitrary multiplicative constant. These n vector equations can have nontrivial solutions only if the n values of λ_i are solutions to the scalar equation

$$|\lambda \mathbf{I}_n - \mathbf{A}| \triangleq \Delta(\lambda) = 0 \tag{2.2-72}$$

$(\lambda \mathbf{I}_n - \mathbf{A})$ is the *characteristic matrix* of $\mathbf{A}$, and $\Delta(\lambda)$ symbolizes the *characteristic polynomial* of $\mathbf{A}$. Because this determinant creates an nth-degree polynomial in λ that can be factored as the product of n binomials,

$$\begin{aligned}\Delta(\lambda) &= \lambda^n + c_{n-1}\lambda^{n-1} + \cdots + c_1\lambda + c_0 \\ &= (\lambda - \lambda_1)(\lambda - \lambda_2)\ldots(\lambda - \lambda_n) = 0\end{aligned} \tag{2.2-73}$$

each of the binomial roots is an eigenvalue of $\Delta(\lambda)$. Real roots are accompanied by real eigenvectors, and complex roots (which always occur in conjugate pairs for real $\mathbf{A}$) are associated with complex-conjugate pairs of eigenvectors (Fig. 2.2-1).* If the roots are *distinct* (i.e., each is different from the others), there are n independent eigenvectors; if some roots are repeated, there may be fewer eigenvectors.

For a distinct eigenvalue λ_i, the eigenvector $\mathbf{e}_i$ is contained in the following:

$$\text{Adj}(\lambda_i \mathbf{I}_n - \mathbf{A}) = [\alpha_1\mathbf{e}_i \quad \alpha_2\mathbf{e}_i \quad \ldots \quad \alpha_n\mathbf{e}_i] \tag{2.2-74}$$

The α_j are arbitrary constants; hence any nonzero column represents $\mathbf{e}_i$. $\text{Adj}(\lambda_i \mathbf{I}_n - \mathbf{A})$ is evaluated for each root ($i = 1, n$), and a single eigenvector is chosen from each evaluation. Together, the n eigenvectors form the columns of the $(n \times n)$ *modal matrix*,

$$\mathbf{E} = [\mathbf{e}_1 \quad \mathbf{e}_2 \quad \ldots \quad \mathbf{e}_n] \tag{2.2-75}$$

where the eigenvectors are scaled so that $|\mathbf{e}_i| = 1$, $i = 1$ to n. The modal

*A *complex scalar number* z has two parts. Written in Cartesian form, $z = x + jy$, with *real* and *imaginary parts*, x and jy. x and y are real numbers, and $j = \sqrt{-1}$. In polar form, $z = re^{j\theta}$, with amplitude r and phase angle θ, both of which are real numbers. The *complex conjugate* of z is $x - jy$ or $re^{-j\theta}$. *Complex vectors and matrices* contain complex scalar elements.

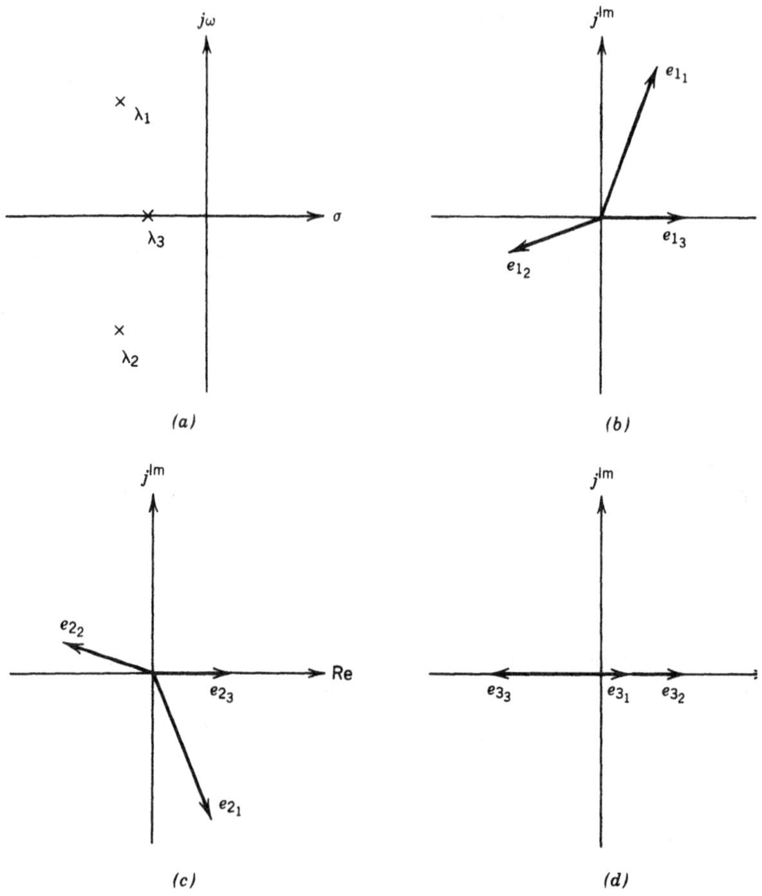

FIGURE 2.2-1 Example of eigenvalues for a third-order system and corresponding eigenvectors. (*a*) Typical eigenvalues; (*b*) eigenvectors of λ_1; (*c*) eigenvectors of λ_2; (*d*) eigenvectors of λ_3.

matrix can be used as a similarity transformation to diagonalize $\mathbf{A}$ (i.e., to transform $\mathbf{A}$ into a diagonal matrix of eigenvalues). From Eq. 2.2-69,

$$\begin{bmatrix} \mathbf{Ae}_1 = \lambda_1 \mathbf{e}_1 \\ \mathbf{Ae}_2 = \lambda_2 \mathbf{e}_2 \\ \vdots \quad\quad \vdots \\ \mathbf{Ae}_n = \lambda_n \mathbf{e}_n \end{bmatrix} \tag{2.2-76a}$$

or

$$\mathbf{AE} = \mathbf{E}\mathbf{\Lambda} \tag{2.2-76b}$$

where Λ is a diagonal matrix of eigenvalues:

$$\Lambda = \begin{bmatrix} \lambda_1 & 0 & \dots & 0 \\ 0 & \lambda_2 & \dots & 0 \\ 0 & 0 & \dots & 0 \\ 0 & 0 & \dots & \lambda_n \end{bmatrix} \tag{2.2-77}$$

$\mathbf{E}$ is nonsingular because the $\mathbf{e}_i$ are linearly independent, so

$$\mathbf{A} = \mathbf{E}\Lambda\mathbf{E}^{-1} \tag{2.2-78}$$

This can be recognized as a similarity transformation (Eq. 2.2-45) from the diagonal matrix of eigenvalues to the original square matrix, and the inverse relationship also holds:

$$\Lambda = \mathbf{E}^{-1}\mathbf{A}\mathbf{E} \tag{2.2-79}$$

If $\mathbf{A}$ is a real, symmetric matrix, the eigenvalues are real, and the eigenvectors are orthogonal to each other. In this case, there are n eigenvectors even if the roots are repeated. The eigenvectors corresponding to distinct eigenvalues are given by Eq. 2.2-74, while the eigenvectors of an eigenvalue λ_i with multiplicity m are the nonzero columns of

$$\frac{d^{m-1}}{d\lambda^{m-1}}[\text{Adj}(\lambda\mathbf{I}_n - \mathbf{A})]\bigg|_{\lambda=\lambda_i} = [\dots \ \mathbf{e}_{i_1} \ \ \mathbf{e}_{i_2} \ \ \dots \ \ \mathbf{e}_{i_m} \ \ \dots] \tag{2.2-80}$$

Modern methods of numerical eigenvalue determination usually involve repeated manipulations of a matrix that transform it into diagonal or almost-upper-triangular form, for which the diagonal or block-diagonal elements *are* the eigenvalues. If $\mathbf{A}$ is a real, symmetric matrix, its eigenvalues are real, and the modal matrix (Eq. 2.2-75) is an orthonormal transformation. Then Eq. 2.2-79 can be written as

$$\Lambda = \mathbf{E}^T\mathbf{A}\mathbf{E} \tag{2.2-81a}$$

or

$$\mathbf{A}\mathbf{E} = \mathbf{E}\Lambda \tag{2.2-81b}$$

Following a method due to Jacobi, Λ is found by constructing an orthonormal matrix from elementary transformations that "annihilate" off-diagonal (or "pivotal") elements term by term, as in (G-9), that is, by constructing $\mathbf{E}$.

If $\mathbf{A}$ is real but not symmetric, it may have complex eigenvalues and may not be reducible to diagonal form; however, it can be written as a "QR product,"

$$\mathbf{A} = \mathbf{QR} \tag{2.2-82}$$

where $\mathbf{Q}$ is a *unitary* matrix (i.e., a complex matrix whose conjugate transpose*, $\mathbf{Q}^H$, is its inverse, $\mathbf{Q}^{-1}$), and $\mathbf{R}$ is an almost-upper-triangular (or block-upper-triangular) matrix, whose diagonal or (2×2) block-diagonal elements are the eigenvalues. The *QR iteration* for finding eigenvalues, as detailed in (W-1), uses a shift variable ρ to develop $\mathbf{A}_k$ as a block-upper-triangular matrix according to the following algorithm:

$$\mathbf{A}_0 = \mathbf{A} \tag{2.2-83}$$

$$\mathbf{A}_k - \rho_k \mathbf{I}_n = \mathbf{Q}_k \mathbf{R}_k \tag{2.2-84}$$

$$\mathbf{A}_{k+1} = \mathbf{R}_k \mathbf{Q}_k + \rho_k \mathbf{I}_n \tag{2.2-85}$$

A similar "double" QR iteration is used in practice to avoid possible numerical difficulties with the basic algorithm (F-2).

In later developments, we will have reason to distinguish between right and left eigenvectors of a matrix as well as right and left modal matrices. The eigenvectors defined earlier are column vectors and, from Eq. 2.2-69, are called *right eigenvectors*, and the *right modal matrix* is defined by Eq. 2.2-75. It also is possible to define *left* (*row*) *eigenvectors*, $\mathbf{d}_i^T$, such that

$$\mathbf{d}_i^T \mathbf{A} = \mathbf{d}_i^T \lambda_i, \qquad i = 1, n \tag{2.2-86}$$

The $\mathbf{d}_i^T$ form the rows of the *left modal matrix*, $\mathbf{D}^T$,

$$\mathbf{D}^T = \begin{bmatrix} \mathbf{d}_1^T \\ \mathbf{d}_2^T \\ \vdots \\ \mathbf{d}_n^T \end{bmatrix} \tag{2.2-87}$$

and $\mathbf{D}^T$ can be scaled so that

$$\mathbf{D}^T = \mathbf{E}^{-1} \tag{2.2-88}$$

It then is possible to perform a *spectral decomposition* of $\mathbf{A}$, that is, to write

*Also called Hermitian transpose. A matrix $\mathbf{B}(j\omega)$ that is identical to its complex-conjugate transpose, $\mathbf{B}^T(-j\omega) = \mathbf{B}^H(j\omega)$, is said to be Hermitian. If $\mathbf{B}^T(-s) = \mathbf{B}(s)$, where $s = \sigma + j\omega$, $\mathbf{B}(s)$ is para-Hermitian.

A as

$$\mathbf{A} = \sum_{i=1}^{n} \lambda_i \mathbf{e}_i \mathbf{d}_i^T = \mathbf{E}\Lambda\mathbf{D}^T \tag{2.2-89}$$

Singular Value Decomposition

Singular values and their associated singular vectors characterize a matrix in much the same way that eigenvalues and eigenvectors do. They have utility in describing the magnitude of a matrix and the relative magnitudes of its components, in determining the conditioning of a matrix, and in constructing generalized inverses of a matrix. If **A** is an $(m \times n)$ complex matrix, it can be put in the form

$$\mathbf{A} = \mathbf{U}\mathbf{S}\mathbf{V}^H \tag{2.2-90}$$

where **S** is an $(m \times n)$ quasidiagonal matrix containing the *singular values* (also called *principal gains* of **A**) of **A**, **U** is an $(m \times m)$ unitary matrix containing the *left singular vectors* of **A**, and **V** is an $(n \times n)$ unitary matrix containing the *right singular vectors* of **A**. (As before, $\mathbf{V}^H$ represents the conjugate transpose of **V**.)

Singular values are defined as the positive square roots of the eigenvalues of **B**, where **B** is defined as

$$\mathbf{B} = \mathbf{A}^H\mathbf{A} \qquad \text{for} \quad m > n \tag{2.2-91}$$

or

$$\mathbf{B} = \mathbf{A}\mathbf{A}^H \qquad \text{for} \quad m < n \tag{2.2-92}$$

Denoting an eigenvalue of **B** as λ_i, the corresponding singular value of **A** is $\sigma_i = \sqrt{\lambda_i}$. The number of nonzero σ_i equals the rank of **A**, and by either definition, **B** has the same rank. The ratio $\sigma_{max}/\sigma_{min}$ is the *condition number* of **A**.

The matrix **S** is partitioned according to the relative magnitudes of m and n. For $m > n$, **S** contains an $(n \times n)$ diagonal submatrix of the singular values, plus $(m - n)$ rows of zeros:

$$\mathbf{S} = \left[\begin{array}{cccc} \sigma_1 & 0 & \dots & 0 \\ 0 & \sigma_2 & \dots & 0 \\ \vdots & \vdots & & \vdots \\ 0 & 0 & \dots & \sigma_n \\ \hdashline 0 & 0 & \dots & 0 \end{array}\right] \begin{array}{l} \\ \\ \\ \\ (m-n) \text{ rows} \end{array} \tag{2.2-93}$$

For $m < n$, $\mathbf{S}$ contains an $(m \times m)$ diagonal submatrix of the σ_i plus $(n - m)$ columns of zeros. The columns of $\mathbf{U}$ are the normalized eigenvectors of $\mathbf{A}\mathbf{A}^H$, and the columns of $\mathbf{V}$ are the normalized eigenvectors of $\mathbf{A}^H\mathbf{A}$.

If $\mathbf{A}$ is a real matrix, then $\mathbf{U}$ and $\mathbf{V}$ are real orthogonal matrices, and

$$\mathbf{A} = \mathbf{U}\mathbf{S}\mathbf{V}^T \tag{2.2-94}$$

If $\mathbf{A}$ is square and nonsingular, its inverse can be computed as

$$\mathbf{A}^{-1} = \mathbf{V}\mathbf{S}^{-1}\mathbf{U}^T \tag{2.2-95}$$

If $\mathbf{A}$ is singular (including the nonsquare case), its pseudoinverse (left or right, as appropriate) is

$$\mathbf{A}^{PI} = \mathbf{V}\mathbf{S}^{PI}\mathbf{U}^T \tag{2.2-96}$$

Numerical methods for performing singular value decomposition can be found in (L-2).

Some Determinant Identities

It will prove useful to manipulate the determinants of nonsingular square matrices in later sections. Given the $(n \times n)$ matrix $\mathbf{A}$

$$|\mathbf{A}^T| = |\mathbf{A}| \tag{2.2-97}$$

because the determinant is a scalar. If $\mathbf{A}$ and $\mathbf{B}$ are both $(n \times n)$ matrices, it is easy to see that

$$|\mathbf{AB}| = |\mathbf{BA}| = |\mathbf{A}||\mathbf{B}| \tag{2.2-98}$$

If $\mathbf{A}$ has dimension $(n \times n)$ and $\mathbf{B}$ has dimension $(n \times m)$, Laplace expansion can be used to verify that

$$\begin{vmatrix} \mathbf{A} & \mathbf{B} \\ \mathbf{0} & \mathbf{I}_m \end{vmatrix} = \begin{vmatrix} \mathbf{A} & \mathbf{0} \\ \mathbf{B}^T & \mathbf{I}_m \end{vmatrix} = |\mathbf{A}| \tag{2.2-99}$$

where $\mathbf{I}_m$ denotes an $(m \times m)$ identity matrix. The last two results combine to show that

$$\begin{aligned} \begin{vmatrix} \mathbf{A}_1 & \mathbf{0} \\ \mathbf{A}_3 & \mathbf{A}_4 \end{vmatrix} &= \left| \begin{bmatrix} \mathbf{A}_1 & \mathbf{0} \\ \mathbf{A}_3 & \mathbf{I}_m \end{bmatrix} \begin{bmatrix} \mathbf{I}_n & \mathbf{0} \\ \mathbf{0} & \mathbf{A}_4 \end{bmatrix} \right| = \begin{vmatrix} \mathbf{A}_1 & \mathbf{0} \\ \mathbf{A}_3 & \mathbf{I}_m \end{vmatrix} \begin{vmatrix} \mathbf{I}_n & \mathbf{0} \\ \mathbf{0} & \mathbf{A}_4 \end{vmatrix} \\ &= |\mathbf{A}_1||\mathbf{A}_4| \end{aligned} \tag{2.2-100}$$

Schur's formula allows the determinant of a partitioned matrix to be written as a product of component determinants, and it makes use of the previous results. Assume that $\mathbf{A}_1$ and $\mathbf{A}_4$ are $(m \times m)$ and $(n \times n)$ matrices, respectively. If $\mathbf{A}_1$ is nonsingular,

$$\begin{vmatrix} \mathbf{A}_1 & \mathbf{A}_2 \\ \mathbf{A}_3 & \mathbf{A}_4 \end{vmatrix} = \begin{vmatrix} \mathbf{I}_m & \mathbf{0} \\ -\mathbf{A}_3\mathbf{A}_1^{-1} & \mathbf{I}_n \end{vmatrix} \begin{vmatrix} \mathbf{A}_1 & \mathbf{A}_2 \\ \mathbf{A}_3 & \mathbf{A}_4 \end{vmatrix} = \begin{vmatrix} \mathbf{A}_1 & \mathbf{A}_2 \\ \mathbf{0} & \mathbf{A}_4 - \mathbf{A}_3\mathbf{A}_1^{-1}\mathbf{A}_2 \end{vmatrix}$$

$$= |\mathbf{A}_1||\mathbf{A}_4 - \mathbf{A}_3\mathbf{A}_1^{-1}\mathbf{A}_2| \tag{2.2-101}$$

If $\mathbf{A}_4$ is nonsingular,

$$\begin{vmatrix} \mathbf{A}_1 & \mathbf{A}_2 \\ \mathbf{A}_3 & \mathbf{A}_4 \end{vmatrix} = \begin{vmatrix} \mathbf{I}_m & -\mathbf{A}_2\mathbf{A}_4^{-1} \\ \mathbf{0} & \mathbf{I}_n \end{vmatrix} \begin{vmatrix} \mathbf{A}_1 & \mathbf{A}_2 \\ \mathbf{A}_3 & \mathbf{A}_4 \end{vmatrix} = \begin{vmatrix} \mathbf{A}_1 - \mathbf{A}_2\mathbf{A}_4^{-1}\mathbf{A}_3 & \mathbf{0} \\ \mathbf{A}_3 & \mathbf{A}_4 \end{vmatrix}$$

$$= |\mathbf{A}_4||\mathbf{A}_1 - \mathbf{A}_2\mathbf{A}_4^{-1}\mathbf{A}_3| \tag{2.2-102}$$

Choosing $\mathbf{A}_1 = \mathbf{I}_m$ and $\mathbf{A}_4 = \mathbf{I}_n$, Schur's formula yields

$$\begin{aligned} \begin{vmatrix} \mathbf{I}_m & \mathbf{A}_2 \\ \mathbf{A}_3 & \mathbf{I}_n \end{vmatrix} &= |\mathbf{I}_m||\mathbf{I}_n - \mathbf{A}_3\mathbf{I}_m^{-1}\mathbf{A}_2| \\ &= |\mathbf{I}_n - \mathbf{A}_3\mathbf{A}_2| \\ &= |\mathbf{I}_n||\mathbf{I}_m - \mathbf{A}_2\mathbf{I}_n^{-1}\mathbf{A}_3| \\ &= |\mathbf{I}_m - \mathbf{A}_2\mathbf{A}_3| \end{aligned} \tag{2.2-103a}$$

leading to the identity

$$|\mathbf{I}_n - \mathbf{A}_3\mathbf{A}_2| = |\mathbf{I}_m - \mathbf{A}_2\mathbf{A}_3| \tag{2.2-103b}$$

Equations 2.2-101 to 2.2-103 find specific application in the evaluation of the asymptotic locations of regulator and estimator eigenvalues (Section 6.4).

2.3 DYNAMIC SYSTEM MODELS AND SOLUTIONS

Nonlinear System Equations

We consider a nonlinear, time-varying dynamic system that can be described by an n-component vector ordinary differential equation representing its dynamics, an r-component algebraic (or transcendental) equation representing its true output, and an r-component equation representing a possibly degraded observation of the true outputs:

Dynamic Equation

$$\frac{d\mathbf{x}(t)}{dt} \triangleq \dot{\mathbf{x}}(t) = \mathbf{f}[\mathbf{x}(t), \mathbf{u}(t), \mathbf{w}(t), \mathbf{p}(t), t] \tag{2.3-1}$$

Output Equation

$$\mathbf{y}(t) = \mathbf{h}[\mathbf{x}(t), \mathbf{u}(t), \mathbf{w}(t), \mathbf{p}(t), t] \tag{2.3-2}$$

Observation Equation

$$\mathbf{z}(t) = \mathbf{j}[\mathbf{y}(t), \mathbf{n}(t), t] \tag{2.3-3}$$

The vector $\mathbf{f}$ contains an element for each element of the state, $\mathbf{x}$, and its dimension is determined by the system's definition. Each element of $\mathbf{f}$ is the appropriate scalar function that defines the *time-rate-of-change* for the corresponding component of $\mathbf{x}$. For example, the vector equation for a three-state model has three scalar components:

$$\begin{aligned} \dot{x}_1(t) &= f_1(x_1, x_2, x_3, \mathbf{u}, \mathbf{w}, \mathbf{p}, t) \\ \dot{x}_2(t) &= f_2(x_1, x_2, x_3, \mathbf{u}, \mathbf{w}, \mathbf{p}, t) \\ \dot{x}_3(t) &= f_3(x_1, x_2, x_3, \mathbf{u}, \mathbf{w}, \mathbf{p}, t) \end{aligned} \tag{2.3-4}$$

The dimension of the output function $\mathbf{h}$ is not governed by the dynamics of the process; it may be larger or smaller than the state dimension, and there is some freedom in its choice. In one system, three first-order differential equations may be required to characterize the dynamics, but perhaps only one state component is accessible as an output:

$$y_1 = x_1 \tag{2.3-5}$$

In another third-order system, four outputs may be available:

$$\begin{aligned} y_1 &= x_1 \\ y_2 &= a_1 x_1 + a_2 x_2 \\ y_3 &= b_0 + b_1 x_3 \\ y_4 &= c_0 x_1^2 + c_1 x_2 x_3 \end{aligned} \tag{2.3-6}$$

For the first system, added sensors could increase the dimension of $\mathbf{y}$, while in the second, the output dimension could be reduced by one without destroying the direct observation of all states.

Whereas Eq. 2.3-5 may represent the ideal output of the system, the actual observation may introduce errors; the measurement corresponding to y_1 could be

$$z_1 = (1 + n_1)x_1 + n_2 \tag{2.3-7}$$

where n_1 and n_2 are unknown multiplicative and additive factors, respectively. If only the latter type occurs, observation error is simply added to the output in Eq. 2.3-3. In either case, z_1 is an imperfect representation of x_1; if the statistical properties of x_1 and of the errors are known, then a *stochastic-optimal estimate* of x_1 could be derived from mathematical processing of z_1. With a measurement vector such as Eq. 2.3-6, subjected to an additive error vector $\mathbf{n}$,

$$\mathbf{z} = \mathbf{y} + \mathbf{n} \tag{2.3-8}$$

it may be desirable not to eliminate a redundant measurement component but to process all measurements in order to "average out" errors. Under what circumstances does this make sense?

Solutions for $\mathbf{x}(t)$ in $[t_0, t_f]$ are found by integrating Eq. 2.3-1. The resulting time histories describe the evolution of motions for given controls, disturbances, and initial conditions; each change in any of these quantities leads to a new state trajectory. While a few low-order classical differential equations possess exact solutions that can be expressed in terms of algebraic or transcendental functions, the most general nonlinear, time-varying equations must be integrated numerically, as described later in the section.

State-space representation of the dynamic equation is central to the methods described in this book because it provides a consistent framework for analyzing systems of any degree of complexity. Many models of physical, chemical, biological, or economic systems are naturally described as assemblages of interconnected first-order differential equations, but others originate as higher-order equations. A scalar nth-order ordinary differential equation can be replaced by n first-order ordinary differential equations or, equivalently, by a single first-order vector equation of dimension n. Groups of scalar equations can be handled in similar fashion.

Two examples make these points. Consider the third-order scalar equation,

$$\dddot{x} + c_3\ddot{x} + c_2\dot{x} + c_1 x = bu \tag{2.3-9}$$

with solution variable x and forcing function u. Defining

$$\mathbf{x} = \begin{bmatrix} x_1 \\ x_2 \\ x_3 \end{bmatrix} = \begin{bmatrix} x \\ \dot{x} \\ \ddot{x} \end{bmatrix} \tag{2.3-10}$$

Eq. 2.3-9 can be written

$$\dot{x}_3 + c_3 x_3 + c_2 x_2 + c_1 x_1 = bu \tag{2.3-11}$$

This equation can be integrated to find x_3 (or $\ddot{x}$), but two additional integrations are required to find x_1 and x_2 (the original x and $\dot{x}$). From Eqs. 2.3-10 and 2.3-11, the three scalar differential equations that must be solved can be combined in a single vector equation,

$$\begin{bmatrix} \dot{x}_1 \\ \dot{x}_2 \\ \dot{x}_3 \end{bmatrix} = \begin{bmatrix} x_2 \\ x_3 \\ (-c_3 x_3 - c_2 x_2 - c_1 x_1 + bu) \end{bmatrix} = \begin{bmatrix} f_1(\mathbf{x}, u) \\ f_2(\mathbf{x}, u) \\ f_3(\mathbf{x}, u) \end{bmatrix} \tag{2.3-12}$$

which has the same form as Eq. 2.3-1.

In the second case, suppose two interconnected second-order scalar differential equations describe the system:

$$\begin{aligned} \ddot{y} - a_2(\dot{z} - \dot{y}) - a_1(z - y)^2 &= b_1 u_1 + b_2 u_2 \\ \ddot{z} - c_2 \dot{z}^2 - c_1(y + z) &= d u_1^2 \end{aligned} \tag{2.3-13}$$

The state vector could be expressed as

$$\mathbf{x} = \begin{bmatrix} x_1 \\ x_2 \\ x_3 \\ x_4 \end{bmatrix} = \begin{bmatrix} y \\ \dot{y} \\ z \\ \dot{z} \end{bmatrix} \tag{2.3-14}$$

yielding the dynamic equation

$$\begin{bmatrix} \dot{x}_1 \\ \dot{x}_2 \\ \dot{x}_3 \\ \dot{x}_4 \end{bmatrix} = \begin{bmatrix} x_2 \\ a_2(x_4 - x_2) + a_1(x_3 - x_1)^2 + b_1 u_1 + b_2 u_2 \\ x_4 \\ c_2 x_4^2 + c_1(x_3 + x_1) + d u_1^2 \end{bmatrix} \tag{2.3-15}$$

The state vector definition is not necessarily unique; for the previous example, we could choose

$$\mathbf{x} = \begin{bmatrix} y \\ \dot{y} \\ (z - y) \\ (\dot{z} - \dot{y}) \end{bmatrix} \tag{2.3-16}$$

The dynamic equation would be adjusted accordingly; this is left as an exercise for the reader.

One characteristic of the dynamic equation that has potential importance is its equilibrium solution. The dynamic system is said to be at *static equilibrium* when the state rates are zero and the state components are unchanging:

$$\dot{\mathbf{x}}(t) = \mathbf{0} = \mathbf{f}[\mathbf{x}(t), \mathbf{u}(t), \mathbf{w}(t), \mathbf{p}(t), t] \tag{2.3-17}$$

This implies that $\mathbf{u}(t)$, $\mathbf{w}(t)$, and $\mathbf{p}(t)$ are constant, although equilibrium might be maintained in the face of varying disturbances by varying control. It may not always be possible (or important) to achieve total static equilibrium, particularly if some state components are pure integrals of others; for example, an automobile can achieve constant velocity on the highway, assuring steadily increasing distance. This is an example of *quasistatic equilibrium*, because those state components that are not actually fixed have steady rates of change.

An equilibrium point is *stable* if small disturbances do not cause the state to diverge; otherwise, it is an unstable equilibrium. One of the strengths of the optimal feedback controller called a linear-quadratic regulator, to be developed in later chapters, is that it can guarantee system stability about an equilibrium under fairly broad conditions.

In the simplest case, the output vector is the state vector, and Eq. 2.3-2 reduces to

$$\mathbf{y} = \mathbf{I}_n \mathbf{x} \tag{2.3-18}$$

The output equation is often more complicated; $\mathbf{y}$ may be a subset of $\mathbf{x}$, it may include redundant measurements of components of $\mathbf{x}$, it may be a linear combination of components, or it may be a nonlinear function of $\mathbf{x}$. Consider the measurement of two components of an aircraft's state variable, speed and altitude, using air pressure measurements. The output vector includes stagnation (total) pressure p_t, which is measured normal to the flow, and static pressure p_s, measured parallel to the flow. At low airspeed,

$$\mathbf{y} = \begin{bmatrix} y_1 \\ y_2 \end{bmatrix} = \begin{bmatrix} p_t \\ p_s \end{bmatrix} = \begin{bmatrix} p_A(h) + \frac{1}{2}\rho(h)v^2 \\ p_A(h) \end{bmatrix} \tag{2.3-19}$$

where p_A is the ambient pressure—a nonlinear function of altitude (h), as is the air density (ρ)—and v is the true airspeed. Assuming different transducers are used to measure each pressure, instrument errors of the two devices are unrelated, allowing the observation vector to be expressed as

$$\mathbf{z} = \begin{bmatrix} z_1 \\ z_2 \end{bmatrix} = \begin{bmatrix} p_t + n_1 \\ p_s + n_2 \end{bmatrix} \tag{2.3-20}$$

In practice, the measurement errors are more complex, but the example illustrates that even simple measurements may have very nonlinear relationships to the state variables.

Local Linearization

Small perturbations from a nominal trajectory can be modeled by linear approximation. To do this, both sides of Eqs. 2.3-1 to 2.3-3 are expanded in Taylor series, and terms beyond the first degree are neglected:

$$\dot{\mathbf{x}}_o(t) + \Delta\dot{\mathbf{x}}(t) = \mathbf{f}[\mathbf{x}_o(t), \mathbf{u}_o(t), \mathbf{w}_o(t), \mathbf{p}_o(t), t] + \Delta\mathbf{f}[\mathbf{x}(t), \mathbf{u}(t), \mathbf{w}(t), \mathbf{p}(t), t] \tag{2.3-21}$$

$$\mathbf{y}_o(t) + \Delta\mathbf{y}(t) = \mathbf{h}[\mathbf{x}_o(t), \mathbf{u}_o(t), \mathbf{w}_o(t), \mathbf{p}_o(t), t] + \Delta\mathbf{h}[\mathbf{x}(t), \mathbf{u}(t), \mathbf{w}(t), \mathbf{p}(t), t] \tag{2.3-22}$$

$$\mathbf{z}_o(t) + \Delta\mathbf{z}(t) = \mathbf{j}[\mathbf{y}_o(t), \mathbf{n}_o(t), t] + \Delta\mathbf{j}[\mathbf{y}(t), \mathbf{n}(t), t] \tag{2.3-23}$$

The zeroth-degree terms generate the nominal solution, and the first-degree terms govern the perturbation solution (Fig. 2.3-1). The *nominal, nonlinear equations* are

$$\dot{\mathbf{x}}_o(t) = \mathbf{f}[\mathbf{x}_o(t), \mathbf{u}_o(t), \mathbf{w}_o(t), \mathbf{p}_o(t), t] \tag{2.3-24}$$

$$\mathbf{y}_o(t) = \mathbf{h}[\mathbf{x}_o(t), \mathbf{u}_o(t), \mathbf{w}_o(t), \mathbf{p}_o(t), t] \tag{2.3-25}$$

$$\mathbf{z}_o(t) = \mathbf{j}[\mathbf{y}_o(t), \mathbf{n}_o(t), t] \tag{2.3-26}$$

Assuming that the parameter time histories are not perturbed, the *associated linear, time-varying equations* that describe perturbations from the nominal solution are

$$\Delta\dot{\mathbf{x}}(t) = \mathbf{F}(t)\Delta\mathbf{x}(t) + \mathbf{G}(t)\Delta\mathbf{u}(t) + \mathbf{L}(t)\Delta\mathbf{w}(t) \tag{2.3-27}$$

$$\Delta\mathbf{y}(t) = \mathbf{H}_{\mathbf{x}}(t)\Delta\mathbf{x}(t) + \mathbf{H}_{\mathbf{u}}(t)\Delta\mathbf{u}(t) + \mathbf{H}_{\mathbf{w}}(t)\Delta\mathbf{w}(t) \tag{2.3-28}$$

$$\Delta\mathbf{z}(t) = \mathbf{J}_{\mathbf{y}}(t)\Delta\mathbf{y}(t) + \mathbf{J}_{\mathbf{n}}(t)\Delta\mathbf{n}(t) \tag{2.3-29}$$

where the Jacobian matrices

$$\mathbf{F}(t) = \frac{\partial \mathbf{f}}{\partial \mathbf{x}} \qquad (n \times n) \tag{2.3-30}$$

$$\mathbf{G}(t) = \frac{\partial \mathbf{f}}{\partial \mathbf{u}} \qquad (n \times m) \tag{2.3-31}$$

$$\mathbf{L}(t) = \frac{\partial \mathbf{f}}{\partial \mathbf{w}} \qquad (n \times s) \tag{2.3-32}$$

$$\mathbf{H}_{\mathbf{x}}(t) = \frac{\partial \mathbf{h}}{\partial \mathbf{x}} \qquad (r \times n) \tag{2.3-33}$$

$$\mathbf{H}_{\mathbf{u}}(t)=\frac{\partial\mathbf{h}}{\partial\mathbf{u}}\qquad(r\times m)\tag{2.3-34}$$

$$\mathbf{H}_{\mathbf{w}}(t)=\frac{\partial\mathbf{h}}{\partial\mathbf{w}}\qquad(r\times s)\tag{2.3-35}$$

$$\mathbf{J}_{\mathbf{y}}(t)=\frac{\partial\mathbf{j}}{\partial\mathbf{y}}\qquad(r\times r)\tag{2.3-36}$$

$$\mathbf{J}_{\mathbf{n}}(t)=\frac{\partial\mathbf{j}}{\partial\mathbf{n}}\qquad(r\times r)\tag{2.3–37}$$

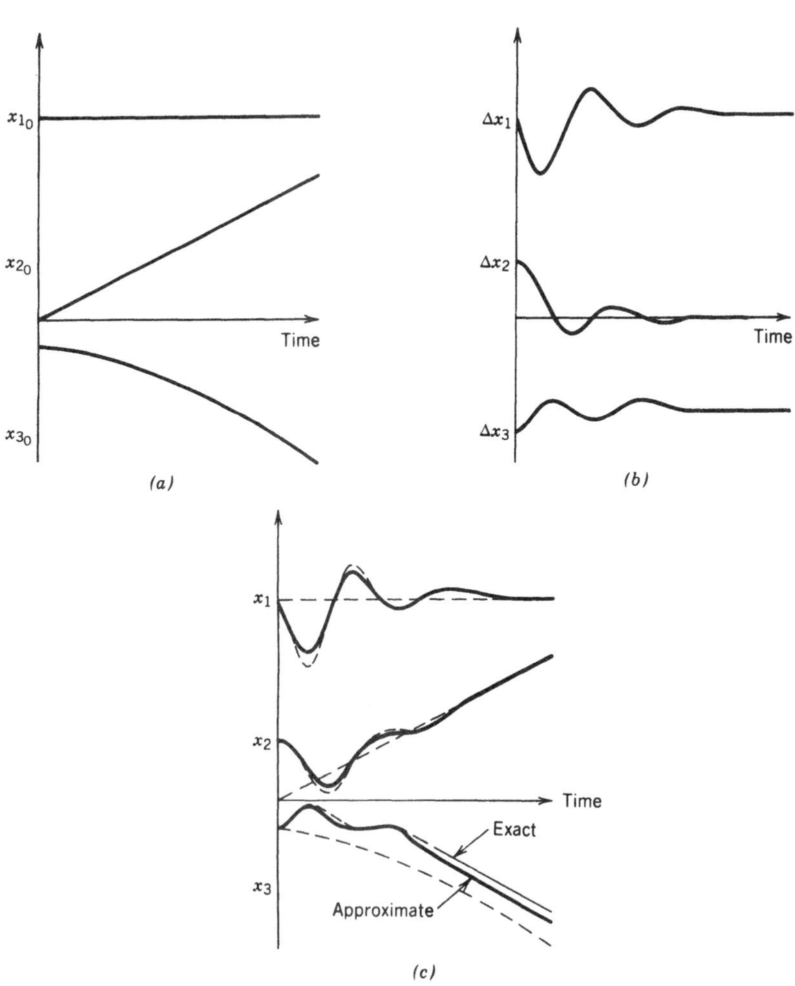

FIGURE 2.3-1 Example of approximation of perturbed solution as the sum of the nominal solution plus a linear perturbation solution. (*a*) Nonlinear nominal solution; (*b*) linear perturbation solution; (*c*) perturbed solution.

are evaluated at $[\mathbf{x}_o(t), \mathbf{u}_o(t), \mathbf{w}_o(t), \mathbf{p}_o(t), t]$. The Jacobian matrices express the linear sensitivity of $\mathbf{f}$ and $\mathbf{h}$ to small perturbations in $\mathbf{x}$, $\mathbf{u}$, and $\mathbf{w}$. Even if $\mathbf{p}_o(t)$ is constant, these matrices may vary in time when the nominal input variables $(\mathbf{u}_o, \mathbf{w}_o)$ and solution variable $(\mathbf{x}_o)$ vary in time. The linear, time-invariant case occurs only with constant nominal inputs, when the state has reached equilibrium, or if there is no functional dependence of the Jacobian matrices on the changing variables.

The two examples used earlier in the section can be used to illustrate local linearization. For the first case (Eq. 2.3-12), $\mathbf{F}$ and $\mathbf{G}$ are

$$\mathbf{F} = \begin{bmatrix} 0 & 1 & 0 \\ 0 & 0 & 1 \\ -c_3 & -c_2 & -c_1 \end{bmatrix} \tag{2.3-38}$$

$$\mathbf{G} = \begin{bmatrix} 0 \\ 0 \\ b \end{bmatrix} \tag{2.3-39}$$

while for the second (Eq. 2.3-15),

$$\mathbf{F} = \begin{bmatrix} 0 & 1 & 0 & 0 \\ -2a_1(x_{3_o} - x_{1_o}) & -a_2 & 2a_1(x_{3_o} - x_{1_o}) & a_2 \\ 0 & 0 & 0 & 1 \\ c_1 & 0 & c_1 & 2c_2 x_{4_o} \end{bmatrix} \tag{2.3-40}$$

$$\mathbf{G} = \begin{bmatrix} 0 & 0 \\ b_1 & b_2 \\ 0 & 0 \\ 2d_1 u_{1_o} & 0 \end{bmatrix} \tag{2.3-41}$$

Keep in mind that the nominal values (x_{1_o} to x_{4_o} and u_{1_o}) can be time-varying. A linearized output equation example (from Eq. 2.3-19) is

$$\Delta\mathbf{y} = \begin{bmatrix} \Delta p_t \\ \Delta p_s \end{bmatrix} = \begin{bmatrix} \rho(h_o)v_o & \left[\dfrac{\partial p_A}{\partial h} + \dfrac{1}{2}\dfrac{\partial \rho}{\partial h} v_o^2\right]\Bigg|_{\mathbf{x}=\mathbf{x}_o} \\ 0 & \dfrac{\partial p_A}{\partial h}\Bigg|_{\mathbf{x}=\mathbf{x}_o} \end{bmatrix} \begin{bmatrix} \Delta v \\ \Delta h \end{bmatrix} \tag{2.3-42}$$

and the corresponding linearized observation equation is (from Eq. 2.3-20)

$$\Delta\mathbf{z} = \begin{bmatrix} \Delta p_t + \Delta n_1 \\ \Delta p_s + \Delta n_2 \end{bmatrix} \tag{2.3-43}$$

Earlier comments about the equilibrium and stability of nonlinear differential equations also apply to linear equations. When the linear equations represent perturbations about a nominal solution, that solution is taken as a reference. If $\Delta\dot{\mathbf{x}}(t) = \mathbf{0}$, the perturbation solution is at equilibrium; this does not prevent $\mathbf{x}_o(t)$ from varying. If the linear system is stable, unforced perturbation solutions do not diverge from the nominal solution with increasing time.

Numerical Integration of Nonlinear Equations

Solving for $\mathbf{x}(t)$ in $[t_0, t_f]$ requires the integration of Eq. 2.3-1; this is expressed symbolically by

$$\mathbf{x}(t) = \mathbf{x}(t_0) + \int_{t_o}^{t} \mathbf{f}[\mathbf{x}(\tau), \mathbf{u}(\tau), \mathbf{w}(\tau), \mathbf{p}(\tau), \tau]d\tau \tag{2.3-44}$$

where $\mathbf{u}(t)$, $\mathbf{w}(t)$, and $\mathbf{p}(t)$ are assumed to be known in the interval, and $\mathbf{x}(t)$ is instantaneously available to allow evaluation of $\mathbf{f}[\cdot]$. If it is sufficient to tabulate $\mathbf{x}(t)$ at discrete instants of time, the integration algorithm could step from one instant to the next, using the previous result as an initial condition for the next time increment:

$$\mathbf{x}(t_k) = \mathbf{x}(t_{k-1}) + \int_{t_{k-1}}^{t_k} \mathbf{f}[\mathbf{x}(t), \mathbf{u}(t), \mathbf{w}(t), \mathbf{p}(t), t]dt \tag{2.3-45}$$

where k is a sampling index. The problem is the same as before, but the time increment $(t_k - t_{k-1})$ has been made arbitrarily small, allowing computational approximations based on multiplication and addition to substitute for the operation of integration.

Algorithms for numerical integration differ from each other in the way that $\mathbf{f}[\cdot]$ is assumed to vary in the interval. There is a trade-off to be made between the length of the time increment and the precision with which the variation in $\mathbf{f}[\cdot]$ is represented. If $\mathbf{f}[\cdot]$ is portrayed precisely in the interval, large time steps can be used; if not, small steps are required.

*Euler** (or *rectangular*) *integration* provides the crudest representation of $\mathbf{f}[\cdot]$, holding it constant at the value that pertains for $t = t_{k-1}$. With $\Delta t = t_k - t_{k-1}$,

$$\mathbf{x}(t_k) = \mathbf{x}(t_{k-1}) + \mathbf{f}[\mathbf{x}(t_{k-1}), \mathbf{u}(t_{k-1}), \mathbf{w}(t_{k-1}), \mathbf{p}(t_{k-1}), t_{k-1}]\Delta t \tag{2.3-46}$$

This algorithm requires a single *function evaluation*—computation of $\mathbf{f}[\cdot]$ with given values of $\mathbf{x}$, $\mathbf{u}$, $\mathbf{w}$, $\mathbf{p}$, and t—and it does not provide for the effects that changing variables may have on $\mathbf{f}[\cdot]$ in the interval.

Modified Euler (or *trapezoidal*) *integration* makes a correction based on

*Leonhard Euler (1707–1783), perhaps the most prolific mathematician of all time, made fundamental contributions in calculus, number theory, geometry, and algebra.

the first function evaluation and the known variations in **u**, **w**, **p**, and t to improve the result. Incremental changes in the state are calculated as

$$\Delta \mathbf{x}_1 = \mathbf{f}[\mathbf{x}(t_{k-1}), \mathbf{u}(t_{k-1}), \mathbf{w}(t_{k-1}), \mathbf{p}(t_{k-1}), t_{k-1})]\Delta t \tag{2.3-47}$$

$$\Delta \mathbf{x}_2 = \mathbf{f}\{[\mathbf{x}(t_{k-1}) + \Delta \mathbf{x}_1], \mathbf{u}(t_k), \mathbf{w}(t_k), \mathbf{p}(t_k), t_k\}\Delta t \tag{2.3-48}$$

and the final computation of $\mathbf{x}(t_k)$ is

$$\mathbf{x}(t_k) = \mathbf{x}(t_{k-1}) + \tfrac{1}{2}(\Delta x_1 + \Delta \mathbf{x}_2) \tag{2.3-49}$$

Two function evaluations are required, effectively doubling the amount of computation per time step, but the accuracy of the result is improved, and the time interval can be increased.

This idea of using interim calculations of $\mathbf{f}[\cdot]$ to improve the fit in the interval is generalized in *Runge–Kutta integration.** For n^{th}-order Runge–Kutta integration, n function evaluations are made to produce n state increments; a weighted sum of the increments is used to update the state. The Euler and modified-Euler algorithms can be considered Runge–Kutta methods of order one and two, respectively. The fourth-order algorithm provides a good trade-off between accuracy and computation; it can be expressed as follows:

$$\Delta \mathbf{x}_1 = \mathbf{f}[\mathbf{x}(t_{k-1}), \mathbf{u}(t_{k-1}), \mathbf{w}(t_{k-1}), \mathbf{p}(t_{k-1}), t_{k-1}]\Delta t \tag{2.3-50}$$

$$\Delta \mathbf{x}_2 = \mathbf{f}\{[\mathbf{x}(t_{k-1}) + \Delta \mathbf{x}_1/2], \mathbf{u}(t_{k-1/2}), \mathbf{w}(t_{k-1/2}), \mathbf{p}(t_{k-1/2}), t_{k-1/2}\}\Delta t \tag{2.3-51}$$

$$\Delta \mathbf{x}_3 = \mathbf{f}\{[\mathbf{x}(t_{k-1}) + \Delta \mathbf{x}_2/2], \mathbf{u}(t_{k-1/2}), \mathbf{w}(t_{k-1/2}), \mathbf{p}(t_{k-1/2}), t_{k-1/2}\}\Delta t \tag{2.3-52}$$

$$\Delta \mathbf{x}_4 = \mathbf{f}\{[\mathbf{x}(t_{k-1}) + \Delta \mathbf{x}_3], \mathbf{u}(t_k), \mathbf{w}(t_k), \mathbf{p}(t_k), t_k\}\Delta t \tag{2.3-53}$$

$$\mathbf{x}(t_k) = \mathbf{x}(t_{k-1}) + \tfrac{1}{6}(\Delta \mathbf{x}_1 + 2\Delta \mathbf{x}_2 + 2\Delta \mathbf{x}_3 + \Delta \mathbf{x}_4) \tag{2.3-54}$$

where $t_{k-1/2}$ symbolizes $t = t_{k-1} + \Delta t/2$.

The integration time step need not be held constant from one interval to the next. In the interests of reducing computation during periods of inactivity while retaining closely spaced computations during fast transients, the relative magnitudes of state increments can be tested. If the magnitudes are small, the time interval can be doubled; if they are large, it can be halved. The doubling/halving process can be repeated, as required.

The Runge–Kutta algorithms described here are called "one-step processes" because each new state computation involves only the last computed state plus an increment. "Multistep processes", also called predictor-

*Carl Runge (1856–1927), considered a mathematician by the physicists and a physicist by the mathematicians, was a professor of applied mathematics at Göttingen. M. Wilhelm Kutta (1867–1944), professor of mathematics at the Technische Hochschule of Stuttgart, is well known for his contributions in aerodynamics and hydrodynamics.

corrector algorithms, involve prior state computations, such as $\mathbf{x}(t_{k-2})$, $\mathbf{x}(t_{k-3})$, and so on. The past values are used to fit a curve that predicts $\mathbf{x}(t_k)$ by extrapolation, and a single function evaluation corrects this prediction. Evidently quite efficient, these algorithms are better than Euler integration only if state variations are relatively smooth. With random inputs, the value of the prediction by extrapolation is minimal. While Runge–Kutta integration is self-starting with given initial conditions, predictor-corrector algorithms must obtain starting values of $\mathbf{x}(t_{k-2})$, and so on, externally; hence a few Runge–Kutta steps may be required to get the predictor-corrector computations started.

Numerical Integration of Linear Equations

The preceding algorithms also apply to linear equations (Eq. 2.3-27), but the principle of superposition can be used to generalize the solution. Because the effects of initial conditions and inputs are additive, the integral

$$\Delta\mathbf{x}(t_k) = \Delta\mathbf{x}(t_{k-1}) + \int_{t_{k-1}}^{t_k} [\mathbf{F}(\tau)\Delta\mathbf{x}(\tau) + \mathbf{G}(\tau)\Delta\mathbf{u}(\tau) + \mathbf{L}(\tau)\Delta\mathbf{w}(\tau)]d\tau \quad (2.3\text{-}55)$$

can be separated into homogeneous (unforced) and nonhomogeneous (forced) solutions. The former describes the system's initial condition response, the latter describes the effects of control and disturbances, and the total response is the sum of the two.

The initial condition response can be expressed in terms of an $(n \times n)$ time-varying *fundamental solution matrix*, $\mathbf{U}(t)$, which characterizes the state time histories resulting from unit initial conditions in each of the state variables (Fig. 2.3-2):

$$\mathbf{U}(t) = [\Delta\mathbf{x}_1(t) \quad \Delta\mathbf{x}_2(t) \ldots \Delta\mathbf{x}_n(t)] \quad (2.3\text{-}56)$$

with

$$\mathbf{U}(t_0) = \mathbf{I}_n \quad (2.3\text{-}57)$$

If each of these unit-initial-condition time histories is known, then, by superposition, the effects of *arbitrary* initial conditions on $\Delta\mathbf{x}(t)$ are easily portrayed as

$$\Delta\mathbf{x}(t) = \mathbf{U}(t)\Delta\mathbf{x}(t_0) \quad (2.3\text{-}58)$$

This is a solution to the homogeneous equation, $\Delta\dot{\mathbf{x}}(t) = \mathbf{F}(t)\Delta\mathbf{x}(t)$.

As a consequence of Eq. 2.3-58, the states at any two times t_1 and t_2 in

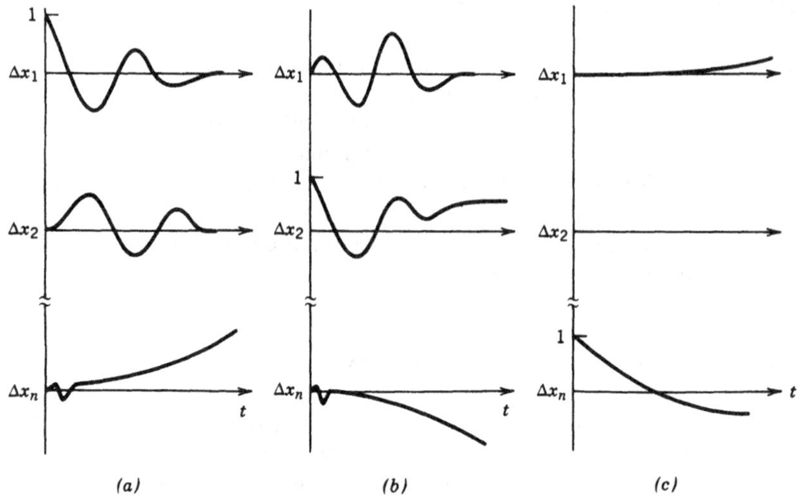

FIGURE 2.3-2 Fundamental solution vectors. State time histories in response to separate unit initial conditions in each component of the state (*a*) Δx_1; (*b*) Δx_2; (*c*) Δx_n.

$[t_0, t_f]$ are directly related, assuming $\mathbf{U}(t)$ is always nonsingular. Since

$$\Delta\mathbf{x}(t_1) = \mathbf{U}(t_1)\Delta\mathbf{x}(t_0) \tag{2.3-59}$$

$$\Delta\mathbf{x}(t_0) = \mathbf{U}^{-1}(t_1)\Delta\mathbf{x}(t_1) \tag{2.3-60}$$

and

$$\begin{aligned}\Delta\mathbf{x}(t_2) &= \mathbf{U}(t_2)\Delta\mathbf{x}(t_0)\\ &= \mathbf{U}(t_2)\mathbf{U}^{-1}(t_1)\Delta\mathbf{x}(t_1)\\ &= \mathbf{\Phi}(t_2, t_1)\Delta\mathbf{x}(t_1)\end{aligned} \tag{2.3-61}$$

$\mathbf{\Phi}(t_2, t_1)$ is the *state transition matrix*, which allows the state at t_2 to be derived from the state at t_1. Then

$$\begin{aligned}\Delta\mathbf{x}(t_1) &= \mathbf{\Phi}^{-1}(t_2, t_1)\Delta x(t_2)\\ &= \mathbf{\Phi}(t_1, t_2)\Delta\mathbf{x}(t_2)\end{aligned} \tag{2.3-62}$$

State transition matrices can be used to propagate the state forward or backward in time, and they have the following easily seen properties:

$$\mathbf{\Phi}(t, t) = \mathbf{I}_n \tag{2.3-63}$$

$$\mathbf{\Phi}(t_2, t_0) = \mathbf{\Phi}(t_2, t_1)\mathbf{\Phi}(t_1, t_0) \tag{2.3-64}$$

Because

$$\Delta\dot{\mathbf{x}}(t) = \dot{\mathbf{\Phi}}(t, t_1)\Delta\mathbf{x}(t_1) = \mathbf{F}(t)\Delta\mathbf{x}(t) = \mathbf{F}(t)\mathbf{\Phi}(t, t_1)\Delta\mathbf{x}(t_1) \tag{2.3-65}$$

the state transition matrix satisfies the original differential equation

$$\dot{\mathbf{\Phi}}(t, t_1) = \mathbf{F}(t)\mathbf{\Phi}(t, t_1) \tag{2.3-66}$$

for any t and t_1. From Eq. 2.3-58, $\mathbf{U}(t)$ also satisfies this equation:

$$\dot{\mathbf{U}}(t) = \mathbf{F}(t)\mathbf{U}(t) \tag{2.3-67}$$

All of the equations up to Eq. 2.3-64 have been presented with the assumption that we know what the fundamental solutions are, allowing $\mathbf{U}(t)$ to be specified, leading in turn to the state transition matrix, $\mathbf{\Phi}(t_2, t_1)$, that relates any two times in the interval. Clearly, we could solve the original homogeneous equation or Eq. 2.3-67 using Runge–Kutta integration to find $\mathbf{U}(t)$, then compute $\mathbf{\Phi}(t_2, t_1)$ as earlier. An alternative is based on *successive approximations* to the solutions for $\Delta\mathbf{x}(t)$. Defining a series of approximate equations

$$\Delta\dot{\mathbf{x}}_1(t) = \mathbf{F}(t)\Delta\mathbf{x}(t_0) \tag{2.3-68a}$$

$$\Delta\dot{\mathbf{x}}_2(t) = \mathbf{F}(t)\Delta\mathbf{x}_1(t_0) \tag{2.3-68b}$$

leads to the following solutions:

$$\begin{aligned}\Delta\mathbf{x}_1(t) &= \Delta\mathbf{x}(t_0) + \int_{t_0}^{t} \mathbf{F}(\tau)\Delta\mathbf{x}(t_0)\,d\tau \\ &= \left[\mathbf{I}_n + \int_{t_0}^{t} \mathbf{F}(\tau)\,d\tau\right]\Delta\mathbf{x}(t_0)\end{aligned} \tag{2.3-69}$$

$$\begin{aligned}\Delta\mathbf{x}_2(t) &= \Delta\mathbf{x}(t_0) + \int_{t_0}^{t} \mathbf{F}(\xi)\Delta\mathbf{x}_1(\xi)\,d\xi \\ &= \Delta\mathbf{x}(t_0) + \int_{t_0}^{t} \mathbf{F}(\xi)\left[\mathbf{I}_n + \int_{t_0}^{\xi} \mathbf{F}(\tau)\,d\tau\right]\Delta\mathbf{x}(t_0)\,d\xi \\ &= \left[\mathbf{I}_n + \int_{t_0}^{t} \mathbf{F}(\tau)\,d\tau + \int_{t_0}^{t} \mathbf{F}(\xi)\int_{t_0}^{\xi} \mathbf{F}(\tau)\,d\tau\,d\xi\right]\Delta\mathbf{x}(t_0)\end{aligned} \tag{2.3-70}$$

These equations take the form of Eq. (2.3-61), so the terms in square brackets are seen to be approximations to $\mathbf{\Phi}(t, t_0)$.

If $\mathbf{F}$ is constant, Eq. 2.3-70 becomes

$$\Delta\mathbf{x}_2(t) = [\mathbf{I}_n + \mathbf{F}(t - t_0) + \tfrac{1}{2}\mathbf{F}^2(t - t_0)^2]\Delta\mathbf{x}(t_0) \tag{2.3-71}$$

Carrying the approximation further, a matrix infinite series for $\mathbf{\Phi}(t, t_0)$ is found:

$$\mathbf{\Phi}(t, t_0) = \mathbf{I}_n + \mathbf{F}(t - t_0) + (1/2!)\mathbf{F}^2(t - t_0)^2 + (1/3!)\mathbf{F}^3(t - t_0)^3 + \cdots \quad (2.3\text{-}72)$$

In the limit, this series represents the matrix exponential function

$$\mathbf{\Phi}(t, t_0) = e^{\mathbf{F}(t - t_0)} \quad (2.3\text{-}73)$$

Only the *difference* between t and t_0 appears in this equation; defining $\Delta t = t - t_0$,

$$\mathbf{\Phi}(t, t_0) = \mathbf{\Phi}(\Delta t) = e^{\mathbf{F}\Delta t} \quad (2.3\text{-}74)$$

and $\mathbf{\Phi}(\Delta t)$ is a constant $(n \times n)$ matrix for fixed $\mathbf{F}$. Then the homogeneous solution for Δx can be propagated from one instant to the next as

$$\Delta\mathbf{x}(t_k) = \mathbf{\Phi}(\Delta t)\Delta\mathbf{x}(t_{k-1}) \quad (2.3\text{-}75)$$

From Eq. 2.3-74,

$$\mathbf{\Phi}(2\Delta t) = \mathbf{\Phi}^2(\Delta t) \quad (2.3\text{-}76)$$

$$\mathbf{\Phi}(n\Delta t) = \mathbf{\Phi}^n(\Delta t) \quad (2.3\text{-}77)$$

These relationships are of particular value in the analysis of linear, time-invariant, discrete-time systems.

Returning to the forced solution of $\Delta\mathbf{x}(t)$, we begin by writing the time-rate-of-change of $\mathbf{U}^{-1}(t)$. From Eqs. 2.2-49 and 2.3-67,

$$\dot{\mathbf{U}}^{-1}(t) = -\mathbf{U}^{-1}(t)\mathbf{F}(t) \quad (2.3\text{-}78)$$

Then

$$\dot{\mathbf{U}}^{-1}(t)\Delta\mathbf{x}(t) = -\mathbf{U}^{-1}(t)\mathbf{F}(t)\Delta\mathbf{x}(t) \quad (2.3\text{-}79)$$

Premultiplying both sides of Eq. 2.3-27 by $\mathbf{U}^{-1}(t)$ and adding Eq. 2.3-79 leads to

$$\mathbf{U}^{-1}(t)\Delta\dot{\mathbf{x}}(t) + \dot{\mathbf{U}}^{-1}(t)\Delta\mathbf{x}(t) = \mathbf{U}^{-1}(t)[\mathbf{F}(t)\Delta\mathbf{x}(t) + \mathbf{G}(t)\Delta\mathbf{u}(t) + \mathbf{L}(t)\Delta\mathbf{w}(t)] - \mathbf{U}^{-1}(t)\mathbf{F}(t)\Delta\mathbf{x}(t) \quad (2.3\text{-}80a)$$

or

$$\frac{d}{dt}[\mathbf{U}^{-1}(t)\Delta\mathbf{x}(t)] = \mathbf{U}^{-1}(t)[\mathbf{G}(t)\Delta\mathbf{u}(t) + \mathbf{L}(t)\Delta\mathbf{w}(t)] \quad (2.3\text{-}80b)$$

Integrating both sides of this equation from t_0 to t yields

$$\mathbf{U}^{-1}(t)\Delta\mathbf{x}(t) - \mathbf{U}^{-1}(t_0)\Delta\mathbf{x}(t_0) = \int_{t_0}^{t} \mathbf{U}^{-1}(\tau))[\mathbf{G}(\tau)\Delta\mathbf{u}(\tau) + \mathbf{L}(\tau)\Delta\mathbf{w}(\tau)]d\tau \tag{2.3-81a}$$

or

$$\begin{aligned}\Delta\mathbf{x}(t) &= \mathbf{U}(t)\mathbf{U}^{-1}(t_0)\Delta\mathbf{x}(t_0) + \int_{t_0}^{t} \mathbf{U}(t)\mathbf{U}^{-1}(\tau)[\mathbf{G}(\tau)\Delta\mathbf{u}(\tau) + \mathbf{L}(\tau)\Delta\mathbf{w}(t)]d\tau \\ &= \mathbf{\Phi}(t, t_0)\Delta\mathbf{x}(t_0) + \int_{t_0}^{t} \mathbf{\Phi}(t, \tau)[\mathbf{G}(\tau)\Delta\mathbf{u}(\tau) + \mathbf{L}(\tau)\Delta\mathbf{w}(\tau)]d\tau \end{aligned} \tag{2.3-81b}$$

which is the total solution for $\Delta\mathbf{x}(t)$. Equation 2.3-81 is the sum of the system's initial condition response and input response. The latter is a *vector convolution integral* that "maps" the inputs into the state. The matrix products, $\mathbf{\Phi}(t, \tau)\mathbf{G}(\tau)$ and $\mathbf{\Phi}(t, \tau)\mathbf{L}(\tau)$, are time-varying *impulse response matrices* (or Green's functions) of dimension $(n \times m)$ and $(n \times s)$, respectively, that weight present and past inputs to determine their effects on the present state.

A scalar impulse is a sudden input to the system; assuming that the impulse occurs at time t, it is idealized by the *Dirac delta function*,* $\delta(t-\tau)$:

$$\delta(t-\tau) = \begin{cases} \infty, & \tau = t \\ 0, & \tau \neq t \end{cases} \tag{2.3-82}$$

$$\lim_{\epsilon \to 0} \int_{t-\epsilon}^{t+\epsilon} \delta(t-\tau)d\tau = 1 \tag{2.3-83}$$

This is called a *unit impulse function*. Although the amplitude of the delta function is instantaneously infinite, the energy contained in the impulse is finite. By way of example, assume that $\mathbf{G}(t) = \mathbf{I}_n$ and $\Delta\mathbf{u}(t) = [\delta(t_0 - \tau) \quad 0 \quad 0 \ldots 0]^T$ with $\Delta\mathbf{w} = \mathbf{0}$. Then the solution of Eq. 2.3-81 is

$$\int_{t_0}^{t} \mathbf{\Phi}(t, \tau) \begin{bmatrix} \delta(t_0 - \tau) \\ 0 \\ \vdots \\ 0 \end{bmatrix} d\tau = \Delta\mathbf{x}_1(t) \tag{2.3-84}$$

as in Fig. 2.3-2 and Eq. 2.3-56. For arbitrary $\mathbf{G}(t)$, the scalar impulse in the

*Cambridge professor Paul Dirac (1902–1984) shared the 1933 Nobel Prize in physics with Erwin Schrödinger for their work in quantum mechanics.

first control component would excite all of the fundamental initial condition responses in direct proportion to the magnitudes of the first column of $\mathbf{G}(t)$. Unit impulses in the remaining control components would excite similar superposed transient responses; hence $\mathbf{\Phi}(t, \tau)\mathbf{G}(\tau)$ represents the $(n \times m)$ state responses at time t that can result from unit impulses in the m control components at time t, with $t \geq \tau$. Unit impulse disturbances act in the same fashion through $\mathbf{\Phi}(t, \tau)\mathbf{L}(\tau)$.

If $\mathbf{F}$, $\mathbf{G}$, $\mathbf{L}$, and $(t_k - t_{k-1})$ are constant, Eq. 2.3-81 provides the basis for a recursive equation to propagate $\Delta\mathbf{x}(t)$ between discrete instants of time. $\mathbf{\Phi}(t, t_0)$ can be replaced by the constant $\mathbf{\Phi}(\Delta t)$, but $\mathbf{\Phi}(t, \tau)$ remains a function of the integration variable τ during the interval:

$$\Delta\mathbf{x}(t_k) = \mathbf{\Phi}(\Delta t)\Delta\mathbf{x}(t_{k-1}) + \int_{t_{k-1}}^{t_k} \mathbf{\Phi}(t_k, \tau)[\mathbf{G}\Delta\mathbf{u}(\tau) + \mathbf{L}\Delta\mathbf{w}(\tau)]d\tau \quad (2.3\text{-}85)$$

If $\Delta\mathbf{u}(t)$ and $\Delta\mathbf{w}(t)$ can be considered constant during the interval, which would be exact for the output of a digital controller and approximate for continuous inputs, further simplification can be made. Using Eq. 2.3-73,

$$\begin{aligned}\Delta\mathbf{x}(t_k) &= \mathbf{\Phi}(\Delta t)\Delta\mathbf{x}(t_{k-1}) + \int_{t_{k-1}}^{t_k} e^{\mathbf{F}(t_k-\tau)}\, d\tau[\mathbf{G}\Delta\mathbf{u}(t_{k-1}) + \mathbf{L}\Delta\mathbf{w}(t_{k-1})] \\ &= \mathbf{\Phi}(\Delta t)\Delta\mathbf{x}(t_{k-1}) + \mathbf{\Phi}(\Delta t)\int_0^{\Delta t} e^{-\mathbf{F}\tau}\, d\tau[\mathbf{G}\Delta\mathbf{u}(t_{k-1}) + \mathbf{L}\Delta\mathbf{w}(t_{k-1})] \quad (2.3\text{-}86a) \\ &= \mathbf{\Phi}(\Delta t)\Delta\mathbf{x}(t_{k-1}) + \mathbf{\Gamma}(\Delta t)\Delta\mathbf{u}(t_{k-1}) + \mathbf{\Lambda}(\Delta t)\Delta\mathbf{w}(t_{k-1}) \quad (2.3\text{-}86b)\end{aligned}$$

where

$$\mathbf{\Gamma}(\Delta t) = \mathbf{\Phi}(\Delta t)[\mathbf{I}_n - \mathbf{\Phi}^{-1}(\Delta t)]\mathbf{F}^{-1}\mathbf{G} = [\mathbf{\Phi}(\Delta t) - \mathbf{I}_n]\mathbf{F}^{-1}\mathbf{G} \quad (2.3\text{-}87)$$

$$\mathbf{\Lambda}(\Delta t) = \mathbf{\Phi}(\Delta t)[\mathbf{I}_n - \mathbf{\Phi}^{-1}(\Delta t)]\mathbf{F}^{-1}\mathbf{L} = [\mathbf{\Phi}(\Delta t) - \mathbf{I}_n]\mathbf{F}^{-1}\mathbf{L} \quad (2.3\text{-}88)$$

Apparently $\mathbf{F}^{-1}$ must exist to define $\mathbf{\Gamma}$ and $\mathbf{\Lambda}$. However, using the series approximation for $e^{\mathbf{F}\tau}$ in Eq. 2.3-86a,

$$\mathbf{\Gamma}(\Delta t) = \mathbf{\Phi}(\Delta t)[\mathbf{I}_n - (1/2)\mathbf{F}\Delta t + (1/3!)\mathbf{F}^2\Delta t^2 - \cdots]\mathbf{G}\Delta t \quad (2.3\text{-}89)$$

$$\mathbf{\Lambda}(\Delta t) = \mathbf{\Phi}(\Delta t)[\cdot]\mathbf{L}\Delta t \quad (2.3\text{-}90)$$

so that $\mathbf{F}^{-1}$ need not exist to compute the discrete-time input matrices.

Representation of Data

The nonlinear functions that appear on the right-hand side of the dynamic equation (Eq. 2.3-1) can be idealized as $\mathbf{f}(\mathbf{x}, \mathbf{u}, \mathbf{w}, \mathbf{p}, t)$, but at some point in the solution of the equation, more explicit dependences on the variables must be defined. While the science or economics of a particular problem specifies the form and dimension of such an equation, there are likely to be important coefficients whose relationships to the variables must be determined experimentally. For example, vertical motions of the spring-mass-damper system shown in Fig. 2.3-3*a* can be represented by

$$m\ddot{x} + d(\dot{x}) + k(x) = g(u) \tag{2.3-91}$$

or, with $x_1 = x$, $x_2 = \dot{x}$,

$$\begin{bmatrix} \dot{x}_1 \\ \dot{x}_2 \end{bmatrix} = \begin{bmatrix} x_2 \\ [-k(x_1) - d(x_2) + g(u)]/m \end{bmatrix} \tag{2.3-92}$$

where m is the mass, $k(x_1)$ is the spring force, $d(x_2)$ is the damping force, and $g(u)$ is a control force. Spring and damping forces often are characterized as linear functions of the state

$$k(x) = k_x x \tag{2.3-93}$$

$$d(\dot{x}) = d_{\dot{x}}\dot{x} \tag{2.3-94}$$

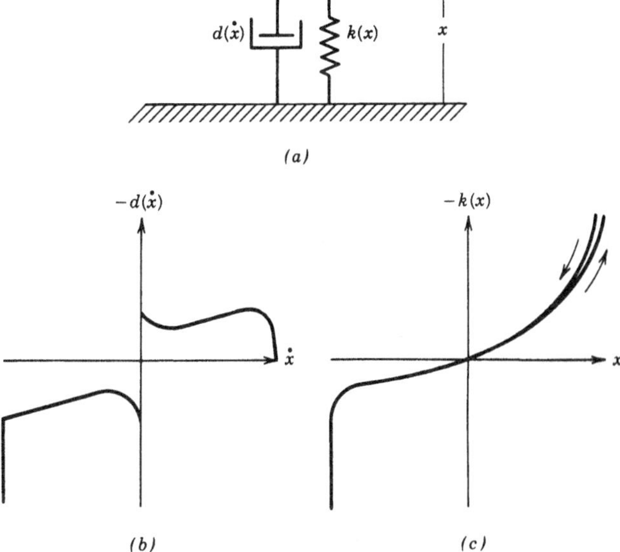

FIGURE 2.3-3 Physical system and experimentally determined forces. (*a*) Spring-mass-damper system; (*b*) damping force; (*c*) spring force.

where k_x and $d_{\dot{x}}$ are evaluated at the nominal operating or equilibrium point. However, experimentation may reveal decidedly nonlinear characteristics, as shown in Figs. 2.3-3*b* and *c*. In addition to curvature, these sketches illustrate limiting, discontinuity, "stiction," asymmetry, and hysteresis in the forces, all of which can occur in practice. Although not shown, it also is possible for the damper to exert a spring force and for the spring to provide damping. Deterministic curve-fitting procedures are required to portray $k(x)$ and $d(\dot{x})$ over the expected ranges of x and $\dot{x}$ for integration of Eq. 2.3-92.

The two most commonly used approaches to deterministic representation of nonlinear functions for digital computation are polynomial and tabular approximation. In the first approach, the original nonlinear function is replaced by an additive series of simpler functions that fits the original function with arbitrarily small error. The function then is evaluated by computing the polynomial. In the second approach, values of the original function are tabulated at given values of the independent variable(s), and the function is evaluated between these points by interpolation. The two approaches can be combined, providing a polynomial fit between tabulated points. Whichever method is used, the instantaneous value of the independent variable becomes the *argument* of the function evaluation. For example, in performing a modified Euler integration of Eq. 2.3-92, $k(x_1)$ should be computed as $k[x_1(t_{k-1})]$ in calculating $\Delta\mathbf{x}_1$ and as $k[x_1(t_{k-1})+\Delta x_1]$ in calculating $\Delta\mathbf{x}_2$.

2.4 RANDOM VARIABLES, SEQUENCES, AND PROCESSES

Scalar Random Variables

A scalar variable x may take different values each time it is sampled; if these values bear no deterministic (or fixed) relationship to the sampling process, x is a *scalar random variable*. Suppose a candy store clerk scoops jelly beans, and x represents the number in one scoop. The number will vary from one scoop (or *event*) to the next, so x is a random variable.

The statistics of x could be determined by careful experiment. Given N events, the frequency of occurrence for specific values of x provides an estimate of the probability that a given result will be obtained in a future sampling:

$$\Pr(x_i) = n_i/N, \qquad i = 1 \text{ to } I \tag{2.4-1}$$

In the example, n_i is the number of scoops containing x_i jelly beans and I is the number of different values of x_i obtained in the experiments. Of course, the probability estimates determined in this way are statistically significant only if N is very large; however, even when N is small, the total number of jelly beans is $\sum_{i=1}^{I} x_i n_i$, and

$$\sum_{i=1}^{I} \Pr(x_i) = \frac{1}{N} \sum_{i=1}^{I} n_i = 1 \tag{2.4-2}$$

By its definition, $\Pr(x_i)$ cannot be negative, and its value is between 0 and 1.

$\Pr(x_i)$, which is also called a *probability mass function*, is the appropriate statistic when x takes discrete values. If x is a continuous variable, the probability mass function must be redefined. Now the clerk is dispensing juice instead of jelly beans, and x represents the volume in each glass. Assume that the device for measuring volume has a resolution of $\pm\Delta x/2$; then n_i represents the number of glasses containing $(x_i \pm \Delta x/2)$ milliliters of juice, and the preceding equations become

$$\Pr\left(x_i \pm \frac{\Delta x}{2}\right) = \frac{n_i}{N}, \qquad i = 1 \text{ to } I \tag{2.4-3}$$

$$\sum_{i=1}^{I} \Pr\left(x_i \pm \frac{\Delta x}{2}\right) = 1 \tag{2.4-4}$$

To this point, it has been assumed that the amplitude of the scalar variable is restricted to a finite set of values (i.e., that it is quantized). The development can be extended to continuous probability distributions by considering what happens as Δx becomes vanishingly small and N becomes arbitrarily large to accommodate the decreasing number of measurements in each interval. First, the *probability density function* for discrete sampling is defined as follows:

$$\text{pr}(x_i) = \frac{\Pr(x_i \pm \Delta x/2)}{\Delta x} \tag{2.4-5a}$$

or

$$\Pr(x_i \pm \Delta x/2) = \text{pr}(x_i)\Delta x \tag{2.4-5b}$$

Then Δx decreases as N and I increase, and in the limit,

$$\sum_{i=1}^{I} \text{pr}(x_i)\Delta x \xrightarrow[\substack{I\to\infty \\ \Delta x\to 0}]{} \int_{x_{min}}^{x_{max}} \text{pr}(x)\,dx = 1 \tag{2.4-6}$$

More generally, x may have no finite limits, so the limits of the summation and the integral are expanded for both continuous and discrete measures of probability:

$$\sum_{i=1}^{\infty} \Pr(x_i) = 1 \tag{2.4-7}$$

$$\int_{-\infty}^{\infty} \text{pr}(x)\, dx = 1 \tag{2.4-8}$$

The average or *expected value* of x is readily determined from $\Pr(x_i)$ or $\text{pr}(x)$. Denoting the expected value by $E(x)$ or $\mathscr{E}(x)$,

$$E(x) = \sum_{i=1}^{\infty} x_i \Pr(x_i) = \bar{x} \tag{2.4-9}$$

for a discrete random variable, and

$$\mathscr{E}(x) = \int_{-\infty}^{\infty} x\, \text{pr}(x)\, dx = \bar{x} \tag{2.4-10}$$

for a continuous random variable. $E(x)$ or $\mathscr{E}(x)$ also is called the *first moment about the origin* or the *mean value* of x, and it is denoted by $\bar{x}$. The mean value has the same units as x. Taking the expected value of a discrete variable is a linear operation because

$$E(ax) = aE(x) \tag{2.4-11}$$

$$E\left[\sum_{k=1}^{K} x_k\right] = \sum_{k=1}^{K} E(x_k) \tag{2.4-12}$$

and the same is true for continuous variables.

Higher moments provide measures of the variability of x, and the nth moments of discrete and continuous variables are defined by similar expressions:

$$E(x^n) = \sum_{i=1}^{\infty} x_i^n \Pr(x_i) \tag{2.4-13}$$

$$\mathscr{E}(x^n) = \int_{-\infty}^{\infty} x^n\, \text{pr}(x)\, dx \tag{2.4-14}$$

Higher moments about the origin reflect not only variation about the mean but variation in the mean value itself. The former taken alone is a more useful measure of variation, leading to the nth *central moments* for discrete and continuous variables:

$$E[(x-\bar{x})^n]=\sum_{i=1}^{\infty}[(x-\bar{x})^n \Pr(x_i)] \tag{2.4-15}$$

$$\mathscr{E}[(x-\bar{x})^n]=\int_{-\infty}^{\infty}(x-\bar{x})^n \operatorname{pr}(x)\,dx \tag{2.4-16}$$

The first central moment is zero in either case, while the second central moment defines the *variance* of x:

$$E[x-E(x)]=E(x-\bar{x})=0 \tag{2.4-17}$$

$$\mathscr{E}[x-\mathscr{E}(x)]=\mathscr{E}(x-\bar{x})=0 \tag{2.4-18}$$

$$E[(x-\bar{x})^2]=\sigma^2 \tag{2.4-19}$$

$$\mathscr{E}[(x-\bar{x})^2]=\sigma^2 \tag{2.4-20}$$

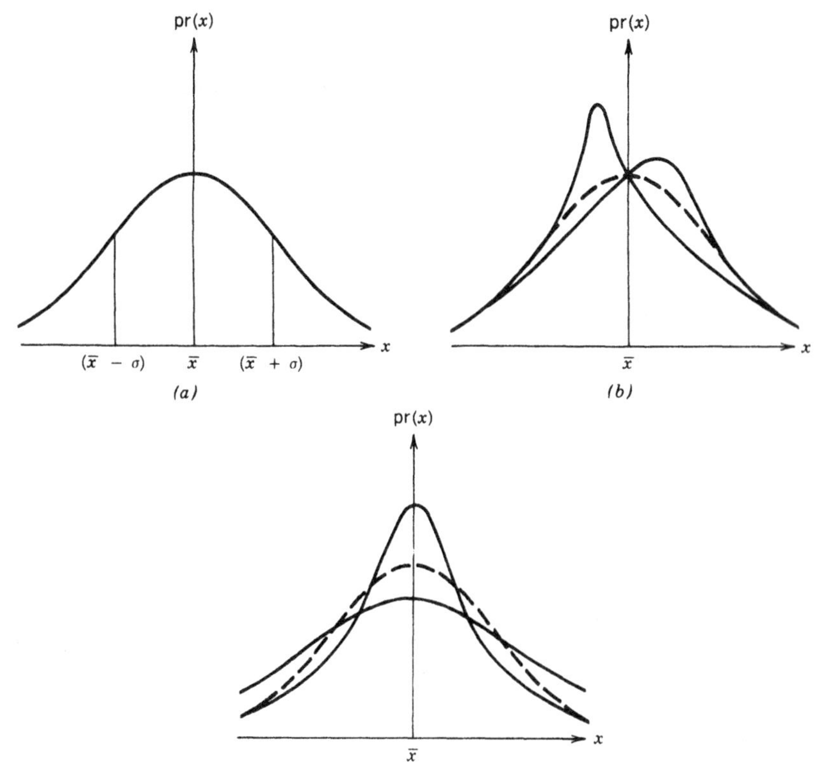

FIGURE 2.4-1 Gaussian and non-Gaussian probability distributions. (*a*) Gaussian distribution; (*b*) skewed distributions; (*c*) distributions with non-Gaussian kurtosis.

The *standard deviation* of x is σ, and it has the same units as x. (Note that the mean and standard deviation of a discrete random variable need not equal one of the discrete values.) It will be seen that the mean and standard deviation are sufficient to define the Gaussian probability distribution (Fig. 2.4-1a) described later. In the remainder, expected values of both discrete and continuous variables will be signified by $E(\cdot)$.

The higher central moments reflect important attributes of non-Gaussian distributions; the third moment indicates *skewness* of the distribution (Fig. 2.4-1b), while the fourth moment (or *kurtosis*) measures the steepness of the distribution's peak (Fig. 2.4-1c). The Gaussian distribution is symmetric about the mean so its third central moment is zero. Its kurtosis is not zero, but the distribution is defined without reference to the fourth moment's value.

Groups of Random Variables

When two random variables are sampled, the probabilities that each takes a certain value may be interrelated. Suppose that the aforementioned jelly beans come in two colors (orange and black), that N scoopfuls are obtained, and that x_i and y_j represent the numbers of each color in each scoopful. The unconditional, individual probabilities can be estimated as before:

$$\Pr(x_i) = n(x_i)/N, \qquad i = 1 \text{ to } I \tag{2.4-21}$$

$$\Pr(y_j) = n(y_j)/N, \qquad j = 1 \text{ to } J \tag{2.4-22}$$

The true probabilities are obtained only as $N \to \infty$, but for any N,

$$\sum_{i=1}^{I} \Pr(x_i) = \frac{1}{N} \sum_{i=1}^{I} n(x_i) = 1 \tag{2.4-23}$$

$$\sum_{j=1}^{J} \Pr(y_j) = \frac{1}{N} \sum_{j=1}^{J} n(y_j) = 1 \tag{2.4-24}$$

Now we are interested in knowing the *joint probability* that x_i and y_j occur simultaneously, signified by $\Pr(x_i, y_j)$. If the orange and black jelly beans were kept in separate barrels and each was sampled in separate scoops, then the probabilities would be unrelated, and the joint probability would be

$$\Pr(x_i, y_j) = \Pr(x_i)\Pr(y_j) \tag{2.4-25}$$

If they were kept in the same barrel and a single scoop was used, then the numbers in a single event would be related, and the joint probability would be estimated by the following:

$$\Pr(x_i, y_j) = \frac{n(x_i)}{N} \text{ and } \frac{n(y_j)}{N}, \qquad i = 1 \text{ to } I, \quad j = 1 \text{ to } J \quad (2.4\text{-}26)$$

Whether or not x_i and y_j are related, the sum over all possible values must equal one:

$$\sum_{i=1}^{I} \sum_{j=1}^{J} \Pr(x_i, y_j) = 1 \qquad (2.4\text{-}27)$$

The *conditional probability* of x_i given y_j is expressed by $\Pr(x_i|y_j)$, and it is simply

$$\Pr(x_i|y_j) = \frac{\Pr(x_i, y_j)}{\Pr(y_j)} \qquad (2.4\text{-}28)$$

Similarly, the conditional probability of y_j given x_i is

$$\Pr(y_j|x_i) = \frac{\Pr(x_i, y_j)}{\Pr(x_i)} \qquad (2.4\text{-}29)$$

If x and y are unrelated (Eq. 2.4-25), then the conditional and unconditional probabilities are the same; that is,

$$\Pr(x_i|y_j) = \Pr(x_i) \qquad (2.4\text{-}30)$$

$$\Pr(y_j|x_i) = \Pr(y_j) \qquad (2.4\text{-}31)$$

Rearranging Eqs. 2.4-28 and 2.4-29,

$$\Pr(x_i|y_j)\Pr(y_i) = \Pr(y_j|x_i)\Pr(x_i) \qquad (2.4\text{-}32)$$

leading to alternate expressions for *Bayes's Rule*:*

$$\Pr(x_i|y_j) = \frac{\Pr(y_j|x_i)\Pr(x_i)}{\Pr(y_j)} \qquad (2.4\text{-}33)$$

$$\Pr(y_j|x_i) = \frac{\Pr(x_i|y_j)\Pr(y_j)}{\Pr(x_i)} \qquad (2.4\text{-}34)$$

*English clergyman Thomas Bayes (1702–1761) established the basis for statistical inference in a brief piece entitled "Essay Towards Solving a Problem in the Doctrine of Chances."

We will have need of such relationships in developing multiple-model estimation algorithms in Section 4.7. Note that the probability of x_i is

$$\Pr(x_i) = \sum_{j=1}^{J} [\Pr(x_i | y_j) \Pr(y_j)] \tag{2.4-35}$$

and if $\Pr(y_j) = 1$, then

$$\Pr(x_i, y_j) = \Pr(x_i | y_j) = \Pr(x_i) \tag{2.4-36}$$

Analogous expressions hold for y_j.

Following the discrete results, conditional and joint probability density functions can be developed for continuous random variables. The two probability density functions must satisfy

$$\int_{-\infty}^{\infty} \mathrm{pr}(x) dx = 1 \tag{2.4-37}$$

$$\int_{-\infty}^{\infty} \mathrm{pr}(y) dy = 1 \tag{2.4-38}$$

and the *joint probability density function* must satisfy

$$\int_{-\infty}^{\infty} \int_{-\infty}^{\infty} \mathrm{pr}(x, y) dx\, dy = 1 \tag{2.4-39}$$

The notations $\mathrm{pr}(x)$, $\mathrm{pr}(y)$, and $\mathrm{pr}(x, y)$ should be read "probability density function of x, of y, and of x and y," respectively, signifying possibly different mathematical functions. The *conditional probability density function* is then defined as

$$\mathrm{pr}(x|y) = \frac{\mathrm{pr}(x, y)}{\mathrm{pr}(y)} \tag{2.4-40}$$

and Bayes's Rule (Eq. 2.4-33) also holds true for probability density functions:

$$\mathrm{pr}(x|y)\Delta x = \frac{\mathrm{pr}(y|x)\Delta y\, \mathrm{pr}(x)\Delta x}{\mathrm{pr}(y)\Delta y} \tag{2.4-41}$$

Canceling the incremental values,

$$\mathrm{pr}(x|y) = \frac{\mathrm{pr}(y|x)\, \mathrm{pr}(x)}{\mathrm{pr}(y)} \tag{2.4-42}$$

It is possible to consider the conditional probabilities of variables with continuous and discrete distributions at the same time. If there are a discrete number of x_i and a continuous distribution of y_j, the previous results imply that

$$\Pr(x_k|y) = \frac{\text{pr}(y|x_k)\Delta y \Pr(x_k)}{\text{pr}(y)\Delta y} = \frac{\text{pr}(y|x_k)\Pr(x_k)}{\text{pr}(y)}$$
$$= \frac{\text{pr}(y|x_k)\Pr(x_k)}{\sum_{i=1}^{I}\text{pr}(y|x_i)\Pr(x_i)} \tag{2.4-43}$$

where k is a particular value in the interval $[1, I]$.

If y is a function of x, its expected value and probability distribution can be determined using the probability distribution of x. We already have seen examples of the expected values of specific functions of x in the discussion of higher moments; given $y = f(x)$, the previous equations suggest that

$$E[f(x)] = \int_{-\infty}^{\infty} f(x)\,\text{pr}(x)\,dx \tag{2.4-44}$$

Assuming that $f(x)$ is a continuous function, y can be expressed as a Taylor series expansion of x,

$$y(x) = y(x_o) + \Delta y = f(x_o) + \left[\frac{\partial f(x)}{\partial x}\right]_{x=x_o} \Delta x + \cdots \tag{2.4-45}$$

so a first-order equation for the perturbation in y due to a variation in x is

$$\Delta y = \left.\frac{\partial f(x)}{\partial x}\right|_{x=x_o} \Delta x = \frac{\partial f(x_o)}{\partial x}\Delta x \tag{2.4-46}$$

Now the probability that y lies in Δy [referenced to $y(x_o)$] must be the same as the probability that x is in Δx; hence

$$\text{pr}[y(x_o)]\Delta y = \text{pr}[y(x_o)]\frac{\partial f(x_o)}{\partial x}\Delta x$$
$$= \text{pr}(x_o)\Delta x \tag{2.4-47a}$$

or

$$\text{pr}[y(x_o)] = \frac{\text{pr}(x_o)}{|\partial f(x_o)/\partial x|} \tag{2.4-47b}$$

This equation assumes that there is a monotonic relationship between y and x (and between x and y), but multivalued functions often occur. In such case, Eq. 2.4-47b must be modified by a suitable multiplier. If

$$y = f(x) = A \sin(\omega t + x) \tag{2.4-48}$$

where x is a random phase angle between 0 and 2π, y takes the same value twice in the range of x. Let x have a *uniform* (or *rectangular*) *distribution*,

$$\text{pr}(x) = \frac{1}{2\pi}, \qquad 0 \le x \le 2\pi \tag{2.4-49}$$

Then the sensitivity of y to variations in x is

$$\begin{aligned} \frac{\partial f}{\partial x} &= A\cos(\omega t + x) = A\sqrt{1 - \sin^2(\omega t + x)} \\ &= \sqrt{A^2 - y^2} \end{aligned} \tag{2.4-50}$$

Multiplying Eq. 2.4-47b by two, the *probability density function of a sine wave* with amplitude A and random phase angle is

$$\text{pr}(y) = \begin{cases} \dfrac{1}{\pi\sqrt{A^2 - y^2}}, & y \le A \\ 0, & y > A \end{cases} \tag{2.4-51}$$

Note that the distribution is independent of the natural frequency ω of the sine wave.

A distinctly different situation arises when the probability distribution of the *sum* of two or more independent random variables must be determined. If z is the sum of independent variables,

$$z = x + y \tag{2.4-52}$$

then the conditional probability density function of z given x is

$$\text{pr}(z|x) = \text{pr}(y) = \text{pr}(z - x) \tag{2.4-53}$$

and the unconditional probability density function of z is

$$\text{pr}(z) = \int_{-\infty}^{\infty} \text{pr}(z|x)\,\text{pr}(x)\,dx = \int_{-\infty}^{\infty} \text{pr}(z - x)\,\text{pr}(x)\,dx \tag{2.4-54}$$

This equation is known as a *convolution* or *superposition integral* (introduced in a different context in Section 2.3); pr(z) results from an integral smoothing of component probability density functions. Furthermore, from Eq. 2.4-12, the mean value and variance of z are

$$\bar{z} = \bar{x} + \bar{y} \tag{2.4-55}$$

$$\sigma_z^2 = \sigma_x^2 + \sigma_y^2 \tag{2.4-56}$$

and these results (Eqs. 2.4-54 to 2.4-56) can be extended for sums of greater numbers of independent variables:

$$z = \sum_{i=1}^{I} a_i x_i \tag{2.4-57}$$

$$\bar{z} = \sum_{i=1}^{I} a_i \bar{x}_i \tag{2.4-58}$$

$$\sigma_z^2 = \sum_{i=1}^{I} a_i^2 \sigma_{x_i}^2 \tag{2.4-59}$$

The *Central Limit Theorem* states that the probability density function of the sum approaches a Gaussian distribution as the number grows, even when the component distributions are markedly non-Gaussian. This finding is fortuitous for estimation problems, which often involve such sums, because the *Gaussian distribution* is specified by only two quantities: its mean value, $\bar{x}$, and its standard deviation, σ_x (or variance, σ_x^2).* If x is a Gaussian random variable, its probability density function is

$$\text{pr}(x) = \frac{1}{(2\pi)^{1/2}\sigma_x} e^{-(x-\bar{x})^2/2\sigma_x^2} \tag{2.4-60}$$

Evaluating $E(x)$ and $E[(x - \bar{x})^2]$ for this distribution will confirm the previous statement.

Scalar Random Sequences and Processes

A *scalar random sequence* consists of a group of related scalar random variables that occur at discrete points in time or space. The variables themselves can have continuous or discrete amplitude distributions, but they are defined at distinct points. In the remainder of the section, it will be assumed that the variables have continuous amplitudes and that the sampling variable represents time increments; hence the random sequence x_1, x_2, x_3 . . . is accompanied by the probability *density* sequence $\text{pr}(x_1)$, $\text{pr}(x_2)$, and so on. A *scalar random process* is continuous in time as well as amplitude, so

*In the multivariate case, these two quantities are the mean value *vector* and the covariance *matrix*, as described later in this section.

$x(t)$ is accompanied by $\text{pr}[x(t)]$. The distinction between processes and sequences becomes important in the latter portion of this section; in the next few paragraphs, the comments on processes apply to both.

A specific "random" process actually is deterministic, in the sense that it happens (or happened) with probability one; however, there is no deterministic way of accurately predicting the future process, and the "random" fluctuations of past measurements may mask an underlying process that really is systematic. In such instance, it is useful to think of the process as one member of an *ensemble of sequences*, each of which characterizes the same phenomenon without having the same time history. The statistics of x derive from estimates of the probability density function across the entire ensemble, and they may vary in time (i.e., they may be *nonstationary*).

An infinity of ensemble members would be required to compute precise statistics for a nonstationary random process (Fig. 2.4-2). In some few cases,

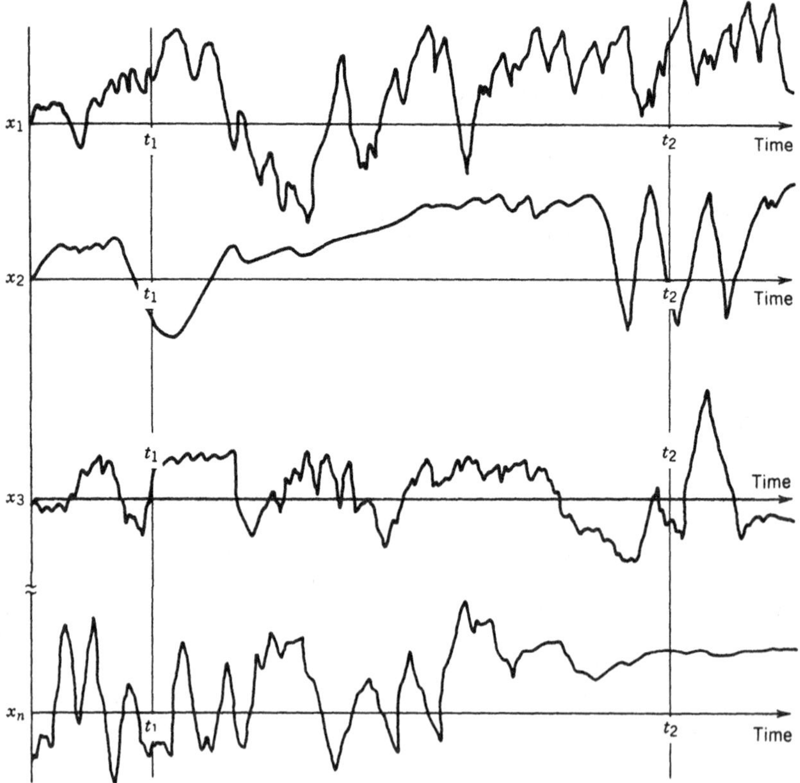

FIGURE 2.4-2 Nonstationary ensemble of random sequences.

enough members of an ensemble are available to provide experimental nonstationary statistics for a random variable; however, nonstationary statistics usually derive from assumptions or from mathematical inferences, such as, optimal estimation algorithms that do exhibit systematic behavior and that model the propagation of the random variable's probability density function in time.

A random process is *stationary* if its ensemble statistics do not vary in time (Fig. 2.4-3). The probability distribution of a stationary process computed at t_1 has the same mean value and central moments as a probability distribution computed at t_2. In both the stationary and nonstationary cases, the probability density function can be defined as

$$\text{pr}[x(t)] = \lim_{\substack{M\to\infty \\ \Delta x\to 0}} \frac{1}{M} \sum_{m=1}^{M} \frac{n[x(t) \pm \Delta x/2]}{\Delta x} \tag{2.4-61}$$

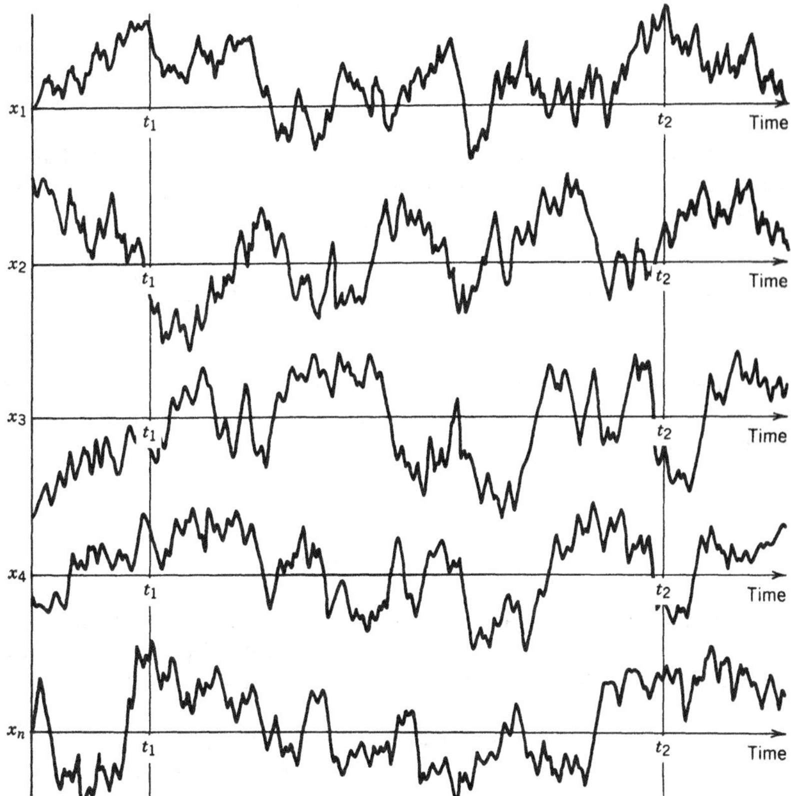

FIGURE 2.4-3 Stationary ensemble of ergodic sequences.

where M is the number of ensemble members and $n[x(t) \pm \Delta x/2]$ is the number of events in the Δx band centered at $x(t)$. For a stationary random process, $\text{pr}[x(t)]$ is invariant with t; for a nonstationary process, it is not.

A stationary random process, $\chi(t)$, can generate a nonstationary random process, $x(t)$, if the former is multiplied by a deterministic, time-varying coefficient, $a(t)$:

$$x(t) = a(t)\chi(t) \tag{2.4-62}$$

Assuming that $\chi(t)$ is a Gaussian random variable with fixed mean χ and standard deviation σ_χ, the *nonstationary Gaussian probability density function* is readily described as

$$\text{pr}[x(t)] = \frac{1}{(2\pi)^{1/2}\sigma_x(t)} e^{-[x(t)-\bar{x}(t)]^2/2\sigma_x^2(t)} \tag{2.4-63}$$

with

$$\bar{x}(t) = a(t)\bar{\chi} \tag{2.4-64}$$

$$\sigma_x^2 = a^2(t)\sigma_\chi^2 \tag{2.4-65}$$

Equation 2.4-61 defines what is sometimes called the *first* probability density function of x, where the *second* probability density function is the *joint* probability density function of $x(t_1)$ and $x(t_2)$, for arbitrary t_1 and t_2:

$$\text{pr}[x(t_1), x(t_2)] = \lim_{\substack{M\to\infty \\ \Delta x\to 0}} \frac{1}{M} \sum_{m=1}^{M} \left\{ \frac{n[x(t_1) \pm \Delta x/2]}{\Delta x^2} \text{ and } \frac{n[x(t_2) \pm \Delta x/2]}{\Delta x^2} \right\} \tag{2.4-66}$$

For a nonstationary process, $\text{pr}[x(t_1), x(t_2)]$ depends on the choice of t_1 and t_2; however, the second probability density function of a stationary process depends only on the *difference* between t_1 and t_2,

$$\begin{aligned} \text{pr}[x(t_1), x(t_2)] &= \text{pr}[x(t_1), x(t_1 + \Delta t)] \\ &= \text{pr}[x(t), x(t + \Delta t)] \end{aligned} \tag{2.4-67}$$

where t and Δt are arbitrary. The nonstationary process's first probability density function (p.d.f.) varies in time, and its second p.d.f. is distributed differently along the $x(t_1)$ and $x(t_2)$ axes (Fig. 2.4-4*a*). The stationary process's first p.d.f. is the same at all times, and its distribution along the $x(t)$ and $x(t+\Delta t)$ axes for any t and Δt are the same (Fig. 2.4-4*b*).

A special case of the stationary process plays an important role in practical application. The probability densities could be calculated for samples of a single random process taken at different times (i.e., using *time*

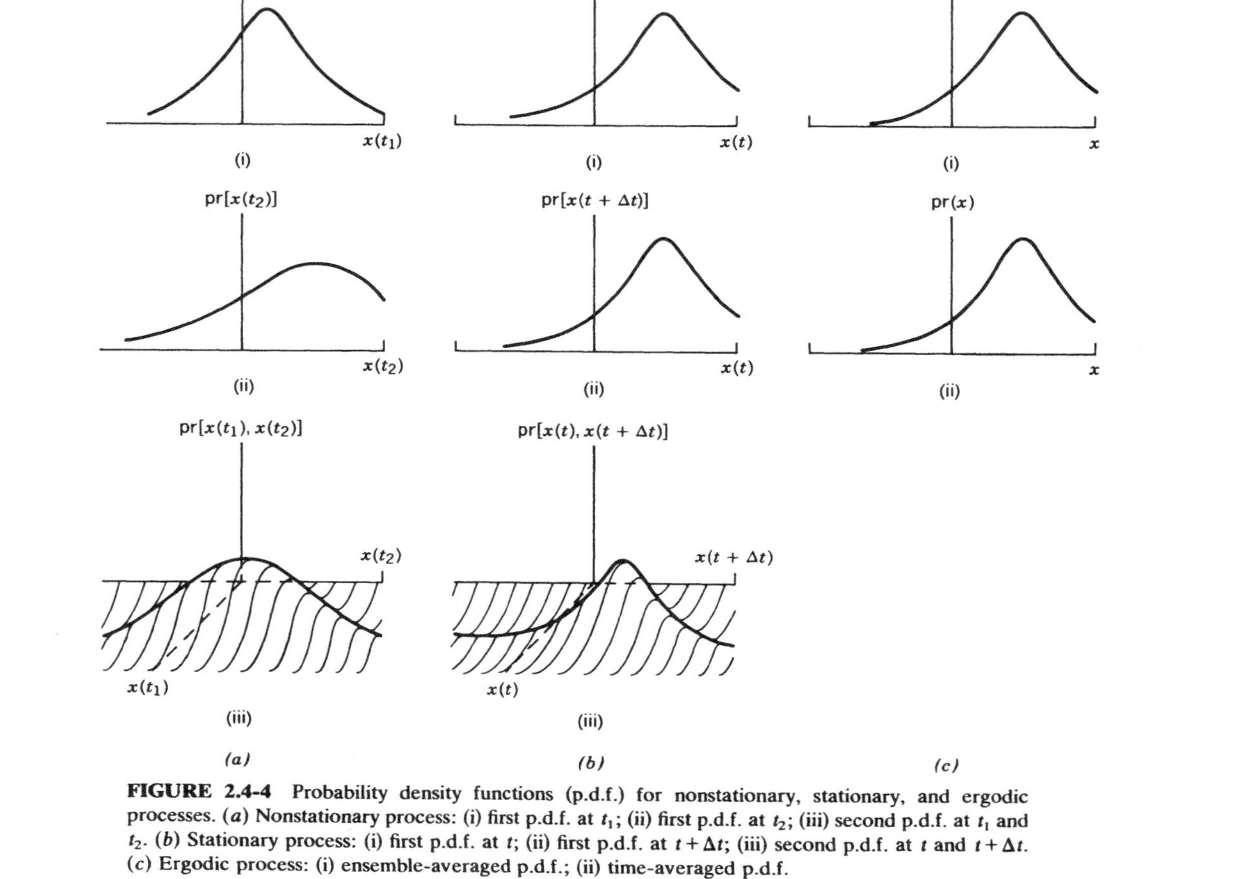

FIGURE 2.4-4 Probability density functions (p.d.f.) for nonstationary, stationary, and ergodic processes. (*a*) Nonstationary process: (i) first p.d.f. at t_1; (ii) first p.d.f. at t_2; (iii) second p.d.f. at t_1 and t_2. (*b*) Stationary process: (i) first p.d.f. at t; (ii) first p.d.f. at $t+\Delta t$; (iii) second p.d.f. at t and $t+\Delta t$. (*c*) Ergodic process: (i) ensemble-averaged p.d.f.; (ii) time-averaged p.d.f.

averaging rather than *ensemble averaging*, see Fig. 2.4-4*c*). If these values are identical to probability densities computed for the ensemble, then the stationary process is said to be *ergodic*. The time-based first and second probability density functions can be calculated as

$$\text{pr}[x(t)] = \text{pr}(x) = \lim_{\substack{M\to\infty \\ \Delta x\to 0}} \frac{1}{M} \sum_{m=1}^{M} \frac{n(x \pm \Delta x/2)}{\Delta x} \tag{2.4-68}$$

$$\text{pr}[x(t), x(t+\Delta t)] = \lim_{\substack{M\to\infty \\ \Delta x\to 0}} \sum_{m=1}^{M} \left\{ \frac{n[x(t) \pm \Delta x/2]}{\Delta x^2} \text{ and } \frac{n[x(t+\Delta t) \pm \Delta x/2]}{\Delta x^2} \right\} \tag{2.4-69}$$

where M is the number of samples taken at different times on a single record. For an ergodic process, the results of these calculations are identical to those derived from Eqs. 2.4-61 and 2.4-67. Examples of stationary, nonergodic processes include ensembles whose members have constant but different mean values and/or standard deviations.

Correlation and Covariance Functions

The second probability density function is used in the definition of a joint expected value of $x(t_1)$ and $x(t_2)$ that is called the *autocorrelation function* of x. If x is defined as having continuous amplitude, the autocorrelation function for a nonstationary random sequence or process is expressed as

$$\begin{aligned} E[x(t_1), x(t_2)] &= \int_{-\infty}^{\infty} \int_{-\infty}^{\infty} x(t_1)x(t_2)\,\text{pr}[x(t_1), x(t_2)]\,dx(t_1)\,dx(t_2) \\ &= \psi[x(t_1), x(t_2)] \end{aligned} \tag{2.4-70}$$

The similarity to higher moments is apparent, and it is sometimes more useful to subtract out mean values, leading to the *autocovariance function*:

$$\begin{aligned} &E\{[x(t_1) - \bar{x}(t_1)][x(t_2) - \bar{x}(t_2)]\} = \phi[\tilde{x}(t_1), \tilde{x}(t_2)] \\ &= \int_{-\infty}^{\infty} \int_{-\infty}^{\infty} [x(t_1) - \bar{x}(t_1)][x(t_2) - \bar{x}(t_2)]\,\text{pr}[x(t_1), x(t_2)]\,dx(t_1)\,dx(t_2) \end{aligned} \tag{2.4-71}$$

with $\tilde{x}(t) \triangleq x(t) - \bar{x}(t)$. When $t_1 = t_2$,

$$\begin{aligned} E\{[x(t_1) - \bar{x}(t_1)][x(t_2) - \bar{x}(t_2)]\} &= E\{[x(t_1) - \bar{x}(t_1)]^2\} \\ &= \sigma_x^2 \end{aligned} \tag{2.4-72}$$

which is the *variance* of x at t_1. The autocorrelation and autocovariance functions are the same when the mean values are zero.

Given two random processes (or sequences), $x(t)$ and $y(t)$, the *cross-correlation* and *cross-covariance functions* are similarly defined. The latter is

$$E\{[x(t_1) - \bar{x}(t_1)][y(t_2) - \bar{y}(t_2)]\} = \phi[\tilde{x}(t_1), \tilde{y}(t_2)]$$

$$= \int_{-\infty}^{\infty}\int_{-\infty}^{\infty} [x(t_1) - \bar{x}(t_1)][y(t_2) - \bar{y}(t_2)] \operatorname{pr}[x(t_1), y(t_2)]\, dx(t_1)\, dy(t_2) \qquad (2.4\text{-}73)$$

with $\tilde{y}(t) \triangleq y(t) - \bar{y}(t)$, and the *covariance* of x and y is given by the value of this equation for $t_1 = t_2$:

$$E\{[x(t_1) - \bar{x}(t_1)][y(t_1) - \bar{y}(t_1)]\} = \sigma_{xy}(t_1) \qquad (2.4\text{-}74)$$

The sample mean x_m of any variable x can be estimated as

$$x_m = \frac{1}{N}\sum_{n=1}^{N} x_n \qquad (2.4\text{-}75)$$

and as $N \rightarrow \infty$, $x_m \rightarrow \bar{x}$. The sample covariance s_{xy} of any two variables x and y can be obtained from

$$s_{xy} = \frac{1}{N-1}\sum_{n=1}^{N} (x_n - x_m)(y_n - y_m) \qquad (2.4\text{-}76)$$

and as $N \rightarrow \infty$, $s_{xy} \rightarrow \sigma_{xy}$. These equations could be used to evaluate the auto- and cross-covariance functions of nonstationary and stationary random processes using ensemble data.

For stationary processes with $\Delta t = t_2 - t_1$ (Fig. 2.4-5*a*),

$$\begin{aligned}\phi[\tilde{x}(t_1), \tilde{x}(t_2)] &= \phi[\tilde{x}(t_1), \tilde{x}(t_1 + \Delta t)] \\ &= \phi_{xx}(\Delta t) \end{aligned} \qquad (2.4\text{-}77)$$

$$\begin{aligned}\phi[\tilde{x}(t_1), \tilde{y}(t_2)] &= \phi[\tilde{x}(t_1), \tilde{y}(t_1 + \Delta t)] \\ &= \phi_{xy}(\Delta t) \end{aligned} \qquad (2.4\text{-}78)$$

Because the ordering of x in Eq. 2.4-77 is immaterial, $\phi_{xx}(\Delta t) = \phi_{xx}(-\Delta t)$, so the autocovariance function is symmetric about the origin. The cross-covariance function is not symmetric because different variables are sampled at the two times separated by Δt. The covariance functions have the following additional properties:

$$\phi_{xx}(0) \geq \phi_{xx}(\Delta t) \qquad (2.4\text{-}79)$$

$$\phi_{yy}(0) \geq \phi_{yy}(\Delta t) \qquad (2.4\text{-}80)$$

$$\phi_{xx}(0)\phi_{yy}(0) \geq [\phi_{xy}(\Delta t)]^2 \qquad (2.4\text{-}81)$$

If the random sequences or processes are ergodic, their covariance functions can be computed from time averages of individual time histories. Let x_i and y_j be the ith and j^{th} samples in ergodic, zero-mean random sequences. Their auto- and cross-covariance functions can be computed as

$$\phi_{xx}(k) = E(x_n x_{n+k}) = \lim_{N\to\infty} \frac{1}{N} \sum_{n=1}^{N} x_n x_{n+k} \tag{2.4-82}$$

$$\phi_{yy}(k) = E(y_n y_{n+k}) = \lim_{N\to\infty} \frac{1}{N} \sum_{n=1}^{N} y_n y_{n+k} \tag{2.4-83}$$

$$\phi_{xy}(k) = E(x_n y_{n+k}) = \lim_{N\to\infty} \frac{1}{N} \sum_{n=1}^{N} x_n y_{n+k} \tag{2.4-84}$$

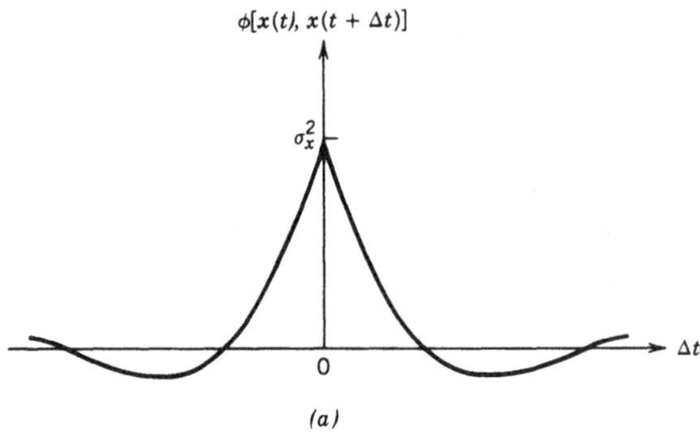

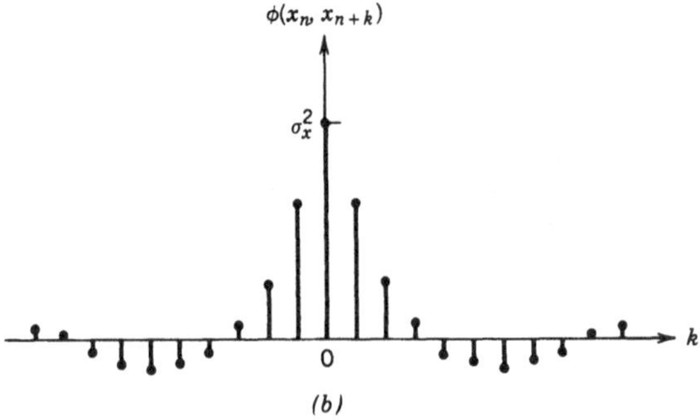

FIGURE 2.4-5 Autocovariance functions of random processes. (*a*) autocovariance function of a stationary random process; (*b*) autocovariance function of a stationary random sequence.

With sampling interval τ, the "lag" Δt is $k\tau$, and the covariance functions are defined only for discrete increments of the time lag (Fig. 2.4-5*b*). As these sequences are stationary, the lower limit of summation could be taken as $-N$, and each sum then would be divided by $2N-1$.

The auto- and cross-covariance functions of ergodic, zero-mean random processes are computed using integration rather than summation:

$$\phi_{xx}(\Delta t) = E[x(t)x(t+\Delta t)]$$
$$= \lim_{T\to\infty} \frac{1}{T}\int_0^T x(t)x(t+\Delta t)\,dt \qquad (2.4\text{-}85)$$
$$\phi_{yy}(\Delta t) = E[y(t)y(t+\Delta t)]$$
$$= \lim_{T\to\infty} \frac{1}{T}\int_0^T y(t)y(t+\Delta t)\,dt \qquad (2.4\text{-}86)$$
$$\phi_{xy}(\Delta t) = E[x(t)y(t+\Delta t)]$$
$$= \lim_{T\to\infty} \frac{1}{T}\int_0^T x(t)y(t+\Delta t)\,dt \qquad (2.4\text{-}87)$$

The covariance functions of random processes are continuous functions of the lag, and they are expressed as *convolution integrals* of the dependent variables. As in the discrete case, the lower limit of integration could be taken as $-T$; then the integral would be divided by $2T$ rather than T.

Two autocovariance functions are of particular interest. The first deals with a variable that is totally uncorrelated with itself from one instant to the next, and it must be handled differently for sequences and processes. If $\phi_{xx}(k)$ or $\phi_{xx}(\Delta t)$ is zero except when its argument is zero, the random sequence or process is called *white noise*, in analogy to the wide-band thermal or "shot" noise produced in electronic amplifiers. As will be seen, "white" connotes equal power at all frequencies in the power spectral density function. For the white random *sequence*, the zero value is the variance of the process:

$$\phi_{xx}(0) = \sigma_x^2 \qquad (2.4\text{-}88)$$

The notion that the autocovariance function of the white random *process* is zero except at zero lag causes a problem, in that Δt can become infinitesimally small, yet $\phi_{xx}(\Delta t \neq 0) = 0$. The problem is resolved by defining the autocovariance function of continuous white noise to be

$$\phi_{xx}(\Delta t) = \phi_{xx}(0)\,\delta(\Delta t) \qquad (2.4\text{-}89)$$

Here $\delta(\Delta t)$ is the Dirac delta function, which has these characteristics:

$$\delta(\Delta t)=\begin{cases}\infty, & \Delta t=0\\ 0, & \Delta t\neq 0\end{cases} \tag{2.4-90}$$

$$\lim_{\Delta t_1\to 0}\int_{-\Delta t_1}^{\Delta t_1}\delta(\Delta t)d(\Delta t)=1 \tag{2.4-91}$$

The value of $\phi_{xx}(0)$ will be deduced in the following section.

A further expression of the independence of white noise derives from the conditional probability density function for sequential samples of the process, which is

$$\text{pr}[x(t)\,|\,x(t+\Delta t)]=\text{pr}[x(t)] \tag{2.4-92}$$

for any Δt. In other words, the amplitudes of contiguous values of x are unrelated, even from a probabilistic standpoint.

The second autocovariance function of interest belongs to the Markov sequence,* which is discussed further in Section 4.2. The Markov sequence provides the simplest example of *colored noise*, with decreasing power at increasing frequencies. Let w_i be a zero-mean white noise sequence with covariance σ_w^2, and assume that w_i is the input to a first-order difference equation with sampling index i:

$$x_{i+1}=ax_i+\sqrt{1-a^2}\,w_i, \qquad 0\leq a\leq 1 \tag{2.4-93}$$

Squaring and taking the expected values,

$$E(x_{i+1}^2)=a^2E(x_i^2)+2a\sqrt{1-a^2}\,E(x_iw_i)+(1-a^2)E(w_i^2) \tag{2.4-94}$$

$E(x_iw_i)=0$ because w_i has no instantaneous effect on x_i. In *stochastic equilibrium*, Eq. 2.4-94 reaches steady state [i.e., $E(x_{i+1}^2)=E(x_i^2)$], leaving

$$E(x_i^2)=\sigma_x^2=\sigma_w^2 \tag{2.4-95}$$

Substituting Eq. 2.4-93 in Eq. 2.4-82,

$$E(x_ix_{i+1})=aE(x_i^2)=a\sigma_x^2 \tag{2.4-96}$$

so the autocovariance function of the Markov sequence (Fig. 2.4-6*a*) is

$$\phi_{xx}(k)=a^{|k|}\sigma_x^2=a^{|k|}\sigma_w^2 \tag{2.4-97}$$

*Together with Chebyshev and Lyapunov, Russian mathematician Andrei Markov (1856–1922) laid the foundations of modern probability theory, studying sequences of variables and developing the methods of moments.

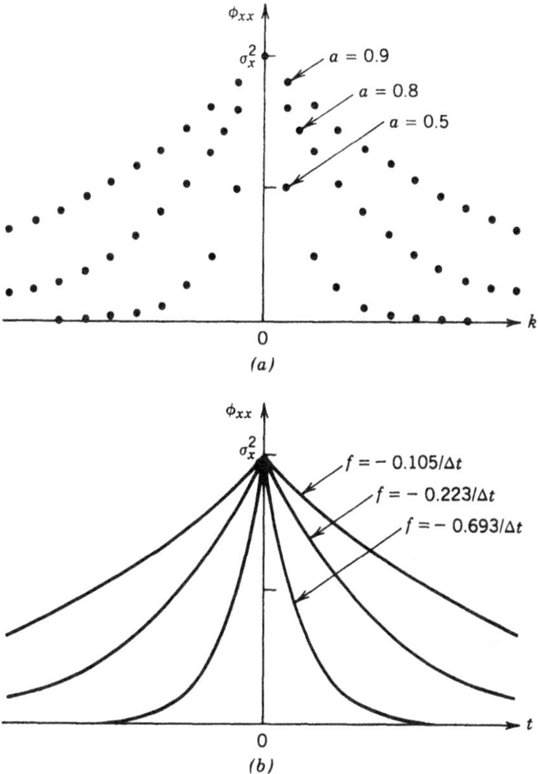

FIGURE 2.4-6 Autocovariance functions of Markov sequences and processes. (Inverse correlation time, $f = -\ln a/\Delta t$.) (*a*) Autocovariance functions of Markov sequences; (*b*) autocovariance functions of Markov processes.

Similar equations can be derived for the Markov process. Using the limiting arguments of Section 2.3, as $\Delta t \to 0$, Eq. 2.4-93 and 2.4-94 become

$$\dot{x}(t) = fx(t) - fw(t), \qquad f < 0 \tag{2.4-98}$$

$$\frac{d[\sigma_x^2(t)]}{dt} = 2f[\sigma_x^2(t) - \sigma_w^2] \tag{2.4-99}$$

where $a = e^{f\Delta t}$ and $\sigma_x^2(t) = E[x^2(t)]$. In stochastic equilibrium, $\sigma_x^2 = \sigma_w^2$. From Eq. 2.3-81b and 2.4-85,

$$\begin{aligned}\phi_{xx}(\Delta t) &= E[x(t)x(t+\Delta t)]\\ &= E\left[e^{f\Delta t}x^2(t) + x(t)\int_t^{t+\Delta t} e^{f(\Delta t-\tau)}fw\,d\tau\right]\\ &= e^{f|\Delta t|}\sigma_x^2 = e^{f|\Delta t|}\sigma_w^2\end{aligned} \tag{2.4-100}$$

The expected value of $x(t)$ times the forcing integral is zero, and the absolute value of Δt accounts for the symmetry of the autocovariance function. In this form, $-1/f$ is called the *correlation time* of the Markov process (Fig. 2.4-6*b*). Even though the Markov process is driven by white noise, it does not suffer from the white noise's singularity problem. The exponential autocorrelation function corresponds to "band limiting," which is presented in greater detail later.

A time-varying Markov process could be defined with $f = f(t)$ and constant σ_w^2. This process would be nonstationary, but its statistics could be derived without reference to entire ensembles (Section 4.2).

Fourier Series and Integrals

Dynamic processes occur during a period of time, and it is most natural to express their governing equations as differential or difference equations which use time as the independent variable. Such equations are said to be written in the *time domain*, and they are the most general mathematical models for physical phenomena, admitting nonlinear and nonstationary effects with relative ease. However, solutions and insight are sometimes difficult to capture with time-domain formulations. The difficulties are reduced somewhat if prior knowledge indicates that solutions can be described by canonical functions, such as exponentials, sinusoids, or other waveforms. Having chosen a set of functions, it remains only to express the solution as some combination of these functions, with suitable multipliers to fit these functions to the problem at hand.

If we choose sinusoids as the canonical functions, then the solutions will be expressed in terms of amplitude, phase angle, and frequency of sine waves, and the equations will be written in the *frequency domain*. Frequency-domain methods are not restricted to stochastic problems, and our review begins by considering the application of Fourier series to strictly periodic signals.* These results are then extended to continuing aperiodic signals (including random processes), culminating in expressions for power spectral density functions, which portray the "frequency content" of signals, and cross-spectral density functions, which indicate the correlation and phasing of two signals as functions of frequency. Although nonstationary methods can be developed, Fourier analysis is most readily applied to stationary processes.

If $x(t)$ is periodic and continuous in time, repeating itself every T sec, then it can be characterized precisely in any time interval $[-T/2, T/2]$ by

*Baron Jean B. J. Fourier (1768–1830) was a professor of mathematics at the Ecole Polytechnique in Paris, later accompanying Napoleon I to Egypt and serving as prefect of the department of Isere in France.

an infinite series of sines and cosines,

$$x(t) = \frac{a(0)}{2} + \sum_{n=1}^{\infty} [a(n\omega_o) \cos n\omega_o t + b(n\omega_o) \sin n\omega_o t] \quad (2.4\text{-}101)$$

where $a(n\omega_o)$ and $b(n\omega_o)$ are chosen to fit $x(t)$. The fundamental frequency, ω_o, is based upon the period T:

$$\omega_o = \frac{2\pi}{T} \text{ (rad/s)} \quad (2.4\text{-}102)$$

$$a(n\omega_o) = \frac{2}{T} \int_{-T/2}^{T/2} x(t) \cos n\omega_o t \, dt, \qquad n = 0 \text{ to } \infty \quad (2.4\text{-}103)$$

$$b(n\omega_o) = \frac{2}{T} \int_{-T/2}^{T/2} x(t) \sin n\omega_o t \, dt, \qquad n = 0 \text{ to } \infty \quad (2.4\text{-}104)$$

This is the basic definition of a *Fourier series*. Using a trigonometric identity, Eq. 2.4-101 can be written in terms of amplitudes $c(n\omega_o)$ and phase angles $\xi(n\omega_o)$,

$$x(t) = \frac{a(0)}{2} + \sum_{n=1}^{\infty} c(n\omega_o) \cos [n\omega_o t + \xi(n\omega_o)] \quad (2.4\text{-}105)$$

with

$$c(n\omega_o) = \sqrt{a^2(n\omega_o) + b^2(n\omega_o)}, \qquad n = 1 \text{ to } \infty \quad (2.4\text{-}106)$$

$$\xi(n\omega_o) = \tan^{-1} \left[\frac{-b(n\omega_o)}{a(n\omega_o)} \right], \qquad n = 1 \text{ to } \infty \quad (2.4\text{-}107)$$

The Fourier series also can be expressed in complex-variable form as

$$x(t) = \frac{1}{T} \sum_{n=-\infty}^{\infty} g(n\omega_o) \, e^{jn\omega_o t} \quad (2.4\text{-}108)$$

where $j = \sqrt{-1}$, g is a complex number,

$$g(n\omega_o) = \begin{cases} T[a(n\omega_o) + jb(n\omega_o)]/2, & 0 \le n \le \infty \\ T[a(-n\omega_o) - jb(-n\omega_o)]/2, & -\infty \le n \le -1 \end{cases} \quad (2.4\text{-}109)$$

and $b(0)$ is zero. The complex coefficients can be written directly, using Eq. 2.4-103 and 2.4-104:

$$g(n\omega_o) = \int_{-T/2}^{T/2} x(t) \, e^{-jn\omega_o t} \, dt \quad (2.4\text{-}110)$$

If $x(t)$ is a sine wave with frequency $m\omega_o$, it is accurately portrayed by the mth terms, and all other coefficients $(n \neq m)$ are zero. If $x(t)$ is a nonsinusoidal periodic function with frequency $m\omega_o$ (e.g., a square or triangular wave), the mth terms of the Fourier series identify the fundamental component of $x(t)$. For $n < m$, the Fourier components are zero, and the appropriate higher harmonic terms $(n > m)$ account for the difference between the actual waveform and a sinusoid. In each case, the discrete frequency content of $x(t)$ is contained in its Fourier series coefficients, which indicate the magnitudes of the real and imaginary components at each frequency or, equivalently, their amplitudes and phase angles.

The Fourier series cannot be applied if $x(t)$ is aperiodic; however, by stretching the assumed period to infinity and replacing the series summation with an equivalent integral, the continuous frequency content of $x(t)$ can be determined. Letting T go to infinity causes ω_o to approach zero. Denoting the vanishing value of ω_o as $\Delta\omega$, Eq. 2.4-108 can be written as

$$x(t) = \frac{1}{2\pi} \sum_{n=-\infty}^{\infty} g(n\Delta\omega)\, e^{jn\Delta\omega t} \Delta\omega \tag{2.4-111}$$

and as T approaches infinity, the limit of this equation is the *Fourier integral*:

$$x(t) = \frac{1}{2\pi} \int_{-\infty}^{\infty} X(\omega)\, e^{j\omega t}\, d\omega \tag{2.4-112}$$

$X(\omega)$ is the complex-valued *Fourier transform* of $x(t)$, defined as the limit of Eq. 2.4-110 as T goes to infinity:

$$X(\omega) = \int_{-\infty}^{\infty} x(t)\, e^{-j\omega t}\, dt \tag{2.4-113}$$

$X(\omega)$ expresses the frequency content of $x(t)$ as a continuous function of ω, for ω in $[-\infty, \infty]$. By implication, the Fourier integral also can be called the *inverse Fourier transform*.

Spectral Density Functions of Random Processes

There are two ways to find the distribution of $x(t)$'s "power" with frequency. The first definition uses the Fourier transform, $X(\omega)$, directly. Letting this complex variable be written as

$$X(\omega) = X_R(\omega) + jX_I(\omega) \tag{2.4-114}$$

its complex conjugate is defined as

$$X^*(\omega) = X_R(\omega) - jX_I(\omega) \tag{2.4-115}$$

Multiplying the transform by its complex conjugate,

$$|X(\omega)|^2 = X(\omega)X^*(\omega)$$
$$= X_R^2(\omega) + X_I^2(\omega)$$
$$= \Psi_{xx}(\omega) \tag{2.4-116}$$

the product specifies the real-valued, scalar *power spectral density function*, $\Psi_{xx}(\omega)$. This definition of $\Psi_{xx}(\omega)$ evidently requires $x(t)$ to be an ergodic random process, as it is defined from a single time history.

The second method of deriving the power spectral density function is based on the autocorrelation (or autocovariance) function. The two constitute a *Fourier transform pair* (i.e., the power spectral density is the Fourier transform of the autocorrelation and the autocorrelation is the inverse Fourier transform of the power spectral density):

$$\Psi_{xx}(\omega) = \int_{-\infty}^{\infty} \psi_{xx}(\tau)e^{-j\omega\tau}\, d\tau \tag{2.4-117}$$

$$\psi_{xx}(\tau) = \frac{1}{2\pi}\int_{-\infty}^{\infty} \Psi_{xx}(\omega)e^{j\omega\tau}\, d\omega \tag{2.4-118}$$

For Eq. 2.4-117 to be true, $x(t)$ must be stationary, but it need not be ergodic.

The power spectral density of the *variation* from the mean is calculated by either method. In the first approach, the Fourier transform of $[x(t) - \bar{x}]$ is calculated, while the autocovariance function replaces the autocorrelation function in the second. Denoting the variational power spectral density function by $\Phi_{xx}(\omega)$ [which should not be confused with the state transition matrix, $\Phi(t_1, t_2)$ used elsewhere],

$$\Phi_{xx}(\omega) = \int_{-\infty}^{\infty} \phi_{xx}(\tau)e^{-j\omega\tau}\, d\tau = 2\int_{0}^{\infty} \phi_{xx}(\tau)\cos\omega\tau\, d\tau \tag{2.4-119}$$

$$\phi_{xx}(\tau) = \frac{1}{2\pi}\int_{-\infty}^{\infty} \Phi_{xx}(\omega)e^{j\omega\tau}\, d\omega = \frac{1}{\pi}\int_{0}^{\infty} \Phi_{xx}(\omega)\cos\omega\tau\, d\omega \tag{2.4-120}$$

Together, these equations are known as the *Wiener theorem*.* Because $\phi_{xx}(0) = \sigma_x^2$, the *variance* of $[x(t) - \bar{x}]$ also can be obtained from Eq. 2.4-120:

$$\frac{1}{\pi}\int_{0}^{\infty} \Phi_{xx}(\omega)\, d\omega = \sigma_x^2 \tag{2.4-121}$$

*Norbert Wiener (1894–1964) studied mathematics at Harvard University and spent most of his career at the Massachusetts Institute of Technology, where he developed fundamental concepts of information theory, signal processing, and cybernetics.

As $\Phi_{xx}(\omega)$ represents "power," its integral over all frequencies, σ_x^2, represents "energy" contained in the process.

The power spectral density of physical system outputs normally "rolls off" (decreases in magnitude) with increasing frequency; otherwise, the variance would be infinite. Recalling the autocovariance function of a *white noise random process* (Eq. 2.4-89), Eq. 2.4-119 indicates that the corresponding power spectral density function is

$$\Phi_{xx}(\omega) = \phi_{xx}(0) \int_{-\infty}^{\infty} \delta(\tau) e^{-j\omega\tau}\, d\tau \tag{2.4-122}$$

which, from Eq. 2.4-91, is

$$\Phi_{xx}(\omega) = \phi_{xx}(0) = \text{constant} \tag{2.4-123}$$

However, Eq. 2.4-121 tells us that the variance of white noise is

$$\frac{1}{\pi} \int_{0}^{\infty} \phi_{xx}(0)\, d\omega = \sigma_x^2 \tag{2.4-124}$$

which can be finite only if $\phi_{xx}(0) = 0$! This is the essential quandary of continuous white noise—it cannot exist in any physical sense.

The problem of characterizing white noise for analytical purposes often can be overcome by heuristic arguments, implying that *band limiting* occurs above the highest frequency of interest (Fig. 2.4-7). Suppose that the power spectral density has a sharp cutoff at a frequency of ω_B:

$$\Phi_{xx}(\omega) = \begin{cases} \Phi, & |\omega| \leq \omega_B \\ 0, & |\omega| > \omega_B \end{cases} \tag{2.4-125}$$

From Eq. 2.4-120 and 2.4-121,

$$\sigma_x^2 = \Phi\omega_B/2\pi \tag{2.4-126}$$

$$\phi_{xx}(\tau) = (\Phi/2\pi\tau) \sin \omega_B\tau \tag{2.4-127}$$

The Markov process provides a more gradual rolloff. If $x(t)$ is a Markov process (Eq. 2.4-100), the corresponding power spectral density function is

$$\Phi_{xx}(\omega) = \frac{2f\sigma_x^2}{\omega^2 + f^2} \tag{2.4-128}$$

This function reaches its "half-power point" at $\omega = f$, and it falls off as $1/\omega^2$ for frequencies above f. Equations 2.4-125 and 2.4-128 often are used to approximate white noise, with large values of ω_B or f.

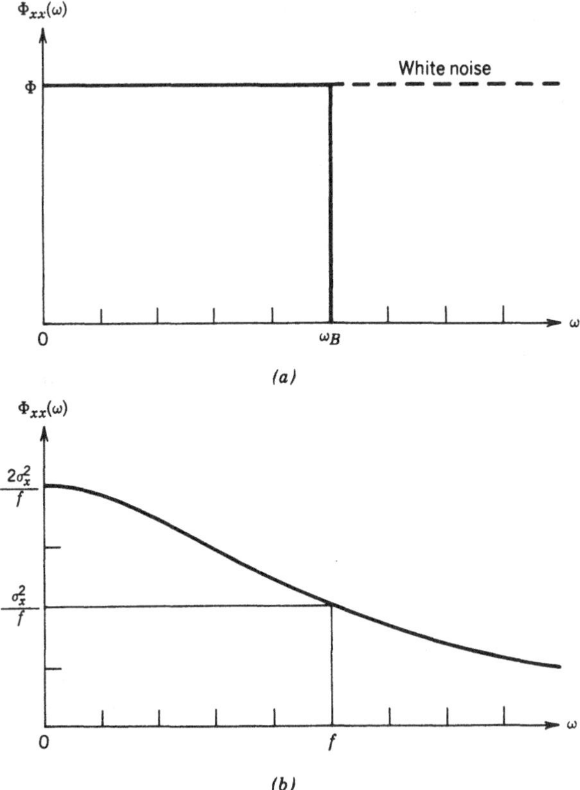

FIGURE 2.4-7 Power spectral density functions for white noise, ideal band limiting, and the Markov process. (*a*) Ideal band limiting and white noise; (*b*) Markov process.

The *cross-spectral density function* for $x(t)$ and $y(t)$ proceeds directly from these results. With $Y(\omega)$ representing the Fourier transform of $y(t)$,

$$\Psi_{xy}(\omega) = X(\omega)\,Y(\omega)^* \tag{2.4-129}$$

Ψ_{xy} is a complex variable, as there is no cancellation of complex conjugates. The variational cross-spectral density function can be written as

$$\Phi_{xy}(\omega) = \int_{-\infty}^{\infty} \phi_{xy}(\tau)e^{-j\omega\tau}\,d\tau \tag{2.4-130}$$

and the inverse Fourier transform is

$$\phi_{xy}(\tau) = \frac{1}{2\pi}\int_{-\infty}^{\infty} \Phi_{xy}(\omega)e^{j\omega\tau}\,d\omega \tag{2.4-131}$$

(The cosine versions cannot be used for lack of symmetry.)

Spectral Functions of Random Sequences

Spectral characteristics of random sequences are defined in much the same way as those for random processes. The principal difference is in the definition of the Fourier transform, and this difference reflects important attributes that must be understood when discrete samples are used to represent continuous-time histories. The *discrete Fourier transform*, $X_D(\omega_D)$, is computed from the sequence x_k as

$$X_D(\omega_D) = \sum_{k=-\infty}^{\infty} x_k e^{-j\omega_D k} \tag{2.4-132}$$

where ω_D is the frequency measured in radians per sampling interval. Assuming that the sampling interval is τ sec, ω_D is related to the time-based frequency, ω_C (rad/s) as

$$\omega_D = \omega_C \tau \tag{2.4-133}$$

and time is related to the sampling index k by

$$t = k\tau \tag{2.4-134}$$

Because $e^{j\theta} = e^{j(\theta+2\pi n)}$, the discrete Fourier transform is periodic in frequency increments of $\Delta\omega_D = 2\pi$ or $\Delta\omega_C = 2\pi/\tau$. Consequently, the *inverse discrete Fourier transform* need be evaluated only in the interval $[-\pi, \pi]$:

$$x_k = \frac{1}{\pi} \int_{-\pi}^{\pi} X_D(\omega_D) e^{j\omega_D k}\, d\omega_D \tag{2.4-135}$$

As in the continuous case, the *power spectrum* can be derived from either the Fourier transform of the autocorrelation function or the Fourier transform of the random sequence. In the first approach,

$$\Psi_{xx}(\omega_D) = \sum_{k=-\infty}^{\infty} \psi_{xx}(k) e^{-j\omega_D k} \tag{2.4-136}$$

Defining the autocorrelation as in Eq. 2.4-82, it can be shown that Eq. 2.4-136 is equivalent to

$$\Psi_{xx}(\omega_D) = \lim_{N\to\infty} \left[\frac{1}{(2N-1)}\right] X_N(\omega_D) X_N^*(\omega_D) \tag{2.4-137}$$

where $X_N(\omega_D)$ is

$$X_N(\omega_D) = \sum_{k=-N}^{N} x_k e^{-j\omega_D k} \tag{2.4-138}$$

The power spectrum is a real, continuous, scalar variable defined in the interval, $-\pi \le \omega_D \le \pi$, and it is periodic outside the interval in 2π increments. Variational spectra can be defined using covariance functions in place of correlation functions.

By analogy, the *cross spectrum* for random sequences x_k and y_k can be computed as

$$\Psi_{xy}(\omega_D) = \sum_{k=-\infty}^{\infty} \psi_{xy}(k)e^{-j\omega_D k} \tag{2.4-139}$$

or

$$\Psi_{xy}(\omega_D) = \lim_{N\to\infty}\left[\frac{1}{(2N-1)}\right] X_N(\omega_D)\,Y_N^*(\omega_D) \tag{2.4-140}$$

The cross spectrum is a complex, continuous, scalar variable that is defined in the interval $-\pi \le \omega_D \le \pi$ and is periodic outside the interval.

If the discrete sequence x_k represents periodic samples of the continuous process, $x(t)$, then the discrete Fourier transform of x_k can be derived from the Fourier transform of $x(t)$. From Eqs. 2.4-112, 2.4-133, and 2.4-134, the sample at the kth instant can be expressed in terms of the continuous Fourier transform,

$$x_k = x(k\tau) = \frac{1}{2\pi}\int_{-\infty}^{\infty} X_C(\omega_C)e^{j\omega_C k\tau}\, d\omega_C \tag{2.4-141}$$

with the continuous transform represented by $X_C(\omega_C)$. The integral of Eq. 2.4-141 can be broken into the summation of integrals over frequency bands of width $2\pi/\tau$:

$$x_k = \frac{1}{2\pi}\sum_{n=-\infty}^{\infty}\int_{(2n-1)\pi/\tau}^{(2n+1)\pi/\tau} X_C(\omega_C)e^{j\omega_C k\tau}\, d\omega_C \tag{2.4-142}$$

Defining

$$\omega_C = \omega + \frac{2\pi n}{\tau} \tag{2.4-143}$$

the inverse Fourier transform can be rewritten as

$$x_k = \frac{1}{2\pi}\sum_{n=-\infty}^{\infty}\int_{-\pi/\tau}^{\pi/\tau} X_C\left(\omega + \frac{2\pi n}{\tau}\right) e^{j\omega k\tau}e^{j2\pi nk}\, d\omega$$

$$\times \frac{1}{2\pi}\int_{-\pi/\tau}^{\pi/\tau}\left[\sum_{n=-\infty}^{\infty} X_C\left(\omega + \frac{2\pi n}{\tau}\right)\right] e^{j\omega k\tau}\, d\omega \tag{2.4-144a}$$

since $e^{j2\pi nk} = 1$ for integer values of n and k. With Eq. 2.4-133, this becomes

$$x_k = \frac{1}{2\pi}\int_{-\pi}^{\pi}\left[\frac{1}{T}\sum_{n=-\infty}^{\infty} X_C\left(\frac{\omega_D}{T}+\frac{2\pi n}{T}\right)\right]e^{j\omega_D k}\,d\omega_D \qquad (2.4\text{-}144\text{b})$$

Comparing Eq. 2.4-144b with Eq. 2.4-135, the discrete Fourier transform is

$$X_D(\omega_C) = \frac{1}{T}\sum_{n=-\infty}^{\infty} X_C\left(\omega+\frac{2\pi n}{T}\right) \qquad (2.4\text{-}145)$$

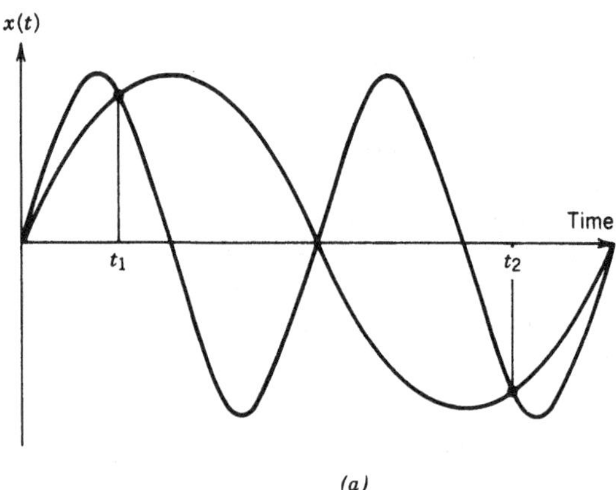

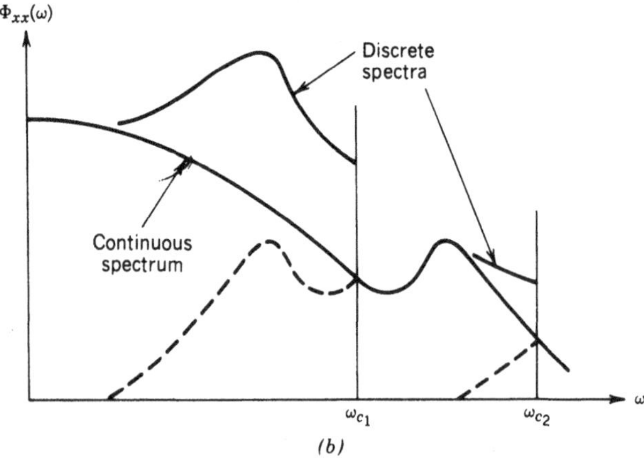

FIGURE 2.4-8 Aliasing and frequency folding caused by sampling continuous signals. (*a*) The two sine waves are indistinguishable if sampled only at t_1 and t_2; (*b*) effects of sampling on discrete power spectral density function.

for ω_C in $[-\pi/\tau, \pi/\tau]$. Since this equation could be used in Eq. 2.4-137, the power spectrum of x_k could be derived from the power spectral density of $x(t)$.

The fact that X_D sums all frequency components of X_C is due to a phenomenon called *aliasing*: signals with frequency magnitudes higher than π/τ cannot be distinguished from signals in the band $\pm\pi/\tau$ using the sampled data alone. Figure 2.4-8*a* illustrates the effect for two sine waves of differing frequencies. If samples are taken only at the points of intersection, there is no way to tell which sine wave has been sampled. Given the power spectral density function Φ_C (Fig. 2.4-8*b*), the summation that generates Φ_D makes Φ_C appear to fold back at $\omega_C = \pi/\tau$. ω_C is called the *Nyquist** or *folding frequency*, and it determines the highest signal frequency that can be captured unambiguously with a sampling interval of τ sec.

One beneficial consequence of this upper limit on signal frequency is that white noise is naturally band-limited at π/τ. Then Eqs. 2.4-125 to 2.4-127 describe the power spectrum, variance, and autocovariance function of the white noise sequence.

Multivariate Statistics

The probability density function for an n-dimensional vector random variable $\mathbf{x}$ with continuous amplitude distribution is a *joint* probability density function of the n elements:

$$\mathrm{pr}(\mathbf{x}) = \mathrm{pr}(x_1, x_2, \ldots, x_n) \tag{2.4-146}$$

It is a scalar function of a vector variable, and following Eq. 2.4-39, the integral under its hypersurface, taken between limits of $\pm\infty$ on all elements, must equal one:

$$\int_{-\infty}^{\infty} \mathrm{pr}(\mathbf{x})\, d\mathbf{x} = \int_{-\infty}^{\infty} \cdots \int_{-\infty}^{\infty} \mathrm{pr}(x_1, x_2, \ldots, x_n)\, dx_1\, dx_2 \ldots dx_n = 1 \tag{2.4-147}$$

Multivariate expected values can be defined to account for the possible values of the n components of $\mathbf{x}$. The *mean value of* $\mathbf{x}$ is an n-dimensional

*Harry Nyquist (1889–1976) was a prolific contributor to control theory and communications engineering, having developed the conditions for stability in negative feedback systems (expressed in the Nyquist diagram), methods for digital switching and coding, and inventions that resulted in 138 American patents.

vector

$$E(\mathbf{x}) = \int_{-\infty}^{\infty} \mathbf{x}\,\text{pr}(\mathbf{x})\,d\mathbf{x}$$

$$= \begin{bmatrix} \int_{-\infty}^{\infty} \cdots \int_{-\infty}^{\infty} x_1\,\text{pr}(x_1, \ldots, x_n)\,dx_1 \ldots dx_n \\ \vdots \\ \int_{-\infty}^{\infty} \cdots \int_{-\infty}^{\infty} x_n\,\text{pr}(x_1, \ldots, x_n)\,dx_1 \ldots dx_n \end{bmatrix}$$

$$= \bar{\mathbf{x}} = \begin{bmatrix} \bar{x}_1 \\ \vdots \\ \bar{x}_n \end{bmatrix} \tag{2.4-148}$$

while the *covariance matrix* is

$$E[(\mathbf{x}-\bar{\mathbf{x}})(\mathbf{x}-\bar{\mathbf{x}})^T] = E(\tilde{\mathbf{x}}\tilde{\mathbf{x}}^T)$$

$$= \int_{-\infty}^{\infty} \tilde{\mathbf{x}}\tilde{\mathbf{x}}^T\,\text{pr}(\mathbf{x})\,d\mathbf{x}$$

$$= \mathbf{P} = \begin{bmatrix} p_{11} & p_{12} & \cdots & p_{1n} \\ p_{21} & p_{22} & \cdots & p_{2n} \\ \vdots & \vdots & & \vdots \\ p_{n1} & p_{n2} & \cdots & p_{nn} \end{bmatrix} \tag{2.4-149}$$

with $\tilde{\mathbf{x}} \triangleq \mathbf{x} - \bar{\mathbf{x}}$, and p_{ij} defined by

$$p_{ij} = \int_{-\infty}^{\infty} \int_{-\infty}^{\infty} \tilde{x}_i \tilde{x}_j\,\text{pr}(x_1, \ldots, x_n)\,dx_1 \ldots dx_n \tag{2.4-150}$$

$\mathbf{P}$ is a symmetric matrix ($p_{ij} = p_{ji}$) because the ordering of $\tilde{x}_i$ and $\tilde{x}_j$ does not matter in evaluating Eq. 2.4-150.

The covariance matrix contains the variances of x_i on its principal diagonal and the covariances of x_i and x_j on its off-diagonal elements:

$$\mathbf{P} = \begin{bmatrix} \sigma_1^2 & \sigma_{12} & \cdots & \sigma_{1n} \\ \sigma_{12} & \sigma_2^2 & \cdots & \sigma_{2n} \\ \vdots & \vdots & & \vdots \\ \sigma_{n1} & \sigma_{n2} & \cdots & \sigma_n^2 \end{bmatrix} \tag{2.4-151a}$$

From prior definitions, the σ_i^2 and σ_{ij} are seen to be the zero-lag components of the corresponding auto- and cross-covariance functions. Defining the *correlation coefficient* ρ_{ij} as $\sigma_{ij}/\sigma_i\sigma_j$, the covariance matrix can be written as

$$P = \begin{bmatrix} \sigma_1^2 & \rho_{12}\sigma_1\sigma_2 & \dots & \rho_{1n}\sigma_1\sigma_n \\ \rho_{12}\sigma_1\sigma_2 & \sigma_2^2 & \dots & \rho_{2n}\sigma_2\sigma_n \\ \vdots & \vdots & & \vdots \\ \rho_{1n}\sigma_1\sigma_n & \rho_{2n}\sigma_2\sigma_n & \dots & \sigma_n^2 \end{bmatrix} \tag{2.4-151b}$$

and from Eq. 2.4-81, $|\rho_{ij}| \leq 1$. It is apparent that cross-covariance and cross-spectral density *matrices* could be formed by repeated application of the scalar results shown previously.

The mean vector and covariance matrix are sufficient to define a *multivariate Gaussian probability density function*, pr(**x**), that satisfies Eq. 2.4-147:

$$\mathrm{pr}(\mathbf{x}) = \frac{1}{(2\pi)^{n/2}|\mathbf{P}|^{1/2}} e^{-(1/2)(\mathbf{x}-\bar{\mathbf{x}})^T\mathbf{P}^{-1}(\mathbf{x}-\bar{\mathbf{x}})} \tag{2.4-152}$$

This probability density function forms the basis for most of the optimal estimation methods discussed in this book.

Conditional density functions also are developed by following the scalar results. The probability density function of **x** given **y** is a scalar variable that plays a major role in optimal estimation.

$$\mathrm{pr}(\mathbf{x}|\mathbf{y}) = \frac{\mathrm{pr}(\mathbf{x}, \mathbf{y})}{\mathrm{pr}(\mathbf{y})} \tag{2.4-153}$$

We will, for example, focus on methods that estimate the *conditional mean value* of **x**, defined in terms of its conditional expected value as

$$E(\mathbf{x}|\mathbf{y}) = \int_{-\infty}^{\infty} \mathbf{x}\,\mathrm{pr}(\mathbf{x}|\mathbf{y})\,d\mathbf{x} \tag{2.4-154}$$

where **y** is fixed with respect to the integration over **x**.

Expected values and probability density functions of functions of **x** follow the reasoning found in Eq. 2.4-44 to 2.4-47. Defining $\mathbf{y} = \mathbf{f}(\mathbf{x})$ as an m-vector, its m-dimensional expected value is

$$E[\mathbf{f}(\mathbf{x})] = \int_{-\infty}^{\infty} \mathbf{f}(\mathbf{x})\,\mathrm{pr}(\mathbf{x})\,d\mathbf{x} \tag{2.4-155}$$

A Taylor series expansion of $\mathbf{y}$ yields

$$\mathbf{y}(\mathbf{x}) = \mathbf{y}_o + \Delta\mathbf{y} = \mathbf{f}(\mathbf{x}_o) + \frac{\partial \mathbf{f}(\mathbf{x}_o)}{\partial \mathbf{x}} \Delta\mathbf{x} + \cdots \qquad (2.4\text{-}156)$$

and the corresponding probability of $\mathbf{y}$ being in $\Delta\mathbf{y}$ is approximately

$$\begin{aligned} \int_{\mathbf{y}_o-\Delta\mathbf{y}}^{\mathbf{y}_o+\Delta\mathbf{y}} \mathrm{pr}(\mathbf{y}_o)\, d\mathbf{y} &= \int_{\mathbf{x}_o-\Delta\mathbf{x}}^{\mathbf{x}_o+\Delta\mathbf{x}} \mathrm{pr}(\mathbf{y}_o) \left[\frac{\partial \mathbf{f}(\mathbf{x}_o)}{\partial \mathbf{x}}\right] d\mathbf{x} \\ &= \int_{\mathbf{x}_o-\Delta\mathbf{x}}^{\mathbf{x}_o+\Delta\mathbf{x}} \mathrm{pr}(\mathbf{x}_o)\, d\mathbf{x} \end{aligned} \qquad (2.4\text{-}157)$$

For small values of $\Delta\mathbf{x}$ and $\Delta\mathbf{y}$, the probability density functions can be considered constant; hence

$$\mathrm{pr}(\mathbf{y}_o) \int \frac{\partial \mathbf{f}}{\partial \mathbf{x}}\, d\mathbf{x} = \mathrm{pr}(\mathbf{x}_o) \int d\mathbf{x} \qquad (2.4\text{-}158)$$

If $m = n$ and $\partial\mathbf{f}/\partial\mathbf{x}$ is nonsingular, both sides of the equation can be multiplied by the inverse of the Jacobian matrix:

$$\begin{aligned} \mathrm{pr}(\mathbf{y}_o) \int \left(\frac{\partial \mathbf{f}}{\partial \mathbf{x}}\right)^{-1} \left(\frac{\partial \mathbf{f}}{\partial \mathbf{x}}\right) d\mathbf{x} &= \mathrm{pr}(\mathbf{y}_o) \int d\mathbf{x} \\ &= \mathrm{pr}(\mathbf{x}_o) \int \left(\frac{\partial \mathbf{f}}{\partial \mathbf{x}}\right)^{-1} d\mathbf{x} \end{aligned} \qquad (2.4\text{-}159)$$

Then, since the integrals are defined to produce scalar values as in Eq. 2.4-147,

$$\begin{aligned} \mathrm{pr}(\mathbf{y}_o) &= \mathrm{pr}(\mathbf{x}_o) \frac{\displaystyle\int \left(\frac{\partial \mathbf{f}}{\partial \mathbf{x}}\right)^{-1} d\mathbf{x}}{\displaystyle\int d\mathbf{x}} \\ &= \mathrm{pr}(\mathbf{x}_o) \left| \left(\frac{\partial \mathbf{f}}{\partial \mathbf{x}}\right)^{-1} \right| \end{aligned} \qquad (2.4\text{-}160)$$

If $m \neq n$, the computation proceeds using the appropriate pseudoinverse of $\partial\mathbf{f}/\partial\mathbf{x}$ (Section 2.2).

Given the function $\mathbf{f}(\mathbf{x}, \mathbf{y})$, its expected value can be expressed as

$$\begin{aligned} E[\mathbf{f}(\mathbf{x}, \mathbf{y})] &= E_{\mathbf{y}}\{E_{\mathbf{x}}[\mathbf{f}(\mathbf{x}, \mathbf{y})|\mathbf{y}]\} \\ &= E_{\mathbf{x}}\{E_{\mathbf{y}}[\mathbf{f}(\mathbf{x}, \mathbf{y})|\mathbf{x}]\} \end{aligned} \qquad (2.4\text{-}161)$$

where the subscripts denote the variables of integration for the expectation operation.

2.5 PROPERTIES OF DYNAMIC SYSTEMS

This section presents some attributes of dynamic system structure and valuation that govern state response and the associated output. Consider a system described as before by an nth-order ordinary differential equation representing dynamic interactions

$$\dot{\mathbf{x}}(t) = \mathbf{f}[\mathbf{x}(t), \mathbf{u}(t), \mathbf{w}(t), \mathbf{p}(t), t] \tag{2.5-1a}$$

and an r-component algebraic equation portraying observable outputs

$$\mathbf{y}(t) = \mathbf{h}[\mathbf{x}(t), \mathbf{u}(t), \mathbf{w}(t), \mathbf{p}(t), t] \tag{2.5-2a}$$

where all variables are defined and dimensioned as in Sections 1.2 and 2.3. Random disturbances, measurement errors, or parameter uncertainties are not considered, and it is sufficient to describe the nonlinear system as

$$\dot{\mathbf{x}}(t) = \mathbf{f}[\mathbf{x}(t), \mathbf{u}(t), \mathbf{w}(t), t] \tag{2.5-1b}$$

$$\mathbf{y}(t) = \mathbf{h}[\mathbf{x}(t), t] \tag{2.5-2b}$$

Corresponding linear models are

$$\dot{\mathbf{x}}(t) = \mathbf{F}(t)\mathbf{x}(t) + \mathbf{G}(t)\mathbf{u}(t) + \mathbf{L}(t)\mathbf{w}(t) \tag{2.5-3}$$

$$\mathbf{y}(t) = \mathbf{H}(t)\mathbf{x}(t) \tag{2.5-4}$$

where $\mathbf{F}(t)$, $\mathbf{G}(t)$, $\mathbf{L}(t)$, and $\mathbf{H}(t)$ may be Jacobian matrices of $\mathbf{f}[\cdot]$ and $\mathbf{h}[\cdot]$, as defined in Section 2.3.

Some structural properties are readily defined. If there is no explicit dependence of equation coefficients or structure on time, these equations are *time-invariant*:

$$\dot{\mathbf{x}}(t) = \mathbf{f}[\mathbf{x}(t), \mathbf{u}(t), \mathbf{w}(t)] \tag{2.5-5}$$

$$\mathbf{y}(t) = \mathbf{h}[\mathbf{x}(t)] \tag{2.5-6}$$

$$\dot{\mathbf{x}}(t) = \mathbf{F}\mathbf{x}(t) + \mathbf{G}\mathbf{u}(t) + \mathbf{L}\mathbf{w}(t) \tag{2.5-7}$$

$$\mathbf{y}(t) = \mathbf{H}\mathbf{x}(t) \tag{2.5-8}$$

Note that a system could be nonlinear but time-invariant, linear but time-varying, and so on.

If the system differential equations are unforced by controls or disturbances, the equations are *homogeneous*:

$$\dot{\mathbf{x}}(t) = \mathbf{f}[\mathbf{x}(t), \mathbf{0}, \mathbf{0}, t] \tag{2.5-9}$$

$$\dot{\mathbf{x}}(t) = \mathbf{F}(t)\mathbf{x}(t) \tag{2.5-10}$$

Homogeneous equations are sufficient to describe the fundamental stability characteristics of the system [i.e., the state's propensity to remain bounded (or not) in the absence of inputs]. If the equations are time-invariant and unforced, the differential equations are said to be *autonomous*:

$$\dot{\mathbf{x}}(t) = \mathbf{f}[\mathbf{x}(t), \mathbf{0}, \mathbf{0}] \tag{2.5-11}$$

$$\dot{\mathbf{x}}(t) = \mathbf{F}\mathbf{x}(t) \tag{2.5-12}$$

Stability assessment is simplified when considering autonomous systems because the state's tendency to converge or diverge does not depend on time variations of inputs or the starting time of the response.

Systems with *feedback control* can be modeled by equivalent homogeneous equations, so their dynamic properties can be evaluated accordingly. For example, if $\mathbf{w}(t)$ equals zero, and the m-dimensional control vector is

$$\mathbf{u}(t) = -\mathbf{C}(t)\mathbf{x}(t) \tag{2.5-13}$$

where $\mathbf{C}(t)$ is any $(m \times n)$ matrix, then Eq. 2.5-3 is equivalent to

$$\begin{aligned}\dot{\mathbf{x}}(t) &= [\mathbf{F}(t) - \mathbf{G}(t)\mathbf{C}(t)]\mathbf{x}(t) \\ &= \mathbf{F}'(t)\mathbf{x}(t)\end{aligned} \tag{2.5-14}$$

Dynamic compensation can be handled by redefining the state variables as well as the fundamental matrix. Suppose the feedback law contains proportional and integral terms

$$\begin{aligned}\mathbf{u}(t) &= -\mathbf{C}_1(t)\mathbf{x}(t) - \mathbf{C}_2(t)\int_0^t \mathbf{H}_\xi \mathbf{x}(\tau)\, d\tau \\ &= -\mathbf{C}_1\mathbf{x}(t) - \mathbf{C}_2(t)\boldsymbol{\xi}(t)\end{aligned} \tag{2.5-15}$$

where $\boldsymbol{\xi}(t)$ is an l-dimensional *integral state* defined by the ordinary differential equation

$$\dot{\boldsymbol{\xi}}(t) = \mathbf{H}_\xi \mathbf{x}(t) \tag{2.5-16}$$

The $(l \times n)$ *selection matrix* $\mathbf{H}_\xi$ selects the state to be integrated by containing ones in the appropriate elements and zeros elsewhere. The equivalent $(n+l)$-dimensional homogeneous equation is

$$\begin{bmatrix}\dot{\mathbf{x}}(t)\\ \dot{\boldsymbol{\xi}}(t)\end{bmatrix}=\begin{bmatrix}[\mathbf{F}(t)-\mathbf{G}(t)\mathbf{C}_1(t)] & -\mathbf{G}(t)\mathbf{C}_2(t)\\ \mathbf{H}_\xi & \mathbf{0}\end{bmatrix}\begin{bmatrix}\mathbf{x}(t)\\ \boldsymbol{\xi}(t)\end{bmatrix} \tag{2.5-17a}$$

or

$$\dot{\boldsymbol{\chi}}(t)=\mathbf{F}'(t)\boldsymbol{\chi}(t) \tag{2.5-17b}$$

with $\boldsymbol{\chi}^T=[\mathbf{x}^T \quad \boldsymbol{\xi}^T]$.

Actuator dynamics can be appended in similar fashion. If the homogeneous equation for the actuators takes the form

$$\dot{\mathbf{u}}(t)=\mathbf{F}_C(t)\mathbf{u}(t) \tag{2.5-18}$$

then the $(n+m)$-dimensional system homogeneous equation is

$$\begin{bmatrix}\dot{\mathbf{x}}(t)\\ \dot{\mathbf{u}}(t)\end{bmatrix}=\begin{bmatrix}\mathbf{F}(t) & \mathbf{G}(t)\\ \mathbf{0} & \mathbf{F}_C(t)\end{bmatrix}\begin{bmatrix}\mathbf{x}(t)\\ \mathbf{u}(t)\end{bmatrix} \tag{2.5-19a}$$

or

$$\dot{\boldsymbol{\chi}}(t)=\mathbf{F}'(t)\boldsymbol{\chi}(t) \tag{2.5-19b}$$

with $\boldsymbol{\chi}^T=[\mathbf{x}^T \quad \mathbf{u}^T]$. It is clear that combinations of actuator dynamics and control compensation can be modeled directly.

Static and Quasistatic Equilibrium

As long as differential equations such as those presented above are satisfied, the system is at *dynamic equilibrium*: forces, kinematic relationships, and state rates of change are balanced. If the state rates are zero, the system is at *static equilibrium* (i.e., $\mathbf{x}(t)$ has a fixed value). Points of static equilibrium represent the *steady-state response of stable systems* (to be defined later), and they serve as reference points in the state space that help organize our understanding of possible trajectories. While constant $\mathbf{x}$ conceivably could be maintained by commanding $\mathbf{u}(t)$ to counter the effects of varying $\mathbf{w}(t)$, static equilibrium normally implies a state of rest for all system variables. Denoting the resting value by $(\cdot)^*$, the static equilibria of nonlinear and linear time-invariant systems then are defined by

$$\mathbf{0}=\mathbf{f}[\mathbf{x}^*,\mathbf{u}^*,\mathbf{w}^*] \tag{2.5-20}$$

$$\mathbf{0}=\mathbf{F}\mathbf{x}^*+\mathbf{G}\mathbf{u}^*+\mathbf{L}\mathbf{w}^* \tag{2.5-21}$$

In either case, n scalar equations are zero simultaneously. Given $\mathbf{u}^*$ and $\mathbf{w}^*$, the values of $\mathbf{x}^*$ that produce static equilibrium of a nonlinear system normally would be defined by iterative solution of Eq. 2.5-20, although closed-form solutions may exist for special cases. If $\mathbf{F}$ is nonsingular, static equilibrium of the linear system is readily defined by

$$\mathbf{x}^* = -\mathbf{F}^{-1}(\mathbf{G}\mathbf{u}^* + \mathbf{L}\mathbf{w}^*) \tag{2.5-22}$$

If $\mathbf{F}$ is singular, the linear system still may possess a *quasistatic equilibrium*, as an example shows. Suppose $\mathbf{F}$ has rank $(n-l)$; reordering state elements if necessary, it may be possible to partition $\mathbf{F}$ as

$$\mathbf{F} = \begin{bmatrix} \mathbf{F}_1 & \mathbf{0} \\ \mathbf{F}_2 & \mathbf{0} \end{bmatrix} \tag{2.5-23}$$

where $\mathbf{F}_1$ is an $(n-l)\times(n-l)$ nonsingular matrix and $\mathbf{F}_2$ is an $l\times(n-l)$ matrix. With corresponding partitioning of the state vector, the system dynamic equation is then

$$\begin{bmatrix} \dot{\mathbf{x}}_1(t) \\ \dot{\mathbf{x}}_2(t) \end{bmatrix} = \begin{bmatrix} \mathbf{F}_1 & \mathbf{0} \\ \mathbf{F}_2 & \mathbf{0} \end{bmatrix} \begin{bmatrix} \mathbf{x}_1(t) \\ \mathbf{x}_2(t) \end{bmatrix} + \begin{bmatrix} \mathbf{G}_1 \\ \mathbf{G}_2 \end{bmatrix} \mathbf{u}(t) + \begin{bmatrix} \mathbf{L}_1 \\ \mathbf{L}_2 \end{bmatrix} \mathbf{w}(t) \tag{2.5-24}$$

$\mathbf{x}_2(t)$ is simply an integral of $\mathbf{x}_1(t)$ and the inputs, and $\mathbf{x}_1(t)$ is independent of $\mathbf{x}_2(t)$; therefore, reduced-order static equilibrium of $\mathbf{x}_1(t)$ is given by

$$\mathbf{x}_1^* = -\mathbf{F}_1^{-1}(\mathbf{G}_1\mathbf{u}^* + \mathbf{L}_1\mathbf{w}^*) \tag{2.5-25}$$

and $\mathbf{x}_2^*(t)$, though by no means fixed, is a well-defined function of the equilibrium solution:

$$\begin{aligned} \mathbf{x}_2^*(t) &= \mathbf{x}_2(0) + \int_0^t (\mathbf{F}_2\mathbf{x}_1^* + \mathbf{G}_2\mathbf{u}^* + \mathbf{L}_2\mathbf{w}^*)\, d\tau \\ &= \mathbf{x}_2(0) + (\mathbf{F}_2\mathbf{x}_1^* + \mathbf{G}_2\mathbf{u}^* + \mathbf{L}_2\mathbf{w}^*)t \end{aligned} \tag{2.5-26}$$

We refer to $\mathbf{x}_2^*(t)$ as being *quasistatic to first degree*, and it is apparent that higher-degree quasistatic solutions could be specified when $\mathbf{F}$ is singular but partitions differently.

As a further example, consider the second-order system

$$\begin{bmatrix} \dot{x}_1(t) \\ \dot{x}_2(t) \end{bmatrix} = \begin{bmatrix} a & b \\ 1 & 0 \end{bmatrix} \begin{bmatrix} x_1(t) \\ x_2(t) \end{bmatrix} + \begin{bmatrix} c \\ 0 \end{bmatrix} u(t) \tag{2.5-27}$$

Static equilibrium is defined by

$$\begin{bmatrix} x_1^* \\ x_2^* \end{bmatrix} = -\begin{bmatrix} a & b \\ 1 & 0 \end{bmatrix}^{-1} \begin{bmatrix} c \\ 0 \end{bmatrix} u^*$$
$$= \frac{1}{b}\begin{bmatrix} 0 & -b \\ -1 & a \end{bmatrix} \begin{bmatrix} c \\ 0 \end{bmatrix} u^*$$
$$= \begin{bmatrix} 0 \\ \dfrac{-cu^*}{b} \end{bmatrix} \tag{2.5-28}$$

If $b = 0$, the fundamental matrix is singular, but the steady-state value of x_1^* is simply

$$x_1^* = -\frac{cu^*}{a} \tag{2.5-29}$$

while $x_2^*(t)$ is a ramp function of x_1^*:

$$x_2^*(t) = x_2(0) - \frac{cu^* t}{a} \tag{2.5-30}$$

Having found a static equilibrium for $\mathbf{x}^*$, given $\mathbf{u}^*$ and $\mathbf{w}^*$, it can be used as a reference point for further study of time-invariant systems. Defining

$$\tilde{\mathbf{x}}(t) = \mathbf{x}(t) - \mathbf{x}^* \tag{2.5-31}$$
$$\tilde{\mathbf{u}}(t) = \mathbf{u}(t) - \mathbf{u}^* = \mathbf{0} \tag{2.5-32}$$
$$\tilde{\mathbf{w}}(t) = \mathbf{w}(t) - \mathbf{w}^* = \mathbf{0} \tag{2.5-33}$$

the nonlinear and linear system differential equations reduce to their respective autononomous versions

$$\dot{\tilde{\mathbf{x}}}(t) = \mathbf{f}'[\tilde{\mathbf{x}}(t), \mathbf{0}, \mathbf{0}] \tag{2.5-34}$$
$$\dot{\tilde{\mathbf{x}}}(t) = \mathbf{F}\tilde{\mathbf{x}}(t) \tag{2.5-35}$$

where $\mathbf{f}'[\cdot]$ is equivalent to $\mathbf{f}[\cdot]$, rewritten to account for the shift in origin. The shift in origin has no effect on the linear system's fundamental matrix. The static equilibrium points for the two cases evidently provide

$$\mathbf{0} = \mathbf{f}'[\mathbf{0}, \mathbf{0}, \mathbf{0}] \tag{2.5-36}$$
$$\mathbf{0} = \mathbf{F}(\mathbf{0}) \tag{2.5-37}$$

Equilibrium points are sometimes called *singular points*, but not because there is anything singular about Eqs. 2.5-36 or 2.5-37. If we choose to plot two components of a state trajectory, $\tilde{x}_i(t)$ and $\tilde{x}_j(t)$, against each other rather than against time, a projection of the trajectory on the $\tilde{x}_i - \tilde{x}_j$ plane is obtained (Fig. 2.5-1). Taking the horizontal axis in the $\tilde{x}_j$ direction, the slope of the resulting curve would be

$$\frac{\partial \tilde{x}_i(t)}{\partial \tilde{x}_j(t)} = \frac{d\tilde{x}_i(t)/dt}{d\tilde{x}_j(t)/dt} = \frac{f_i[\tilde{x}(t)]}{f_j[\tilde{x}(t)]} \tag{2.5-38}$$

However, at an equilibrium point, $f_i[\cdot]$ and $f_j[\cdot]$ are zero, and the slope is indeterminate; hence the point is singular in the $\tilde{x}_i - \tilde{x}_j$ plane. Projections of the trajectory are analogous to the *phase-plane plots* of second-order systems (described later), and they often provide insights regarding the nature of motions.

If the state is perturbed from an equilibrium point, its tendency to remain in the neighborhood of the point is governed by the system's stability, which is detailed in the next section. A system can have more than one equilibrium point $\mathbf{x}^*$ for a given control setting and disturbance input, $\mathbf{u}^*$ and $\mathbf{w}^*$, and the number of equilibria may depend on the values of $\mathbf{u}^*$ and $\mathbf{w}^*$. If a given equilibrium point (for fixed $\mathbf{u}^*$ and $\mathbf{w}^*$) is stable, then adjacent equilibrium points are unstable, and vice versa. Which equilibrium point attracts or repels the state is dependent on the initial conditions; however, if $\mathbf{u}^*$ or $\mathbf{w}^*$ is changing, the dynamic state may jump abruptly from the "sphere of influence" of one equilibrium to another, giving rise to the name *catastrophe theory* for the description of such behavior. It can be concluded that multiple equilibria can lead to multiple locally optimal solutions.

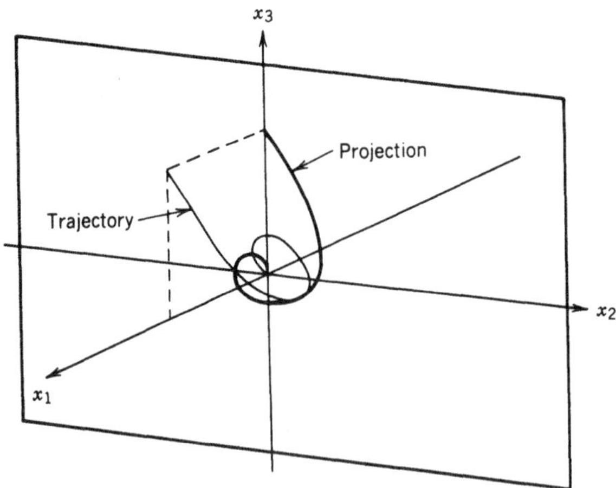

FIGURE 2.5-1 Projection of a multidimensional trajectory on a plane.

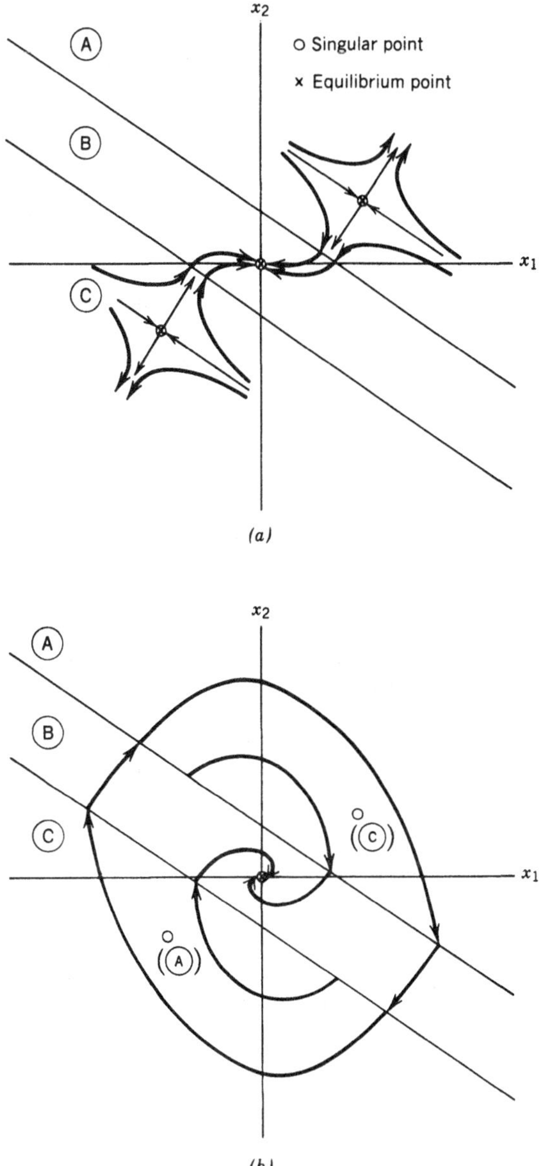

FIGURE 2.5-2 Phase-plane plots for two piecewise-linear systems with different dynamic characteristics. (*a*) All singular points are equilibrium points; (*b*) only one singular point is an equilibrium point.

It is possible to have singular points that are not equilibrium points. Suppose, for example, that a linear system has a saturating feedback control law such that the state space is separated into three distinct regions (A, B, C) with differing dynamic characteristics (Fig. 2.5-2). Each region has one singular point defined by Eq. 2.5-37 and 2.5-38. Consider two different piecewise-linear systems. If a region's singular point is contained within itself (Fig. 2.5-2*a*), that singular point is an equilibrium point (H-1). If a region's singular point lies within a different region (Fig. 2.5-2*b*), it is not an equilibrium point because the trajectory at that point is unaffected by the singularity. Nevertheless, the singular points affect the phase-plane trajectories in their respective regions, as shown below.

Stability

A system response's tendency to grow or decay in time characterizes the system's stability. In the most general sense, stability refers to characteristics of the time variation of *response*. From a practical viewpoint, we may prefer to determine the stability of the *system* which produces that response. In the first instance, we specify criteria for the time evolution of the state, (that is, for the stability of *trajectories*). In the second instance, we determine criteria for the *structures and parameters* of the system that can be related to these trajectories.

In order to decide whether the response is growing or decaying, a measure of magnitude or length—a norm in the parlance of Section 2.1—must be defined, and its value in some time interval must be known. The *Euclidean norm* introduced in an earlier section (Eq. 2.1-29),

$$\begin{aligned} \|\tilde{\mathbf{x}}(t)\| &= |\tilde{\mathbf{x}}^T(t)\tilde{\mathbf{x}}(t)|^{1/2} \\ &= [\tilde{x}_1^2 + \tilde{x}_2^2 + \cdots + \tilde{x}_n^2]^{1/2} \end{aligned} \tag{2.5-39}$$

is a useful measure for multivariable systems because it is a positive definite scalar function, being zero only if all components are zero and diverging if any single component diverges. For stability analysis, it is reasonable to form the Euclidean norm of the perturbation from static equilibrium, $\tilde{\mathbf{x}}(t)$ (Eq. 2.5-31), as we might expect a stable trajectory to remain in the neighborhood of the equilibrium point, if not actually converge to it, as time passes.

There are numerous definitions of stability that depend on pragmatic as well as rigorous goals; it is most convenient to restrict our attention to autonomous systems and to begin by considering *uniform stability*: an equilibrium point is stable if arbitrary bounds on a trajectory can be related to bounds on initial conditions, independent of the starting time t_0. More specifically, if, for every $\epsilon > 0$, we can define $\delta(\epsilon)$ such that

$$\|\tilde{\mathbf{x}}(t_0)\| \leq \delta, \qquad \delta > 0 \tag{2.5-40}$$

leads to

$$\|\tilde{\mathbf{x}}(t)\| \leq \epsilon, \qquad \epsilon \geq \delta > 0, \quad t \geq t_0 \tag{2.5-41}$$

for all t beyond the starting time, then the origin is a stable point. This notion is sketched in Fig. 2.5-3, where stable and unstable trajectories and norms are shown. This definition allows the state to oscillate forever about the equilibrium.

The static equilibrium point evidences *local asymptotic stability* if Eq. 2.5-40 and 2.5-41 are satisfied *and* if

$$\lim_{t \to \infty} \|\tilde{\mathbf{x}}(t)\| = 0 \tag{2.5-42}$$

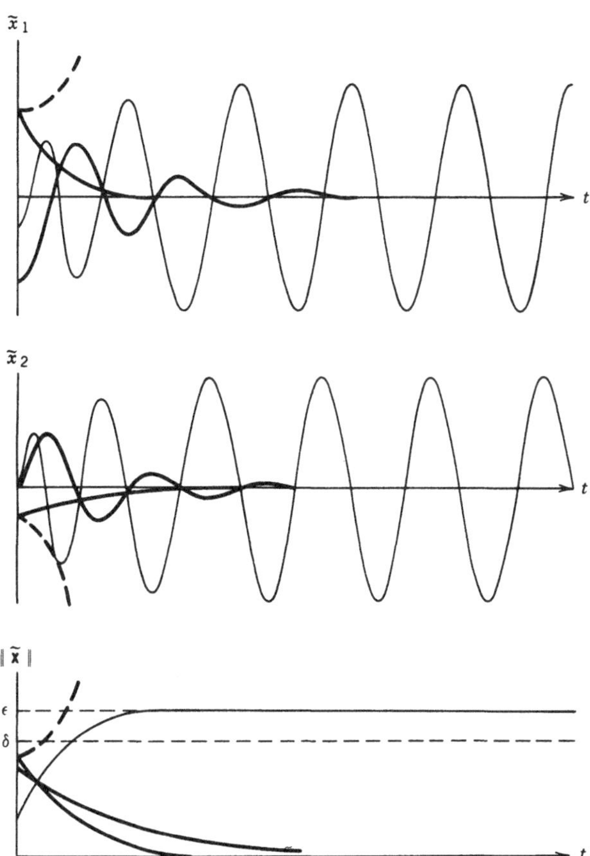

FIGURE 2.5-3 Uniformly stable (solid line) and unstable (dashed line) trajectories and norms.

In other words, it is insufficient for the response to be bounded; it must decay to zero as time passes. This definition of stability is somewhat more appealing, in that it guarantees that transients will decay and that the response will approach a steady state. Here Eq. 2.5-41 serves only to limit transient peaks to arbitrarily large (or small) values.

If Eqs. 2.5-40 to 2.5-42 are satisfied for *any* choice of ϵ and lim $\delta(\epsilon)$ approaches ∞ as ϵ approaches ∞, the equilibrium point possesses *uniform asymptotic stability in the whole*. This can occur in nonlinear systems that become more stable with increased $\|\tilde{\mathbf{x}}\|$, as well as linear systems, whose stability characteristics are independent of amplitude. In the latter case, local asymptotic stability guarantees asymptotic stability in the whole.

Meeting the criterion for *exponential asymptotic stability* assures a given rate of convergence to the equilibrium. If we replace Eq. 2.5-41 by

$$\|\tilde{\mathbf{x}}(t)\| \le ke^{-\alpha t}\|\tilde{\mathbf{x}}(0)\|, \qquad k, \alpha > 0 \tag{2.5-43}$$

then the norm of $\tilde{\mathbf{x}}(t)$ is guaranteed to lie within a decaying envelope whose convergence rate is determined by α. Because

$$\begin{aligned}\int_0^\infty e^{-\alpha t}\,dt &= -\frac{1}{\alpha}e^{-\alpha t}\Big|_0^\infty \\ &= -\frac{1}{\alpha}(0-1) \\ &= \frac{1}{\alpha}\end{aligned} \tag{2.5-44}$$

integrals of the norm of $\tilde{\mathbf{x}}(t)$ are bounded:

$$\begin{aligned}\int_0^\infty \|\tilde{\mathbf{x}}(t)\|\,dt &= \int_0^\infty [\tilde{\mathbf{x}}^T(t)\tilde{\mathbf{x}}(t)]^{1/2}\,dt \\ &\le \frac{k}{\alpha}\|\tilde{\mathbf{x}}(0)\|\end{aligned} \tag{2.5-45}$$

$$\begin{aligned}\int_0^\infty \|\tilde{\mathbf{x}}(t)\|^2\,dt &= \int_0^\infty [\tilde{\mathbf{x}}^T(t)\tilde{\mathbf{x}}(t)]\,dt \\ &\le (k^2/2\alpha)\|\tilde{\mathbf{x}}(0)\|^2\end{aligned} \tag{2.5-46}$$

Consequently, if we define

$$\tilde{\chi}(t) = \mathbf{D}\tilde{\mathbf{x}}(t) \tag{2.5-47}$$

such that

$$\mathbf{Q} = \mathbf{D}^T\mathbf{D} > \mathbf{0} \tag{2.5-48}$$

then an integral cost function of the form

$$\int_0^\infty [\tilde{\chi}^T(t)\tilde{\chi}(t)]\,dt = \int_0^\infty [\tilde{\mathbf{x}}^T(t)\mathbf{D}^T\mathbf{D}\tilde{\mathbf{x}}(t)]\,dt$$

$$= \int_0^\infty [\tilde{\mathbf{x}}^T(t)\mathbf{Q}\tilde{\mathbf{x}}(t)]\,dt \tag{2.5-49}$$

is bounded for stable $\tilde{\mathbf{x}}(t)$. The converse also is true; a bounded integral of the norm implies stability.

This result has practical significance in the development of the linear-quadratic regulator, which is an optimal control law for a time-invariant system with quadratic cost function and infinite final time (Section 5.4). This control law is expressed as

$$\tilde{\mathbf{u}}(t) = -\mathbf{C}\tilde{\mathbf{x}}(t) \tag{2.5-50}$$

where $\mathbf{C}$ is a constant optimal gain matrix. From Section 2.3, the closed-loop initial condition response of a linear, time-invariant system is

$$\tilde{\mathbf{x}}(t) = \mathbf{\Phi}_{\mathrm{CL}}(t, 0)\tilde{\mathbf{x}}(0)$$

$$= e^{\mathbf{F}_{\mathrm{CL}}t}\tilde{\mathbf{x}}(0) \tag{2.5-51}$$

where $\mathbf{F}_{\mathrm{CL}}$ is the closed-loop fundamental matrix $(\mathbf{F} - \mathbf{GC})$, and $\mathbf{\Phi}_{\mathrm{CL}}$ is the corresponding state transition matrix. If a minimized cost function of the form

$$J = \frac{1}{2}\int_0^\infty [\tilde{\mathbf{x}}^T(t)\mathbf{Q}\tilde{\mathbf{x}}(t) + \tilde{\mathbf{u}}^T(t)\mathbf{R}\tilde{\mathbf{u}}(t)]\,dt$$

$$= \frac{1}{2}\int_0^\infty [\tilde{\mathbf{x}}^T(t)(\mathbf{Q} + \mathbf{C}^T\mathbf{R}\mathbf{C})\tilde{\mathbf{x}}(t)]\,dt, \qquad (\mathbf{Q} + \mathbf{C}^T\mathbf{R}\mathbf{C}) > \mathbf{0} \tag{2.5-52}$$

has finite bound, then $\tilde{\mathbf{x}}(t)$ must be exponentially asymptotically stable, as

$$\|\tilde{\mathbf{x}}(t)\| = \|e^{\mathbf{F}_{\mathrm{CL}}t}\tilde{\mathbf{x}}(0)\|$$

$$\le k'e^{-\alpha t}\|\tilde{\mathbf{x}}(0)\| \tag{2.5-53}$$

where α represents the slowest response mode of the system. (More is said about response modes later.)

A. M. Lyapunov* established two stability theorems that underlie much

*Alexandr Lyapunov (1857–1918), professor of mathematics at Kharkov, first published his stability criteria for physical systems in 1892. He is well known for his contributions to probability theory (method of characteristic functions) as well as stability theory. Lyapunov cut his own life short on the day his wife died of tuberculosis.

of modern stability analysis (L-4). We will restrict our attention to autonomous systems, although the results can be applied to time-varying systems as well. Given the nonlinear system

$$\dot{\tilde{\mathbf{x}}}(t) = \mathbf{f}[\tilde{\mathbf{x}}(t)] \tag{2.5-54}$$

with state equilibrium at the origin, consider the corresponding linear perturbation model

$$\Delta\dot{\tilde{\mathbf{x}}}(t) = \frac{\partial \mathbf{f}}{\partial \mathbf{x}} \Delta\tilde{\mathbf{x}}(t) \triangleq \mathbf{F}\Delta\tilde{\mathbf{x}}(t) \tag{2.5-55}$$

where the fundamental matrix is evaluated at the origin. *Lyapunov's first theorem* says that the nonlinear system (Eq. 2.5-54) is asymptotically stable at the origin if the linear model is asymptotically stable at the origin. No assurances can be given if the linear model is only neutrally stable, in the sense of Eqs. 2.5-40 and 2.5-41, for then at least some elements of the response may be determined exclusively by nonlinear effects. Furthermore, the theorem gives no clues regarding the extent of the nonlinear system's stable region; if Eq. 2.5-55 is asymptotically stable, it is globally asymptotically stable and that may not be the case for Eq. 2.5-54. Nevertheless, the first theorem makes a powerful statement about the perturbation stability of nonlinear systems.

Lyapunov's second theorem (or *direct method*) is a generalization and extension of Eqs. 2.5-40 to 2.5-42 that provides a formal means of assessing stability in terms of the system structure and parameters. Suppose $\mathcal{V}[\tilde{\mathbf{x}}(t)]$ is a positive definite scalar function of $\tilde{\mathbf{x}}(t)$ whose time-derivative is

$$\begin{aligned} \dot{\mathcal{V}}[\tilde{\mathbf{x}}(t)] &= \left\{\frac{\partial \mathcal{V}[\tilde{\mathbf{x}}(t)]}{\partial \tilde{\mathbf{x}}}\right\} \frac{d\tilde{\mathbf{x}}(t)}{dt} \\ &= \mathcal{V}_{\mathbf{x}}[\tilde{\mathbf{x}}(t)]\mathbf{f}[\tilde{\mathbf{x}}(t)] \end{aligned} \tag{2.5-56}$$

for the autononomous system described by Eq. 2.5-54. If $\dot{\mathcal{V}}[\tilde{\mathbf{x}}(t)] \leq 0$ in the neighborhood of the equilibrium point, then $\mathcal{V}[\tilde{\mathbf{x}}(t)]$ is a *Lyapunov function*, and the neighborhood is a *stable* region. If $\dot{\mathcal{V}}[\tilde{\mathbf{x}}(t)] < 0$ in the neighborhood, then it is an *asymptotically stable* region. As before, the result is conservative, in that failure to find a Lyapunov function does not preclude stability. Furthermore, for a given Lyapunov function, regions in which $\dot{\mathcal{V}}[\tilde{\mathbf{x}}(t)]$ is not negative still may be stable.

The principal difficulty with applying the direct method is that no comprehensive guidance for choosing Lyapunov functions can be given: different Lyapunov functions may identify different regions of stability. At this point, it may not be surprising that a quadratic form often is chosen, as it can be specified to be positive definite with a symmetric defining matrix,

P, and its derivative is continuous. Choosing

$$\mathcal{V}[\tilde{\mathbf{x}}(t)] = \tilde{\mathbf{x}}^T(t)\mathbf{P}\tilde{\mathbf{x}}(t) \tag{2.5-57}$$

the scalar time-derivative of $\mathcal{V}[\cdot]$ is

$$\begin{aligned}\dot{\mathcal{V}}[\tilde{\mathbf{x}}(t)] &= 2\tilde{\mathbf{x}}^T(t)\mathbf{P}\dot{\tilde{\mathbf{x}}}(t)\\ &= 2\tilde{\mathbf{x}}^T(t)\mathbf{P}\mathbf{f}[\tilde{\mathbf{x}}(t)]\end{aligned} \tag{2.5-58}$$

Given $\mathbf{f}[\cdot]$, $\dot{\mathcal{V}}[\cdot]$ can be mapped as a function of $\tilde{\mathbf{x}}(t)$ and the parameters of $\mathbf{f}[\cdot]$ to estimate the stable regions of response (in which $\dot{\mathcal{V}} \leq 0$). However, a suitable definition of **P** may be difficult to obtain.

For the linear system (Eq. 2.5-55), $\dot{\mathcal{V}}[\cdot]$ is

$$\begin{aligned}\dot{\mathcal{V}}[\tilde{\mathbf{x}}(t)] &= 2\tilde{\mathbf{x}}^T(t)\mathbf{P}\mathbf{F}\tilde{\mathbf{x}}(t)\\ &= \tilde{\mathbf{x}}^T(t)[\mathbf{P}\mathbf{F} + \mathbf{F}^T\mathbf{P}]\tilde{\mathbf{x}}(t)\end{aligned} \tag{2.5-59}$$

and, although the vector-matrix product is a scalar, it is advantageous to replace the asymmetric $2\mathbf{PF}$ with the symmetric $(\mathbf{PF} + \mathbf{F}^T\mathbf{P})$. We wish to define **P** so that

$$(\mathbf{PF} + \mathbf{F}^T\mathbf{P}) = -\mathbf{Q} \tag{2.5-60}$$

where **Q** is a positive definite matrix, thus assuring that $\dot{\mathcal{V}}[\cdot]$ is negative definite. Conversely, we may define **Q** as an easily specified positive definite matrix (e.g., a diagonal matrix), then solve Eq. 2.5-60 for **P**. This equation is known as a *Lyapunov equation*, and although it is linear in **P**, it normally requires an iterative solution [as in (B-3) or (G-6)].

Having found a satisfactory **P** for the linear model, it can be inferred from Lyapunov's first theorem that this is a satisfactory defining function for the nonlinear system, at least in the neighborhood of equilibrium. Of course, **F** may not be stable, in which case Eq. 2.5-60 cannot be solved with positive definite **P** and **Q**. In (N-1), it is shown that finding the necessary **P** and **Q** is equivalent to satisfying the well-known Hurwitz and Nyquist stability criteria for linear systems.

There is a striking similarity between Lyapunov's direct method and the principle of optimality used in dynamic programming (Chapters 3 and 5). Both are based on positive definite (often quadratic) performance measures which must decrease uniformly from the initial to the final times. Furthermore, stability and optimality are, in some cases, synonymous [Section 5.4 and (K-3)]. These parallels remain to be exploited.

Finally, we mention bounded-input/bounded-output (BIBO) stability, which describes observable response (Eq. 2.5-2) to bounded inputs in Eq. 2.5-1 (J-2). Satisfying the preceding criteria assures BIBO stability;

however, BIBO stability also admits the possibility of internal instability that is undetectable in the output.

Modes of Motion for Linear, Time-Invariant Systems

The initial-condition responses of linear, time-invariant systems always consist of linear combinations of classical waveforms, including exponentials and sinusoids as well as powers of time. These waveforms establish the *modes of motion* of constant-coefficient systems, and they have general application in control system design. Forced responses also can be described as superposition (or convolution) integrals of the response modes and the forcing input; hence the *input-output characteristics* of linear, time-invariant systems reflect these modes of motion. As this section is an extension of the previous section on stability, attention is focused on initial-condition response, with input–output characteristics reappearing in Section 2.6.

While the Fourier transform introduced in Section 2.4 has obvious application for describing modes of motion, the associated integrals (Eqs. 2.4-112 and 2.4-113) cannot be defined uniquely for some common functions (e.g., a unit step input) unless a convergence factor is incorporated. Therefore, we replace the imaginary variable, $j\omega$, by the complex variable, $s = \sigma + j\omega$, before evaluating the integrals. The *Laplace* operator*, s, contains the convergence factor σ as well as the imaginary frequency $j\omega$, and it has units of frequency (e.g., rad/s). Furthermore, if we restrict our attention to scalar time histories, $x(t)$, that begin at $t = 0$, we arrive at definitions for the one-sided *Laplace transform* and its inverse, as follows:

$$\mathfrak{L}[x(t)] \triangleq x(s) = \int_0^{\infty} x(t)e^{-st}\,dt \tag{2.5-61}$$

$$\mathfrak{L}^{-1}[x(s)] = x(t) = \frac{1}{2\pi j}\int_{\sigma-j\infty}^{\sigma+j\infty} x(s)e^{st}\,ds \tag{2.5-62}$$

We shall need to evaluate very few of these integrals here; for now, it is sufficient to know that $x(s)$ exists for any $x(t)$ of exponential order, that is, for which

$$\lim_{t\to\infty} x(t)e^{kt} = 0, \qquad k = \text{a real number} \tag{2.5-63}$$

We are more concerned with selected properties of the Laplace transform.

*Pierre Simon, Marquis de Laplace (1749–1827), was a French astronomer and mathematician who studied the mechanics of the solar system as well as the theory of probability.

Because it is a linear operator,

$$\mathfrak{L}[x_1(t)+x_2(t)]=\mathfrak{L}[x_1(t)]+\mathfrak{L}[x_2(t)]=x_1(s)+x_2(s) \qquad (2.5\text{-}64)$$

$$\mathfrak{L}[ax(t)]=a\mathfrak{L}[x(t)]=ax(s) \qquad (2.5\text{-}65)$$

Laplace transforms of vectors and matrices are defined similarly; for the n-vector $\mathbf{x}(t)$ and the $(n\times r)$-matrix $\mathbf{F}(t)$,

$$\mathfrak{L}[\mathbf{x}(t)]\triangleq\mathbf{x}(s)=\begin{bmatrix} x_1(s)\\ x_2(s)\\ \vdots\\ x_n(s)\end{bmatrix} \qquad (2.5\text{-}66)$$

$$\mathfrak{L}[\mathbf{F}(t)]\triangleq\mathbf{F}(s)=\begin{bmatrix} f_{11}(s) & f_{12}(s) & \dots & f_{1r}(s)\\ f_{21}(s) & f_{22}(s) & \dots & f_{2r}(s)\\ \vdots & \vdots & & \vdots\\ f_{n1}(s) & f_{n2}(s) & \dots & f_{nr}(s)\end{bmatrix} \qquad (2.5\text{-}67)$$

The Laplace transform of the time-derivative of $\mathbf{x}(t)$ is

$$\mathfrak{L}[\dot{\mathbf{x}}(t)]=s\mathbf{x}(s)-\mathbf{x}(0) \qquad (2.5\text{-}68)$$

where the initial conditon $\mathbf{x}(0)$ is the numerical value of $\mathbf{x}(t)$ immediately before the imposition of any starting singularity [e.g., impulse or step function] (K-1).

These equations are adequate to develop the major relationships between time-domain and frequency-domain state-space models. Given the linear, time-invariant system and observation equations

$$\dot{\mathbf{x}}(t)=\mathbf{F}\mathbf{x}(t)+\mathbf{G}\mathbf{u}(t)+\mathbf{L}\mathbf{w}(\mathrm{t}) \qquad (2.5\text{-}69)$$

$$\mathbf{y}(t)=\mathbf{H}\mathbf{x}(t) \qquad (2.5\text{-}70)$$

the corresponding Laplace transforms are

$$s\mathbf{x}(s)-\mathbf{x}(0)=\mathbf{F}\mathbf{x}(s)+\mathbf{G}\mathbf{u}(s)+\mathbf{L}\mathbf{w}(s) \qquad (2.5\text{-}71)$$

$$\mathbf{y}(s)=\mathbf{H}\mathbf{x}(s) \qquad (2.5\text{-}72)$$

The frequency-domain state equation then can be manipulated as follows:

$$s\mathbf{x}(s)-\mathbf{F}\mathbf{x}(s)=\mathbf{x}(0)+\mathbf{G}\mathbf{u}(s)+\mathbf{L}\mathbf{w}(s)$$

$$(s\mathbf{I}_n-\mathbf{F})\mathbf{x}(s)=\mathbf{x}(0)+\mathbf{G}\mathbf{u}(s)+\mathbf{L}\mathbf{w}(s)$$

$$\mathbf{x}(s)=(s\mathbf{I}_n-\mathbf{F})^{-1}[\mathbf{x}(0)+\mathbf{G}\mathbf{u}(s)+\mathbf{L}\mathbf{w}(s)] \qquad (2.5\text{-}73)$$

This expresses the Laplace transform of the state in terms of the initial condition, the constant-coefficient input-effect matrices (**G** and **L**), the Laplace transforms of the inputs [$\mathbf{u}(s)$ and $\mathbf{w}(s)$], and $(s\mathbf{I}_n - \mathbf{F})^{-1}$, where

$$(s\mathbf{I}_n - \mathbf{F})^{-1} = \frac{\text{Adj}(s\mathbf{I}_n - \mathbf{F})}{|s\mathbf{I}_n - \mathbf{F}|} \tag{2.5-74}$$

as defined in Section 2.2. Recall that $\text{Adj}(s\mathbf{I}_n - \mathbf{F})$ is an $(n \times n)$ matrix, while $|s\mathbf{I}_n - \mathbf{F}|$ is a scalar variable. Both of these contain polynomials in s as a consequence of their definitions (Eqs. 2.2-17 and 2.2-13 to 2.2-15).

The Laplace transform of the system's initial-condition response is specified by Eq. 2.5-73, with $\mathbf{u}(s) = \mathbf{w}(s) = \mathbf{0}$,

$$\mathbf{x}(s) = (s\mathbf{I}_n - \mathbf{F})^{-1}\mathbf{x}(0) \tag{2.5-75}$$

while Eqs. 2.3-61 and 2.3-73 give the time response of the linear, time-invariant system in terms of the state transition matrix:

$$\mathbf{x}(t) = \mathbf{\Phi}(t, 0)\mathbf{x}(0) = e^{\mathbf{F}t}\mathbf{x}(0) \tag{2.5-76}$$

Because $\mathbf{x}(0)$ is a constant, it can be concluded that $(s\mathbf{I}_n - \mathbf{F})^{-1}$ is the Laplace transform of $e^{\mathbf{F}t}$:

$$\mathcal{L}(e^{\mathbf{F}t}) = (s\mathbf{I}_n - \mathbf{F})^{-1} \triangleq \mathbf{A}(s) \tag{2.5-77}$$

As the elements of the matrix Laplace transform are the Laplace transforms of the elements, and

$$\mathcal{L}^{-1}[(s\mathbf{I}_n - \mathbf{F})^{-1}] = e^{\mathbf{F}t} \triangleq \mathbf{\Phi}(t, 0) \tag{2.5-78}$$

this provides a straightforward method of determining the state transition matrix; the elements are the inverse Laplace transforms of the corresponding elements of $(s\mathbf{I}_n - \mathbf{F})^{-1}$.

The denominator of $(s\mathbf{I}_n - \mathbf{F})^{-1}$ is a determinant, and it is identical to Eq. 2.2-72; hence $|s\mathbf{I}_n - \mathbf{F}|$ can be expanded as

$$|s\mathbf{I}_n - \mathbf{F}| \triangleq \Delta(s) = s^n + c_{n-1}s^{n-1} + \cdots + c_1 s + c_0 \tag{2.5-79}$$

and this can be factored as

$$\Delta(s) = (s - \lambda_1)(s - \lambda_2)\ldots(s - \lambda_n) \tag{2.5-80}$$

Therefore, the roots of the denominator are the *eigenvalues* of **F**. These roots may be real or complex, but if they are the latter, they occur in complex-conjugate pairs for real-valued **F**.

The elements of $\text{Adj}(s\mathbf{I}_n - \mathbf{F})$ are cofactors of $(s\mathbf{I}_n - \mathbf{F})$, which in turn are determinants of dimension $(n-1)\times(n-1)$; therefore, they are polynomials in s of degree no greater than $(n-1)$. Consequently, an element of $\mathbf{A}(s) = (s\mathbf{I}_n - \mathbf{F})^{-1}$ could take the form

$$a_{jk}(s) = \frac{k_{jk}(s^{n-1} + b_{n-2}s^{n-2} + \cdots + b_1 s + b_0)}{\Delta(s)}$$

$$= \frac{k_{jk}(s-\beta_1)(s-\beta_2)\ldots(s-\beta_{n-1})}{(s-\lambda_1)(s-\lambda_2)\ldots(s-\lambda_n)} \tag{2.5-81}$$

where the coefficients b_i are specific to the jkth term, and the numerator roots β_i are called *zeros* of $a_{jk}(s)$. There are *at most* $(n-1)$ zeros, although there may be less.

From Eq. 2.5-75, the jth element of the state vector's Laplace transform is

$$x_j(s) = a_{j1}(s)x_1(0) + a_{j2}(s)x_2(0) + \cdots + a_{jn}(s)x_n(0)$$

$$= \frac{n_j(s)}{\Delta(s)} \tag{2.5-82}$$

with $n_j(s)$ representing the linear combination of numerators in the jth row of $\mathbf{A}(s)$, each weighted by the corresponding initial condition. With distinct roots, this can be expressed as a partial-fraction expansion

$$x_j(s) = d_0 + \frac{d_1}{(s-\lambda_1)} + \frac{d_2}{(s-\lambda_2)} + \cdots + \frac{d_n}{(s-\lambda_n)} \tag{2.5-83}$$

where the d_i are specific to x_j and are defined as

$$d_i = (s-\lambda_i)\left.\frac{n_j(s)}{\Delta(s)}\right|_{s=\lambda_i} \tag{2.5-84}$$

For the moment, assume that the λ_i are real and distinct; then the d_i are real, and $x_j(t)$ is the sum of d_0 plus n inverse Laplace transforms of the $d_i/(s-\lambda_i)$. The inverse Laplace transform of this function is

$$\mathfrak{L}^{-1}\left(\frac{d_i}{s-\lambda_i}\right) = d_i e^{\lambda_i t} \tag{2.5-85}$$

and $x_j(t)$ is the weighted superposition of n modes of motion, each of which grows or decays exponentially as determined by the magnitude and sign of λ_i . If the magnitude of λ_i is large, the mode is "fast." If the sign of λ_i is positive, the mode is divergent. Consequently, all λ_i must be negative for

$x_j(t)$ to have an exponentially stable response. By *Routh's criterion*, it is necessary (but not sufficient) that all the coefficients of Eq. 2.5-79 have the same sign for this to be true; as the coefficient of s^n is $+1$, the c_i must be positive.

If λ_i is complex and distinct, then another root, say λ_{i+1}, is its complex conjugate, because $x(t)$ is a real variable, and the net contribution of the two complex modes also must be real. Then d_{i+1} is the complex conjugate of d_i, and $(\lambda_i, \lambda_{i+1})$ together define a single, real mode of motion:

$$\frac{d_i}{(s-\lambda_i)} + \frac{d_{i+1}}{(s-\lambda_{i+1})} = \frac{(s-\lambda_i^*)d_i + (s-\lambda_i)d_i^*}{[s^2 - (\lambda_i + \lambda_i^*)s + \lambda_i\lambda_i^*]}$$

$$= \frac{2a[s-(\mu + \nu b/a)]}{s^2 - 2\mu s + (\mu^2 + \nu^2)} \tag{2.5-86}$$

where $(\cdot)^*$ denotes the complex conjugate, and

$$d_i = a + jb \tag{2.5-87}$$

$$\lambda_i = \mu + j\nu \tag{2.5-88}$$

The inverse Laplace transform of Eq. 2.5-86 is

$$\mathfrak{L}^{-1}[\cdot] = \frac{2a}{\nu}\left[\left(2\mu + \frac{\nu b}{a}\right)^2 + \nu^2\right]^{1/2} e^{\mu t} \sin(\nu t + \psi) \tag{2.5-89}$$

with

$$\psi = \tan^{-1}\left[\frac{a}{(-b)}\right] \tag{2.5-90}$$

Clearly, the oscillatory mode diverges if μ is positive.

As a consequence, there can be no more than n and no fewer than $n/2$ modes of real motion in an nth-order system. In the former case, all the modes are exponential; in the latter, all are sinusoidal. Repeated (non-distinct) roots complicate the partial-fraction expansion as detailed, for example, in (D-2); however, the same conclusions can be reached about requirements for asymptotic stability: there must be no eigenvalues with positive real parts.

Following Section 2.2, the eigenvalues can be used to specify corresponding eigenvectors $\mathbf{e}_i$, which in turn allow us to form a modal matrix $\mathbf{D}$. With distinct λ_i, $\mathbf{D}$ can be used to diagonalize $\mathbf{F}$, that is, to create a diagonal matrix $\mathbf{\Lambda}$ such that

$$\mathbf{\Lambda} = \mathbf{D}^{-1}\mathbf{F}\mathbf{D}$$

$$= \begin{bmatrix} \lambda_1 & 0 & \dots & 0 \\ 0 & \lambda_2 & \dots & 0 \\ \vdots & \vdots & & \vdots \\ 0 & 0 & \dots & \lambda_n \end{bmatrix} \qquad (2.5\text{-}91)$$

Hence there is a diagonalized dynamic system corresponding to Eq. 2.5-69 that can be expressed as

$$\dot{\mathbf{q}}(t) = \mathbf{\Lambda}\mathbf{q}(t) + \mathbf{D}^{-1}[\mathbf{G}\mathbf{u}(t) + \mathbf{L}\mathbf{w}(t)] \qquad (2.5\text{-}92)$$

where $\mathbf{q}(t)$ is the *normal-mode state vector*. This is called the *normal form* of the state equation. The initial-condition responses of the elements of $\mathbf{q}(t)$ are uncoupled, but the input responses are not (in general). If the λ_i are real, then all elements of $\mathbf{q}(t)$ are real, but complex λ_i lead to complex elements of $\mathbf{q}(t)$ and $\mathbf{D}$. Because complex λ_i occur in conjugate pairs, the corresponding elements of $\mathbf{\Lambda}$ and $\mathbf{q}(t)$ can be manipulated to provide equivalent real terms. The columns of $\mathbf{D}$ corresponding to $\lambda_{i,\,i+1}$ are $\mathbf{e}_{i,\,i+1}$, and their elements take the form $\mathbf{e}_{i,\,i+1} = \mathbf{a}_i \pm j\mathbf{b}_i$. If these columns are replaced by $\mathbf{e}_i = \mathbf{a}_i$, $\mathbf{e}_{i+1} = \mathbf{b}_i$, then Eq. 2.2-69 indicates that complex (2×2) blocks in $\mathbf{\Lambda}$ like

$$\begin{bmatrix} \lambda_i & 0 \\ 0 & \lambda_{i+1} \end{bmatrix}$$

can be replaced by

$$\begin{bmatrix} \mu_i & \nu_i \\ -\nu_i & \mu_i \end{bmatrix}$$

where $\lambda_{i,\,i+1} = \mu_i \pm j\nu_i$. With repeated roots, diagonalization may not be possible unless $\mathbf{F}$ is symmetric, and modal decomposition requires use of the upper-block-triangular or nondiagonal Jordan canonical form [as in (D-2)].

When the eigenvalues are distinct, any component of the normal-mode state vector is coupled with no more than one other component, and their related motions can be portrayed in two-dimensional *phase-plane plots*. As before, these plots are projections of the trajectory when the state dimension is greater than two; however, there is no coupling of the projected initial-condition response with motions out of the plane. Trajectories for linear, second-order systems have been classified and are illustrated in Fig.

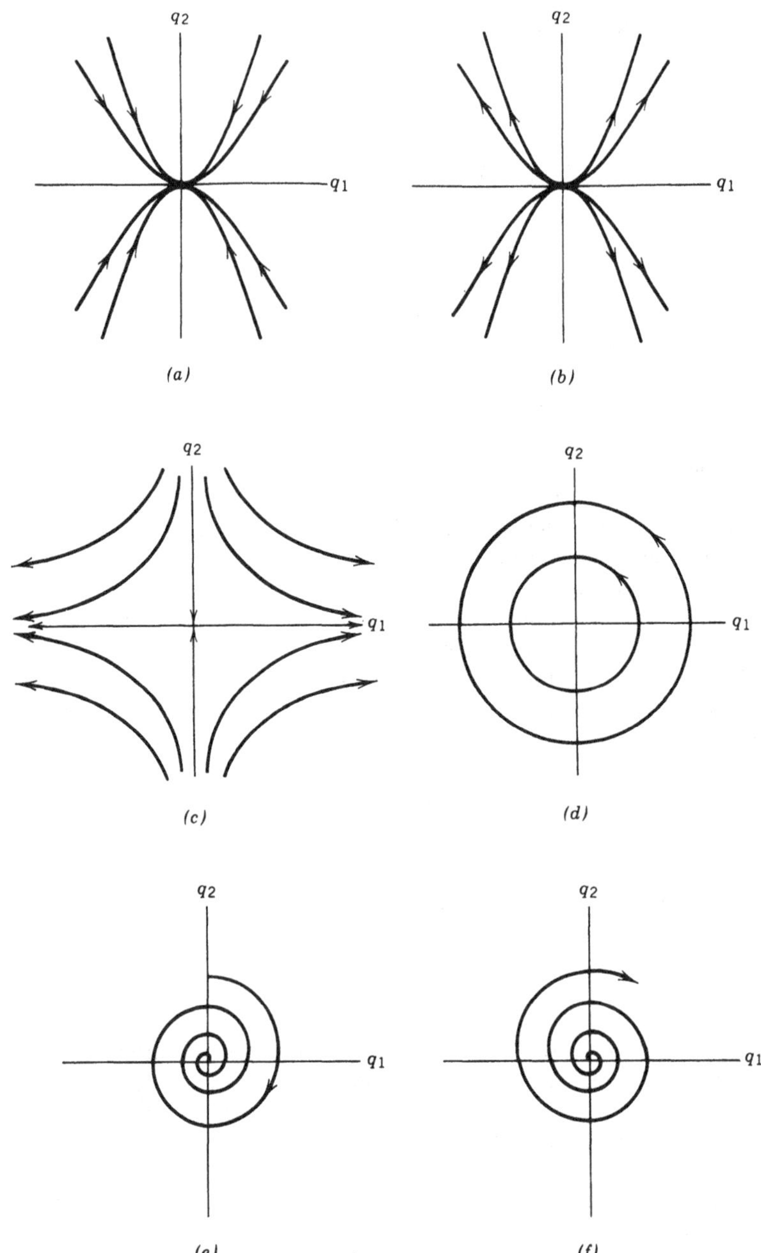

FIGURE 2.5-4 Classification of second-order singular points. (*a*) Stable node ($\lambda_1, \lambda_2 < 0$); (*g*) unstable node ($\lambda_1, \lambda_2 > 0$); (*c*) saddle point ($\lambda_1 < 0, \lambda_2 > 0$); (*d*) center ($\lambda_1 = j\nu, \lambda_2 = -j\nu$); (*e*) stable focus ($\lambda_{1,2} = \mu \pm j\nu, \mu < 0$); (*f*) unstable focus ($\lambda_{1,2} = \mu \pm j\nu, \mu > 0$).

2.5-4. If the two eigenvalues are real and stable (unstable), the singular point is a stable (unstable) *node*, and trajectories either decay to the origin or depart from it. If one root is stable and the other unstable, the singular point is a *saddle point*, and it is reached only if the initial condition in the unstable mode is zero. If λ_1 and λ_2 are purely imaginary conjugates, the singular point is a *center*, and trajectories orbit the point without growth or decay.* When λ_1 and λ_2 are complex conjugates with nonzero real parts, the origin is a stable or unstable *focus*, and trajectories spiral into or out of it.

Reachability, Controllability, and Stabilizability

In designing control systems, it is important to know whether or not *any* control law can be effective, optimal or not. For this preliminary evaluation, the control might be considered effective if it can force the state from $\mathbf{x}(t_0)$ to $\mathbf{x}(t_f)$, where $(t_f - t_0)$ is a finite interval, or if it can assure system stability, as defined earlier. The issue is not the structure and parameters of the control law but the structure of the dynamic system that allows the control to be felt with adequate force by all state components that need control (Section 1.2).

From prior developments, it is reasonable to expect the control effectiveness of a nonlinear system to vary from one region of the state space to another. For a linear system, control effectiveness remains the same throughout the state space because dynamic properties do not change with $\mathbf{x}$, and there are no limits (in principle) on the magnitude of $\mathbf{u}$. In parallel with the stability results, the control effectiveness along a nominal trajectory of a nonlinear system can be approximated by the control effectiveness of a locally linearized model of the system. Therefore, we will devote most of our attention to criteria for linear systems.

For linear, continuous-time systems there is no distinction between reachability and controllability, but for nonlinear systems the difference may be significant. For fixed values of t_0 and t_f, a *reachable region* is defined as the envelope of $\mathbf{x}(t_f)$ that can be reached from $\mathbf{x}(t_0) = \mathbf{0}$ through the use of control. The *controllable region* is defined as the envelope of $\mathbf{x}(t_0)$ from which $\mathbf{x}(t_f) = \mathbf{0}$ can be reached through the use of control. Figure 2.5-5 shows the difference between the two for a simple example [from (S-3)]: the system is a double integrator with bounded control ($\ddot{x} = u$, $|u| \leq 1$). Although the two regions overlap, they are clearly not the same. This system is nonlinear only because u is bounded; if u were unbounded, the reachable and controllable regions would be unbounded and, therefore, the same.

Local controllability of the nonlinear system can be deduced by examin-

*These are not limit cycles, for unlike the "locked-in" oscillations of the nonlinear system, their amplitudes, depend on the initial conditions. Furthermore, the oscillations are sinusoidal, whereas limit cycles are periodic but not sinusoidal.

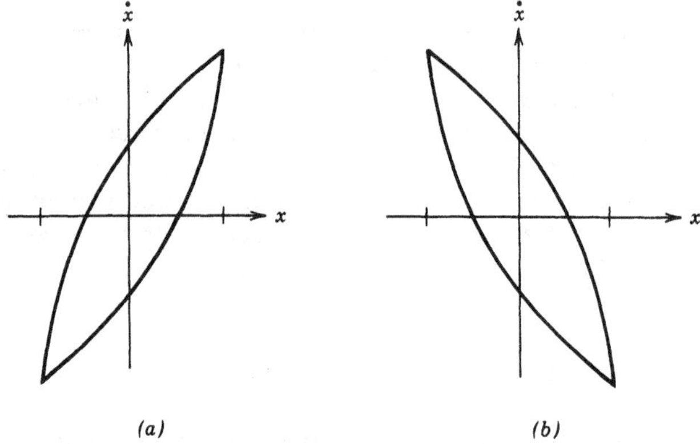

FIGURE 2.5-5 Reachable and controllable regions of an example system [from (S-3)].

ing the effects of control on small perturbations from a nominal trajectory, $\mathbf{x}_o(t)$ in $[t_0, t_f]$. From Eq. 2.3-81b, neighboring trajectories, $\mathbf{x}(t) = \mathbf{x}_o(t) + \Delta\mathbf{x}(t)$, could be approximated as

$$\mathbf{x}(t) = \mathbf{x}_o(t) + \mathbf{\Phi}(t, t_0)\Delta\mathbf{x}(t_0) + \int_{t_0}^{t} \mathbf{\Phi}(t, \tau)\mathbf{G}(\tau)\Delta\mathbf{u}(\tau)\, d\tau \tag{2.5-93}$$

where $\mathbf{\Phi}(t, \cdot)$ and $\mathbf{G}(\tau)$ are evaluated along the nominal trajectory and disturbance effects are neglected $[\Delta\mathbf{w}(t) = \mathbf{0}]$. $\mathbf{x}(t)$ is an n-vector, and $\mathbf{u}(t)$ is an m-vector. The system is locally controllable in the interval $[t_0, t_f]$ [i.e., $x(t_f)$ can be driven to zero] if the $(n \times n)$ *controllability Grammian matrix* $\mathcal{M}(t_f, t_0)$, defined as

$$\mathcal{M}(t_f, t_0) = \int_{t_0}^{t_f} \mathbf{\Phi}(t_f, \tau)\mathbf{G}(\tau)\mathbf{G}^T(\tau)\mathbf{\Phi}^T(t_f, \tau)\, d\tau \tag{2.5-94}$$

is nonsingular. Then, if we choose

$$\Delta\mathbf{u}(t) = \mathbf{G}^T(t)\mathbf{\Phi}^T(t_f, t)\mathcal{M}^{-1}(t_f, t_0)[-\mathbf{x}_o(t_f) - \mathbf{\Phi}(t_f, t_0)\Delta\mathbf{x}(t_0)] \tag{2.5-95}$$

where the term in square brackets, denoted by $[\cdot]$, is evaluated as a constant, Eq. 2.5-93 can be rearranged to yield

$$[-\mathbf{x}_o(t_f) - \mathbf{\Phi}(t_f, t_0)\Delta\mathbf{x}(t_0)] = \int_{t_0}^{t_f} \mathbf{\Phi}(t_f, \tau)\mathbf{G}(\tau)\mathbf{G}^T(\tau)\mathbf{\Phi}^T(t_f, \tau)\, d\tau\mathcal{M}^{-1}(t_f, t_0)[\cdot] \tag{2.5-96a}$$

or

$$[\cdot] = [\cdot] \tag{2.5-96b}$$

providing a criterion for *local complete controllability of the nonlinear system*. Taking $\mathbf{x}_o(t) = \mathbf{0}$, this also provides a global criterion for *complete controllability of a linear, time-varying system*.

Development of the criterion for complete controllability of a linear, time-invariant system also begins with Eq. 2.3-81b, which can be written as

$$\mathbf{x}(t_f) = \mathbf{0} = e^{\mathbf{F}t_f}\mathbf{x}(0) + \int_0^{t_f} e^{\mathbf{F}(t_f-\tau)}\,\mathbf{G}\mathbf{u}(\tau)\,d\tau$$

$$= e^{\mathbf{F}t_f}\left[\mathbf{x}(0) + \int_0^{t_f} e^{-\mathbf{F}\tau}\mathbf{G}\mathbf{u}(\tau)\,d\tau\right] \tag{2.5-97}$$

The matrix exponential can be expanded as a power series

$$e^{-\mathbf{F}\tau} = \mathbf{I}_n - \mathbf{F}\tau + \frac{\mathbf{F}^2\tau^2}{2} - \frac{\mathbf{F}^3\tau^3}{6} + \cdots \tag{2.5-98}$$

Using the Cayley–Hamilton theorem [e.g., (B-6)], it can be expressed as a matrix polynomial with, at most, $(n-1)$ terms,

$$e^{-\mathbf{F}\tau} = r_0\mathbf{I}_n + r_1\mathbf{F} + r_2\mathbf{F}^2 + \cdots + r_{n-1}\mathbf{F}^{n-1} \tag{2.5-99}$$

where the r_i are scalar functions of τ and the eigenvalues of $\mathbf{F}$. (The r_i need not be evaluated in the present instance; it is sufficient to know that they exist.) Then

$$-\mathbf{x}(0) = \mathbf{G}\int_0^{t_f} r_0(\tau)\mathbf{u}(\tau)\,d\tau + \mathbf{F}\mathbf{G}\int_0^{t_f} r_1(\tau)\mathbf{u}(\tau)\,d\tau + \cdots$$

$$+ \mathbf{F}^{n-1}\mathbf{G}\int_0^{t_f} r_{n-1}(\tau)\mathbf{u}(\tau)\,d\tau$$

$$= [\mathbf{G}\quad \mathbf{F}\mathbf{G}\ldots\mathbf{F}^{n-1}\mathbf{G}]\int_0^{t_f}\begin{bmatrix}\mathbf{v}_0(\tau)\\ \mathbf{v}_1(\tau)\\ \vdots\\ \mathbf{v}_{n-1}(\tau)\end{bmatrix}d\tau$$

$$= \mathbb{C}\int_0^{t_f}\begin{bmatrix}\mathbf{v}_0(\tau)\\ \mathbf{v}_1(\tau)\\ \vdots\\ \mathbf{v}_{n-1}(\tau)\end{bmatrix}d\tau \tag{2.5-100}$$

where $\mathbf{v}_i(\tau) = r_i(\tau)\mathbf{u}(\tau)$ and $\mathbb{C}$ is the $(n \times nm)$ *controllability matrix*:

$$\mathbb{C} = [\mathbf{G}\quad \mathbf{F}\mathbf{G}\ldots\mathbf{F}^{n-1}\mathbf{G}] \tag{2.5-101}$$

If the n rows of $\mathbb{C}$ are linearly independent (i.e., if $\mathbb{C}$ has rank n), the n components of $\mathbf{x}(t_f)$ can be driven to zero independently by suitable choice of $\mathbf{u}(t)$. This defines *complete controllability of the linear, time-invariant system*. Practically, it may be easier to establish the rank of $\mathbb{C}$ by finding n independent columns, as it may not be necessary to compute the rightmost products of $\mathbf{F}$ and $\mathbf{G}$. For example, suppose $m = n$ and $\mathbf{G}$ has full rank; then controllability is assured without further evaluation of $\mathbb{C}$. Conversely, if $m = 1$, then all $(n-1)$ products must be evaluated to define the $(n \times n)$ matrix $\mathbb{C}$ before determining its rank.

If a linear, time-invariant system has distinct roots, it can be described in normal form by Eq. 2.5-92. Every component of the normal-mode state vector is directly affected by $\mathbf{u}(t)$ if there are no zero rows in $\mathbf{D}^{-1}\mathbf{G}$. This is the criterion for *complete modal controllability of a linear, time-invariant system*. Complete modal controllability assures that all the eigenvalues (also called *poles*) of the system can be placed at arbitrary positions, because a feedback law such as Eq. 2.5-13 can affect each modal coordinate independently.

If it is necessary only to assure that there is sufficient control to stabilize unstable modes of a system, incomplete controllability may be satisfactory (i.e., stable modes need no further control). A criterion for *complete stabilizability* of a linear, time-invariant system with distinct roots is that the nonzero rows of $\mathbf{D}^{-1}\mathbf{G}$ correspond to the normal-mode state components of unstable modes.

Controllability should not be taken for granted, as the following example shows. Complete controllability may be forfeited by the introduction of *redundant* integral compensation in the control law (S-6).* Although the dynamic system itself may remain completely controllable, identical compensation-state components are linearly dependent and are driven by a common input; bias errors can cause such components to grow without bound unless the compensators are individually asymptotically stable. Even then, the offset in parallel control outputs can cause problems in fault-detection logic.

Constructability, Observability, and Detectability

Observation of a system is the dual of its control, so there are equivalent criteria for determining the n-dimensional state, $\mathbf{x}(t)$, from the r-dimensional output, $\mathbf{y}(t)$. Because the effects of control, $\mathbf{u}(t)$, can be reconstructed using the system model, it is adequate to consider the observation of response to initial conditions alone. Nonlinear observations arise when the output equation (Eq. 2.5-2) is nonlinear; the measurements may be effective only within a limited region of the state space, whereas the effectiveness of linear observations is consistent throughout the state space.

*Redundancy of control logic is sometimes implemented to increase reliability.

Local observation effectiveness of the nonlinear system can be assessed using the appropriate linear model.

For fixed values of t_0 and t_f, a *constructable region* is the envelope of $\mathbf{x}(t_f)$ that can be determined from $\mathbf{y}(t)$ in $[t_0, t_f]$. The *observable region* is the envelope of $\mathbf{x}(t_0)$ that can be determined from $\mathbf{y}(t)$ in $[t_0, t_f]$. The two regions are the same for linear, continuous-time systems, but they may not be the same for nonlinear systems.

Local observability of the nonlinear system can be determined from the equivalent linear, time-varying system, whose homogeneous equation is

$$\mathbf{x}(\tau) = \mathbf{\Phi}(\tau, t_0)\mathbf{x}(t_0) \tag{2.5-102}$$

with corresponding output

$$\mathbf{y}(\tau) = \mathbf{H}(\tau)\mathbf{\Phi}(\tau, t_0)\mathbf{x}(t_0) \tag{2.5-103}$$

Setting $\tau = t_f$, observability could be proved if $\mathbf{H}(t_f)\mathbf{\Phi}(t_f, t_0)$ could be inverted, but this is rarely the case, and it certainly is not possible with $r < n$. We can, however, multiply both sides of the last equation by $\mathbf{\Phi}^T(\tau, t_0)\mathbf{H}^T(\tau)$ and integrate from t_0 to t_f. Because $\mathbf{x}(t_0)$ is a constant, this can be written as

$$\begin{aligned}\int_{t_0}^{t_f} \mathbf{\Phi}^T(\tau, t_0)\mathbf{H}^T(\tau)\mathbf{y}(\tau)\, d\tau &= \int_{t_0}^{t_f} \mathbf{\Phi}^T(\tau, t_0)\mathbf{H}^T(\tau)\mathbf{H}(\tau)\mathbf{\Phi}(\tau, t_0)\, d\tau\mathbf{x}(t_0) \\ &= \mathbb{N}(t_f, t_0)\mathbf{x}(t_0) \end{aligned} \tag{2.5-104}$$

where the *observability Grammian matrix* is

$$\mathbb{N}(t_f, t_0) = \int_{t_0}^{t_f} \mathbf{\Phi}^T(\tau, t_0)\mathbf{H}^T(\tau)\mathbf{H}(\tau)\mathbf{\Phi}(\tau, t_0)\, d\tau \tag{2.5-105}$$

Consequently, $\mathbf{x}(t_0)$ is observable from $\mathbf{y}(t)$ in $[t_0, t_f]$ if $\mathbb{N}(t_f, t_0)$ is nonsingular,

$$\mathbf{x}(t_0) = \mathbb{N}^{-1}(t_f, t_0)\int_{t_0}^{t_f} \mathbf{\Phi}^T(\tau, t_0)\mathbf{H}^T(\tau)\mathbf{y}(\tau)\, d\tau \tag{2.5-106}$$

establishing the *observability criterion for linear, time-varying systems.*

This result for $\mathbb{N}(t_f, t_0)$ is the *dual* of the $\mathbb{M}(t_f, t_0)$ criterion for controllability (Eq. 2.5-94); hence it suggests that the remaining observability criteria for linear, time-invariant systems can be shown by applying the controllability criteria to a system of the form

$$\dot{\mathbf{x}}(t) = -\mathbf{F}^T\mathbf{x}(t) + \mathbf{H}^T\mathbf{u}(t) \tag{2.5-107}$$

since $\mathbf{\Phi}^T(-\tau, 0) = e^{-\mathbf{F}^T\tau}$. We then define the $(n \times nr)$ *observability matrix* as

$$\mathbb{O} = [\mathbf{H}^T \quad \mathbf{F}^T\mathbf{H}^T \ldots (\mathbf{F}^T)^{n-1}\mathbf{H}^T] \tag{2.5-108}$$

and require the rank of $\mathbb{O}$ to be n for complete observability of the linear, time-invariant system. ($\mathbb{O}$ is frequently defined as the transpose of this, but the result is the same.) The modal matrix of Eq. 2.5-106 is $\mathbf{D}^{-T}$, so the observability criterion for a linear, time-invariant system with distinct roots is that $\mathbf{D}^T\mathbf{H}^T$ have no zero rows (or that $\mathbf{HD}$ have no zero columns). If it is necessary to observe only those modes that are unstable, the nonzero columns of $\mathbf{HD}$ must correspond to the unstable normal-mode components, establishing the criterion for complete detectability of the system with distinct roots.

Discrete-Time Systems

Our principal interest in discrete-time systems is as sampled-data models of continuous-time systems. In that regard, we have seen (Section 2.3) that discrete-time models of nonlinear continuous-time systems are algorithmic (e.g., Runge–Kutta integration); hence their dynamic properties are complicated. To the extent that the algorithm should be a good portrayal of the continuous-time system, it may suffice to approximate the discrete-time properties by the continuous-time properties. We will, however, consider the properties of linear, discrete-time models, principally noting the differences between them and their continuous-time counterparts.

The system equations for a linear, time-varying, discrete-time model are

$$\mathbf{x}_{k+1} = \mathbf{\Phi}_k\mathbf{x}_k + \mathbf{\Gamma}_k\mathbf{u}_k + \mathbf{\Lambda}_k\mathbf{w}_k \tag{2.5-109}$$

$$\mathbf{y}_k = \mathbf{H}_k\mathbf{x}_k \tag{2.5-110}$$

and they can be considered sampled-data models of the linear, continuous-time equations (Eqs. 2.5-3 and 2.5-4) with the definitions of $\mathbf{\Phi}$, $\mathbf{\Gamma}$, $\mathbf{\Lambda}$, and $\mathbf{H}$ introduced in Section 2.3 for fixed sampling interval T. The corresponding time-invariant equations are

$$\mathbf{x}_{k+1} = \mathbf{\Phi}\mathbf{x}_k + \mathbf{\Gamma}\mathbf{u}_k + \mathbf{\Lambda}\mathbf{w}_k \tag{2.5-111}$$

$$\mathbf{y}_k = \mathbf{H}\mathbf{x}_k \tag{2.5-112}$$

while the autonomous difference equation is

$$\mathbf{x}_{k+1} = \mathbf{\Phi}\mathbf{x}_k \tag{2.5-113}$$

Feedback control could be incorporated as before; for example, the proportional controller

$$\mathbf{u}_k = -\mathbf{C}_k \mathbf{x}_k \tag{2.5-114}$$

could be substituted in Eq. 2.5-109 to form a homogeneous equation (with $\mathbf{w}_k = \mathbf{0}$):

$$\begin{aligned} \mathbf{x}_{k+1} &= (\boldsymbol{\Phi}_k - \boldsymbol{\Gamma}_k \mathbf{C}_k)\mathbf{x}_k \\ &= \boldsymbol{\Phi}' \mathbf{x}_k \end{aligned} \tag{2.5-115}$$

Compensation and actuator dynamics also could be added as in the continuous-time case.

In static equilibrium, $\mathbf{x}_{k+1} = \mathbf{x}_k = \mathbf{x}^*$, where $(\cdot)^*$ denotes a fixed value, so Eq. 2.5-111 becomes

$$\begin{aligned} \mathbf{x}^* &= \boldsymbol{\Phi}\mathbf{x}^* + \boldsymbol{\Gamma}\mathbf{u}^* + \boldsymbol{\Lambda}\mathbf{w}^* \\ (\mathbf{I}_n - \boldsymbol{\Phi})\mathbf{x}^* &= \boldsymbol{\Gamma}\mathbf{u}^* + \boldsymbol{\Lambda}\mathbf{w}^* \\ \mathbf{x}^* &= (\mathbf{I}_n - \boldsymbol{\Phi})^{-1}(\boldsymbol{\Gamma}\mathbf{u}^* + \boldsymbol{\Lambda}\mathbf{w}^*) \end{aligned} \tag{2.5-116}$$

provided $(\mathbf{I}_n - \boldsymbol{\Phi})^{-1}$ exists. ($\boldsymbol{\Lambda}$ should not be confused with the diagonalized version of $\mathbf{F}$. Here it represents the discrete-time disturbance effect matrix.) Because $\boldsymbol{\Gamma}$ and $\boldsymbol{\Lambda}$ have been defined previously as

$$\begin{aligned} \boldsymbol{\Gamma} &= \boldsymbol{\Phi}(\mathbf{I}_n - \boldsymbol{\Phi}^{-1})\mathbf{F}^{-1}\mathbf{G} \\ &= (\boldsymbol{\Phi} - \mathbf{I}_n)\mathbf{F}^{-1}\mathbf{G} \end{aligned} \tag{2.5-117}$$

$$\begin{aligned} \boldsymbol{\Lambda} &= \boldsymbol{\Phi}(\mathbf{I}_n - \boldsymbol{\Phi}^{-1})\mathbf{F}^{-1}\mathbf{L} \\ &= (\boldsymbol{\Phi} - \mathbf{I}_n)\mathbf{F}^{-1}\mathbf{L} \end{aligned} \tag{2.5-118}$$

Eq. 2.5-116 reduces to the continuous-time relation (Eq. 2.5-22),

$$\begin{aligned} \mathbf{x}^* &= (\mathbf{I}_n - \boldsymbol{\Phi})^{-1}(\boldsymbol{\Phi} - \mathbf{I}_n)\mathbf{F}^{-1}(\mathbf{G}\mathbf{u}^* + \mathbf{L}\mathbf{w}^*) \\ &= -\mathbf{F}^{-1}(\mathbf{G}\mathbf{u}^* + \mathbf{L}\mathbf{w}^*) \end{aligned} \tag{2.5-119}$$

provided $\mathbf{F}^{-1}$ exists.

Quasistatic equilibrium is analogous to that of the continuous-time case. The state transition matrix for the singular example of $\mathbf{F}$ given by Eq. 2.5-23 takes the partitioned form

$$\boldsymbol{\Phi} = \begin{bmatrix} \boldsymbol{\Phi}_1 & \mathbf{0} \\ \boldsymbol{\Phi}_2 & \mathbf{I} \end{bmatrix} \tag{2.5-120}$$

Hence $(\mathbf{I}_n - \mathbf{\Phi})$ is singular for the same reason that $\mathbf{F}$ is singular. Given the system

$$\begin{bmatrix} \mathbf{x}_1 \\ \mathbf{x}_2 \end{bmatrix}_{k+1} = \begin{bmatrix} \mathbf{\Phi}_1 & \mathbf{0} \\ \mathbf{\Phi}_2 & \mathbf{I} \end{bmatrix} \begin{bmatrix} \mathbf{x}_1 \\ \mathbf{x}_2 \end{bmatrix}_k + \begin{bmatrix} \mathbf{\Gamma}_1 \\ \mathbf{\Gamma}_2 \end{bmatrix} \mathbf{u}_k + \begin{bmatrix} \mathbf{\Lambda}_1 \\ \mathbf{\Lambda}_2 \end{bmatrix} \mathbf{w}_k \tag{2.5-121}$$

the reduced-order equilibrium of $\mathbf{x}_1$ is

$$\mathbf{x}_1^* = (\mathbf{I} - \mathbf{\Phi}_1)^{-1}(\mathbf{\Gamma}_1 \mathbf{u}^* + \mathbf{\Lambda}_1 \mathbf{w}^*) \tag{2.5-122}$$

and the corresponding quasistatic value of $\mathbf{x}_{2_{k+1}}$ is

$$\mathbf{x}_{2_{k+1}}^* = \mathbf{\Phi}_2 \mathbf{x}_1^* + \mathbf{x}_{2_k}^* + \mathbf{\Gamma}_2 \mathbf{u}^* + \mathbf{\Lambda}_2 \mathbf{w}^* \tag{2.5-123}$$

With fixed control and disturbance inputs, the autonomous discrete-time dynamic equation can be expressed in terms of $\tilde{\mathbf{x}}_k = \mathbf{x}_k - \mathbf{x}^*$,

$$\tilde{\mathbf{x}}_{k+1} = \mathbf{\Phi} \tilde{\mathbf{x}}_k = e^{\mathbf{F}T} \tilde{\mathbf{x}}_k \tag{2.5-124}$$

$$\mathbf{0} = \mathbf{\Phi}(\mathbf{0}) \tag{2.5-125}$$

where T is the interval between samples. Singular points, phase-plane trajectories, and stability concepts are analogous to their continuous-time counterparts, except that trajectories take discrete jumps from one sample to the next; intersample behavior of the continuous-time system is not contained in the discrete-time model. Note that Eq. 2.5-124 is the form used to describe exponential asymptotic stability of the continuous-time system (Eq. 2.5-51). The discrete-time state is asymptotically stable if

$$\|\tilde{\mathbf{x}}_{k+1}\| = \|\mathbf{\Phi} \tilde{\mathbf{x}}_k\| \le c e^{-k\alpha} \|\tilde{\mathbf{x}}_0\|, \qquad c, \alpha > 0 \tag{2.5-126}$$

where α is a positive constant.

Summations of the norm and the norm-squared are bounded with stable $\tilde{\mathbf{x}}_k$ for reasons mentioned previously, and discrete-time Lyapunov functions play the same role as before. Defining the quadratic Lyapunov function

$$\mathcal{V}_k = \mathcal{V}(\tilde{\mathbf{x}}_k) = \tilde{\mathbf{x}}_k^T \mathbf{P} \tilde{\mathbf{x}}_k \tag{2.5-127}$$

where $\mathbf{P}$ is a positive definite defining matrix, $\mathcal{V}_{k+1} \le \mathcal{V}_k$ indicates stability, and $\mathcal{V}_{k+1} < \mathcal{V}_k$ indicates asymptotic stability. From Eq. 2.5-124,

$$\mathcal{V}_{k+1} = \tilde{\mathbf{x}}_{k+1}^T \mathbf{P} \tilde{\mathbf{x}}_{k+1} = \tilde{\mathbf{x}}_k^T \mathbf{\Phi}^T \mathbf{P} \mathbf{\Phi} \tilde{\mathbf{x}}_k \tag{2.5-128}$$

Then

$$\mathcal{V}_{k+1} - \mathcal{V}_k = \tilde{\mathbf{x}}_k^T (\mathbf{\Phi}^T \mathbf{P} \mathbf{\Phi} - \mathbf{P}) \tilde{\mathbf{x}}_k \triangleq -\tilde{\mathbf{x}}_k^T \mathbf{Q} \tilde{\mathbf{x}}_k \tag{2.5-129}$$

and the system is asymptotically stable if $\mathbf{Q}$ is a positive definite matrix. Consequently,

$$\boldsymbol{\Phi}^T \mathbf{P} \boldsymbol{\Phi} - \mathbf{P} = -\mathbf{Q} \tag{2.5-130}$$

is a *discrete-time Lyapunov equation*.

Modal properties of discrete-time systems are described using the z *transform*, which is the Laplace transform of a periodic sequence of weighted impulse (Dirac delta) functions (Eqs. 2.3-82 and 2.3-83). Given the scalar sequence $x_k = x(t_k) = x(kT)$ beginning at zero time, the Laplace transform of $x(kT)\delta(t-kT)$ is

$$\begin{aligned} \mathfrak{L}[x(kT)\delta(t-kT)] &= \int_0^\infty x(kT)\delta(t-kT)e^{-sT}\, ds \\ &= \sum_{k=0}^{\infty} x(kT)e^{-kTs} \end{aligned} \tag{2.5-131}$$

Defining the z transform operator

$$z \triangleq e^{Ts} \tag{2.5-132}$$

the z transform of x_k is

$$x(z) \triangleq \mathfrak{Z}(x_k) = \mathfrak{Z}[x(kT)] = \sum_{k=0}^{\infty} x_k z^{-k} \tag{2.5-133}$$

The inverse z transform of $x(z)$ is defined by a contour integral

$$x_k = \mathfrak{Z}^{-1}[x(z)] = \frac{1}{2\pi j} \oint x(z) z^{k-1}\, dz \tag{2.5-134}$$

where the region enclosed by the contour contains the singularities of $x(z)$. Given the power series representation of $\mathfrak{Z}(x_k)$, the inverse z transform can be found by inspection, as the x_k are the coefficients of z^{-k}:

$$\mathfrak{Z}(x_k) = x_0 + x_1 z^{-1} + x_2 z^{-2} + \cdots + x_k z^{-k} + \cdots \tag{2.5-135}$$

Consequently, z can be viewed as a *time-shift operator* in sampling the x_k sequence.

The z transform is a linear operator, so

$$\mathfrak{Z}(x_k + y_k) = \mathfrak{Z}(x_k) + \mathfrak{Z}(y_k) = x(z) + y(z) \tag{2.5-136}$$

$$\mathfrak{Z}(ax_k) = a\mathfrak{Z}(x_k) = ax(z) \tag{2.5-137}$$

and vectors and matrices are transformed term by term. The z transform of the sequence shifted ahead by one sampling interval can be derived from the unshifted transform and the initial condition:

$$\begin{aligned}\mathcal{Z}[x(t_k+T)] &= \mathcal{Z}[x(t_{k+1})] \\ &= z\mathcal{Z}[x(t_k)] - zx(0) = zx(z) - zx(0)\end{aligned} \tag{2.5-138}$$

These properties can be combined to write the z transforms of the system equations (Eqs. 2.5-111, 2.5-112):

$$z\mathbf{x}(z) - z\mathbf{x}(0) = \mathbf{\Phi}\mathbf{x}(z) + \mathbf{\Gamma}\mathbf{u}(z) + \mathbf{\Lambda}\mathbf{w}(z) \tag{2.5-139}$$

$$\mathbf{y}(z) = \mathbf{H}\mathbf{x}(z) \tag{2.5-140}$$

The difference equation is then manipulated to solve for $\mathbf{x}(z)$,

$$\mathbf{x}(z) = (z\mathbf{I}_n - \mathbf{\Phi})^{-1}[z\mathbf{x}(0) + \mathbf{\Gamma}\mathbf{u}(z) + \mathbf{\Lambda}\mathbf{w}(z)] \tag{2.5-141}$$

which is nearly identical in form to the Laplace transform of the continuous-time system, with z replacing s and $(\mathbf{\Phi}, \mathbf{\Gamma}, \mathbf{\Lambda})$ replacing $(\mathbf{F}, \mathbf{G}, \mathbf{L})$. [The only structural difference is that $\mathbf{x}(0)$ is multiplied by z.] Consequently, the previous developments of eigenvalues, eigenvectors, and partial-fraction expansions apply here as well.

The z transform of initial-condition response is

$$\begin{aligned}\mathbf{x}(z) &= (z\mathbf{I}_n - \mathbf{\Phi})^{-1} z\mathbf{x}(0) \\ &= \frac{\mathrm{Adj}(z\mathbf{I}_n - \mathbf{\Phi})}{|z\mathbf{I}_n - \mathbf{\Phi}|} z\mathbf{x}(0) \triangleq \mathbf{A}(z)\mathbf{x}(0)\end{aligned} \tag{2.5-142}$$

The characteristic equation is

$$\begin{aligned}|z\mathbf{I}_n - \mathbf{\Phi}| &\triangleq \Delta(z) = z^n + c_{n-1}z^{n-1} + \cdots + c_1 z + c_0 \\ &= (z-\lambda_1)(z-\lambda_2)\ldots(z-\lambda_n) = 0\end{aligned} \tag{2.5-143}$$

and the jkth element of $\mathbf{A}(z)$ is

$$a_{jk} = \frac{k_{jk} z(z-\beta_1)(z-\beta_2)\ldots(z-\beta_{n-1})}{\Delta(z)} \tag{2.5-144}$$

although there may be fewer than $(n-1)$ zeros. With distinct roots, the partial-fraction expansion of $x_j(z)$ is

$$x_j(z) = d_0 + \frac{d_1 z}{(z-\lambda_1)} + \frac{d_2 z}{(z-\lambda_2)} + \cdots + \frac{d_n z}{(z-\lambda_n)} \tag{2.5-145}$$

and with real roots, the inverse z transforms of each element are

$$x_{ij}(kT) = \mathfrak{Z}^{-1}\left[\frac{d_i z}{(z-\lambda_i)}\right] = d_i(\lambda_i)^k \tag{2.5-146}$$

Consequently, each element represents an exponential response mode that diverges for $|\lambda_i| > 1$ and converges for $|\lambda_i| < 1$. Note that the response oscillates for negative λ_i even though this is a first-order mode.

As in the continuous-time case, a complex λ_i must be accompanied by its complex conjugate, λ_{i+1}, for real $\mathbf{x}_k$. Assuming that the d_i are real and that $d_i = d_{i+1}$,

$$\frac{d_i z}{(z-\lambda_i)} + \frac{d_{i+1} z}{(z-\lambda_{i+1})} = \frac{2d_i z(z-\mu)}{z^2 - 2\mu z + (\mu^2+\nu^2)} \tag{2.5-147}$$

where

$$\lambda_i = \mu + j\nu \tag{2.5-148}$$

Defining

$$\omega_n = \frac{1}{T}\cos^{-1}\left[\frac{\mu}{\sqrt{\mu^2+\nu^2}}\right] \tag{2.5-149}$$

$$\zeta = -\frac{\ln(\mu^2+\nu^2)}{2\omega_n T} \tag{2.5-150}$$

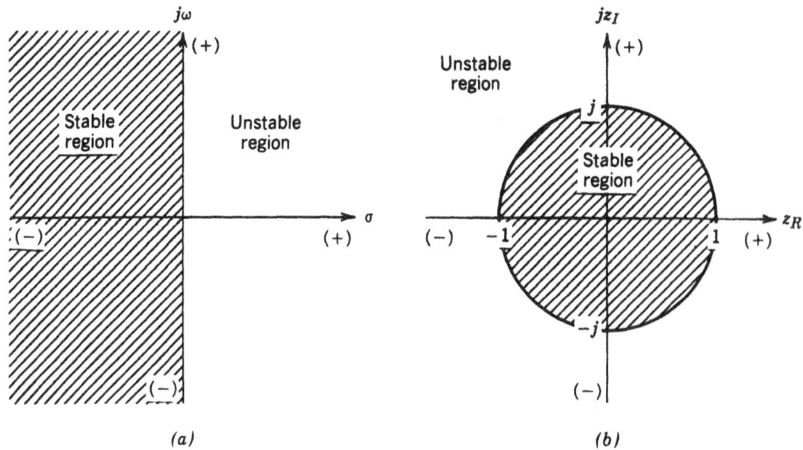

FIGURE 2.5-6 Regions of stability and instability for linear, time-invariant system eigenvalues. (*a*) Complex *s* plane for continuous-time systems; (*b*) complex *z* plane for discrete-time systems.

the inverse z transform of Eq. 2.5-147 is

$$\mathcal{Z}^{-1}[\cdot] = 2d_i e^{-2\zeta\omega_n kT} \cos \omega_n kT \tag{2.5-151}$$

Whether λ_i is real or complex, its magnitude must be less than one for convergence. Therefore, asymptotic stability of the discrete-time system requires that the roots of $|z\mathbf{I}_n - \mathbf{\Phi}|$ lie within the unit circle of the complex z plane. By comparison, the roots of $|s\mathbf{I}_n - \mathbf{F}|$ must lie in the left half of the complex s plane for continuous system stability (Fig. 2.5-6). (This assumes that real parts are plotted along the horizontal axis and imaginary parts along the vertical axis.)

Controllability and observability results are analogous to the continuous-time results with one exception. The criteria are as follows:

Controllability

1. Linear, Time-Varying System:

$$\mathbb{M}(k_f, 0) = \sum_{k=0}^{k_f} \mathbf{\Phi}(k_f, k)\mathbf{\Gamma}(k)\mathbf{\Gamma}^T(k)\mathbf{\Phi}^T(k_f, k)$$

   is nonsingular, k_f finite.
2. Linear, Time-Invariant System:

   $\mathbb{C} = [\mathbf{\Gamma} \quad \mathbf{\Phi\Gamma} \ldots \mathbf{\Phi}^{n-1}\mathbf{\Gamma}]$ has rank n.
3. Linear, Time-Invariant System with Distinct Roots:
   No zero rows of $\mathbf{D}^{-1}\mathbf{\Gamma}$, where $\mathbf{D}$ is the modal matrix of $\mathbf{\Phi}$.

Observability

1. Linear, Time-Varying System:

$$\mathbb{N}(k_f, 0) = \sum_{k=0}^{k_f} \mathbf{\Phi}^T(k, 0)\mathbf{H}^T(k)\mathbf{H}(k)\mathbf{\Phi}(k, 0)$$

   is nonsingular, k_f finite.
2. Linear, Time-Invariant System:
   $\mathbb{O} = [\mathbf{H}^T \quad \mathbf{\Phi}^T\mathbf{H}^T \ldots (\mathbf{\Phi}^T)^{n-1}\mathbf{H}^T]$ has rank n.
3. Linear, Time-Invariant System with Distinct Roots:
   No zero columns of $\mathbf{HD}$.

If $\mathbf{\Phi}$ is nonzero but $\mathbf{\Phi}^k = \mathbf{0}$, $\mathbf{\Phi}$ is a *nilpotent* matrix of index k; it allows $\mathbf{x}_k = \mathbf{\Phi}^k\mathbf{x}_0 = \mathbf{0}$ without use of control. Thus the system could be considered "controllable" by the earlier definition even if $\mathbb{C}$ does not have rank n.

However, the system would not be "reachable" unless $\mathbb{C}$ had rank n, so the criterion still provides a valid test of control effectiveness. Similar distinctions can be made regarding observability and constructability (K-1).

2.6 FREQUENCY DOMAIN MODELING AND ANALYSIS

Frequency domain descriptions of linear, time-invariant systems are useful in analysis of stability and response. Modal properties are expressed directly in system transfer functions, the steady-state response to slow and fast oscillatory inputs is characterized easily, and the computationally difficult convolution integrals of the time domain are replaced by equivalent multiplications of transforms and transfer functions in the frequency domain. The principal complication is that the variables of frequency domain analysis are complex-valued even though the variables of the original time domain model are usually real-valued; however, this complication proves helpful in allowing input–output relationships to be described not only by amplitude ratios but by phase angles, which characterize the relative "lead" or "lag" between oscillatory inputs and outputs.

Frequency domain descriptions are based on the Laplace transform for continuous-time systems and the z transform for discrete-time systems (Section 2.5). Given the time domain description of a continuous-time system,

$$\dot{\mathbf{x}}(t) = \mathbf{F}\mathbf{x}(t) + \mathbf{G}\mathbf{u}(t) \tag{2.6-1}$$

$$\mathbf{y}(t) = \mathbf{H}\mathbf{x}(t) \tag{2.6-2}$$

the corresponding frequency domain description is

$$s\mathbf{x}(s) = \mathbf{F}\mathbf{x}(s) + \mathbf{G}\mathbf{u}(s) + \mathbf{x}(0) \tag{2.6-3}$$

$$\mathbf{y}(s) = \mathbf{H}\mathbf{x}(s) \tag{2.6-4}$$

s is the complex variable, $\sigma + j\omega$, and all matrices, vectors, and transforms are defined as in previous sections. For the discrete-time system,

$$\mathbf{x}_{k+1} = \mathbf{\Phi}\mathbf{x}_k + \mathbf{\Gamma}\mathbf{u}_k \tag{2.6-5}$$

$$\mathbf{y}_k = \mathbf{H}\mathbf{x}_k \tag{2.6-6}$$

the frequency domain description is

$$z\mathbf{x}(z) = \mathbf{\Phi}\mathbf{x}(z) + \mathbf{\Gamma}\mathbf{u}(z) + z\mathbf{x}(0) \tag{2.6-7}$$

$$\mathbf{y}(z) = \mathbf{H}\mathbf{x}(z) \tag{2.6-8}$$

where z is the complex variable $\mu + j\nu$.

It is evident that these equations yield comparable state and output response transforms for given inputs; neglecting initial conditions,

$$\mathbf{x}(s) = (s\mathbf{I}_n - \mathbf{F})^{-1}\mathbf{G}\mathbf{u}(s) \tag{2.6-9}$$

$$\mathbf{y}(s) = \mathbf{H}(s\mathbf{I}_n - \mathbf{F})^{-1}\mathbf{G}\mathbf{u}(s) \tag{2.6-10}$$

$$\mathbf{x}(z) = (z\mathbf{I}_n - \mathbf{\Phi})^{-1}\mathbf{\Gamma}\mathbf{u}(z) \tag{2.6-11}$$

$$\mathbf{y}(z) = \mathbf{H}(z\mathbf{I}_n - \mathbf{\Phi})^{-1}\mathbf{\Gamma}\mathbf{u}(z) \tag{2.6-12}$$

where $\mathbf{x}$, $\mathbf{u}$, and $\mathbf{y}$ have dimensions n, m, and r in both cases. In this section, attention is directed toward systems with scalar inputs and outputs (i.e., *single-input/single-output systems*). Although the state vector may have many components, m and r are assumed to be one. Consequently, the *input–output transfer functions* can be written as

$$\frac{y(s)}{u(s)} = \mathbf{H}(s\mathbf{I}_n - \mathbf{F})^{-1}\mathbf{G} = \frac{\mathbf{H}\,\mathrm{Adj}(s\mathbf{I}_n - \mathbf{F})\mathbf{G}}{|s\mathbf{I}_n - \mathbf{F}|} = Y(s) \tag{2.6-13}$$

$$\frac{y(z)}{u(z)} = \mathbf{H}(z\mathbf{I}_n - \mathbf{\Phi})^{-1}\mathbf{\Gamma} = \frac{\mathbf{H}\,\mathrm{Adj}(z\mathbf{I}_n - \mathbf{\Phi})\mathbf{\Gamma}}{|z\mathbf{I}_n - \mathbf{\Phi}|} = Y(z) \tag{2.6-14}$$

The continuous- and discrete-time transfer functions are ratios of scalar polynomials,

$$\begin{aligned} Y(s) &= \frac{b_q s^q + b_{q-1}s^{q-1} + \cdots + b_1 s + b_0}{s^n + a_{n-1}s^{n-1} + \cdots + a_1 s + a_0} \\ &= \frac{b_q(s-\beta_1)(s-\beta_2)\ldots(s-\beta_q)}{(s-\alpha_1)(s-\alpha_2)\ldots(s-\alpha_n)} \end{aligned} \tag{2.6-15}$$

$$\begin{aligned} Y(z) &= \frac{d_q z^q + d_{q-1}z^{q-1} + \cdots + d_1 z + d_0}{z^n + c_{n-1}z^{n-1} + \cdots + c_1 z + c_0} \\ &= \frac{d_q(z-\delta_1)(z-\delta_2)\ldots(z-\delta_q)}{(z-\gamma_1)(z-\gamma_2)\ldots(z-\gamma_n)} \end{aligned} \tag{2.6-16}$$

β_i and δ_i are the *zeros*, and α_j and γ_j are the *poles* (or eigenvalues) of their respective transfer functions.

As shown in Fig. 2.6-1, the zeros and poles can be plotted in the s or z complex planes. For the continuous-time system, left-half-plane poles are stable, and right-half-plane poles are unstable (Fig. 2.6-1*a*). s-plane locations of the zeros do not affect open-loop stability, although they do affect the stability of systems in which $y(s)$ is fed back to $u(s)$ through a control system. Similar observations can be made for the discrete-time system, with the unit circle serving as the demarcation between stable and unstable poles in the z plane.

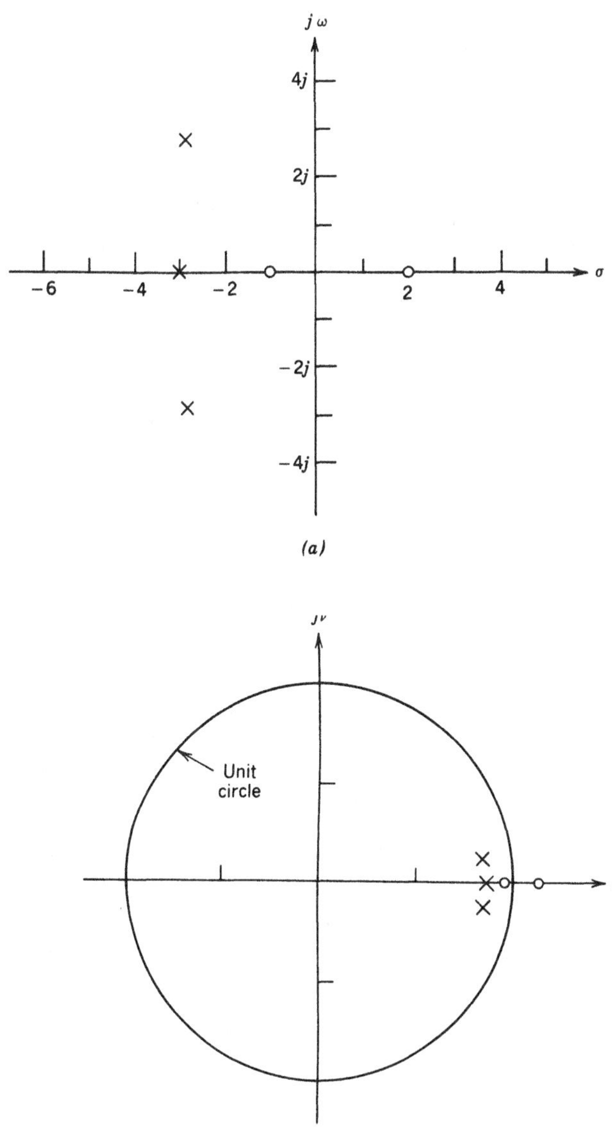

FIGURE 2.6-1 Zeros (○) and poles (×) of input-output transfer functions. (*a*) *s*-plane plot of the continuous-time transfer function:

$$Y(s) = [k(s+1)(s-2)]/\{(s+3)[s^2 + 2(0.707)(4)s + (4^2)]\};$$

(*b*) *z*-plane plot of the discrete-time transfer function:

$$Y(z) = [k(z-0.951)(z-1.105)]/[(z-0.861)(z^2 - 1.719z + 0.754)].$$

If the state-transition and control-effect matrices are defined by

$$\mathbf{\Phi} = e^{\mathbf{F}T} \tag{2.6-17}$$

$$\mathbf{\Gamma} = (\mathbf{\Phi} - \mathbf{I}_n)\mathbf{F}^{-1}\mathbf{G} \tag{2.6-18}$$

then Eq. 2.6-14 is a sampled-data equivalent of Eq. 2.6-13 for the sampling interval T sec. It is not surprising, therefore, that correspondences can be drawn between the zero and pole locations in the s and z planes. Recalling the definition of the z transform operator,

$$z = e^{sT} \tag{2.6-19}$$

real singularities, λ, are transformed from the s to the z plane by

$$\lambda_z = e^{\lambda_s T} \tag{2.6-20}$$

while complex singularities, $\sigma + j\omega$, transform as

$$\begin{aligned} \mu + j\nu &= e^{(\sigma + j\omega)T} = e^{\sigma T} e^{j\omega T} \\ &= e^{\sigma T}(\cos \omega T + j \sin \omega T) \end{aligned} \tag{2.6-21}$$

Figure 2.6-1b illustrates such a transformation of Fig. 2.6-1a, with $T =$ 0.05 sec. Note the similarity in pole-zero configurations, with the $(1, 0)$ point of the z plane taking the place of the s-plane origin. More comprehensive mappings are presented in (F-3).

EXAMPLE 2.6-1 TRANSFER FUNCTIONS FOR PITCHING MOTION OF AN AIRCRAFT

The pitching motions of an aircraft flying at constant speed V are largely described by its *short period mode*. This response mode is well described by the following second-order equation:

$$\begin{bmatrix} \dot{q} \\ \dot{\alpha} \end{bmatrix} = \begin{bmatrix} M_q & M_\alpha \\ 1 & -L_\alpha/V \end{bmatrix} \begin{bmatrix} q \\ \alpha \end{bmatrix} + \begin{bmatrix} M_{\delta E} & M_{\delta F} \\ -L_{\delta E}/V & -L_{\delta F}/V \end{bmatrix} \begin{bmatrix} \delta E \\ \delta F \end{bmatrix} - \begin{bmatrix} M_\alpha \\ -L_\alpha/V \end{bmatrix} \alpha_w$$

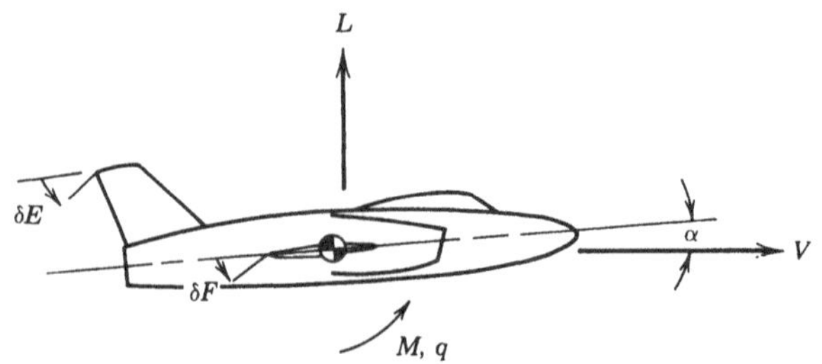

As illustrated in the figure, q represents the pitching rate, α is the angle of attack, the elevator and flap angles are δE and δF, and the angle-of-attack variation due to vertical wind input is α_w. $M_{(\cdot)}$ and $L_{(\cdot)}$ are pitching and lift acceleration sensitivities to the subscript variables.

There are six scalar transfer functions relating δE, δF, and α_w to q and α. Choosing the $\alpha(s)/\delta E(s)$ transfer function for illustration,

$$\frac{\alpha(s)}{\delta E(s)} = [0 \quad 1]\begin{bmatrix}(s - M_q) & -M_\alpha \\ -1 & (s + L_\alpha/V)\end{bmatrix}^{-1}\begin{bmatrix}M_{\delta E} \\ -L_{\delta E}/V\end{bmatrix}$$

$$= [0 \quad 1]\frac{\begin{bmatrix}(s + L_\alpha/V) & M_\alpha \\ 1 & (s - M_q)\end{bmatrix}\begin{bmatrix}M_{\delta E} \\ -L_{\delta E}/V\end{bmatrix}}{s^2 + (L_\alpha/V - M_q)s - M_\alpha}$$

$$= \frac{-(L_{\delta E}/V)[s - (M_q + VM_{\delta E}/L_{\delta E})]}{s^2 + 2\zeta\omega_n s + \omega_n^2}$$

For conventional aircraft, $M_{(\cdot)} < 0$ and $L_{(\cdot)} > 0$, resulting in a stable short period mode and a left-half-plane zero. Furthermore, since $(L_\alpha/V - M_q)$ usually is less than $2\sqrt{-M_\alpha}$, the mode normally is oscillatory.

Root Locus

It often is necessary to determine the effect of a scalar multiplier k on the roots of a characteristic equation such as

$$\begin{aligned}&s^n + a_{n-1}s^{n-1} + \cdots + a_1 s + a_0 + k(b_q s^q + b_{q-1}s^{q-1} + \cdots + b_1 s + b_0) \\ &\quad = (s - \alpha_1)(s - \alpha_2)\ldots(s - \alpha_n) + k(s - \beta_1)(s - \beta_2)\ldots(s - \beta_q) = 0 \quad (2.6\text{-}22)\end{aligned}$$

The root locus technique of Evans provides a graphical means of evaluating this effect (E-2). The *locus of roots* is the s-plane plot of root locations as k varies in some range of values [e.g., $(0, \infty)$ or $(-\infty, \infty)$]. One common application is for the determination of the closed-loop roots of a negative-unity-feedback, continuous-time control system, as shown in Fig. 2.6-2. Performing block-diagram algebra, the transfer function between $y_c(s)$ and $y(s)$ is seen to be

$$\frac{y(s)}{y_c(s)} = \frac{Y(s)}{1 + Y(s)} \tag{2.6-23}$$

and the closed-loop roots are solutions to the equation

$$1 + Y(s) = 1 + \frac{kn(s)}{\Delta(s)} = 0 \tag{2.6-24}$$

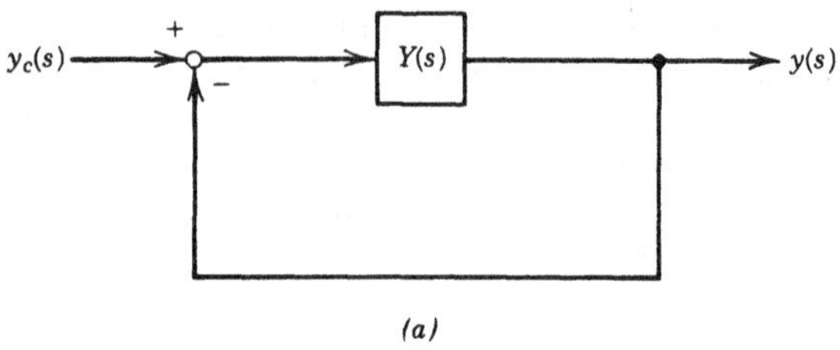

(a)

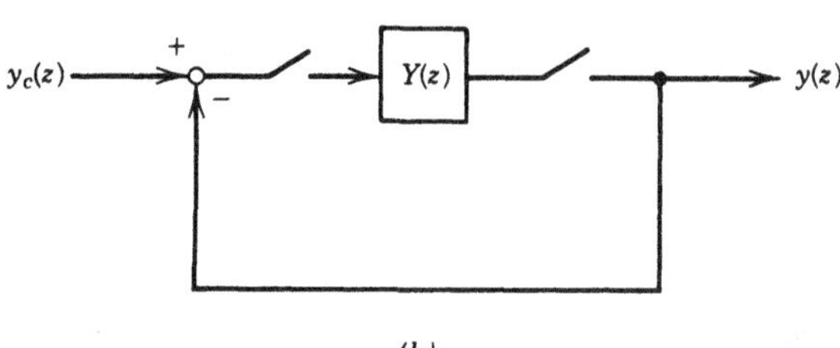

(b)

FIGURE 2.6-2 Block diagrams of unity-feedback control systems. (*a*) Continuous-time system; (*b*) discrete-time system.

$[1 + Y(s)]$ is called the *return difference function*, and k is the system's *loop gain*. Defining $Y(s)$ as in Eq. 2.6-15, this is seen to yield the same equation as Eq. 2.6-22 after multiplying by $\Delta(s)$. Furthermore, since the closed-loop roots of a discrete-time control system are similarly defined by solving the equation

$$1 + Y(z) = 1 + \frac{kn(z)}{\Delta(z)} = 0 \tag{2.6-25}$$

the root locus technique also can be applied to such systems using Eq. 2.6-16.

Although graphical constructions provide adequate accuracy for many analyses, the advent of digital computers and root-finding algorithms has largely preempted the detailed application of Evans's method. Nevertheless, the principal rules of the method are easily remembered, and together with free-hand sketches, they provide a quick way to gain insights into the effects of loop gain variations. The most significant rules of construction are summarized below for the continuous-time case.

First, it is apparent that Eq. 2.6-22 must have n roots for any value of k; hence *there are n branches to the root locus*. For $k = 0$, *the roots of the equation are the poles of* $Y(s)$, α_i, $i = 1$ to n. Put another way, they are the zeros of the characteristic polynomial $\Delta(s)$, since Eq. 2.6-22 becomes

$$\Delta(s) = (s - \alpha_1)(s - \alpha_2) \ldots (s - \alpha_n) = 0 \tag{2.6-26}$$

For k approaching $\pm\infty$, the second term of Eq. 2.6-22 predominates, so q *closed-loop roots approach the zeros of* $Y(s)$, which also are the zeros of $n(s)$. Alternately, dividing by k,

$$0 = \frac{1}{k}[(s-\alpha_1)\ldots(s-\alpha_n)] + (s-\beta_1)\ldots(s-\beta_q) \xrightarrow[k\to\infty]{} (s-\beta_1)\ldots(s-\beta_q) \tag{2.6-27}$$

However, there are n roots, so the locations of the remaining $(n-q)$ roots must be determined. Dividing by $\Delta(s)$, Eq. 2.6-24 takes the form

$$1 + k(b_q s^{q-n} + b'_{q-1} s^{q-n-1} + \cdots) = 0 \tag{2.6-28}$$

Multiplying by s^{n-q}, this equation could be approximated by

$$s^{n-q} + kb_q = 0 \tag{2.6-29}$$

for large values of k, so it is apparent that *the remaining* $(n-q)$ *roots go to infinity as* k *goes to* $\pm$ *infinity*. The roots radiate outward along linear asymptotes that emanate from the root locus "center of gravity" (c.g.).

As shown, for example, in (N-2), the term b'_{q-1} determines the *center-of-gravity location*, which is computed as

$$\text{c.g.} = \frac{\sum_{i=1}^{n} \alpha_i - \sum_{j=1}^{q} \beta_j}{n-q} \tag{2.6-29a}$$

Because all complex zeros and poles must be accompanied by their complex conjugates, *the* c.g. *lies on the real axis*, and Eq. 2.6-29a is equivalent to

$$\text{c.g.} = \frac{\sum_{i=1}^{n} \text{Re}(\alpha_i) - \sum_{j=1}^{q} \text{Re}(\beta_j)}{n-q} \tag{2.6-29b}$$

The angles that the asymptotes make with the real axis are computed using the *phase angle criterion* that derives from Eq. 2.6-24. Because

$$\frac{kn(s)}{\Delta(s)} = -1 \tag{2.6-30}$$

the net phase angle of $n(s)/\Delta(s)$ must be 180° for $kb_q > 0$ and 0° for $kb_q < 0$. There is symmetry about the real axis, and the roots that approach the q zeros of $n(s)$ have no effect on the asymptotic angles of the radiating $(n-q)$ roots; therefore, the $(n-q)$ asymptotes must themselves be symmetric about the real axis. The Laplace operator s is a complex variable that can be written in polar form, $re^{j\phi}$, so Eq. 2.6-29 can be written as

$$(re^{j\phi})^{n-q} = r^{n-q}e^{j(n-q)\phi}$$
$$= -kb_q = \begin{cases} |kb_q|e^{180^\circ}, & kb_q > 0 \\ |kb_q|e^{0^\circ}, & kb_q < 0 \end{cases} \tag{2.6-31}$$

The asymptotic angles are defined by

$$\phi = \frac{(2m+1)180^\circ}{n-q}, \qquad m = 0, 1, 2, \ldots (n-q)-1 \tag{2.6-32}$$

for positive loop gain kb_q and

$$\phi = \frac{2m(180^\circ)}{n-q}, \qquad m = 0, 1, 2, \ldots (n-q)-1 \tag{2.6-33}$$

for negative loop gain.

The phase angle criterion also clearly indicates that *branches of the root locus lie on the real axis*. For the 180° criterion, there is a root locus branch on any segment which has an odd number of zeros or poles to the right. Those segments that do not (do) have root locus branches with the 180° criterion do (do not) have branches for the 0° criterion. Nevertheless, branches can only begin ($k = 0$) at poles and end ($k = \pm\infty$) at zeros; if there is a branch between adjacent real poles, it must split at a "breakaway point" to form complex-conjugate roots with increasing gain. Conversely, a branch between adjacent zeros must be entered at an interior point by complex-conjugate roots originating elsewhere. The locus for a single complex-conjugate pair necessarily leaves or strikes the real axis at angles of ±90°.

Angles of departure from poles and arrival at zeros also are governed by the phase angle criterion. The angle of pole departure (zero arrival) is established by subtracting the angles to the remaining poles (zeros), summing the angles to the zeros (poles), and subtracting the criterion phase angle (180° or 0°). Some examples of root locus plots are given in Fig. 2.6-3. With conventional feedback (positive loop gain with negative feedback), high gain leads to instability when $(n-q) \geq 3$. With negative loop gain and negative feedback (resulting in positive feedback), high gain leads to instability in every case.

As an exercise, sketch the root loci for the continuous- and discrete-time configurations shown in Fig. 2.6-1. Your results should show that conventional feedback leads to oscillatory instability at a moderate level of loop gain in both cases. As the gain is increased further, the divergent oscillatory mode becomes two real modes, one of which eventually returns to the left

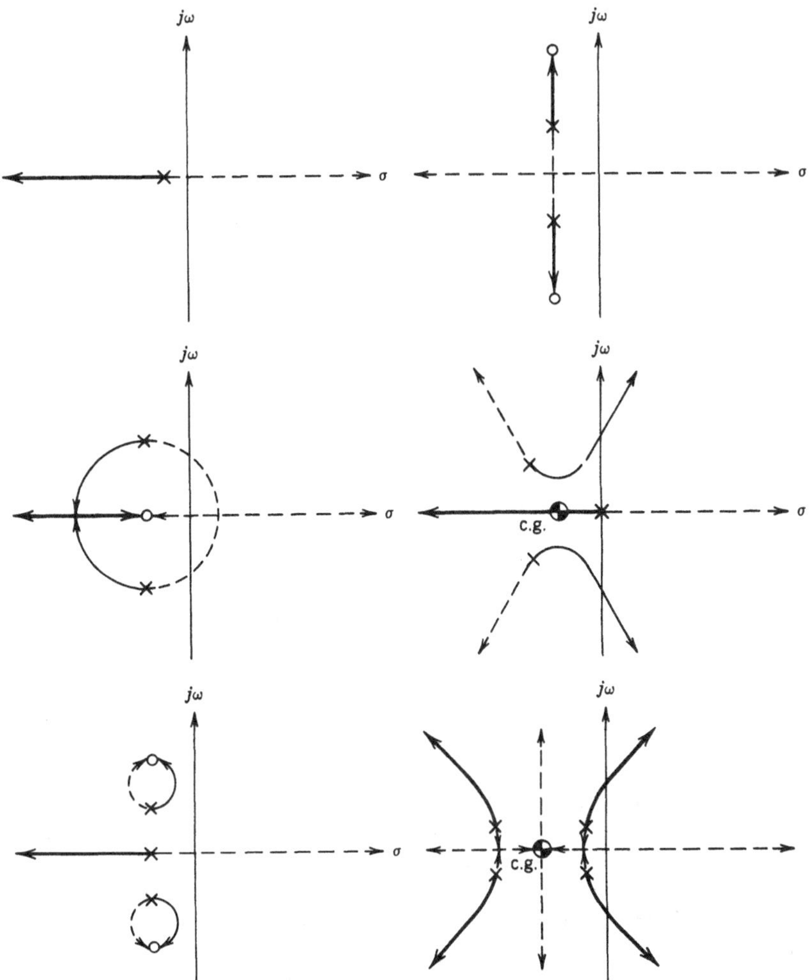

FIGURE 2.6-3 Examples of root-locus plots for positive and negative loop gains. (Solid line, positive; dashed line, negative.)

half plane. The real continuous-time mode that begins at $\alpha = -3$ becomes increasingly stable with increasing k; the corresponding discrete-time mode is ultimately destabilized by high gain.

Frequency-Response Function and Bode Plot

If the control input is a sinusoidal oscillation of frequency ω_o, then the output of a linear, time-invariant system also is a sinusoidal oscillation, differing at most from the input in amplitude and phase angle. For the input

$$u(t) = u_o \sin \omega_o t \tag{2.6-34}$$

the steady-state oscillatory response takes the form

$$y(t) = a(\omega_o) \sin [\omega_o t + \phi(\omega_o)] \tag{2.6-35}$$

where the amplitude ratio, $a(\omega_o)/u_o$, and phase angle, $\phi(\omega_o)$, depend upon the input–output transfer function, $Y(s)$. That dependence, which is called the *frequency response* at ω_o, is determined by setting $s = j\omega_o$ in $Y(s)$ and evaluating the result. More generally, given the scalar continuous-time transfer function,

$$Y(s) = \mathbf{H}(s\mathbf{I}_n - \mathbf{F})^{-1}\mathbf{G} \tag{2.6-36a}$$

the *frequency response function* is

$$Y(j\omega) = \mathbf{H}(j\omega\mathbf{I}_n - \mathbf{F})^{-1}\mathbf{G} \tag{2.6-36b}$$

where the driving frequency ω can take all values in $(-\infty, \infty)$. From Eq. 2.6-15, $Y(j\omega)$ can be written as

$$Y(j\omega) = \frac{b_q(j\omega - \beta_1)(j\omega - \beta_2)\ldots(j\omega - \beta_q)}{(j\omega - \alpha_1)(j\omega - \alpha_2)\ldots(j\omega - \alpha_n)} \tag{2.6-37}$$

This is equivalent to evaluating $Y(s)$ on a contour defined as the $j\omega$ axis of the s plane. It is sufficient to evaluate frequency response in the range $(0, \infty)$.

$Y(j\omega)$ can be precisely zero or infinity only at zeros or poles that lie on the $j\omega$ axis. The frequency response is finite at all other points, and it is always finite if all the zeros and poles have nonzero real parts. Nevertheless, the frequency response is measurable only if a steady-state response to oscillatory inputs exists (i.e., if all modes are stable).

A graph of the frequency response function, called the *Bode plot,** can be constructed readily. $Y(j\omega)$ is complex-valued and can be expressed in the form

$$Y(j\omega) = \text{Re}(\omega) + j\,\text{Im}(\omega) = A(\omega)e^{j\phi(\omega)} \tag{2.6-38}$$

where $A(\omega)$ is the *amplitude ratio* and $\phi(\omega)$ is the *phase angle*. The natural

*Hendrik W. Bode (1905–1982) developed techniques for network analysis and feedback design at the Bell Laboratories.

logarithm and corresponding base-10 logarithm are

$$\ln Y(j\omega) = \ln A(\omega) + j\phi(\omega) \tag{2.6-39a}$$

$$\log Y(j\omega) = \log A(\omega) + 0.434j\phi(\omega) \tag{2.6-39b}$$

The Bode plot consists of separate curves for $20 \log A(\omega)$, which is measured in decibels (dB), and $\phi(\omega)$, in deg. Both are plotted against $\log \omega$ for $(\omega_{min}, \omega_{max})$, as shown for a typical transfer function in Fig. 2.6-4.

Because the logarithms of products and divisions are sums and

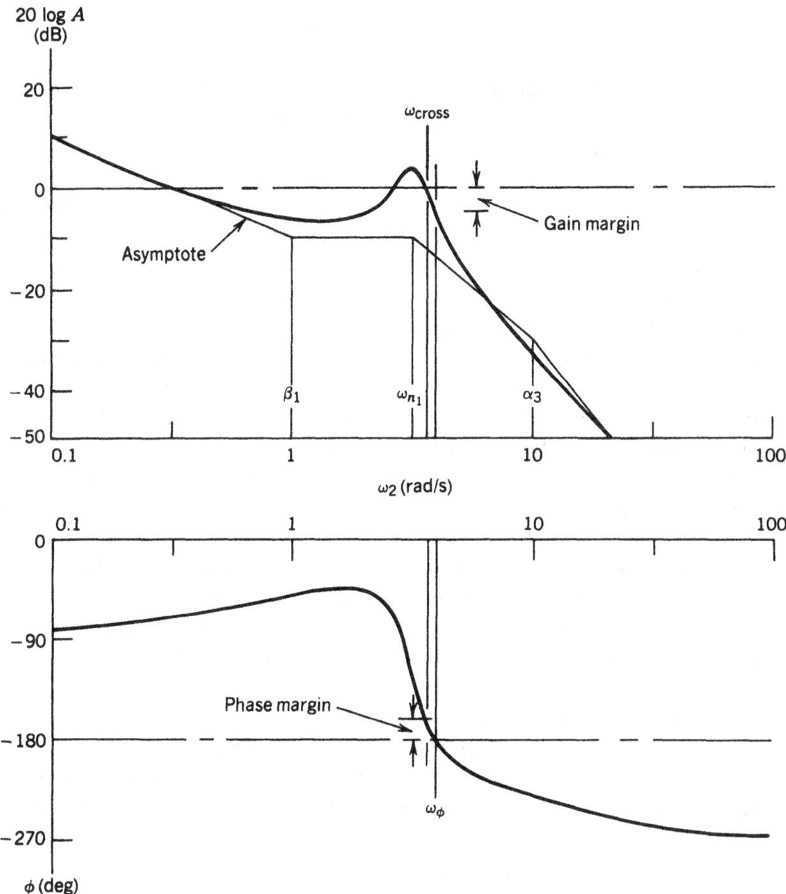

FIGURE 2.6-4 Bode plot of the transfer function:

$$Y(s) = k(s + \beta_1)/\{s[s^2 + 2(0.1)\omega_{n1}s + \omega_{n1}^2](s + \alpha_3)\}.$$

differences of logarithms,

$$\log\left(\frac{ab}{cd}\right) = \log a + \log b - \log c - \log d \tag{2.6-40}$$

the contributions of each zero and pole to $20 \log A(\omega)$ and $\phi(\omega)$ can simply be added or subtracted. Complex pairs of zeros and poles are written in second-order form:

$$\begin{aligned}(j\omega - \lambda_i)(j\omega - \lambda_i^*) &= [j\omega - (\sigma_i + j\omega_i)][j\omega - (\sigma_i - j\omega_i)] \\ &= (j\omega)^2 + 2\sigma_i(j\omega) + (\sigma_i^2 + \omega_i^2) \\ &= (j\omega)^2 + 2\zeta_i\omega_{n_i}(j\omega) + \omega_{n_i}^2 \end{aligned} \tag{2.6-41}$$

The locations of the singularities then are noted ($\omega = |\beta_i|$ and $|\alpha_i|$ for real zeros and poles or $\omega = \omega_{n_i}$ for complex pairs), with identical zeros and poles canceling each other.

In the first step, *amplitude-ratio asymptotes* are laid out between the singularities of the transfer function. In the second step, *amplitude-ratio differences* (or departures) defined for each singularity are superimposed on the asymptotes to yield the actual amplitude ratios. *Phase-shift deviations* are plotted in a single step.

The amplitude-ratio asymptotes are straight lines whose slopes depend upon adjacent singularities, beginning at $\omega = 0$ and working toward higher frequency. (Of course, $\omega = 0$ does not appear on the plot due to the logarithmic scale.) The initial slope is either $20n_z$ dB/dec or $-20n_p$ dB/dec,* depending on the net number of zeros (n_z) or poles (n_p) at the s-plane origin. With increasing frequency, the slope of the asymptote increases (decreases) by 20 dB/dec each time a real zero (pole) is encountered. Each time a complex pair of zeros (poles) is reached, the slope increases (decreases) by 40 dB/dec. The vertical positioning of the asymptotes is fixed by evaluating $20 \log A(\omega)$ at some easily described point in $(\omega_{\min}, \omega_{\max})$ that is not too close to a zero or a pole. The amplitude-ratio differences, shown in Fig. 2.6-5, then are added to the asymptotes to complete the plot. The differences have the same shapes for zeros and poles; however, they are applied with opposite sign.

There is no advantage in using asymptotes to define $\phi(\omega)$. It is easier to work directly with the phase-shift deviation plots of Fig. 2.6-6. The phase angle increases (decreases) with increasing frequency as left-half-plane zeros (poles) are encountered; the opposite occurs with right-half-plane zeros and poles. Left-half-plane zeros are called *minimum-phase zeros*

*A factor of 10 in frequency is a *decade*.

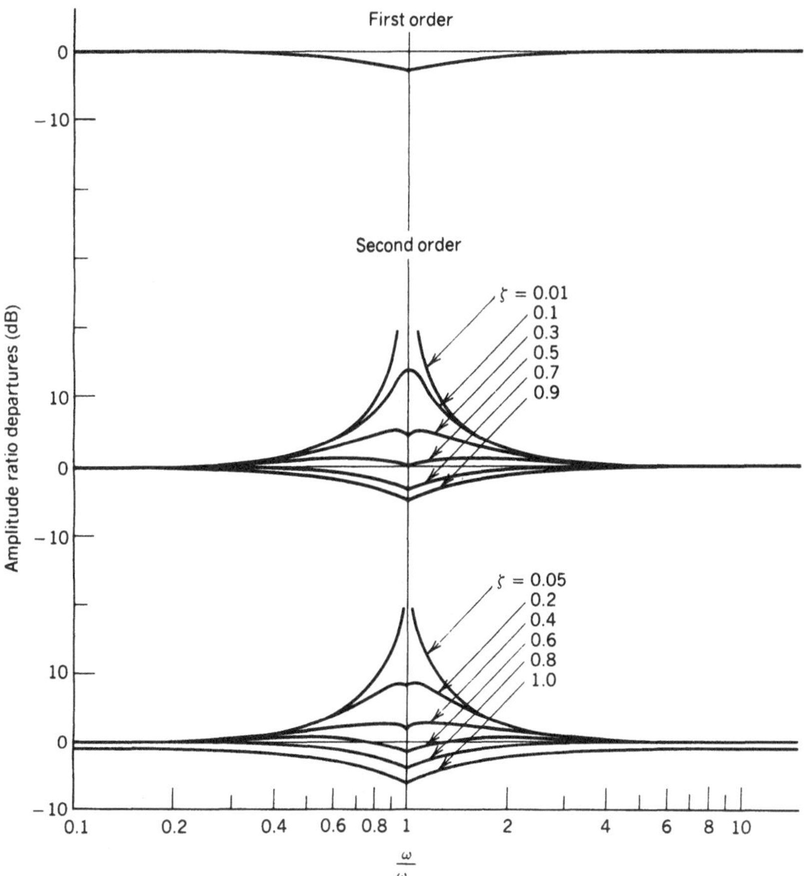

FIGURE 2.6-5 Amplitude ratio departures for first- and second-order poles [from (M-4)].

because a stable, proper* transfer function's net phase shift is less (as ω varies from zero to infinity) than it would be with an equal number of right-half-plane (non-minimum-phase) zeros.

With increasing frequency, the total phase shift due to a single, real, minimum-phase zero, β_i, is $+90°$. The phase *lead* is 5.7° when ω is one decade below $|\beta_i|$, 45° when $\omega = |\beta_i|$, and 84.3° when ω is a decade above $|\beta_i|$. Similarly, the phase lead for a complex pair of zeros is $+90°$ when $\omega = \omega_{n_i}$, and there is a total shift of $+180°$ in $(0, \infty)$. The phase lead at other

*A *proper* transfer function has no more zeros than poles; it is *strictly proper* if there are fewer zeros than poles. Transfer functions derived from Eq. 2.6-36 are strictly proper if **H**, **F**, and **G** are real matrices.

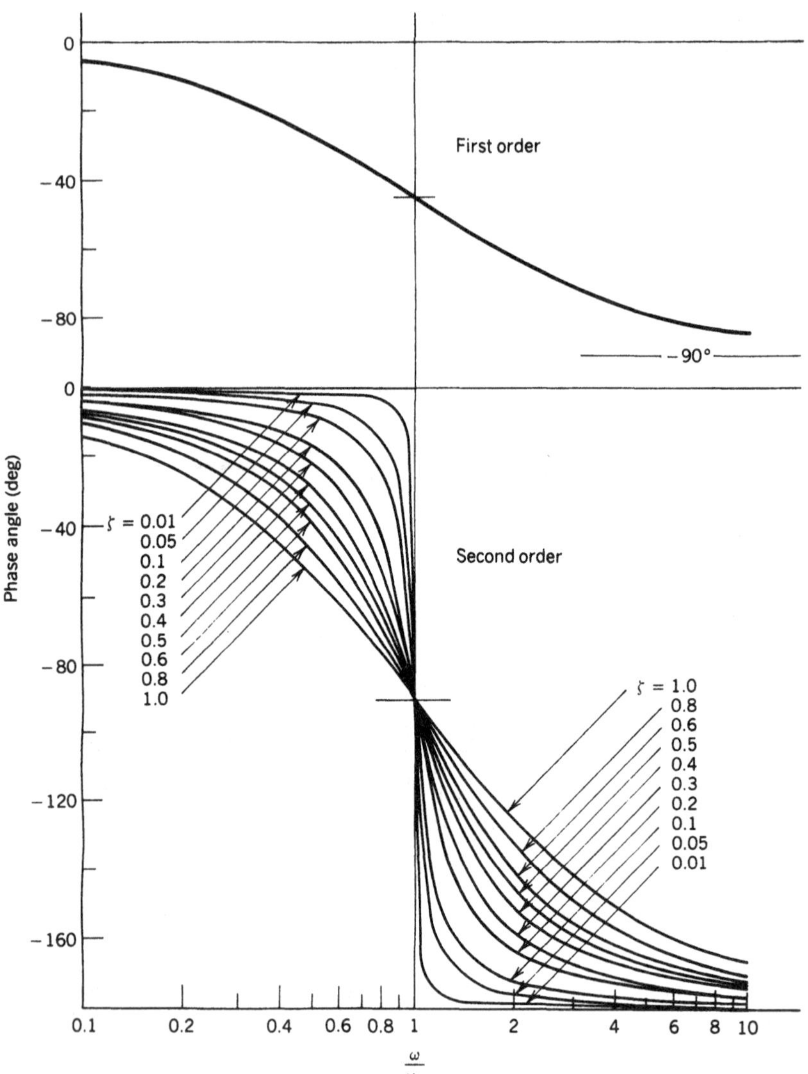

FIGURE 2.6-6 Phase angles for first- and second-order poles [from (M-4)].

frequencies depends on ζ_i. The shift as ω passes through ω_{n_i} is most abrupt for low values of ζ_i (Fig. 2.6-6).

A unity-feedback controller's stability and stability margins are governed by the relationship between amplitude ratio and phase angle. Feedback has a regenerative effect; input oscillations produce output oscillations. If the latter are amplified and fed back in phase with the former, the oscillations will grow without bound. In a *negative* feedback system, an output with $\pm 180°$ phase angle is actually in phase with the input. Therefore, if there is unity feedback around $Y(s)$ (Fig. 2.6-2), $A(\omega)$ should be less than one when $\phi(\omega)$ is $\pm 180°$.

Two important stability-margin characteristics for unity-feedback systems are portrayed by the Bode plot. The *gain margin* (GM) is the amplitude-ratio increment between $A(\omega_\phi)$ and one, where ω_ϕ is the frequency at which $\phi = -180°$. GM expresses the loop gain required to produce neutral stability, and it can be read directly (in dB) from the Bode plot (Fig. 2.6-4). The *phase margin* (PM) is the difference between $\phi(\omega_{\text{cross}})$ and $-180°$ where ω_{cross} is the *crossover frequency*, at which the amplitude ratio is one. PM indicates the additional phase lag that would cause neutral stability. Classical control design procedures typically call for phase margins of 40° and gain margins of 50% (6 dB) or more.

$|Y(s)|$ can be plotted as a *contoured surface* in the third dimension above and below the s plane, and $|Y(j\omega)|$ is represented by the contour height for a cut along the $j\omega$ axis. Converting to logarithmic scales, this contour becomes the Bode plot. Cutting the surface along the σ axis produces a corresponding *sigma plot* (M-4). Although constructed in a manner similar to the Bode plot, the sigma plot has no relationship to open-loop frequency response. The sigma plot is useful in determining root locus breakaway points, closed-loop frequency response asymptotes, and relationships between loop gain and real root location.

Nyquist Plot and Stability Criterion

$Y(j\omega)$ is a complex variable that can be characterized by a polar plot (or Argand diagram) as ω varies from $-\infty$ to $+\infty$. The resulting *Nyquist plot*, is, in effect, a cross plot of the information contained in the Bode plot; the amplitude ratio, $A(\omega)$, forms the magnitude of a vector whose angle is $\phi(\omega)$, and the head of the vector is plotted over the entire frequency range (Fig. 2.6-7). Frequency is an implicit variable that can only be signified by tick marks on the curve. Consequently, frequency response is not easily determined from the Nyquist plot; however, this condensed curve presents information that is essential for determining the closed-loop stability of a unity-feedback controller containing $Y(s)$ (Fig. 2.6-2).

The Nyquist plot for a strictly proper transfer function begins and ends at the origin. If we imagine a point in the s plane traveling along the $j\omega$ axis from $-\infty$ to $+\infty$, the Nyquist plot is a *conformal mapping* of the s-plane contour into the $Y(s)$ plane. Even though the initial and final points of the $Y(s)$ contour are the same, the mapping represents an open s-plane contour unless some formal closure is specified. The *Nyquist* $\mathbb{D}$ *contour* provides closure along a semicircle of radius R, $R \to \infty$, in the right half of the s plane.

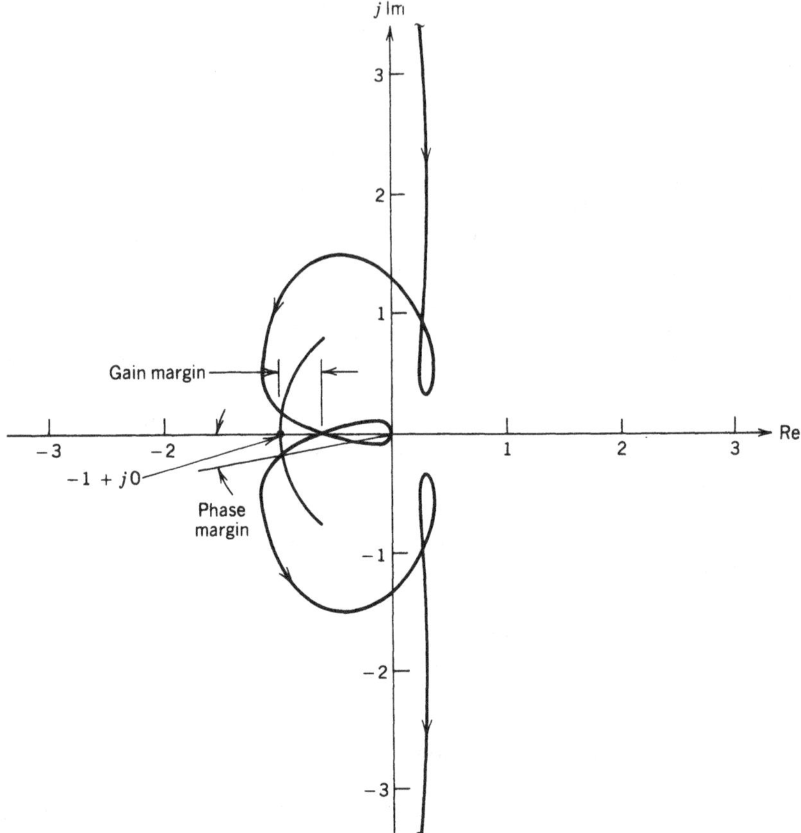

FIGURE 2.6-7 Nyquist plot of the transfer function:

$$Y(s) = k(s+\beta_1)/\{s[s^2+2(0.1)\omega_{n1}s+\omega_{n1}^2](s+\alpha_3)\}.$$

If $Y(s)$ contains no poles on the $j\omega$ axis, the $\mathbb{D}$ contour consists of the $j\omega$ axis plus the right-half-plane semicircle (Fig. 2.6-8a). If there are purely imaginary poles of $Y(s)$, the $\mathbb{D}$ contour avoids each by indenting into the right half plane along a semicircle of vanishingly small radius (so as to avoid neighboring singularities in the right half plane) (Fig. 2.6-8b). Cauchy's *principle of the argument* then relates the number of poles and zeros within the $\mathbb{D}$ contour to the number of encirclements of the origin made by the return difference function $[1+Y(s)]$ as the $\mathbb{D}$ contour is traversed (Fig. 2.6-8c). This is equivalent to the number of encirclements of the point $(-1+j0)$ made by $Y(s)$. Specifically, as s varies from $-j\infty$ to $+j\infty$ back to $-j\infty$ (traversing the $\mathbb{D}$ contour in a clockwise rotation), the corresponding number of counterclockwise encirclements is $N = P - Z$. P (or Z) is the number of poles (zeros) of the return difference function that lie within the

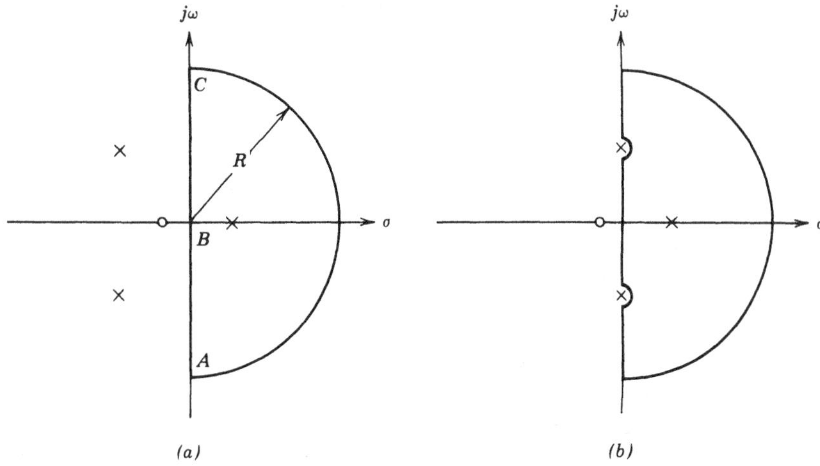

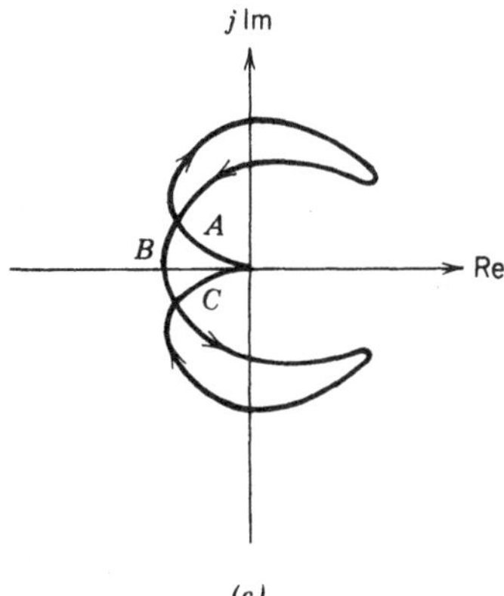

FIGURE 2.6-8 Examples of Nyquist $\mathbb{D}$ contours and plot. (a) Nyquist $\mathbb{D}$ contour (s plane); (b) Nyquist $\mathbb{D}$ contour with imaginary roots (s plane); (c) Nyquist plot corresponding to (a) [$Y(s)$ plane].

$\mathbb{D}$ contour (i.e., in the right half plane).* Since

$$1 + Y(s) = 1 + \frac{kn(s)}{\Delta(s)}$$

$$= \frac{\Delta(s) + kn(s)}{\Delta(s)} \tag{2.6-42}$$

*To apply Cauchy's principle, it is necessary only that the $\mathbb{D}$ contour enclose all right-half-plane singularities; if these singularities have finite magnitude, the $\mathbb{D}$ contour can be defined with finite radius R.

Z is the number of right-half-plane zeros of the closed-loop characteristic polynomial, and P is the number of right-half-plane zeros of the open-loop characteristic polynomial. If the closed-loop system is stable, $N = P$, which is the number of unstable poles in the open-loop system, $Y(s)$. This is the *Nyquist stability criterion* for closed-loop systems. If the open-loop system is stable, $N = P = 0$.

Stability margins are readily shown on the Nyquist plot. The gain margin is determined at the point where $Y(j\omega)$ crosses the negative real axis [i.e., where the phase angle is 180° (Fig. 2.6-7)]. GM is the distance between $(-1 + j0)$ and $|Y(j\omega)|$ at that point. The phase margin is measured as the angle between the negative real axis and the point for which $|Y(j\omega)| = 1$.

As k varies, $Y(j\omega)$ expands or shrinks without changing its shape, so the stability margins change too. A graphical alternative to replotting $Y(j\omega)$ for each new value of k is to plot $n(j\omega)/\Delta(j\omega)$ instead, evaluating stability margins with respect to the point $(-1/k + j0)$. GM is then k times the distance between the $-1/k$ point and the contour, while PM is evaluated at $|n(j\omega)/\Delta(j\omega)| = 1/k$.

Effects of Sampling

Spectral characteristics of continuous-signal sampling were discussed in Section 2.4, and discrete-time equivalents to continuous-time systems are presented throughout the book. This section provides a brief qualitative review of additional effects that periodic sampling can have on control and estimation.

It already has been noted that sampling has an "aliasing" effect on continuous-time signals. Frequency components higher than π/T, where T is the sampling interval in sec, are "folded" back to lower frequencies, becoming indistinguishable from true low-frequency components. (As a practical matter, continuous signals containing significant frequency components above π/T should be subjected to low-pass filtering prior to sampling.) There is an analogous aliasing effect on a system's transfer characteristics (Fig. 2.6-9a). The frequency response of a z-domain transfer function can be determined by replacing z with e^{sT} and letting $s = j\omega$ (Eq. 2.6-19). At frequencies much below π/T, the discrete transfer function's frequency response is quite similar to that of an equivalent continuous transfer function; however, near π/T the frequency response is markedly different, and no signals are passed at π/T and above. [If a discrete transfer function is to be generated from a given continuous transfer function, aliasing effects can be reduced by using a bilinear (or Tustin) transform rather than a z transform, as, for example, in (F-3) or (A-1)].

The frequency response of a hybrid system consisting of a digital filter, sampled-data hold, and continuous process is more complex. The sampled-data hold that converts the discrete output of the digital filter to an "analog" signal has a frequency response characteristic of its own, and the

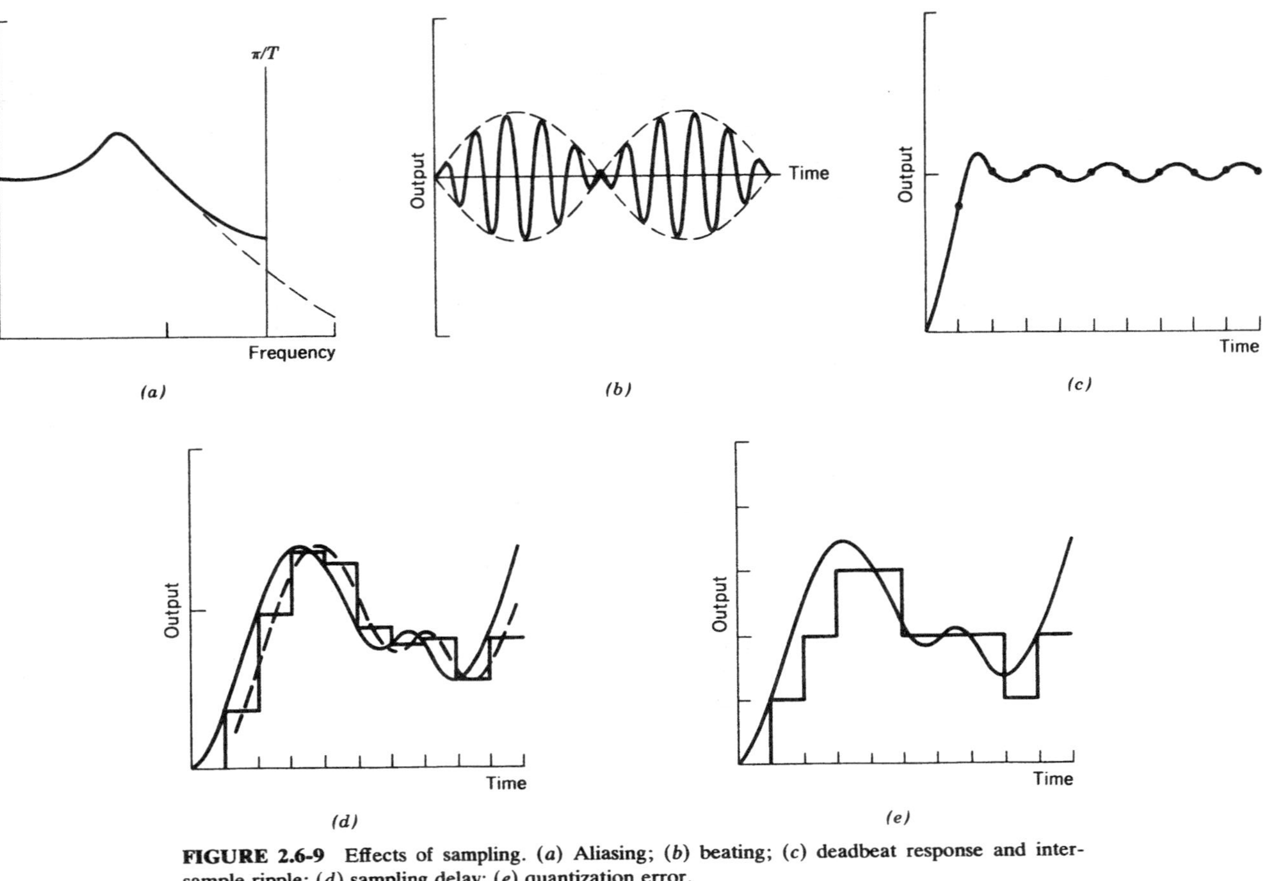

FIGURE 2.6-9 Effects of sampling. (*a*) Aliasing; (*b*) beating; (*c*) deadbeat response and inter-sample ripple; (*d*) sampling delay; (*e*) quantization error.

succession of "staircased" inputs excites harmonic response of the continuous process at frequencies above π/T. The latter can be expressed as response to the *sidebands* of the sampled, periodic input; given an input frequency of ω and a sampling frequency of π/T, the sampled, periodic input has spectral power at frequencies of $k\pi/T \pm \omega$, $k = 0$ to ∞. Interference between the sidebands and the fundamental frequency also gives rise to the phenomenon of *beating*, in which the amplitude envelope of the system's periodic output is itself periodic (Fig. 2.6-9*b*).

Linear sampled-data systems can achieve *deadbeat response*, in which steady-state response to a constant input is obtained in finite time and with no overshoot. A corresponding continuous system cannot achieve the steady state in finite time unless it possesses some degree of overshoot, because its response is a sum of exponentials. A deadbeat controller can be designed by choosing the control gain matrix $\mathbf{C}$ such that the closed-loop state transition matrix, $(\mathbf{\Phi} - \mathbf{\Gamma C})$, is nilpotent (Section 2.5), that is, so that $(\mathbf{\Phi} - \mathbf{\Gamma C})^k = \mathbf{0}$ for some $k > 1$. Nevertheless, satisfactory response is not assured because achieving the steady state at sampling instants does not assure that the response is well-behaved between sampling instants (Fig. 2.6-9*c*). Lightly damped continuous processes can experience *intersample ripple*, and unstable processes tend to diverge between samples. Qualitatively better response may be obtained by varying the sampling interval or by using sampled-data linear-quadratic theory to design the controller (Section 3.7).

Sampling introduces an inherent lag in the processing of continuous input signals; once a sample has been taken via analog-to-digital (A/D) conversion, there is no new information until the next sampling instant. Measurements taken at one instant can be considered "clamped" and unchanging in the interval. Digital-to-analog (D/A) conversion of computer outputs using a zero-order hold produces the same result. In both cases, it is fairly easy to show that the sampling process effectively delays reconstruction of a continually varying signal by one-half the sampling interval (Fig. 2.6-9*d*). A discrete step input can be delayed up to a full sampling interval while a discrete impulse can be missed entirely if it occurs between sampling instants. Equivalent phase lag can be reduced somewhat by predictive techniques (e.g., the first-order hold), but it cannot be eliminated entirely.

Another source of pure delay is the nonzero execution time of digital computers. Control and estimation algorithms require some degree of sequential processing, and each machine instruction takes time. Consequently, there is unavoidable delay between the acceptance of measurement inputs and the delivery of command outputs. Computation time sets a lower limit on the minimum sampling interval, although it is possible to achieve an input–output delay that is less than the total time required for control and estimation computations. Control and estimation algorithms should be arranged to minimize the amount of computation that occurs between the input and output times. For example, the Kalman filter (Section 4.3) consists of propagation and update steps; the propagation should be calculated before the measurement takes place, leaving only the update calculations for postmeasurement execution. In a multivariable controller, "multiplexing" of inputs and outputs introduces time skewing; hence the delays between various input–output pairs may be different.

Digital computers operate with finite word lengths, so there is granularity (or quantization) in the expression of all numbers (Fig. 2.6-9*e*). Error models for quantization depend upon implementational specifics, such as floating- versus fixed-point arithmetic, rounding versus truncation, and so on. These effects are nonlinear, further complicating analysis (P-3).

PROBLEMS

Section 2.1

1. Find the stationary points of the following and determine whether they are maxima, minima, or inflection points:
   (a) $J(u) = 10 + 15u + 5u^2$
   (b) $J(u) = 3 + 4u - 6u^2 + 10u^3$
   (c) $J(u) = (u-1)(u+2)(u-3)$
   (d) $J(u) = e^u + e^{-u}$

2. Find the stationary points of the following and determine whether they are maximum, minimum, saddle, or inflection points:
   (a) $J(u) = (u_1^2 + 3u_1 - 4)(u_2^2 - u_2 + 6)$
   (b) $J(u)$ as above, subject to the following equality constraint:

$$f(u) = u_1 - 2u_2 = 0$$

3. Find the optimal values of x and u for the following cost function and equality constraint:

$$J = \tfrac{1}{2}\begin{bmatrix} x_1 & x_2 \end{bmatrix}\begin{bmatrix} 2 & 0 \\ 0 & 4 \end{bmatrix}\begin{bmatrix} x_1 \\ x_2 \end{bmatrix} + u^2$$

$$\mathbf{f} = \begin{bmatrix} 0 & 1 \\ -1 & -1 \end{bmatrix}\begin{bmatrix} x_1 \\ x_2 \end{bmatrix} + \begin{bmatrix} 0 \\ 1 \end{bmatrix} u = 0$$

4. Repeat Problem 3 with the following change in J:

$$J = \tfrac{1}{2}\begin{bmatrix} x_1 & x_2 \end{bmatrix}\begin{bmatrix} 2 & 2 \\ 2 & 4 \end{bmatrix}\begin{bmatrix} x_1 \\ x_2 \end{bmatrix} + u^2$$

5. Compute the partial-derivative vectors and matrices (gradient, Hessian, and Jacobian) in symbolic form for the following problem:

$$J = (a + bu^2)x_2^2$$

$$\mathbf{f} = \begin{bmatrix} (a + bu^2)cx_2^2 - x_1 \\ dux_2 - e/x_2 \end{bmatrix} = \mathbf{0}$$

6. Compute a numerical solution for Problem 5 with $a = 1$, $b = 0.1$, $c = 10$, $d = 1$, and $e = 10$.

7. Find the minimizing solution for the following problem:

$$J = x^2 + 5u^2$$

$$f(x, u) = x - 3u - 3 \leq 0$$

8. Suppose that

$$x(t) = 1 - e^{-\zeta\omega_n t}\cos(\omega_n t + \phi), \qquad t \geq 0$$

What is the maximum value of $x(t)$, and at what time does it occur if $\omega_n = 1$, $\phi = 0$, given:

(a) $\zeta = 0.5$

(b) $\zeta = 1$

(c) $\zeta = 2$

Section 2.2

1. Compute the determinant of the following matrices by Laplace expansion and pivotal condensation:

(a) $\mathbf{F} = \begin{bmatrix} a & b & 0 & 0 \\ c & d & 0 & 0 \\ 0 & 0 & e & 0 \\ 0 & 0 & 0 & f \end{bmatrix}$

(b) $\mathbf{F} = \begin{bmatrix} 1 & 0 & 0 & 0 \\ 2 & 3 & 0 & 0 \\ 4 & 5 & 6 & 0 \\ 7 & 8 & 9 & 10 \end{bmatrix}$

(c) $\mathbf{F} = \begin{bmatrix} 0 & -1 & 2 & -3 \\ 1 & 0 & -3 & 4 \\ -2 & 3 & 0 & -5 \\ 3 & -4 & 5 & 0 \end{bmatrix}$

2. Compute the inverses of the matrices in Problem 1 if they exist.

3. Determine the "definiteness" of the following matrices:

(a) $\mathbf{Q} = \begin{bmatrix} 1 & 0 & 0 \\ 0 & 2 & 0 \\ 0 & 0 & 3 \end{bmatrix}$

(b) $\mathbf{Q} = \begin{bmatrix} 1 & 0 & 0 \\ 0 & -2 & 0 \\ 0 & 0 & 0 \end{bmatrix}$

(c) $\mathbf{Q} = \begin{bmatrix} 1 & 0 & 0 \\ 0 & 2 & 0 \\ 0 & 0 & 0 \end{bmatrix}$

(d) $\mathbf{Q} = \begin{bmatrix} 1 & 100 \\ 0.01 & 1 \end{bmatrix}$

4. Compute the appropriate (right or left) pseudoinverses of the following:

(a) $\mathbf{H} = \begin{bmatrix} 1 & 1 & 1 \\ 1 & 1 & 1 \end{bmatrix}$

(b) $\mathbf{H} = \begin{bmatrix} 1 & 0 \\ 0 & 1 \\ 1 & 0 \end{bmatrix}$

(c) $\mathbf{H} = \begin{bmatrix} 1 & 0 & 0 & 0 \\ 0 & 2 & 0 & 0 \\ 0 & 0 & 3 & 0 \end{bmatrix}$

5. Suppose that

$$\begin{bmatrix} \dot{x}_1 \\ \dot{x}_2 \\ \dot{x}_3 \end{bmatrix} = \begin{bmatrix} -1 & 0 & 0 \\ 0 & -10 & 0 \\ 0 & 0 & -100 \end{bmatrix} \begin{bmatrix} x_1 \\ x_2 \\ x_3 \end{bmatrix} + \begin{bmatrix} 1 & 0 \\ 0 & 1 \\ 1 & 0 \end{bmatrix} \begin{bmatrix} u_1 \\ u_2 \end{bmatrix}$$

and

$$\begin{bmatrix} y_1 \\ y_2 \\ y_3 \end{bmatrix} = \begin{bmatrix} 3x_1 + x_2 \\ x_2 + 3x_3 \\ x_1 + x_3 \end{bmatrix}$$

What is the corresponding differential equation for $\mathbf{y}$?

6. Show that

$$(\mathbf{I}_n - \mathbf{A}\mathbf{B}^{-1}\mathbf{C})^{-1} = \mathbf{I}_n - \mathbf{A}(\mathbf{C}\mathbf{A} - \mathbf{B})^{-1}\mathbf{C}$$

7. Find the eigenvalues and a set of eigenvectors for the matrix

$$\mathbf{A} = \begin{bmatrix} -2 & 1 & 1 \\ 0 & 5 & 1 \\ 0 & 0 & 3 \end{bmatrix}$$

What are the corresponding modal and diagonalized matrices?

8. Perform a singular value decomposition of the matrix in Problem 7.

Section 2.3

1. Use rectangular, trapezoidal, and fourth-order Runge–Kutta algorithms to integrate the following equation, and compare results to the exact value of the integral:

$$\dot{x}(t) = \cos 2\pi t, \qquad t = 0 \text{ to } 1 \text{ in } 0.1 \text{ increments}$$

2. Perform a local linearization of the following dynamic system:

$$\dot{x}_1 = a_1 x_1 + a_2 x_1^2 x_2^3 + a_3 x_3 \sin x_2 + b_1 u^2$$

$$\dot{x}_2 = (x_1 + x_3)^2/2 + b_2 u$$

$$\dot{x}_3 = a_4 x_1 - a_4 x_3$$

3. Van der Pol's equation is described by

$$\ddot{x} + a(1 - x^2)\dot{x} + bx = c$$

(a) Integrate van der Pol's equation with the following parameters and initial conditions:

$$a = 1, \quad b = 1, \quad c = 0, \quad x = 0.1, \quad \dot{x} = 0$$

(b) Repeat with the following initial conditions:

$$x = 0.2, \qquad \dot{x} = 0$$

(c) Perform a local linearization of van der Pol's equation about the nominal equilibrium condition.

(d) Integrate the linearized perturbation equation with the initial conditions, $\Delta x = 0.1$, $\Delta \dot{x} = 0$; add the perturbation solution to the solution obtained in (a); and compare this result with the solution obtained in (b).

4. Compute state transition and control effect matrices, $\mathbf{\Phi}$ and $\mathbf{\Gamma}$, for the following differential equation, with $\Delta t = 0.1$ s:

$$\begin{bmatrix} \dot{x}_1 \\ \dot{x}_2 \end{bmatrix} = \begin{bmatrix} -0.5 & 1 \\ 0 & -0.7 \end{bmatrix} \begin{bmatrix} x_1 \\ x_2 \end{bmatrix} + \begin{bmatrix} 0 \\ 1 \end{bmatrix} u$$

Propagate x_1 and x_2 by fourth-order integration of the differential equations and by the recursive expression

$$\mathbf{x}(t_k) = \mathbf{\Phi}\mathbf{x}(t_{k-1}) + \mathbf{\Gamma}\mathbf{u}(t_{k-1})$$

with zero initial conditions and $u = 1$, for $t = 0$ to 10 s.

5. The linearized equations of motion for an orbiting satellite spinning with nominal angular rate p_o about the x axis are

$$\Delta \dot{p} = \frac{M_x}{I_x}$$

$$\Delta \dot{q} = \frac{p_o(I_x - I_z)\Delta r + M_y}{I_y}$$

$$\Delta \dot{r} = \frac{p_o(I_y - I_x)\Delta q + M_z}{I_z}$$

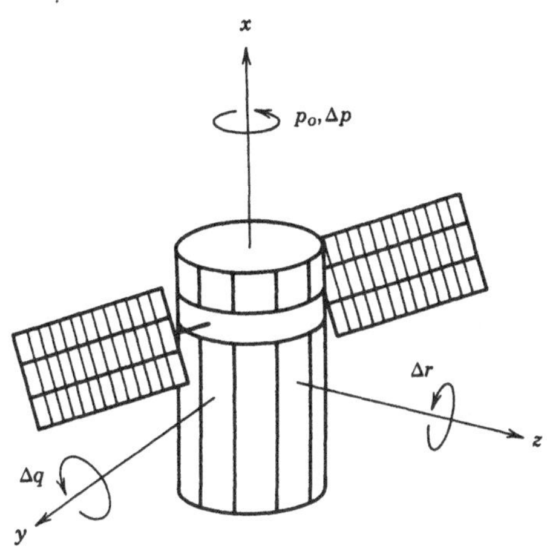

where I_x, I_y, and I_z are the moments of inertia about the roll, pitch, and yaw axes; M_x, M_y, and M_z are the corresponding control and disturbance torques; and Δp, Δq, and Δr are perturbations in rolling, pitching, and yawing rates.

(a) Compute the state transition and control effect matrices, assuming that $I_y = 750$ kg m^2, $I_z = 1000$ kg m^2, $I_x = 500$ kg m^2, $p_0 = 20$ rad/s, and $\Delta t = 0.1$ s.

(b) Compute the state history for 5 s, assuming zero inputs, $\Delta q(0) = 0.1$ rad/s, and $\Delta p(0) = \Delta r(0) = 0$.

Section 2.4

1. Using the random-number-generating routine on a personal computer, generate a sequence of 10, 100, and 1000 random numbers.

   (a) Calculate the sample means and standard deviations of each sequence and compare with the average value and spread assumed in calling the routine.

   (b) Compute histograms (sampled estimates of the probability distribution) for each sequence. What is the probability distribution used by the random number generator?

   (c) Let x_1, x_2, and x_3 each be 1000-number sequences generated by the random number routine. What is the histogram of $(x_1 + x_2)/2$? Of $(x_1 + x_2 + x_3)/3$?

2. A rectangular probability density function is represented by

$$\mathrm{pr}(x) = \frac{1}{2a}, \qquad -a \le x \le a$$

$$\mathrm{pr}(x) = 0, \qquad x \text{ outside this interval}$$

   Calculate the mean and variance of this distribution.

3. The probability density functions of x_i, $i = 1$–3, are

$$\mathrm{pr}(x_i) = \begin{cases} 0, & x_i < -1 \\ \tfrac{1}{2}, & -1 \le x_i \le 1 \\ 0, & x_i > 1 \end{cases}$$

   Using a discrete approximation, compute the probability density function for $z = x_1 + x_2 + x_3$.

4. The Rayleigh probability distribution is described by the following density function:

$$\mathrm{pr}(x) = (x/r)e^{-x^2/2r}, \qquad x \ge 0$$

$$\mathrm{pr}(x) = 0, \qquad x < 0$$

(a) Verify that the area under pr(x), integrating from minus to plus infinity, is one.

(b) Compute the mean and standard deviation of the Rayleigh distribution.

5. Suppose that x is a Gaussian random variable with mean and standard deviation each equal to one. What is the probability density function of $f(x) = x^2$ in the range (0, 2)?

6. Use a random-number-generating routine to calculate the 1000-number sequence u_k, and compute

$$x_{k+1} = 0.8x_k + 0.6u_k, \quad k = 1\text{-}1000, \quad x_0 = 0.$$

(a) Compute the auto- and cross-covariance functions of u_k and x_k. Comment on the results.

(b) Compute and compare the histograms of u_k and x_k.

(c) Calculate the discrete Fourier transforms, spectra, and cross-spectra of u_k and x_k.

Section 2.5

1. Compute the steady-state response x_1^* of the system

$$\begin{bmatrix} \dot{x}_1 \\ \dot{x}_2 \\ \dot{x}_3 \end{bmatrix} = \begin{bmatrix} -1 & 1 & 0 \\ 1 & -2 & 2 \\ 0 & 2 & -4 \end{bmatrix} \begin{bmatrix} x_1 \\ x_2 \\ x_3 \end{bmatrix} + \begin{bmatrix} 1 & 0 \\ 1 & 1 \\ 0 & 1 \end{bmatrix} \begin{bmatrix} u_1 \\ u_2 \end{bmatrix}$$

for

(a) $\mathbf{u}^* = [1 \quad 0]^T$

(b) $\mathbf{u}^* = [0 \quad 1]^T$

(c) $\mathbf{u}^* = [1 \quad 1]^T$, with the (3, 3) element of the first matrix changed from -4 to 0.

2. Van der Pol's equation is

$$\ddot{x} + a(1 - x^2)\dot{x} + bx = c$$

(a) Generate a phase plane plot of this equation's response for various initial conditions and $a = -1$, $b = 1$, and $c = 0$.

(b) Determine the stability of the equilibrium point and of any limit cycles that may occur.

(c) Define a quadratic Lyapunov function for the system. How does the region of stability that it predicts compare with the actual stable region?

3. A classical control description of the system presented in Problem 1 can be useful for analysis.

(a) Compute the system's Laplace transform.

(b) Express $\mathbf{x}(s)$ as a function of the initial condition, $\mathbf{x}(0)$, assuming no control inputs.

(c) Express $\mathbf{x}(s)$ as a function of the control input transform, $\mathbf{u}(s)$, for zero initial condition.

(d) Compute the roots of the system's characteristic equation.

(e) Using partial-fraction expansion, express the time response, $\mathbf{x}(t)$, for unit initial condition.

4. Determine whether or not the following $(\mathbf{F}, \mathbf{G})$ pairs are completely controllable:

(a) $\mathbf{F} = \begin{bmatrix} -1 & 0 \\ -1 & 0 \end{bmatrix}; \quad \mathbf{G} = \begin{bmatrix} 0 \\ 1 \end{bmatrix}$

(b) $\mathbf{F} = \begin{bmatrix} -1 & 0 \\ 0 & -1 \end{bmatrix}; \quad \mathbf{G} = \begin{bmatrix} 0 \\ 1 \end{bmatrix}$

(c) $\mathbf{F} = \begin{bmatrix} -1 & -1 \\ 0 & 0 \end{bmatrix}; \quad \mathbf{G} = \begin{bmatrix} 0 \\ 1 \end{bmatrix}$

(d) $\mathbf{F} = \begin{bmatrix} -1 & 0 \\ 0 & -1 \end{bmatrix}; \quad \mathbf{G} = \begin{bmatrix} 1 \\ 1 \end{bmatrix}$

(e) $\mathbf{F} = \begin{bmatrix} -1 & 1 \\ 1 & -1 \end{bmatrix}; \quad \mathbf{G} = \begin{bmatrix} 1 \\ 1 \end{bmatrix}$

(f) $\mathbf{F} = \begin{bmatrix} -1 & 0 \\ 0 & -2 \end{bmatrix}; \quad \mathbf{G} = \begin{bmatrix} 1 \\ 1 \end{bmatrix}$

5. Express the observability duals of Problem 4, defining $\mathbf{H}$ as $\mathbf{G}^T$.

6. Define examples of (2×2) matrices that are nilpotent with indices of 2, 3, and 4.

Section 2.6

1. The short-period motion of a jet fighter aircraft can be described by the equations of Example 2.6-1. The following are typical values of the equation coefficients for a modern open-loop-unstable configuration flying at an altitude of 5,000 ft and an airspeed of 650 ft/s (neglecting flap effects):

$$M_q = -0.568$$

$$M_\alpha = 17.98$$

$$\frac{L_\alpha}{V} = 1.237$$

$$M_{\delta E} = 0.175$$

$$\frac{L_{\delta E}}{V} = 0.001$$

(a) Plot the root locus for pitch-rate feedback to the elevator with both positive and negative feedback gains.

(b) Repeat (a) for angle-of-attack feedback to the elevator.

(c) Plot the open-loop frequency responses of pitch rate and angle of attack to elevator inputs.

(d) Generate the corresponding Nyquist plots.

2. Suppose that the elevator actuator has a dynamic lag ($\tau = 0.5$ s) that is modeled as

$$\dot{u}(t) = -\frac{1}{\tau}\,u(t) + \frac{1}{\tau}\,u_c(t)$$

where $u_c(t)$ is the commanded elevator angle. Determine the effects that the lag has on the results of Problem 1.

3. Simulate a digitally controlled aircraft with angle-of-attack feedback on a personal computer. Use Runge-Kutta integration with time steps that are at least five times smaller than the assumed time delay of 0.1 s. Compute the step responses for several cases with decreasing stability margins (due to increasing feedback gain). Comment on the observed intersample behavior.

4. Single-axis motions of a flexible robot arm could be modeled by a fourth-order system that accounts for rigid-body motion plus the first bending mode:

$$\dot{x}_1 = x_2$$

$$\dot{x}_2 = -\,a_{21}\cos x_1 + a_{23}x_3 + b_2 u$$

$$\dot{x}_3 = x_4$$

$$\dot{x}_4 = -\,a_{41}\cos x_1 - \omega_n^2 x_3 - 2\zeta\omega_n x_4 + b_4 u$$

$$y = x_1 + x_3$$

x_1 and x_2 represent the angular position and rate of the arm's rigid-body mode, while x_3 and x_4 represent the angular position and rate deflections due to bending. x_1 and x_3 add to produce the net position of the end effector y. The bending mode's natural frequency and damping

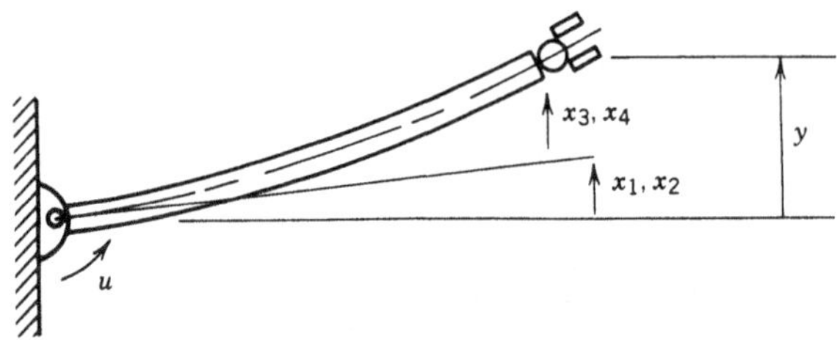

ratio are ω_n and ζ. A control torque u is applied at the robot arm's hinge point; b_2 and b_4 characterize the forcing of the rigid-body and bending modes by this torque. Neglecting gravitational and coupling effects, the coefficients for this model are as follows:

$$a_{21} = a_{23} = a_{41} = 0$$

$$b_2 = 1, \quad b_4 = -0.05, \quad \omega_n = 62.8 \text{ rad/s}, \quad \zeta = 0.2$$

(a) Plot the root loci for individual negative feedback of x_1, x_2, x_3, x_4, and y to u.
(b) Plot the open-loop frequency response $y(\omega)/u(\omega)$.
(c) Generate the corresponding Nyquist plot.

5. The robot arm is to be commanded by a digital computer that introduces a pure delay in the feedback loop. Using root locus and Nyquist plots, examine the effects that delays of 0.001, 0.01, and 0.1 s have on closed-loop stability with varying negative feedback gain from y to u.

REFERENCES

A-1 Åström, K. J. and Wittenmark, B., *Computer Controlled Systems: Theory and Design*, Prentice-Hall, Englewood Cliffs, NJ, 1984.

A-2 Athans, M. The Matrix Minimum Principle, *Information and Control*, **11**, (5, 6), November 1967. 592–606.

B-1 Barnett, S., *Matrices in Control Theory*, Van Nostrand Reinhold, London, 1971.

B-2 Barnett, S., *Introduction to Mathematical Control Theory*, Clarendon Press, Oxford, 1975.

B-3 Bartels, R. H. and Stewart, G. W., Solution of the Matrix Equation AX + XB = C, *Communications of the ACM*, **15**, 820–826, September 1972.

B-4 Bendat, J. S. and Piersol, A. G., *Measurement and Analysis of Random Data*, Wiley, New York, 1966.

B-5 Ben-Israel, A. and Greville, T.N.E., *Generalized Inverses: Theory and Application*, Wiley, New York, 1974.

B-6 Brogan, W. L., *Modern Control Theory*, Prentice-Hall, Englewood Cliffs, NJ, 1985.

B-7 Bryson, A. E., Jr. and Ho, Y. C., *Applied Optimal Control*, Hemisphere Publishing Co., Washington, D.C., 1975.

D-1 Davenport, W. B., Jr., *Probability and Random Processes*, McGraw-Hill, New York, 1970.

D-2 DeRusso, P. M., Roy, R. J. and Close, C. M., *State Variables for Engineers*, Wiley, New York, 1965.

D-3 Dorf, R. C., *Modern Control Systems*, Addison-Wesley, Reading, MA, 1974.

D-4 Dorny, N. C., *A Vector Space Approach to Models and Optimization*, Wiley, New York, 1975.

E-1 El-Hawary, M. E., *Control System Engineering*, Reston Publishing Co., Reston, VA, 1984.

E-2 Evans, W. R., Graphical Analysis of Control Systems, *Transactions of the American Institute of Electrical Engineers*, **67**, 547–551, 1948.

F-1 Fel'dbaum, A. A., *Optimal Control Systems*, Academic, New York, 1965.

F-2 Francis, J. G. F., The QR Transformation—Parts I and II, *Computer Journal*, **4**, 265–271, 332–345, 1961.

F-3 Franklin, G. F. and Powell, J. D., *Digital Control of Dynamic Systems*, Addison-Wesley, Reading, MA, 1980.

F-4 Froberg, C-E., *Introduction to Numerical Analysis*, Addison-Wesley, Reading, MA, 1965.

G-1 Gantmacher, F. R., *The Theory of Matrices*, Chelsea, New York, 1959.

G-2 Gardner, M. F. and Barnes, J. L., *Transients in Control Systems*, Wiley, New York, 1942.

G-3 Gibson, J. E., *Nonlinear Automatic Control*, McGraw-Hill, New York, 1963.

G-4 Gilbert, E. G., Controllability and Observability in Multivariable Control Systems, *SIAM Journal on Control*, Series A, **2**(1), 128–151, 1963.

G-5 Gilmore, R., *Catastrophe Theory for Scientists and Engineers*, Wiley, New York, 1981.

G-6 Golub, G. H., Nash, S. and Van Loan, C., A Hessenburg-Schur Method for the Problem $AX + XB = C$, *IEEE Transactions on Automatic Control*, **AC-24**(6), 909–913, December 1979.

G-7 Graham, D. and McRuer, D., *Analysis of Nonlinear Control Systems*, Wiley, New York, 1961.

G-8 Greenberg, M. D., *Foundations of Applied Mathematics*, Prentice-Hall, Englewood Cliffs, NJ, 1978.

G-9 Greenstadt, J., The Determination of the Characteristic Roots of a Matrix by the Jacobi Method in *Mathematical Methods for Digital Computers*, A. Ralston and H. S. Wilf, eds., Wiley, New York, 1965.

H-1 Hanson, G. D. and Stengel, R. F., Effects of Displacement and Rate Saturation on the Control of Statically Unstable Aircraft, *AIAA Journal of Guidance, Control, and Dynamics*, **7**(2), 197–205, March–April 1984.

H-2 Hermes, H., Controllability and the Singular Value Problem, *SIAM Journal of Control*, Series A, **2**(2), 241–260, 1965.

H-3 Hildebrand, F. B., *Advanced Calculus for Applications*, Prentice-Hall, Englewood Cliffs, NJ, 1976.

J-1 Jacquot, R. G., *Modern Digital Control Systems*, Marcel Dekker, New York, 1981.

J-2 James, H. M., Nichols, N. B. and Phillips, R. S., eds., *Theory of Servomechanisms*, McGraw-Hill, New York, 1947.

K-1 Kailath, T., *Linear Systems*, Prentice-Hall, Englewood Cliffs, NJ, 1980.

K-2 Kalman, R. E., On the General Theory of Control Systems, *Automatic and Remote Control*, Vol. I, J. F. Coales, ed., Butterworths, London, 1961, pp. 481–491.

K-3 Kalman, R. E., When Is a Linear Control System Optimal?, *Transactions of the ASME, Journal of Basic Engineering*, **86**, March 1964, pp. 51–60.

K-4 Kaplan, W., *Advanced Mathematics for Engineers*, Addison-Wesley, Reading, MA, 1981.

K-5 Klema, V. C. and Laub, A. J., The Singular Value Decomposition: Its Computation and Some Applications, *IEEE Transactions on Automatic Control*, **AC-25**(2), 164–176, April 1980.

L-1 Laning, J. H., Jr. and Battin, R. H., *Random Processes in Automatic Control*, McGraw-Hill, New York, 1956.

L-2 Lawson, C. L. and Hanson, R. J., *Solving Least Squares Problems*, Prentice-Hall, Englewood Cliffs, NJ, 1974.

L-3 Lee, Y. W., *Statistical Theory of Communication*, Wiley, New York, 1960.

L-4 Lyapunov, A. M., The General Motion Stability Problem, Kharkov, 1892. (Also *Annals of Mathematical Studies*, No. 17, Princeton University Press, Princeton, NJ, 1947.)

M-1 Mallinckrodt, A. J., Aliasing Errors in Sampled Data Systems, AGARD Report 316, Paris, 1961.

M-2 Maybeck, P. S., *Stochastic Models, Estimation, and Control*, Vol. 1, Academic, New York, 1979.

M-3 McClamroch, N. H., *State Models of Dynamic Systems*, Springer-Verlag, New York, 1980.

M-4 McRuer, D., Ashkenas, I. and Graham, D., *Aircraft Dynamics and Automatic Control*, Princeton University Press, Princeton, NJ, 1973.

M-5 Miller, I. and Freund, J. E., *Probability and Statistics for Engineers*, Prentice-Hall, Englewood Cliffs, NJ, 1977.

N-1 Narendra, K. S. and Taylor, J. H., *Frequency Domain Criteria for Absolute Stability*, Academic, New York, 1973.

N-2 Netushil, A., ed., *Theory of Automatic Control*, MIR Publishers, Moscow, 1978.

O-1 Oppenheim, A. V. and Schafer, R. W., *Digital Signal Processing*, Prentice-Hall, Englewood Cliffs, NJ, 1975.

P-1 Papoulis, A., *The Fourier Integral and Its Applications*, McGraw-Hill, New York, 1962.

P-2 Papoulis, A., *Probability, Random Variables, and Stochastic Processes*, McGraw-Hill, New York, 1965.

P-3 Phillips, C. L. and Nagle, H. T., Jr., *Digital Control System Analysis and Design*, Prentice-Hall, Englewood Cliffs, NJ, 1984.

P-4 Popov, E. P., *The Dynamics of Automatic Control Systems*, Addison-Wesley, Reading, MA, 1962.

R-1 Richards, R. J., *An Introduction to Dynamics and Control*, Longman Group, London, 1979.

S-1 Shinners, S. M., *Modern Control System Theory and Application*, Addison-Wesley, Reading, MA, 1973.

S-2 Smith, J. M., *Mathematical Modeling and Digital Simulation*, Wiley, New York, 1977.

S-3 Snow, D. R., Determining Reachable Regions and Optimal Controls, in *Advances in Control Systems: Theory and Applications*, Vol. 5, C. T. Leondes, ed., Academic, New York, 1967.

S-4 Sokolnikoff, I. S. and Redheffer, R. M., *Mathematics of Physics and Modern Engineering*, McGraw-Hill, New York, 1958.

S-5 Stengel, R. F., Equilibrium Response of Flight Control Systems, *Automatica*, **18**(3), 343–348, May 1982.

S-6 Stengel, R. F., Some Effects of Bias Errors in Redundant Flight Control Systems, *AIAA Journal of Aircraft*, **10**(3), 150–156, March 1973.

T-1 Takahashi, Y., Rabins, M. J., and Auslander, D. M., *Control and Dynamic Systems*, Addison-Wesley, Reading, MA, 1972.

T-2 Thomas, G. B., Jr., *Calculus and Analytic Geometry*, Addison–Wesley, Reading, MA, 1956.

T-3 Thompson, J.M.T., *Instabilities and Catastrophes in Science and Engineering*, Wiley, New York, 1982.

W-1 Wilkinson, J. H., *The Algebraic Eigenvalue Problem*, Clarendon, Oxford, 1965.

W-2 Wonham, W. M., On Pole Assignment in Multi-Input Controllable Linear Systems, *IEEE Transactions on Automatic Control*, **AC-12**(6), 660–665, December 1967.

3 OPTIMAL TRAJECTORIES AND NEIGHBORING-OPTIMAL SOLUTIONS

The general problem of optimal control is to find a history of the control vector [$\mathbf{u}(t)$ for $t_0 \leq t \leq t_f$] which forces the state from its initial value to its final value (along a *trajectory*) and, at the same time, maximizes or minimizes a cost function. The resulting state history [$\mathbf{x}(t)$ for $t_0 \leq t \leq t_f$] is an *optimal trajectory*. Once an optimal trajectory has been defined, a small perturbation in initial value, final value, system parameters, or external disturbances could destroy the optimality of the trajectory; however, it may be possible to recompute the control history, achieving a *neighboring-optimal trajectory*.

The methods of finding optimal control histories for dynamic systems are analogous to the static optimizations discussed in Sections 1.3 and 2.1, but the computations are necessarily more complex. The need for a control history rather than a fixed value of control is one obvious reason, and the possible dependence of the control at some arbitrary time on past and future states of the system is another. In application, the information used to compute the optimal trajectory may be imprecise, so we might well wonder if there is any hope of defining optimal controls in an actual situation. Practical solutions are possible because:

- Certain combinations of cost functions and dynamic system models have analytic (or "exact") optimal control solutions;
- In a wide class of problems, the requisite control histories can be expressed as optimal control *laws*, that is, feedback relationships between the measured (or estimated) states and the optimal controls; and
- Contemporary digital and analog computers are capable of performing the needed calculations in a reasonable amount of time for many systems of interest.

The remainder of the chapter deals primarily with optimal trajectory computation for systems described by nonlinear differential equations. After stating the problem, the types of cost functions that can be optimized and the conditions that define optimality are presented (Sections 3.1 and 3.2). It will be seen that not only the cost function but the dynamic constraint imposed by the system model shapes the optimal control history. Parametric optimization finds utility when the shape of the control function is specified prior to design (Section 3.3), although there is no guarantee that the result is truly optimal. Necessary and sufficient conditions for optimality are discussed, drawing principally on the Calculus of Variations and showing its relationship to the Minimum Principle and the Hamilton–Jacobi–Bellman Equation of Dynamic Programming (Section 3.4). The effects of various constraints on the optimal trajectory are presented, as is an introduction to solutions of the singular optimal control problem, for which the usual necessary and sufficient conditions are not satisfied (Section 3.5). While some combinations of cost function and dynamic system model are simple enough that an exact optimal control solution can be found, optimal control solutions for complicated systems require the use of numerical (generally iterative) computations (Section 3.6). Neighboring-optimal trajectories can be computed by perturbing the system and recomputing the optimal control as before, but if the perturbations are sufficiently small, there is a simpler way. The equations of linear neighboring-optimal control allow a family of perturbed optimal trajectories to be specified by a single feedback control law (Section 3.7). The linear control law may be specified either as a continuous-time controller or as a discrete-time controller; the latter is more suitable for implementation in a digital computer.

3.1 STATEMENT OF THE PROBLEM

In this chapter, we consider optimal control strategies for dynamic system models whose parameters, disturbances, and initial conditions are known without error. By convention, the control history, $\mathbf{u}^*(t)$, $t_0 \le t \le t_f$, that *minimizes* a cost function J is to be found. (Seeking to *maximize* the same cost function but with opposite sign is another way of expressing the same problem; hence the distinction between optimizing to achieve a maximum or a minimum is mathematically trivial.) The cost function considered here consists of two parts, a scalar algebraic function of the final state and time, and a scalar integral function of the state and control:

$$J = \phi[\mathbf{x}(t_f), \mathbf{w}(t_f), \mathbf{p}(t_f), t_f] + \int_{t_0}^{t_f} \mathscr{L}[\mathbf{x}(t), \mathbf{u}(t), \mathbf{w}(t), \mathbf{p}(t), t]\, dt \quad (3.1\text{-}1a)$$

As in Section 1.2, the dimensions of $\mathbf{x}$, $\mathbf{u}$, $\mathbf{w}$, and $\mathbf{p}$ are, respectively, n, m, s, and l. Since disturbances and parameters, $\mathbf{w}(t)$ and $\mathbf{p}(t)$, are assumed to be known functions of time, the cost function can be expressed as

$$J = \phi[\mathbf{x}(t_f), t_f] + \int_{t_0}^{t_f} \mathscr{L}[\mathbf{x}(t), \mathbf{u}(t), t]\, dt \tag{3.1-1b}$$

without loss of generality.

The choice of the final state penalty $\phi[\cdot]$ and the integrand $\mathscr{L}[\cdot]$ (sometimes called the *Lagrangian**) dictates the nature of the optimizing solution, and there is some flexibility in expressing a particular cost as a terminal or integral function. Of course, it need not contain both parts. A cost function containing only a terminal penalty is of the *Mayer type* (Fig. 3.1-1*a*), while one with just the integral term is of the *Lagrange type* (Fig. 3.1-1*b*). Equation 3.1-1 represents the *Bolza type* (Fig. 3.1-1*c*), as it has both terms.† The form is chosen largely as a matter of convenience, for it is easily shown that a cost function of one type can be transformed into either of the other two types. The Bolza form is used here because it demonstrates the roles played by both terminal and integral costs.

The optimizing control acts upon the dynamic system, whose trajectory is modeled by integrating the ordinary differential equation

$$\dot{\mathbf{x}}(t) = \mathbf{f}[\mathbf{x}(t), \mathbf{u}(t), \mathbf{w}(t), \mathbf{p}(t), t] \tag{3.1-2a}$$

for given initial conditions

$$\mathbf{x}(t_0) = \mathbf{x}_0 \tag{3.1-3}$$

Then the trajectory is described by

$$\mathbf{x}(t) = \mathbf{x}_0 + \int_{t_0}^{t} \mathbf{f}[\mathbf{x}(\tau), \mathbf{u}(\tau), \mathbf{w}(\tau), \mathbf{p}(\tau), \tau]\, d\tau \qquad t_0 \le t \le t_f \tag{3.1-4}$$

When $\mathbf{u}(t) = \mathbf{u}^*(t)$, the state history produced by Eq. 3.1-4 is optimized, that is, $\mathbf{x}(t) = \mathbf{x}^*(t)$. As before, it can be assumed that $\mathbf{w}(t)$ and $\mathbf{p}(t)$ are known during the solution interval, so Eq. 3.1-2a can be expressed as

$$\dot{\mathbf{x}}(t) = \mathbf{f}[\mathbf{x}(t), \mathbf{u}(t), t] \tag{3.1-2b}$$

*Lagrange derived methods for solving problems with isoperimetric (constant-perimeter) constraints, which can be expressed as integrals of functions such as $\mathscr{L}$. In classical mechanics, the Lagrangian represents the difference between the kinetic and potential energies of a dynamic system.

†Christian Mayer (1839–1908) was professor of mathematics at Heidelberg. Oscar Bolza (1857–1942) spent nearly 20 years teaching in the United States, returning to Germany in 1910. Disillusioned by "the war to end all wars", he devoted his later years to religious psychology and Sanskrit.

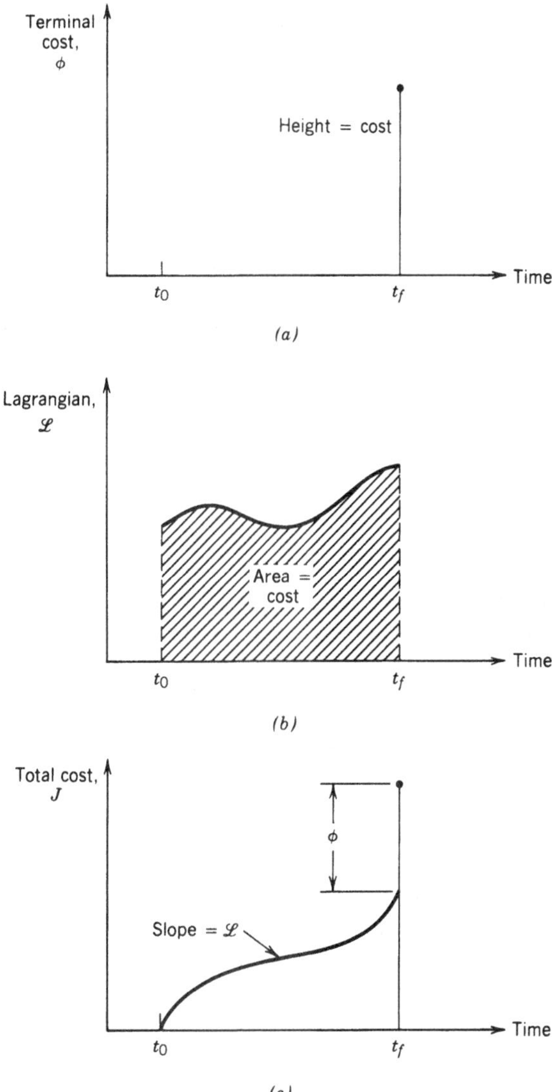

FIGURE 3.1-1 Three types of cost function for dynamic systems. (*a*) Mayer cost function; (*b*) Lagrange cost function; (*c*) Bolza cost function.

Because Eq. 3.1-2 must be satisfied while J is being minimized, it produces a *dynamic constraint* on Eq. 3.1-1.

3.2 COST FUNCTIONS

A cost function measures the penalty that must be paid as a consequence of the dynamic system's trajectory. The "cost" may reflect actual financial expenditures and be expressed in monetary units; it may indicate deviation from some ideal physical situation and be expressed in engineering units; or it simply may represent the passage of time in going from initial to final values of the state. A positive "cost" could be considered a negative "benefit," or vice versa.

For the *minimum-time problem,*

$$J = \int_{t_0}^{t_f} (1)\, dt = t_f - t_0 \tag{3.2-1}$$

is the appropriate cost function, as minimizing J minimizes the time interval, but one must ask "minimum time for what purpose"? A reasonable answer is "the minimum time interval to cause one or more components of the state to achieve desired final values". In other words, not only must $\mathbf{x}_0$ be specified, but some combination of the components of $\mathbf{x}(t_f)$ must be specified as well. Classical examples include the minimum time for an airplane to climb from one altitude and velocity to another or the minimum time to reorient the attitude of an Earth satellite using reaction control thrusters. State components that are not specified at the final time are said to be *free*.

The minimum-time problem is a special case of the *open-end-time problem*. The final time is analogous to a control variable because its value has direct effect on the cost function and can be altered in the optimization.

The cost function for a *minimum-fuel problem* is

$$J = \int_{t_0}^{t_f} \dot{m}(t)\, dt = m(t_f) - m(t_0) \tag{3.2-2}$$

where $\dot{m}(t)$ is the mass flow rate of the fuel. It is reasonable to assume that $\dot{m}(t)$ is a function of at least one control (e.g., throttle position), and it may depend on the state as well. The throttle in a reciprocating engine carburetor controls fuel flow indirectly; the actual mass flow rate depends on air flow, which depends on engine rpm, air density (in turn, a function of altitude and temperature), and so on. Therefore, fuel flow rate can be expressed as

$$\dot{m}(t) = \mathscr{L}[\mathbf{x}(t), \mathbf{u}(t), t] \tag{3.2-3}$$

leading to the cost function

$$J = \int_{t_0}^{t_f} \mathscr{L}[\mathbf{x}(t), \mathbf{u}(t), t]\, dt \tag{3.2-4}$$

Alternatively, fuel mass itself can be considered a state component. (It *must* be considered a state component if the remaining dynamic equations depend on fuel mass, as in the long-range cruise of an aircraft.) If fuel mass is the ith component of $\mathbf{x}$, then $f_i[\mathbf{x}(t), \mathbf{u}(t), t] = \mathscr{L}[\mathbf{x}(t), \mathbf{u}(t), t]$, and the cost function can be written in either integral or final-state form:

$$\begin{aligned} J &= \int_{t_0}^{t_f} f_i[\mathbf{x}(t), \mathbf{u}(t), t]\, dt \\ &= x_i(t_f) - x_i(t_0) \\ &\triangleq \phi[\mathbf{x}(t_f)] \end{aligned} \tag{3.2-5}$$

The final time can be fixed or free, but some combination of final values of other state components must be specified in the latter case, either as constraints or as terminal costs.

A *minimum-cost problem* could be formulated from the previous results. Suppose the *cost of time* (operator salaries, time-based rental costs, penalties for holding up the production line, etc.) is k_1 \$/hr and the *cost of fuel* is k_2 \$/kg. Then a suitable cost function is

$$\begin{aligned} J &= \int_{t_0}^{t_f} [k_1 + k_2 f_i(\cdot)]\, dt \\ &= \int_{t_0}^{t_f} \mathscr{L}[\mathbf{x}(t), \mathbf{u}(t), t]\, dt \end{aligned} \tag{3.2-6}$$

and both t_f and $\mathbf{u}(t)$ are varied to achieve a minimum. Depending on the relative magnitudes of k_1 and k_2, the optimizing control solution tends to the minimum-time or minimum-fuel solution (Fig. 3.2-1). In the limit ($k_1 = 0$ or $k_2 = 0$), minimum-cost control *is* minimum-fuel or minimum-time control.

A *minimum-cost mixing problem* could be posed by defining the cost function

$$J = \int_{t_0}^{t_f} \mathbf{k}^T \mathbf{f}_m(\cdot)\, dt \tag{3.2-7}$$

Here, $\mathbf{f}_m(\cdot)$ represents the mass flow dynamics of the components, and $\mathbf{k}$ contains the unit costs of the components. Once again, some purpose for the mixing must be specified, either by constraints or additional costs. For example, the mixing of a rocket engine's fuel and oxidizer to achieve a net energy conversion at minimum cost would use Eq. 3.2-7 plus a specification of the energy required.

Errors in achieving a desired final state $\mathbf{x}_D$ can be expressed as $\phi\{[\mathbf{x}(t_f)-\mathbf{x}_D], t_f\}$, with the scalar function $\phi\{\cdot\}$ chosen to achieve a minimum when $\mathbf{x}=\mathbf{x}_D$. Clearly, a linear function such as

$$\phi[\mathbf{x}(t_f)-\mathbf{x}_D]=\mathbf{k}^T[\mathbf{x}(t_f)-\mathbf{x}_D] \tag{3.2-8}$$

simply becomes more negative as $\mathbf{x}(t_f)$ becomes more negative. An absolute-value function such as

$$\phi[\mathbf{x}(t_f)-\mathbf{x}_D]=|\mathbf{k}^T[\mathbf{x}(t_f)-\mathbf{x}_D]| \tag{3.2-9}$$

establishes a minimum when $\mathbf{x}(t_f)=\mathbf{x}_D$, but it may introduce solution difficulties near the minimum (where the function slope, or gradient with respect to $\mathbf{x}$, abruptly changes sign). A weighted quadratic function like those discussed in Section 2.1

$$\phi[\mathbf{x}(t_f)-\mathbf{x}_D]=[\mathbf{x}(t_f)-\mathbf{x}_D]^T\mathbf{Q}[\mathbf{x}(t_f)-\mathbf{x}_D] \tag{3.2-10}$$

establishes a minimum at $\mathbf{x}=\mathbf{x}_D$, and it has the advantage of a smooth gradient variation near the minimum. In fact, the gradient is precisely zero at the minimum. The quadratic terminal cost function has intuitive appeal, not only because it represents the second term in a Taylor series expansion of more general functions, but because it penalizes large errors more heavily than small errors. A missile that misses its target by a small amount may still be effective, while doubling its error may reduce its effect by more than a factor of two.

Quadratic forms have similar utility in integral cost functions. Consider the control of forward speed to minimize passenger discomfort as a railroad train negotiates curving track. Lateral acceleration tends to be uncomfortable, and doubling the acceleration intensity more than doubles discomfort. Furthermore, the amount of discomfort is proportional to the length of exposure. Denoting acceleration intensity and discomfort sensitivity as $a(t)$ and k, respectively, the cost function

$$J=\int_{t_0}^{t_f} ka^2(t)\,dt \tag{3.2-11}$$

captures the "cost" of curving from the passenger's point of view. A more

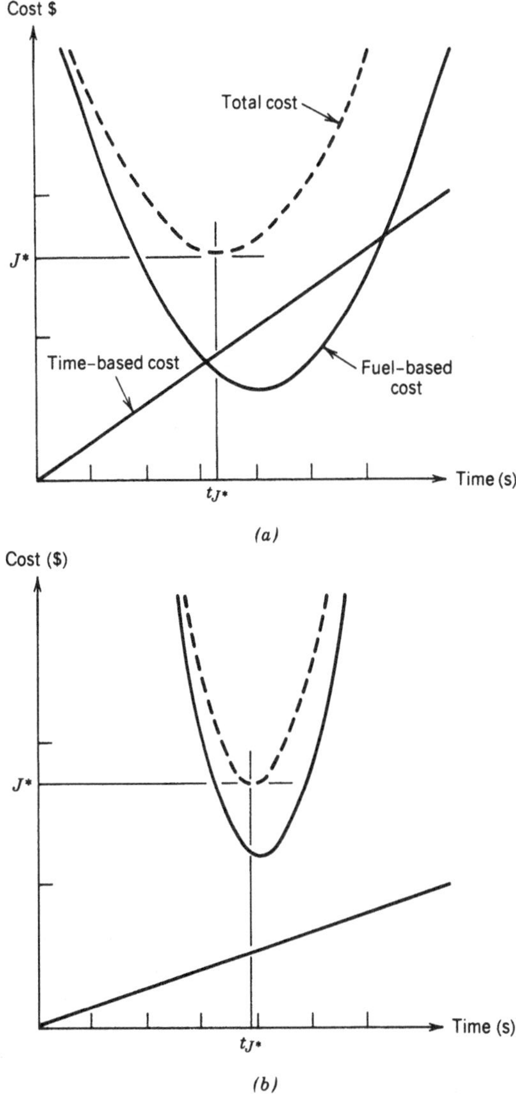

FIGURE 3.2-1 Cost minimization with trade-off between time-based and fuel-based costs. (*a*) Nominal minimization; (*b*) minimization with increased fuel cost and decreased cost of time.

general quadratic cost function

$$J = \int_{t_0}^{t_f} \left\{ [\mathbf{x}^T(t)\mathbf{u}^T(t)] \begin{bmatrix} \mathbf{Q} & \mathbf{M} \\ \mathbf{M}^T & \mathbf{R} \end{bmatrix} \begin{bmatrix} \mathbf{x}(t) \\ \mathbf{u}(t) \end{bmatrix} \right\} dt$$

$$= \int_{t_0}^{t_f} [\mathbf{x}^T(t)\mathbf{Q}\mathbf{x}(t) + 2\mathbf{x}^T(t)\mathbf{M}\mathbf{u}(t) + \mathbf{u}^T(t)\mathbf{R}\mathbf{u}(t)]\, dt \qquad (3.2\text{-}12)$$

penalizes not only state excursions but control excursions and state-control products as well. This general form will be seen to have major significance for the design of linear-optimal controllers.

Finally, it should be observed that cost functions of various types can be combined to achieve multiple objectives simultaneously. In such a case, there may be a trade-off between conflicting demands, as in conducting a cost-benefit analysis. We have seen that some optimizing objectives (e.g., minimizing time or fuel) *must* be supplemented by an additional statement of purpose. This may be expressed as a "hard" constraint on the terminal state, in which the final state must be achieved without error, or it may be specified by a "soft" constraint, such as the terminal cost function of Eq. 3.2-10. In the latter case, the terminal and integral cost functions are added, creating a cost function in the form of Eq. 3.1-1. The aggregate cost function then is minimized by one of the following optimizing methods.

3.3 PARAMETRIC OPTIMIZATION

In many problems, the *admissible control histories*, $\mathbf{u}(t)$, $t_0 \leq t \leq t_f$, are assumed to have known shapes but unknown magnitudes. Control parameters $\mathbf{k}$, which are distinct from the system parameters contained in $\mathbf{p}(t)$, can be chosen to minimize J in much the same way that ordinary minima are found (Section 2.1). There is no guarantee that a parametric-optimal control history is truly optimal or, for that matter, that it is unique in minimizing J. However, it is a feasible solution (by definition), and it may be the best that can be done for the admissible control functions.

The parametric optimization problem can be summarized as follows: The cost function is

$$J = \phi[\mathbf{x}(t_f), t_f] + \int_{t_0}^{t_f} \mathscr{L}[\mathbf{x}(t), \mathbf{u}(t), t]\, dt \qquad (3.3\text{-}1)$$

the dynamic system model during $[t_0, t_f]$ is

$$\dot{\mathbf{x}}(t) = \mathbf{f}[\mathbf{x}(t), \mathbf{u}(t), t], \qquad \mathbf{x}(t_0) = \mathbf{x}_0 \qquad (3.3\text{-}2)$$

and the admissible control is

$$\mathbf{u}(t) = \mathbf{u}(\mathbf{k}, t), \qquad t_0 \leq t \leq t_f \tag{3.3-3}$$

The specified control functions are given by $\mathbf{u}(\mathbf{k}, t)$, and the control parameters are contained in the constant vector $\mathbf{k}$, whose dimension normally would be greater than or equal to the control dimension m. The objective is to find the value of $\mathbf{k}$ that minimizes J while satisfying the dynamic constraint (Eq. 3.3-2).

Control functions that are reasonable for one application may not be reasonable for another; a number of examples are given to indicate the breadth of possibilities. If control settings must be constant during $[t_0, t_f]$, the admissible controls are described by

$$\mathbf{u}(\mathbf{k}, t) = \mathbf{k} \tag{3.3-4}$$

If they "ramp" up or down (increase or decrease linearly),

$$\mathbf{u}(\mathbf{k}, t) = \mathbf{k}_1 + \mathbf{k}_2 t \tag{3.3-5}$$

where $\mathbf{k}^T = [\mathbf{k}_1^T\, \mathbf{k}_2^T]$ and has dimension $2m$. Higher powers of t could be added, leading to the truncated power series,

$$\mathbf{u}(\mathbf{k}, t) = \mathbf{k}_1 + \mathbf{k}_2 t + \mathbf{k}_3 t^2 + \cdots \tag{3.3-6}$$

with an increased number of control parameters. Similarly, one could employ trigonometric functions, such as

$$\mathbf{u}(\mathbf{k}, t) = \mathbf{k} \sin \pi \left[\frac{t}{(t_f - t_0)}\right] \tag{3.3-7}$$

truncated Fourier series,

$$\mathbf{u}(\mathbf{k}, t) = \sum_{i=1}^{N} \left[\mathbf{k}_{1i} \sin \frac{i\pi t}{(t_f - t_0)} + \mathbf{k}_{2i} \cos \frac{i\pi t}{(t_f - t_0)}\right] \tag{3.3-8}$$

or other orthogonal functions. It should be clear that the parametric-optimal solutions obtained in each case would be very different, and no general guidelines to function selection can be given, other than to use functions that could be implemented easily in a control computer or that are "natural" for the problem at hand. As a corollary, it often is desirable to use as few control parameters as possible to minimize computations, yet increasing the number of parameters may improve the optimality of the result.

It will be seen that there is an interesting correspondence between parametric optimization and the techniques of numerical optimization presented in Section 3.6. In the latter case, the variational correction to the control history could be considered the "given" control function and the optimal iterative step size could be interpreted as a control parameter.

Having chosen the functional form of $\mathbf{u}(\mathbf{k}, t)$, the optimizing value of $\mathbf{k}$ must be determined. From Section 2.1, a necessary condition for $\mathbf{k}$ to minimize J is that

$$\frac{\partial J}{\partial \mathbf{k}} = \mathbf{0} \tag{3.3-9}$$

If, in addition,

$$\frac{\partial^2 J}{\partial \mathbf{k}^2} > \mathbf{0} \tag{3.3-10}$$

the two equations provide necessary and sufficient conditions for a *local minimum*. Finding one value of $\mathbf{k}$ that satisfies these equations does not, by itself, preclude the existence of other minimizing values of $\mathbf{k}$. The *global minimum* is defined by the locally minimizing $\mathbf{k}$ that produces the lowest value of J. The problem description may give enough additional information to determine by inspection if a local minimum is global. In simple cases, Eq. 3.3-1 could be evaluated as Eq. 3.3-2 is integrated, yielding a closed-form solution for $J(\mathbf{k})$; then $\partial J(\mathbf{k})/\partial \mathbf{k}$ would be evaluated symbolically and set equal to zero to find the minimum. For more difficult problems, numerical evaluations and the iterative methods of Section 3.6 would be used.

Example 3.3-1 Cart on a Track, Part 1

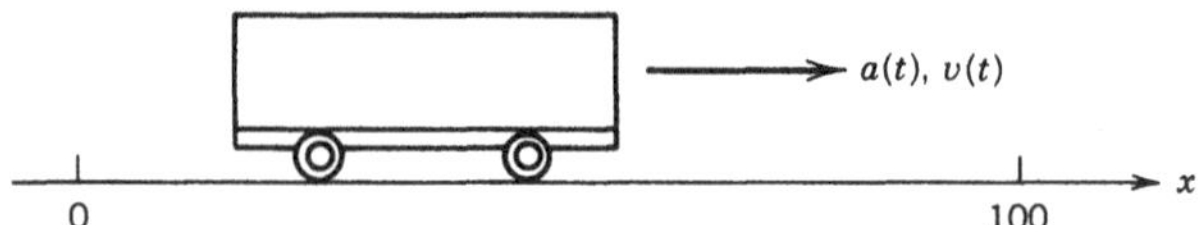

a) A cart is free to move on a track under the influence of a constant control force f. We want the cart to move 100 m in 10 s, but we also want to use as little force as possible. The initial velocity v is zero, the final velocity is free, the mass of the cart is m, and all forces other than the control are zero. What is a suitable value of f?

The state equations are

$$\dot{x} = v$$

$$\dot{v} = \frac{f}{m}$$

Letting $x = x_1$, $v = x_2$, $u = f/m$,

$$\dot{x}_1 = x_2$$

$$\dot{x}_2 = u = k$$

with initial values

$$x_{1_0} = 0$$

$$x_{2_0} = 0$$

and desired final values (at $t_f = 10$ s)

$$x_{1_D} = 100 \text{ m}$$

$$x_{2_D} = \text{free} .$$

Since the state equations can be integrated to obtain $x_1(t) = kt^2/2$ and $x_2(t) = kt$, there is only one value of k that causes $x_1(10)$ to be 100 m ($k = 2$). Our aim of using as little force as possible is contradictory, requiring $k = 0$, so there is no admissible control that satisfies both goals precisely.

Nevertheless, we can trade one objective against the other by minimizing the quadratic cost function

$$J = q(x_{1_f} - 100)^2 + r \int_0^{t_f} u^2 \, dt$$

where q and r reflect the relative importance of satisfying each objective. In this simple example,

$$J = q(50k - 100)^2 + 10rk^2$$

and

$$\frac{\partial J}{\partial k} = k(500q + 2r) - 1000q$$

Setting $\partial J/\partial k = 0$, the parametric-optimal control is

$$k = 1000q/(500q + 2r) = 1000(q/r)/[500(q/r) + 2]$$

In the limit, as $q/r \rightarrow \infty$, the final-position objective dominates, and $k = 2$; conversely, as $q/r \rightarrow 0$, $k \rightarrow 0$. The following chart provides results for three choices of cost function weighting factors:

Cost function	q	100	1	1
weighting	r	1	1	100
Parametric-optimal control	k	1.999	1.992	1.429
Final state	x_{1_f}	99.996	99.602	71.429
components	x_{2_f}	19.999	19.920	14.286
Control penalty	$\int u^2\,dt$	39.997	39.682	20.408
Cost	J	39.999	39.840	2857.1

As might be expected, the actual final positions are less than the desired values in all cases, and reduced control penalty can be obtained at the expense of increased final position error. Also note that increasing q/r beyond one has negligible effect on the optimizing control and the resulting performance, indicating a "knee" in the curve that relates cost function weighting and performance. This is a typical characteristic of cost functions that establish trade-offs: individual components can dominate the optimal solution (asymptotically) as their relative weightings become large with respect to the other weightings.

b) Assuming that the admissible control is

$$u = k_1 + k_2 t$$

what are the minimizing values of k_1 and k_2?

In this case,

$$\dot{x}_1 = x_2$$

$$\dot{x}_2 = k_1 + k_2 t$$

and

$$x_1(t) = \frac{k_1 t^2}{2} + \frac{k_2 t^3}{6}$$

$$x_2(t) = k_1 t + \frac{k_2 t^2}{2}$$

Here an infinity of control parameters could cause the position at $t = 10$ s to be 100 m as long as

$$100 = 50 k_1 + 166.7 k_2$$

or

$$k_1 = \frac{100 - 166.7k_2}{50}$$

This equation is called *Constraint A* in the remainder of the problem. Then

$$J = q[(50k_1 + 166.7k_2) - 100]^2 + r(10k_1^2 + 100k_1k_2 + 333.3k_2^2)$$

and the gradient is expressed by the two equations

$$\frac{\partial J}{\partial k_1} = k_1(500q + 2r) + k_2(1666.7q + 10r) - 1000q$$

$$\frac{\partial J}{\partial k_2} = k_1(1666.7q + 10r) + k_2(5555.6q + 66.7r) - 3333.3q$$

The minimizing control is defined by $\partial J/\partial \mathbf{k} = \mathbf{0}$, which requires that $\partial J/\partial k_1$ and $\partial J/\partial k_2$ simultaneously equal zero. This leads to the equation

$$\begin{bmatrix} (500q + 2r) & (1666.7q + 10r) \\ (1666.7q + 10r) & (5555.6q + 66.7r) \end{bmatrix} \begin{bmatrix} k_1 \\ k_2 \end{bmatrix} = \begin{bmatrix} 1000 \\ 3333.3 \end{bmatrix} q$$

which takes the form

$$\mathbf{Ak} = \mathbf{B}q$$

and has the solution

$$\mathbf{k} = \mathbf{A}^{-1}\mathbf{B}q$$

Expressing the solution symbolically is arduous, but calculating numerical values of optimal-control settings and performance measures for example values of q and r is relatively easy:*

Cost function	q	100	1	1
weighting	r	1	1	100
Parametric-optimal	k_1	3.000	2.991	2.308
control	k_2	−0.300	−0.299	−0.231
Final state	x_{1_f}	99.997	99.701	76.923
components	x_{2_f}	15.000	14.955	11.538
Control penalty	$\int u^2\, dt$	29.998	29.821	17.751
Cost	J	32.794	29.923	2307.7

*The gradient equations are expressed to one decimal place, but the calculations used 10 significant digits.

Comparing these results to Part a, it is clear that the increased flexibility afforded by two control parameters provides a qualitatively better solution, in the sense that low terminal error can be achieved with reduced control penalty. Even though the final position could have been achieved precisely, the parametric optimization of the cost function allows a small error in final position in order to achieve a smaller control penalty. How might the position error be forced to zero? In at least two conceptually different ways: either by choosing q much larger than r (infinitely larger in the limit) or by forcing k_1 and k_2 to be related by Constraint A. Then, choosing one parameter defines the other, so the optimization need generate only a single control parameter.

c) For the previous conditions, we wish to bring the final velocity close to zero as well as to minimize the final position error and the integrated control penalty. What are the minimizing values of k_1 and k_2?

The state equations are unchanged, and, for $t_f = 10$ s, zero final velocity could be obtained by setting

$$0 = 10k_1 + 50k_2$$

or

$$k_1 = -5k_2$$

This requires the acceleration (i.e., the control) to go to zero at the midpoint of the trajectory, and the final acceleration is equal and opposite to the initial acceleration. Put another way, a control that is antisymmetric about the midpoint (i.e., an odd function) results in zero final velocity. We shall call this equation *Constraint B*. Together with Constraint A, k_1 and k_2 could be determined uniquely, without regard to the control penalty. However, if we choose to trade final state errors against control usage, a cost function of the following form is appropriate:

$$J = q_1(x_{1_f} - x_{1_D})^2 + q_2(x_{2_f})^2 + r\int_{t_0}^{t_f} u^2\,dt$$

$$= q_1\left[\left(\frac{k_1 t_f^2}{2} + \frac{k_2 t_f^3}{6}\right) - x_{1_D}\right]^2$$

$$+ q_2\left(k_1 t_f + \frac{k_2 t_f^2}{2}\right)^2 + r\left[k_1^2 t_f + k_1 k_2 t_f^2 + \frac{k_2^2 t_f^3}{3}\right]$$

Setting the gradient equal to zero provides the following control parameter equations:

$$\begin{bmatrix} k_1 \\ k_2 \end{bmatrix} = \begin{bmatrix} [q_1 t_f^3/2 + 2q_2 t_f + 2r] & [q_1 t_f^4/6 + q_2 t_f^2 + r t_f] \\ [q_1 t_f^3/6 + q_2 t_f + r] & [q_1 t_f^4/18 + q_2 t_f^2/2 + 2r t_f/3] \end{bmatrix}^{-1} \begin{bmatrix} 1 \\ \frac{1}{3} \end{bmatrix} 100 q_1 t_f$$

Numerical examples shed light on the relative importance of cost function weighting:

Cost function	q_1	100	1	1	1
weighting	q_2	1	100	1	1
	r	1	1	1	100
Parametric-optimal	k_1	5.142	5.917	5.094	2.325
control	k_2	−0.943	−1.182	−0.934	−0.244
Final state	x_{1_f}	99.991	99.818	99.033	75.646
components	x_{2_f}	4.285	0.059	4.240	11.070
Control penalty	$\int u^2\,dt$	75.903	116.480	74.494	17.197
cost	J	95.211	118.271	93.407	733.15

Once again, the effect of relative cost function weighting is apparent. For example, heavy x_{1_f} error weighting causes k_1 and k_2 to approximate Constraint A, while heavy x_{2_f} weighting causes them to approximate Constraint B. Heavy control penalty produces nearly the same control parameters in both parts b and c. It is emphasized that the *relative* weights define the optimal control. Choosing (q_1, q_2, r) to be (1000, 1000, 1000) would provide the same control parameters as (1, 1, 1). (Verify this by substitution in the control parameter equation.)

d) Suppose that the admissible control is

$$u = k_1 \cos \omega_1 t + k_2 \sin \omega_2 t$$

where $\omega_1 = \pi/10$ and $\omega_2 = 2\omega_1$. What are the optimizing values of k_1 and k_2?

The control function is odd (about the midpoint) in the time interval [0, 10]; hence the final velocity is constrained to be zero for all values of k_1 and k_2. The time history of velocity is described by

$$x_2(t) = \frac{k_1}{\omega_1} \sin \omega_1 t + \frac{k_2}{\omega_2}(1 - \cos \omega_2 t)$$

while the position is

$$x_1(t) = k_1\left[\frac{1}{\omega_1^2}(1 - \cos \omega_1 t)\right] + k_2\left[\frac{t}{\omega_2} - \frac{1}{\omega_2^2} \sin \omega_2 t\right]$$

At the final time,

$$x_1(10) = k_1 \frac{200}{\pi^2} + k_2 \frac{50}{\pi}$$

$$x_2(10) = 0$$

Using the cost function of Part c,

$$J = q_1\left(k_1\frac{200}{\pi^2} + k_2\frac{50}{\pi} - 100\right)^2 + q_2(0)^2 + r\int_0^{t_f}(k_1\cos\omega_1 t + k_2\sin\omega_2 t)^2\,dt$$

As in the previous parts, setting $\partial J/\partial \mathbf{k} = \mathbf{0}$ yields an equation for the parametric-optimal control parameters:

$$\begin{bmatrix} k_1 \\ k_2 \end{bmatrix} = \begin{bmatrix} [(80000/\pi^4)q_1 + 10r] & [(20000/\pi^3)q_1 + (80/3\pi)r] \\ [(20000/\pi^3)q_1 + (80/3\pi)r] & [(5000/\pi^2)q_1 + 10r] \end{bmatrix}^{-1} \cdot \begin{bmatrix} 40000/\pi^2 \\ 10000/\pi \end{bmatrix} q_1$$

The solution is independent of q_2, and it provides the following numerical examples:

Cost function	q_1	100	1	1
weighting	r	1	1	100
Parametric-optimal	k_1	3.462	3.426	1.688
control	k_2	1.874	1.855	0.914
Final state	x_{1_f}	99.989	99.960	48.766
components	x_{2_f}	0	0	0
Control penalty	$\int u^2\,dt$	105.041	102.889	24.985
cost	J	105.053	102.891	5123.4

The more complex control function does not necessarily improve the optimality of the solution. In fact, Section 3.4 shows that the simpler control function assumed for Parts b and c provides the exact optimal solution.

The preceding examples make the point that parametric optimization produces only relative optimality in the control of dynamic systems, for the "goodness" of the result is dependent on the choice of admissible control functions. In many problems, a good, feasible solution may be all that is needed, and parametric-optimal control can come arbitrarily close to the exact optimal control as the number of parameters is increased (in the same way that power series and Fourier series converge to an arbitrary nonlinear function). Because parametric-optimal controls are less than optimum, the possibility of *multiple solutions* (i.e., more than one locally optimal control history) exists. The exact optimal control is not guaranteed, so it is necessary to develop methods that are not constrained by prior choice of the control function.

The preceding discussion pertains to optimization of the parameters of control *time* functions; given a feedback control structure in which the control is some function of the dynamic state, such as

$$\mathbf{u}(t) = \boldsymbol{\psi}[\mathbf{k}, \mathbf{x}(t), t] \tag{3.3-11}$$

the parameters of the control *law*, $\boldsymbol{\psi}[\cdot]$ could be chosen to minimize the cost. The previous problem is called *open-loop parametric optimization*, while this is called *closed-loop parametric optimization*, for obvious reasons.

In Section 3.7, it will be seen that optimal closed-loop control policies require *full state feedback* to all controls. Even the simplest constant-coefficient linear-optimal control law (Section 5.4)

$$\mathbf{u}(t) = -\mathbf{C}\mathbf{x}(t) \tag{3.3-12}$$

has an $(m \times n)$ gain matrix $\mathbf{C}$ to weight the information flow from the n state components to the m control components. Many of these gains may be inconsequential in practice, and additional knowledge about the system (inherent stability, weak coupling between its elements, large differences in speeds of response, etc.) may suggest a control structure with fewer than mn scalar gains. At best, the performance of a *reduced-order parametric optimal feedback controller* approaches unconstrained optimal performance as a limit. Multiple optimizing solutions can be obtained, and numerous aspects of optimality, stability, and robustness remain issues for further research.

3.4 CONDITIONS FOR OPTIMALITY

Optimal trajectories of dynamic systems are defined by equations that establish necessary and sufficient conditions for minimum cost. Assuming that there are no limits on the control, *necessary conditions* must be satisfied for the cost function to be *stationary* with respect to the control (and with respect to the final time, if that is free to vary). Taken alone, however, they do not guarantee optimality. Having satisfied necessary conditions, *sufficient conditions* guarantee optimality, but particular sufficient conditions may not be satisfied on trajectories that are known to be optimal by satisfying other conditions.

Given an optimal control history, small perturbations in the control should have vanishingly small effect on the cost—more precisely, necessary conditions usually indicate that the cost function's linear sensitivity to control variations about the optimum is zero. A trajectory defined by the necessary conditions alone is called an *extremal*, because it remains to be determined if the trajectory minimizes or maximizes the cost function.

If control perturbations of either sign occurring at any time can only increase the cost, the stationary solution is minimizing. An optimal trajectory's higher-order sensitivity to control variations normally is not zero;

having attained cost stationarity, optimality is determined by evaluating this higher-order sensitivity. With adequate prior knowledge about the system and the cost function, assuring cost stationarity may guarantee optimality. As an example, if a linear dynamic system with a positive–definite quadratic cost function has a stationary control solution, then sufficient conditions for a minimum are satisfied.

Other criteria for optimality can be applied. If the control magnitude is bounded (or constrained), optimality within the constraints is defined by the variational necessary and sufficient conditions described above; however, the minimizing control also could be constrained to the boundary at points that violate these conditions. The Minimum Principle provides a necessary and sufficient condition in such instance. Dynamic Programming takes an entirely different approach, using a partial differential equation that is sufficient to establish optimality. The equation is not a necessary condition for optimality, as control histories that do not satisfy the equation may, nevertheless, be optimizing.

Necessary Conditions for Optimality

We seek a set of equations for the control history that causes the cost function (Eq. 3.1-1b) to be stationary with respect to an unconstrained control vector, $\mathbf{u}(t)$, subject to the dynamic constraint imposed by Eq. 3.1-2b and the initial state (Eq. 3.1-3). The necessary conditions for optimality are derived using the Calculus of Variations, the branch of mathematics which deals with extreme values of functions of continuous variables. Unlike the previous section, the admissible controls include any finite, piecewise-continuous time histories of proper dimension. To assure that the dynamic constraint is considered in the minimization, it is added (or *adjoined*) to the cost function in such a way that the numerical value of the cost function is unchanged at its optimum; nevertheless, the *augmented* cost function's sensitivity to state and control variations reflects admissible interplay within the dynamic system. Recall that in static minimization (Section 2.1), an equality constraint of the form

$$\mathbf{f}(\mathbf{x}, \mathbf{u}) = \mathbf{0} \tag{3.4-1}$$

can be adjoined to the cost function J using Lagrange multipliers. These multipliers are contained in the constant adjoint vector $\boldsymbol{\lambda}$. Then the *augmented cost function* J_A

$$J_A = J + \boldsymbol{\lambda}^T \mathbf{f}(\mathbf{x}, \mathbf{u}) \tag{3.4-2}$$

is numerically identical to J at the minimum [because $\mathbf{f}(\mathbf{x}, \mathbf{u})$ must equal zero], but its first variation is altered by the state and control dependencies of the equality constraint (Eqs. 2.1-55 and 2.1-56). Equations 2.1-57 to

2.1-59 are solved to find the constant values of $\mathbf{x}$, $\mathbf{u}$, and $\boldsymbol{\lambda}$ that minimize J_A, and, therefore, J.

The dynamic minimization proceeds in similar fashion; however, the equality constraint is derived from the system's differential equation,

$$\dot{\mathbf{x}}(t) = \mathbf{f}[\mathbf{x}(t), \mathbf{u}(t), t], \qquad \mathbf{x}(t_0) = \mathbf{x}_0 \tag{3.4-3a}$$

The equivalent constraint,

$$\mathbf{f}[\mathbf{x}(t), \mathbf{u}(t), t] - \dot{\mathbf{x}}(t) = \mathbf{0} \tag{3.4-3b}$$

is time-varying, as are the optimal values of $\mathbf{x}(t)$, $\mathbf{u}(t)$, and $\boldsymbol{\lambda}(t)$. Furthermore, the equality constraint now depends on $\dot{\mathbf{x}}(t)$ as well. Because Eq. 3.4-3 is to be satisfied over the entire time interval $[t_0, t_f]$, it is adjoined to the *integrand* of J,

$$J_A = \phi[\mathbf{x}(t_f), t_f] + \int_{t_0}^{t_f} [\mathscr{L}[\mathbf{x}(t), \mathbf{u}(t), t] + \boldsymbol{\lambda}^T(t)\{\mathbf{f}[\mathbf{x}(t), \mathbf{u}(t), t] - \dot{\mathbf{x}}(t)\}] \, dt \tag{3.4-4}$$

and the integrand itself must satisfy necessary conditions throughout the interval to provide stationarity of J_A. (From this point, the augmented cost function will be written without the subscript A, as it is numerically identical to the original cost function when the dynamic equations are satisfied.)

The necessary conditions are most readily found if the integrand of Eq. 3.4-4 is recast in terms of the *Hamiltonian** $\mathscr{H}$ where

$$\mathscr{H}[\mathbf{x}(t), \mathbf{u}(t), \boldsymbol{\lambda}(t), t] = \mathscr{L}[\mathbf{x}(t), \mathbf{u}(t), t] + \boldsymbol{\lambda}^T(t)\mathbf{f}[\mathbf{x}(t), \mathbf{u}(t), t] \tag{3.4-5}$$

The cost function then is written as

$$J = \phi[\mathbf{x}(t_f), t_f] + \int_{t_0}^{t_f} \{\mathscr{H}[\mathbf{x}(t), \mathbf{u}(t), \boldsymbol{\lambda}(t), t] - \boldsymbol{\lambda}^T(t)\dot{\mathbf{x}}(t)\} \, dt \tag{3.4-6a}$$

or

$$J = \phi[\cdot] + \int_{t_0}^{t_f} \mathscr{H}[\cdot] \, dt - \int_{t_0}^{t_f} \boldsymbol{\lambda}^T(t)\dot{\mathbf{x}}(t) \, dt \tag{3.4-6b}$$

*Sir William Hamilton (1805–1865) derived the "principle of varying action," first applied to optics, then to problems of dynamics. In classical mechanics, the Hamiltonian represents the sum of the potential and kinetic energies of a dynamic system.

Using integration by parts,

$$\int u\,dv = uv - \int v\,du \tag{3.4-7}$$

the second integral in Eq. 3.4-6b is

$$\int_{t_0}^{t_f} \boldsymbol{\lambda}^T(t)\dot{\mathbf{x}}(t)\,dt = \boldsymbol{\lambda}^T(t)\mathbf{x}(t)\Big|_{t_0}^{t_f} - \int_{t_0}^{t_f} \dot{\boldsymbol{\lambda}}^T(t)\mathbf{x}(t)\,dt \tag{3.4-8}$$

so the cost function becomes

$$J = \phi[\mathbf{x}(t_f), t] + [\boldsymbol{\lambda}^T(t_0)\mathbf{x}(t_0) - \boldsymbol{\lambda}^T(t_f)\mathbf{x}(t_f)] + \int_{t_0}^{t_f} \{\mathscr{H}[\mathbf{x}(t), \mathbf{u}(t), \boldsymbol{\lambda}(t), t] + \dot{\boldsymbol{\lambda}}^T(t)\mathbf{x}(t)\}\,dt \tag{3.4-6c}$$

For stationarity, we require that the first-order effect of control variations on the cost function be zero throughout the time interval. Because control perturbations generally lead to state perturbations at or beyond the time of application, *first variations* of the cost function's components can be expressed as

$$\Delta(\cdot) = \frac{\partial(\cdot)}{\partial \mathbf{u}}\Delta\mathbf{u} + \frac{\partial(\cdot)}{\partial \mathbf{x}}\Delta\mathbf{x}(\Delta\mathbf{u}) \tag{3.4-9}$$

where $\Delta\mathbf{x}(\Delta\mathbf{u})$ denotes state perturbations arising from control perturbations at or before the time of interest. [$\Delta\mathbf{x}(t)$ depends on $\Delta\mathbf{u}$ in (t_0, t), and it is called a *functional*, i.e., a function of the function, $\Delta\mathbf{u}$]. Assuming that the final time is fixed, the first variation of J due to control is

$$\begin{aligned}\Delta J &= \left\{\left[\frac{\partial \phi}{\partial \mathbf{x}} - \boldsymbol{\lambda}^T\right]\Delta\mathbf{x}(\Delta\mathbf{u})\right\}\Bigg|_{t=t_f} + [\boldsymbol{\lambda}^T\Delta\mathbf{x}(\Delta\mathbf{u})]\Big|_{t=t_0} \\ &\quad + \int_{t_0}^{t_f} \left\{\frac{\partial \mathscr{H}}{\partial \mathbf{u}}\Delta\mathbf{u} + \left[\frac{\partial \mathscr{H}}{\partial \mathbf{x}} + \dot{\boldsymbol{\lambda}}^T\right]\Delta\mathbf{x}(\Delta\mathbf{u})\right\}dt \\ &\triangleq \Delta J(t_f) + \Delta J(t_0) + \Delta J(t_0, t_f)\end{aligned} \tag{3.4-10}$$

where $\Delta\mathbf{u}(t)$, $t_0 \le t \le t_f$, is an arbitrary (presumably small) function. Unless there are implied constraints not expressed in Eq. 3.4-10, the three parts of ΔJ must equal zero separately in the vicinity of the optimal trajectory. It can be assumed that the initial control has no effect on the initial state conditions, so $\Delta J(t_0) = 0$. We wish to choose the adjoint vector time history such that ΔJ is insensitive to arbitrary (nonzero) values of $\Delta\mathbf{u}$ and $\Delta\mathbf{x}(\Delta\mathbf{u})$ in the interval; consequently $\boldsymbol{\lambda}(t)$ should satisfy the differential equation

$$\dot{\boldsymbol{\lambda}}^T(t) = \frac{-\partial \mathscr{H}[\mathbf{x}(t), \mathbf{u}(t), \boldsymbol{\lambda}(t), t]}{\partial \mathbf{x}} \qquad (3.4\text{-}11a)$$

or

$$\dot{\boldsymbol{\lambda}}(t) = \left\{\frac{-\partial \mathscr{H}[\cdot]}{\partial \mathbf{x}}\right\}^T, \qquad t_0 \leq t \leq t_f \qquad (3.4\text{-}11b)$$

subject to the terminal conditions

$$\boldsymbol{\lambda}^T(t_f) = \left.\frac{\partial \phi[\mathbf{x}(t), t]}{\partial \mathbf{x}}\right|_{t=t_f} \qquad (3.4\text{-}12a)$$

or

$$\boldsymbol{\lambda}(t_f) = \left\{\frac{\partial \phi[\cdot]}{\partial \mathbf{x}}\right\}^T \qquad (3.4\text{-}12b)$$

$\Delta J(t_f)$ is zero by Eq. 3.4–12; for $\Delta J(t_0, t_f)$ to be zero, Eq. 3.4-11 and the following must be satisfied:

$$\frac{\partial \mathscr{H}[\mathbf{x}(t), \mathbf{u}(t), \boldsymbol{\lambda}(t), t]}{\partial \mathbf{u}} = \mathbf{0} \qquad (3.4\text{-}13a)$$

This optimality condition also can be expressed as the column vector

$$\left\{\frac{\partial \mathscr{H}[\cdot]}{\partial \mathbf{u}}\right\}^T = \mathbf{0} \qquad (3.4\text{-}13b)$$

Equations 3.4-11 to 3.4-13 are the *Euler–Lagrange equations*,* the three necessary conditions for optimality when the final time is fixed. Because these conditions must apply along a specific trajectory, they are *local* conditions. They indicate that the Hamiltonian is stationary with respect to infinitesimal control variations, but they do not preclude the existence of other optimizing paths; therefore, they are not *global* criteria.

Optimization of the integral cost function subject to a dynamic constraint is a *two-point boundary value problem*, as constants of integration for the state (Eq. 3.4-3a) and its adjoint (Eq. 3.4-12) are specified at opposite ends

*Some sources include Eqs. 3.4-11 and 3.4-13 among the "Weierstrass necessary conditions for an extremal." The third Weierstrass condition is $\dot{\mathbf{x}} = (\partial \mathscr{H}/\partial \boldsymbol{\lambda})^T$. Does this follow from our definition of the Hamiltonian?

(or boundaries) of the time interval. An optimal trajectory has been achieved when Eqs. 3.4-3 and 3.4-11 to 3.4-13 have been satisfied simultaneously (with these boundary conditions) throughout the interval $[t_0, t_f]$. Integration of the state equation forward in time from t_0 to t_f and the adjoint equation back from t_f to t_0 is complicated, because $\mathbf{u}(t)$ (a function of $\mathbf{x}$ and $\boldsymbol{\lambda}$) appears in the state equation and $\mathbf{x}(t)$ appears in the adjoint equation.

It is helpful to relate the Euler–Lagrange equations to the integrand of the original cost function and the dynamic equation. Because $\boldsymbol{\lambda}$ is not a direct function of either $\mathbf{x}$ or $\mathbf{u}$, the gradients of the Hamiltonian can be expressed as follows:

$$\frac{\partial \mathscr{H}}{\partial \mathbf{x}} = \frac{\partial \mathscr{L}}{\partial \mathbf{x}} + \boldsymbol{\lambda}^T \frac{\partial \mathbf{f}}{\partial \mathbf{x}} \tag{3.4-14a}$$

or

$$\left(\frac{\partial \mathscr{H}}{\partial \mathbf{x}}\right)^T = \left(\frac{\partial \mathscr{L}}{\partial \mathbf{x}}\right)^T + \mathbf{F}^T \boldsymbol{\lambda} \tag{3.4-14b}$$

and

$$\frac{\partial \mathscr{H}}{\partial \mathbf{u}} = \frac{\partial \mathscr{L}}{\partial \mathbf{u}} + \boldsymbol{\lambda}^T \frac{\partial \mathbf{f}}{\partial \mathbf{u}} \tag{3.4-15a}$$

or

$$\left(\frac{\partial \mathscr{H}}{\partial \mathbf{u}}\right)^T = \left(\frac{\partial \mathscr{L}}{\partial \mathbf{u}}\right)^T + \mathbf{G}^T \boldsymbol{\lambda} \tag{3.4-15b}$$

where the Jacobian matrices, $\partial \mathbf{f}/\partial \mathbf{x}$ and $\partial \mathbf{f}/\partial \mathbf{u}$, are symbolized by $\mathbf{F}$ and $\mathbf{G}$, as in Section 2.3. Thus the adjoint vector is obtained by integrating the *linear, time-varying differential equation*

$$\dot{\boldsymbol{\lambda}}(t) = -\mathbf{F}^T(t)\boldsymbol{\lambda}(t) - \left[\frac{\partial \mathscr{L}(t)}{\partial \mathbf{x}}\right]^T \tag{3.4-16}$$

back in time from the terminal condition

$$\boldsymbol{\lambda}(t_f) = \frac{\partial \phi[\mathbf{x}(t), t]}{\partial \mathbf{x}}^T \bigg|_{t=t_f} \tag{3.4-17}$$

Having defined $\boldsymbol{\lambda}(t)$ for $t_0 \leq t \leq t_f$, the optimal control history is computed by solving the equation

$$\left[\frac{\partial \mathscr{L}(t)}{\partial \mathbf{u}}\right]^T + \mathbf{G}^T(t)\boldsymbol{\lambda}(t) = \mathbf{0} \tag{3.4-18}$$

for $\mathbf{u}(t)$.

Let's reflect for a moment on the role played by the adjoint vector in the optimization. $\boldsymbol{\lambda}(t)$ adjoins the dynamic constraint to the cost function integrand, so it represents the cost function's *sensitivity to dynamic effects*. This sensitivity is specified by $\partial\phi/\partial\mathbf{x}$ at the end time (Eq. 3.4-17), providing a boundary condition for the solution of $\boldsymbol{\lambda}(t)$. Hence $\boldsymbol{\lambda}(t_f)$ scales the effects of final state variations on the cost function. Once the Euler–Lagrange equations have been satisfied, Eq. 3.4-10 indicates that first variations in the cost function at some arbitrary time t are given by

$$\Delta J(t) = -\boldsymbol{\lambda}^T(t)\Delta\mathbf{x}(t) \tag{3.4-19}$$

It is clear that $\boldsymbol{\lambda}(t)$ represents the cost sensitivity to state perturbations on the optimal trajectory at time t; it is an *influence function* that can be written as the gradient (row vector),

$$\frac{\partial J(t)}{\partial \mathbf{x}} = -\boldsymbol{\lambda}^T(t) \tag{3.4-20}$$

During the interval $[t_0, t_f]$, the adjoint vector's influence is felt in two ways. In Eq. 3.4-16, $\boldsymbol{\lambda}(t)$ "feeds back" on itself, helping govern its own value as a function of $\mathbf{F}^T$ in a differential equation that is forced by $(\partial\mathscr{L}/\partial\mathbf{x})^T$. Once its value is known throughout the interval, $\boldsymbol{\lambda}(t)$ is proportional to the amount of control required to bring $\mathscr{H}_\mathbf{u}$ to zero (Eq. 3.4-18). Nevertheless, because the boundary condition on $\boldsymbol{\lambda}(t)$ is specified at t_f, the instantaneous optimal control (at time t) requires prior computation of $\boldsymbol{\lambda}(t)$ in the *future* interval $[t, t_f]$.

The optimization problem is similar to progressing through a maze; in order to decide which way to turn, it helps to know the effects of future turns on the resulting path. In optimization, the adjoint vector provides this information. If the entire maze can be observed, as on a video game, the optimal strategy is most easily deduced by tracing paths back from the destination. With no view of the future, a decidedly suboptimal strategy results. This concept of optimizing by working back from the final state is implicit in the computation of the adjoint vector, and it forms the basis for the principle of Dynamic Programming, which is discussed at the end of this section.

For simplicity, the various preceding partial derivatives (i.e., the gradients and Jacobian matrices) have been written without regard to the nominal values of $\mathbf{x}$ and $\mathbf{u}$ at which they are evaluated. As discussed in Section 2.3, a partial derivative such as $\partial\mathbf{f}(\mathbf{x}, \mathbf{u})/\partial\mathbf{x}$ can only be evaluated for specified (nominal) values, $\mathbf{x} = \mathbf{x}_o$ and $\mathbf{u} = \mathbf{u}_o$. In keeping with our concern for the behavior of first variations about the optimal trajectory, $\mathbf{x}_o(t)$ and $\mathbf{u}_o(t)$ must represent the *optimal* state and control histories, $\mathbf{x}^*(t)$ and $\mathbf{u}^*(t)$, which are the quantities to be found through optimization. In the most general nonlinear case, the only hope for finding $\mathbf{x}^*(t)$ and $\mathbf{u}^*(t)$ is to use a method of successive approximations, in which a reasonable guess at the final answer is postulated and an iterative numerical algorithm converges (hopefully) to the correct result (as in Section 3.6). At the other end of the scale, we shall see that the linear dynamic system with quadratic cost function does not have this problem because all of the needed partial derivatives are independent of the state and control perturbations (Section 3.7 and Chapter 6).

EXAMPLE 3.4-1 CART ON A TRACK, PART 2

a) The problem of Example 3.3-1b is reconsidered, but this time there are no restrictions on the control's functional form and the Euler–Lagrange equations are employed to define the stationary (minimizing) solution. Once again, the objective is to move the cart 100 m in 10 s, using as little control as possible. The state vector and state equation are

$$\mathbf{x}(t) = \begin{bmatrix} x_1(t) \\ x_2(t) \end{bmatrix}$$

$$\dot{\mathbf{x}}(t) = \begin{bmatrix} \dot{x}_1(t) \\ \dot{x}_2(t) \end{bmatrix} = \mathbf{f} = \begin{bmatrix} x_2(t) \\ u(t) \end{bmatrix}$$

while the scalar control, $u(t)$, $0 \le t \le t_f$, which minimizes the following cost function is to be found:

$$J = \phi(\mathbf{x}_f) + \int_{t_0}^{t_f} \mathcal{L}(t)\, dt$$

with

$$\phi = q(x_{1_f} - 100)^2, \qquad \mathcal{L}(t) = ru^2(t)$$

The linear sensitivity matrices derived from the dynamic vector $\mathbf{f}$ are

$$\mathbf{F} = \begin{bmatrix} 0 & 1 \\ 0 & 0 \end{bmatrix}, \qquad \mathbf{G} = \begin{bmatrix} 0 \\ 1 \end{bmatrix}$$

Because $\mathbf{f}$ is not an explicit function of time and $\mathbf{F}$ and $\mathbf{G}$ are constant, it is apparent that the original dynamic equation is linear and time-invariant. The final value of the adjoint vector is

$$\boldsymbol{\lambda}(t_f) = \left(\frac{\partial \phi}{\partial \mathbf{x}}\right)^T \bigg|_{t=t_f} = \begin{bmatrix} 2q(x_{1_f} - 100) \\ 0 \end{bmatrix} = \begin{bmatrix} \lambda_1(t_f) \\ \lambda_2(t_f) \end{bmatrix}$$

while the differential equation that describes the adjoint vector's time history is

$$\begin{aligned} \dot{\boldsymbol{\lambda}}(t) &= -\mathbf{F}^T \boldsymbol{\lambda}(t) - \left[\frac{\partial \mathscr{L}(t)}{\partial \mathbf{x}}\right]^T \\ &= -\begin{bmatrix} 0 & 0 \\ 1 & 0 \end{bmatrix} \begin{bmatrix} \lambda_1(t) \\ \lambda_2(t) \end{bmatrix} - \begin{bmatrix} 0 \\ 0 \end{bmatrix} \end{aligned}$$

or

$$\begin{bmatrix} \dot{\lambda}_1(t) \\ \dot{\lambda}_2(t) \end{bmatrix} = -\begin{bmatrix} 0 \\ \lambda_1(t) \end{bmatrix}$$

This equation can be integrated from t_f to t_0, yielding

$$\begin{aligned} \lambda_1(t) &= \lambda_1(t_f) + \int_{t_f}^{t} (0)\, dt \\ &= \lambda_1(t_f) = 2q(x_{1_f} - 100) \end{aligned}$$

$$\begin{aligned} \lambda_2(t) &= \lambda_2(t_f) - \int_{t_f}^{t} \lambda_1(t)\, dt \\ &= (0) + \lambda_1(t_f)(t_f - t) \\ &= 2q(x_{1_f} - 100)(t_f - t) \end{aligned}$$

The control history is defined by solving the equation

$$\frac{\partial \mathscr{L}(t)}{\partial u} + \mathbf{G}^T \boldsymbol{\lambda}(t) = 0$$

which leads to

$$2ru(t) + [0 \quad 1] \begin{bmatrix} 2q(x_{1_f} - 100) \\ 2q(x_{1_f} - 100)(t_f - t) \end{bmatrix} = 0$$

or

$$u(t) = \frac{q}{r}(x_{1_f} - 100)(t - t_f)$$

The control could be expressed as

$$u(t) = k_1 + k_2 t$$

where

$$k_1 = -\frac{q}{r}(x_{1_f} - 100)t_f$$

$$k_2 = \frac{q}{r}(x_{1_f} - 100)$$

Consequently, the exact optimal solution takes the same form as the parametric-optimal control function considered in Example 3.3-1b. From this example, we know that the final position can be written as

$$x_{1_f} = \frac{k_1 t_f^2}{2} + \frac{k_2 t_f^3}{6}$$

With $t_f = 10$ s, we substitute for k_1 and k_2 to find the optimal final position:

$$x_{1_f} = \frac{100}{1 + (r/q)(3/1000)}$$

The values of x_{1_f}, k_1, and k_2 derived from this result are identical to those of Example 3.3-1b.

b) As in Example 3.3-1c, we repeat the process with a penalty on final velocity x_{2_f}. Then

$$\phi = q_1(x_{1_f} - 100)^2 + q_2(x_{2_f})^2$$

and

$$\boldsymbol{\lambda}(t_f) = \left(\frac{\partial \phi}{\partial \mathbf{x}}\right)^T = \begin{bmatrix} 2q_1(x_{1_f} - 100) \\ 2q_2 x_{2_f} \end{bmatrix}$$

The differential equation for $\boldsymbol{\lambda}(t)$ is unchanged, but the final condition $\boldsymbol{\lambda}(t_f)$

is modified by the velocity penalty. Then

$$\lambda_1(t) = 2q_1(x_{1_f} - 100)$$
$$\lambda_2(t) = 2[q_2 x_{2_f} + q_1(x_{1_f} - 100)(t_f - t)]$$

and setting $\partial\mathcal{H}/\partial u = 0$ leads to

$$\begin{aligned} u(t) &= \frac{-\lambda_2(t)}{2r} \\ &= \frac{-[q_2 x_{2_f} + q_1(x_{1_f} - 100)(t_f - t)]}{r} \\ &= k_1 + k_2 t \end{aligned}$$

where

$$k_1 = -\frac{[q_2 x_{2_f} + q_1(x_{1_f} - 100)t_f]}{r}$$

$$k_2 = \frac{q_1(x_{1_f} - 100)}{r}$$

Note that the constant term (k_1) alone accounts for the final velocity penalty, while the final position penalty affects both k_1 and k_2. The final state can be found as in Example 3.3-1:

$$x_{1_f} = \frac{k_1 t_f^2}{2} + \frac{k_2 t_f^3}{6}$$

$$x_{2_f} = k_1 t_f + \frac{k_2 t_f^2}{2}$$

and these relationships can be substituted above to provide explicit equations for k_1 and k_2:

$$\begin{bmatrix} k_1 \\ k_2 \end{bmatrix} = \begin{bmatrix} (q_1 t_f^3/3 + q_2 t_f + r) & (q_1 t_f^4/6 + q_2 t_f^2/2) \\ q_1 t_f^2/2 & (q_1 t_f^3/6 - r) \end{bmatrix}^{-1} \begin{bmatrix} t_f \\ 1 \end{bmatrix} 100 q_1$$

Although not apparent from a simple comparison, this provides the same gains and terminal values as were found in Example 3.3-1c.

For many problems, the Hamiltonian is constant on the optimal trajectory, providing an alternate necessary condition. From Eq. 3.4-5, the time-rate-of-change of $\mathscr{H}$ is

$$\frac{d\mathscr{H}}{dt}[\mathbf{x}(t), \mathbf{u}(t), \boldsymbol{\lambda}(t), t] = \frac{d}{dt}\{\mathscr{L}[\mathbf{x}(t), \mathbf{u}(t), t] + \boldsymbol{\lambda}^T(t)\mathbf{f}[\mathbf{x}(t), \mathbf{u}(t), t]\}$$

$$= \left[\frac{\partial \mathscr{L}}{\partial t} + \boldsymbol{\lambda}^T \frac{\partial \mathbf{f}}{\partial t} + \left(\frac{\partial \boldsymbol{\lambda}}{\partial t}\right)^T \mathbf{f}\right] + \left(\frac{\partial \mathscr{L}}{\partial \mathbf{x}} + \boldsymbol{\lambda}^T \mathbf{F}\right)\frac{d\mathbf{x}}{dt}$$

$$+ \left(\frac{\partial \mathscr{L}}{\partial \mathbf{u}} + \boldsymbol{\lambda}^T \mathbf{G}\right)\frac{d\mathbf{u}}{dt} \tag{3.4-21}$$

Because Eqs. 3.4-16 and 3.4-18 are satisfied on the optimal trajectory, and $d\mathbf{x}/dt = \mathbf{f}$, this equation reduces to

$$\frac{d\mathscr{H}}{dt} = \frac{\partial \mathscr{L}}{\partial t} + \boldsymbol{\lambda}^T \frac{\partial \mathbf{f}}{\partial t} \tag{3.4-22}$$

If neither $\mathscr{L}$ nor $\mathbf{f}$ is an *explicit* function of time [i.e., if their time variations are due solely to time variations in $\mathbf{x}(t)$ and $\mathbf{u}(t)$], then $d\mathscr{H}/dt = 0$; therefore,

$$\mathscr{H} = \text{constant} \tag{3.4-23}$$

on the optimal trajectory. Equation 3.4-23 is not satisfied if $\mathbf{f}$ contains time-varying parameters $\mathbf{p}(t)$, as in Section 1.2, or if the penalty weights in the Lagrangian vary in time.

Having considered the fixed-end-time optimization problem, we must consider the effect of *open end time*, in which the final time of the trajectory can be chosen to further minimize the cost function. The final time t_f behaves as an additional control in this case, so it is natural to anticipate an additional necessary condition for optimality.

Returning to Eq. 3.4-10, it can be concluded that the first variation in J at the final time, $\Delta J(t_f)$, still must equal zero; however, its form should be altered to account for cost variations that might arise from changing values of t_f. Then

$$\Delta J(t_f)_{\text{open}} = \Delta J(t_f)\Big|_{\text{fixed}} + \frac{dJ}{dt}\Big|_{t=t_f} \Delta t_f \tag{3.4-24}$$

From Eq. 3.4-6a,

$$\frac{dJ}{dt}\Big|_{t=t_f} = \left\{\left[\frac{\partial \phi}{\partial t} + \frac{\partial \phi}{\partial \mathbf{x}}\dot{\mathbf{x}}\right] + (\mathscr{H} - \boldsymbol{\lambda}^T\dot{\mathbf{x}})\right\}\Big|_{t=t_f} \tag{3.4-25a}$$

Equation 3.4-12 indicates that $\partial\phi/\partial\mathbf{x} = \boldsymbol{\lambda}^T$ when $t = t_f$; hence the additional necessary condition for optimality with open end time is that

$$\left.\frac{dJ}{dt}\right|_{t=t_f} = \left.\left(\frac{\partial \phi}{\partial t} + \mathcal{H}\right)\right|_{t=t_f} = 0 \tag{3.4-25b}$$

This allows an additional characteristic of the Hamiltonian to be identified. If the terminal cost is independent of the final time,

$$\left.\frac{dJ}{dt}\right|_{t=t_f} = \left.\mathcal{H}\right|_{t=t_f} = 0 \tag{3.4-26}$$

If $\mathcal{H}$ also does not depend explicitly on time, it is constant (Eq. 3.4-23); then, $\mathcal{H} = 0$ over the entire interval.

A fundamental problem arises when Eq. 3.4-18 provides no mechanism for determining the control history. Having taken the required partial derivatives, suppose $\mathbf{u}(t)$ (or some of its components) does not appear in the equation; how can the control be found? If the control is bounded, the answer may lie in the Minimum Principle, which provides an alternate necessary condition and is discussed later in this section. Otherwise, a singular control solution may exist (Section 3.5).

Sufficient Conditions for Optimality

If the necessary conditions are met and any change in the control can only increase the cost, then these facts are sufficient to guarantee an optimal (minimum-cost) trajectory. If changes in the system's initial conditions would cause an unambiguous redefinition of the optimal control history and a corresponding neighboring-optimal (nonintersecting) path, then the original optimal trajectory is unique. In many practical control problems, there is sufficient knowledge of the system and flexibility in the choice of cost functions that cost stationarity may imply optimality. Consequently, the present discussion of sufficient conditions is brief, introducing three common tests that normally would assure optimality for a system with continuous parameters, controls, and cost gradients.

The first, and most important, is the "strengthened" *Legendre–Clebsch* (or *convexity*) *condition*:*

$$\frac{\partial^2 \mathcal{H}}{\partial \mathbf{u}^2}[\mathbf{x}^*(t), \mathbf{u}^*(t), \boldsymbol{\lambda}^*(t), t] = \mathcal{H}_{\mathbf{uu}} > \mathbf{0} \tag{3.4-27}$$

*Adrien-Marie Legendre (1752–1833) was the first mathematician to publish the "method of least squares" (Section 4.1), although it had been developed earlier by Gauss. Rudolph Clebsch (1833–1872), a student of Hesse, worked on the theory of invariants, as well as the calculus of variations.

for all t in $[t_0, t_f]$. This says that the second derivative matrix of the scalar Hamiltonian with respect to the control (i.e., its Hessian matrix) is positive definite for the optimal trajectory-control pair, $[\mathbf{x}^*(t), \mathbf{u}^*(t)]$, throughout the solution time interval. In the vicinity of the optimal trajectory, the quadratic form $\Delta\mathbf{u}^T(t)\mathcal{H}_{\mathbf{uu}}(t)\Delta\mathbf{u}(t)$ then would describe a multidimensional (possibly time-varying) bowl which is convex when viewed from below. Justification of this criterion follows that for ordinary maxima and minima in Section 2.1.

A so-called weakened version of the condition allowing null values of the Hessian,

$$\mathcal{H}_{\mathbf{uu}} \geq \mathbf{0} \tag{3.4-28}$$

may be admissible if additional criteria are satisfied. As in Section 2.1, convexity could be established by a higher-order derivative such as $\partial^4\mathcal{H}/\partial\mathbf{u}^4$, even if $\partial^2\mathcal{H}/\partial\mathbf{u}^2$ is zero (Fig. 3.4-1).

For controllable systems, satisfying Eq. 3.4-27 often is sufficient to guarantee optimality, but it should be remembered that the convexity condition is a local criterion. $\mathcal{H}_{\mathbf{uu}}(t)$ is proportional to the curvature of $\mathcal{H}(t)$ with respect to $\mathbf{u}$ along the optimal trajectory. By implication, the condition

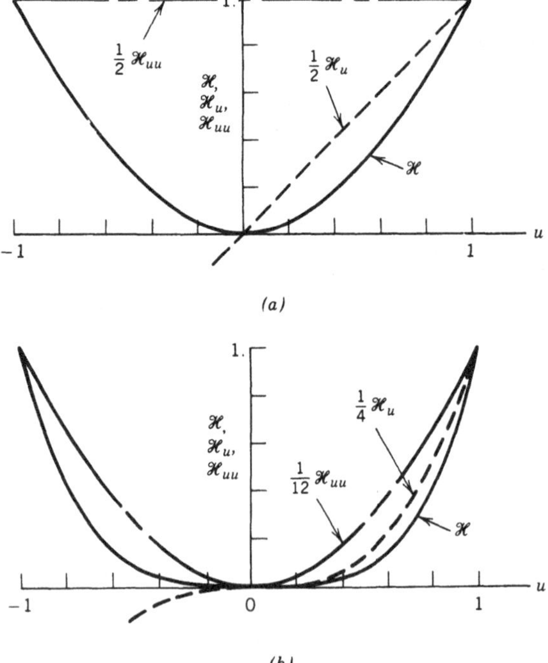

FIGURE 3.4-1 Two convex Hamiltonians, and their first and second derivatives. (*a*) $\mathcal{H}(u) = u^2$; (*b*) $\mathcal{H}(u) = u^4$.

applies in the immediate neighborhood of $\mathbf{x}^*(t)$ and $\mathbf{u}^*(t)$; however, in strongly nonlinear systems, $\mathcal{H}_{\mathbf{uu}}$ may not remain positive definite for large excursions from the optimal path. This could be of particular concern in numerical optimization (Section 3.6), where rapid convergence to the optimal path may depend on positive-definite $\mathcal{H}_{\mathbf{uu}}$ in an extended region. (Recall that the control gradient $\mathcal{H}_{\mathbf{u}} = \mathbf{0}$ on the optimal path only. It is nonzero for any excursion from the optimum, and this fact is put to use in numerical optimization.)

The very existence of neighboring-optimal solutions can be considered a sufficiency criterion, and while a discussion of such solutions is saved for Section 3.7, at least the philosophy can be discussed here. First, let us distinguish between a neighboring-optimal path and a nonoptimal neighboring path. Suppose that an optimal trajectory and control, $[\mathbf{x}_a^*(t), \mathbf{u}_a^*(t)]$ have been computed for the initial condition, $\mathbf{x}_a(0)$. If the initial condition is changed by a small amount to $\mathbf{x}_b(0)$ with the control history fixed as $\mathbf{u}_a^*(t)$, then a new state history, $\mathbf{x}_b(t)$, can be derived from Eq. 3.4-3, but there is no reason to expect the cost function (Eq. 3.4-4) to be minimized. Therefore, $[\mathbf{x}_b(t), \mathbf{u}_a^*(t)]$ is a nonoptimal pair. If, on the other hand, we satisfy Eqs. 3.4-11 to 3.4-13 while integrating Eq. 3.4-3 with the new initial condition, a new optimal control history, $\mathbf{u}_b^*(t)$, that minimizes J locally is generated, and $[\mathbf{x}_b^*(t), \mathbf{u}_b^*(t)]$ is a neighboring-optimal pair. It is easy to contemplate other ways of producing nonoptimal neighboring trajectories (e.g., by imposing disturbances on the system or changing physical constraints without accounting for them in the optimization). In each case, it would be necessary to redefine the control history using the Euler–Lagrange equations to achieve the neighboring-optimal solution.

The *normality condition* expresses the circumstances under which neighboring-optimal trajectories exist and terminal constraints can be satisfied. Mathematical expressions for normality can be found in (B-8) and Section 3.5. It requires that the system be controllable, in the sense of Sections 1.2 and 2.3 and that the neighboring-optimal perturbation equations of Section 3.7 be amenable to solution. This also assumes that the strengthened Legendre–Clebsch condition (Eq. 3.4–27) is satisfied.

If optimal trajectories intersect, then there is an indeterminacy about the optimal solution. The minimum distance from the North Pole to the equator is well defined, but the minimizing path is indeterminate, because any meridian of longitude could be followed to achieve the same distance. The north pole is called a *conjugate point*, a point where minimizing paths are joined together (Fig. 3.4-2).* The third sufficient condition, the *Jacobi condition*, is that there be no conjugate points on the optimal path. If there are none, then the locally optimal path is unique. Returning to the example,

*This is also called a *focal point*, from the application of variational calculus in optics. Reflected or refracted rays of light follow time-minimizing paths; where they converge, their image is indeterminate.

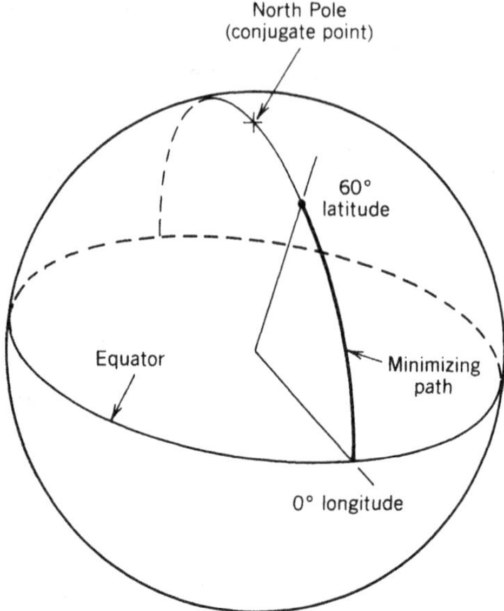

FIGURE 3.4-2 Example of conjugate point and its effect on the minimizing path.

starting from a point below the North Pole—say 0° longitude and 60°N latitude—the implied Hamiltonian is stationary if we proceed either north or south, but only the latter path is minimizing.

From the standpoint of neighboring-optimal solutions (Section 3.7), which assume that Eq. 3.4-27 is satisfied, the Jacobi condition requires that a finite optimal feedback relationship between state and control perturbations be maintained (B-8). If a finite state perturbation commands an infinite control at any point, that point is a conjugate point.

The Minimum Principle

We have seen that the first and second derivatives of the Hamiltonian

$$\frac{\partial \mathscr{H}}{\partial \mathbf{u}}[\mathbf{x}^*(t), \mathbf{u}^*(t), \boldsymbol{\lambda}^*(t), t] = \mathbf{0} \tag{3.4-29}$$

and

$$\frac{\partial^2 \mathscr{H}}{\partial \mathbf{u}^2}[\mathbf{x}^*(t), \mathbf{u}^*(t), \boldsymbol{\lambda}^*(t), t] \geq \mathbf{0} \tag{3.4-30}$$

provide necessary and sufficient conditions for optimality of $\mathbf{x}^*(t)$ and

$\mathbf{u}^*(t)$ in the interval $[t_0, t_f]$, if the system is normal and the optimal path contains no conjugate points.* Treating $\mathbf{x}^*(t)$ and $\boldsymbol{\lambda}^*(t)$ as fixed parameters at each time t and referring to Section 2.1, it can be seen that these are precisely the conditions for an ordinary minimum of $\mathcal{H}$ with respect to $\mathbf{u}$. Therefore

$$\mathcal{H}^* = \mathcal{H}[\mathbf{x}^*(t), \mathbf{u}^*(t), \boldsymbol{\lambda}^*(t), t] \leq \mathcal{H}[\mathbf{x}^*(t), \mathbf{u}(t), \boldsymbol{\lambda}^*(t), t] \tag{3.4-31}$$

in $[t_0, t_f]$, where $\mathbf{u}(t)$ is any admissible, neighboring (nonoptimal) control history. This equation states the Minimum Principle for optimality developed by Pontryagin† and his co-workers in the 1950s. Equation 3.4-31 is a necessary and sufficient condition for optimality. If $\mathcal{H}$ is stationary and convex, the minimum principle is satisfied. If it is one but not the other, Eq. 3.4-31 is not satisfied *unless* there is a constraint on the admissible controls.

Actually, a stronger case for the minimum principle can be made. Suppose that $J^*[\mathbf{u}^*(t)]$ is the minimum value of a cost function described by Eq. 3.4-4, and $J[\mathbf{u}^*(t) + \Delta\mathbf{u}(t)]$ is the same cost function evaluated with a perturbed (nonoptimal) control. The control perturbation history, $\Delta\mathbf{u}(t)$, $t_0 \leq t \leq t_f$, is arbitrary and "small." Then $J[\mathbf{u}^*(t) + \Delta\mathbf{u}(t)]$ must be greater than (or, at best, equal to) $J^*[\mathbf{u}^*(t)]$; eliminating like terms,

$$\begin{aligned} &J[\mathbf{u}^*(t) + \Delta\mathbf{u}(t)] - J^*[\mathbf{u}^*(t)] \\ &\quad = \int_{t_0}^{t_f} \{\mathcal{H}[\mathbf{x}^*(t), \mathbf{u}^*(t) + \Delta\mathbf{u}(t), \boldsymbol{\lambda}^*(t), t] - \mathcal{H}[\mathbf{x}^*(t), \mathbf{u}^*(t), \boldsymbol{\lambda}^*(t), t]\}\, dt \\ &\quad \geq \mathbf{0} \end{aligned} \tag{3.4-32}$$

For this equation to be true with completely arbitrary $\Delta\mathbf{u}(t)$, Eq. 3.4-31 must be satisfied.

If the controls are *bounded* by inequality constraints (Section 3.5), the minimum principle can demonstrate optimality where the variational equations fail. (Recall the example of a nonstationary local maximum of an algebraic function at the boundary of $\mathbf{u}$, presented in Section 2.1). A comparison of $\mathcal{H}$ computed on the boundary with any other $\mathcal{H}$ establishes optimality, and the optimizing control is determined directly from $\mathcal{H}$ (rather than $\partial\mathcal{H}/\partial\mathbf{u}$). Nevertheless, while the minimum principle is more comprehensive than the variational equations (one equality does the work of two equations in stating necessary and sufficient conditions), it is less specific

*Optimal values of variables are denoted by asterisks.

†Academician Lev Pontryagin (1908–) headed the differential controls section of the USSR Mathematics Institute of the Academy of Sciences at the time. In this work, $\mathcal{H}$ is defined as $\lambda_0\mathcal{L} + \boldsymbol{\lambda}^T\mathbf{f}$, where λ_0 is a *negative* constant; hence J is minimized when $\mathcal{H}$ is maximized, and the corresponding convexity condition is $\partial^2\mathcal{H}/\partial\mathbf{u}^2 \leq \mathbf{0}$. What is described here as the "minimum" principle originated as the "maximum" principle.

about the means of solution when the control constraints are not in effect. Then, the minimum principle must be expressed by Eqs. 3.4-29 and 3.4-30, resulting in a "classical" calculus-of-variations statement of the problem and solution.

The minimum principle appears to provide a global criterion for optimality. Given two locally optimal trajectories with identical boundary conditions, the one with lower $\mathcal{H}^*$ should have lower cost. If, however, $\mathcal{H}^*$ is time-varying, a lower value at one time may be offset by a higher value at another time, as discussed below.

The Hamilton–Jacobi–Bellman Equation

Consider the minimization of a *value function* (which is related to the cost function as shown below) during the reduced time interval $[t_1, t_f]$, where $t_0 \le t_1 \le t_f$, t_f is fixed, and the control is unbounded. Having found the optimal control, $\mathbf{u}(t) = \mathbf{u}^*(t)$ in $[t_1, t_f]$, the minimized value function could be expressed as

$$\begin{aligned} V^*[\mathbf{x}^*(t_1), t_1] &= \phi[\mathbf{x}^*(t_f), t_f] + \int_{t_1}^{t_f} \mathcal{L}[\mathbf{x}^*(t), \mathbf{u}^*(t), t]\, dt \\ &= \phi[\mathbf{x}^*(t_f), t_f] - \int_{t_f}^{t_1} \mathcal{L}[\mathbf{x}^*(t), \mathbf{u}^*(t), t]\, dt \\ &= \min_{\mathbf{u}} \left\{ \phi[\mathbf{x}^*(t_f), t_f] - \int_{t_f}^{t_1} \mathcal{L}[\mathbf{x}^*(t), \mathbf{u}(t), t]\, dt \right\} \end{aligned} \tag{3.4-33}$$

where it is assumed that the dynamic constraint is satisfied:

$$\dot{\mathbf{x}}^*(t) = \mathbf{f}[\mathbf{x}^*(t), \mathbf{u}^*(t), t], \qquad \mathbf{x}^*(t_1) = \mathbf{x}_1 \tag{3.4-34}$$

The total derivative of the value function with respect to time is

$$\left.\frac{dV^*}{dt}\right|_{t=t_1} = -\mathcal{L}[\mathbf{x}^*(t_1), \mathbf{u}^*(t_1), t_1] \tag{3.4-35}$$

The negative sign is appropriate because the minimal value of V^* occurs at t_f and the maximal value at t_0; dV^*/dt reflects changes in the *lower* limit of the integral in the first line of Eq. 3.4-33.

A distinction between the value function V and the cost function J is that V takes its maximum value at t_0 and its minimum value at t_f, while J does just the opposite. Both are positive numbers, $V_{max} = J_{max}$ for a given trajectory, and the two are related as follows (Fig. 3.4-3):

$$V(t) = \begin{cases} J_{max} - J(t), & t < t_f \\ V_{min} = \phi[\mathbf{x}(t_f), t_f], & t = t_f \end{cases} \tag{3.4-36}$$

$$J(t) = \begin{cases} V_{\max} - V(t), & t < t_{f^-} \\ V_{\max} - V_{\min}, & t = t_{f^-} \\ V_{\max}, & t = t_{f^+} \end{cases} \tag{3.4-37}$$

The total derivative is composed of partial derivatives of V^* with respect to t and $\mathbf{x}$ (but not $\mathbf{u}$, because the first-order sensitivity to $\mathbf{u}$ is zero on

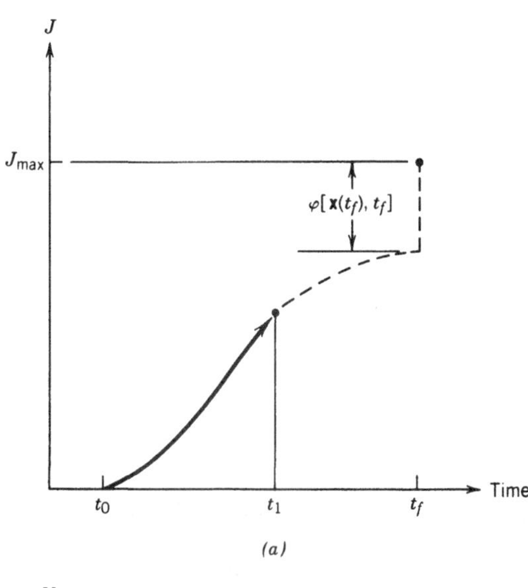

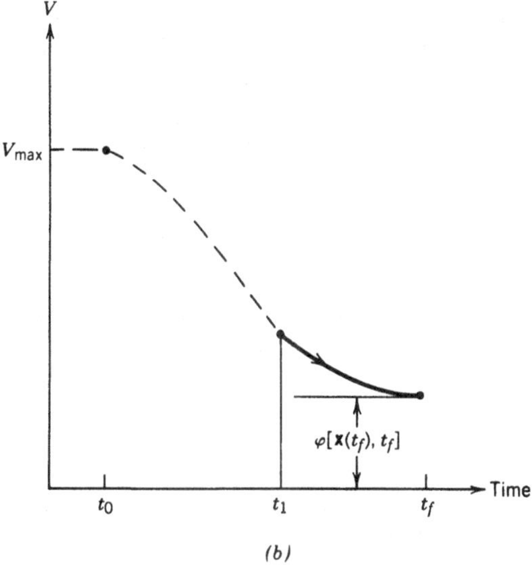

FIGURE 3.4-3 Comparison of cost and value functions.

the optimal path):

$$\left.\frac{dV^*}{dt}\right|_{t=t_1} = \left.\left(\frac{\partial V^*}{\partial t} + \frac{\partial V^*}{\partial \mathbf{x}}\dot{\mathbf{x}}\right)\right|_{t=t_1}$$
$$= \left.\left(\frac{\partial V^*}{\partial t} + \frac{\partial V^*}{\partial \mathbf{x}}\mathbf{f}\right)\right|_{t=t_1} \tag{3.4-38}$$

Equating the two,

$$\frac{\partial V^*}{\partial t}[\mathbf{x}^*(t_1), t_1] = -\mathscr{L}[\mathbf{x}^*(t_1), \mathbf{u}^*(t_1), t_1] - \frac{\partial V^*}{\partial \mathbf{x}}[\mathbf{x}^*(t_1), t_1]\mathbf{f}[\mathbf{x}^*(t_1), \mathbf{u}^*(t_1), t_1] \tag{3.4-39a}$$

Defining a Hamiltonian in which $\mathbf{f}$ is adjoined to $\mathscr{L}$ by $\partial V/\partial \mathbf{x}$, this can be restated as

$$\frac{\partial V^*}{\partial t}[\mathbf{x}^*(t_1), t_1] = -\mathscr{H}\left\{\mathbf{x}^*(t_1), \mathbf{u}^*(t_1), \frac{\partial V^*}{\partial \mathbf{x}}[\mathbf{x}^*(t_1), t_1]\right\}$$
$$= -\min_{\mathbf{u}} \mathscr{H}\left\{\mathbf{x}^*(t_1), \mathbf{u}(t_1), \frac{\partial V^*}{\partial \mathbf{x}}[\mathbf{x}^*(t_1), t_1]\right\} \tag{3.4-39b}$$

This *partial differential equation* is known as the Hamilton–Jacobi–Bellman* (HJB) equation. It provides the rule for defining optimal controls of continuous-time systems using Dynamic Programming. The similarity between the Hamiltonians of Eqs. 3.4-39 and 3.4-5 is apparent; from Eq. 3.4-20, it can be seen that they are *identical*, because the transpose of the adjoint vector, $\boldsymbol{\lambda}^{*T}(t)$, equals the value function's sensitivity to state perturbations, $\partial V^*/\partial \mathbf{x}$, on the optimal trajectory. Nevertheless, they are formulated using two very different approaches. The finding that

$$\frac{\partial V^*}{\partial t}(t) = -\min_{\mathbf{u}} \mathscr{H}(t) \tag{3.4-40}$$

on an optimal trajectory is in concert with the minimum principle, since the lowest values of the positive-semidefinite $\mathscr{H}(t)$ will result in the lowest net cost.

*Richard Bellman (1920–1984), a dominant figure in the development of modern control theory and systems analysis, was the first to recognize that the Hamilton–Jacobi equation of classical mechanics had application in optimal control. The "HJB" appellation makes note of this.

The HJB equation is a sufficient condition for local optimality, but it is not a necessary condition; a value function might fail to satisfy differentiability and continuity conditions that are required to solve the partial differential condition yet still be optimal. Sufficiency is verified when the minimum value of $\mathscr{H}$ with respect to all $\mathbf{u}$ for all $\mathbf{x}$ and t is found to be unique. In principle, Eq. 3.4-39 could be solved subject to the terminal boundary condition

$$V^*[\mathbf{x}(t_f), t_f] = \phi[\mathbf{x}^*(t_f), t_f] \tag{3.4-41}$$

and the dynamic constraint (Eq. 3.4-34) to yield the value function, $V^*(\mathbf{x}_1, t_1)$, at some arbitrary time, $t_0 \le t_1 \le t_f$, and "place," $\mathbf{x}(t_1)$.* Progressing back from $[\mathbf{x}^*(t_1), t_1]$ to $[\mathbf{x}^*(t_0), t_0]$ does not change the nature of the optimal trajectory from $[\mathbf{x}^*(t_1), t_1]$ to $[\mathbf{x}^*(t_f), t_f]$, so the latter is a *segment* of the optimal trajectory between $[\mathbf{x}^*(t_0), t_0]$ and $[\mathbf{x}^*(t_f), t_f]$. This is a reflection of the *Principle of Optimality* stated by Bellman:†

> An optimal policy has the property that whatever the initial state and initial decision are, the remaining decisions must constitute an optimal policy with regard to the state resulting from the first decision.

Consequently, the value function $V^*(\mathbf{x}, t)$, describes a *hypersurface of minimum cost* in the possible range of $\mathbf{x}$ during the solution interval $[t_0, t_f]$. Each entry for the value function is associated with a specific optimal trajectory and control history; thus the state-control space contains a *family* (or *field*) *of extremals* corresponding to each set of possible initial conditions. If the normality and Jacobi conditions are satisfied, the neighboring extremals are nonintersecting and unique.

Dynamic programming justifies the existence of a *nonlinear feedback control law* to minimize a cost function such as Eq. 3.1-1. For each permissible $\mathbf{x}(t_1)$ and t_1, there is a minimized value function, $V^*[\mathbf{x}(t_1), t_1]$, and an optimal state-control history, $[\mathbf{x}^*(t), \mathbf{u}^*(t)]$, $t_1 \le t \le t_f$. Therefore, the state and time could be fed back as arguments for an optimal control function

$$\mathbf{u}(t) = \mathbf{u}^*[\mathbf{x}(t), t] \tag{3.4-42}$$

that minimizes the instantaneous value function, on the assumption that the remainder of the state and control history will adhere to optimizing values.

It could occur that two locally minimizing paths connect the same initial and final points (hence they are exterior conjugate points), and one would

*The value function is sometimes called the optimal return function or "cost to go," because it measures the cost of completing the trajectory.

†In (K-1), it is noted that similar statements were made by Jakob and Johann Bernoulli (1697, 1706), Leonhard Euler (1744), and C. Caratheodory (1935).

like to define the *global optimum* on the basis of minimum $\mathscr{H}^*$ alone. This is possible if $\mathscr{H}$ has no explicit dependence on time; then $\mathscr{H}^*$ is constant (Eq. 3.4-23), and the lower $\mathscr{H}^*$ does indeed yield lower cost. If $\mathscr{H}^*$ is not constant, the test could fail, because $\mathscr{H}_1^*(t_1) < \mathscr{H}_2^*(t_1)$ may not preclude $\mathscr{H}_1^*(t_2) > \mathscr{H}_2^*(t_2)$, and V^* is the time integral of the Hamiltonian. By analogy, consider two towns in a mountainous region connected by two valleys. If the objective is to travel from one town to the other with minimum hillclimbing, either valley provides a feasible path. If it is known that the valley floors have fixed height, then the choice is simple (and global). If the floor heights are known to vary, the global minimum cannot be found without additional information.

With few exceptions, the HJB equation is more difficult to solve than the Euler–Lagrange equations, and it has been suggested that the HJB equation's principal value is as a check for sufficiency (A-3). One important exception is that the HJB equation provides a straightforward solution for the linear-quadratic controller (Section 3.7), and it gives insight regarding the nature of optimal solutions. It would appear that the equation provides an advantage for determining global optimality, in that the value function is evaluated as part of the optimization procedure. This advantage may be minimal, as direct evaluation of Eq. 3.1-1 while solving the Euler–Lagrange equations or satisfying the minimum principle is possible and may, in fact, be a crucial element of efficient numerical solution. Dynamic programming "comes into its own" in the optimization of discrete, multi-stage systems such as the maze mentioned earlier, which have not been considered here, and it provides the only general formulation for stochastic optimal control of nonlinear systems, as presented in Chapter 5.

The classical variational equations, the minimum principle, and the HJB equation are complementary tools for optimization. They are not entirely redundant, although it is reassuring that they reduce to the same conditions for many problems. One formulation may be better suited than the others for solving a particular problem.

3.5 CONSTRAINTS AND SINGULAR CONTROL

To this point, we have considered optimal trajectories with unrestricted state and control vectors and cost functions that are reasonably well-behaved. The current section introduces a number of complicating factors, including terminal state constraints, path constraints on the state and control, and "singular arcs," along which the necessary and sufficient conditions of the previous section fail to provide enough information to determine the optimal control history.

As noted in Section 1.3, constraints may fall into two categories. Equality constraints require that some function of variables be maintained at a constant value throughout the solution interval (or at the endpoint), and inequality constraints prohibit a function of variables from exceeding some limit during the interval. Equality and inequality constraints can be viewed as "surfaces"* in the state-control space. For the former, the constrained variables always lie on the surface; for the latter, they lie either to one side of the surface or on the surface itself.

Singularity arises when second-order conditions for "convexity" are not met (i.e., when the Hamiltonian's Hessian matrix is zero). Optimal singular arcs must be defined by additional criteria that characterize higher-order behavior of the cost function and its associated Hamiltonian.

Terminal State Equality Constraints

It may be desirable to assure that one or more state variables satisfy some condition precisely at the end of the trajectory. Perhaps a state component should be driven to zero or to some specified value. One approach is simply to form a convex error function and weight the difference between achieved and desired final states heavily in a Mayer- or Bolza-type cost function; this is referred to as a *penalty function method* of satisfying the constraint. Rather than accepting even a small terminal error, as in Examples 3.3-1 and 3.4-1, we might specify that the final position or velocity error be exactly zero.† A combination of variables representing a final reaction product in a chemical process, an energy level in a missile trajectory, or a physical boundary in machine control may be the primary objective for control. All of these could be described as *terminal state equality constraints*, and this section deals with the conditions that must be satisfied to achieve the zero-error objective.

If the desired end condition can be written as the *scalar* equality constraint

$$\psi[\mathbf{x}(t_f), t_f] = 0 \tag{3.5-1}$$

the problem can be solved using a Lagrange multiplier, as in Section 2.1 (B-8). The unconstrained cost function that is to be minimized while satisfying Eq. 3.5-1 is

$$J_0 = \phi[\mathbf{x}(t_f), t_f] + \int_{t_0}^{t_f} \mathscr{L}[\mathbf{x}(t), \mathbf{u}(t), t]\, dt \tag{3.5-2}$$

*If the constraint is a function of three scalar variables, the "surface" *is* a surface in the conventional sense. For a one-variable constraint, the "surface" is a *point*. For two variables, it is a (possibly curved) *line*. For four or more variables, it is a *hypersurface*, the "hyper-" indicating that the surface is beyond visualization in our three-dimensional world.

†Constraints that must be satisfied without error are sometimes called "hard" constraints; those that are enforced by penalty functions and which, therefore, tolerate small errors are called "soft" constraints.

subject to the dynamic constraint

$$\dot{\mathbf{x}}(t) = \mathbf{f}[\mathbf{x}(t), \mathbf{u}(t), t], \qquad \mathbf{x}(t_0) = \mathbf{x}_0 \tag{3.5-3}$$

The cost function is augmented by the terminal constraint

$$J_c = J_0 + \mu J_1 \tag{3.5-4}$$

where $J_1 = \psi[\mathbf{x}(t_f), t_f]$, and μ is a scalar Lagrange multiplier. The terminal constraint should not conflict with the terminal cost; if $\psi[\mathbf{x}(t_f)]$ implies one value of $\mathbf{x}(t_f)$ and $\phi[\mathbf{x}(t_f)]$ implies another, the terminal cost cannot be minimized while satisfying the constraint.

For the moment, let's consider the necessary conditions for finding an optimal (zero) value of the constraint alone. Since the corresponding augmented cost function is

$$J_1 = \psi + \int_{t_0}^{t_f} \boldsymbol{\lambda}_1^T(\mathbf{f} - \dot{\mathbf{x}})\, dt = \psi + \int_{t_0}^{t_f} (\mathcal{H}_1 - \boldsymbol{\lambda}_1^T \dot{\mathbf{x}})\, dt \tag{3.5-5}$$

the first two Euler–Lagrange equations become

$$\dot{\boldsymbol{\lambda}}_1(t) = -\left(\frac{\partial \mathcal{H}_1}{\partial \mathbf{x}}\right)^T = -\mathbf{F}^T(t)\boldsymbol{\lambda}_1(t) \tag{3.5-6}$$

$$\boldsymbol{\lambda}_1(t_f) = \left(\frac{\partial \psi}{\partial \mathbf{x}}\right)^T \bigg|_{t=t_f} \tag{3.5-7}$$

These equations specify $\boldsymbol{\lambda}_1(t)$ throughout the interval. From Eq. 3.4-10,

$$\Delta J_1 = \int_{t_0}^{t_f} \left(\frac{\partial \mathcal{H}_1}{\partial \mathbf{u}}\right) \Delta \mathbf{u}(t)\, dt = \int_{t_0}^{t_f} \boldsymbol{\lambda}_1^T(t)\mathbf{G}(t)\Delta\mathbf{u}(t)\, dt \tag{3.5-8}$$

when there are no perturbations in $\mathbf{x}_0$ or $\mathbf{x}_f$. We could assure that $\Delta J_1 = 0$ (which is necessary at its optimum) by satisfying the third Euler–Lagrange equation (Eq. 3.4-18), but there is another way. Suppose that we choose $\Delta\mathbf{u}(t)$ such that (1) the integral (Eq. 3.5-8) is zero and (2) J_c is minimized concurrently. Then the terminal constraint is satisfied, and J_0 is minimized subject to the constraint.

Neglecting the terminal constraint, the adjoint equation for J_0 is found by integrating

$$\dot{\boldsymbol{\lambda}}_0 = -\left(\frac{\partial \mathcal{H}_0}{\partial \mathbf{x}}\right)^T = -\mathbf{F}^T \boldsymbol{\lambda}_0 - \left(\frac{\partial \mathcal{L}}{\partial \mathbf{x}}\right)^T \tag{3.5-9}$$

with the terminal condition

$$\boldsymbol{\lambda}_0(t_f) = \left(\frac{\partial \phi}{\partial \mathbf{x}}\right)^T\Bigg|_{t=t_f} \tag{3.5-10}$$

The first variation of the constrained cost function is

$$\begin{aligned}\Delta J_c &= \Delta J_0 + \mu \Delta J_1 \\ &= \int_{t_0}^{t_f} \left[\frac{\partial \mathcal{H}_0}{\partial \mathbf{u}} + \mu \frac{\partial \mathcal{H}_1}{\partial \mathbf{u}}\right] \Delta \mathbf{u}(t)\, dt \\ &= \int_{t_0}^{t_f} \frac{\partial \mathcal{H}_c}{\partial \mathbf{u}} \Delta \mathbf{u}(t)\, dt \end{aligned} \tag{3.5-11}$$

and it must equal zero on the optimal trajectory, because the gradient itself must equal zero:

$$\frac{\partial \mathcal{H}_c}{\partial \mathbf{u}} = \left(\frac{\partial \mathcal{L}}{\partial \mathbf{u}} + \boldsymbol{\lambda}_0^T \mathbf{G}\right) + \mu \boldsymbol{\lambda}_1^T \mathbf{G} = \mathbf{0} \tag{3.5-12}$$

If we were to choose $\Delta \mathbf{u}(t)$ as

$$\begin{aligned}\Delta \mathbf{u}(t) &= \epsilon \left[\frac{\partial \mathcal{H}_c}{\partial \mathbf{u}}(t)\right]^T \\ &= \epsilon \left\{\left[\frac{\partial \mathcal{L}}{\partial \mathbf{u}}(t)\right]^T + \mathbf{G}^T(\boldsymbol{\lambda}_0 + \mu \boldsymbol{\lambda}_1)\right\}\end{aligned} \tag{3.5-13}$$

where ϵ is an arbitrary small constant, then ΔJ_c is doubly zero since $\partial \mathcal{H}_c / \partial \mathbf{u} = \mathbf{0}$. Nevertheless, the Lagrange multiplier μ remains to be found. An equation for μ results from substituting Eq. 3.5-13 in Eq. 3.5-8. Then

$$\begin{aligned}\Delta J_1 &= \epsilon \int_{t_0}^{t_f} \boldsymbol{\lambda}_1^T \mathbf{G}\left[\left(\frac{\partial \mathcal{L}}{\partial \mathbf{u}}\right)^T + \mathbf{G}^T \boldsymbol{\lambda}_0\right] dt + \epsilon\mu \int_{t_0}^{t_f} \boldsymbol{\lambda}_1^T \mathbf{G}\mathbf{G}^T \boldsymbol{\lambda}_1\, dt \\ &\triangleq \epsilon(a + \mu b)\end{aligned} \tag{3.5-14}$$

and the required value of μ is

$$\mu = -\frac{a}{b} \tag{3.5-15}$$

The second integral must be nonzero for μ to exist, assuring adequate *controllability* to meet the terminal constraint. If b is zero, the *normality condition* mentioned in Section 3.4 is not satisfied.

Note that two sets of adjoint variables must be computed. $\boldsymbol{\lambda}_1(t)$ represents a vector *influence function* that brings the state to the terminal constraint through Eq. 3.5-8, and it is independent of J_0 and the minimization of J_c. The unconstrained adjoint vector, $\boldsymbol{\lambda}_0(t)$, is not influenced by J_1. Both adjoint vectors are necessary to find μ and $\partial\mathcal{H}_c/\partial\mathbf{u}$, and $\mathbf{u}(t)$ is chosen to force $\partial\mathcal{H}_c/\partial\mathbf{u}$ to zero.

The scalar *isoperimetric constraint*

$$\int_{t_0}^{t_f} c[\mathbf{x}(t), \mathbf{u}(t), t]\, dt = \text{constant} = c_f \tag{3.5-16}$$

can be treated as a terminal state constraint if we define a new state variable component x_{n+1}, whose differential equation is

$$\dot{x}_{n+1}(t) = c[\mathbf{x}(t), \mathbf{u}(t), t] \tag{3.5-17}$$

and set $\psi = [x_{n+1}(t_f) - c_f]$. Equation 3.5-17 is added to Eq. 3.5-3, and the solution proceeds as above. This is analogous to transforming a Lagrange cost function to a Bolza type, the only difference being that $c[\cdot]$ and ψ are treated as constraints rather than as costs.

In the event that more than one terminal constraint must be satisfied, the preceding results can be extended to the *vector*

$$\mathbf{J}_T = \boldsymbol{\psi}[\mathbf{x}(t_f), t_f] = \begin{bmatrix} \psi_1 \\ \cdot \\ \cdot \\ \cdot \\ \psi_r \end{bmatrix} = \mathbf{0} \tag{3.5-18}$$

Given r constraints, the associated r adjoint vectors are computed as

$$\dot{\boldsymbol{\lambda}}_i = -\mathbf{F}^T \boldsymbol{\lambda}_i \tag{3.5-19}$$

$$\boldsymbol{\lambda}_i^T(t_f) = \frac{\partial \psi_i}{\partial \mathbf{x}}, \qquad i = 1 \text{ to } r \tag{3.5-20}$$

and the following r variations must be zero:

$$\Delta J_i = \int_{t_0}^{t_f} \boldsymbol{\lambda}_i^T \mathbf{G} \Delta\mathbf{u}(t)\, dt, \qquad i = 1 \text{ to } r \tag{3.5-21}$$

The constrained cost function is found using an r-component Lagrange multiplier $\boldsymbol{\mu}$:

$$J_c = J_0 + \boldsymbol{\mu}^T \mathbf{J}_T \tag{3.5-22}$$

The first variation of the cost function is

$$\Delta J_c = \Delta J_0 + \boldsymbol{\mu}^T \boldsymbol{\Delta J}_T \tag{3.5-23}$$

and the minimizing gradient of the Hamiltonian is

$$\begin{aligned}\frac{\partial \mathcal{H}_c}{\partial \mathbf{u}} &= \frac{\partial \mathcal{H}_0}{\partial \mathbf{u}} + \boldsymbol{\mu}^T \begin{bmatrix} \partial \mathcal{H}_1/\partial \mathbf{u} \\ \cdot \\ \cdot \\ \cdot \\ \partial \mathcal{H}_r/\partial \mathbf{u} \end{bmatrix} \\ &= \left(\frac{\partial \mathcal{L}}{\partial \mathbf{u}} + \boldsymbol{\lambda}_0^T \mathbf{G}\right) + (\mu_1 \boldsymbol{\lambda}_1^T + \cdots + \mu_r \boldsymbol{\lambda}_r^T)\mathbf{G} \\ &= \mathbf{0}\end{aligned} \tag{3.5-24}$$

Choosing $\Delta\mathbf{u} = \epsilon(\partial \mathcal{H}_c/\partial \mathbf{u})^T$ and substituting in Eq. 3.5-21, ΔJ_i is zero when

$$\int_{t_0}^{t_f} \boldsymbol{\lambda}_i^T \mathbf{G}\left[\left(\frac{\partial \mathcal{L}}{\partial \mathbf{u}}\right)^T + \mathbf{G}^T \boldsymbol{\lambda}_0\right] dt + \mu_1 \int_{t_0}^{t_f} \boldsymbol{\lambda}_i^T \mathbf{G}\mathbf{G}^T \boldsymbol{\lambda}_1 \, dt + \cdots + \mu_r \int_{t_0}^{t_f} \boldsymbol{\lambda}_i^T \mathbf{G}\mathbf{G}^T \boldsymbol{\lambda}_r \, dt = 0 \tag{3.5-25a}$$

which takes the form

$$a_i + \mu_1 b_{i1} + \cdots + \mu_r b_{ir} = 0 \tag{3.5-25b}$$

Repeating this for $i = 1$ to r provides r equations to be solved for the Lagrange multipliers:

$$\mathbf{B}\boldsymbol{\mu} = -\mathbf{A} \tag{3.5-26a}$$

or

$$\boldsymbol{\mu} = -\mathbf{B}^{-1}\mathbf{A} \tag{3.5-26b}$$

where

$$\mathbf{A} = \begin{bmatrix} a_1 \\ \cdot \\ \cdot \\ \cdot \\ a_r \end{bmatrix} \tag{3.5-27}$$

$$\mathbf{B} = \begin{bmatrix} b_{11} & \cdots & b_{1r} \\ \cdot & & \cdot \\ \cdot & & \cdot \\ \cdot & & \cdot \\ b_{r1} & \cdots & b_{1r} \end{bmatrix} \tag{3.5-28}$$

Hence **B** must be nonsingular to assure sufficient controllability to satisfy the r terminal constraints. Having found $\boldsymbol{\mu}$, the minimizing control is chosen to satisfy Eq. 3.5-24.

Before leaving terminal equality constraints, a simplified solution for a special problem is noted. Consider the open-end-time minimization of

$$J = \phi[\mathbf{x}(t_f)] + \int_{t_0}^{t_f} \mathscr{L}\, dt \tag{3.5-29}$$

subject to the dynamic constraint

$$\dot{\mathbf{x}}(t) = \mathbf{f}[\mathbf{x}(t), \mathbf{u}(t)], \qquad \mathbf{x}(t_0) = \mathbf{x}_0 \tag{3.5-30}$$

and the scalar terminal constraint

$$\psi[\mathbf{x}(t_f)] = \psi_f = 0 \tag{3.5-31}$$

For illustration, any explicit time dependences in the system or constraints are neglected. The problem is made difficult by the open end time and the terminal constraint, both of which increase the number of conditions that must be satisfied to achieve optimality. Although it may not be of obvious concern, Eq. 3.5-31 could be evaluated throughout the interval $[t_0, t_f]$. If $\psi[\mathbf{x}(t)]$ is likely to be a monotonic (continually increasing or decreasing) function of time, then $d\psi/dt \neq 0$ in the interval and Eqs. 3.5-29 and 3.5-30 could be replaced by

$$J = \phi[\mathbf{x}(\psi_f)] + \int_{\psi_0}^{\psi_f} \mathscr{L}\, d\psi \tag{3.5-32}$$

$$\begin{aligned} \frac{\partial \mathbf{x}(\psi)}{\partial \psi} &= \frac{\mathbf{f}[\mathbf{x}(t), \mathbf{u}(t)]}{\dfrac{d\psi}{dt}} \\ &= \mathbf{f}_\psi[\mathbf{x}(\psi), \mathbf{u}(\psi)], \qquad \mathbf{x}(\psi_0) = \mathbf{x}_0 \end{aligned} \tag{3.5-33}$$

Consequently, the open end condition of the old independent variable t is replaced by the fixed end condition of the new independent variable ψ, and the terminal constraint (Eq. 3.5-31) is satisfied automatically. A fixed-range, minimum-fuel aircraft trajectory provides an obvious example, in which time could be replaced by range as the independent variable.

A third benefit may accrue if ψ is an uncomplicated function of the states. Suppose for example, that $\psi(\mathbf{x}) = x_n$; then $\partial x_n/\partial \psi = 1$ throughout the interval, and the order of Eq. 3.5-33 and the dimension of the associated differential equation for $\boldsymbol{\lambda}$ can be reduced from n to $(n-1)$.

EXAMPLE 3.5-1 TERMINAL CONSTRAINTS ON A ROBOT ARM

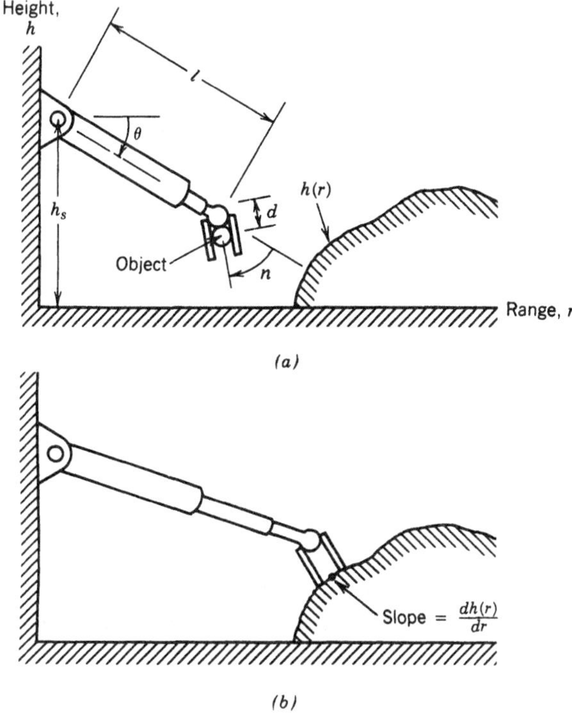

(1) Overall layout; (2) terminal contraint.

The motions of the extendable robot arm and wrist shown in Diagrams 1 and 2 could be modeled by the sixth-order differential equation,

$$\dot\theta = q$$

$$\dot q = \Big[u_1 + g\{(m_{\text{arm}}/2)l\cos\theta + m_{\text{object}}(l^2 + d^2 + 2ld\cos n)^{1/2} \times \cos[\theta + d\sin n/(l^2 d^2 + 2ld\cos n)^{1/2}]\}\Big] / [m_{\text{arm}} l^2/12 + m_{\text{object}}(l^2 + d^2 + 2ld\cos n)]$$

$$\dot l = v$$

$$\dot v = \frac{u_2}{(m_{\text{fore}} + m_{\text{object}})} + g\sin\theta$$

$$\dot n = p$$

$$\dot p = \Big[u_3 + g(m_{\text{hand}} + m_{\text{object}})d\cos(\theta + n)\Big] / (m_{\text{hand}}d^2/12 + m_{\text{object}}d^2)$$

or

$$\dot{\mathbf{x}} = \mathbf{f}(\mathbf{x}, \mathbf{u})$$

where g is gravitational acceleration

$$\begin{aligned} x^T &= [\theta \quad q \quad l \quad v \quad n \quad p] \\ &= [x_1 \ldots x_6] \end{aligned}$$

and

$$\mathbf{u} = \begin{bmatrix} u_1 \\ u_2 \\ u_3 \end{bmatrix} = \begin{bmatrix} \text{shoulder control torque} \\ \text{length control force} \\ \text{wrist control torque} \end{bmatrix}$$

The mass of the total arm assembly is m_{arm} while the masses of the forearm (including the hand), the hand, and the held object are m_{fore}, m_{hand}, and m_{object}, respectively. The arm length is l and the hand length is d. Common approximations for weight distribution and moment of inertia have been employed.

It might be desired to minimize accelerations in moving the hand from starting to final position in $[t_0, t_f]$; hence the following is a suitable cost function:

$$J = \int_{t_0}^{t_f} (k_1\dot{q}^2 + k_2\dot{v}^2 + k_3\dot{p}^2)\, dt$$

The k_i provide relative weightings of accelerations, and the accelerations are expressed in terms of **x** and **u** using the right sides of the second, fourth, and sixth equations. A number of terminal constraints might be desirable, and they could be expressed as follows:

a. Zero terminal rates:

$$\boldsymbol{\psi}[\mathbf{x}(t_f)] = \begin{bmatrix} x_2(t_f) \\ x_4(t_f) \\ x_6(t_f) \end{bmatrix}$$

b. Desired terminal angles and length:

$$\boldsymbol{\psi}[\mathbf{x}(t_f)] = \begin{bmatrix} x_1(t_f) - \theta_d \\ x_3(t_f) - l_d \\ x_5(t_f) - n_d \end{bmatrix}$$

c. Desired terminal wrist height and range:

$$\boldsymbol{\psi}[\mathbf{x}(t_f)] = \begin{bmatrix} [h_s - x_3(t_f)\sin x_1(t_f)] - h_d \\ x_3(t_f)\cos x_1(t_f) - r_d \end{bmatrix}$$

d. Desired terminal position with hand on and perpendicular to an arbitrary object defined by $h_d = h(r_d)$:

$$\boldsymbol{\psi}[\mathbf{x}(t_f)] = \begin{bmatrix} \{h_s - a\sin[x(t_f) + b]\} - h_d \\ a\cos[x_1(t_f) + b] - r_d \\ [x_1(t_f) + x_5(t_f)] - \tan^{-1}(dh_d/dr) \end{bmatrix}$$

where

$$a = [x_3^2(t_f) + d^2 + 2x_3(t_f)d\cos x_5(t_f)]^{1/2}$$

$$b = \frac{d\sin x_5(t_f)}{a}$$

Equality Constraints on the State and Control

The state constraints discussed in the last section become important when the end of the trajectory is reached, but equality constraints that must be satisfied at every instant of the trajectory also can occur. It may be possible to *redefine the problem* so that the constraints are satisfied explicitly, and since this normally would allow the dimension of the state and/or the control to be reduced, this is a desirable alternative for problem solution. Another alternative is to augment the Lagrangian with a *penalty function* that attributes high cost to deviations from the constraint. The third alternative is to use *Lagrange multipliers* to adjoin equality constraints to the cost function. This approach can be employed in situations where the explicit substitution of the constraint is not practical, and it is a helpful antecedent to the discussion of inequality constraints.

The cost function

$$J = \phi[\mathbf{x}(t_f), t_f] + \int_{t_0}^{t_f} \mathscr{L}[\mathbf{x}(t), \mathbf{u}(t), t]\, dt \tag{3.5-34}$$

is to be minimized subject to the n-dimensional dynamic constraint

$$\dot{\mathbf{x}}(t) = \mathbf{f}[\mathbf{x}(t), \mathbf{u}(t), t], \qquad \mathbf{x}(t_0) = \mathbf{x}_0 \tag{3.5-35}$$

and the r-dimensional equality constraint

$$\mathbf{0} = \mathbf{c}[\mathbf{x}(t), \mathbf{u}(t), t] \tag{3.5-36}$$

in the interval $[t_0, t_f]$. As before, $\mathbf{x}$ and $\mathbf{u}$ have the dimensions n and m, respectively. The n-dimensional adjoint vector $\boldsymbol{\lambda}(t)$ can be associated with the dynamic constraint, while the r-dimensional Lagrange multiplier $\boldsymbol{\mu}(t)$ is associated with the equality constraint. The augmented cost function becomes

$$J = \phi + \int_{t_0}^{t_f} [\mathscr{L} + \boldsymbol{\lambda}^T(\mathbf{f} - \dot{\mathbf{x}}) + \boldsymbol{\mu}^T \mathbf{c}]\, dt \tag{3.5-37}$$

Consequently, the Hamiltonian can be written as

$$\mathscr{H} = \mathscr{L} + \boldsymbol{\lambda}^T \mathbf{f} + \boldsymbol{\mu}^T \mathbf{c} \tag{3.5-38}$$

and the corresponding Euler–Lagrange equations are

$$\begin{aligned}\dot{\boldsymbol{\lambda}} &= -\left(\frac{\partial \mathscr{H}}{\partial \mathbf{x}}\right)^T \\ &= -\left(\frac{\partial \mathscr{L}}{\partial \mathbf{x}}\right)^T - \mathbf{F}^T\boldsymbol{\lambda} - \left(\frac{\partial \mathbf{c}}{\partial \mathbf{x}}\right)^T \boldsymbol{\mu} \qquad (n \text{ scalar equations})\end{aligned} \tag{3.5-39}$$

$$\boldsymbol{\lambda}(t_f) = \left(\frac{\partial \phi}{\partial \mathbf{x}}\right)^T\bigg|_{t=t_f} \tag{3.5-40}$$

$$\begin{aligned}\mathbf{0} &= \left(\frac{\partial \mathscr{H}}{\partial \mathbf{u}}\right)^T \\ &= \left(\frac{\partial \mathscr{L}}{\partial \mathbf{u}}\right)^T + \mathbf{G}^T\boldsymbol{\lambda} + \left(\frac{\partial \mathbf{c}}{\partial \mathbf{u}}\right)^T \boldsymbol{\mu} \qquad (m \text{ scalar equations})\end{aligned} \tag{3.5-41}$$

There are enough equations here to solve for the n components of $\boldsymbol{\lambda}(t)$ and the m components of $\mathbf{u}(t)$, but the r components of $\boldsymbol{\mu}(t)$ also must be determined. The constraint equation (Eq. 3.5-36) provides r scalar equations which, together with Eqs. 3.5-39 to 3.5-41, allows $\mathbf{u}(t)$, $\boldsymbol{\lambda}(t)$, and $\boldsymbol{\mu}(t)$ to be found simultaneously.

Of course, $\boldsymbol{\mu}(t)$ does not appear in Eq. 3.5-36, so the implication is that one of the other equations [e.g., Eq. 3.5-41], provides the solution for $\boldsymbol{\mu}(t)$. The appropriate manipulations are problem-dependent, but we can get an idea of the methodology by considering a case in which $r = m$ and $\partial \mathbf{c}/\partial \mathbf{u}$ is nonsingular throughout the interval. Then the control $\mathbf{u}(t)$ is specified entirely by Eq. 3.5-36 (without regard for optimality), and Eq. 3.5-41 can be rearranged to find $\boldsymbol{\mu}(t)$:

$$\boldsymbol{\mu} = -\left[\left(\frac{\partial \mathbf{c}}{\partial \mathbf{u}}\right)^T\right]^{-1}\left[\left(\frac{\partial \mathscr{L}}{\partial \mathbf{u}}\right)^T + \mathbf{G}^T\boldsymbol{\lambda}\right] \tag{3.5-42}$$

If $r < m$, $(m-r)$ elements of $\mathbf{u}(t)$ can be obtained from the optimality condition (Eq. 3.5-41), the remaining control components are specified by the constraint (Eq. 3.5-36), and the r components of μ are found from

$$\boldsymbol{\mu} = -\left[\left(\frac{\partial \mathbf{c}_R}{\partial \mathbf{u}}\right)^T\right]^{-1}\left[\left(\frac{\partial \mathscr{L}}{\partial \mathbf{u}}\right)^T + \mathbf{G}^T\boldsymbol{\lambda}\right] \tag{3.5-43}$$

where $\partial \mathbf{c}_R/\partial \mathbf{u}$ is a nonsingular $(r \times r)$ partition of the $(r \times m)$ $\partial \mathbf{c}/\partial \mathbf{u}$. $\boldsymbol{\mu}(t)$ also might be found by Gaussian elimination or by using the left pseudoinverse (Section 2.2).

If the constraint is a function of the control alone,

$$\mathbf{0} = \mathbf{c}[\mathbf{u}(t), t] \tag{3.5-44}$$

it is again apparent that r must be less than m for the optimality conditions to have any effect on the control. Because $\partial \mathbf{c}/\partial \mathbf{x} = \mathbf{0}$, the control has no effect on the integration of the adjoint equation (Eq. 3.5-39), but the remainder of the preceding discussion regarding the determination of $\mathbf{u}$ and $\boldsymbol{\mu}$ applies.

An important aspect of these two equality constraints (Eqs. 3.5-36 and 3.5-44) is that both contain control mechanisms for maintaining the constraint. For example, at any given time on a time-invariant state-control equality constraint,

$$\mathbf{0} = \frac{\partial \mathbf{c}}{\partial \mathbf{x}}\Delta\mathbf{x} + \frac{\partial \mathbf{c}}{\partial \mathbf{u}}\Delta\mathbf{u} \tag{3.5-45}$$

so a state perturbation $\Delta\mathbf{x}$ could be offset by a control perturbation $\Delta\mathbf{u}$. For both constraints, a change in one control variable Δu_i could be opposed by another control variable Δu_j. If the r-dimensional constraint is a function of the state only,

$$\mathbf{0} = \mathbf{c}^{(0)}[\mathbf{x}(t), t] \tag{3.5-46}$$

where the meaning of the superscript is made clear below, then $\partial \mathbf{c}^{(0)}/\partial \mathbf{u} = \mathbf{0}$, and there is no immediate control over the constraint. Nevertheless, if $\mathbf{c}^{(0)} = \mathbf{0}$ throughout the interval, then $d\mathbf{c}^{(0)}/dt$, and all higher time-derivatives of the constraint must be zero as well. From the chain rule, the first derivative of Eq. 3.5-46 is

$$\frac{d\mathbf{c}^{(0)}}{dt} = \frac{\partial \mathbf{c}^{(0)}}{\partial t} + \frac{\partial \mathbf{c}^{(0)}}{\partial \mathbf{x}}\dot{\mathbf{x}}(t) = \mathbf{0} \tag{3.5-47a}$$

Using Eq. 3.5-35, this is

$$\frac{d\mathbf{c}^{(0)}}{dt} = \frac{\partial \mathbf{c}^{(0)}}{\partial t} + \frac{\partial \mathbf{c}^{(0)}}{\partial \mathbf{x}} \mathbf{f}[\mathbf{x}(t), \mathbf{u}(t), t]$$
$$\triangleq \mathbf{c}^{(1)}[\mathbf{x}(t), \mathbf{u}(t), t] \tag{3.5-47b}$$

where the superscript indicates the number of differentiations required to generate the constraint (B-8). In other words, the first derivative of the state constraint apparently provides a state-control constraint of the proper form.*

For the first-order state equality constraint, there are $2r$ equations to be satisfied. $\mathbf{c}^{(1)}$ is treated like the equality constraint of Eq. 3.5-36, and $\mathbf{c}^{(0)}$ either replaces r components of the state equation (Eq. 3.5-35) or establishes the initial conditions for r states as functions of the remaining $(n-r)$ states. In the latter case, the constrained-optimal control forces the integration to satisfy both Eqs. 3.5-46 and 3.5-47.

EXAMPLE 3.5-2 EQUALITY CONSTRAINTS ON TAKEOFF AND LANDING

The differential equations that describe the low-speed longitudinal motions† of an aircraft's center of mass can be written as

$$\dot{v} = \frac{T(\rho, v, \delta T)\cos\alpha - C_D(\alpha, \delta F)\rho v^2 S/2}{m} - g\sin\gamma$$
$$= f_1(\mathbf{x}, \mathbf{u})$$
$$\dot{\gamma} = \frac{T(\rho, v, \delta T)\sin\alpha + C_L(\alpha, \delta F)\rho v^2 S/2}{mv} - \frac{g}{v}\cos\gamma$$
$$= f_2(\mathbf{x}, \mathbf{u})$$
$$\dot{h} = v\sin\gamma = f_3(\mathbf{x}, u)$$
$$\dot{r} = v\cos\gamma = f_4(\mathbf{x}, \mathbf{u})$$
$$\dot{m} = f_5(\rho, v, \delta T) = f_5(\mathbf{x}, \mathbf{u})$$

*Due to the specific structures of $\mathbf{c}^{(0)}$ and $\mathbf{f}$, it may occur that the product $(\partial \mathbf{c}^{(0)}/\partial \mathbf{x})\,\mathbf{f}$ does not contain the control in each of its r elements. In such case, the differentiation is repeated (row by row) until the control appears. If q differentiations are required to bring out the control dependence, a qth-*order state equality constraint* $\mathbf{c}^{(q)}$ results. If the control never appears in the derivatives, the $(\mathbf{c}^{(0)}, \mathbf{f})$ pair lacks adequate controllability and is, therefore, abnormal.

†Motions that occur in a vertical plane fixed by the local vertical and the aircraft's fore-aft axis, with wings level.

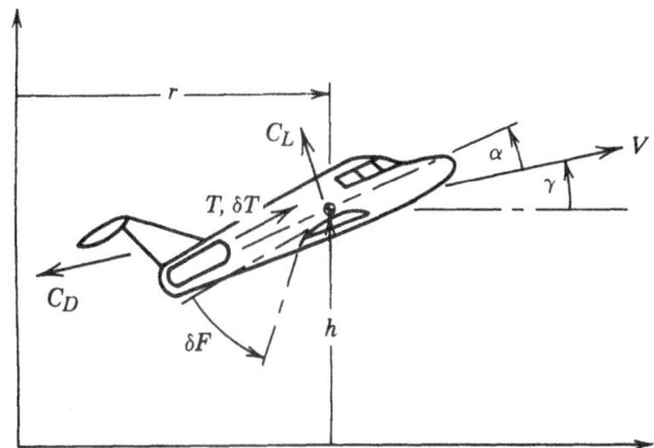

where the velocity (v), flight path angle (γ), height (h), and range (r) are defined in the sketch, and m is the aircraft's mass. The thrust is T, the gravitational acceleration is g, air density is ρ, the aircraft's reference area is S, and the lift and drag coefficients are C_L and C_D, respectively. The equations take the form

$$\dot{\mathbf{x}} = \mathbf{f}(\mathbf{x}, \mathbf{u})$$

with

$$\mathbf{x}^T = [v \quad \gamma \quad h \quad r \quad m]$$

and

$$\mathbf{u} = \begin{bmatrix} u_1 \\ u_2 \\ u_3 \end{bmatrix} = \begin{bmatrix} \alpha \\ \delta T \\ \delta F \end{bmatrix} = \begin{matrix} \text{angle of attack} \\ \text{throttle setting} \\ \text{flap setting} \end{matrix}$$

The aircraft's pitching motion about the center of mass is assumed to be well controlled, allowing the angle between the centerline and the velocity vector, α, to be used as a control variable.

Minimizing the fuel used always is a concern, and if the airport is near a population center, the noise N of takeoff and landing—as perceived at various points near the approach and departure paths—should be minimized as well. Therefore, the following is a feasible cost function for minimization:

$$J = \int_{t_0}^{t_f} \left\{ k_0 \dot{m} + \sum_{i=1}^{n} k_i \, N_i[T(\rho, v, \delta T), h, r] \right\} dt$$

The relative importances of fuel used and noise perceived at n points are weighted by the k_i.

Examples of equality constraints can be given:

a) Flaps and throttle held constant during a trajectory segment:

$$\mathbf{c}(\mathbf{u}) = \begin{bmatrix} (\delta T - \delta T_c) \\ (\delta F - \delta F_c) \end{bmatrix} = \mathbf{0}$$

b) Constant flight path angle rate during the "flare," which matches the glide slope and the touchdown point:

$$c(\mathbf{x}, \mathbf{u}) = \frac{T(\rho, v, \delta T)\sin\alpha + C_L(\alpha, \delta F)\rho v^2 S/2}{mv} - \frac{g}{v}\cos\gamma - \dot{\gamma}_c = 0$$

c) Constant flight path angle on the glide slope:

$$c(\mathbf{x}) = \gamma - \gamma_c = 0$$

plus

$$\begin{aligned}\frac{\partial c}{\partial t} &= \dot{\gamma} - 0 \\ &= \frac{T(\rho, v, \delta T)\sin\alpha + C_L(\alpha, \delta F)\rho v^2 S/2}{mv} - \frac{g}{v}\cos\gamma = 0\end{aligned}$$

d) Constant height in a holding pattern:

$$c(\mathbf{x}) = h - h_c = 0$$

plus

$$\begin{aligned}\frac{\partial c}{\partial t} &= \dot{h} - 0 \\ &= v\sin\gamma = 0 \Rightarrow \gamma = 0 \text{ (because } v \neq 0)\end{aligned}$$

plus

$$\frac{\partial^2 c}{\partial t^2} = \ddot{h}$$

$$= \dot{v} \sin \gamma + v\dot{\gamma} \cos \gamma$$

$$= v\dot{\gamma} \text{ (because } \gamma = 0)$$

$$= \frac{T(\rho, v, \delta T) \sin \alpha + C_L(\alpha, \delta F)\rho v^2 S/2}{m} - g \cos \gamma = 0$$

e) Constant dynamic pressure ($\bar{q} \triangleq \rho v^2/2$) during climb or descent, with air density a function of height:

$$c(\mathbf{x}) = \frac{\rho v^2}{2} - \bar{q}_c = 0$$

plus

$$\frac{\partial c}{\partial t} = \frac{\dot{\rho} v^2}{2} + \rho v \dot{v}$$

$$= \frac{1}{2} \frac{\partial \rho}{\partial h} \dot{h} v^2 + \rho v \dot{v}$$

$$= \frac{1}{2} \frac{\partial \rho}{\partial h} (v^3 \sin \gamma)$$

$$+ \rho v \left[\frac{T(\rho, v, \delta T) \cos \alpha - C_D(\alpha, \delta F)\rho v^2 S/2}{m} - g \sin \gamma \right] = 0$$

The five examples include control, state-control, and state equality constraints. Examples c and e are first-order constraints, and Example d is a second-order constraint.

Inequality Constraints on the State and Control

In practice, the state and control variables of *all* control problems are bounded by physical or economic considerations. For example, a control valve has maximum and minimum openings, an airplane's altitude is limited by the earth's surface on one end and by its operating ceiling on the other, the budget for fuel is fixed, and so on. As long as the variations in $\mathbf{x}(t)$ and $\mathbf{u}(t)$ remain within "reasonable limits" that do not challenge these boundaries, they can be neglected during optimization. However, it often occurs that the bounds intrude upon a system's intended region of operation, so it is necessary to develop methods for handling these constraints. Furthermore, if the cost function is not *convex* (i.e., if it does not possess a stationary minimum in the conventional sense), control bounds may establish part or all of the optimal control history.

The general form of the r-component inequality constraint is

$$\mathbf{c}[\mathbf{x}(t), \mathbf{u}(t), t] \leq \mathbf{0} \tag{3.5-48}$$

in the interval $[t_0, t_f]$, and while it indicates that $\mathbf{c}$ is negative semidefinite and is bounded on one side, the values of $\mathbf{x}$ and $\mathbf{u}$ within the constraint can be positive or negative and bounded on two sides if appropriate. Considering scalar constraints on a scalar control for illustration,

$$(u - u_{max}) \leq 0, \qquad u_{max} > 0 \tag{3.5-49}$$

allows positive values of u with a one-sided boundary,

$$|u| - |u_{max}| \leq 0 \tag{3.5-50}$$

allows positive and negative values of u with a symmetric two-sided boundary, and

$$(u - u_{max})(u - u_{min}) \leq 0 \tag{3.5-51}$$

provides an asymmetric two-sided boundary, with positive or negative values of u (depending on the choice of u_{max} and u_{min}). These are "hard" constraints; a "soft" constraint could be defined using a penalty function, such as

$$\mathscr{L}_c(u) = \begin{cases} k(u - u_{max})^2, & u \geq u_{max} \\ 0, & u_{min} < u < u_{max} \\ k(u - u_{min})^2, & u \leq u_{min} \end{cases} \tag{3.5-52}$$

and adding it to the cost function's Lagrangian. Looking ahead to our interest in *stochastic* optimal control, which must tolerate random variations in system variables, soft constraints may prove to be more useful than hard constraints, which could be applied sensibly only for deterministic limits.

Clearly, there are two possibilities for solution: the state and control are on one or more constraining boundaries, or they are not. In the first case, the corresponding equality constraints apply, and the optimizing criteria of the previous section can be employed. Therefore, at least the first derivatives of $\mathbf{c}[\cdot]$ should be well-behaved on the boundary. In the second case, the constraints are ignored. Consequently, the specifics of $\mathbf{c}[\cdot]$ are immaterial in the interior region. The only additional problem is that the constrained and unconstrained optimal control solutions must be matched at those points where the solution transitions from one to the other, suggesting that the derivatives of $\mathbf{c}[\cdot]$ be well-behaved in the interior neighborhood of the boundary. Furthermore, additional necessary conditions have to be satisfied at the junctures of the constrained and unconstrained trajectory segments (commonly called "arcs") to assure continuity of the solution.

Two additional differences between equality and inequality constraints should be noted. Consider an r-dimensional constraint. If it is an equality constraint, all components are in force all the time, but if it is an inequality constraint, from zero to r constraints may be invoked, depending on the trajectory. Figure 3.5-1 provides an example of the possible effects of a three-dimensional state inequality constraint on two optimal trajectories, both beginning at $\mathbf{x}_0$ and ending at either $\mathbf{x}_{f_1}$ or $\mathbf{x}_{f_2}$. In the first case, c_1 clips the trajectory and is the only component of $\mathbf{c}[\mathbf{x}(t), t]$ to come into play. The trajectory consists of two unconstrained arcs separated by a constrained arc. In the second case, $c_1 \leq 0$ acts first, then $c_2 \leq 0$ constrains the trajectory. Note that the two scalar constraints act concurrently at only one point, and $c_3 \leq 0$ never has any effect. Because the constraint may or may not be affecting the control at any given time, the simple expedient of redefining the problem to satisfy the constraint explicitly will not work.

A second dissimilarity between equality and inequality constraints regards the number of allowable constraints. A control equality constraint

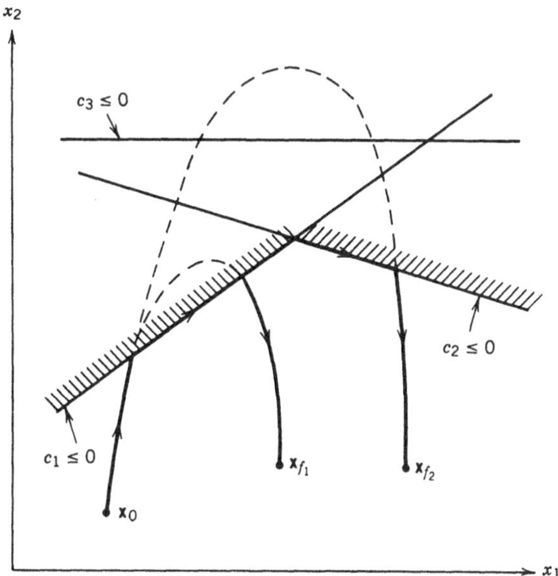

FIGURE 3.5-1 Effects of state inequality constraints on two trajectories.

of the same dimension as the control is overly restrictive, preventing optimization. However, an inequality constraint can have the same dimension as the control and still allow optimization because (1) the control need not stay on the boundary, and (2) there may be maximum and minimum boundaries, implying more than one level of admissible control (Fig. 3.5-2).

From the previous discussion and results, it can be concluded that the necessary conditions for optimality subject to the *state-control inequality constraint* (Eq. 3.5-48) are as follows. The Hamiltonian is defined as

$$\mathcal{H} = \mathcal{L} + \boldsymbol{\lambda}^T \mathbf{f} + \boldsymbol{\mu}_{\text{eff}}^T \mathbf{c}_{\text{eff}} \begin{cases} \boldsymbol{\mu}_{\text{eff}} = \mathbf{0} \text{ if } \mathbf{c}_{\text{eff}} < \mathbf{0} \\ \boldsymbol{\mu}_{\text{eff}} \geq \mathbf{0} \text{ if } \mathbf{c}_{\text{eff}} = \mathbf{0} \end{cases} \tag{3.5-53}$$

where $\mathbf{c}_{\text{eff}}[\mathbf{x}(t), \mathbf{u}(t), t]$ is the vector of the components of $\mathbf{c}$ in effect on the boundary, and $\boldsymbol{\mu}_{\text{eff}}(t)$ is the corresponding Lagrange multiplier. (The dimension of $\mathbf{c}_{\text{eff}}$ is assumed to be less than or equal to r.) The adjoint equation is

$$\dot{\boldsymbol{\lambda}} = \begin{cases} -\left(\dfrac{\partial \mathcal{L}}{\partial \mathbf{x}}\right)^T - \mathbf{F}^T \boldsymbol{\lambda}, & \mathbf{c}_{\text{eff}} < \mathbf{0} \\ -\left(\dfrac{\partial \mathcal{L}}{\partial \mathbf{x}}\right)^T - \mathbf{F}^T \boldsymbol{\lambda} - \left(\dfrac{\partial \mathbf{c}_{\text{eff}}}{\partial \mathbf{x}}\right)^T \boldsymbol{\mu}_{\text{eff}}, & \mathbf{c}_{\text{eff}} = \mathbf{0} \end{cases} \tag{3.5-54}$$

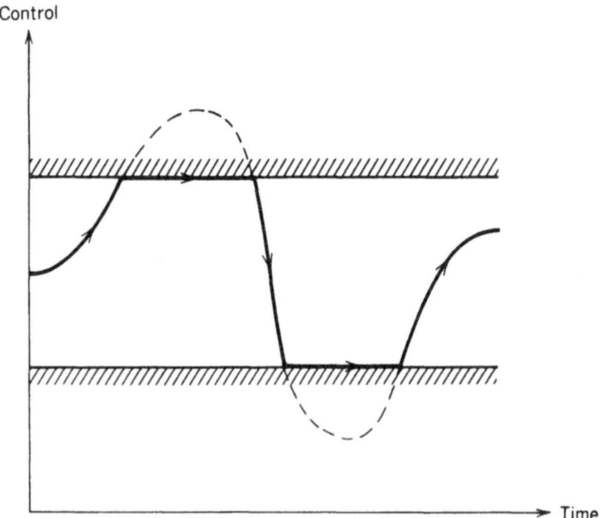

FIGURE 3.5-2 Effect of a single inequality constraint on a single control.

Equation 3.5-54 has the end condition

$$\boldsymbol{\lambda}(t_f) = \left(\frac{\partial \phi}{\partial \mathbf{x}}\right)^T \bigg|_{t=t_f} \tag{3.5-55}$$

and the condition for control optimality is

$$\mathbf{0} = \begin{cases} \left(\dfrac{\partial \mathscr{L}}{\partial \mathbf{u}}\right)^T + \mathbf{G}^T \boldsymbol{\lambda}, & \mathbf{c}_{\text{eff}} < \mathbf{0} \\[2ex] \left(\dfrac{\partial \mathscr{L}}{\partial \mathbf{u}}\right)^T + \mathbf{G}^T \boldsymbol{\lambda} + \left(\dfrac{\partial \mathbf{c}_{\text{eff}}}{\partial \mathbf{u}}\right)^T \boldsymbol{\mu}_{\text{eff}}, & \mathbf{c}_{\text{eff}} = \mathbf{0} \end{cases} \tag{3.5-56}$$

If the constraint is independent of the state

$$\mathbf{c}[\mathbf{u}(t), t] \leq \mathbf{0} \tag{3.5-57}$$

then

$$\frac{\partial \mathbf{c}}{\partial \mathbf{x}} = \mathbf{0} \tag{3.5-58}$$

but the solution is otherwise, as above (Eqs. 3.5-51 to 3.5-56).

It has been tacitly assumed in the previous discussion that the cost function has a well-defined minimum because the Hamiltonian is convex; thus the inequality constraint merely plays the role of limiting the admissible control. However, it is possible that $\mathscr{H}$ is not convex, so the constraint is called upon to establish the minimizing control. *Minimum-time* and *minimum-fuel* problems often fall in this category and always do so if the dynamic system to be controlled is linear. These are reasonable objectives for optimal control, but they do not fit the mold. Fortunately, the minimum principle provides an easy solution.

As a demonstration of the optimality conditions enforced by bounded control, consider the linear dynamic system with constant coefficients

$$\dot{\mathbf{x}}(t) = \mathbf{F}\mathbf{x}(t) + \mathbf{G}u(t), \qquad \mathbf{x}(0) = \mathbf{x}_0 \tag{3.5-59}$$

whose scalar control magnitude is constrained to remain less than one:

$$|u(t)| - 1 \leq 0 \tag{3.5-60}$$

(It can be assumed that the physical scaling of the control is absorbed in **G**, so that u represents the full-scale control motion, ±100%.) The time to reach a terminal state constraint $[\psi(t_f) = 0]$ is to be minimized, so the cost function is

$$J = \int_{t_0}^{t_f} (1)\, dt \tag{3.5-61}$$

and the Hamiltonian is

$$\mathcal{H}(t) = 1 + \boldsymbol{\lambda}^T(t)[\mathbf{F}\mathbf{x}(t) + \mathbf{G}u(t)] \tag{3.5-62}$$

Then the adjoint equation is

$$\begin{aligned} \dot{\boldsymbol{\lambda}}(t) &= -\left[\frac{\partial \mathcal{H}(t)}{\partial \mathbf{x}}\right]^T \\ &= -\mathbf{F}^T \boldsymbol{\lambda}(t) \end{aligned} \tag{3.5-63}$$

subject to

$$\boldsymbol{\lambda}^T(t_f) = \left(\frac{\partial \psi}{\partial \mathbf{x}}\right)\Big|_{t=t_f} \tag{3.5-64}$$

$$\mathcal{H}(t_f) = 0 \tag{3.5-65}$$

The first and second derivatives of $\mathcal{H}$ with respect to u are

$$\frac{\partial \mathcal{H}}{\partial u} = \boldsymbol{\lambda}^T \mathbf{G} \tag{3.5-66}$$

$$\frac{\partial^2 \mathcal{H}}{\partial u^2} = 0 \tag{3.5-67}$$

The control variable does not appear in $\partial \mathcal{H}/\partial u$, so Eq. 3.5-66 does not reveal the optimal u, and $\mathcal{H}$ is not convex because $\partial^2 \mathcal{H}/\partial u^2$ is zero. Nevertheless, an optimal control exists because u is bounded.

The minimum principle and Eq. 3.5-62 tell us that

$$1 + \boldsymbol{\lambda}^{*T}(t)[\mathbf{F}\mathbf{x}^*(t) + \mathbf{G}u^*(t)] \le 1 + \boldsymbol{\lambda}^{*T}(t)[\mathbf{F}\mathbf{x}^*(t) + \mathbf{G}u(t)] \tag{3.5-68}$$

where $u^*(t)$ is the minimizing control and $u(t)$ is any other control. Eliminating like elements on both sides of the equation,

$$\boldsymbol{\lambda}^{*T}(t)\mathbf{G}u^*(t) \le \boldsymbol{\lambda}^{*T}(t)\mathbf{G}u(t) \tag{3.5-69}$$

and the minimum-$\mathcal{H}$ control is that which makes $\boldsymbol{\lambda}^{*T}(t)\mathbf{G}u(t)$ most negative at every time t:

$$u^*(t) = \begin{cases} +1 \text{ if } \boldsymbol{\lambda}^{*T}(t)\mathbf{G} < 0 \\ -1 \text{ if } \boldsymbol{\lambda}^{*T}(t)\mathbf{G} > 0 \end{cases} \tag{3.5-70}$$

Consequently, $u^*(t)$ is specified not by setting $\partial\mathcal{H}/\partial u$ to zero but by minimizing $\mathcal{H}$. It turns out that $\partial\mathcal{H}/\partial u$ is never zero for any length of time in this problem; it is zero only at the *switching points*, when $u^*(t)$ instantaneously reverses sign. This is an example of "bang-bang" or "maximum-effort" control, in which optimality requires full use of the control available.

Some degree of continuity must be assured in making the transition between optimal arcs with differing constraints, and, therefore, differing conditions for optimality. If the constraint is a function of the control, the *Weierstrass–Erdmann conditions*,* derived elsewhere [e.g., (C-1)], require that

$$\boldsymbol{\lambda}(t_1-) = \boldsymbol{\lambda}(t_1+) \tag{3.5-71}$$

$$\mathcal{H}(t_1-) = \mathcal{H}(t_1+) \tag{3.5-72}$$

where t_1 is the time of transition, t_1- is infinitesimally earlier than the transition, and t_1+ is infinitesimally later. Furthermore, optimality requires that

$$\mathcal{H}_{\mathbf{u}}(t_1-) = \mathcal{H}_{\mathbf{u}}(t_1+) = \mathbf{0} \tag{3.5-73}$$

In other words, $\boldsymbol{\lambda}$, $\mathcal{H}$, and $\mathcal{H}_{\mathbf{u}}$ are continuous, although the control may be discontinuous, as in the case of "bang-bang" control. These *transversality conditions*† apply not only to transitions between constrained and unconstrained arcs but to points where the trajectory goes from one set of constraints to another.

As long as $\partial\mathbf{c}_{\text{eff}}/\partial\mathbf{x}$ and $\boldsymbol{\mu}_{\text{eff}}$ are finite, the integral of Eq. 3.5-54 is continuous, so Eq. 3.5-71 is satisfied with the transition to the state-control equality constraint at $t = t_1$. Equation 3.5-72 is satisfied at $t = t_1$ by inspection of Eq. 3.5-53. If $\partial\mathbf{c}_{\text{eff}}/\partial\mathbf{u}$ is finite and $\boldsymbol{\mu}_{\text{eff}}$ is continuous at $t = t_1$ (i.e.,

*Karl Weierstrass (1815–1897), professor at the University of Berlin, was a central figure among nineteenth century mathematicians. His lectures on the calculus of variations (1865–1890) were profound but were not formally published, entering the literature primarily through the work of his students. G. Erdmann derived the corner conditions independently in 1877, 12 years after Weierstrass first presented them.

†In the most general sense, transversality conditions must be satisfied where two arcs are "transverse", i.e., where they intersect. The term also is applied to end conditions (e.g., terminal constraints), which apply to the intersection of the trajectory with the terminal surfaces specified by either Eq. 3.5-1 or 3.5-18.

if $\boldsymbol{\mu}_{\text{eff}}$ does not jump instantaneously to a nonzero value), then the control history $\mathbf{u}(t)$, derived from Eq. 3.5-56 also is continuous,

$$\mathbf{u}(t_1-) = \mathbf{u}(t_1+) \tag{3.5-74}$$

implying that

$$\dot{\mathbf{x}}(t_1-) = \dot{\mathbf{x}}(t_1+) \tag{3.5-75}$$

If the control is *not* continuous, this implies that the state rate jumps at $t = t_1$, in which case the juncture between the constrained and unconstrained arcs is called a *corner*. With a simple discontinuity in $\dot{\mathbf{x}}(t_1)$, the state itself remains continuous:

$$\mathbf{x}(t_1-) = \mathbf{x}(t_1+) \tag{3.5-76}$$

Higher-order rate discontinuities (which imply infinite acceleration) are required to force the state to be discontinuous. Although physically unlikely, such jumps could occur in reduced-order (approximate) models of complex dynamic systems (e.g., those derived from singular perturbation theory or "residualization").

The transition from an unconstrained arc to a state-constrained arc is made difficult by the lack of control dependency in $\mathbf{c}[\mathbf{x}(t), t] \leq \mathbf{0}$; however, the problem is clarified if we consider the unconstrained arc as a separate *open-end-time problem with terminal equality constraint*. For illustration, assume that the constraint is scalar. The time of the transition to the constrained arc t_1 takes the role of the final time, and the terminal constraint for the unconstrained arc is specified by

$$\boldsymbol{\psi}(t_1) = \begin{bmatrix} c^{(0)}(t_1) \\ c^{(1)}(t_1) \\ \vdots \\ c^{(q-1)}(t_1) \end{bmatrix} = \mathbf{0} \tag{3.5-77}$$

where the control explicitly occurs in the qth time derivative of the constraint $c^{(q)}$. The constraint $\boldsymbol{\psi}$ can be adjoined to the cost function through a corresponding constant Lagrange multiplier of appropriate dimension $\boldsymbol{\nu}$ (Eq. 3.5-22), with $\boldsymbol{\nu}$ specified as in Eq. 3.5-26. The Weierstrass–Erdmann conditions are equivalent to the open-end-time conditions on $\boldsymbol{\lambda}$ and $\mathscr{H}$ (Eqs. 3.5-20 and 3.4-25b):

$$\boldsymbol{\lambda}^T(t_1-) = \boldsymbol{\lambda}^T(t_1+) + \boldsymbol{\nu}^T \frac{\partial \boldsymbol{\psi}(t_1)}{\partial \mathbf{x}} \tag{3.5-78}$$

$$\mathscr{L}(t_1-)+\boldsymbol{\lambda}^T(t_1-)\mathbf{f}(t_1-)+\boldsymbol{\nu}^T\frac{\partial\boldsymbol{\psi}(t_1)}{\partial t}=\mathscr{L}(t_1+)+\boldsymbol{\lambda}^T(t_1+)\mathbf{f}(t_1+)+\mu c^{(q)}(t_1+) \tag{3.5-79}$$

The Lagrange multiplier μ adjoins the effective state-control constraint $c^{(q)}$ to the Hamiltonian for the constrained arc, and it is calculated following Eq. 3.5-43. The term $\boldsymbol{\nu}^T\partial\boldsymbol{\psi}/\partial t$ adjusts the terminal Hamiltonian of the unconstrained arc. Consequently, $\boldsymbol{\lambda}$, $\mathscr{H}$, and $\mathscr{H}_\mathbf{u}$ may be discontinuous at the entry point of a constrained arc. There is the additional requirement that

$$c^{(0)}(t_1+)=\cdots=c^{(q-1)}(t_1+)=0 \tag{3.5-80}$$

providing enough entry conditions so that maintaining $c^{(q)}=0$ assures that $c^{(0)}=0$ on the constrained arc. Having matched the arcs at the entry point, the Weierstrass–Erdmann conditions are satisfied automatically at the exit point, as the juncture forms an initial condition for the unconstrained arc that follows. Transitions between differing constrained arcs also must satisfy the optimality condition (Eq. 3.5-73) on both sides of the juncture. Alternate and additional necessary conditions for problems with state-inequality constraints are presented in (J-4) and (K-6).

Example 3.5-3 Inequality Constraints on the San Diego Freeway

The differential equations that describe the motion of an automobile on a horizontal road, assuming a "no-slip condition" between the tires and the

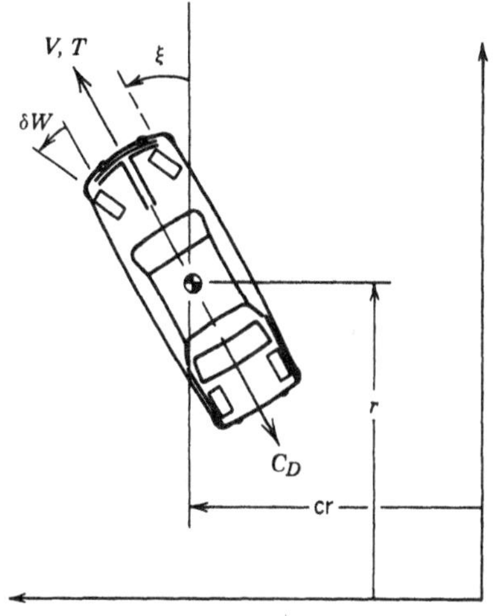

road, are approximately

$$\dot{v} = \frac{T(v, \delta T, \delta B)\cos^2 \delta W - C_D \rho v^2 S/2}{m}$$

$$\dot{\xi} = 2\left(\frac{v}{l}\right)\sin\left(\frac{\delta W}{2}\right)$$

$$\dot{r} = v \cos \xi$$

$$\dot{cr} = v \sin \xi$$

$$\dot{m} = f_5(v, \delta T)$$

As shown in the sketch, the state variables are forward velocity (v), heading angle (ξ), range (r), and crossrange (cr), plus the vehicle's mass. The net thrust produced by the engine in high gear is proportional to the engine's torque, which is related to throttle setting δT and to forward velocity through the engine rpm. The car's wheelbase is l, and the equations assume rear-wheel drive. Brake deflection δB reduces the thrust, as does the aerodynamic drag, $C_D \rho v^2 S/2$. The equations take the form

$$\dot{\mathbf{x}} = \mathbf{f}(\mathbf{x}, \mathbf{u})$$

with

$$\mathbf{x}^T = [v \quad \xi \quad r \quad cr \quad m]$$

and

$$\mathbf{u} = \begin{bmatrix} \delta W \\ \delta T \\ \delta B \end{bmatrix} \begin{matrix} \text{front wheel angle} \\ \text{throttle setting} \\ \text{brake setting} \end{matrix}$$

If we choose to minimize the fuel used per unit distance n while maintaining a desired average speed v_d, then

$$J = \int_{t_0}^{t_f} \left[\frac{k_1 f_5}{v} + k_2(v - v_d)^2\right] dt$$

is a suitable cost function. k_1 and k_2 weight the relative importance of fuel use and distance traveled in the time interval.

Examples of inequality constraints can be given:

a. Maximum and minimum control deflections:

$$\mathbf{c}(\mathbf{u}) = \begin{bmatrix} (\delta W - \delta W_{max})(\delta W - \delta W_{min}) \\ \delta T - 1 \\ \delta B - 1 \end{bmatrix} \leq \mathbf{0}$$

b. Maximum acceleration and deceleration:

$$c(\mathbf{x}, \mathbf{u}) = (\dot{v} - \dot{v}_{\text{max}})(\dot{v} - \dot{v}_{\text{min}}) \leq 0$$

where $\dot{v}$ is defined as above.

c. Staying within the speed limit:

$$c(\mathbf{x}) = v - v_{\text{max}} \leq 0$$

plus

$$\frac{\partial c}{\partial t} = \dot{v} = \frac{[T(v, \delta T, \delta B)\cos^2 \delta W - C_D \rho v^2 S/2]}{m} \leq 0$$

d. Staying within a straight lane (aligned with r):

$$c(\mathbf{x}) = (cr - cr_{\text{max}})(cr - cr_{\text{min}}) \leq 0$$

plus

$$\frac{\partial c}{\partial t} = 2cr\,\dot{cr} - (cr_{\text{max}} + cr_{\text{min}})\dot{cr}$$

$$= (2cr - cr_{\text{max}} - cr_{\text{min}})v \sin \xi \leq 0$$

plus

$$\frac{\partial^2 c}{\partial t^2} = 2\dot{cr}\, v \sin \xi + (2cr - cr_{\text{max}} - cr_{\text{min}})(\dot{v} \sin \xi + v\dot{\xi} \cos \xi) \leq 0$$

with the derivatives defined as above.

The four examples illustrate control, state-control, first-order state, and second-order state inequality constraints.

Singular Control

Optimal singular control becomes a possibility when the necessary and sufficient conditions described in the last section fail to define the control history and the corresponding trajectory. Failure of these conditions during a finite interval does not necessarily preclude optimality, nor does it guarantee that the trajectory contains an optimizing singular arc. It means that other criteria for optimality must be satisfied. Singular control is a subject of ongoing research with many problems and special cases, and only major points are summarized here.

A typical situation leading to the singular arc is that stationarity is achieved,

$$\frac{\partial \mathscr{H}}{\partial \mathbf{u}} = \mathscr{H}_{\mathbf{u}} = \mathbf{0} \tag{3.5-81}$$

but convexity is not:

$$\frac{\partial^2 \mathscr{H}}{\partial \mathbf{u}^2} = \mathscr{H}_{\mathbf{uu}} = \mathbf{0} \tag{3.5-82}$$

Thus the weakened convexity condition is satisfied, but the strengthened condition is not.

To get a notion of why we call this condition "singular," consider the minimization of a quadratic cost function,

$$J = \mathbf{x}^T(t_f)\mathbf{P}\mathbf{x}(t_f) + \frac{1}{2}\int_{t_0}^{t_f} [\mathbf{x}^T(t)\mathbf{Q}\mathbf{x}(t) + \mathbf{u}^T(t)\mathbf{R}\mathbf{u}(t)]\, dt \tag{3.5-83}$$

where $\mathbf{Q}$ and $\mathbf{R}$ are symmetric constant matrices. The minimization is subject to a linear, time-invariant dynamic constraint:

$$\dot{\mathbf{x}}(t) = \mathbf{F}\mathbf{x}(t) + \mathbf{G}\mathbf{u}(t), \qquad \mathbf{x}(t_0) = \mathbf{x}_0 \tag{3.5-84}$$

with open end time. The Hamiltonian is

$$\mathscr{H}(t) = \frac{1}{2}[\mathbf{x}^T(t)\mathbf{Q}\mathbf{x}(t) + \mathbf{u}^T(t)\mathbf{R}\mathbf{u}(t)] + \boldsymbol{\lambda}^T[\mathbf{F}\mathbf{x}(t) + \mathbf{G}\mathbf{u}(t)] \tag{3.5-85}$$

and its first and second derivatives with respect to $\mathbf{u}$ are

$$\mathscr{H}_{\mathbf{u}}(t) = \mathbf{u}^T(t)\mathbf{R} + \boldsymbol{\lambda}^T(t)\mathbf{G} = [\mathbf{R}\mathbf{u}(t) + \mathbf{G}^T\boldsymbol{\lambda}(t)]^T \tag{3.5-86}$$

$$\mathscr{H}_{\mathbf{uu}}(t) = \mathbf{R} \tag{3.5-87}$$

The extremal control, found from Eqs. 3.5-81 and 3.5-86, is

$$\mathbf{u}(t) = -\mathbf{R}^{-1}\mathbf{G}^T\boldsymbol{\lambda}(t) \tag{3.5-88}$$

and it is minimizing as long as $\mathbf{R}$ ($= \mathscr{H}_{\mathbf{uu}}$) is positive definite. If, however, $\mathbf{R}$ is singular, then $\mathbf{R}^{-1}$ does not exist, and the control history can not be found from Eq. 3.5-88.

Suppose that the cost is independent of the control:

$$J = \frac{1}{2}\int_{t_0}^{t_f} [\mathbf{x}^T(t)\mathbf{Q}\mathbf{x}(t)]\, dt \tag{3.5-89}$$

Setting $\mathbf{R}=\mathbf{0}$ removes the control from the Lagrangian, leaving a Hamiltonian (Eq. 3.5-85) that is linear in $\mathbf{u}$, the most common cause of a singular arc. Interestingly enough, the problem chosen as an example does have a singular optimal control, while the minimum-time problem discussed earlier (with $\mathcal{H}$ linear in u) does not.

Viewing Eq. 3.5-89 as a *penalty function that enforces a state equality constraint* provides insight into possible solutions. Just as time-derivatives of $c[\mathbf{x}(t), t]$ may bring out the control dependency in state equality constraints, time-derivatives of $\mathcal{H}_{\mathbf{u}}$ may identify the singular control. Because $\mathcal{H}_{\mathbf{u}}=\mathbf{0}$ throughout, all its derivatives with respect to time also should be zero. In the present case, the Hamiltonian's time dependence is implicit, and the optimal value of $\mathcal{H}$ is constant throughout the interval. This, together with Eq. 3.5-86 (and $\mathbf{R}=\mathbf{0}$), and

$$\frac{d}{dt}(\mathcal{H}_{\mathbf{u}})=\dot{\boldsymbol{\lambda}}^T\mathbf{G}=-\left(\frac{\partial\mathcal{H}}{\partial\mathbf{x}}\right)\mathbf{G}$$

$$=-(\mathbf{x}^T\mathbf{Q}+\boldsymbol{\lambda}^T\mathbf{F})\mathbf{G}=\mathbf{0} \tag{3.5-90}$$

$$\frac{d^2}{dt^2}(\mathcal{H}_{\mathbf{u}})=-[(\mathbf{Fx}+\mathbf{Gu})^T\mathbf{Q}-(\mathbf{x}^T\mathbf{Q}+\boldsymbol{\lambda}^T\mathbf{F})\mathbf{F}]\mathbf{G}=\mathbf{0} \tag{3.5-91}$$

provides enough equations to define the singular control.

The issue of how to reach the singular arc from arbitrary initial conditions must be addressed. With unbounded control and no penalty for its use, infinite pulses of infinitesimal width are admissible for matching both initial and final constraints. The resulting discontinuous jumps in the state are obtained with no increase in cost because the integration time is zero.

Realistically, the control must be bounded; the minimum principle would command saturated control for a nonzero time interval to reach the singular arc and to leave it for the terminal conditions. The cost increment would be minimal (but nonzero) on these nonsingular arcs. Because $\mathcal{H}_{\mathbf{u}}=\mathbf{0}$ on the singular arc and $\mathcal{H}_{\mathbf{u}}\neq\mathbf{0}$ on the control-constrained arcs, a transversality condition must be satisfied at the junctures. While a smooth transition is possible for some problems, in others it is not, requiring infinite control impulses to match the arcs.

Derivatives with respect to time figure in many of the necessary conditions that have been stated for singular control, and the *order of the singularity* can be related to the number of derivatives that are required for identification of the singular control. A *generalized convexity condition* can be expressed in two parts (J-3):

$$\frac{\partial}{\partial\mathbf{u}}\left[\frac{d^{(q)}}{dt^{(q)}}\left(\frac{\partial\mathcal{H}}{\partial\mathbf{u}}\right)\right]=\mathbf{0} \tag{3.5-92}$$

$$(-1)^{p/2}\frac{\partial}{\partial\mathbf{u}}\left[\frac{d^{(p)}}{dt^{(p)}}\left(\frac{\partial\mathcal{H}}{\partial\mathbf{u}}\right)\right]\geq\mathbf{0} \tag{3.5-93}$$

for q odd and p zero or even (0, 2, 4, . . .). The order of the singularity for each scalar control component u_i is determined by the first value of p for which

$$\frac{d^p}{dt^p}\left(\frac{\partial \mathcal{H}}{\partial u_i}\right) = a_i(\mathbf{x}, \boldsymbol{\lambda}, t) + u_i b_i(\mathbf{x}, \boldsymbol{\lambda}, t), \qquad i = 1 \text{ to } m \tag{3.5-94}$$

allowing u_i to be defined by a_i and b_i. Convexity is established by the additional differentiation with respect to the control (Eq. 3.5-93). Note that all control components need not have the same order of singularity, so it is conceivable that an optimization problem could be singular with respect to some controls and nonsingular with respect to others.

An alternative necessary condition for the singular second variation to be nonnegative is offered in (J-3). This condition does more than test the instantaneous convexity of $\mathcal{H}$, as it minimizes the second variation of the cost function (a so-called *accessory minimum problem*). The following equations must be satisfied on the singular arc:

$$\frac{\partial}{\partial \mathbf{u}}\left[\frac{\partial \mathcal{H}(t)}{\partial \mathbf{x}}\right]\mathbf{G}(t) + \mathbf{G}^T(t)\mathbf{P}(t)\mathbf{G}(t) \geq \mathbf{0} \tag{3.5-95}$$

$\mathbf{P}(t)$ is the solution to a linear differential equation

$$\dot{\mathbf{P}}(t) = -\left[\frac{\partial^2 \mathcal{H}(t)}{\partial \mathbf{x}^2} + \mathbf{F}^T(t)\mathbf{P}(t) + \mathbf{P}(t)\mathbf{F}(t)\right] \tag{3.5-96}$$

subject to the following terminal condition (at the end of the arc, t_2):

$$\mathbf{P}(t_2) = \frac{\partial^2 \phi}{\partial \mathbf{x}^2}[\mathbf{x}(t), t]_{t=t_2} \tag{3.5-97}$$

Investigation of accessory minimum problems in the neighborhood of steady singular arcs has led to the possibility of *periodic controls* providing lower cost than the steady solutions (B-6, S-4). Minimum-fuel cruising flight of aircraft has provided a prototype for study; a succession of physical models have been subjected to the tests described earlier, as well as to frequency domain tests of second-order variations. Optimal periodic solutions, in themselves, may be nonsingular and the result of inequality constraints (a generalization and repetition of "bang-bang" control over a long time interval), while the singular arcs may exist but be nonoptimal.

Hamiltonians that satisfy the minimum principle but fail the strengthened convexity condition (Eq. 3.4-27) could occur. For example, if there is a fourth-degree penalty on the scalar control of a linear dynamic system, the

Hamiltonian is

$$\mathcal{H} = \tfrac{1}{2}(\mathbf{x}^T\mathbf{Q}\mathbf{x} + ru^4) + \boldsymbol{\lambda}^T(\mathbf{F}\mathbf{x} + \mathbf{G}u) \tag{3.5-98}$$

If the problem is stated such that $u^*(t) = 0$ for any finite time interval, then $\mathcal{H}^*_{uu}(t) = 12ru^*(t)^2 = 0$ even though $\mathcal{H}^*$ is convex throughout. The problem is not singular, but it might appear to be, as a consequence of the chosen test for optimality. In a similar vein, singular problems sometimes can be made nonsingular by a *transformation of variables* or by *regularization*, in which a small convex control penalty such as ϵu^2, $\epsilon > 0$, is added to the cost function.

The need to use computers and iterative algorithms for all but the simplest nonlinear optimization problems has given rise to alternate solution techniques. A *generalized gradient method* interchanges the roles of selected components of the state and control along constrained or singular arcs (M-3). This can have the effect of redefining the arc as unconstrained or nonsingular. An *indirect numerical method* iterates components of the initial adjoint vector to enforce the transversality conditions, while satisfying Eqs. 3.5-81 and 3.5-90 to 3.5-93 on the singular arc (A-1). In the final analysis, numerical methods for singular control optimization must be chosen to fit the problem because there is no single definition of what the singular problem *is*; the singular problem is defined by what it *isn't*.

Example 3.5-4 Biological Population Dynamics

A predator-prey biological system can be modeled as

$$\dot{x}_1 = x_1(1 - x_2) - 0.5[e^{-a(t-t_i)} + b]x_1 u$$

$$\dot{x}_2 = x_2(x_1 - 1)$$

where x_1 represents the prey population, x_2 represents the predator population, and u represents a pesticide that kills only the prey (H-1). The pesticide has its strongest effect when it is applied at $t = t_i$; then its effect decays with time constant $(1/a)$ to its steady-state value b. Without the pesticide, the predator-prey population oscillates about the equilibrium point $x_1 = x_2 = 1$. It is desired to bring this population to its equilibrium value through minimum application of the pesticide. The pesticide can be applied at a rate in the range $0 < u < u_{\max}$. A cost function that penalizes the use of pesticide alone is

$$J = \int_{t_0}^{t_f} u\,dt$$

Write the Hamiltonian and Euler–Lagrange equations for the system. What control law is indicated by the minimum principle? Simulate a numerical-

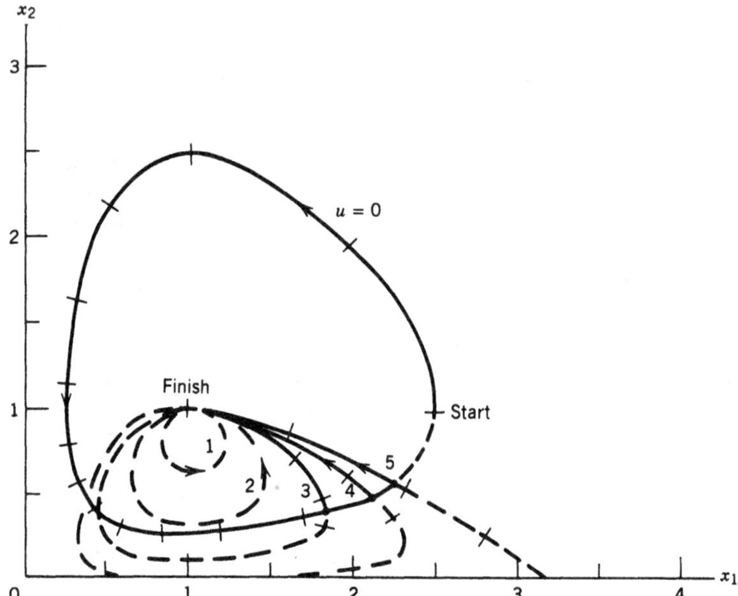

Phase plot of predator (x_2) versus prey (x_1) interchange and the effect of pesticide.

optimal solution for the following conditions:

$$a = 0.3, \quad b = 0.4, \quad x_1(0) = 2.5, \quad x_2(0) = 1\,,$$

$$t_0 = 0, \quad t_f = \text{open}, \quad x_1(t_f) = x_2(t_f) = 1\,, \quad u_{\max} = 1, 2, 3, 4, 5$$

The Hamiltonian is

$$\mathcal{H} = u + \lambda_1[x_1(1 - x_2) - 0.5(e^{-a(t-t_i)} + b)x_1 u] + \lambda_2[x_2(x_1 - 1)]$$

and the Euler–Lagrange equations are

$$\dot{\boldsymbol{\lambda}}(t) = -\left(\frac{\partial \mathcal{H}}{\partial \mathbf{x}}\right)^T$$

$$= \begin{bmatrix} \lambda_1[1 - x_2 - 0.5(e^{-a(t-t_1)} + b)u] \\ \lambda_2(x_1 - 1) - \lambda_1 x_1 \end{bmatrix}$$

$$\boldsymbol{\lambda}(t_f) = \left\{\frac{\partial \phi}{\partial \mathbf{x}}[\mathbf{x}(t_f)] = \mathbf{0}\right\}^T$$

$$\frac{\partial \mathcal{H}}{\partial u} = 1 - \lambda_1\,[0.5(e^{-a(t-t_1)} + b)\,x_1] = 0$$

This is a "bang-bang" control problem in that the control does not appear in $\mathcal{H}_u$; hence convexity cannot be established, and setting $\mathcal{H}_u = 0$ does not lead to a control solution. The minimum principle tells us that u should be chosen to minimize $\mathcal{H}$, and since u is bounded in $(0, u_{max})$, it takes one of the bounding values. If λ_1 is negative, choose $u = u_{max}$; if it is positive, choose $u = 0$. Therefore, $\mathcal{H}$ provides a *switching function*, which is the needed control law. Switching functions normally are continuous functions of the form $\sigma(\mathbf{x}, t)$. If $\sigma(\mathbf{x}, t) < 0$, the control is set to one limit; if $\sigma(\mathbf{x}, t) > 0$, it is set to the other.

Finding the adjoint variables $\lambda_1(t)$ and $\lambda_2(t)$ for this problem's switching function is an involved process, because the end time is open, there is a two-component terminal constraint, and the transversality conditions must be satisfied at all switching points. As an alternative, the switching function can be determined directly from solutions to this system's dynamic equations, with u set to its limiting values. Specifically, we must generate all trajectories passing through the initial and final conditions for which u equals 0 or u_{max}. The most direct way of doing this is to integrate *forward* from the initial conditions and *backward* from the terminal conditions, noting where the families of trajectories intersect.

Since there is *no inherent damping* in this system (as indicated by the eigenvalues of $\partial\mathbf{f}/\partial\mathbf{x}$), unforced trajectories are *closed contours* about the equilibrium point. For a range of u_{max}, forced trajectories also become closed contours about a shifted equilibrium point as t becomes large and the starting transient effect of the pesticide decays. It can be concluded that the control could be 0 or u_{max} at the starting point, but it *must* be at u_{max} at the terminal point. Consequently, the terminal switching function is determined by the unique forced trajectory that ends at the equilibrium point.

Phase plots (x_2 vs. x_1) of the solutions are shown in the accompanying figure. The pesticide's starting transient effect was not modeled ($a = 0$)—more about that later. With no pesticide application ($u = 0$), the starting condition represents an overabundance of the prey that leads to a rapid growth in the predator population (at the expense of the prey). Having "overgrazed," the predator population diminishes to a point that the prey population can be reestablished, and the cycle begins again.

While various application strategies for achieving a "balanced indigenous population" could be considered, the objective is to do so with a minimum amount of pesticide. (u actually represents the *rate* of application and/or uptake, so the total quantity is the time integral.) A family of trajectories with u_{max} set at values from 1 to 5 represents a range of maximum application rates. For the indicated initial conditions, it can be seen that application rates of about 2.2 or less are inadequate to solve the problem because there are no intersections of the starting and final trajectories. For higher rates, the goal can be accomplished with applications of shorter duration and later starting times (relative time is indicated by tick marks on the trajectories).

To account for the pesticide's starting transient effect, it would be necessary to iterate, making a range of assumptions about $(t_f - t_i)$ until the proper value matched the time required to go from the unforced contour to the equilibrium point. In general, this would lead to later starting times for application and smaller amounts of pesticide.

3.6 NUMERICAL OPTIMIZATION

One of the virtues of vector-matrix notation is that it allows broad statements to be made about the *nature* of the solutions to complex problems; its corresponding vice is that it gives little preparation for the *computational difficulties* that can be encountered in achieving numerical solutions to these problems. Bellman aptly identified the "curse of dimensionality" that accompanies the optimization of high-order systems, as the number of computations typically grows much faster than the dimensions of the state and control. In addition, formidable problems are associated with nonlinearities in a system's differential equations, particularly those that "map" the state and control into system-dependent forcing terms. These often must be represented by multidimensional polynomials or tables of discrete points. Integrating such equations, evaluating the partial derivatives required for the corresponding Euler–Lagrange equations, and finding the optimizing solution to this two-point boundary-value problem then becomes a challenging task, one that can be accomplished only with the help of a computing machine.

When we speak of a "computing machine" of a type that currently exists and is adequate for the optimizing task, we mean a digital or hybrid (digital-analog) computer that performs at least some (if not all) of its functions in a discrete, sequential fashion. Consequently, in order to achieve a numerical solution, the machine must solve equations by *iteration*, "stepping" rather than "sweeping" from start to finish. Actually, several independent iterations may be called for in generating a single optimal trajectory. For all-digital computation, the system differential equation typically is *iterated in time increments* from beginning to end, the adjoint differential equation is iterated in time increments from end to beginning, and the optimization step is concluded with the computation of a control history perturbation that is supposed to reduce the cost. Then the control history is *iterated in optimization steps*, each of which includes the state, adjoint, and control computations described earlier.

In this section, iteration is taken to be synonymous with iteration in optimization steps, the iterative solution of ordinary differential equations having been discussed in Section 2.3. Furthermore, we direct our attention to the numerical optimization of well-posed fixed-end-time Bolza problems without constraints or singularities, noting that

- Open-end-time problems effectively add another control component (time),
- Constraints may be represented by penalty functions, and
- Singular control often requires problem-specific special treatment.

Iterative techniques for finding the optimal control history, $\mathbf{u}^*(t)$, $t_0 \leq t \leq t_f$, and the corresponding state trajectory, $\mathbf{x}(t)$, must provide solutions of the n-component differential equation

$$\dot{\mathbf{x}}(t) = \mathbf{f}[\mathbf{x}(t), \mathbf{u}(t), t], \qquad \mathbf{x}(t_0) \text{ given} \tag{3.6-1}$$

subject to initial conditions for the state $\mathbf{x}(t_0)$ while minimizing the cost function

$$J = \phi[\mathbf{x}(t_f), t_f] + \int_{t_0}^{t_f} \mathscr{L}[\mathbf{x}(t), \mathbf{u}(t), t]\, dt \tag{3.6-2}$$

The calculus-of-variations approach requires that the Euler–Lagrange equations be satisfied. The adjoint vector history is obtained by integrating

$$\dot{\boldsymbol{\lambda}}(t) = -\left\{\frac{\partial \mathscr{H}[\mathbf{x}(t), \mathbf{u}(t), \boldsymbol{\lambda}(t), t]}{\partial \mathbf{x}}\right\}^T \tag{3.6-3}$$

subject to the terminal condition $\boldsymbol{\lambda}(t_f)$, where

$$\boldsymbol{\lambda}(t_f) = \left[\frac{\partial \phi(t)}{\partial \mathbf{x}}\right]^T\Bigg|_{t=t_f} \tag{3.6-4}$$

Finally, the optimality condition is computed,

$$\mathbf{0} = \frac{\partial \mathscr{H}[\mathbf{x}(t), \mathbf{u}(t), \boldsymbol{\lambda}(t), t]}{\partial \mathbf{u}} \tag{3.6-5}$$

and sufficient conditions are verified. Numerical solutions to the two-point boundary-value problem can be found using parametric optimization (Section 3.3), penalty functions, extremal fields (dynamic programming), neighboring extremals, quasilinearization, and gradient methods. The principal distinctions between these methods are which equations are nominally satisfied by the problem formulation itself and which must be forced to solution by iteration, as summarized in Table 3.6-1.

TABLE 3.6-1 Comparison of Numerical Optimization Methods

	Optimality of Solution	Solution Method $\mathbf{x}(t)$	$\boldsymbol{\lambda}(t)$	$\mathbf{u}(t)$	Iteration Variables	Order of ODE[a] Solution
Parametric	approximate	ODE[a]	–	I^b	$\mathbf{u}(\mathbf{k}_u, t)$	n
Penalty function	approximate	I	–	I	$\mathbf{x}(\mathbf{k}_x, t), \mathbf{u}(\mathbf{k}_u, t)$	none
Dynamic programming	exact	ODE	PDE[c]	I	$\mathbf{u}(t)$	n
Neighboring extremal	exact	ODE	ODE	$\mathscr{H}_\mathbf{u} = \mathbf{0}$	$\boldsymbol{\lambda}(t_0)$	$2n$
Quasilinearization	exact	I	I	$\mathscr{H}_\mathbf{u} = \mathbf{0}$	$\mathbf{x}(t), \boldsymbol{\lambda}(t)$	$2n^d$
Gradient	exact	ODE	ODE	I	$\mathbf{u}(t)$	$2n$

[a] ODE: ordinary differential equation.

[b] Iteration.

[c] PDE: Partial differential equation; HJB equation; one dependent variable (V), $(n+1)$ independent variables $(\mathbf{x}, t)$, $\partial V/\partial \mathbf{x}$ corresponds to $\boldsymbol{\lambda}^T$.

[d] Perturbation equation for $\Delta\mathbf{x}(t)$ and $\Delta\boldsymbol{\lambda}(t)$.

Penalty Function Method

Penalty function methods such as Balakrishnan's ϵ technique adjoin the dynamic constraint to the cost function (B-1),

$$J = \phi[\mathbf{x}(t_f), t_f] + \int_{t_0}^{t_f} \Big\{ \mathscr{L}[\mathbf{x}(t), \mathbf{u}(t), t] + \frac{1}{\epsilon}[\dot{\mathbf{x}}(t) - \mathbf{f}[\mathbf{x}(t), \mathbf{u}(t), t]]^T [\dot{\mathbf{x}}(t) - \mathbf{f}[\mathbf{x}(t), \mathbf{u}(t), t]] \Big\} dt \tag{3.6-6}$$

with ϵ small and positive, and with $\mathbf{x}(t)$ and $\mathbf{u}(t)$ initially chosen as known functions of time that satisfy end conditions precisely:

$$\mathbf{x}(t) = \mathbf{x}(\mathbf{k}_x, t) \tag{3.6-7}$$

$$\mathbf{u}(t) = \mathbf{u}(\mathbf{k}_u, t) \tag{3.6-8}$$

$\mathbf{k}_x$ and $\mathbf{k}_u$ are parameter vectors, as in Section 3.3. For example, the starting values of $\mathbf{x}(t)$ and $\mathbf{u}(t)$ could be linear in t, with specified $[\mathbf{x}(t_0), \mathbf{u}(t_0)]$ and $[\mathbf{x}(t_f), \mathbf{u}(t_f)]$. Since $\mathbf{x}(t)$ is known at each optimization step, the corresponding $\dot{\mathbf{x}}(t)$ can be calculated, and Eq. 3.6-6 can be evaluated without integrating Eq. 3.6-1; however, the state equation is not guaranteed to be satisfied. The iteration proceeds by varying $\mathbf{k}_x$ and $\mathbf{k}_u$ using an algorithm such as Newton–Raphson iteration until J is minimized. If the value of ϵ is concurrently made smaller, the dynamic constraint error may be reduced to a small value; however, J is truly optimized only if Eqs. 3.6-7 and 3.6-8 admit the exact optimal solution.

Dynamic Programming

The dynamic programming solution uses the HJB equation (Eq. 3.4-39) rather than the Euler–Lagrange equations (Eqs. 3.6-3 to 3.6-5) in iteration; hence the integral of a scalar partial differential equation for J must be minimized subject to the dynamic constraint of the n ordinary differential equations for $\mathbf{x}$. Although the partial differential equation has a single solution variable, it has $(n+1)$ independent variables $(\mathbf{x}, t)$, and it is difficult to solve numerically.

An alternate definition of dynamic programming is sometimes given. If a field of extremal trajectories and control histories has been established by *any* solution technique, an optimal trajectory corresponding to particular initial conditions could be generated by *interpolating* between the computed trajectories to define the optimizing control history (Eq. 3.4-42). This would be an expensive method of producing a single optimal trajectory, but once the field of extremals has been obtained, the interpolation itself could be extremely fast. If there is uncertainty regarding the initial conditions that will be in effect when the optimal control is applied, or if there is a continuing need to generate many optimal trajectories under essentially similar dynamic conditions, then the use of dynamic programming to define an optimal nonlinear feedback control law could be quite practical.

Neighboring Extremal Method

In the neighboring extremal (or variation-of-extremal) approach, starting values of the adjoint vector $\boldsymbol{\lambda}(t_0)$ are guessed, and Eqs. 3.6-1 and 3.6-3 are integrated from t_0 to t_f, with the control history specified by Eq. 3.6-5:

$$\mathbf{u}(t) = \mathbf{u}[\mathbf{x}(t), \boldsymbol{\lambda}(t), t] \tag{3.6-9}$$

Obtaining a closed-form solution for $\mathbf{u}(t)$ from Eq. 3.6-5 could require extensive calculation; it could be necessary to use an *iterative zero-finding algorithm* to identify the vector $\mathbf{u}(t)$ that makes each component of $\mathcal{H}_{\mathbf{u}}$ zero. The result is an extremal trajectory (because it satisfies the Euler–Lagrange equations) that does not necessarily satisfy the end conditions on $\boldsymbol{\lambda}(t_f)$ and $\mathbf{x}(t_f)$. The starting values of $\boldsymbol{\lambda}$ are perturbed by some function of the terminal errors, and the optimization step is repeated. This is called a *shooting method* of numerical optimization because all components of $\mathbf{x}$ and $\boldsymbol{\lambda}$ are projected from the starting condition to the "target" condition. Small variations in $\boldsymbol{\lambda}(t_0)$ often lead to large variations in $\boldsymbol{\lambda}(t)$ due to the adjoint equation's stability properties. State equations that are stable with time running forward imply adjoint equations that are stable with time running in reverse (Chapter 5). In such instance, the adjoint equations are unstable when they are integrated forward in time.

As an alternative, $\mathbf{x}$ and $\boldsymbol{\lambda}$ could be specified at t_f rather than t_0, and both Eqs. 3.6-1 and 3.6-3 would be integrated back in time. Solution stability problems still might exist for the variable whose solution is stable in forward time; this would be of secondary concern in generating a field of extremals for a dynamic-programming control law.

Quasilinearization Method

The method of quasilinearization requires a starting guess not only for the state history $\mathbf{x}(t)$ and the initial adjoint vector $\boldsymbol{\lambda}(t_0)$, but for the entire adjoint history $\boldsymbol{\lambda}(t)$ as well. Given $\mathbf{x}_o(t)$ and $\boldsymbol{\lambda}_o(t)$ in $[t_0, t_f]$, the corresponding value of control $\mathbf{u}_o(t)$ can be computed from Eq. 3.6-9. Then the state equation can be expressed as

$$\begin{aligned}\dot{\mathbf{x}}(t) &= \mathbf{f}[\mathbf{x}(t), \mathbf{u}_o(t), t] \\ &= \mathbf{f}'[\mathbf{x}(t), \boldsymbol{\lambda}_o(t), t]\end{aligned} \tag{3.6-10}$$

where $\mathbf{f}'$ is distinguished from $\mathbf{f}$ by the substitution of Eq. 3.6-9 for $\mathbf{u}(t)$. Nevertheless, there is no guarantee that Eq. 3.6-10 is satisfied by $\mathbf{x}_o(t)$ and $\boldsymbol{\lambda}_o(t)$, which were chosen arbitrarily to start the procedure. Similarly, the adjoint equation, with Eq. 3.6-9 substituted for $\mathbf{u}$, may not be satisfied. If we were lucky and chose $\mathbf{x}_o(t)$ and $\boldsymbol{\lambda}_o(t)$ close to the optimal values, then the difference between $[\mathbf{x}_o(t), \boldsymbol{\lambda}_o(t)]$ and $[\mathbf{x}^*(t), \mathbf{u}^*(t)]$ could be modeled using local linearization. Expanding both sides of Eqs. 3.6-1 and 3.6-3 about $\mathbf{x}_o(t)$ and $\boldsymbol{\lambda}_o(t)$, retaining only first-order variations, and collecting nominal and perturbation terms on opposite sides of the "about equal" sign,

$$\dot{\mathbf{x}}_o(t) - \mathbf{f}'[\mathbf{x}_o(t), \boldsymbol{\lambda}_o(t), t] \cong -\Delta\dot{\mathbf{x}}_o(t) + \left[\left(\frac{\partial \mathbf{f}'}{\partial \mathbf{x}}\right)\Delta\mathbf{x}_o(t) + \frac{\partial \mathbf{f}'}{\partial \boldsymbol{\lambda}}\Delta\boldsymbol{\lambda}_o(t)\right]_{\mathbf{x}=\mathbf{x}_o,\, \boldsymbol{\lambda}=\boldsymbol{\lambda}_o} \tag{3.6-11}$$

$$\begin{aligned}&\dot{\boldsymbol{\lambda}}_o(t) + \mathscr{H}_{\mathbf{x}}^T[\mathbf{x}_o(t), \boldsymbol{\lambda}_o(t), t] \cong \\ &\qquad -\Delta\dot{\boldsymbol{\lambda}}_o(t) - \left[\frac{\partial}{\partial \mathbf{x}}\mathscr{H}_{\mathbf{x}}^T\Delta\mathbf{x}_o(t) + \frac{\partial}{\partial \boldsymbol{\lambda}}\mathscr{H}_{\mathbf{x}}^T\Delta\boldsymbol{\lambda}_o(t)\right]_{\mathbf{x}=\mathbf{x}_o,\, \boldsymbol{\lambda}=\boldsymbol{\lambda}_o}\end{aligned} \tag{3.6-12}$$

where $\mathscr{H}$ is the Hamiltonian for $\mathscr{L}$ and $\mathbf{f}'$; $\mathscr{H}_{\mathbf{x}}^T$ is its gradient with respect to $\mathbf{x}$. Having evaluated the necessary partial derivatives at $[\mathbf{x}_o(t), \boldsymbol{\lambda}_o(t)]$ throughout the interval, equations of the form

$$\Delta\dot{\mathbf{x}}_o(t) = \mathbf{F}(t)\Delta\mathbf{x}_o(t) + \mathbf{G}(t)\Delta\boldsymbol{\lambda}_o(t) \tag{3.6-13}$$

$$\Delta\dot{\boldsymbol{\lambda}}_o(t) = \mathbf{H}(t)\Delta\mathbf{x}_o(t) + \mathbf{K}(t)\Delta\boldsymbol{\lambda}_o(t) \tag{3.6-14}$$

could be solved to obtain $\Delta\mathbf{x}(t)$ and $\Delta\boldsymbol{\lambda}(t)$ in $[t_0, t_f]$, subject to appropriate boundary conditions. (Similar equations are used to compute neighboring-optimal trajectories in Section 3.7.)

An iterative procedure for finding $\mathbf{x}^*(t)$, $\mathbf{u}^*(t)$, and $\boldsymbol{\lambda}^*(t)$ can be based on a succession of first-order improvements to the state and adjoint histories,

$$\begin{bmatrix} \mathbf{x}_{k+1}(t) \\ \boldsymbol{\lambda}_{k+1}(t) \end{bmatrix} = \begin{bmatrix} \mathbf{x}_k(t) \\ \boldsymbol{\lambda}_k(t) \end{bmatrix} + \epsilon \begin{bmatrix} \Delta\mathbf{x}_k(t) \\ \Delta\boldsymbol{\lambda}_k(t) \end{bmatrix} \tag{3.6-15}$$

plus the control solution from Eq. 3.6-5

$$\mathbf{u}_{k+1}(t) = \mathbf{u}[\mathbf{x}_{k+1}(t), \boldsymbol{\lambda}_{k+1}(t), t] \tag{3.6-16}$$

where $\epsilon < 1$ to aid convergence in early iterations, and k is an index of optimization steps. $\mathbf{F}(t)$, $\mathbf{G}(t)$, $\mathbf{H}(t)$, and $\mathbf{K}(t)$ are reevaluated on each optimization step to account for the changing values of $\mathbf{x}$ and $\boldsymbol{\lambda}$.

Gradient Methods

A particular appeal of the gradient methods is that the dynamic system equation is solved exactly on each iteration, with the control being perturbed from step to step to "home in" on the optimal solution. In other words, the algorithm *simulates* the system's dynamic response with varying control histories from one iteration to the next, and the physical effects of the optimization are quite visible at each step. In the remainder of this section, we will concentrate on gradient methods for solving Eqs. 3.6-1 to 3.6-5.

A schematic of gradient optimization is shown in Fig. 3.6-1. The successive approximation requires a step-by-step progression of solutions from the starting trajectory to the final trajectory, with the control history being adjusted on each step to further reduce the cost. The process begins with the specification of initial conditions and a nominal control history. The simulation program integrates Eq. 3.6-1 from t_0 to t_f to obtain $\mathbf{x}_0(t)$. At the same time, the cost function, system sensitivity matrices [$\mathbf{F}(t)$ and $\mathbf{G}(t)$], and Lagrangian gradients [$\mathscr{L}_{\mathbf{x}}(t)$ and $\mathscr{L}_{\mathbf{u}}(t)$] are evaluated. The adjoint vector is integrated back from the final point to obtain $\boldsymbol{\lambda}_0(t)$, allowing $\mathscr{H}_{\mathbf{u}}(t)$ to be determined. In early iterations, $\mathscr{H}_{\mathbf{u}}(t)$ may not be close to zero because the trajectory is not optimal. The control history is perturbed as some function of $\mathscr{H}_{\mathbf{u}}(t)$, and the process is repeated until the solution is perceived to be sufficiently close to the optimum. In most cases, sufficiency need not be verified on every iteration, so computation of $\mathscr{H}_{\mathbf{uu}}$ may be optional.

The choice of a stopping condition for this iterative process is somewhat arbitrary. If the minimizing value of J is known (e.g., it could be zero by definition), the iteration can stop when the computed cost is acceptably close to its known minimum value. Otherwise, the iteration can stop when

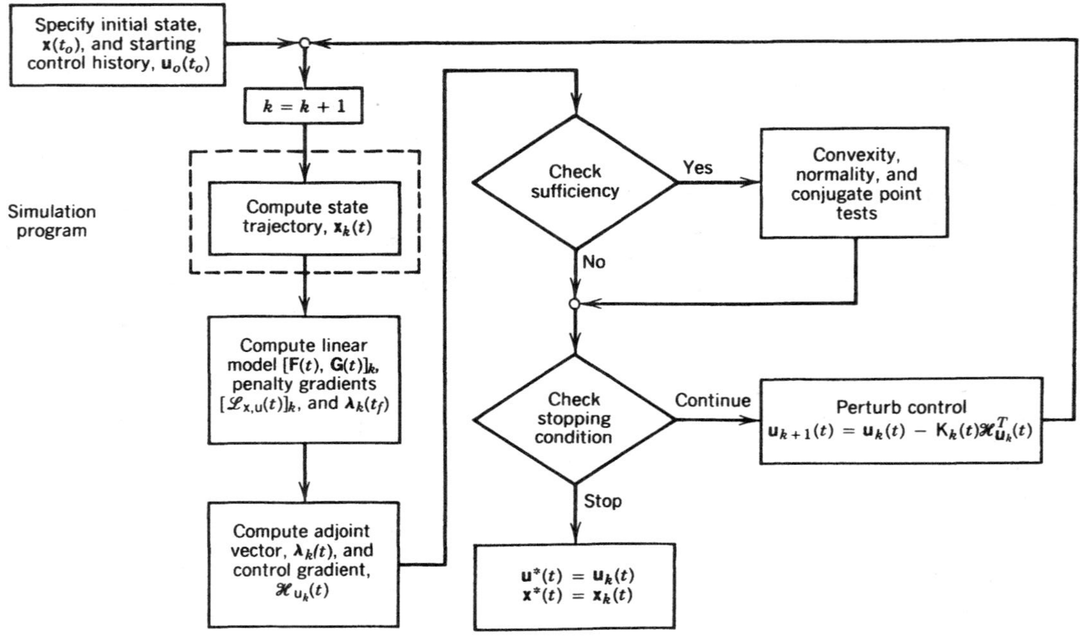

FIGURE 3.6-1 Gradient method of numerical optimization.

some scalar function of the control gradient, $\mathscr{H}_{\mathbf{u}}(t)$ (e.g., the integral of $\mathscr{H}_{\mathbf{u}}^T(t)\mathscr{H}_{\mathbf{u}}(t)$ over the time interval) is acceptably close to zero.

As shown in Fig. 3.6-1, the control history is perturbed and the partial derivatives are calculated on each iteration. One particularly simple control adjustment scheme derives the control perturbation for the $(k+1)$st iteration from the control gradient computed on the kth iteration:

$$\mathbf{u}_{k+1}(t) = \mathbf{u}_k(t) - \epsilon_k(t)\left[\frac{\partial \mathscr{H}(t)}{\partial \mathbf{u}}\right]_k^T \tag{3.6-17}$$

where $\epsilon_k(t)$ is a scalar control-gradient weighting function. This is known as a *steepest-descent algorithm*. Visualizing a bowl-shaped Hamiltonian $\mathscr{H}(\mathbf{u})$, the gradient $\mathscr{H}_{\mathbf{u}}$ defines the magnitude and direction of the local slope of the function. Perturbing the control by some function of $\mathscr{H}_{\mathbf{u}}$ moves the control toward the bottom of the bowl. Proper selection of the step size is critical for rapid convergence to the minimizing control. If $\epsilon_k(t)$ is too small, convergence may require a large number of iterations; if it is too large, convergence may not occur at all.

The steepest-descent concept is illustrated in Fig. 3.6-2 for a two-control minimization of the Hamiltonian $\mathscr{H}(t)$ at a given t. The direction of the line from 0 to 1 is established by the relative magnitudes of the components of $\mathscr{H}_{\mathbf{u}}$, and the distance along the line—the step size—is determined by $\epsilon\mathscr{H}_{\mathbf{u}}$. If ϵ is chosen correctly, the point 1 is at a local minimum. Having completed the first step, point 1 is the starting point for the next step. Note that the second control correction (from 1 to 2) tends to be orthogonal to the first

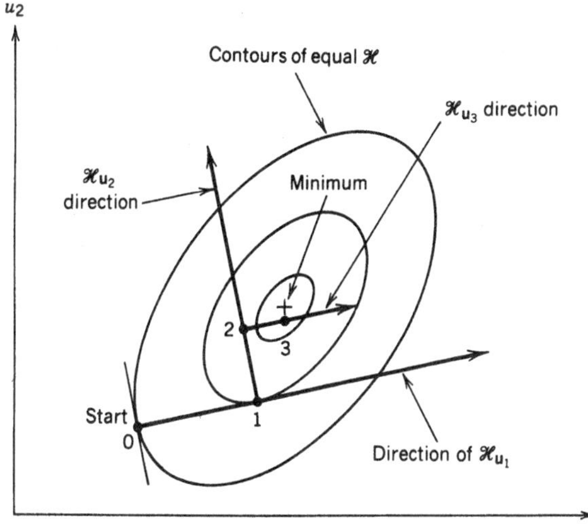

FIGURE 3.6-2 Illustration of efficient iterative steepest-descent approach to a minimum in the Hamiltonian.

correction if ϵ has been chosen well. If ϵ is too large or too small, more steps will be required to reach the minimum. On reaching the minimum, the optimizing values of $\mathbf{u}^*(t)$ are found, and $\mathscr{H}_\mathbf{u}$ is zero throughout the interval; hence this is a "zero-finding algorithm" of the sort mentioned earlier.

The steepest-descent algorithm can be applied to an entire trajectory by performing control corrections at discrete instants in $[t_0, t_f]$. Having chosen the first control correction at time t, the first control correction at the next sampling instant t_1 is computed before computing the second correction at time t. As a result, when we return to make the second control correction at time t, the estimated Hamiltonian contours are likely to have moved, and we cannot be sure that the second correction is orthogonal to the first, even if ϵ was properly chosen in the first instance.

Often, small changes in the control that occur late in the trajectory cause large variations in the terminal conditions. To allow large control perturbations early in the trajectory without upsetting the end condition, $\epsilon_k(t)$ can ramp from maximum to minimum value as t progresses from t_0 to t_f.

In the neighborhood of the optimal trajectory, $\mathscr{H}_{\mathbf{uu}}(t)$ should be greater than zero (by convexity); if the control correction is defined as

$$\mathbf{u}_{k+1}(t) = \mathbf{u}_k(t) - \mathscr{H}_{\mathbf{uu}_k}^{-1}(t)[\mathscr{H}_{\mathbf{u}_k}(t)]^T \tag{3.6-18}$$

the numerical method becomes a *Newton–Raphson* or *second-order gradient algorithm*. If $\mathscr{H}$ is quadratic in $\mathbf{u}$, Eq. 3.6-18 should converge to the minimum in two iterations. In practice, the symmetric matrix $\mathscr{H}_{\mathbf{uu}}(t)$ can be difficult to calculate precisely, $\mathscr{H}$ may not be quadratic in $\mathbf{u}$, and the calculated value may not be positive definite during early iterations. Consequently, more than two iterations to the minimum can be expected. It then may be hard to assure that each step actually reduces the cost, and several systematic but *ad hoc* modifications to the algorithm can be made. Simply attenuating the control perturbation by some positive factor ϵ, whose magnitude is less than one, can prevent overshoots, although it slows progress toward the minimum. Because imprecision in the off-diagonal elements of $\mathscr{H}_{\mathbf{uu}}$ is most likely to prevent positive definiteness, using just the diagonal elements of $\mathscr{H}_{\mathbf{uu}}$ during early iterations (assuming each diagonal element is positive) should provide improving control.

In the *generalized gradient algorithm*,

$$\mathbf{u}_{k+1}(t) = \mathbf{u}_k(t) - \mathbf{K}_k[\mathscr{H}_{\mathbf{u}_k}(t)]^T \tag{3.6-19}$$

efficient constant values of the $(m \times m)$ matrix $\mathbf{K}_k$ can be found by numerical search. The key to success in using a numerical search is to choose the elements of $\mathbf{K}_k$ so that the control perturbation yields the maximum possible reduction in J for the given $\mathscr{H}_{\mathbf{u}_k}(t)$; this implies not only that Eq. 3.6-2 must be evaluated on each optimization step but that several "substeps" must be taken within each step. The additional iteration can be viewed as an internal parameter optimization, in which the elements of $\mathbf{K}_k$ serve as parameters that are chosen to minimize J on each step. By comparison to using parameter optimization alone to define the trajectory (as in Section 3.3), the advantage of the gradient method is that no prior control function need be assumed. The system equation (Eq. 3.6-1) and cost function (Eq. 3.6-2) must be evaluated several times on each optimization step with varying control perturbations; because the same $\mathscr{H}_{\mathbf{u}_k}(t)$ is used for each of these, Eqs. 3.6-3 to 3.6-5 are evaluated only once per major step.

To add substance to the concept, consider Fig. 3.6-3, which portrays equal-cost contours in a space defined by the elements of a (2×2) diagonal matrix $\mathbf{K}$. The constant values of k_{11} and k_{22} that provide the most improvement in J for the calculated history of $\mathscr{H}_{\mathbf{u}}(t)$ in $[t_0, t_f]$ are to be found. Not knowing the best search direction, we might begin by picking equal values of k_{11} and k_{22}, as shown. The state equation and cost function first are evaluated for three control histories, denoted by (0), (1), and (2). With prior results from the kth iteration, these are

- $J_{(0)}$ and $\mathbf{u}_{(0)}(t) = \mathbf{u}_k(t)$
- $J_{(1)}$ and $\mathbf{u}_{(1)}(t) = \mathbf{u}_k(t) - \mathbf{K}_k \mathscr{H}^T_{\mathbf{u}_k}(t)$
- $J_{(2)}$ and $\mathbf{u}_{(2)}(t) = \mathbf{u}_k(t) - 2\mathbf{K}_k \mathscr{H}^T_{\mathbf{u}_k}(t)$

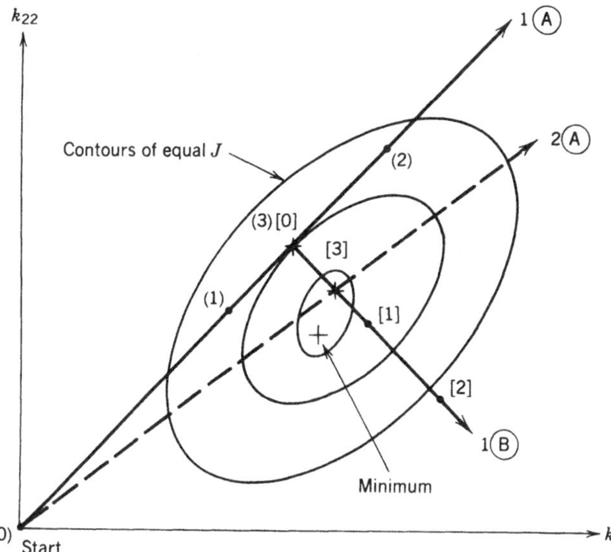

FIGURE 3.6-3 Illustration of a numerical search to find efficient values for the diagonal elements of K_k.

Fitting a quadratic function to these three points along line 1A and finding its minimum leads to

- $J_{(3)}$ and $\mathbf{u}_{(3)}(t) = \mathbf{u}_k(t) - \alpha\mathbf{K}_k\mathscr{H}^T_{\mathbf{u}_k}(t)$

where $\alpha\mathbf{K}_k$ corresponds to the minimizing gradient weighting. $\mathbf{u}_{(3)}$ is taken as the starting point $\mathbf{u}_{[0]}$ for a search in a direction that is orthogonal to the original search direction (line 2A). Once again, three cost function evaluations and control histories establish a local minimum, and the optimization step is completed with the definition of

$$\begin{aligned}\mathbf{u}_{[3]} &= \mathbf{u}_{k+1}(t)\\ &= \mathbf{u}_k(t) - \begin{bmatrix} k_{11_{[3]}} & 0 \\ 0 & k_{22_{[3]}} \end{bmatrix} \mathscr{H}^T_{\mathbf{u}_k}(t)\end{aligned} \tag{3.6-20}$$

The process is repeated for the next major optimization step. It may be desirable to initiate the next search along the line 2A, particularly as the optimum becomes near. The diagonal $\mathbf{K}_k$ could be considered an approximation to the average $\mathscr{H}_{\mathbf{uu}_k}$, which would appear constant for locally quadratic J. The numerical search can be extended to an m-component control vector with $(m \times m)$ diagonal $\mathbf{K}$. *Gram–Schmidt orthogonalization* can be used to identify the search directions which follow the initial search (within a major step).* It also should be noted that more than three evaluations along a line could be required if the test values of $\mathbf{K}_k$ do not capture the minimizing values of $\mathbf{K}$'s elements.

Various accelerated-gradient methods have been proposed to speed convergence, the most common of which is called the *conjugate-gradient method*. An interpretation of this method is that if the direction established by $\mathscr{H}_{\mathbf{u}}(t)$ on successive iterations changes, then this curvature can be enlisted to improve convergence. For the first iteration, the preceding numerical search is used to define

$$\mathbf{u}_1(t) = \mathbf{u}_0(t) - \mathbf{K}_0\mathscr{H}^T_{\mathbf{u}_0}(t) \tag{3.6-21}$$

The second iteration begins with the computation of a new gradient, $\mathscr{H}_{\mathbf{u}_1}(t)$. An average ratio of the square of the new and old gradients can be defined as

$$b = \frac{\displaystyle\int_{t_0}^{t_f} \mathscr{H}_{\mathbf{u}_1}\mathscr{H}^T_{\mathbf{u}_1}\,dt}{\displaystyle\int_{t_0}^{t_f} \mathscr{H}_{\mathbf{u}_0}\mathscr{H}^T_{\mathbf{u}_0}\,dt} \tag{3.6-22}$$

An effective gradient that combines $\mathscr{H}_{\mathbf{u}_1}$ and $\mathscr{H}_{\mathbf{u}_0}$,

$$\mathscr{H}'_{\mathbf{u}_1}(t) = \mathscr{H}_{\mathbf{u}_1}(t) - b\mathscr{H}_{\mathbf{u}_0}(t) \tag{3.6-23}$$

*See Section 4.3 for a weighted Gram–Schmidt orthogonalization procedure. In the present case, the weighting matrix would be set to the identity matrix.

is used in the second numerical search. This yields the following control history:

$$\mathbf{u}_2(t) = \mathbf{u}_1(t) - \mathbf{K}_1 \mathscr{H}'^T_{\mathbf{u}_1}(t)$$
$$= \mathbf{u}_1(t) - \mathbf{K}_1[\mathscr{H}_{\mathbf{u}_1}(t) - b\mathscr{H}_{\mathbf{u}_0}(t)]^T \tag{3.6-24}$$

The recursion is repeated until convergence is satisfactory. For a linear dynamic system and quadratic cost function, the conjugate-gradient algorithm converges in m steps, where m is the dimension of the control vector; however, such rapid convergence is not assured for more complex systems.

It should be recognized that the amount of computation associated with iterative numerical optimization is immense, and it rarely will be practical to solve these equations in "real time", (i.e., during the actual application of control to the real dynamic system). Indeed, even if the computing "power" to effect such a solution is available, uncertainties regarding disturbances during $[t_0, t_f]$ and variations in the target time and state may invalidate the optimality of the instantaneous control shortly after it is applied.

The principal value of such solutions, to date, has been to identify nominal trajectories and control histories prior to application. In addition to providing valuable planning information for designing the process (or vehicle) and its control system, the precalculated optimal trajectories and control histories can determine time-varying set points for feedback controllers (e.g., the neighboring-optimal control laws described in the next section). An interesting alternative is to use dynamic programming to implement a nonlinear feedback control law (S-8).

Example 3.6-1 Optimal Space Shuttle Reentry

The terminal phase of the space shuttle's return from orbit is governed by the following set of differential equations (S-8):

$$\dot{v} = \frac{-C_D k e^{-\beta h} v^2}{2} - g \sin \gamma$$

$$\dot{\gamma} = C_L k e^{-\beta h} \frac{v}{2} \cos \phi - \frac{g}{v} \cos \gamma$$

$$\dot{h} = v \sin \gamma$$

$$\dot{r} = v \cos \gamma \cos \xi$$

$$\dot{\xi} = \frac{-C_L k e^{-\beta h} \frac{v}{2} \sin\phi}{\cos \gamma}$$

$$\dot{c} = v \cos \gamma \sin \xi$$

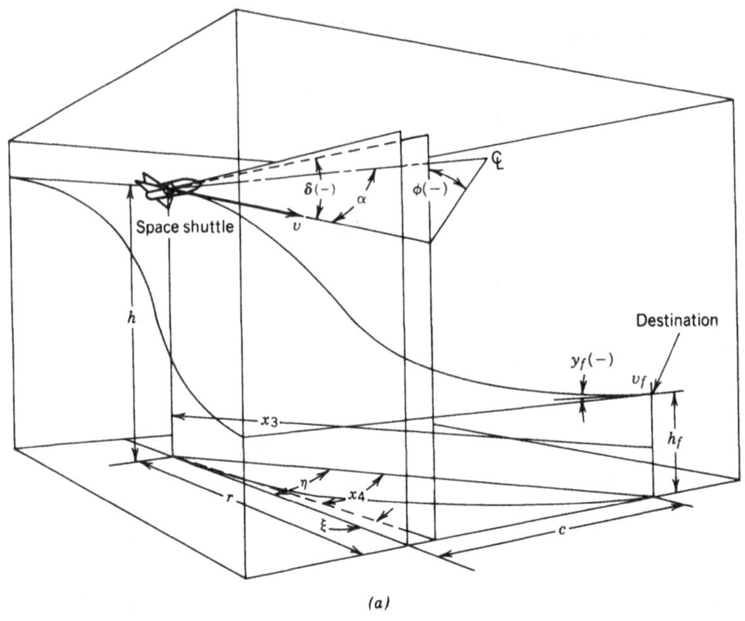

(a)

(1) Coordinate system for the space shuttle reentry.

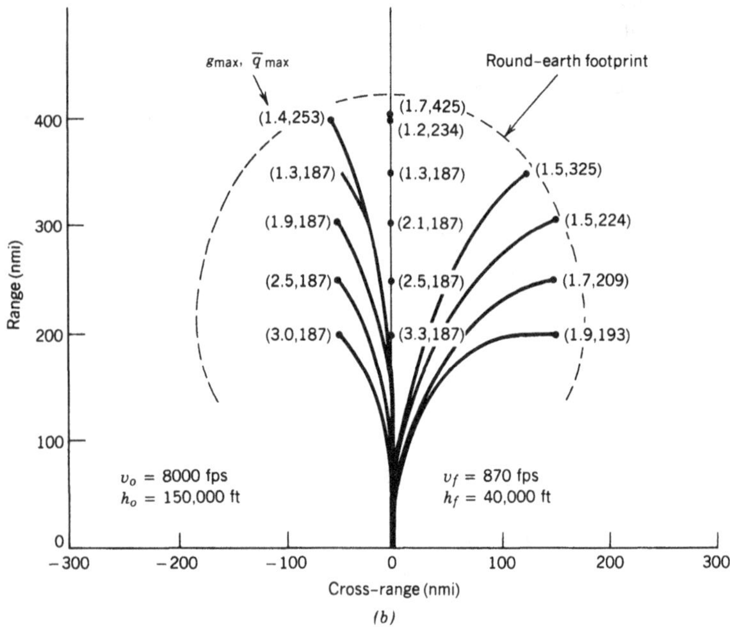

(b)

(2) Ground tracks of 15 transition trajectories calculated with flat-Earth assumptions. Maximum load factor (g's) and dynamic pressure (psf) are shown in parentheses next to each terminal point.

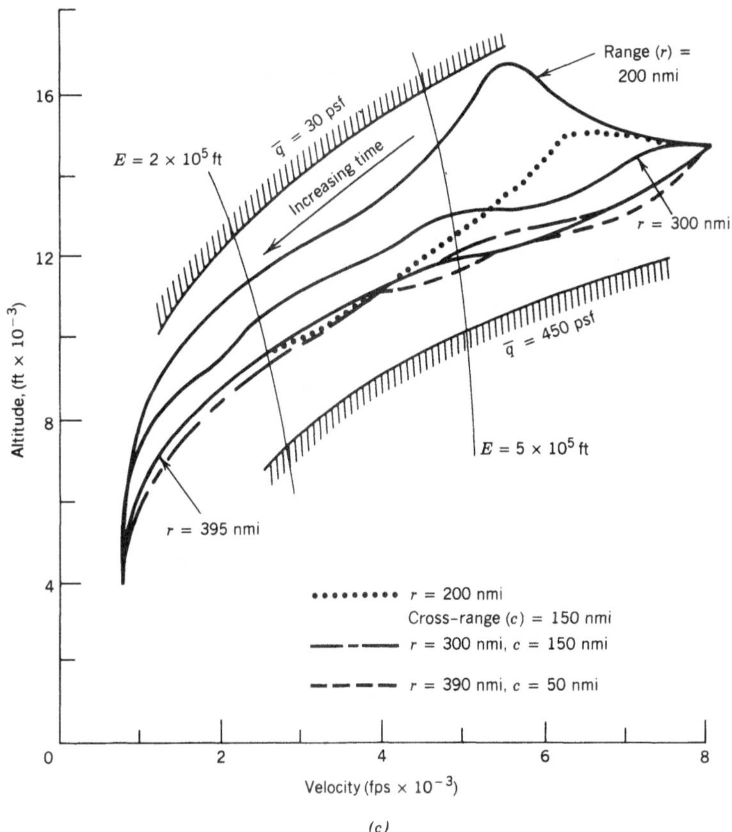

(3) Altitude-velocity profiles for several transition trajectories. Short-range trajectories require early deceleration and, therefore, high α. This leads to an initial increase in altitude. Long path-length trajectories require high kinetic energy at a fixed level of specific energy; hence dynamic pressure is higher.

The six state variables are velocity (v), vertical flight path angle (γ), height (h), range (r), horizontal flight path angle (ξ), and crossrange (c), as shown in Diagram 1. The control variables are roll angle (ϕ) and angle of attack (α), which affects the aerodynamic lift and drag coefficients, C_L and C_D. Given a set of initial conditions high in the atmosphere, a reasonable control objective is to minimize a terminal error cost at lower altitude and open end time, subject to integral costs that penalize deviations from a smooth reentry trajectory.

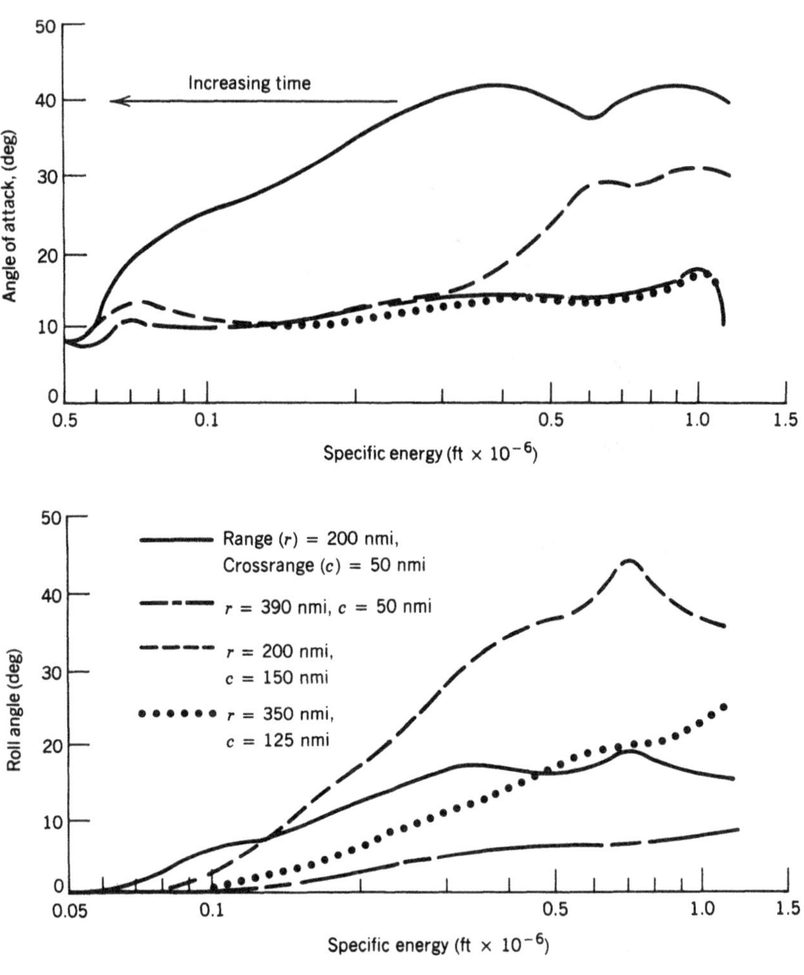

(d)

(4) Control histories for four transition trajectories. Angle-of-attack (α) trends can be related to the h-v trends seen in the previous figure. Reference to the Diagram 5 indicates that roll angle (ϕ) is a straightforward function of azimuth-to-go.

Transformation of the independent variable reduces the model order and converts the optimization to a fixed-end-point problem. The total specific energy measured relative to the earth's surface is the sum of potential and kinetic energies

$$E = K + H = \frac{v^2}{2g} + H$$

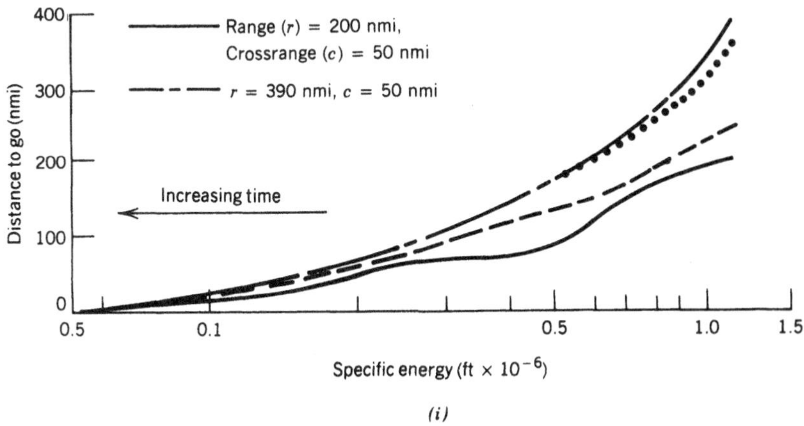

(i)

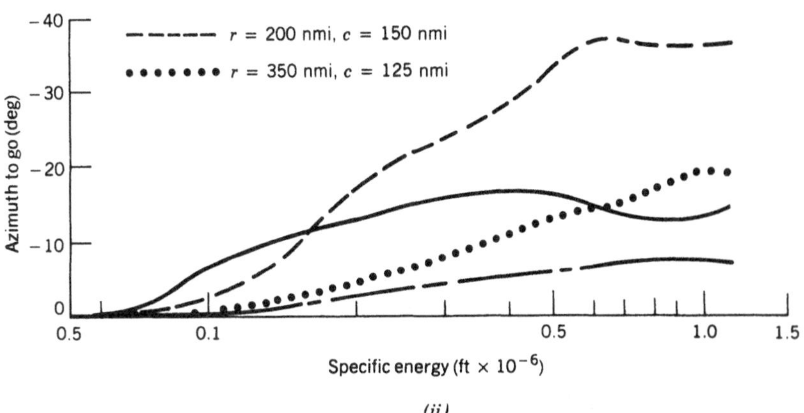

(ii)

(5*a*) Control histories for four transition trajectories. Angle-of-attack (α) trends can be related to the h-v trends seen in Diagram 4. Reference to Diagram 5*b* indicates that roll angle (ϕ) is a straightforward function of azimuth-to-go. (5*b*) Position histories for four transition trajectories. Distance-to-go (x_3) trends are similar to the h-v and α trends of the preceding diagrams. The convergence of azimuth-to-go (x_4) indicates a "straight-in" final approach to the destination.

and it decreases monotonically through the effect of aerodynamic drag in going from initial to final altitude and velocity (i.e., from E_0 and E_f). Using total specific energy rather than time as the independent variable leads to the fixed-end-point cost function

$$J = \phi[\mathbf{x}(E_f)] + \int_{E_0}^{E_f} \{\mathscr{L}[\mathbf{x}(E), \mathbf{u}(E)] + \boldsymbol{\lambda}^T(E)[\mathbf{f}(\mathbf{x}(E), \mathbf{u}(E)) - \mathbf{x}'(E)]\}\, dE$$

and the fifth-order dynamic equation

$$\mathbf{x}'(E) \triangleq \frac{d[\mathbf{x}(E)]}{dE} = \mathbf{f}[\mathbf{x}(E), \mathbf{u}(E)]$$

It is convenient to redefine the state variables as specific kinetic energy (x_1), vertical flight path angle (x_2), distance to go (x_3), horizontal flight path angle (x_4), and azimuth angle to the final destination.

In this example, $\phi[\mathbf{x}(E_f)]$ expresses a quadratic cost for terminal errors in x_1, x_2, and x_3. The Lagrangian provides trajectory damping by weighting the rate of change of dynamic pressure $\bar{q}$, which is half the air density times the velocity squared.

Optimal trajectories were computed using a steepest-descent algorithm with

$$\mathbf{u}(E)_{k+1} = \mathbf{u}_k(E) - \epsilon \frac{(E_f - E)}{(E_f - E_0)} \begin{bmatrix} \mathcal{H}_\alpha(E) \\ \sigma \mathcal{H}_\phi(E) \end{bmatrix}$$

where ϵ and σ were chosen to maximize the cost reduction on each step by successive direct searches of J. Examples of computational results are given in Sketches *b–e*.

3.7 NEIGHBORING-OPTIMAL SOLUTIONS

Once an optimal trajectory has been computed, it may be desirable to generate other optimal trajectories in the neighborhood of the original. Variations in initial conditions, end conditions, parameters, or deterministic disturbances would alter the optimal state and control histories, and while the original computations could be repeated with changes made, there is a simpler way to investigate the effects of "small" variations. Locally linearized dynamic models can be used in conjunction with a quadratic cost function derived from the second variation of the original cost function. The neighboring-optimal solution is then approximated as the sum of the original plus the linear-optimal solution.

The necessary linear formulations are derived in the remainder of this section. We begin with the continuous-time case, using both the calculus of variations and dynamic programming to achieve the solution. This is followed by the discrete-time case, again using two optimizing techniques. It will be seen that the linear-optimal control *histories* are expressed as the outputs of linear-optimal *feedback control laws*. These developments form the basis not only for the neighboring-optimal trajectory calculation but for the presentation of linear-quadratic optimal control theory to follow.

Continuous Neighboring-Optimal Control

The linear perturbation models introduced in Sections 2.3 and 3.6 can be put to use to examine the effects of small variations in initial conditions and terminal cost on the optimal trajectory and controls. Denoting the optimal solution by $(\cdot)^*$, the cost function can be expanded as

$$J[\mathbf{x}^*(t)+\Delta\mathbf{x}(t)] \cong J[\mathbf{x}^*(t)] + \Delta^2 J[\Delta\mathbf{x}(t)] \tag{3.7-1}$$

(Optimality guarantees that the first variation of J is zero on the optimal path.)

The second variation can be expressed as

$$\Delta^2 J = \tfrac{1}{2}\Delta\mathbf{x}^T(t_f)\phi_{\mathbf{xx}}(t_f)\Delta\mathbf{x}(t_f) + \frac{1}{2}\int_{t_0}^{t_f}\left\{[\Delta\mathbf{x}^T(t)\ \Delta\mathbf{u}^T(t)]\begin{bmatrix}\mathscr{L}_{\mathbf{xx}}(t) & \mathscr{L}_{\mathbf{xu}}(t)\\ \mathscr{L}_{\mathbf{ux}}(t) & \mathscr{L}_{\mathbf{uu}}(t)\end{bmatrix}\begin{bmatrix}\Delta\mathbf{x}(t)\\ \Delta\mathbf{u}(t)\end{bmatrix}\right\}dt \tag{3.7-2}$$

subject to the linear dynamic constraint

$$\Delta\dot{\mathbf{x}}(t) = \mathbf{F}(t)\Delta\mathbf{x}(t) + \mathbf{G}(t)\Delta\mathbf{u}(t), \qquad \Delta\mathbf{x}(t_0)\ \text{given} \tag{3.7-3}$$

$\mathbf{F}(t)$ and $\mathbf{G}(t)$ are defined as $\partial\mathbf{f}/\partial\mathbf{x}$ and $\partial\mathbf{f}/\partial\mathbf{u}$, evaluated along the optimal path. Perturbations from the nominal optimal solution are

$$\Delta\mathbf{x}(t) = \mathbf{x}(t) - \mathbf{x}^*(t) \qquad t_0 \le t \le t_f \tag{3.7-4}$$

$$\Delta\mathbf{u}(t) = \mathbf{u}(t) - \mathbf{u}^*(t) \tag{3.7-5}$$

A notational change is introduced in the quadratic cost function:

$$\Delta^2 J \triangleq J = \tfrac{1}{2}\Delta\mathbf{x}^T(t_f)\mathbf{P}(t_f)\Delta\mathbf{x}(t_f) + \frac{1}{2}\int_{t_0}^{t_f}\left\{[\Delta\mathbf{x}^T(t)\ \Delta\mathbf{u}^T(t)]\begin{bmatrix}\mathbf{Q}(t) & \mathbf{M}(t)\\ \mathbf{M}^T(t) & \mathbf{R}(t)\end{bmatrix}\begin{bmatrix}\Delta\mathbf{x}(t)\\ \Delta\mathbf{u}(t)\end{bmatrix}\right\}dt \tag{3.7-6}$$

The corresponding Hamiltonian can be written as

$$\mathscr{H}(t) = \tfrac{1}{2}[\Delta\mathbf{x}^T(t)\mathbf{Q}(t)\Delta\mathbf{x}(t) + 2\Delta\mathbf{x}^T(t)\mathbf{M}(t)\Delta\mathbf{u}(t) + \Delta\mathbf{u}^T\mathbf{R}(t)\Delta\mathbf{u}(t)] + \Delta\boldsymbol{\lambda}^T(t)[\mathbf{F}(t)\Delta\mathbf{x}(t) + \mathbf{G}(t)\Delta\mathbf{u}(t)] \tag{3.7-7}$$

The Euler–Lagrange equations, Eqs. 3.4-11 to 3.4-13, are

$$\Delta\dot{\boldsymbol{\lambda}}(t) = -\mathbf{Q}(t)\Delta\mathbf{x}(t) - \mathbf{M}(t)\Delta\mathbf{u}(t) - \mathbf{F}^T(t)\Delta\boldsymbol{\lambda}(t) \tag{3.7-8}$$

$$\Delta\boldsymbol{\lambda}(t_f) = \mathbf{P}(t_f)\Delta\mathbf{x}(t_f) \tag{3.7-9}$$

$$\mathbf{0} = \mathbf{M}^T(t)\Delta\mathbf{x}(t) + \mathbf{R}(t)\Delta\mathbf{u}(t) + \mathbf{G}^T(t)\Delta\boldsymbol{\lambda}(t) \tag{3.7-10}$$

Equation 3.7-10 can be rearranged to solve for the control,

$$\Delta\mathbf{u}(t) = -\mathbf{R}^{-1}(t)[\mathbf{M}^T(t)\Delta\mathbf{x}(t) + \mathbf{G}^T(t)\Delta\boldsymbol{\lambda}(t)] \tag{3.7-11}$$

and this relationship can be substituted in Eqs. 3.7-3 and 3.7-8 to express the linear, two-point boundary value problem as

$$\begin{aligned}\Delta\dot{\mathbf{x}}(t) = {} & [\mathbf{F}(t) - \mathbf{G}(t)\mathbf{R}^{-1}(t)\mathbf{M}^T(t)]\Delta\mathbf{x}(t) \\ & - \mathbf{G}(t)\mathbf{R}^{-1}(t)\mathbf{G}^T(t)\Delta\boldsymbol{\lambda}(t), \qquad \Delta\mathbf{x}(t_0) \text{ given}\end{aligned} \tag{3.7-12}$$

$$\begin{aligned}\Delta\dot{\boldsymbol{\lambda}}(t) = {} & [-\mathbf{Q}(t) + \mathbf{M}(t)\mathbf{R}^{-1}(t)\mathbf{M}^T(t)]\Delta\mathbf{x}(t) \\ & - [\mathbf{F}(t) - \mathbf{G}(t)\mathbf{R}^{-1}(t)\mathbf{M}^T(t)]^T\Delta\boldsymbol{\lambda}(t), \qquad \Delta\boldsymbol{\lambda}(t_f) \text{ given}\end{aligned} \tag{3.7-13}$$

Because $\Delta\mathbf{x}(t)$ and $\Delta\boldsymbol{\lambda}(t)$ are adjoint, Eq. 3.7-9 applies not only at the final time but during the entire interval:

$$\Delta\boldsymbol{\lambda}(t) = \mathbf{P}(t)\Delta\mathbf{x}(t), \qquad t_0 \le t \le t_f \tag{3.7-14}$$

Then Eq. 3.7-11 can be expressed as the *optimal feedback control law*,

$$\begin{aligned}\Delta\mathbf{u}^*(t) &= -\mathbf{R}^{-1}(t)[\mathbf{G}^T(t)\mathbf{P}(t) + \mathbf{M}^T(t)]\Delta\mathbf{x}(t) \\ &= -\mathbf{C}(t)\Delta\mathbf{x}(t)\end{aligned} \tag{3.7-15}$$

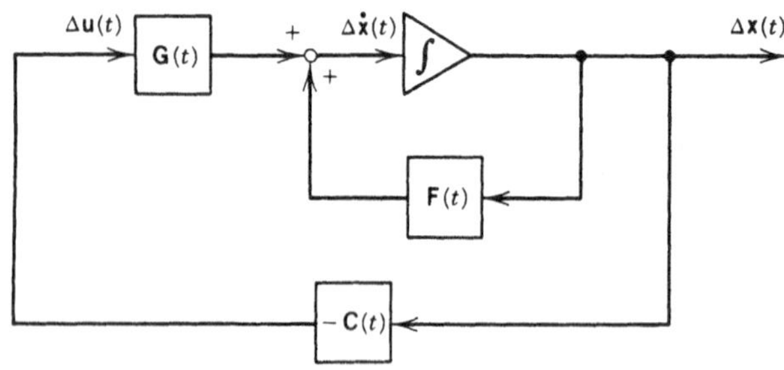

FIGURE 3.7-1 Block diagram of the linear-quadratic control law for a continuous-time system.

where $\mathbf{C}(t)$ is the time-varying $(m \times n)$ *control gain matrix*. Equation 3.7-15 describes a *linear-quadratic* (*LQ*) *control law*, so called because the dynamic constraint is linear and the cost function is quadratic. A schematic drawing of the control law is shown in Fig. 3.7-1.

The $(n \times n)$ matrix $\mathbf{P}(t)$ remains to be found. Differentiating Eq. 3.7-14, incorporating Eqs. 3.7-12 to 3.7-14, and rearranging terms, a nonlinear differential equation for $\mathbf{P}(t)$ can be derived. Dropping the (t) for clarity,

$$\begin{aligned}\dot{\mathbf{P}}\Delta\mathbf{x} &= \Delta\dot{\boldsymbol{\lambda}} - \mathbf{P}\Delta\dot{\mathbf{x}} \\ &= [(-\mathbf{Q} + \mathbf{M}\mathbf{R}^{-1}\mathbf{M}^T)\Delta\mathbf{x} - (\mathbf{F} - \mathbf{G}\mathbf{R}^{-1}\mathbf{M}^T)^T\Delta\boldsymbol{\lambda}] \\ &\quad - \mathbf{P}[(\mathbf{F} - \mathbf{G}\mathbf{R}^{-1}\mathbf{M}^T)\Delta\mathbf{x} - \mathbf{G}\mathbf{R}^{-1}\mathbf{G}^T\Delta\boldsymbol{\lambda}] \\ &= [(-\mathbf{Q} + \mathbf{M}\mathbf{R}^{-1}\mathbf{M}^T) - (\mathbf{F} - \mathbf{G}\mathbf{R}^{-1}\mathbf{M}^T)^T\mathbf{P}]\Delta\mathbf{x} \\ &\quad - \mathbf{P}[(\mathbf{F} - \mathbf{G}\mathbf{R}^{-1}\mathbf{M}^T) - \mathbf{G}\mathbf{R}^{-1}\mathbf{G}^T\mathbf{P}]\Delta\mathbf{x} \end{aligned} \tag{3.7-16}$$

Canceling $\Delta\mathbf{x}$ on both sides of the equation leads to a *matrix Riccati equation** for $\mathbf{P}$; hence we shall refer to $\mathbf{P}(t)$ as a *Riccati matrix*, whose differential equation is

$$\begin{aligned}\dot{\mathbf{P}} &= -(\mathbf{F} - \mathbf{G}\mathbf{R}^{-1}\mathbf{M}^T)^T\mathbf{P} - \mathbf{P}(\mathbf{F} - \mathbf{G}\mathbf{R}^{-1}\mathbf{M}^T) + \mathbf{P}\mathbf{G}\mathbf{R}^{-1}\mathbf{G}^T\mathbf{P} \\ &\quad - \mathbf{Q} + \mathbf{M}\mathbf{R}^{-1}\mathbf{M}^T, \qquad \mathbf{P}(t_f) = \phi_{\mathbf{xx}}(t_f)\end{aligned} \tag{3.7-17}$$

The solution for $\mathbf{P}(t)$ and, therefore, for $\mathbf{C}(t)$ is seen to be independent of $\Delta\mathbf{x}(t)$. Variations in $\Delta\mathbf{x}(t_0)$ or $\Delta\mathbf{x}(t_f)$ have no effect on $\mathbf{C}(t)$, although the *linear-optimal control history* obviously is affected by state perturbations.

From Eq. 3.7-5, the total control is formed as the sum of the nominal and perturbation optimal controls:

$$\begin{aligned}\mathbf{u}(t) &= \mathbf{u}^*(t) + \Delta\mathbf{u}^*(t) \\ &= \mathbf{u}^*(t) - \mathbf{C}(t)[\mathbf{x}(t) - \mathbf{x}^*(t)]\end{aligned} \tag{3.7-18}$$

The *prototype neighboring-optimal control law* is illustrated in Fig. 3.7-2. This diagram reflects Eq. 3.7-18, with all nominal and perturbation variables continuous functions of time. It also would be possible to implement a discrete-time version of this control law, using methods developed below.

The diagram introduces the notion that an alternative to time scheduling of $\mathbf{x}^*$, $\mathbf{u}^*$, and $\mathbf{C}$ can be considered. When t_f is not fixed and is of secondary importance, it may be desirable to choose a scalar function η of the state with fixed endpoint as the scheduling variable. This scalar function must be monotonic in time, so as not to introduce singularities in the scheduling. For

*Jacopo Francesco, Count Riccati (1676–1754), published a similar scalar equation in 1724.

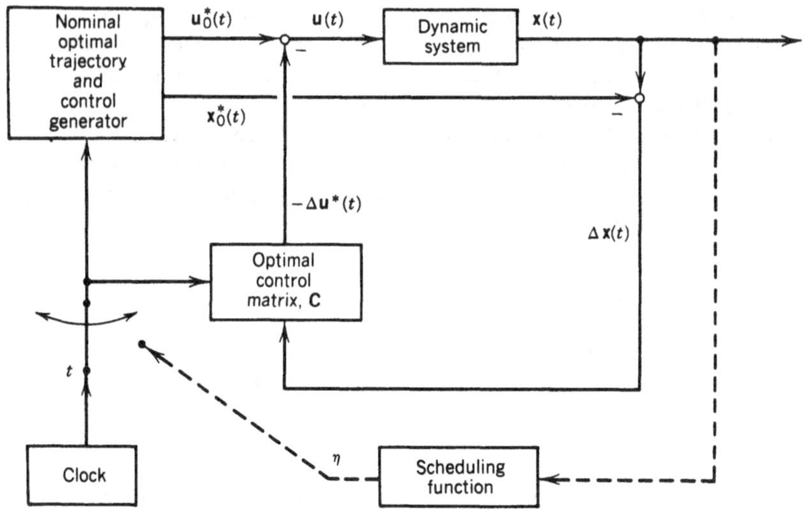

FIGURE 3.7-2 Prototype neighboring-optimal control law.

example, "time to go," $(t_f - t)$, would be a better scheduling variable than $(t - t_0)$ for a terminal controller; then "range to go," $(r_f - r)$ (or some equally useful measure of the "distance" to the end of the trajectory) could be substituted for $(t_f - t)$. Using a state-dependent scheduling variable that is single-valued as the objective is approached allows the control to adapt to unexpected variations in t_f, as might result from the effects of disturbances or of a maneuvering target.

Dynamic Programming Solution for Continuous Linear-Quadratic Control

The LQ law for continuous-time control is derived easily using the HJB equation. We begin by restating the control problem as the minimization of a quadratic value function (Eq. 3.7-6),

$$V = \tfrac{1}{2}\Delta\mathbf{x}^T(t_f)\phi_{\mathbf{xx}}(t_f)\Delta\mathbf{x}(t_f) + \frac{1}{2}\int_{t_0}^{t_f}\left\{[\Delta\mathbf{x}^T(t)\ \Delta\mathbf{u}^T(t)]\begin{bmatrix}\mathbf{Q}(t) & \mathbf{M}(t)\\ \mathbf{M}^T(t) & \mathbf{R}(t)\end{bmatrix}\begin{bmatrix}\Delta\mathbf{x}(t)\\ \Delta\mathbf{u}(t)\end{bmatrix}\right\}dt \tag{3.7-19}$$

subject to the dynamic constraint

$$\Delta\dot{\mathbf{x}}(t) = \mathbf{F}(t)\Delta\mathbf{x}(t) + \mathbf{G}(t)\Delta\mathbf{u}(t), \qquad \Delta\mathbf{x}_0 \text{ given} \tag{3.7-20}$$

The HJB equation (Eq. 3.4-39) provides the optimality criterion

$$\frac{\partial V^*}{\partial t}[\Delta\mathbf{x}^*(t), t] = -\min_{\mathbf{u}}\left\{\mathscr{L}[\Delta\mathbf{x}^*(t), \Delta\mathbf{u}(t), t] + \frac{\partial V^*}{\partial \Delta\mathbf{x}}[\Delta\mathbf{x}^*(t), t][\mathbf{F}(t)\Delta\mathbf{x}^*(t) + \mathbf{G}(t)\Delta\mathbf{u}(t)]\right\}$$

$$= -\min_{\mathbf{u}} \mathscr{H}[\Delta\mathbf{x}^*(t), \Delta\mathbf{u}(t), t] \tag{3.7-21}$$

where

$$\mathscr{L}[\Delta\mathbf{x}(t), \Delta\mathbf{u}(t), t] = \tfrac{1}{2}[\Delta\mathbf{x}^T(t)\ \Delta\mathbf{u}^T(t)]\begin{bmatrix}\mathbf{Q}(t) & \mathbf{M}(t)\\ \mathbf{M}^T(t) & \mathbf{R}(t)\end{bmatrix}\begin{bmatrix}\Delta\mathbf{x}(t)\\ \Delta\mathbf{u}(t)\end{bmatrix}$$
$$= \tfrac{1}{2}[\Delta\mathbf{x}^T(t)\mathbf{Q}(t)\Delta\mathbf{x}(t) + 2\Delta\mathbf{x}^T(t)\mathbf{M}(t)\Delta\mathbf{u}(t) + \Delta\mathbf{u}^T(t)\mathbf{R}(t)\Delta\mathbf{u}(t)] \tag{3.7-22}$$

It is plausible that the optimal value function is a quadratic function of $\Delta\mathbf{x}^*(t)$,

$$V^*[\Delta\mathbf{x}^*(t), t] = \tfrac{1}{2}\Delta\mathbf{x}^{*T}(t)\mathbf{P}(t)\Delta\mathbf{x}^*(t) \tag{3.7-23}$$

where $\mathbf{P}(t)$ is a positive definite, symmetric matrix, and this assumption on our part is formally derived in (A-2). The left-hand side of Eq. 3.7-21 is

$$\frac{\partial V^*}{\partial t}[\Delta\mathbf{x}^*(t), t] = \tfrac{1}{2}\Delta\mathbf{x}^{*T}(t)\dot{\mathbf{P}}(t)\Delta\mathbf{x}^*(t) \tag{3.7-24}$$

and the partial derivative of V^* with respect to $\Delta\mathbf{x}$ is

$$\frac{\partial V^*}{\partial \Delta\mathbf{x}}[\Delta\mathbf{x}^*(t), t] = \Delta\mathbf{x}^{*T}(t)\mathbf{P}(t) \tag{3.7-25}$$

The right side of Eq. 3.7-21 is found by taking the derivative of $\mathscr{H}[\cdot]$ with respect to $\Delta\mathbf{u}$, setting it equal to zero to find $\Delta\mathbf{u}$, and substituting for $\Delta\mathbf{u}$ in the equation. The differentiation produces

$$\Delta\mathbf{x}^T(t)\mathbf{M}(t) + \Delta\mathbf{u}^T(t)\mathbf{R}(t) + \Delta\mathbf{x}^T(t)\mathbf{P}(t)\mathbf{G}(t) = \mathbf{0} \tag{3.7-26}$$

which defines the *linear-optimal control law*:

$$\Delta\mathbf{u}(t) = -\mathbf{R}^{-1}(t)[\mathbf{G}^T(t)\mathbf{P}(t) + \mathbf{M}^T(t)]\Delta\mathbf{x}(t) \tag{3.7-27}$$

Substituting in Eq. 3.7-21 and suppressing explicit reference to the time dependence,

$$\Delta\mathbf{x}^T\dot{\mathbf{P}}\Delta\mathbf{x} = -\Delta\mathbf{x}^T[(\mathbf{F}-\mathbf{G}\mathbf{R}^{-1}\mathbf{M}^T)^T\mathbf{P}+\mathbf{P}(\mathbf{F}-\mathbf{G}\mathbf{R}^{-1}\mathbf{M}^T) \\ -\mathbf{P}\mathbf{G}\mathbf{R}^{-1}\mathbf{G}^T\mathbf{P}+\mathbf{Q}-\mathbf{M}\mathbf{R}^{-1}\mathbf{M}^T]\Delta\mathbf{x} \tag{3.7-28}$$

The solution must hold for arbitrary $\Delta\mathbf{x}(t)$, so these terms can be canceled, leaving the *matrix Riccati equation* found earlier:

$$\dot{\mathbf{P}} = -(\mathbf{F}-\mathbf{G}\mathbf{R}^{-1}\mathbf{M}^T)\mathbf{P}^T-\mathbf{P}(\mathbf{F}-\mathbf{G}\mathbf{R}^{-1}\mathbf{M}^T) \\ +\mathbf{P}\mathbf{G}\mathbf{R}^{-1}\mathbf{G}^T\mathbf{P}-\mathbf{Q}+\mathbf{M}\mathbf{R}^{-1}\mathbf{M}^T \tag{3.7-29}$$

As the value function must equal the terminal cost at $t = t_f$,

$$\mathbf{P}(t_f) = \phi_{\mathbf{xx}}(t_f) \tag{3.7-30}$$

Because the Riccati equation was derived earlier from a necessary condition and is now derived from a sufficient condition, it can be concluded that existence of solutions to this equation are necessary and sufficient conditions for optimality in the linear-quadratic case.

Discrete Neighboring-Optimal Control

An important alternative to continuous neighboring-optimal control results when a digital computer implements control system logic. In such a case, the nominal trajectory may have been generated by solving a continuous-time nonlinear optimal control problem, but small corrections are to be made at discrete instants of time (or sampling instants). An appropriate *discrete-time neighboring-optimal control law* is derived by minimizing Eq. 3.7-6 with the assumption that the control is piecewise-constant between sampling instants, which normally are assumed to occur at equal intervals.

The neighboring-optimal cost function is expressed as a summation rather than an integral by first noting that Eq. 3.7-6 could be written as

$$\Delta^2 J = \tfrac{1}{2}\Delta\mathbf{x}^T(t_f)\mathbf{P}(t_f)\Delta\mathbf{x}(t_f) \\ +\frac{1}{2}\sum_{k=0}^{k_f-1}\int_{t_k}^{t_{k+1}}\left\{[\Delta\mathbf{x}^T(t)\ \Delta\mathbf{u}^T(t)]\begin{bmatrix}\mathbf{Q}(t) & \mathbf{M}(t)\\ \mathbf{M}^T(t) & \mathbf{R}(t)\end{bmatrix}\begin{bmatrix}\Delta\mathbf{x}(t)\\ \Delta\mathbf{u}(t)\end{bmatrix}\right\}dt \tag{3.7-31}$$

The index k_f corresponds to t_f, and the constant time increment Δt is t_f/k_f. We wish to minimize the cost subject to a dynamic constraint. With piecewise-constant control, it is appropriate to express this constraint using the difference-equation equivalent of Eq. 3.7-3:

$$\Delta\mathbf{x}(t_{k+1}) = \mathbf{\Phi}(t_{k+1}, t_k)\Delta\mathbf{x}(t_k)+\mathbf{\Gamma}(t_{k+1}, t_k)\Delta\mathbf{u}(t_k), \qquad \Delta\mathbf{x}(0) = \Delta\mathbf{x}_0 \tag{3.7-32}$$

with $\mathbf{\Phi}(t_{k+1}, t_k)$ and $\mathbf{\Gamma}(t_{k+1}, t_k)$ defined as in Section 2.3. At any arbitrary time t during the interval (t_k, t_{k+1}), the state perturbation can be computed as

$$\Delta\mathbf{x}(t) = \mathbf{\Phi}(t, t_k)\Delta\mathbf{x}(t_k) + \mathbf{\Gamma}(t, t_k)\Delta\mathbf{u}(t_k) \tag{3.7-33}$$

so the integral of Eq. 3.7-31 can be written

$$\begin{aligned}\int_{t_k}^{t_{k+1}} \{\cdot\}\, dt &= \int_{t_k}^{t_{k+1}} \left\{[\Delta\mathbf{x}^T(t_k)\ \Delta\mathbf{u}^T(t_k)]\begin{bmatrix}\mathbf{\Phi}^T\mathbf{Q}\mathbf{\Phi} & \mathbf{\Phi}^T(\mathbf{Q}\mathbf{\Gamma}+\mathbf{M})\\ (\mathbf{Q}\mathbf{\Gamma}+\mathbf{M})^T\mathbf{\Phi} & \mathbf{R}'\end{bmatrix}\begin{bmatrix}\Delta\mathbf{x}(t_k)\\ \Delta\mathbf{u}(t_k)\end{bmatrix}\right\} dt\\ &= [\Delta\mathbf{x}^T(t_k)\ \Delta\mathbf{u}^T(t_k)]\int_{t_k}^{t_{k+1}}\left\{\begin{bmatrix}\mathbf{\Phi}^T\mathbf{Q}\mathbf{\Phi} & \mathbf{\Phi}^T(\mathbf{Q}\mathbf{\Gamma}+\mathbf{M})\\ (\mathbf{Q}\mathbf{\Gamma}+\mathbf{M})^T\mathbf{\Phi} & \mathbf{R}'\end{bmatrix} dt\right\}\begin{bmatrix}\Delta\mathbf{x}(t_k)\\ \Delta\mathbf{u}(t_k)\end{bmatrix}\end{aligned} \tag{3.7-34}$$

where the arguments of $\mathbf{\Phi}$, $\mathbf{\Gamma}$, $\mathbf{Q}$, $\mathbf{M}$, and $\mathbf{R}$ have been suppressed for clarity, and

$$\mathbf{R}' = \mathbf{R} + \mathbf{\Gamma}^T\mathbf{M} + \mathbf{M}^T\mathbf{\Gamma} + \mathbf{\Gamma}^T\mathbf{Q}\mathbf{\Gamma} \tag{3.7-35}$$

$\Delta\mathbf{x}(t_k)$ and $\Delta\mathbf{u}(t_k)$ are constant in the integration interval and can be brought outside the integral. Denoting variables at the sampling instant t_k by the subscript k, the quadratic cost function (Eq. 3.7-31) becomes

$$\begin{aligned}\Delta^2 J &\triangleq J_{\text{SD}}\\ &= \tfrac{1}{2}\Delta\mathbf{x}_f^T\mathbf{P}_f\Delta\mathbf{x}_f + \frac{1}{2}\sum_{k=0}^{k_f-1}[\Delta\mathbf{x}_k^T\ \Delta\mathbf{u}_k^T]\begin{bmatrix}\hat{\mathbf{Q}}_k & \hat{\mathbf{M}}_k\\ \hat{\mathbf{M}}_k^T & \hat{\mathbf{R}}_k\end{bmatrix}\begin{bmatrix}\Delta\mathbf{x}_k\\ \Delta\mathbf{u}_k\end{bmatrix}\end{aligned} \tag{3.7-36}$$

The sampled-data cost weighting matrices are defined as

$$\hat{\mathbf{Q}}_k = \int_{t_k}^{t_{k+1}} \mathbf{\Phi}^T(t, t_k)\mathbf{Q}(t)\mathbf{\Phi}(t, t_k)\, dt \tag{3.7-37}$$

$$\hat{\mathbf{M}}_k = \int_{t_k}^{t_{k+1}} \mathbf{\Phi}^T(t, t_k)[\mathbf{Q}(t)\mathbf{\Gamma}(t, t_k) + \mathbf{M}(t)]\, dt \tag{3.7-38}$$

$$\hat{\mathbf{R}}_k = \int_{t_k}^{t_{k+1}} [\mathbf{\Gamma}^T(t, t_k)\mathbf{Q}(t)\mathbf{\Gamma}(t, t_k) + \mathbf{\Gamma}^T(t, t_k)\mathbf{M}(t) + \mathbf{M}^T(t)\mathbf{\Gamma}(t, t_k) + \mathbf{R}(t)]\, dt \tag{3.7-39}$$

Note that $\mathbf{\Phi}$ and $\mathbf{\Gamma}$ are changing during the sampling interval, and it is imperative to account for these changes (D-3). Lagrangians for continuous-time, discrete-time, and sampled-data problems are compared in Fig. 3.7-3. In the first case, the Lagrangian changes smoothly, whereas in the second, it steps from one discrete value to the next. The sampled-data formulation estimates the smooth variation of the continuous-time case between sampling instants.

The sampled-data cost function is seen to contain cross-product weighting, $\hat{\mathbf{M}}_k$, whether or not $\mathbf{M}(t)$ appears in the continuous cost function. If

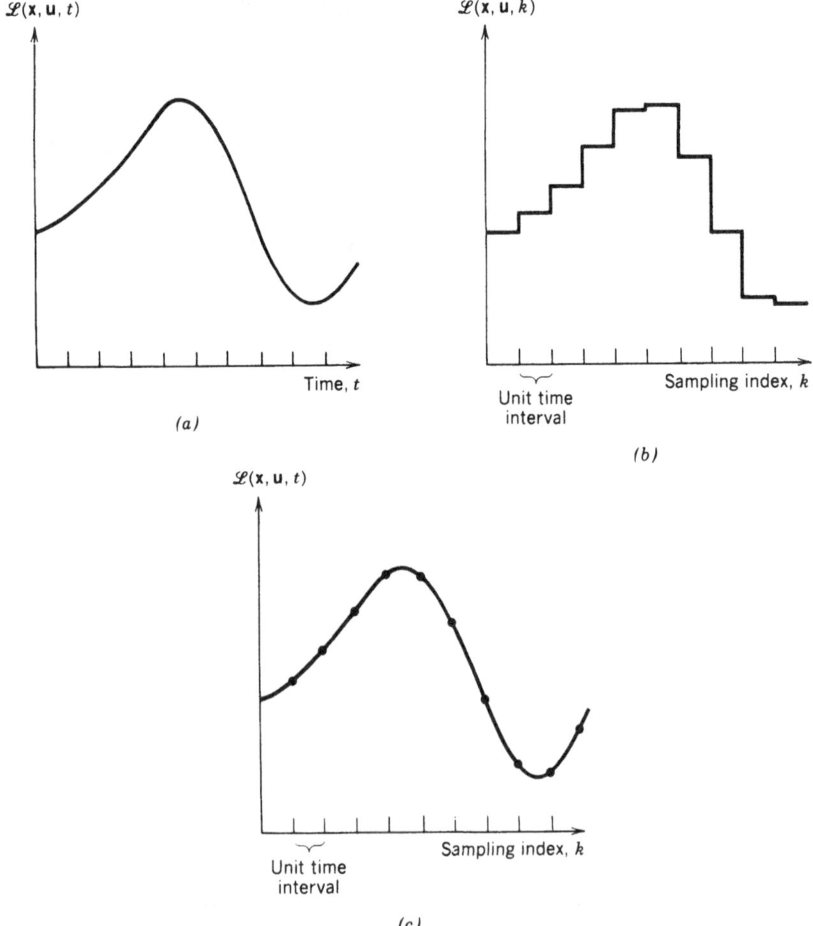

FIGURE 3.7-3 Comparison of Lagrangians for (*a*) continuous-time, (*b*) discrete-time, and (*c*) sampled-data cost functions. In each case, the "integrated" cost is the area under the curve.

desired, state and control vectors can be transformed to eliminate $\hat{\mathbf{M}}_k$, following the development for continuous systems in Section 5.4.

Minimization of the *sampled-data cost function* (Eq. 3.7-36) proceeds as in the continuous-time case. The dynamic constraint

$$\Delta \mathbf{x}_{k+1} = \mathbf{\Phi}_k \, \Delta \mathbf{x}_k + \mathbf{\Gamma}_k \, \Delta \mathbf{u}_k \tag{3.7-40}$$

is adjoined to the cost function

$$J_{\mathrm{SD}} = \tfrac{1}{2} \Delta \mathbf{x}_f^T \mathbf{P}_f \, \Delta \mathbf{x}_f + \sum_{k=0}^{k_f - 1} (\mathcal{H}_k - \Delta \boldsymbol{\lambda}_{k+1}^T \, \Delta \mathbf{x}_{k+1}) \tag{3.7-41}$$

The Hamiltonian is defined as

$$\begin{aligned} \mathcal{H}_k &= \tfrac{1}{2} [\Delta \mathbf{x}_k^T \; \Delta \mathbf{u}_k^T] \begin{bmatrix} \hat{\mathbf{Q}}_k & \hat{\mathbf{M}}_k \\ \hat{\mathbf{M}}_k^T & \hat{\mathbf{R}}_k \end{bmatrix} \begin{bmatrix} \Delta \mathbf{x}_k \\ \Delta \mathbf{u}_k \end{bmatrix} + \Delta \boldsymbol{\lambda}_{k+1}^T (\mathbf{\Phi}_k \, \Delta \mathbf{x}_k + \mathbf{\Gamma}_k \, \Delta \mathbf{u}_k) \\ &= \tfrac{1}{2} (\Delta \mathbf{x}_k^T \hat{\mathbf{Q}}_k \, \Delta \mathbf{x}_k + 2 \Delta \mathbf{x}_k^T \hat{\mathbf{M}}_k \, \Delta \mathbf{u}_k + \Delta \mathbf{u}_k^T \hat{\mathbf{R}}_k \, \Delta \mathbf{u}_k) \\ &\quad + \Delta \boldsymbol{\lambda}_{k+1}^T \mathbf{\Phi}_k \, \Delta \mathbf{x}_k + \Delta \boldsymbol{\lambda}_{k+1}^T \mathbf{\Gamma}_k \, \Delta \mathbf{u}_k \end{aligned} \tag{3.7-42}$$

The limits of summation and the index of the adjoint term are modified,

$$J_{\mathrm{SD}} = \tfrac{1}{2} \Delta \mathbf{x}_f^T \mathbf{P}_f \, \Delta \mathbf{x}_f + \sum_{k=1}^{k_f - 1} (\mathcal{H}_k - \Delta \boldsymbol{\lambda}_k^T \, \Delta \mathbf{x}_k) + \mathcal{H}_0 - \Delta \boldsymbol{\lambda}_f^T \, \Delta \mathbf{x}_f \tag{3.7-43}$$

and the first variation of J_{SD}, following Section 3.4, is

$$\begin{aligned} \Delta J_{\mathrm{SD}} &= (\Delta \mathbf{x}_f^T \mathbf{P}_f - \Delta \boldsymbol{\lambda}_f^T) \delta \mathbf{x}_f + \frac{\partial \mathcal{H}_0}{\partial \Delta \mathbf{x}_0} \delta \mathbf{x}_0 + \frac{\partial \mathcal{H}_0}{\partial \Delta \mathbf{u}_0} \delta \mathbf{u}_0 \\ &\quad + \sum_{k=1}^{k_f - 1} \left[\left(\frac{\partial \mathcal{H}_k}{\partial \Delta \mathbf{x}_k} - \Delta \boldsymbol{\lambda}_k^T \right) \delta \mathbf{x}_k + \frac{\partial \mathcal{H}_k}{\partial \Delta \mathbf{u}_k} \delta \mathbf{u}_k \right] \end{aligned} \tag{3.7-44}$$

$\delta \mathbf{x}$ and $\delta \mathbf{u}$ are differential changes in the perturbation variables $\Delta \mathbf{x}$ and $\Delta \mathbf{u}$. The sensitivity to differential changes in the state is minimized by choosing

$$\Delta \boldsymbol{\lambda}_k^T = \frac{\partial \mathcal{H}_k}{\partial \Delta \mathbf{x}_k} = \Delta \mathbf{x}_k^T \hat{\mathbf{Q}}_k + \Delta \mathbf{u}_k^T \hat{\mathbf{M}}_k^T + \Delta \boldsymbol{\lambda}_{k+1}^T \mathbf{\Phi}_k \tag{3.7-45}$$

$$\Delta \boldsymbol{\lambda}_f^T = \Delta \mathbf{x}_f^T \mathbf{P}_f \tag{3.7-46}$$

The necessary condition for optimality is that control perturbations have no first-order effect on the Hamiltonian:

$$\mathbf{0} = \frac{\partial \mathcal{H}_k}{\partial \Delta \mathbf{u}_k} = \Delta \mathbf{x}_k^T \hat{\mathbf{M}}_k + \Delta \mathbf{u}_k^T \hat{\mathbf{R}}_k + \Delta \boldsymbol{\lambda}_{k+1}^T \mathbf{\Gamma}_k \tag{3.7-47}$$

The optimal feedback control law is obtained from Eq. 3.7-47. Because $\Delta\boldsymbol{\lambda}$ is the adjoint of $\Delta\mathbf{x}$, Eq. 3.7-46 is satisfied at all sampling instants:

$$\Delta\boldsymbol{\lambda}_k = \mathbf{P}_k\,\Delta\mathbf{x}_k \tag{3.7-48a}$$

$$\Delta\boldsymbol{\lambda}_{k+1} = \mathbf{P}_{k+1}\,\Delta\mathbf{x}_{k+1} \tag{3.7-48b}$$

Substituting first Eq. 3.7-40 in Eq. 3.7-48b, then Eq. 3.7-48b in Eq. 3.7-47,

$$\mathbf{0} = (\hat{\mathbf{R}}_k + \boldsymbol{\Gamma}_k^T\mathbf{P}_{k+1}\boldsymbol{\Gamma}_k)\,\Delta\mathbf{u}_k + (\hat{\mathbf{M}}_k^T + \boldsymbol{\Gamma}_k^T\mathbf{P}_{k+1}\boldsymbol{\Phi}_k)\,\Delta\mathbf{x}_k \tag{3.7-49a}$$

or

$$\begin{aligned}\Delta\mathbf{u}_k^* &= -(\hat{\mathbf{R}}_k + \boldsymbol{\Gamma}_k^T\mathbf{P}_{k+1}\boldsymbol{\Gamma}_k)^{-1}(\hat{\mathbf{M}}_k^T + \boldsymbol{\Gamma}_k^T\mathbf{P}_{k+1}\boldsymbol{\Phi}_k)\,\Delta\mathbf{x}_k \\ &= -\mathbf{C}_k\,\Delta\mathbf{x}_k \end{aligned} \tag{3.7-49b}$$

where $\mathbf{C}_k$ is the $(m \times n)$ *optimal sampled-data feedback gain matrix.*

The adjoining matrix $\mathbf{P}_k$ must be calculated for k running backwards from $(k_f - 1)$ to 1, with the starting condition defined by $\mathbf{P}_{k+1} = \mathbf{P}_{k_f} \triangleq \mathbf{P}_f$. Substituting Eqs. 3.7-40, 3.7-48, and 3.7-49 in Eq. 3.7-45,

$$\begin{aligned}\mathbf{P}_k\,\Delta\mathbf{x}_k &= \hat{\mathbf{Q}}_k\,\Delta\mathbf{x}_k + \boldsymbol{\Phi}_k^T\mathbf{P}_{k+1}\boldsymbol{\Phi}_k\,\Delta\mathbf{x}_k \\ &\quad - (\hat{\mathbf{M}}_k^T + \boldsymbol{\Gamma}_k^T\mathbf{P}_{k+1}\boldsymbol{\Phi}_k)^T(\hat{\mathbf{R}}_k + \boldsymbol{\Gamma}_k^T\mathbf{P}_{k+1}\boldsymbol{\Gamma}_k)^{-1}(\hat{\mathbf{M}}_k^T + \boldsymbol{\Gamma}_k^T\mathbf{P}_{k+1}\boldsymbol{\Phi}_k)\,\Delta\mathbf{x}_k\end{aligned} \tag{3.7-50}$$

The common factor $\Delta\mathbf{x}_k$ can be eliminated from all terms, leaving the *discrete-time Riccati equation*

$$\begin{aligned}\mathbf{P}_k &= \hat{\mathbf{Q}}_k + \boldsymbol{\Phi}_k^T\mathbf{P}_{k+1}\boldsymbol{\Phi}_k \\ &\quad - (\hat{\mathbf{M}}_k + \boldsymbol{\Phi}_k^T\mathbf{P}_{k+1}\boldsymbol{\Gamma}_k)(\hat{\mathbf{R}}_k + \boldsymbol{\Gamma}_k^T\mathbf{P}_{k+1}\boldsymbol{\Gamma}_k)^{-1}(\hat{\mathbf{M}}_k^T + \boldsymbol{\Gamma}_k^T\mathbf{P}_{k+1}\boldsymbol{\Phi}_k)\end{aligned} \tag{3.7-51}$$

As in the continuous case, the total control is the sum of the nominal and perturbation components:

$$\begin{aligned}\mathbf{u}(t) &= \mathbf{u}^*(t) + \Delta\mathbf{u}_k^* \\ &= \mathbf{u}^*(t) - \mathbf{C}_k[\mathbf{x}(t_k) - \mathbf{x}^*(t_k)]\end{aligned} \tag{3.7-52}$$

This is expressed as a *hybrid control law*, with the nominal control changing continuously and the perturbation control changing at the sampling instants. In practice, the nominal control would be sampled as well. The continuous $\mathbf{u}^*(t)$ would be transformed to the sampled $\mathbf{u}^*(t_k)$ either by performing the nonlinear optimization with piecewise-constant constraints on the control or by *ad hoc* means, as presented in Chapter 5.

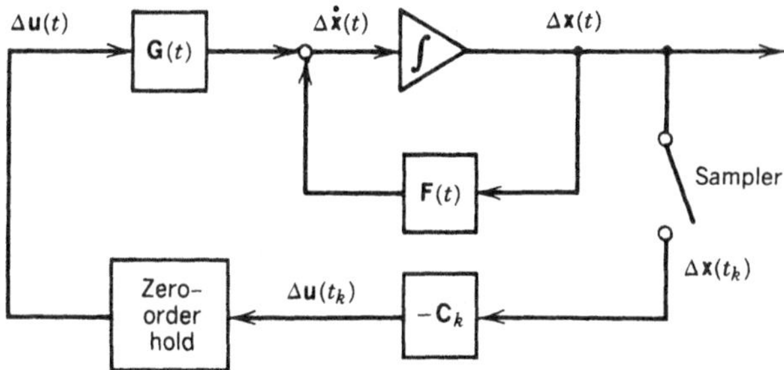

FIGURE 3.7-4 Block diagram of the sampled-data control law.

The sampled-data control law (Eq. 3.7-49b) can be described by a block diagram (Fig. 3.7-4) that includes a continuous-time perturbation dynamic model, where $\mathbf{F}(t)$ and $\mathbf{G}(t)$ are related to the $\mathbf{\Phi}_k$ and $\mathbf{\Gamma}_k$ used in control design. The principal differences between this and the continuous-time controller (Fig. 3.7-2) are that all measurements are sampled, the control path to the dynamic system is passed through a time delay or "zero-order hold" (which clamps the command from one sampling instant to the next), and the control gain values account for the sample-and-hold process.

Dynamic Programming Solution for Discrete Linear-Quadratic Control

The principle of optimality provides an alternate means of deriving the discrete-time LQ controller (Section 3.4). A quadratic value function

$$V_0 = \tfrac{1}{2}\Delta\mathbf{x}_f^T \phi_{\mathbf{xx}_f} \Delta\mathbf{x}_f + \frac{1}{2}\sum_{k=0}^{k_f-1}\left\{[\Delta\mathbf{x}_k^T\ \Delta\mathbf{u}_k^T]\begin{bmatrix}\hat{\mathbf{Q}}_k & \hat{\mathbf{M}}_k\\ \hat{\mathbf{M}}_k^T & \hat{\mathbf{R}}_k\end{bmatrix}\begin{bmatrix}\Delta\mathbf{x}_k\\ \Delta\mathbf{u}_k\end{bmatrix}\right\} \tag{3.7-53}$$

is to be minimized subject to the dynamic constraint

$$\Delta\mathbf{x}_{k+1} = \mathbf{\Phi}_k\,\Delta\mathbf{x}_k + \mathbf{\Gamma}_k\,\Delta\mathbf{u}_k, \qquad \Delta\mathbf{x}_0 = \Delta\mathbf{x}(0) \tag{3.7-54}$$

Recalling that the optimal trajectory from t_k to t_{k_f} contains future segments that are themselves optimal trajectories, the optimal value function at t_k is expressed as the sum of the optimal value function evaluated from $(k+1)$ to k_f plus the incremental change in value during the $(k, k+1)$ interval, minimized with respect to the control:

$$V_k^* = \min_{\Delta\mathbf{u}_k}\,[\mathscr{L}(\Delta\mathbf{x}_k^*, \Delta\mathbf{u}_k, k) + V_{k+1}^*]$$

$$= \min_{\Delta\mathbf{u}_k}\,\mathscr{H}_k \tag{3.7-55}$$

For the linear-quadratic problem described by Eqs. 3.7-53 and 3.7-54, the discrete Lagrangian is

$$\mathscr{L}(\Delta\mathbf{x}_k, \Delta\mathbf{u}_k, k) = \tfrac{1}{2}(\Delta\mathbf{x}_k^T\hat{\mathbf{Q}}_k\,\Delta\mathbf{x}_k + \Delta\mathbf{x}_k^T\hat{\mathbf{M}}\,\Delta\mathbf{u}_k + \Delta\mathbf{u}_k^T\hat{\mathbf{M}}_k^T\,\Delta\mathbf{x}_k + \Delta\mathbf{u}_k^T\hat{\mathbf{R}}_k\,\Delta\mathbf{u}_k) \tag{3.7-56}$$

As in the continuous-time development, it is plausible for the optimal value function to be a quadratic function of $\Delta\mathbf{x}^*$,

$$V_k^* = \tfrac{1}{2}\,\Delta\mathbf{x}_k^{*T}\mathbf{P}_k\,\Delta\mathbf{x}_k^* \tag{3.7-57a}$$

$$V_{k+1}^* = \tfrac{1}{2}\,\Delta\mathbf{x}_{k+1}^{*T}\mathbf{P}_{k+1}\,\Delta\mathbf{x}_{k+1}^* \tag{3.7-57b}$$

with $\mathbf{P}$ being a positive definite, symmetric matrix. Using Eqs. 3.7-54 and 3.7-57b to define V_{k+1}, the discrete Hamiltonian is

$$\begin{aligned}\mathscr{H}_k = \tfrac{1}{2}[(&\Delta\mathbf{x}_k^T\hat{\mathbf{Q}}_k\,\Delta\mathbf{x}_k + \Delta\mathbf{x}_k^T\hat{\mathbf{M}}_k\,\Delta\mathbf{u}_k + \Delta\mathbf{u}_k^T\hat{\mathbf{M}}_k^T\,\Delta\mathbf{x}_k\\ &+ \Delta\mathbf{u}_k^T\hat{\mathbf{R}}_k\,\Delta\mathbf{u}_k) + (\Delta\mathbf{x}_k^T\mathbf{\Phi}_k^T + \Delta\mathbf{u}_k^T\mathbf{\Gamma}_k^T)\mathbf{P}_{k+1}(\mathbf{\Phi}_k\,\Delta\mathbf{x}_k + \mathbf{\Gamma}_k\,\Delta\mathbf{u}_k)]\end{aligned} \tag{3.7-58}$$

This is to be minimized with respect to the control $\Delta\mathbf{u}_k$, and the minimum occurs when the gradient (a row vector) is equal to zero:

$$\begin{aligned}\frac{\partial\mathscr{H}_k}{\partial\Delta\mathbf{u}_k} &= \Delta\mathbf{x}_k^T\hat{\mathbf{M}}_k + \Delta\mathbf{u}_k^T\hat{\mathbf{R}}_k + (\Delta\mathbf{x}_k^T\mathbf{\Phi}_k^T + \Delta\mathbf{u}_k^T\mathbf{\Gamma}_k^T)\mathbf{P}_{k+1}\mathbf{\Gamma}_k\\ &= \Delta\mathbf{x}_k^T(\hat{\mathbf{M}}_k + \mathbf{\Phi}_k^T\mathbf{P}_{k+1}\mathbf{\Gamma}_k) + \Delta\mathbf{u}_k(\hat{\mathbf{R}}_k + \mathbf{\Gamma}_k^T\mathbf{P}_{k+1}\mathbf{\Gamma}_k)\\ &= \mathbf{0}\end{aligned} \tag{3.7-59}$$

Transposing and solving for $\Delta\mathbf{u}_k$, the *linear-optimal control law* is found to be

$$\begin{aligned}\Delta\mathbf{u}_k &= -(\hat{\mathbf{R}}_k + \mathbf{\Gamma}_k^T\mathbf{P}_{k+1}\mathbf{\Gamma}_k)^{-1}(\hat{\mathbf{M}}_k^T + \mathbf{\Gamma}_k^T\mathbf{P}_{k+1}\mathbf{\Phi}_k)\,\Delta\mathbf{x}_k\\ &= -\mathbf{C}_k\,\Delta\mathbf{x}_k\end{aligned} \tag{3.7-60}$$

Substituting for V_k and $\Delta\mathbf{u}_k$ in the optimality equation (Eq. 3.7-55) leads to a relation that is quadratic in $\Delta\mathbf{x}$:

$$\begin{aligned}\Delta\mathbf{x}^T\mathbf{P}_k\,\Delta\mathbf{x} = \Delta\mathbf{x}^T\hat{\mathbf{Q}}\,\Delta\mathbf{x} - \Delta\mathbf{x}^T\hat{\mathbf{M}}\mathbf{C}\,\Delta\mathbf{x} - \Delta\mathbf{x}^T\mathbf{C}^T\hat{\mathbf{M}}^T\,\Delta\mathbf{x} + \Delta\mathbf{x}^T\mathbf{C}^T\hat{\mathbf{R}}\mathbf{C}\,\Delta\mathbf{x}\\ + (\Delta\mathbf{x}^T\mathbf{\Phi}^T - \Delta\mathbf{x}^T\mathbf{C}^T\mathbf{\Gamma}^T)\mathbf{P}_{k+1}(\mathbf{\Phi}\,\Delta\mathbf{x} - \mathbf{\Gamma}\mathbf{C}\,\Delta\mathbf{x})\end{aligned} \tag{3.7-61}$$

The $\Delta\mathbf{x}$ terms can be canceled if the solution is to hold for arbitrary $\Delta\mathbf{x}$.

After manipulation,

$$\begin{aligned}\mathbf{P}_k = \hat{\mathbf{Q}} + \boldsymbol{\Phi}^T\mathbf{P}_{k+1}\boldsymbol{\Phi} + \mathbf{C}^T(\hat{\mathbf{R}} + \boldsymbol{\Gamma}^T\mathbf{P}_{k+1}\boldsymbol{\Gamma})\mathbf{C}\\ - \mathbf{C}^T(\boldsymbol{\Gamma}^T\mathbf{P}_{k+1}\boldsymbol{\Phi} + \hat{\mathbf{M}}^T) - (\boldsymbol{\Phi}^T\mathbf{P}_{k+1}\boldsymbol{\Gamma} + \hat{\mathbf{M}})\mathbf{C}\end{aligned} \tag{3.7-62}$$

and replacing $\mathbf{C}$ by its equivalent (Eq. 3.7-60), this can be condensed to

$$\mathbf{P}_k = \hat{\mathbf{Q}} + \boldsymbol{\Phi}^T\mathbf{P}_{k+1}\boldsymbol{\Phi} - (\hat{\mathbf{M}} + \boldsymbol{\Phi}^T\mathbf{P}_{k+1}\boldsymbol{\Gamma})(\hat{\mathbf{R}} + \boldsymbol{\Gamma}^T\mathbf{P}_{k+1}\boldsymbol{\Gamma})^{-1}(\hat{\mathbf{M}}^T + \boldsymbol{\Gamma}^T\mathbf{P}_{k+1}\boldsymbol{\Phi}) \tag{3.7-63}$$

which is identical to Eq. 3.7-51. From Eqs. 3.7-53 and 3.7-57,

$$\mathbf{P}_{k_f} = \phi_{\mathbf{xx}_f} \tag{3.7-64}$$

For reasons mentioned earlier, Eq. 3.7-63 is necessary and sufficient for the discrete-time LQ solution.

Small Disturbances and Parameter Variations

Equations 3.7-18 and 3.7-52 indicate that the total control approaches its nominal value as the actual state approaches its optimal value. This result apparently assures that the total control is optimal (i.e., that the dynamic system is forced to follow the optimal trajectory). If, however, there are disturbances not modeled in Eq. 3.1-2 or the actual system differs from the dynamic model, Eq. 3.7-18 may not be up to the task, and $\mathbf{C}(t)$ may not be optimal.

The optimal treatment of disturbances and system variations depends on whether they are deterministic (certain) or random (uncertain) and, in the latter case, whether they are random constants or random processes (continually changing in time). From Eq. 2.3-27, disturbances, $\Delta\mathbf{w}(t)$, that are certain and therefore known can be opposed by perturbing the control setting. A rocket launch through a known vertical wind profile could be treated in this fashion. Known parameter variations should be taken into account in calculating $\mathbf{C}(t)$.

Disturbances and parameter variations that are random processes pose a different problem because the cost functions presented earlier cannot be minimized with certainty. At best, we can hope to minimize some statistical measure of the cost, and this implies that statistical characteristics of the random processes are known. A reasonable approach is to minimize the expected value of the cost, conditioned on the statistics of the random processes. Random disturbances and parameter variations can be treated in the same manner i.e., both can be modeled as "process noise"), and adaptive filtering (Section 4.7) can be used to estimate both disturbance levels and parameters. Although conditions for stochastic optimality of nonlinear systems have been developed, the most useful results apply to linear systems with zero-mean Gaussian process noise. For this case, the LQ control law defined by Eq. 3.7-15 is optimal without qualification (Chapter 5).

Solving the problem of random *constant* disturbances or parameter variations is more involved because the induced error does not average zero during the interval. Two possible solutions are estimation of the constant (Section 4.7) and integral compensation. Assuming that the disturbance or parameter is observable, the state can be augmented to include the random constant, and an estimate of the constant can be based upon observation of the system. The estimate then is used in the same way as its deterministic equivalent. Recognizing that the random constant would prevent $[\mathbf{x}(t)-\mathbf{x}^*(t)]$ from reaching null, an integral of the state error could be added to Eq. 3.7-18, seeking to provide asymptotic convergence of the actual and optimal states.

Figure 3.7-1 implies direct feedback of the state in the neighboring-optimal control, but it should be recalled that the state may not be measured directly and that the available measurements may be corrupted by error (Fig. 1.2-1). In such instance, Eq. 3.7-18 must use an *estimate* of the state, $\hat{\mathbf{x}}(t)$, which is derived from an observation, $\mathbf{z}(t)$. If the statistics of the measurement errors are known, the estimate can be *optimal* in some sense. For example, estimators that minimize the root-mean-square error between $\mathbf{x}(t)$ and $\hat{\mathbf{x}}(t)$ or which maximize the probability that $\mathbf{x}(t)$ and $\hat{\mathbf{x}}(t)$ are the same can be found. (These two estimators are identical for a wide class of problems.) By the separation theorem, discussed in Chapter 5, the controller and the estimator often can be optimized separately; however, contrary to early suppositions, the stochastic optimal control law that results is *not* guaranteed to have satisfactory performance and robustness. These issues are addressed in Chapter 6, while Chapter 4 presents a number of important aspects of optimal state estimation.

PROBLEMS

Section 3.2

1. Identify the important state, control, and output variables of the chemical reactor problem described in Section 1.2 and formulate a suitable cost function to be minimized by the application of control.

2. A robot arm that is free to rotate and translate about three axes must pick small screws out of a tray and insert them in a device. The screws must be properly oriented and tightened in the process. The robot-arm components are essentially rigid, and each degree of freedom is controlled by a separate electric motor. Identify the important state,

control, and output variables of the problem, and formulate a suitable cost function to be minimized by the application of control.

3. A high-speed impact printer for computer output contains 138 print hammers that must be actuated quickly and precisely to engage the correct typeslugs as they pass by. The typeslugs are mounted on a continuous chain that moves at constant speed; to make an imprint, the hammer strikes the paper, driving it into the ribbon when the proper character is in position behind the ribbon. The hammer is actuated by sending a current to a moving coil mounted on the hammer, which is in the field of a strong permanent magnet. For a single print hammer mechanism, identify the important state, control, and output variables of the problem, and formulate a suitable cost function to be minimized by the application of control.

4. A military helicopter must fly from point A to point B through an area of hills and valleys without being seen and as quickly as possible. The control system must provide terrain avoidance as well as proper guidance to the destination, and it must minimize abrupt accelerations and angular motions for crew effectiveness and comfort. The helicopter's controls consist of commands that vary the pitch and roll angles of the main rotor relative to the helicopter body, main-rotor-blade incidence angle, and tail-rotor blade incidence angle. Identify the important state, control, and output variables of the problem, and formulate a suitable cost function to be minimized by the application of control.

5. A government agency is the principle source of funds for power generation facilities in an isolated region of the country. The ability to generate power at low cost affects the region's ability to compete with other regions, the net growth of the region's population, and the quality of life of those who live there. Costs of constructing new, more efficient generating facilities must be weighed against the costs of maintaining old generating equipment. The current quality of life must be weighed against the future quality of life when the agency reviews its 5-year investment policy each year. Identify the important state, control, and output variables of the problem, and formulate a suitable cost function to be minimized by the application of control.

Section 3.3

1. Do the parametric optimization problem described in Example 3.3-1c with a desired final value of 10 rather than zero.

2. The velocity (v), flight path angle (γ), altitude (h), and mass (m) of an

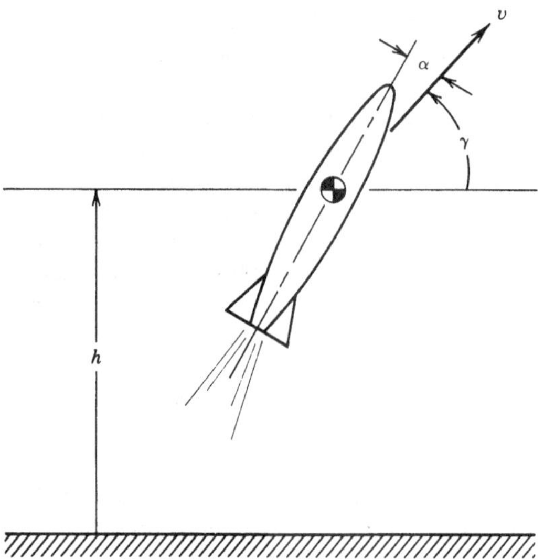

ascending rocket can be found by integrating the following equations:

$$\dot{v} = \frac{1}{m}(T \cos \alpha - D - g \sin \gamma)$$

$$\dot{\gamma} = \frac{1}{mv}(T \sin \alpha + L - g \cos \gamma)$$

$$\dot{h} = v \sin \gamma$$

$$\dot{m} = f(T)$$

The thrust T has a constant value, and the angle of attack α is the control variable. D and L represent aerodynamic drag and lift, while $f(T)$ is the mass flow rate of the rocket motor. A cost function of the form

$$J = \frac{1}{2}\left\{[\mathbf{x}(t_f) - \mathbf{x}_d]^T \mathbf{Q}[\mathbf{x}(t_f) - \mathbf{x}_d] + r\int_0^{t_f} \alpha^2(t)\, dt\right\}$$

is to be minimized by a control history that is linear in time:

$$\alpha(t) = \alpha_0 + \alpha_1 t \text{ rad}$$

$\mathbf{Q}$ is a (3×3) positive definite, diagonal matrix, and

$$\mathbf{x}^T = (v \quad \gamma \quad h)$$

(a) What are the two scalar equations that must be satisfied to find the constant α_0 and α_1 that minimize J?

(b) Find the optimal trajectory, the minimizing α_0 and α_1, and the cost J for the following specifications:

$$T = 10{,}000 \text{ lb}, \qquad D = L = 0, \qquad f(T) = 0, \qquad t_f = 10 \text{ s},$$

$$g = 32 \text{ ft/s}^2, \qquad v(0) = 100 \text{ ft/s}, \qquad v_d \text{ open}$$

$$\gamma(0) = \pi/2 \qquad \gamma_d = \pi/3$$

$$h(0) = 0 \text{ ft} \qquad h_d \text{ open}$$

$$m(0) = 20 \text{ slugs} \qquad m(10) = 20 \text{ slugs}$$

$$\mathbf{Q} = \text{Diag}(0, 1, 0), \qquad r = 1$$

(c) Repeat (b) with $\mathbf{Q} = \text{Diag}(0, 10, 0)$.

(d) Repeat (c) with $f(T) = 1$ [hence $m(10) = 10$ slugs] and $D = 0.0004v^2$.

(e) repeat (d) with $h_d = 20{,}000$ ft and $\mathbf{Q} = \text{Diag}(0, 10, 0.01)$

3. Repeat Problem 2c, assuming that the control takes the following form:

$$\alpha(t) = \alpha_0 + \alpha_1 \sin \omega t + \alpha_2 \cos \omega t$$

$$\omega = 2\pi / t_f$$

Minimizing values of α_0, α_1, and α_2 must be found.

4. Assume that the cost function in Problem 2 is revised to read

$$J = \tfrac{1}{2}\left\{[\mathbf{x}(t_f) - \mathbf{x}_d]^T \mathbf{Q}[\mathbf{x}(t_f) - \mathbf{x}_d] + \int_0^{t_f} [r_1 \alpha^2(t) + r_2 T^2(t)\, dt\right\}$$

while the two controls are constant in the interval:

$$\alpha(t) = \alpha$$

$$T(t) = T$$

Find the optimal trajectory, the minimizing controls, and the cost for the following:

$$v_d = 5000 \text{ ft/s}$$

$$\gamma_d = \pi/3 \text{ rad}$$

$$h_d = 20{,}000 \text{ ft}$$

$$\mathbf{Q} = \text{Diag}(0.1, 10, 0.01), \qquad r_1 = r_2 = 1$$

5. A second-order system described by

$$\begin{bmatrix}\dot{x}_1(t)\\ \dot{x}_2(t)\end{bmatrix}=\begin{bmatrix}0 & 1\\ -\omega_n^2 & -2\zeta\omega_n\end{bmatrix}\begin{bmatrix}x_1(t)\\ x_2(t)\end{bmatrix}+\begin{bmatrix}0\\ \omega_n^2\end{bmatrix}u(t)$$

is to be controlled by a control *law* of the form,

$$u(t)=-c_1[x_1(t)-x_{1_d}]-c_2x_2(t)$$

in order to minimize the cost function

$$J=\tfrac{1}{2}\left\{q[x_1(t_f)-x_{1_d}]^2+\int_0^{t_f} ru^2(t)\,dt\right\}$$

(a) Find the optimal trajectory, the minimizing constant control gains, c_1 and c_2, and the cost function for the following:

$$\omega_n=6.28,\qquad \zeta=0.707,\qquad t_f=4\text{ s}$$

$$q=1,\qquad r=1$$

$$x_1(0)=0,\qquad x_{1_d}=1$$

$$x_2(0)=0,\qquad x_{2_d}\text{ open}$$

(b) Repeat (a), assuming that only x_1 is used for control, (i.e., $c_2=0$).

Section 3.4

1. It has been proposed that the rate of pollution increase be modeled by a first-order equation

$$\dot{x}(t)=-ax(t)+bu(t),\qquad x(0)\text{ given}$$

where $x(t)$ is the amount of pollution, $u(t)$ is the amount of consumption, a represents the decay of pollution due to natural and regulatory action, and b provides the proportionality between consumption and the rate of pollution (F-1). A *utility function* that expresses the relative values of pollution and consumption has been expressed as the following Lagrangian:

$$\mathscr{L}=[1-e^{-u/u_o}]\left[1-\left(\frac{x}{x_o}\right)^n\right]e^{-ct}$$

u_o is a nominal consumption level, x_o is a tolerable level of pollution, n is a factor expressing the sharpness of concern for pollution, and c is a factor that "de-weights" concern for the future in comparison to the present. The utility function is to be maximized by choice of control. Write the Euler–Lagrange equations for a fixed-end-time optimization of this problem.

2. Using numerical methods, solve Problem 1 with $u_o = 1$, $x_o = 0.5$, $n = 2$, $a = 0.015$, $b = 0.01$, $c = 0$, $x(0) = 0.4$, and $t_f = 10$.

3. Many mechanical systems can be modeled by the following second-order differential equation:

$$\begin{bmatrix} \dot{x}_1 \\ \dot{x}_2 \end{bmatrix} = \begin{bmatrix} 0 & 1 \\ -\omega_n^2 & -2\zeta\omega_n \end{bmatrix} \begin{bmatrix} x_1 \\ x_2 \end{bmatrix} + \begin{bmatrix} 0 \\ \omega_n^2 \end{bmatrix} u$$

Write the Hamiltonian and Euler–Lagrange equations for the system, assuming that the cost function is quadratic in $\mathbf{x}$ and $\mathbf{u}$.

4. A high temperature oven can be modeled by

$$\begin{aligned} \dot{x} &= -k_1(x - T_s) - k_2(x^4 - T_s^4) + k_3 u \\ &= -k_1 x - k_2 x^4 + k_3 u + k_4 \end{aligned}$$

where x is the oven temperature, T_s is the temperature of the air surrounding the oven, and u is the controlled input heat transfer rate. This model accounts for convective (linear) and radiative (quartic) losses as scaled by the coefficients k_i. The oven is to be brought to a desired final temperature x_d at the fixed final time t_f subject to the cost function,

$$J = q[x(t_f) - x_d] + \int_0^{t_f} ru^2\, dt$$

where q and r are constants. Find the minimizing control *law* for this problem. (Hint: The Hamiltonian is not an explicit function of time; therefore, it is constant in the interval.)

5. A first-order model for chemical reactor concentration response $x(t)$ to flow rate $u(t)$ can be written as

$$\dot{x}(t) = ax^n(t) + bu(t)$$

where n is the reaction order, a is a reaction constant, and b is a constant relating the inlet flow rate to the outlet concentration. We wish to minimize a value function

$$V=[x(t_f)-x_d]^2+\int_0^{t_f}\{[x(t)-x_d]^2+u^2\}\,dt$$

using the HJB equation, with the assumption that the optimal value function takes the general form

$$V^*[x(t),t]=p(t)x^2(t), \qquad p(t)>0$$

Find the optimal trajectory beginning at $x(0)=0.1$ for $a=-2$ $n=2$, $b=1$, and $x_d=1$ using the principle of optimality.

Section 3.5

1. Rotations of a nonspinning satellite about a single axis can be represented by the second-order differential equation,

$$\ddot{\theta}=u$$

Assuming $u\le u_{\max}$, find the minimum-time solution from $\theta(0)=1$, $\dot{\theta}(0)=0$, to $\theta(t_f)=\dot{\theta}(t_f)=0$.

2. Repeat Example 3.5-4 assuming that the pesticide kills only the predator and not the prey.

3. The temperature in a one-room building can be modeled by the following first-order differential equation:

$$\dot{x}_1=\frac{1}{c_p m}[k(x_1-x_{\text{air}})+u]$$

The temperature in the room is x_1, the temperature of the outside air is x_{air}, the mass of air in the room is m, the heat capacity is c_p, and the heat-rate input from an electric furnace is u. The equation can be written as

$$\dot{x}_1=-\frac{1}{T}(x_1-x_{\text{air}})+k_1 u$$

where T is a time constant (in hours) and k_1 is a proportionality constant. A cost function that penalizes the direct use of power while meeting a hard terminal constraint can be expressed as

$$J=\int_{t_0}^{t_f}|u|\,dt$$

$$\psi=x_1(t_f)-x_d=0$$

Suppose that $x_1(t_0) = x_{\text{air}} = 45°\text{F}$, $T = 4$ hr, $k_1 = 1$, $(t_f - t_0) = 12$ hr, and $x_d = 70°\text{F}$. Find the minimizing control history, and plot or tabulate the resulting temperature rise in the room.

4. A cost function that trades the cost of achieving the final temperature against the cost of providing the necessary heat rate can be expressed as

$$J = c_1[x_1(t_f) - x_d]^2 + \int_{t_0}^{t_f} c_2 u^2\, dt$$

Suppose $c_1 = 5c_2$. Find the minimizing control history, and plot or tabulate the resulting temperature rise in the room.

5. Change the Lagrangian in Problem 4 to

$$\mathscr{L} = c_1[x_1(t) - x_d]^2 + c_2 u^2$$

and repeat the problem.

6. A two-room building heated by a furnace in the first room has the following differential equations:

$$\dot{x}_1 = -\frac{1}{T_1}(x_1 - x_{\text{air}}) + k_1(x_2 - x_1) + k_2 u$$

$$\dot{x}_2 = -\frac{1}{T_2}(x_2 - x_{\text{air}}) + k_1(x_1 - x_2)$$

Write the Euler–Lagrange equations for this problem using the Lagrangian of Problem 5. Find the minimizing control history assuming that $k_2 = k_1$ and $T_2 = T_1$.

7. Simplified equations of motion for a robot arm can be expressed as follows:

$$\ddot{\theta} = \frac{M}{l^2}$$

$$\ddot{l} = F$$

The arm is free to rotate and change its length, but it is constrained to move to the right of a vertical wall, as shown in the following sketch.

This state inequality constraint could be expressed as

$$c[x(t)] = x - x_0 \geq 0$$

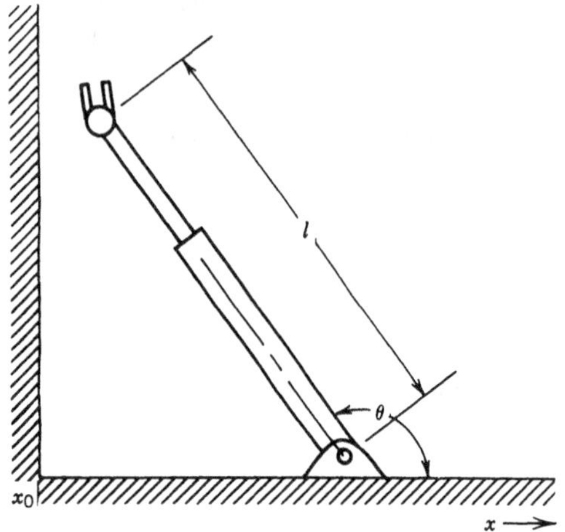

Write the qth-order state-control inequality constraint that must be satisfied on the wall.

8. The trajectory of a vertical sounding rocket vehicle traveling over a flat earth in the absence of an atmosphere could be computed from the following equations:

$$\dot{h} = v$$

$$\dot{v} = -g + a(t)$$

Here g is the acceleration due to gravity and $a(t)$ is the acceleration due to the rocket motor. The total impulse I of the rocket motor is limited, so

$$I = \int_0^{t_f} a(t)\, dt = \text{constant}$$

We would like to choose $a(t)$ so that the rocket vehicle's peak altitude is maximized, in other words, so that

$$\max_{a(t)} J = h_f$$

(a) Write the necessary conditions for optimality, assuming that there is no constraint on $a(t)$, and comment on the optimal solution.

(b) Assume that $|a(t)| \leq a_{\max}$. Does this make a difference? Explain.

9. A "two-compartment" model of the ingestion and metabolism of a drug is given by [M-1]

$$\dot{x}_1 = -k_1 x_1 + u$$
$$\dot{x}_2 = k_1 x_1 - k_2 x_2$$
$$z = x_2 + n$$

where k_1 and k_2 are positive constants and

$x_1 =$ mass of drug in gastrointestinal tract $\le x_{1\,\max}$

$x_2 =$ mass of drug in bloodstream

$u =$ ingestion rate of the drug $\le u_{\max}$

(a) What are the eigenvalues of the system?
(b) What does the time response of x_2 look like for $u =$ constant, $x_1 = x_2 = 0$?
(c) How would you compute the time-optimal control law for achieving $x_2 = x_{2_{\text{desired}}}$?

Section 3.6

1. Find the optimal control $\alpha(t)$ for the rocket problem described in Problem 2e, Section 3.3, using the penalty-function method. In this case, $\alpha(t)$ is not constrained to vary linearly in time; it is an arbitrary function to be determined in the optimization procedure.
2. Solve Problem 1 using the neighboring-extremal method.
3. Solve Problem 1 using the quasilinearization method.
4. Solve Problem 1 using the steepest-descent method.
5. Using the steepest-descent method, find the minimizing control for the following problem:

$$\begin{bmatrix} \dot{x}_1(t) \\ \dot{x}_2(t) \end{bmatrix} = \begin{bmatrix} 0 & 1 \\ 9 & -9 \end{bmatrix} \begin{bmatrix} x_1(t) \\ x_2(t) \end{bmatrix} + \begin{bmatrix} 0 \\ -9 \end{bmatrix} u(t)$$

$$J = \tfrac{1}{2}\left\{[x_1^2(10) + x_2^2(10)] + \int_0^{10} [5x_1^2(t) + x_2^2(t) + u^2(t)]\,dt\right\}$$

$x_1(10) = 10, \qquad x_2(10) = 0$

Section 3.7

1. A first-order continuous-time system can be described by the following differential equation:

$$\dot{x} = ax + bu$$

We want to minimize the quadratic cost function

$$J = \lim_{T \to \infty} \frac{1}{2T} \int_0^T (qx^2 + ru^2)\, dt$$

with respect to the control u. This leads to a control *law* of the form

$$u = -\left(\frac{1}{r}\right) bsx = -cx$$

where s is the steady-state solution of a Riccati equation.

(a) Express the control gain c as a function of a, q, and r, assuming that $b = 1$ and $a > 0$. What are the limiting values as q (or r) becomes very large (small) or small (large)?

(b) Repeat, assuming that $a < 0$. What observations can be made about the ranges of control gain and about the resulting closed-loop system characteristics?

2. A second-order spring-mass-damper system can be expressed as

$$\begin{bmatrix} \dot{x}_1 \\ \dot{x}_2 \end{bmatrix} = \begin{bmatrix} 0 & 1 \\ a & b \end{bmatrix} \begin{bmatrix} x_1 \\ x_2 \end{bmatrix} + \begin{bmatrix} 0 \\ -a \end{bmatrix} u$$

and the corresponding quadratic cost function is minimized by solving a matrix Riccati equation. The steady-state solution corresponds to an infinite final time, as above.

(a) Express the constant control gain matrix for the following cost function weighting matrices:

$$\mathbf{Q} = \begin{bmatrix} q_{11} & 0 \\ 0 & 0 \end{bmatrix}, \qquad \mathbf{R} = r$$

(b) Repeat with the following:

$$\mathbf{Q} = \begin{bmatrix} 0 & 0 \\ 0 & q_{22} \end{bmatrix}, \qquad \mathbf{R} = r$$

3. Do the discrete-time equivalent of Problem 2. Begin by finding the state transition and discrete control-effect matrices, $\mathbf{\Phi}$ and $\mathbf{\Gamma}$, assuming that the time sampling interval is 0.1 s.
   (a) Assume $\hat{\mathbf{Q}} = \mathbf{Q}$ and $\hat{\mathbf{R}} = \mathbf{R}$ (i.e., neglect sampled-data effects on the discrete cost function).
   (b) Repeat, accounting for sampled-data effects on $\hat{\mathbf{Q}}$ and $\hat{\mathbf{R}}$.
4. Repeat Problem 5, Section 3.6, using the linear-optimal control law defined by Eqs. 3.7-15 and 3.7-17 to generate the control history. Compare the results from the two approaches.
5. Generate the equivalent sampled-data dynamic model and cost function for Problem 4. Compute the optimal sampled-data control law and the corresponding trajectory.

REFERENCES

A-1 Anderson, G. M., An Indirect Numerical Method for the Solution of a Class of Optimal Control Problems with Singular Arcs, *IEEE Transactions on Automatic Control*, **AC-17**(3) 363–365, June 1972.

A-2 Anderson, B.D.O. and Moore, J. B., *Linear Optimal Control*, Prentice-Hall, Englewood Cliffs, NJ, 1971.

A-3 Athans, M. and Falb, P. L., *Optimal Control*, McGraw-Hill, New York, 1966.

B-1 Balakrishnan, A. V., On a New Computing Technique in Optimal Control, *SIAM Journal of Control*, **6**, 149–173, 1968.

B-2 Bell, D. J. and Jacobson, D. H., *Singular Optimal Control Problems*, Academic, New York, 1975.

B-3 Bellman, R., *Dynamic Programming*, Princeton University Press, Princeton, NJ, 1957.

B-4 Bellman, R. and Kalaba, R., Dynamic Programming and Adaptive Processes: Mathematical Foundations. *IRE Transactions on Automatic Control*, **AC-5**(1), January 1960, pp. 5–10.

B-5 Bellmann, R. and Kalaba, R., eds., *Selected Papers on Mathematical Trends in Control Theory*, Dover, New York, 1964.

B-6 Bernstein, D. S. and Gilbert, E. G., Optimal Periodic Control: The Pi Test Revisited, *IEEE Transactions on Automatic Control*, **AC-25**(4), 673–684, August 1980.

B-7 Bryson, A. E., Jr., Denham, W. F., and Dreyfus, S. E., Optimal Programming Problems with Inequality Constraints. I: Necessary Conditions for Extremal Solutions, *AIAA Journal*, **1**(11), 2544–2550, November 1963.

B-8 Bryson, A. E., Jr. and Ho, Y. C., *Applied Optimal Control*, Hemisphere Publishing, Washington, D.C., 1975.

C-1 Citron, S. J., *Elements of Optimal Control*, Holt, Rinehart & Winston, New York, 1969.

C-2 Clements, D. J. and Anderson, B.D.O., *Singular Optimal Control Theory: The Linear-Quadratic Problem*, Springer-Verlag, Berlin, 1978.

C-3 Contensou, P., Étude Théoretique des Trajectories Optimales dans un Champ de

Gravitation. Application au Cas d'un Centre d'Attraction Unique, *Astronautica Acta*, **8**, 134–150, 1962.

D-1 Davis, M.H.A., *Linear Estimation and Stochastic Control*, Chapman and Hall, London, 1977.

D-2 Denham, W. F. and Bryson, A. E., Jr., Optimal Programming Problems with Inequality Constraints. II: Solution by Steepest-Ascent, *AIAA Journal*, **2**(1), 25–34, January, 1964.

D-3 Dorato, P. and Levis, A., Optimal Linear Regulators: The Discrete-Time Case, *IEEE Transactions on Automatic Control*, **AC-16**(6), 613–620, December 1971.

D-4 Dyer, P. and McReynolds, S. R., *The Computation and Theory of Optimal Control*, Academic, New York, 1970.

F-1 Fleming, R. N. and Pantell, R. H., The Conflict Between Consumption and Pollution, *IEEE Transactions on Systems, Man, and Cybernetics*, **SMC-4**(3), 204–208, March 1974.

H-1 Hirata, H., Optimal Control of Nonlinear Population Dynamics, *IEEE Transactions on Systems, Man, and Cybernetics*, **SMC-10**(1), 32–38, January 1980.

H-2 Holley, W. and Bryson, A. E., Jr., Multi-Input, Multi-Output Regulator Design for Constant Disturbances and Non-Zero Set Points with Application to Automatic Landing in a Crosswind, Stanford University SUDAAR No. 465, April 1973.

J-1 Jacob, H. G., An Engineering Optimization Method with Application to STOL-Aircraft Approach and Landing Trajectories, NASA TN D-6978, September, 1972.

J-2 Jacobson, D. H. and Mayne, D. Q., *Differential Dynamic Programming*, American Elsevier, New York, 1970.

J-3 Jacobson, D. H., Totally Singular Quadratic Minimization Problems, *IEEE Transactions on Automatic Control*, **AC-16**(6), 651–658, December 1971.

J-4 Jacobson, D. H., Lele, M. M. and Speyer, J. L., New Necessary Conditions of Optimality for Control Problems with State-Variable Inequality Constraints, *Journal of Mathematical Analysis and Applications*, **35**(2), 255–284, August 1971.

J-5 Johnson, C. D. and Gibson, J. E., Singular Solutions in Problems of Optimal Control, *IEEE Transactions on Automatic Control* **AC-8**(1), 4–15, January 1963.

K-1 Kailath, T., *Linear Systems*, Prentice-Hall, Englewood Cliffs, NJ, 1980.

K-2 Kelley, H. J., Method of Gradients, Chapter 6 in *Optimization Techniques*, Academic, New York, 1962.

K-3 Kelley, H. J., Kopp, R. E., and Moyer, H. G., Singular Extremals, Chapter 3 in *Topics in Optimization*, Academic, New York, 103–155, 1967.

K-4 Kirk, D. E., *Optimal Control Theory*, Prentice-Hall, Englewood Cliffs, NJ, 1970.

K-5 Kopp, R. E. and Moyer, H. G., Trajectory Optimization Techniques, in *Advances in Control Systems*, Vol. 4, C. T. Leondes, ed., Academic, New York, 1966.

K-6 Kreindler, E., Additional Necessary Conditions for Optimal Control with State-Variable Inequality Constraints, *Journal of Optimization Theory and Applications*, **38**(2), 241–250, October 1982.

L-1 Larson, R. E., *State Increment Dynamic Programming*, American Elsevier, New York, 1968.

L-2 Lasdon, L. S., Mitter, S. K. and Waren, A. D., The Conjugate Gradient Method for Optimal Control Problems, *IEEE Transactions on Automatic Control*, **AC-12**(2), 132–138, April 1967.

L-3 Lee, E. B. and Markus, L., *Foundations of Optimal Control Theory*, Wiley, New York, 1967.

L-4 Leitmann, G. *An Introduction to Optimal Control*, McGraw-Hill, New York, 1966.

L-5 Leitmann, G., *The Calculus of Variations and Optimal Control: An Introduction*, Plenum, New York, 1981.

L-6 Levis, A. H., Schleuter, R. A. and Athans, M., On the Behaviour of Optimal Linear Sampled-Data Regulators, *International Journal of Control*, **13**(2), 343–361, February 1971.

M-1 McClamrock, N. H., *State Models of Dynamic Systems*, Springer-Verlag, New York, 1980.

M-2 McIntyre, J. and Paiewonsky, B., "On Optimal Control with Bounded State Variables," in *Advances in Control Systems*, Vol. 5, C. T. Leondes, ed., Academic, New York, 1967, pp. 389–419.

M-3 Mehra, R. K. and Davis, R. E., "A Generalized Gradient Method for Optimal Control Problems with Inequality Constraints and Singular Arcs," *IEEE Transactions on Automatic Control*, **AC-17**(1), 69–79, February 1972.

M-4 Miele, A., Pritchard, R. E., and Damoulakis, J. N., Sequential Gradient-Restoration Algorithm for Optimal Control Algorithms, *Journal of Optimization Theory and Applications*, **5**(4), April 1970, pp. 235–282.

M-5 Miele, A., Recent Advances in Gradient Algorithms for Optimal Control Problems, *Journal of Optimization Theory and Applications*, **17**(5/6), December 1975, pp. 361–430.

O-1 Oldenburger, R., *Optimal and Self-Optimizing Control*, MIT Press, Cambridge, MA, 1966.

P-1 Polak, E., An Historical Survey of Computation Methods in Optimal Control, *SIAM Review*, **15**(2), 553–584, April 1973.

P-2 Pontryagin, L. S., Boltyanskii, V. G., Gamkrelidze, R. V. and Mishenko, E. F., *The Mathematical Theory of Optimal Processes* (translated by D. E. Brown), Macmillan, New York, 1964.

P-3 Psiaki, M. L. and Stengel, R. F., Optimal Flight Paths Through Microburst Wind Profiles, *Proceedings of the AIAA 12th Atmospheric Flight Mechanics Conference*, Snowmass, Colorado, August 1985, pp. 494–506.

R-1 Robbins, H. M., Optimality of Intermediate-Thrust Arcs of Rocket Trajectories, *AIAA Journal* **3**(6), 1094–1098, June 1965.

S-1 Schultz, R. L., Fuel Optimality of Cruise, *AIAA Journal of Aircraft*, **11**(9), 586–587, September 1974.

S-2 Speyer, J. L. and Jacobson, D. H., Necessary and Sufficient Conditions for Optimality for Singular Control Problems: A Transformation Approach, *Journal of Mathematical Analysis & Applications*, **33**, 163–187, 1971.

S-3 Speyer, J. L., On the Fuel Optimality of Cruise, *AIAA Journal of Aircraft*, **10**(12), 763–765, December 1973.

S-4 Speyer, J. L., Nonoptimality of the Steady-State Cruise for Aircraft, *AIAA Journal of Aircraft*, **14**(11), 1604–1610, November 1976.

S-5 Stengel, R. F. and Marcus, F. J., Energy Management Techniques for Fuel Conservation in Military Transport Aircraft, AFFDL-TR-156, February 1976.

S-6 Stengel, R. F., Optimal Transition from Entry to Cruising Flight, *AIAA Journal of Spacecraft and Rockets*, **8**(11), 1126–1132, November 1971.

S-7 Stengel, R. F., Strategies for Control of the Space Shuttle Transition, *AIAA Journal of Spacecraft and Rockets*, **10**(1), 77–84, January 1973.

S-8 Stengel, R. F., Optimal Guidance for the Space Shuttle Transition, *AIAA Journal of Spacecraft and Rockets*, **11**(3), 173–179, March 1974.

S-9 Stengel, R. F., An Introduction to Stochastic Optimal Control Theory, *Theory and Applications of Optimal Control in Aerospace Systems*, **AGARD-AG-251**, 3–1 to 3–33, July 1981.

T-1 Tait, T. S., Singular Problems in Optimal Control, Ph.D. thesis, Harvard University, 1965.

W-1 Wonham, W. M. and Johnson, C. D., Optimal Bang-Bang Control with Quadratic Performance Index, *Transactions of the American Society of Mechanical Engineers*, Series D, **86**(1), 107–115, March 1964.

4 OPTIMAL STATE ESTIMATION

When a control law depends on state feedback to provide stability, optimality, or other desired response characteristics, dynamic measurements must be taken. As noted earlier, these measurements may not yield the entire state vector directly, and they may be subject to significant errors. Therefore, it is necessary to consider ways in which *estimates* of the state vector can be derived from available measurements. It is consistent with our design objectives that these state estimates contain minimal error (i.e., that they be optimal in some sense). This, in turn, requires

- The ability to define a state-estimate error metric (to be minimized in estimation)
- A knowledge of measurement error statistics, dynamic system models, and system input statistics, and
- Algorithms for using this information to compute minimum-error state estimates

This chapter addresses the design of the necessary state estimators.

We begin the development of *dynamic* state estimators by first considering *static* state estimators [i.e., algorithms that compute the elements of a constant vector from redundant observations of the vector (Section 4.1)]. Given several (k) equally valid but "noisy" measurements of a constant *scalar* quantity, it would be reasonable to simply average the measurements to produce an estimate of the constant. Such an estimate is optimal in the sense that it minimizes the sum of the squared errors between the measurements and a linear function of the estimated constant, (i.e., it is a "*least-squares*" *estimate*). The same holds true for estimating a constant *vector*. Prior knowledge of the measurement errors may indicate that some measurements are better than others; this could be taken into account in the estimate, and the better measurements could be weighted more heavily in the averaging process. The proper weighting factors could be determined by penalizing the poorer measurements in the squared-error criterion, leading to a *weighted least-squares estimate* of the constant vector. An additional vector of measurements could be incorporated in the estimate by repeating the computation with $(k+1)$ terms, but it is possible to get an identical estimate using a *recursive weighted least-squares algorithm* that sums the prior estimate and the measurement update with suitable weighting.

In all these cases, the estimated vector can be considered a state vector that has been modeled as a constant. If, however, the state vector is varying in time as a random sequence, this algorithm can be extended to account for the passage of time and for an underlying linear, stochastic dynamic model. Because the estimator is recursive, it is logical to express the dynamic model as a difference equation (i.e., as a discrete-time model with finite time steps; Section 4.2). This model performs two functions: it is used directly to propagate the time-varying mean value of the state vector estimate, and it enters into the calculation of the coefficients (or gains) of the recursive least-squares estimator. Given "white" Gaussian inputs and measurement errors, the optimal filter, presented in Section 4.3, is called the discrete-time *Kalman filter*. Problems of numerical stability can arise in the computation of estimator gains. These can be reduced with alternate algorithms, notably using the square root of the state covariance or "U–D factorizations" rather than the covariance itself. Filters for correlated disturbances and measurement errors and for time-correlated measurement errors are presented in Section 4.4.

The equivalent continuous-time *Kalman–Bucy filter** is most readily obtained by letting the time interval between samples approach zero (Section 4.5). In the limit, the difference equations become differential equations, but while the proper estimation equations result from this procedure, there are certain theoretical objections to the necessary characterization of random inputs and observation errors as continuous-time "white noise." The mathematical formalisms that overcome these objections are reviewed in Section 4.2 and applied to the estimator in Section 4.5.

The Kalman filter provides the optimal (minimum-variance) state estimate when the dynamic system is linear and the random inputs and errors are Gaussian. Consequently, it would be appropriate for use in neighboring-optimal control as described in Section 3.7. However, there could be cases in which nonlinear and/or non-Gaussian effects are significant. Fairly

*Rudolf Kalman (1930–   ), professor of mathematics at the University of Florida and the Swiss Federal Institute of Technology, has made numerous fundamental contributions to linear systems theory. Richard Bucy (1935–   ), professor of mathematics and aerospace engineering at the University of Southern California, also has made many contributions in the field. The two men derived the linear-optimal filter equations independently, then documented their work jointly.

straightforward extensions of the Kalman filter address the first issue, while higher-order filters address the second (Section 4.6). When the parameters of the system model and the statistical properties of the inputs and the observation errors are unknown, these must be estimated along with the state, as in Section 4.7.

4.1 LEAST-SQUARES ESTIMATES OF CONSTANT VECTORS

Constant vectors can be treated as easily as constant scalars, so we begin with the linear observation equation

$$\begin{aligned} \mathbf{z} &= \mathbf{H}\mathbf{x} + \mathbf{n} \\ &= \mathbf{y} + \mathbf{n} \end{aligned} \tag{4.1-1}$$

where the constant state vector $\mathbf{x}$ has dimension n, the measurement and error-free output vectors, $\mathbf{z}$ and $\mathbf{y}$, have dimension $k_1 \geq n$, $\mathbf{H}$ is a $(k_1 \times n)$ observation matrix, and the error vector $\mathbf{n}$ has dimension k_1. This equation is analogous to Eqs. 2.3-28 and 2.3-29, with $\mathbf{H} = \mathbf{H_x}$, $\mathbf{H_u} = \mathbf{H_w} = \mathbf{0}$, and $\mathbf{J_y} = \mathbf{J_n} = \mathbf{I}$.

Our objective is to compute an estimate of the state, denoted by $\hat{\mathbf{x}}$, from the measurement $\mathbf{z}$. Having obtained $\hat{\mathbf{x}}$, an estimate of the output can be constructed:

$$\hat{\mathbf{y}} = \mathbf{H}\hat{\mathbf{x}} \tag{4.1-2}$$

Then we can define the state's residual error (also called the *state residual*) as

$$\boldsymbol{\epsilon}_\mathbf{x} \triangleq \mathbf{x} - \hat{\mathbf{x}} \tag{4.1-3}$$

and the *measurement residual* as

$$\boldsymbol{\epsilon}_\mathbf{z} \triangleq \mathbf{z} - \hat{\mathbf{y}} \tag{4.1-4}$$

These residuals are conceptually very different from each other. $\boldsymbol{\epsilon}_\mathbf{x}$ reflects the difference between the actual (error-free) state and the state estimate, while $\boldsymbol{\epsilon}_\mathbf{z}$ portrays the difference between the corrupted measurement and a reconstruction of the output.

Least-Squares Estimator

We would like the state estimate to be optimal, and for reasons described in earlier chapters, *minimizing a quadratic error function* is a suitable objective. A quadratic cost function of the state residual

$$J(\mathbf{x}) = \tfrac{1}{2}\boldsymbol{\epsilon}_\mathbf{x}^T \boldsymbol{\epsilon}_\mathbf{x} = \tfrac{1}{2}(\mathbf{x}-\hat{\mathbf{x}})^T(\mathbf{x}-\hat{\mathbf{x}}) \tag{4.1-5}$$

would be most meaningful, but the reason for estimation is that $\mathbf{x}$ is unknown. Hence $\mathbf{x}$ normally is unavailable for cost function evaluation and for formulation of the estimation equations. On the other hand, the measurements are available, and with the reasonable assumption that $\mathbf{z}$ bears some systematic relationship to $\mathbf{x}$,

$$\begin{aligned} J(\mathbf{z}) &= \tfrac{1}{2}\boldsymbol{\epsilon}_\mathbf{z}^T \boldsymbol{\epsilon}_\mathbf{z} = \tfrac{1}{2}(\mathbf{z}-\hat{\mathbf{y}})^T(\mathbf{z}-\hat{\mathbf{y}}) \\ &= \tfrac{1}{2}(\mathbf{z}-\mathbf{H}\hat{\mathbf{x}})^T(\mathbf{z}-\mathbf{H}\hat{\mathbf{x}}) \end{aligned} \tag{4.1-6a}$$

is a useful cost function. It can be evaluated without prior knowledge of $\mathbf{x}$ or $\mathbf{n}$, and it can be minimized to derive $\hat{\mathbf{x}}$ from $\mathbf{z}$. If the mean value of $\mathbf{n}$ is zero, then minimizing $J(\mathbf{z})$ minimizes $J(\mathbf{x})$ in the limit as $k_1 \to \infty$.

Expanding $J(\mathbf{z})$,

$$J(\mathbf{z}) = \tfrac{1}{2}(\mathbf{z}^T\mathbf{z} - \mathbf{z}^T\mathbf{H}\hat{\mathbf{x}} - \hat{\mathbf{x}}^T\mathbf{H}^T\mathbf{z} + \hat{\mathbf{x}}^T\mathbf{H}^T\mathbf{H}\hat{\mathbf{x}}) \tag{4.1-6b}$$

Its minimum with respect to $\hat{\mathbf{x}}$ is found by setting the gradient to zero. Recalling that we have defined the derivative of a scalar with respect to a vector as a row vector (Section 2.1),

$$\frac{\partial J(\mathbf{z})}{\partial \hat{\mathbf{x}}} = (\mathbf{H}^T\mathbf{H}\hat{\mathbf{x}} - \mathbf{H}^T\mathbf{z})^T = \mathbf{0} \tag{4.1-7}$$

Solving for $\hat{\mathbf{x}}$, the *least-squares estimator* is found to be

$$\begin{aligned} \hat{\mathbf{x}} &= (\mathbf{H}^T\mathbf{H})^{-1}\mathbf{H}^T\mathbf{z} \\ &= \mathbf{H}^L\mathbf{z} \end{aligned} \tag{4.1-8}$$

where $\mathbf{H}^L$ is the left pseudoinverse of $\mathbf{H}$ (Section 2.2). No reference to the statistical properties of $\mathbf{z}$ is made in this equation. A solution may be found with $k_1 = n$; however, for averaging to occur, k_1 must be greater than n. Note that the second derivative of J with respect to $\hat{\mathbf{x}}$ is the $(n \times n)$ Hessian matrix:

$$\frac{\partial^2 J(\mathbf{z})}{\partial \hat{\mathbf{x}}^2} = \mathbf{H}^T\mathbf{H} \tag{4.1-9}$$

It is positive definite when $\mathbf{H}$ has rank n, providing the sufficient condition for a minimum.

An alternate way of defining the minimum is called *completing the square*. Adding and subtracting the term $\frac{1}{2}\mathbf{z}^T\mathbf{H}(\mathbf{H}^T\mathbf{H})^{-1}\mathbf{H}^T\mathbf{z}$ to $J(\mathbf{z})$ allows the following factorization:

$$\begin{aligned} J(\mathbf{z}) &= \tfrac{1}{2}\{[\hat{\mathbf{x}}^T\mathbf{H}^T\mathbf{H}\hat{\mathbf{x}} - \mathbf{z}^T\mathbf{H}\hat{\mathbf{x}} - \hat{\mathbf{x}}^T\mathbf{H}^T\mathbf{z} + \mathbf{z}^T\mathbf{H}(\mathbf{H}^T\mathbf{H})^{-1}\mathbf{H}^T\mathbf{z}] \\ &\quad + \mathbf{z}^T\mathbf{z} - \mathbf{z}^T\mathbf{H}(\mathbf{H}^T\mathbf{H})^{-1}\mathbf{H}^T\mathbf{z}\} \\ &= \tfrac{1}{2}\{[\hat{\mathbf{x}} - (\mathbf{H}^T\mathbf{H})^{-1}\mathbf{H}^T\mathbf{z}]^T(\mathbf{H}^T\mathbf{H})[\hat{\mathbf{x}} - (\mathbf{H}^T\mathbf{H})^{-1}\mathbf{H}^T\mathbf{z}] \\ &\quad + \mathbf{z}^T\mathbf{z} - \mathbf{z}^T\mathbf{H}(\mathbf{H}^T\mathbf{H})^{-1}\mathbf{H}^T\mathbf{z}\} \end{aligned} \tag{4.1-6c}$$

The minimizing choice of $\hat{\mathbf{x}}$ is seen to be given by Eq. 4.1-8.

Example 4.1-1 Estimating a Scalar Constant

To illustrate that Eq. 4.1-8 averages the measurements, assume that x is a scalar and that the k_1-vector of measurements, $\mathbf{z}$, is degraded by the error $\mathbf{n}$. In scalar notation, Eq. 4.1-1 is

$$z_k = x + n_k, \qquad k = 1 \text{ to } k_1$$

Then $\mathbf{H}$ is

$$\mathbf{H} = \begin{bmatrix} 1 \\ 1 \\ \cdot \\ \cdot \\ \cdot \\ 1 \end{bmatrix}$$

and

$$\begin{aligned} \hat{\mathbf{x}} &= (\mathbf{H}^T\mathbf{H})^{-1}\mathbf{H}^T\mathbf{z} \\ &= \frac{1}{k_1}[1 \quad 1 \quad \ldots \quad 1]\mathbf{z} \\ &= \frac{1}{k_1}\sum_{k=1}^{k_1} z_k \end{aligned}$$

The product $\mathbf{H}^T\mathbf{z}$ merely sums the measurements, which are divided by k_1 to produce the estimate.

Example 4.1-2 Estimation of Cubic Polynomial Coefficients

Suppose that a number of measurements, z_k, $k = 1$ to k_1, are taken at different times, t_k, with the intent of fitting a cubic polynomial to the data:

$$y(t) = x_1 + x_2 t + x_3 t^2 + x_4 t^3$$

How can the vector of polynomial coefficients, $\mathbf{x}$, be estimated?

The measurements represent

$$\begin{aligned} \mathbf{z} &= \mathbf{y} + \mathbf{n} \\ &= \mathbf{H}\mathbf{x} + \mathbf{n} \end{aligned}$$

or

$$\begin{bmatrix} z_1 \\ z_2 \\ \vdots \\ \vdots \\ z_{k_1} \end{bmatrix} = \begin{bmatrix} 1 & t_1 & t_1^2 & t_1^3 \\ 1 & t_2 & t_2^2 & t_2^3 \\ \vdots & \vdots & \vdots & \vdots \\ \vdots & \vdots & \vdots & \vdots \\ 1 & t_{k_1} & t_{k_1}^2 & t_{k_1}^3 \end{bmatrix} \begin{bmatrix} x_1 \\ x_2 \\ x_3 \\ \\ x_4 \end{bmatrix} + \begin{bmatrix} n_1 \\ n_2 \\ \vdots \\ \vdots \\ n_{k_1} \end{bmatrix}$$

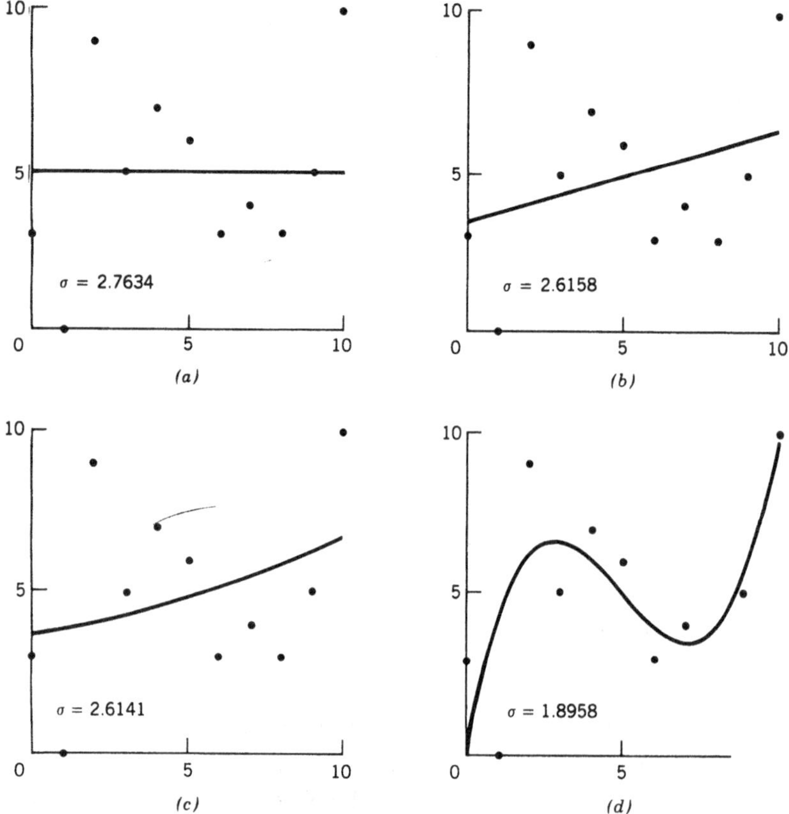

Least-squares curve fitting for polynomials of varying degree. σ = root-mean-square fit error. (*a*) Zeroth degree; (*b*) first degree; (*c*) second degree; (*d*) third degree.

and Eq. 4.1-8 can be applied to find $\hat{\mathbf{x}} = [\hat{x}_1 \quad \hat{x}_2 \quad \hat{x}_3 \quad \hat{x}_4]^T$. Examples of least-squares polynomial fits of zeroth to third degree are shown in diagrams 1-4. The data points are the same in all cases.

EXAMPLE 4.1-3 TRANSFORMATION AND ESTIMATION OF A CONSTANT LINEAR VELOCITY

Example 2.2-2 presented a single velocity vector in two Cartesian coordinate systems, $\mathbf{v}_B$ and $\mathbf{v}_E$:

$$\mathbf{v}_B = \begin{bmatrix} v_x \\ v_y \\ v_z \end{bmatrix}_B$$

$$\mathbf{v}_E = \begin{bmatrix} v_x \\ v_y \\ v_z \end{bmatrix}_E$$

$\mathbf{v}_B$ could be calculated from $\mathbf{v}_E$ using the orthogonal transformation matrix, $\mathbf{H}_{BE}(\phi, \theta, \psi)$,

$$\mathbf{v}_B = \mathbf{H}_{BE}\mathbf{v}_E$$

and the inverse relationship also holds true:

$$\mathbf{v}_E = \mathbf{H}_{EB}(\psi, \theta, \phi)\mathbf{v}_B$$

Given the Euler angles $\phi = 0°$, $\theta = 30°$, and $\psi = 45°$, plus four noisy measurements of the velocity vector (measured in body axes),

$$\mathbf{v}_{B_1} = [10 \quad 15 \quad 20]^T, \qquad \mathbf{v}_{B_2} = [12 \quad 14 \quad 23]^T$$
$$\mathbf{v}_{B_3} = [\ 9 \quad 17 \quad 15]^T, \qquad \mathbf{v}_{B_4} = [\ 9 \quad 14 \quad 22]^T$$

estimate $\mathbf{v}_E$ using Eq. 4.1-8.

Define

$$\mathbf{z} = [\mathbf{v}_{B_1}^T \quad \mathbf{v}_{B_2}^T \quad \mathbf{v}_{B_3}^T \quad \mathbf{v}_{B_4}^T]^T, \qquad \mathbf{x} = \mathbf{v}_E,$$

$$\mathbf{H} = \begin{bmatrix} H_{BE} \\ \vdots \\ H_{BE} \\ H_{BE} \end{bmatrix}$$

$$\mathbf{H}_{BE}(0, 30, 45) = \begin{bmatrix} 0.613 & 0.613 & 0.5 \\ -0.707 & 0.707 & 0 \\ 0.354 & 0.354 & 0.867 \end{bmatrix}$$

Then, because $\mathbf{H}_{EB} = (\mathbf{H}_{BE})^{-1} = (\mathbf{H}_{BE})^T$

$$(\mathbf{H}^T\mathbf{H})^{-1} = (4\mathbf{H}_{EB}\mathbf{H}_{BE})^{-1} = \tfrac{1}{4}\mathbf{H}_{EB}\mathbf{H}_{BE}$$
$$= \begin{bmatrix} 0.25 & 0 & 0 \\ 0 & 0.25 & 0 \\ 0 & 0 & 0.25 \end{bmatrix}$$

and

$$\mathbf{H}^L = (\mathbf{H}^T\mathbf{H})^{-1}\mathbf{H}^T = \begin{bmatrix} 0.25 & 0 & 0 \\ 0 & 0.25 & 0 \\ 0 & 0 & 0.25 \end{bmatrix} [\mathbf{H}_{EB} \quad \mathbf{H}_{EB} \quad \mathbf{H}_{EB} \quad \mathbf{H}_{EB}]$$
$$= (0.25)[\mathbf{H}_{EB} \quad \mathbf{H}_{EB} \quad \mathbf{H}_{EB} \quad \mathbf{H}_{EB}]$$

leading to

$$\hat{\mathbf{v}}_E = \hat{\mathbf{x}} = \mathbf{H}^L\mathbf{z} = \begin{bmatrix} 2.588 \\ 23.801 \\ 12.321 \end{bmatrix}$$

No doubt you have recognized that this simple example could have been solved more easily by averaging the body-axis measurements and then performing the transformations; however, it serves to illustrate the least-squares procedure in a verifiable fashion.

The minimum value of $J(\mathbf{z})$ and the corresponding value of $J(\mathbf{x})$ can be calculated if $\mathbf{n}$ is known. Substituting Eq. 4.1-8 in Eq. 4.1-6

$$J(\mathbf{z}) = \tfrac{1}{2}[\mathbf{z}^T\mathbf{z} - \mathbf{z}^T\mathbf{H}(\mathbf{H}^T\mathbf{H})^{-1}\mathbf{H}^T\mathbf{z}] \qquad (4.1\text{-}10)$$

Using Eq. 4.1-1, this becomes

$$J(\mathbf{z}) = \tfrac{1}{2}\{\mathbf{n}^T[\mathbf{I}_{k_1} - \mathbf{H}(\mathbf{H}^T\mathbf{H})^{-1}\mathbf{H}^T]\mathbf{n}\} \qquad (4.1\text{-}11)$$

Substituting Eqs. 4.1-1 and 4.1-8 in Eq. 4.1-5,

$$J(\mathbf{x}) = \tfrac{1}{2}[\mathbf{n}^T\mathbf{H}(\mathbf{H}^T\mathbf{H})^{-1}(\mathbf{H}^T\mathbf{H})^{-1}\mathbf{H}^T\mathbf{n}] \qquad (4.1\text{-}12)$$

Some basic properties of estimation error are illustrated in the following example.

EXAMPLE 4.1-4 ESTIMATION ERRORS OF A SCALAR CONSTANT

Consider again the measurement of a scalar constant as in Example 4.1-1. First, let $k_1 = 2$. Then, $\hat{x} = (z_1 + z_2)/2$, and

$$J(\mathbf{z}) = \tfrac{1}{2}[n_1 \quad n_2]\left\{\begin{bmatrix}1 & 0\\0 & 1\end{bmatrix} - \begin{bmatrix}1\\1\end{bmatrix}\tfrac{1}{2}[1 \quad 1]\right\}\begin{bmatrix}n_1\\n_2\end{bmatrix}$$
$$= \tfrac{1}{4}(n_1 - n_2)^2$$

while $J(x)$ is

$$J(x) = \tfrac{1}{2}[n_1 \quad n_2]\begin{bmatrix}1\\1\end{bmatrix}\tfrac{1}{4}[1 \quad 1]\begin{bmatrix}n_1\\n_2\end{bmatrix}$$
$$= \tfrac{1}{8}(n_1 + n_2)^2$$

When $n_1 = n_2$, n is called a *bias error*, and all measurements are wrong by an additive constant. $J(\mathbf{z}) = 0$, but $J(x) = n_1^2/2$, indicating that although $\hat{x}$ indeed minimizes $J(\mathbf{z})$, there is a bias in the estimate. If $n_1 = -n_2$, the error mean value is zero; while $J(\mathbf{z})$ is minimized, it equals n_1^2 and not zero. However, $J(x)$ is identically zero, and the estimate matches the actual value of $\mathbf{x}$.

$\mathbf{H}^T\mathbf{H} = k_1$ and $\mathbf{H}\mathbf{H}^T$ is a full matrix of "ones" in this example; for arbitrary k_1,

$$J(\mathbf{z}) = \tfrac{1}{2}\mathbf{n}^T\left[\mathbf{I}_{k_1} - \frac{1}{k_1}\mathbf{H}\mathbf{H}^T\right]\mathbf{n}$$
$$= \sum_{k=1}^{k_1} n_k^2 - \frac{1}{k_1}\sum_{i=1}^{k_1}\left[n_i \sum_{j=1}^{k_1} n_j\right]$$

and

$$J(x) = \frac{1}{2k_1^2}\mathbf{n}^T\mathbf{H}\mathbf{H}^T\mathbf{n}$$
$$= \frac{1}{2k_1^2}\left(\sum_{k=1}^{k_1} n_k\right)^2$$

As before, a bias error goes undetected in $J(\mathbf{z})$. If the error mean value is zero, the measurement residual cost is finite, but the state residual cost is zero.

It has been assumed that the relationship between the measurement and the state is linear (Eq. 4.1-1). When there is a nonlinear relationship, $\mathbf{z} = \mathbf{h}(\mathbf{x}) + \mathbf{n}$, a closed-form least-squares solution for $\hat{\mathbf{x}}$ normally cannot be found. $J(\mathbf{z})$ must be minimized by iteration, using, for example, the *Newton–Raphson algorithm* presented in Section 2.1.

Weighted Least-Squares Estimator

The measurement error may vary from one point to the next, in which case it is desirable to weight the good data more heavily than the poor data in the estimator. (In the previous examples, all measurements were given equal weights.) A weighted least-squares estimator can be derived by minimizing a quadratic cost function of a normalized measurement residual. From Eqs. 4.1-1 and 4.1-2, the measurement residual for the kth (scalar) data point is

$$\begin{aligned} \epsilon_{z_k} &= (z_k - \hat{y}_k) = \mathbf{H}_k(\mathbf{x} - \hat{\mathbf{x}}) + n_k \\ &= \mathbf{H}_k \boldsymbol{\epsilon}_x + n_k \end{aligned} \tag{4.1-13}$$

where $\mathbf{z}$ and $\mathbf{y}$ have k_1 elements, $\mathbf{x}$ has n components, and $\mathbf{H}_k$ is the kth row of the $(k_1 \times n)$ matrix $\mathbf{H}$. If $\boldsymbol{\epsilon}_{x_k}$ and n_k were known, a normalized scalar residual ϵ'_{z_k} could be formed,

$$\epsilon'_{z_k} = \frac{\epsilon_{z_k}}{\mathbf{H}_k \boldsymbol{\epsilon}_x + n_k} = \frac{\epsilon_{z_k}}{v_k} \tag{4.1-14}$$

and the corresponding normalized residual vector could be written as,

$$\begin{aligned} \boldsymbol{\epsilon}'_z &= \mathbf{N}^{-1} \boldsymbol{\epsilon}_z \\ &= \mathbf{N}^{-1}(\mathbf{z} - \mathbf{H}\hat{\mathbf{x}}) \end{aligned} \tag{4.1-15}$$

with

$$\mathbf{N} = \begin{bmatrix} v_1 & 0 & \dots & 0 \\ 0 & v_2 & \dots & 0 \\ \vdots & \vdots & & \vdots \\ 0 & 0 & \dots & v_{k_1} \end{bmatrix} \tag{4.1-16}$$

Then the weighted quadratic cost function becomes

$$\begin{aligned} J(\mathbf{z}') &= \tfrac{1}{2}(\mathbf{z} - \mathbf{H}\hat{\mathbf{x}})^T \mathbf{N}^{-T} \mathbf{N}^{-1} (\mathbf{z} - \mathbf{H}\hat{\mathbf{x}}) \\ &= \tfrac{1}{2}(\mathbf{z} - \mathbf{H}\hat{\mathbf{x}})^T \mathbf{S}^{-1} (\mathbf{z} - \mathbf{H}\hat{\mathbf{x}}) \end{aligned} \tag{4.1-17}$$

where the diagonal weighting matrix $\mathbf{S}$ reflects the square of the deterministic measurement errors:

$$\mathbf{S} = \mathbf{N}^T\mathbf{N} \tag{4.1-18}$$

The *weighted least-squares estimator* is derived by setting $\partial J(\mathbf{z}')/\partial\hat{\mathbf{x}}$ to zero,

$$\partial J(\mathbf{z}')/\partial\hat{\mathbf{x}} = (\mathbf{H}^T\mathbf{S}^{-1}\mathbf{H}\hat{\mathbf{x}} - \mathbf{H}^T\mathbf{S}^{-1}\mathbf{z})^T = \mathbf{0} \tag{4.1-19}$$

and solving for $\hat{\mathbf{x}}$:

$$\begin{aligned}\hat{\mathbf{x}} &= (\mathbf{H}^T\mathbf{S}^{-1}\mathbf{H})^{-1}\mathbf{H}^T\mathbf{S}^{-1}\mathbf{z} \\ &= \mathbf{H}^{WL}\mathbf{z}\end{aligned} \tag{4.1-20}$$

$\mathbf{H}^{WL}$ is the weighted left pseudoinverse of $\mathbf{H}$ (Section 2.2). When $\mathbf{S}$ is a diagonal matrix of equal constants, this equation provides the same estimate as Eq. 4.1-8.

Trouble is, $\boldsymbol{\epsilon}_\mathbf{x}$ and $\mathbf{n}$ normally are not known deterministically; otherwise there would be no need to estimate $\mathbf{x}$. When $\boldsymbol{\epsilon}_\mathbf{x}$ and $\mathbf{n}$ are random variables, a cost function based on the *expected value* of the error can be formed, leading to a *stochastic estimator* for the constant vectors $E(\hat{\mathbf{x}})$ and $E(\mathbf{x})$. We should first note that if the state residual contains a known bias, $\mathbf{m}_\mathbf{x}$,

$$E(\boldsymbol{\epsilon}_\mathbf{x}) = E(\mathbf{x} - \hat{\mathbf{x}}) = E(\mathbf{x}) - E(\hat{\mathbf{x}}) = \mathbf{m}_\mathbf{x} \tag{4.1-21}$$

then the expected value of the actual state is

$$\begin{aligned}E(\mathbf{x}) &= \mathbf{m}_\mathbf{x} + E(\hat{\mathbf{x}}) \\ &= \mathbf{m}_\mathbf{x} + \hat{\mathbf{x}}\end{aligned} \tag{4.1-22}$$

Similarly, if there is a known bias in the measurement,

$$E(\mathbf{n}) = \mathbf{m}_\mathbf{n} \tag{4.1-23}$$

If we interpret the $\mathbf{S}$ of Eq. 4.1-17 as a symmetric matrix of expected stochastic errors rather than a diagonal matrix of deterministic errors, then Eq. 4.1-20 becomes a stochastic estimator with no further modification, and $E(\hat{\mathbf{x}}) = \hat{\mathbf{x}}$. The expected value of the state is

$$E(\mathbf{x}) = \mathbf{m}_\mathbf{x} + \mathbf{H}^{WL}(\mathbf{z} - \mathbf{m}_\mathbf{n}) \tag{4.1-24}$$

At least two definitions of $\mathbf{S}$ can be considered. The expected *measure-*

ment error covariance provides a suitable weighting matrix:

$$\begin{aligned}\mathbf{S}_1 &= E[(\mathbf{z}-\mathbf{y})(\mathbf{z}-\mathbf{y})^T] \\ &= E[(\mathbf{z}-\mathbf{Hx})(\mathbf{z}-\mathbf{Hx})^T] \\ &= E(\mathbf{nn}^T) = \mathbf{R}\end{aligned} \tag{4.1-25}$$

An alternate definition of $\mathbf{S}$ is based on the expected *measurement residual covariance*:

$$\begin{aligned}\mathbf{S}_2 &= E[(\mathbf{z}-\mathbf{H\hat{x}})(\mathbf{z}-\mathbf{H\hat{x}})^T] \\ &= E[(\mathbf{H}\boldsymbol{\epsilon}_\mathbf{x}+\mathbf{n})(\mathbf{H}\boldsymbol{\epsilon}_\mathbf{x}+\mathbf{n})^T] \\ &= E(\mathbf{H}\boldsymbol{\epsilon}_\mathbf{x}\boldsymbol{\epsilon}_\mathbf{x}^T\mathbf{H}^T+\mathbf{H}\boldsymbol{\epsilon}_\mathbf{x}\mathbf{n}^T+\mathbf{n}\boldsymbol{\epsilon}_\mathbf{x}^T\mathbf{H}^T+\mathbf{nn}^T) \\ &= \mathbf{H}E(\boldsymbol{\epsilon}_\mathbf{x}\boldsymbol{\epsilon}_\mathbf{x}^T)\mathbf{H}^T+\mathbf{H}E(\boldsymbol{\epsilon}_\mathbf{x}\mathbf{n}^T)+E(\mathbf{n}\boldsymbol{\epsilon}_\mathbf{x}^T)\mathbf{H}^T+E(\mathbf{nn}^T) \\ &= \mathbf{HPH}^T+\mathbf{HM}+\mathbf{M}^T\mathbf{H}^T+\mathbf{R}\end{aligned} \tag{4.1-26}$$

Here $\mathbf{R}$ is the $(k_1 \times k_1)$ measurement error covariance matrix (defined above). $\mathbf{P}$ is the $(n \times n)$ state residual covariance matrix, and $\mathbf{M}$ is an $(n \times k_1)$ matrix coupling the state residual and the measurement error:

$$\mathbf{P} = E[(\mathbf{x}-\hat{\mathbf{x}})(\mathbf{x}-\hat{\mathbf{x}})^T] \tag{4.1-27}$$

$$\mathbf{M} = E[(\mathbf{x}-\hat{\mathbf{x}})\mathbf{n}^T] \tag{4.1-28}$$

From Eqs. 4.1-1 and 4.1-20, the state estimate can be expressed as the sum of the actual state and the effective estimation error:

$$\begin{aligned}\hat{\mathbf{x}} &= (\mathbf{H}^T\mathbf{S}^{-1}\mathbf{H})^{-1}\mathbf{H}^T\mathbf{S}^{-1}\mathbf{z} \\ &= \mathbf{x}+(\mathbf{H}^T\mathbf{S}^{-1}\mathbf{H})^{-1}\mathbf{H}^T\mathbf{S}^{-1}\mathbf{n}\end{aligned} \tag{4.1-29}$$

Substituting in Eqs. 4.1-27 and 4.1-28, $\mathbf{P}$ and $\mathbf{M}$ can be calculated as

$$\mathbf{P} = (\mathbf{H}^T\mathbf{S}^{-1}\mathbf{H})^{-1}\mathbf{H}^T\mathbf{S}^{-1}\mathbf{R}\mathbf{S}^{-1}\mathbf{H}(\mathbf{H}^T\mathbf{S}^{-1}\mathbf{H})^{-1} \tag{4.1-30}$$

$$\mathbf{M} = -(\mathbf{H}^T\mathbf{S}^{-1}\mathbf{H})^{-1}\mathbf{H}^T\mathbf{S}^{-1}\mathbf{R} \tag{4.1-31}$$

With $\mathbf{S} = \mathbf{S}_1 = \mathbf{R}$, $\mathbf{P}$ and $\mathbf{M}$ reduce to

$$\mathbf{P} = (\mathbf{H}^T\mathbf{R}^{-1}\mathbf{H})^{-1} \tag{4.1-32}$$

$$\mathbf{M} = -(\mathbf{H}^T\mathbf{R}^{-1}\mathbf{H})^{-1}\mathbf{H}^T \tag{4.1-33}$$

With $\mathbf{S} = \mathbf{S}_2$, the values of $\mathbf{P}$, $\mathbf{M}$, and $\mathbf{S}$ are dependent on one another, so a simultaneous solution of Eqs. 4.1-26, 4.1-30, and 4.1-31 is required. The following iteration can be considered. Substituting Eq. 4.1-30 and 4.1-31 in Eq. 4.1-26 leads to an equation of the form

$$\mathbf{S}_2 = \mathbf{L}\mathbf{R}\mathbf{L}^T \tag{4.1-34}$$

where

$$\mathbf{L} = \mathbf{H}(\mathbf{H}^T\mathbf{S}_2^{-1}\mathbf{H})^{-1}\mathbf{H}^T\mathbf{S}_2^{-1} - \mathbf{I}_{k_1} = \mathbf{L}^T \tag{4.1-35}$$

This provides a basis for iteration to find $\mathbf{S}_2$:

$$\mathbf{S}_{2_{i+1}} = \mathbf{L}(\mathbf{S}_{2_i})\mathbf{R}\mathbf{L}^T(\mathbf{S}_{2_i}),\ \mathbf{S}_{2_0} = \mathbf{R} \qquad i = 1, 2, \ldots \tag{4.1-36}$$

The iteration is independent of the actual values of $\mathbf{z}$ and $\mathbf{x}$, depending only on the error statistics $\mathbf{R}$ and the observation matrix $\mathbf{H}$.

Example 4.1-5 A Least-Squares Parameter Estimation Problem

The parameters of a second-order system

$$\begin{bmatrix} \dot{x}_1 \\ \dot{x}_2 \end{bmatrix} = \begin{bmatrix} 0 & 1 \\ -\omega_n^2 & -2\zeta\omega_n \end{bmatrix} \begin{bmatrix} x_1 \\ x_2 \end{bmatrix} + \begin{bmatrix} 0 \\ g \end{bmatrix} u$$

are to be estimated from time history measurements of displacement (x_1), rate (x_2), acceleration ($\dot{x}_2$), and control (u), taken at the sampling instants, $k = 1$ to K. The parameter vector can be defined as

$$\mathbf{p} = \begin{bmatrix} p_1 \\ p_2 \\ p_3 \end{bmatrix} = \begin{bmatrix} \omega_n^2 \\ 2\zeta\omega_n \\ g \end{bmatrix}$$

where

$$\omega_n^2 = p_1$$

$$\zeta = \frac{p_2}{2\sqrt{p_1}}$$

Assuming that the measurements of x_1, x_2, and u are precise while those of $\dot{x}_2$ are noisy, a scalar observation equation can be formed as

$$z_k = \mathbf{H}\mathbf{p} + v_k, \qquad k = 1 \text{ to } K$$

where

$$z_k = \dot{x}_{2_k}$$

and

$$\mathbf{H} = \begin{bmatrix} -x_{1_1} & -x_{2_1} & u_1 \\ -x_{1_2} & -x_{2_2} & u_2 \\ \vdots & \vdots & \vdots \\ -x_{1_n} & -x_{2_n} & u_n \end{bmatrix}$$

With equally noisy measurements of $\dot{x}_2$, the estimate is formed from Eq. 4.1-8:

$$\hat{\mathbf{p}} = (\mathbf{H}^T\mathbf{H})^{-1}\mathbf{H}^T\mathbf{z}$$

If the measurement uncertainty is varying, a suitable weighting matrix $\mathbf{S}$ is defined, and Eq. 4.1-20 indicates that

$$\hat{\mathbf{p}} = (\mathbf{H}^T\mathbf{S}^{-1}\mathbf{H})^{-1}\mathbf{H}^T\mathbf{S}^{-1}\mathbf{z}$$

The roles of state and parameter vectors have been interchanged in the example; although a dynamic system is involved, this is a static estimation problem. When the measurements of displacement, rate, and control are noisy, a much more complicated problem arises, for then x_1, x_2, and u must be appended to the state vector, producing a nonlinear, dynamic state estimation problem (Section 4.6).

Recursive Least-Squares Estimator

The previous estimators are "batch processing" algorithms, in that all measurements are processed together to provide the estimate of a constant vector. To use these equations when new measurements become available, it would be necessary to append the new data to $\mathbf{z}$ and repeat the entire process. As an alternative, the prior estimate can be used as the starting point for a sequential estimation algorithm that assigns proper relative weighting to the old and new data.

Given k_1 measurements, $\mathbf{z}_1$, the corresponding output and error matrices, $\mathbf{H}_1$ and $\mathbf{S} = \mathbf{R}_1$, and the resulting estimate $\hat{\mathbf{x}}_1$:

$$\mathbf{z}_1 = \mathbf{H}_1\mathbf{x} + \mathbf{n}_1 \tag{4.1-37}$$

$$\hat{\mathbf{x}}_1 = (\mathbf{H}_1^T\mathbf{R}_1^{-1}\mathbf{H}_1)^{-1}\mathbf{H}_1^T\mathbf{R}_1^{-1}\mathbf{z}_1 \tag{4.1-38}$$

The *new* measurement $\mathbf{z}_2$ with dimension k_2, is

$$\mathbf{z}_2 = \mathbf{H}_2\mathbf{x} + \mathbf{n}_2 \tag{4.1-39}$$

$\mathbf{R}_2$ is a $(k_2 \times k_2)$ matrix containing the expected squared errors in the new measurement. The cost function for all $(k_1 + k_2)$ measurements

$$\mathbf{z} = \begin{bmatrix} \mathbf{z}_1 \\ \mathbf{z}_2 \end{bmatrix} \tag{4.1-40}$$

can be partitioned as

$$J(\mathbf{z}_1, \mathbf{z}_2) = [(\mathbf{z}_1 - \mathbf{H}_1\hat{\mathbf{x}}_2)^T (\mathbf{z}_2 - \mathbf{H}_2\hat{\mathbf{x}}_2)^T] \begin{bmatrix} \mathbf{R}_1^{-1} & \mathbf{0} \\ \mathbf{0} & \mathbf{R}_2^{-1} \end{bmatrix} \begin{bmatrix} (\mathbf{z}_1 - \mathbf{H}_1\hat{\mathbf{x}}_2) \\ (\mathbf{z}_2 - \mathbf{H}_2\hat{\mathbf{x}}_2) \end{bmatrix} \tag{4.1-41}$$

where $\hat{\mathbf{x}}_2$ is the state estimate obtained by using all the data. Taking the derivative of $J(\mathbf{z}_1, \mathbf{z}_2)$ and setting it equal to zero provides the least-squares estimate $\hat{\mathbf{x}}_2$:

$$\hat{\mathbf{x}}_2 = (\mathbf{H}_1^T\mathbf{R}_1^{-1}\mathbf{H}_1 + \mathbf{H}_2^T\mathbf{R}_2^{-1}\mathbf{H}_2)^{-1}(\mathbf{H}_1^T\mathbf{R}_1^{-1}\mathbf{z}_1 + \mathbf{H}_2^T\mathbf{R}_2^{-1}\mathbf{z}_2) \tag{4.1-42}$$

From Eq. 4.1-32, we know that

$$\mathbf{H}_1^T\mathbf{R}_1^{-1}\mathbf{H}_1 = \mathbf{P}_1^{-1} \tag{4.1-43}$$

The matrix inversion lemma (Eq. 2.2-67b) allows the following substitution to be made:

$$(\mathbf{P}_1^{-1} + \mathbf{H}_2^T\mathbf{R}_2^{-1}\mathbf{H}_2)^{-1} = \mathbf{P}_1 - \mathbf{P}_1\mathbf{H}_2^T(\mathbf{H}_2\mathbf{P}_1\mathbf{H}_2^T + \mathbf{R}_2)^{-1}\mathbf{H}_2\mathbf{P}_1 \tag{4.1-44}$$

(For comparison with the original equation, $\mathbf{A}_1 = -\mathbf{R}_2$, $\mathbf{A}_2 = \mathbf{H}_2$, and $\mathbf{A}_4 = \mathbf{P}_1^{-1}$.) Equation 4.1-42 can be written as

$$\begin{aligned}
\hat{\mathbf{x}}_2 &= \mathbf{P}_1\mathbf{H}_1^T\mathbf{R}_1^{-1}\mathbf{z}_1 - \mathbf{P}_1\mathbf{H}_2^T(\mathbf{H}_2\mathbf{P}_1\mathbf{H}_2^T + \mathbf{R}_2)^{-1}\mathbf{H}_2\mathbf{P}_1\mathbf{H}_1^T\mathbf{R}_1^{-1}\mathbf{z}_1 \\
&\quad + \mathbf{P}_1\mathbf{H}_2^T\mathbf{R}_2^{-1}\mathbf{z}_2 - \mathbf{P}_1\mathbf{H}_2^T(\mathbf{H}_2\mathbf{P}_1\mathbf{H}_2^T + \mathbf{R}_2)^{-1}\mathbf{H}_2\mathbf{P}_1\mathbf{H}_2^T\mathbf{R}_2^{-1}\mathbf{z}_2 \\
&= \hat{\mathbf{x}}_1 - \mathbf{P}_1\mathbf{H}_2^T(\mathbf{H}_2\mathbf{P}_1\mathbf{H}_2^T + \mathbf{R}_2)^{-1}\mathbf{H}_2\hat{\mathbf{x}}_1 \\
&\quad + \mathbf{P}_1\mathbf{H}_2^T[\mathbf{I}_{k_2} - (\mathbf{H}_2\mathbf{P}_1\mathbf{H}_2^T + \mathbf{R}_2)^{-1}\mathbf{H}_2\mathbf{P}_1\mathbf{H}_2^T]\mathbf{R}_2^{-1}\mathbf{z}_2
\end{aligned} \tag{4.1-45}$$

Noting that

$$(\mathbf{H}_2\mathbf{P}_1\mathbf{H}_2^T + \mathbf{R}_2)^{-1}(\mathbf{H}_2\mathbf{P}_1\mathbf{H}_2^T + \mathbf{R}_2) = \mathbf{I}_{k_2} \tag{4.1-46}$$

and substituting for $\mathbf{I}_{k_2}$ in Eq. 4.1-45,

$$\begin{aligned}
\hat{\mathbf{x}}_2 &= \hat{\mathbf{x}}_1 + \mathbf{P}_1\mathbf{H}_2^T(\mathbf{H}_2\mathbf{P}_1\mathbf{H}_2^T + \mathbf{R}_2)^{-1}(\mathbf{z}_2 - \mathbf{H}_2\hat{\mathbf{x}}_1) \\
&= \hat{\mathbf{x}}_1 + \mathbf{K}_2(\mathbf{z}_2 - \mathbf{H}_2\hat{\mathbf{x}}_1)
\end{aligned} \tag{4.1-47}$$

where $\mathbf{K}_2$ is the *recursive weighted-least-squares estimator gain matrix*:

$$\mathbf{K}_2 = \mathbf{P}_1\mathbf{H}_2^T(\mathbf{H}_2\mathbf{P}_1\mathbf{H}_2^T + \mathbf{R}_2)^{-1} \tag{4.1-48}$$

Although the derivation is complex, the result is quite simple, because the new estimate is based on the old estimate plus a gain matrix times the residual.

Equation 4.1-47 looks like a digital filter, and measurements taken over a period of time could update the estimate as they occur. Redefining k as a *time index* and letting the observation vector at time k have r components, the *recursive mean-value estimator* is

$$\hat{\mathbf{x}}_k = \hat{\mathbf{x}}_{k-1} + \mathbf{K}_k(\mathbf{z}_k - \mathbf{H}_k\hat{\mathbf{x}}_{k-1}) \tag{4.1-49}$$

with

$$\mathbf{K}_k = \mathbf{P}_{k-1}\mathbf{H}_k^T(\mathbf{H}_k\mathbf{P}_{k-1}\mathbf{H}_k^T + \mathbf{R}_k)^{-1} \tag{4.1-50}$$

and

$$\mathbf{P}_k = (\mathbf{P}_{k-1}^{-1} + \mathbf{H}_k^T\mathbf{R}_k^{-1}\mathbf{H}_k)^{-1} \tag{4.1-51}$$

$\mathbf{K}_k$ is an $(n \times r)$ gain matrix, while $\mathbf{P}_k$ is an $(n \times n)$ matrix that represents the estimation error at the kth sampling instant.

EXAMPLE 4.1-6 RECURSIVE ESTIMATION OF A SCALAR CONSTANT

Returning to Example 4.1-1, assume that k_1 measurements $\mathbf{z}_1$ have been used to derive the estimate $\hat{x}_1$ with equal weighting of all measurements ($\mathbf{R}_1 = \mathbf{I}_{k_1}$). $\mathbf{H}_1$ is a column vector of ones, and

$$P_1 = (\mathbf{H}_1^T\mathbf{R}_1^{-1}\mathbf{H}_1)^{-1} = \frac{1}{k_1}$$

$$\hat{x}_1 = \mathbf{P}_1\mathbf{H}_1^T\mathbf{R}_1^{-1}\mathbf{z}_1 = \frac{1}{k_1}\sum_{k=1}^{k_1} z_k$$

A single new measurement with the same relative accuracy is made. What are H_2, R_2, P_2, K_2, and $\hat{x}_2$?

H_2 and R_2 are both unit scalars, so

$$P_2 = (P_1^{-1} + H_2R_2^{-1}H_2)^{-1}$$
$$= (k_1 + 1)^{-1} = 1/(k_1 + 1)$$

Then

$$\begin{aligned} K_2 &= P_1 H_2 (H_2 P_1 H_2 + R_2)^{-1} \\ &= \frac{1}{k_1}\left(\frac{1}{k_1} + 1\right)^{-1} \\ &= \frac{1}{k_1}\left(\frac{k_1 + 1}{k_1}\right)^{-1} \\ &= \frac{1}{k_1 + 1} \end{aligned}$$

and

$$\begin{aligned} \hat{x}_2 &= \hat{x}_1 + K_2(z_2 - \hat{x}_1) \\ &= \hat{x}_1 + \frac{1}{k_1 + 1}(z_2 - \hat{x}_1) \end{aligned}$$

As $k_1 \rightarrow \infty$, $K_2 \rightarrow 0$.

Keep in mind that these recursive equations estimate a constant vector, and the gain matrix normally gets smaller as more measurements are made. If least-squares estimation is to be applied to the states of a dynamic system over a period of time, the estimated quantities may not be constant, and a vanishing gain matrix would tend to ignore information contained in the later measurements. There must be a mechanism for including dynamic effects of the system model and its inputs (both controlled and uncontrolled) on $\hat{\mathbf{x}}_k$ and $\mathbf{P}_k$. Given the uncertainty in the measurement-based estimates, algorithms that optimally combine these estimates with model-based estimates can be formed. The system's dynamic model provides the basis for propagating the state and its corresponding uncertainty. The combined optimal estimate provides lower state estimate uncertainty than could be obtained from either estimate alone.

4.2 PROPAGATION OF THE STATE ESTIMATE AND ITS UNCERTAINTY

The state of a dynamic system varies in time, and state estimators must account for such motion. If the initial conditions and inputs are known without error, then the methods of Section 2.3 can be used to calculate a time history of the state for t in $[t_0, t_f]$ from the system model, without any "updating" by measurements of the state. If the initial conditions and inputs

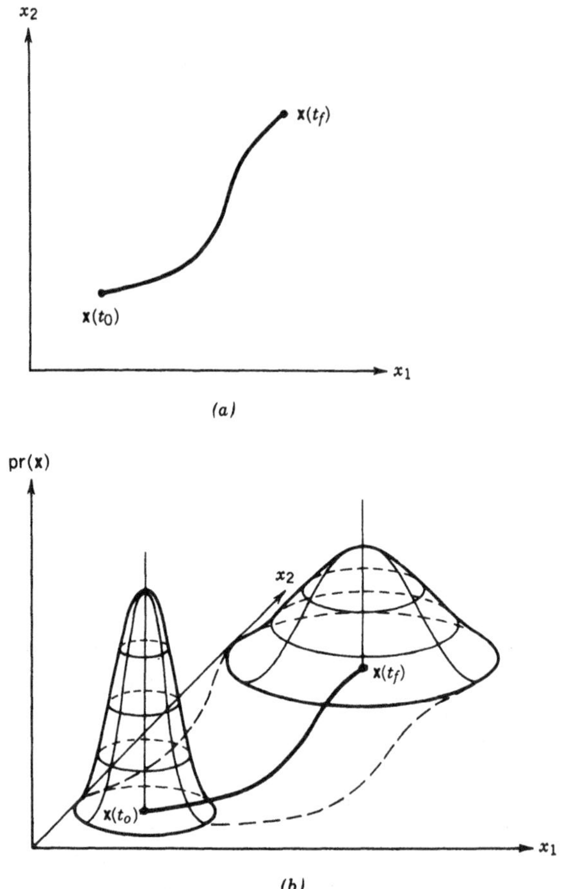

FIGURE 4.2-1 Comparison of deterministic and stochastic state propagation.

are random, then the state cannot be propagated with certainty, and the state itself must be considered a random variable. This section will illustrate how the system model equations, in combination with the statistics of the initial conditions and inputs, can be used to propagate the statistics of the state without considering measurements.

Deterministic and stochastic state propagation are compared in Fig. 4.2-1. In the former case, the deterministic state travels from $\mathbf{x}(t_0)$ to $\mathbf{x}(t_f)$ without ambiguity, while in the latter case, the trajectory represents the *most likely* path, as indicated by the state's probability density function. The most likely value and the uncertainty in the *random* variable, $\mathbf{x}(t)$, are completely specified by its joint probability density function or likelihood function, $\text{pr}[\mathbf{x}(t)]$. Consequently, if the multivariate function $\text{pr}(\mathbf{x})$ can be

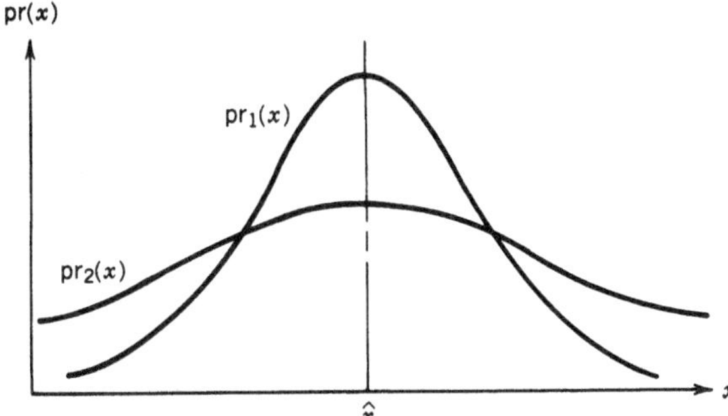

FIGURE 4.2-2 Identical estimates of x derived from probability density functions with different uncertainties.

propagated for all values of $\mathbf{x}$ in the time interval $[t_0, t_f]$, then a stochastic state estimate can be made. The estimate of $\mathbf{x}(t)$ is prescribed by the *mean*, *mode*, or *median* of $\text{pr}[\mathbf{x}(t)]$,* and the uncertainty of this estimate is reflected in the spread of the density function. As illustrated in Fig. 4.2-2, alternate probability density functions could provide the same estimate of $\mathbf{x}$ with different uncertainties.

Although $\text{pr}[\mathbf{x}(t)]$ is a scalar function of a vector variable, providing a time history of the entire distribution normally would be a most difficult and computer-intensive task. (Recall that a scalar function of a vector variable plots as a hypersurface when the vector's dimension exceeds two. The objective is a time history of the entire hypersurface.) Knowing the probability distributions for the initial conditions and the inputs, a numerical experiment called *Monte Carlo simulation* could be performed. The system equations would be integrated from t_0 to t_f many times, each time using a different initial condition and input time history, in turn defined by their respective probability distributions. The frequencies of occurrence of all values of $\mathbf{x}$ at selected times in $[t_0, t_f]$ would be tabulated to approximate $\text{pr}[\mathbf{x}(t)]$ by multivariate histograms at each time. Not a "real-time" process suitable for feedback control, this example indicates the enormity of the problem that we have set out to solve.

Fortunately, there are alternate ways of propagating the essential features of the state's probability density function that are particularly appropriate when the initial conditions and inputs are Gaussian and the dynamic system is linear. Because a Gaussian multivariate distribution is completely defined by its n-dimensional mean value $\mathbf{m}$ and symmetric $(n \times n)$ covariance matrix $\mathbf{P}$, the number of variables to be propagated is

*All three are the same for symmetric, unimodal probability density functions. For nonsymmetric or multimodal distributions, one of these must be chosen as a basis for state estimation. The initial sections of this chapter specifically deal with mean value estimators.

tractable, being $n(n+3)/2$. The principle of superposition for linear systems assures that Gaussian distributions remain Gaussian, although their means and covariances may change. When the random variables are non-Gaussian or the system is nonlinear, the procedures described below can be applied with minor modifications. They are not optimal, but they may provide useful suboptimal estimates.

Discrete-Time Systems

We begin by considering the propagation of the state and its uncertainty in linear discrete-time systems, which can be modeled deterministically as

$$\mathbf{x}_k = \mathbf{\Phi}_{k-1}\mathbf{x}_{k-1} + \mathbf{\Gamma}_{k-1}\mathbf{u}_{k-1} + \mathbf{\Lambda}_{k-1}\mathbf{w}_{k-1}, \qquad \mathbf{x}_0 \text{ given} \tag{4.2-1}$$

with $k = 1$ to $k_{\max}$. The dimensions of $\mathbf{x}$, $\mathbf{u}$, and $\mathbf{w}$ are n, m, and s, respectively. It is assumed that $\mathbf{\Phi}_{k-1}$, $\mathbf{\Gamma}_{k-1}$, $\mathbf{\Lambda}_{k-1}$, and $\mathbf{u}_{k-1}$ are known without error; however, the initial condition $\mathbf{x}_0$ is a Gaussian random variable prescribed by its mean value and covariance matrix

$$E(\mathbf{x}_0) = \mathbf{m}_0 \tag{4.2-2}$$

$$E[(\mathbf{x}_0 - \mathbf{m}_0)(\mathbf{x}_0 - \mathbf{m}_0)^T] = \mathbf{P}_0 \tag{4.2-3}$$

while the disturbance input (sometimes called "process noise") is a zero-mean Gaussian random sequence:

$$E(\mathbf{w}_k) = \mathbf{0} \tag{4.2-4}$$

$$E(\mathbf{w}_k\mathbf{w}_k^T) = \mathbf{Q}'_k \tag{4.2-5}$$

$$k = 0 \text{ to } k_{\max} - 1$$

The disturbance input covariance matrix, $\mathbf{Q}'_k$ could be symmetric, implying instantaneous cross-correlation of disturbances, but if $\mathbf{Q}'_k$ is diagonal, then components of the disturbance input are uncorrelated with each other. If, in addition, $E(\mathbf{w}_k\mathbf{w}_{k-l}^T) = \mathbf{0}$ for nonzero values of l, disturbances at one instant are completely uncorrelated with those at any other instant, and the sequence can be considered discrete "white noise."

The expected value of the state at the kth instant is

$$\begin{aligned}E(\mathbf{x}_k) = \mathbf{m}_k &= E[\mathbf{\Phi}_{k-1}\mathbf{x}_{k-1} + \mathbf{\Gamma}_{k-1}\mathbf{u}_{k-1} + \mathbf{\Lambda}_{k-1}\mathbf{w}_{k-1}] \\ &= \mathbf{\Phi}_{k-1}E(\mathbf{x}_{k-1}) + \mathbf{\Gamma}_{k-1}\mathbf{u}_{k-1} + \mathbf{\Lambda}_{k-1}E(\mathbf{w}_{k-1}), \qquad E(\mathbf{x}_0) \text{ given} \\ &= \mathbf{\Phi}_{k-1}\mathbf{m}_{k-1} + \mathbf{\Gamma}_{k-1}\mathbf{u}_{k-1} + \mathbf{\Lambda}_{k-1}(\mathbf{0}), \qquad \mathbf{m}_0 \text{ given}\end{aligned} \tag{4.2-6}$$

In this equation, known mean values are indistinguishable from deterministic values. The equation propagates the expected value of the state, with the understanding that random perturbations from the expected value are likely to occur.

The covariance of the state perturbations can be propagated in a similar way. From Eqs. 4.2-1 and 4.2-6, the outer product of the state perturbations at the kth instant is

$$(\mathbf{x}_k - \mathbf{m}_k)(\mathbf{x}_k - \mathbf{m}_k)^T = [\boldsymbol{\Phi}_{k-1}(\mathbf{x}_{k-1} - \mathbf{m}_{k-1}) + \boldsymbol{\Gamma}_{k-1}(\mathbf{0}) + \boldsymbol{\Lambda}_{k-1}\mathbf{w}_{k-1}] \cdot [\boldsymbol{\Phi}_{k-1}(\mathbf{x}_{k-1} - \mathbf{m}_{k-1}) + \boldsymbol{\Gamma}_{k-1}(\mathbf{0}) + \boldsymbol{\Lambda}_{k-1}\mathbf{w}_{k-1}]^T \quad (4.2\text{-}7)$$

Taking the expected values of both sides,

$$\begin{aligned} E[(\mathbf{x}_k - \mathbf{m}_k)(\mathbf{x}_k - \mathbf{m}_k)^T] &= \mathbf{P}_k \\ &= E[\boldsymbol{\Phi}_{k-1}(\mathbf{x}_{k-1} - \mathbf{m}_{k-1})(\mathbf{x}_{k-1} - \mathbf{m}_{k-1})^T\boldsymbol{\Phi}_{k-1}^T \\ &\quad + \boldsymbol{\Lambda}_{k-1}\mathbf{w}_{k-1}\mathbf{w}_{k-1}^T\boldsymbol{\Lambda}_{k-1}^T + \boldsymbol{\Phi}_{k-1}(\mathbf{x}_{k-1} - \mathbf{m}_{k-1})\mathbf{w}_{k-1}^T\boldsymbol{\Lambda}_{k-1}^T \\ &\quad + \boldsymbol{\Lambda}_{k-1}\mathbf{w}_{k-1}(\mathbf{x}_{k-1} - \mathbf{m}_{k-1})^T\boldsymbol{\Phi}_{k-1}^T] \\ &= \boldsymbol{\Phi}_{k-1}\mathbf{P}_{k-1}\boldsymbol{\Phi}_{k-1}^T + \boldsymbol{\Lambda}_{k-1}\mathbf{Q}'_{k-1}\boldsymbol{\Lambda}_{k-1}^T \\ &\quad + \boldsymbol{\Phi}_{k-1}\mathbf{M}_{k-1}\boldsymbol{\Lambda}_{k-1}^T + \boldsymbol{\Lambda}_{k-1}\mathbf{M}_{k-1}^T\boldsymbol{\Phi}_{k-1}^T, \qquad \mathbf{P}_0 \text{ given} \end{aligned} \quad (4.2\text{-}8)$$

where the cross-correlation between state perturbations and disturbance input,

$$\mathbf{M}_{k-1} = E[(\mathbf{x}_{k-1} - \mathbf{m}_{k-1})\mathbf{w}_{k-1}^T] \quad (4.2\text{-}9)$$

is zero if $\mathbf{w}_k$ is a white-noise sequence.

EXAMPLE 4.2-1 URBAN DYNAMICS

Consider an isolated first-order model of the number of unskilled workers x in an urban area as measured on a quarterly basis (K-1). The important variables in the model include the number of workers arriving in the area during the quarter, a, the number leaving the area, b, the number who become skilled and therefore leave the cohort, c, and the net gain or loss due to age and health, d. A suitable deterministic model of the underlying process is

$$x_k = x_{k-1} + a_{k-1} + b_{k-1} + c_{k-1} + d_{k-1}$$

The number of unskilled workers is the scalar state variable in this problem, and the remaining variables are the inputs.

Many factors will influence the rates of arrival and departure, including seasonal effects, the local economy, the attractiveness of the area as perceived by those within and those without, the group's motivation, ability, and opportunity to become skilled, and so on. It is plausible that each input variable will have a component that is proportional to the number of unskilled workers in the region at the beginning of the quarter, a deterministic component representing seasonal and long-term trends, and a random component representing uncertainty in the modeling process. These effects could be modeled as

$$a_{k-1} = a_x x_{k-1} + u_{a_{k-1}} + w_{a_{k-1}}$$

$$b_{k-1} = b_x x_{k-1} + u_{b_{k-1}} + w_{b_{k-1}}$$

$$c_{k-1} = c_x x_{k-1} + u_{c_{k-1}} + w_{c_{k-1}}$$

$$d_{k-1} = d_x x_{k-1} + u_{d_{k-1}} + w_{d_{k-1}}$$

allowing the dynamic equation to be written as

$$\begin{aligned} x_k &= (1 + a_x + b_x + c_x + d_x)x_{k-1} + (u_a + u_b + u_c + u_d)_{k-1} \\ &\quad + (w_a + w_b + w_c + w_d)_{k-1} \\ &= \phi x_{k-1} + u_{k-1} + w_{k-1} \end{aligned}$$

Now assume that ϕ, m_0, p_0, and u_k are known, while

$$E[w_k] = 0, \quad E[w_k^2] = q, \quad E[(x_k - m_k)w_k] = 0$$

The mean number of unskilled workers, as well as the corresponding uncertainty can be estimated using Eqs. 4.2-6 and 4.2-8:

$$m_k = \phi m_{k-1} + u_{k-1}, \qquad m_0 \text{ given}$$

$$p_k = \phi^2 p_{k-1} + q, \qquad p_0 \text{ given}$$

For illustration, assume that

$$m_0 = 1000 \text{ people}$$

$$p_0 = 0 \text{ (number of people)}^2$$

$$u_{k-1} = 25 \text{ people (constant)}$$

$$q = 100 \text{ (number of people)}^2$$

Consider four cases: (a) No dependence on number of unskilled workers ($\phi = 1$), (b) 10% amplification factor per quarter ($\phi = 1.1$), (c) 10% attenuation factor per quarter ($\phi = 0.9$), and (d) 25% attenuation factor ($\phi = 0.75$). Results for 20 quarters (5 years) follow, with $p_k^{1/2}$ rather than p_k plotted.

With no feedback of the number of local unskilled workers, the quarterly mean value increases linearly, while the uncertainty $p_k^{1/2}$ grows with decreasing slope (Case a). The amplification of Case b leads to divergent growth in both the mean value and the uncertainty, whereas the attenuation of Case c leads to a stabilized decay in the mean value with a bounded

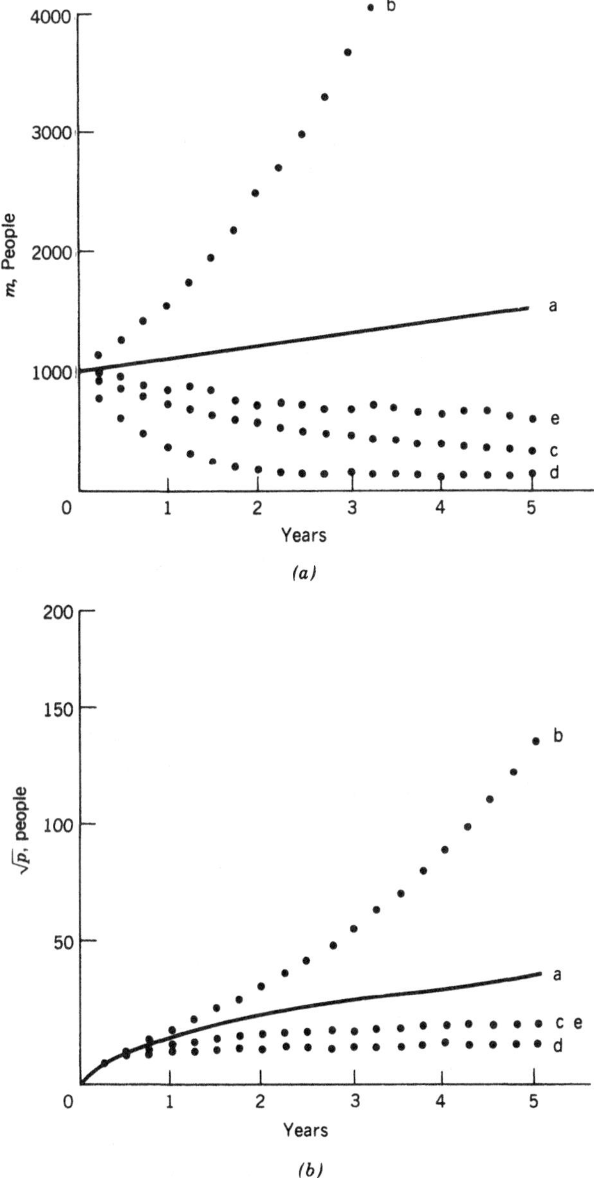

Estimates of unskilled workers in an urban area for various models. (*a*) Mean value; (*b*) uncertainty.

uncertainty. The rate of decay is increased in Case d, which also shows reduced uncertainty. [Equilibria, if they exist, occur when $(\cdot)_k = (\cdot)_{k-1}$—what equilibrium values of m and $p^{1/2}$ are predicted for these four cases?]

Although the dynamics of m and p are related through ϕ, the actual solutions are uncoupled. This is illustrated by Case e, which has the same values of ϕ and q as Case c, but which has the following seasonal forcing function:

$$u_{k-1} = \begin{cases} 100, & k = 1, 5, 9, 13, 17 \\ 50, & k = 2, 6, 10, 14, 18 \\ 25, & k = 3, 7, 11, 15, 19 \\ 50, & k = 4, 8, 12, 16, 20 \end{cases}$$

Note that there is a periodicity in the decaying mean value for Case e, but its uncertainty is the same as that of Case c.

Together, Eqs. 4.2-6 and 4.2-8 specify the conditional probability density function of the random sequence $\mathbf{x}$ at the kth instant:

$$\mathrm{pr}(\mathbf{x}_k \mid \mathbf{x}_{k-1}, \mathbf{x}_{k-2}, \ldots, \mathbf{x}_0) = \frac{1}{(2\pi)^{n/2}|\mathbf{P}_k|^{1/2}} e^{-(1/2)(\mathbf{x}_k - \mathbf{m}_k)^T \mathbf{P}_k^{-1}(\mathbf{x}_k - \mathbf{m}_k)} \quad (4.2\text{-}10)$$

The density function is conditioned by all prior knowledge of the sequence, because $\mathbf{m}_k$ and $\mathbf{P}_k$ are related to $\mathbf{m}_0$ and $\mathbf{P}_0$ (and all intermediate values) by their respective difference equations. This conditioning is expressed in the equation

$$\mathrm{pr}(\mathbf{x}_k) = \mathrm{pr}(\mathbf{x}_k \mid \mathbf{x}_{k-1}, \mathbf{x}_{k-2}, \ldots, \mathbf{x}_0)\, \mathrm{pr}(\mathbf{x}_{k-1}, \mathbf{x}_{k-2}, \ldots, \mathbf{x}_0) \quad (4.2\text{-}11)$$

However, substituting Eqs. 4.2-6 and 4.2-8 for $\mathbf{m}_k$ and $\mathbf{P}_k$ in Eq. 4.2-10 shows that the conditional relationship explicitly depends only on the previous instant. In other words, conditioning prior to the $(k-1)$st instant affects $\mathbf{x}_k$ and $\mathbf{P}_k$ only through its effect on the values of $\mathbf{x}_{k-1}$ and $\mathbf{P}_{k-1}$:

$$\mathrm{pr}(\mathbf{x}_k \mid \mathbf{x}_{k-1}, \mathbf{x}_{k-2}, \ldots, \mathbf{x}_0) = \mathrm{pr}(\mathbf{x}_k \mid \mathbf{x}_{k-1}) \quad (4.2\text{-}12)$$

As a consequence, the conditional density functions have the property of *transition functions*, allowing the probability density function to be written as follows:

$$\begin{aligned} \mathrm{pr}(\mathbf{x}_k) &= \mathrm{pr}(\mathbf{x}_k \mid \mathbf{x}_{k-1})\, \mathrm{pr}(\mathbf{x}_{k-1} \mid \mathbf{x}_{k-2}), \ldots, \mathrm{pr}(\mathbf{x}_1 \mid \mathbf{x}_0)\, \mathrm{pr}(\mathbf{x}_0) \\ &= \prod_{i=1}^{k} \mathrm{pr}(\mathbf{x}_i \mid \mathbf{x}_{i-1})\, \mathrm{pr}(\mathbf{x}_0) \end{aligned} \quad (4.2\text{-}13)$$

Equation 4.2-12 describes the *Markov property*. $\mathbf{x}$ is, therefore, a *Markov sequence*. Furthermore, as its probability density function is Gaussian, $\mathbf{x}$ is a *Gauss–Markov sequence*.

Summarizing the propagation equations for $\mathbf{m}$ and $\mathbf{P}$ with Gaussian initial conditions and white Gaussian inputs,

$$\mathbf{m}_k = \mathbf{\Phi}_{k-1}\mathbf{m}_{k-1} + \mathbf{\Gamma}_{k-1}\mathbf{u}_{k-1}, \qquad \mathbf{m}_0 \text{ given} \tag{4.2-14}$$

$$\mathbf{P}_k = \mathbf{\Phi}_{k-1}\mathbf{P}_{k-1}\mathbf{\Phi}_{k-1}^T + \mathbf{\Lambda}_{k-1}\mathbf{Q}'_{k-1}\mathbf{\Lambda}_{k-1}^T, \qquad \mathbf{P}_0 \text{ given} \tag{4.2-15a}$$

It can be seen that both equations are linear and uncoupled from each other. The first equation propagates the most likely value of the state, as represented by its mean, and the second propagates the spread of the probability density function. Defining $\mathbf{Q}_k = \mathbf{\Lambda}_k\mathbf{Q}'_k\mathbf{\Lambda}_k^T$, the latter can be written as

$$\mathbf{P}_k = \mathbf{\Phi}_{k-1}\mathbf{P}_{k-1}\mathbf{\Phi}_{k-1}^T + \mathbf{Q}_{k-1} \tag{4.2-15b}$$

Because $\mathbf{P}_0$ and $\mathbf{Q}$ are symmetric and Eq. 4.2-15 itself is symmetric, $\mathbf{P}_k$ is symmetric; therefore, $p_{ij} = p_{ji}$, and only the upper (*or* lower) triangular block of this $(n \times n)$ equation need be solved. The equation could be rewritten as

$$\boldsymbol{\xi}_k = \mathbf{\Psi}_{k-1}\boldsymbol{\xi}_{k-1} + \mathbf{\Xi}_{k-1}\boldsymbol{\eta}_{k-1} \tag{4.2-16}$$

with the $[n(n \times 1)/2 \times 1]$ state covariance *vector*,

$$\boldsymbol{\xi} = [p_{11}\ p_{12}\ \ \dots\ \ p_{22}\ p_{23}\ \ \dots\ \ p_{nn}]^T \tag{4.2-17}$$

the $(n \times 1)$ input covariance *vector*,

$$\boldsymbol{\eta} = [q'_{11}\ q'_{22}\ \ \dots\ \ q'_{nn}]^T \tag{4.2-18}$$

and $\mathbf{\Psi}$ and $\mathbf{\Xi}$ defined accordingly. For example, when $n = 2$, expanding and reforming Eq. 4.2-15 leads to

$$\begin{bmatrix} p_{11} \\ p_{12} \\ p_{22} \end{bmatrix}_k = \begin{bmatrix} \phi_{11}^2 & 2\phi_{11}\phi_{12} & \phi_{12}^2 \\ \phi_{11}\phi_{21} & (\phi_{11}\phi_{22} + \phi_{12}\phi_{21}) & \phi_{12}\phi_{22} \\ \phi_{21}^2 & 2\phi_{21}\phi_{22} & \phi_{22}^2 \end{bmatrix}_{k-1} \begin{bmatrix} p_{11} \\ p_{12} \\ p_{22} \end{bmatrix}_{k-1} + \begin{bmatrix} \lambda_{11}^2 & \lambda_{12}^2 \\ \lambda_{11}\lambda_{21} & \lambda_{12}\lambda_{22} \\ \lambda_{21}^2 & \lambda_{22}^2 \end{bmatrix}_{k-1} \begin{bmatrix} q'_{11} \\ q'_{22} \end{bmatrix}_{k-1} \tag{4.2-19}$$

This alternate form of the covariance propagation equation illustrates the reduced dimensionality of the solution due to symmetry in $\mathbf{P}$, and it has potential use for examining the stability and input response of the covariance solution.

If the value of the disturbance input at one time is correlated with the value at another time, then the disturbance is not white; the time-correlation often can be taken into account by modeling the input as a Gauss–Markov sequence. Suppose that the s-dimensional disturbance $\mathbf{w}_k$ is described by

$$E(\mathbf{w}_k) = \mathbf{0} \tag{4.2-20}$$

$$E(\mathbf{w}_k \mathbf{w}_k^T) = \mathbf{W}_k \tag{4.2-21}$$

and

$$E(\mathbf{w}_k \mathbf{w}_{k-1}^T) = \mathbf{V}_k \tag{4.2-22}$$

Here, the present value of the disturbance input $\mathbf{w}_k$ is correlated with the past value $\mathbf{w}_{k-1}$, suggesting that $\mathbf{w}_k$ be modeled as

$$\mathbf{w}_k = \mathbf{A}_{k-1}\mathbf{w}_{k-1} + \boldsymbol{\eta}_{k-1} \tag{4.2-23}$$

where $\boldsymbol{\eta}_k$ is a white-noise sequence,

$$E(\boldsymbol{\eta}_k) = \mathbf{0} \tag{4.2-24}$$

$$E(\boldsymbol{\eta}_k \boldsymbol{\eta}_l^T) = \begin{cases} \mathbf{Q}_k, & k = l \\ \mathbf{0}, & k \neq l \end{cases} \tag{4.2-25}$$

From Eqs. 4.2-21 to 4.2-23,

$$E(\mathbf{w}_k \mathbf{w}_{k-1}^T) = E[(\mathbf{A}_{k-1}\mathbf{w}_{k-1} + \boldsymbol{\eta}_{k-1})\mathbf{w}_{k-1}^T] \tag{4.2-26a}$$

or

$$\mathbf{V}_k = \mathbf{A}_{k-1}\mathbf{W}_{k-1} + (\mathbf{0}) \tag{4.2-26b}$$

since $E(\boldsymbol{\eta}_{k-1}\mathbf{w}_{k-1}^T) = \mathbf{0}$. Then the transition matrix for the time-correlated disturbance input is,

$$\mathbf{A}_{k-1} = \mathbf{V}_k \mathbf{W}_{k-1}^{-1} \tag{4.2-27}$$

provided the inverse exists. Furthermore, from Eqs. 4.2-21 and 4.2-23,

$$\mathbf{W}_k = \mathbf{A}_{k-1}\mathbf{W}_{k-1}\mathbf{A}_{k-1}^T + \mathbf{Q}_{k-1} \tag{4.2-28}$$

so the covariance of the disturbance input sequence is

$$\mathbf{Q}_{k-1} = \mathbf{W}_k - \mathbf{A}_{k-1}\mathbf{W}_{k-1}\mathbf{A}_{k-1}^T \tag{4.2-29}$$

Thus the time-correlated disturbance is itself a dynamic model that can be added to the original model (Eq. 4.2-1). Defining an augmented state vector $\mathbf{x}_A$ of dimension $(n+s)$,

$$E(\mathbf{x}_{A_k}) = E(\boldsymbol{\Phi}_{A_{k-1}}\mathbf{x}_{A_{k-1}} + \boldsymbol{\Gamma}_{A_{k-1}}\mathbf{u}_{k-1} + \boldsymbol{\Lambda}_{A_{k-1}}\boldsymbol{\eta}_{n-1}) \tag{4.2-30a}$$

or

$$E\begin{bmatrix}\mathbf{x}\\ \mathbf{w}\end{bmatrix}_k = E\left\{\begin{bmatrix}\boldsymbol{\Phi} & \boldsymbol{\Lambda}\\ \mathbf{0} & \mathbf{A}\end{bmatrix}_{k-1}\begin{bmatrix}\mathbf{x}\\ \mathbf{w}\end{bmatrix}_{k-1} + \begin{bmatrix}\boldsymbol{\Gamma}\\ \mathbf{0}\end{bmatrix}_{k-1}\mathbf{u}_{k-1}\right\} \tag{4.2-30b}$$

The corresponding covariance equation is

$$\begin{bmatrix}\mathbf{P} & \mathbf{0}\\ \mathbf{0} & \mathbf{W}\end{bmatrix}_k = \begin{bmatrix}\boldsymbol{\Phi} & \boldsymbol{\Lambda}\\ \mathbf{0} & \mathbf{A}\end{bmatrix}_{k-1}\begin{bmatrix}\mathbf{P} & \mathbf{0}\\ \mathbf{0} & \mathbf{W}\end{bmatrix}_{k-1}\begin{bmatrix}\boldsymbol{\Phi} & \boldsymbol{\Lambda}\\ \mathbf{0} & \mathbf{A}\end{bmatrix}^T_{k-1} + \begin{bmatrix}\mathbf{0}\\ \mathbf{I}_s\end{bmatrix}\mathbf{Q}_{k-1}[\mathbf{0} \quad \mathbf{I}_s] \tag{4.2-31}$$

Equation 4.2-23 is sometimes called a shaping filter, as it gives shape to the autocorrelation function of the disturbance input. The filter's output is called "colored noise" because its output spectrum varies with frequency. If $\mathbf{Q}_k$ and $\mathbf{A}_k$ are constant, and if $\mathbf{A}_k$ represents a stable transition (Section 2.5), then Eq. 4.2-28 can reach a steady-state condition, in which $\mathbf{W}_k = \mathbf{W}_{k-1}$. Of course, not all components of $\mathbf{w}_k$ need be correlated at once; if only l terms are correlated, an l-dimensional subset of Eqs. 4.2-22 to 4.2-29 should be considered, with corresponding changes in Eqs. 4.2-30 and 4.2-31.

The preceding example of a time-correlated disturbance input is called a *first-order Gauss–Markov sequence*, in that the model equation (Eq. 4.2-23) is a first-order difference equation. It is possible to model higher-order time-correlation and still satisfy the Markov property (Eq. 4.2-12). A *second-order Gauss–Markov sequence* would be expressed by

$$\begin{bmatrix}\mathbf{w}_1\\ \mathbf{w}_2\end{bmatrix}_k = \begin{bmatrix}\mathbf{A} & \mathbf{B}\\ \mathbf{C} & \mathbf{D}\end{bmatrix}_{k-1}\begin{bmatrix}\mathbf{w}_1\\ \mathbf{w}_2\end{bmatrix}_{k-1} + \begin{bmatrix}\mathbf{0}\\ \boldsymbol{\eta}\end{bmatrix}_{k-1} \tag{4.2-32}$$

where $\mathbf{w}_1$ is the $(s \times 1)$ disturbance input acting on the system (the original $\mathbf{w}$), $\mathbf{w}_2$ is its $(s \times 1)$ derivative, and $\boldsymbol{\eta}$ is an $(s \times 1)$ white Gaussian sequence. For a time-invariant sampled-data model (discussed later), the transition matrix might represent the canonical second-order form for the sampling interval Δt,

$$\begin{bmatrix}\mathbf{A} & \mathbf{B}\\ \mathbf{C} & \mathbf{D}\end{bmatrix} = \exp\begin{bmatrix}\mathbf{0} & \mathbf{I}_s\\ -\mathbf{M} & -\mathbf{N}\end{bmatrix}\Delta t \tag{4.2-33}$$

with the diagonal matrices $\mathbf{M}$ and $\mathbf{N}$ representing spring and damping terms. Since this could be expressed by the $2s$-dimensional equation

$$\mathbf{w}'_k = \mathbf{A}'_{k-1}\mathbf{w}'_{k-1} + \boldsymbol{\eta}'_{k-1} \tag{4.2-34}$$

the previous developments also relate to the second-order model. Models of arbitrary dimension could be expressed by expanding the definitions of $\mathbf{w}'$ and $\mathbf{A}'$ to higher dimension to account for higher derivatives, still maintaining the Markov property for the probability density function of $\mathbf{w}'$.

Sampled-Data Representation of Continuous-Time Systems

The previous section addressed linear discrete-time systems without any particular relationship to an equivalent continuous-time system. Here we discuss the matrix definitions that represent continuous-time systems sampled at periodic instants. The distinction has no effect on the form of the previous equations; it merely relates the state transition properties to the corresponding differential equations and the covariance matrices of random sequences to the spectral density matrices of random processes.

Given the linear, time-varying differential equation model for a dynamic system,

$$\dot{\mathbf{x}}(t) = \mathbf{F}(t)\mathbf{x}(t) + \mathbf{G}(t)\mathbf{u}(t) + \mathbf{L}(t)\mathbf{w}(t), \qquad \mathbf{x}(0) \text{ given} \tag{4.2-35}$$

the corresponding discrete-time model is (from Eq. 2.3-81b),

$$\mathbf{x}(t_k) = \boldsymbol{\Phi}(t_k, t_{k-1})\mathbf{x}(t_{k-1}) + \int_{t_{k-1}}^{t_k} \boldsymbol{\Phi}(t_k, \tau)[\mathbf{G}(\tau)\mathbf{u}(\tau) + \mathbf{L}(\tau)\mathbf{w}(\tau)]\, d\tau, \qquad \mathbf{x}(0) \text{ given} \tag{4.2-36}$$

which, for piecewise constant inputs and periodic sampling ($\Delta t = t_k - t_{k-1}$), can be written as

$$\mathbf{x}_k = \boldsymbol{\Phi}_{k-1}\mathbf{x}_{k-1} + \boldsymbol{\Gamma}_{k-1}\mathbf{u}_{k-1} + \boldsymbol{\Lambda}_{k-1}\mathbf{w}_{k-1}, \qquad \mathbf{x}(0) \text{ given} \tag{4.2-37}$$

with $\boldsymbol{\Phi}$, $\boldsymbol{\Gamma}$, and $\boldsymbol{\Lambda}$ defined as in Section 2.3. If $\mathbf{F}$, $\mathbf{G}$, and $\mathbf{L}$ are constant and the sampling interval is Δt,

$$\boldsymbol{\Phi}(\Delta t) = e^{\mathbf{F}\Delta t} \tag{4.2-38}$$

$$\boldsymbol{\Gamma}(\Delta t) = \boldsymbol{\Phi}(\Delta t)[\mathbf{I}_n - \boldsymbol{\Phi}^{-1}(\Delta t)]\mathbf{F}^{-1}\mathbf{G} \tag{4.2-39}$$

$$\boldsymbol{\Lambda}(\Delta t) = \boldsymbol{\Phi}(\Delta t)[\mathbf{I}_n - \boldsymbol{\Phi}^{-1}(\Delta t)]\mathbf{F}^{-1}\mathbf{L} \tag{4.2-40}$$

With piecewise-constant inputs, both the mean-value and covariance propagation equations of the discrete-time problem (Eqs. 4.2-6 and 4.2-8) can be applied to the sampled-data problem without change. If, however, the inputs cannot be modeled as constant in the interval $[t_{k-1}, t_k]$, the matrices $\mathbf{\Gamma}_{k-1}$ and $\mathbf{\Lambda}_{k-1}$ are not readily identified, so it is necessary to treat the integrated input effects explicitly in propagating the mean and the covariance. Once again, it is assumed that

$$E(\mathbf{x}_0) = \mathbf{m}_0 \tag{4.2-41}$$

$$E[(\mathbf{x}_0 - \mathbf{m}_0)(\mathbf{x}_0 - \mathbf{m}_0)^T] = \mathbf{P}_0 \tag{4.2-42}$$

and that $\mathbf{u}(t)$ is known without error:

$$E[\mathbf{u}(t)] = \mathbf{u}(t) \tag{4.2-43}$$

$$E[\mathbf{u}(t)\mathbf{u}^T(\tau)] = \mathbf{0} \tag{4.2-44}$$

The continuous disturbance input $\mathbf{w}(t)$ is assumed to be a white noise random process with zero mean and known *spectral density matrix* $\mathbf{Q}'_C$,

$$E[\mathbf{w}(t)] = \mathbf{0} \tag{4.2-45}$$

$$E[\mathbf{w}(t)\mathbf{w}^T(\tau)] = \mathbf{Q}'_C(t)\delta(t-\tau) \tag{4.2-46}$$

where $\delta(t-\tau)$ is the Dirac delta function. The relationship between spectral density and covariance matrices is discussed below.

Since the disturbance input averages zero, the mean value of the state is propagated as

$$\mathbf{m}_k = \mathbf{\Phi}_{k-1}\mathbf{m}_{k-1} + \int_{t_{k-1}}^{t_k} \mathbf{\Phi}(t_k, \tau)\mathbf{G}(\tau)\mathbf{u}(\tau)\, d\tau \tag{4.2-47}$$

without further regard for the disturbance input. The expected value of the state covariance is

$$\begin{aligned}
\mathbf{P}_k &= \mathbf{\Phi}_{k-1}\mathbf{P}_{k-1}\mathbf{\Phi}_{k-1}^T \\
&\quad + E\left\{\left[\int_{t_{k-1}}^{t_k} \mathbf{\Phi}(t_k, \tau)\mathbf{L}(\tau)\mathbf{w}(\tau)\, d\tau\right]\left[\int_{t_{k-1}}^{t_k} \mathbf{\Phi}(t_k, \alpha)\mathbf{L}(\alpha)\mathbf{w}(\alpha)\, d\alpha\right]^T\right\} \\
&= \mathbf{\Phi}_{k-1}\mathbf{P}_{k-1}\mathbf{\Phi}_{k-1}^T \\
&\quad + E\left\{\int_{t_{k-1}}^{t_k} \int_{t_{k-1}}^{t_k} \mathbf{\Phi}(t_k, \tau)\mathbf{L}(\tau)\mathbf{w}(\tau)\mathbf{w}^T(\alpha)\mathbf{L}^T(\alpha)\mathbf{\Phi}^T(t_k, \alpha)\, d\alpha\, dt\right\}
\end{aligned} \tag{4.2-48}$$

Here, two dummy variables α and τ initially are needed to distinguish between the integrated effects of the disturbance and its transpose in the interval. Because there is nothing uncertain about $\mathbf{\Phi}$ or $\mathbf{L}$, the expectation can be brought inside the integral:

$$\mathbf{P}_k = \mathbf{\Phi}_{k-1}\mathbf{P}_{k-1}\mathbf{\Phi}_{k-1}^T + \int_{t_{k-1}}^{t_k}\int_{t_{k-1}}^{t_k} \mathbf{\Phi}(t_k, \tau)\mathbf{L}(\tau)E[\mathbf{w}(\tau)\mathbf{w}^T(\alpha)]\mathbf{L}^T(\alpha)\mathbf{\Phi}^T(t_k, \alpha)\, d\alpha\, d\tau \quad (4.2\text{-}49)$$

From Eq. 4.2-46,

$$E[\mathbf{w}(\tau)\mathbf{w}^T(\alpha)] = \mathbf{Q}'_C(\tau)\delta(\tau - \alpha) \quad (4.2\text{-}50)$$

Integrating over α, this allows Eq. 4.2-49 to be written as

$$\begin{aligned}\mathbf{P}_k &= \mathbf{\Phi}_{k-1}\mathbf{P}_{k-1}\mathbf{\Phi}_{k-1}^T + \int_{t_{k-1}}^{t_k} \mathbf{\Phi}(t_k, \tau)\mathbf{L}(\tau)\mathbf{Q}'_C(\tau)\mathbf{L}^T(\tau)\mathbf{\Phi}^T(t_k, \tau)\, d\tau \\ &= \mathbf{\Phi}_{k-1}\mathbf{P}_{k-1}\mathbf{\Phi}_{k-1}^T + \mathbf{Q}_{k-1} \quad (4.2\text{-}51)\end{aligned}$$

as in Eq. 4.2-15b. $\mathbf{Q}_{k-1}$ represents the integrated effect of the continuous disturbance during the interval, and although the forms of the discrete-time and sampled-data equations are identical, the disturbance input matrices, $\mathbf{Q}'_{k-1}$ and $\mathbf{Q}'_C(t)$, are, of course, numerically different. ($\mathbf{Q}'_{k-1}$ is a covariance matrix, and $\mathbf{Q}'_C(t)$ is a spectral density matrix for a white-noise process.) Even if $\mathbf{Q}'_C$ and $\mathbf{L}$ are diagonal, indicating zero cross-correlation in the driving terms, coupling in $\mathbf{\Phi}$ could cause the covariance matrix, $\mathbf{Q}_{k-1}$, to be nondiagonal (but symmetric), cross-correlating the disturbance effect over the interval. Note that $\mathbf{Q}_{k-1}$ and $\mathbf{Q}'_C$ need not have the same dimension: $\mathbf{Q}_{k-1}$ is $(n \times n)$, while $\mathbf{Q}'_C$ is $(s \times s)$.

The relationship between the covariance matrix, $\mathbf{Q}_{k-1}$, and the spectral density matrix, $\mathbf{Q}'_C(t)$, depends on the dynamic system model, as

$$\mathbf{Q}_{k-1} = \int_{t_{k-1}}^{t_k} \mathbf{\Phi}(t_k, \tau)\mathbf{L}(\tau)\mathbf{Q}'_C(\tau)\mathbf{L}^T(\tau)\mathbf{\Phi}^T(t_k, \tau)\, d\tau \quad (4.2\text{-}52)$$

This arises purely from the sampling process and is unrelated to colored process noise or the effects of an input shaping filter. Given symmetric $\mathbf{Q}'_C$, $\mathbf{Q}_{k-1}$ also is symmetric. For vanishingly small values of $\Delta t = t_k - t_{k-1}$,

$\Phi(t, t) = \mathbf{I}_n$, and with slowly varying (or constant) values of $\mathbf{L}(t)$ and $\mathbf{Q}'_C(t)$,

$$\mathbf{Q}_{k-1} \cong \mathbf{L}(t_k)\mathbf{Q}'_C(t_k)\mathbf{L}^T(t_k)\,\Delta t \tag{4.2-53}$$

For larger values of Δt, the entire integral must be evaluated. As an example, define

$$\mathbf{A}(t, \tau) = \Phi(t, \tau)\mathbf{L}(\tau) \tag{4.2-54}$$

and consider the integrand of Eq. 4.2-52 with $n = s = 2$ and diagonal spectral density matrix:

$$\mathbf{A} = \begin{bmatrix} a_{11} & a_{12} \\ a_{21} & a_{22} \end{bmatrix} \tag{4.2-55}$$

$$\mathbf{Q}'_C = \begin{bmatrix} q_{11} & 0 \\ 0 & q_{22} \end{bmatrix} \tag{4.2-56}$$

The integrand is

$$\mathbf{A}\mathbf{Q}'_C\mathbf{A}^T = \begin{bmatrix} (a_{11}^2 q_{11} + a_{12}^2 q_{22}) & (a_{11}a_{21}q_{11} + a_{12}a_{22}q_{22}) \\ (a_{11}a_{21}q_{11} + a_{12}a_{22}q_{22}) & (a_{21}^2 q_{11} + a_{22}^2 q_{22}) \end{bmatrix} \tag{4.2-57}$$

illustrating the coupling mentioned earlier. $\mathbf{Q}_{k-1}$ would be diagonal only if $a_{11}a_{21}q_{11} = -a_{12}a_{22}q_{22}$, and it could be surmised that diagonal $\mathbf{Q}_{k-1}$ would rarely occur in sampled-data models. With coupling in Φ and $\mathbf{L}$, a diagonal $\mathbf{Q}_{k-1}$ implies a nondiagonal $\mathbf{Q}'_C$.

Example 4.2-2 Disturbance Covariance for First- and Second-Order Systems

Consider a first-order system,

$$\dot{x}(t) = fx(t) + w(t)$$

with state transition function

$$\phi(\Delta t) = e^{f\Delta t}$$

and disturbance spectral density q_C such that

$$E[w(t)^2] = q_C\delta(t - \tau)$$

From Eq. 4.2-52, the disturbance covariance for a sampled-data model of this system would be

$$q = \int_0^{\Delta t} e^{2f(\Delta t - \tau)} q_C \, d\tau$$

$$= \frac{q_C}{2f}(e^{2f\Delta t} - 1)$$

Examples of the disturbance covariance are tabulated below for $q_C = 1$, $f = \pm 1$, and several values of Δt. The approximation of Eq. 4.2-53 is seen to be close for $|f\Delta t| \le 0.01$. The value of q diverges as $f\Delta t$ becomes increasingly positive, while it converges to 0.5 as $f\Delta t$ becomes increasingly negative.

Δt (sec)	$f = 1$ q	$f = -1$ q
10^{-6}	10^{-6}	10^{-6}
0.001	0.001	0.001
0.01	0.0101	0.0099
0.1	0.1107	0.0906
1.	3.195	0.4323
10.	2.5×10^8	0.499

Next, a second-order system with scalar forcing is considered:

$$\begin{bmatrix} \dot{x}_1 \\ \dot{x}_2 \end{bmatrix} = \begin{bmatrix} 0 & 1 \\ a & b \end{bmatrix} \begin{bmatrix} x_1 \\ x_2 \end{bmatrix} + \begin{bmatrix} 0 \\ w \end{bmatrix}$$

$$E[w(t)^2] = q_C \delta(t - \tau)$$

Equation 4.2-52 has been evaluated to find $\mathbf{Q}$, dividing the interval Δt into 100 steps and using rectangular integration. (The integrand takes the form of Eq. 4.2-57, with q_{11} and the first column of $\mathbf{A}$ equal to zero.) The elements of $\mathbf{Q}$ are shown in the following list for several values of a, b, and Δt, with $q_C = 1$. In all cases, the correlation, which is reflected by the nonzero values of q_{11} and q_{12}, increases as the sampling interval (Δt) increases. q_{22} is reasonably well approximated by $q_C \Delta t$ when $\Delta t = 0.01$, but not when the sampling interval is larger. When either a or b is positive in these examples, the system is unstable (as defined in Section 6.2); effects of system divergence on $\mathbf{Q}$ are most notable at the longest tabulated time interval ($\Delta t = 1$ s).

a	b	Δt	q_{11}	q_{12}	q_{22}
−1	−2	0.01	3×10^{-7}	5×10^{-5}	0.0098
		0.1	3×10^{-4}	0.0041	0.0823
		1.	0.0802	0.0677	0.2212
−1	−2	0.01	3×10^{-7}	5×10^{-5}	0.0098
		0.1	3×10^{-4}	0.0039	0.0786
		1.	0.013	0.0025	0.2269
−1	2	0.01	3×10^{-7}	5×10^{-5}	0.0102
		0.1	4×10^{-4}	0.0057	0.1162
		1.	0.0769	0.1073	2.137
−1	2	0.01	3×10^{-7}	5×10^{-5}	0.0102
		0.1	4×10^{-4}	0.006	0.1223
		1.	1.561	3.621	8.844
1	2	0.01	3×10^{-7}	5×10^{-5}	0.0102
		0.1	4×10^{-4}	0.0061	0.1231
		1.	2.428	6.739	19.07
1	−2	0.01	3×10^{-7}	5×10^{-5}	0.0098
		0.1	3×10^{-4}	0.0041	0.0828
		1.	0.1109	0.1259	0.2958

The use of white noise and delta functions is not entirely rigorous, although the resulting equations for $\mathbf{m}_k$ and $\mathbf{P}_k$ are the same as those derived by a more rigorous procedure. The latter begins by defining a *Brownian motion** (or Wiener or diffusion) process $\boldsymbol{\beta}(t)$ modeled after the irregular motions of small particles suspended in a fluid. Such particles are observed to have random velocities whose magnitudes and directions are uncorrelated with previous velocities. Given two arbitrary times, t_1 and t_2, and treating $\boldsymbol{\beta}(t)$ as an indicator of particle position

$$E[\boldsymbol{\beta}(t_2)-\boldsymbol{\beta}(t_1)]=\mathbf{0} \tag{4.2-58}$$

$$E\{[\boldsymbol{\beta}(t_2)-\boldsymbol{\beta}(t_1)][\boldsymbol{\beta}(t_2)-\boldsymbol{\beta}(t_1)]^T\}=\int_{t_1}^{t_2}\mathbf{Q}'_C(t)\,dt \tag{4.2-59}$$

where $\mathbf{Q}'_C(t)$ is called the *diffusion* of the process. This is said to be a process with "independent increments," of duration (t_2-t_1), in that motions in one time increment are assumed to be entirely unrelated to those in the

*Scottish botanist Robert Brown (1773–1858) was the first to study this effect.

next increment. Because Eq. 4.2-58 could be rewritten

$$E[\boldsymbol{\beta}(t_2)] = E[\boldsymbol{\beta}(t_1)] \tag{4.2-60}$$

the best estimate of the present mean is the estimate of the past mean; hence the process has a stationary mean value. The process $\boldsymbol{\beta}(t)$ is Gaussian and continuous but nondifferentiable, because its velocity is free to change instantaneously.

Nevertheless, if we define a white-noise process as the derivative of the Brownian motion process,

$$\mathbf{w}(t) = \frac{d\boldsymbol{\beta}(t)}{dt} \tag{4.2-61}$$

then Eq. 4.2-35 could be expressed as a stochastic differential state equation. Neglecting the control input and multiplying both sides by dt,

$$d\mathbf{x}(t) = \mathbf{F}(t)\mathbf{x}(t)\,dt + \mathbf{L}(t)\,d\boldsymbol{\beta}(t), \qquad \mathbf{x}(t_0) \text{ given} \tag{4.2-62}$$

with the corresponding integral equation

$$\mathbf{x}_k = \boldsymbol{\Phi}_k\mathbf{x}_{k-1} + E\left[\int_{\boldsymbol{\beta}(t_{k-1})}^{\boldsymbol{\beta}(t_k)} \boldsymbol{\Phi}(t_k, \tau)\mathbf{L}(\tau)\,d\boldsymbol{\beta}(\tau)\right] \tag{4.2-63}$$

The integral is zero by Eq. 4.2-58, as could be deduced by replacing the integration with summation over arbitrarily small independent increments. The corresponding covariance equation is

$$\begin{aligned}\mathbf{P}_k &= \boldsymbol{\Phi}_{k-1}\mathbf{P}_{k-1}\boldsymbol{\Phi}_{k-1}^T \\ &\quad + E\left\{\left[\int_{\boldsymbol{\beta}(t_{k-1})}^{\boldsymbol{\beta}(t_k)} \boldsymbol{\Phi}(t_k, \tau)\mathbf{L}(\tau)\,d\boldsymbol{\beta}(\tau)\right]\left[\int_{\boldsymbol{\beta}(t_{k-1})}^{\boldsymbol{\beta}(t_k)} \boldsymbol{\Phi}(t_k, \alpha)\mathbf{L}(\alpha)\,d\boldsymbol{\beta}(\alpha)\right]^T\right\} \\ &= \boldsymbol{\Phi}_{k-1}\mathbf{P}_{k-1}\boldsymbol{\Phi}_{k-1}^T \\ &\quad + E\left[\int_{\boldsymbol{\beta}(t_{k-1})}^{\boldsymbol{\beta}(t_k)} \int_{\boldsymbol{\beta}(t_{k-1})}^{\boldsymbol{\beta}(t_k)} \boldsymbol{\Phi}(t_k, \tau)\mathbf{L}(\tau)\,d\boldsymbol{\beta}(\tau)\,d\boldsymbol{\beta}^T(\alpha)\mathbf{L}^T(\alpha)\boldsymbol{\Phi}^T(t_k, \alpha)\right]\end{aligned} \tag{4.2-64}$$

[Here, the term $\mathbf{L}^T(\alpha)\boldsymbol{\Phi}^T(t_k, \alpha)$ is part of the integrand; however, it is written after $d\boldsymbol{\beta}^T$ to preserve proper, conformable ordering of vector-matrix products.] As before, the expectation operator can be brought inside the integral. Letting t_2 become arbitrarily close to t_1, writing $\Delta t = t_2 - t_1$, and using the assumption of independent increments, Eq. 4.2-58 leads to

$$\begin{aligned}\lim_{\Delta t \to dt} E[d\boldsymbol{\beta}(\tau)\, d\boldsymbol{\beta}^T(\alpha)] &= E[d\boldsymbol{\beta}(\tau)\, d\boldsymbol{\beta}^T(\alpha)] \\ &= E[d\boldsymbol{\beta}(\tau)\, d\boldsymbol{\beta}^T(\tau)] \\ &= \lim_{\Delta t \to dt} \int_0^{\Delta t} \mathbf{Q}'_C\, d\tau \\ &= \mathbf{Q}'_C(t)\, dt \end{aligned} \tag{4.2-65}$$

This allows Eq. 4.2-64 to be written as

$$\mathbf{P}_k = \boldsymbol{\Phi}_{k-1}\mathbf{P}_{k-1}\boldsymbol{\Phi}_{k-1}^T + \int_{t_{k-1}}^{t_k} \boldsymbol{\Phi}(t_k, \tau)\mathbf{L}(\tau)\mathbf{Q}'_C(\tau)\mathbf{L}^T(\tau)\boldsymbol{\Phi}^T(t_k, \tau)\, d\tau \tag{4.2-66}$$

the same result as Eq. 4.2-51. In effect, the assumption of continuous-time white noise (which rigorously cannot exist) is acceptable because only its integral (which rigorously does exist) is of concern in evaluating $\mathbf{P}_k$.

EXAMPLE 4.2-3 WEATHERVANE ANGLE AND RATE UNCERTAINTY

Assume that the angular motion of a weathervane is modeled by

$$\begin{bmatrix} \dot{x}_1 \\ \dot{x}_2 \end{bmatrix} = \begin{bmatrix} 0 & 1 \\ -\omega_n^2 & -2\zeta\omega_n \end{bmatrix} \begin{bmatrix} x_1 \\ x_2 \end{bmatrix} + \begin{bmatrix} 0 \\ \omega_n^2 \end{bmatrix} w$$

where x_1 represents angular position, x_2 represents angular rate, and w is the random angular direction of the wind. With a wind speed that provides $\zeta = 0.3$, $\omega_n = 6.28$ rad/s, and a wind angle spectral density of 1 deg²/s, compute the state covariance matrix with a 0.1-s sampling interval. Assume that $\mathbf{P}_0 = \mathbf{0}$.

Using Eq. 4.2-52, the covariance matrix is

$$\mathbf{Q}(0.1) = \omega_n^4 \begin{bmatrix} 0.0002 & 0.003 \\ 0.003 & 0.063 \end{bmatrix}$$

while the state transition matrix for $\Delta t = 0.1$ s is

$$\boldsymbol{\Phi}(0.1) = \begin{bmatrix} 0.8309 & 0.078 \\ -3.075 & 0.5372 \end{bmatrix}$$

The state covariance matrix elements (normalized by ω_n^4) are computed by

$$\mathbf{P}_k = \boldsymbol{\Phi}\mathbf{P}_{k-1}\boldsymbol{\Phi}^T + \mathbf{Q}$$

and are plotted below. The diagonal elements (p_{11} and p_{22}) reach stochastic

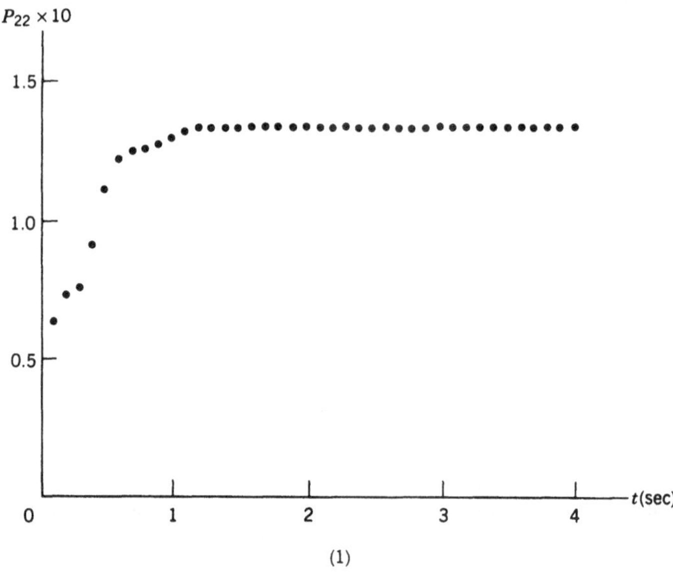

(1)

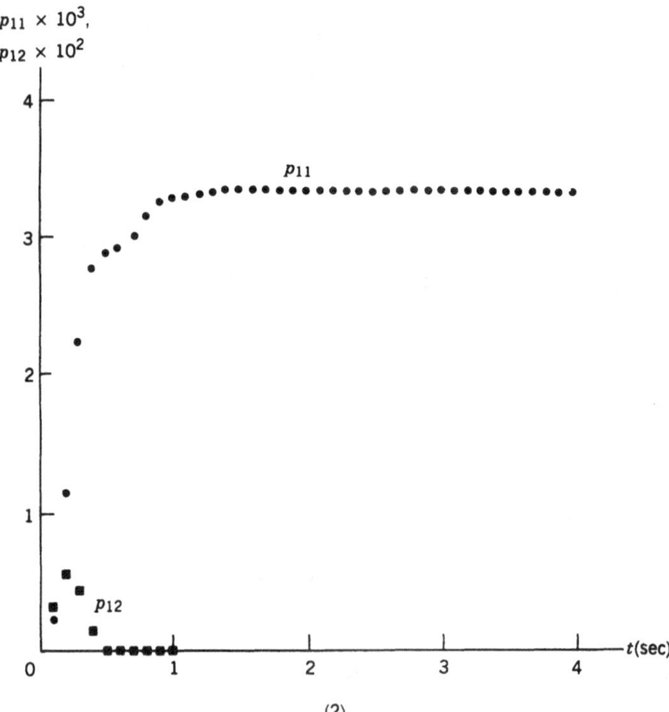

(2)

(1) Covariance element p_{22}; (2) covariance elements p_{11} and p_{12}; (3) p_{22} computed with shorter sampling interval.

equilibrium in a little more than 1 s, representing standard deviations, $\omega_n^2(p_{11})^{1/2}$ and $\omega_n^2(p_{22})^{1/2}$, of 2.3 deg and 14.4 deg/s, respectively. The cross-correlation term (p_{12}) briefly rises, then drops to zero, indicating no correlation between instantaneous angle and angular rate at stochastic equilibrium.

More detail of the **P** time history could be achieved by using a shorter

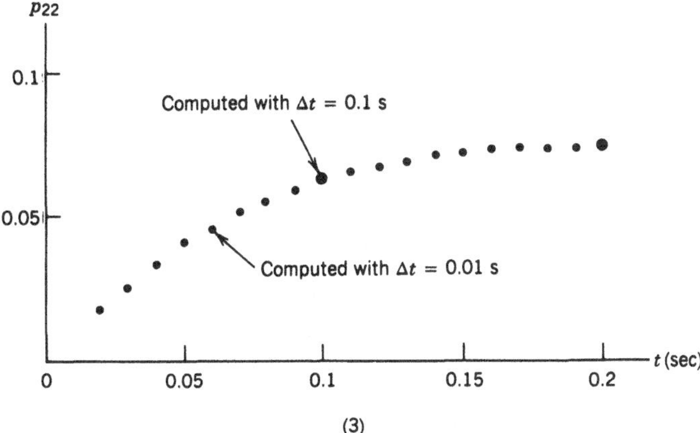

(3)

sampling interval, as illustrated in Diagram 3. Here, $\Delta t = 0.01$ s, leading tc

$$\mathbf{Q}(0.01) = \omega_n^4 \begin{bmatrix} 3.2 \times 10^{-7} & 4.8 \times 10^{-5} \\ 4.8 \times 10^{-5} & 0.0096 \end{bmatrix}$$

$$\mathbf{\Phi}(0.01) = \begin{bmatrix} 0.9981 & 0.0098 \\ -0.3868 & 0.9611 \end{bmatrix}$$

The estimates of p_{11} for $t = 0.1$ and 0.2 s computed from Eq. 4.2-51 are the same as before, but intermediate points are presented as well. In the limit as $\Delta t \rightarrow 0$, a continuous time history would be provided.

Continuous-Time Systems

Sampled-data systems represent continuous-time systems at discrete instants of time, so their propagation equations provide a starting point for the equations that represent the continuous-time system in continuous time, that is, throughout the interval $[t_0, t_f]$. What remains is the derivation of differential equations from difference equations using a conventional limiting procedure. The previous discussion of the reality of white noise applies here, and the conclusions are the same: the assumption of white noise leads to the proper results. The corresponding integral equations then are defined as in Section 2.3.

We can simply begin with Eq. 4.2-35, replacing deterministic values with expected values:

$$E[\dot{\mathbf{x}}(t)] = \mathbf{F}(t)E[\mathbf{x}(t)] + \mathbf{G}(t)E[\mathbf{u}(t)] + \mathbf{L}(t)E[\mathbf{w}(t)], \qquad E[\mathbf{x}(0)] = \mathbf{m}_0 \tag{4.2-67a}$$

Using Eqs. 4.2-43 and 4.2-45, and with

$$E[\mathbf{x}(t)] = \mathbf{m}(t) \tag{4.2-68}$$

$$E\{[\mathbf{x}(t) - \mathbf{m}(t)][\mathbf{x}(t) - \mathbf{m}(t)]^T\} = \mathbf{P}(t) \tag{4.2-69}$$

the differential equation describing the mean-value dynamics is

$$\dot{\mathbf{m}}(t) = \mathbf{F}(t)\mathbf{m}(t) + \mathbf{G}(t)\mathbf{u}(t), \qquad \mathbf{m}(0) = \mathbf{m}_0 \tag{4.2-67b}$$

The same result could have been obtained by applying a limiting process to the equivalent sampled-data equation (Eq. 4.2-47). Letting $\Delta t = t_k - t_{k-1} \ll 1$,

$$\mathbf{\Phi}_{k-1} \simeq \mathbf{I}_n + \mathbf{F}_{k-1}\,\Delta t \tag{4.2-70}$$

as could be deduced from Eq. 2.3-72. Then

$$\mathbf{m}_k \simeq (\mathbf{I}_n + \mathbf{F}_{k-1}\Delta t)\mathbf{m}_{k-1} + \int_{t_{k-1}}^{t_k} \mathbf{\Phi}(t_k - \tau)\mathbf{G}(\tau)\mathbf{u}(\tau)\, d\tau \tag{4.2-71}$$

or

$$\frac{(\mathbf{m}_k - \mathbf{m}_{k-1})}{\Delta t} \simeq \mathbf{F}_{k-1}\mathbf{m}_{k-1} + \frac{1}{\Delta t}\int_{t_{k-1}}^{t_k} \mathbf{\Phi}(t_k - \tau)\mathbf{G}(\tau)\mathbf{u}(\tau)\, d\tau \tag{4.2-72}$$

As $\Delta t \to 0$, $t_k = t_{k-1} = t$, and Eq. 4.2-72 becomes

$$\dot{\mathbf{m}}(t) = \mathbf{F}(t)\mathbf{m}(t) + \mathbf{G}(t)\mathbf{u}(t) \tag{4.2-73}$$

In the limit, the mean-value differential becomes the derivative, values at t_{k-1} are evaluated at t, and the derivative of the integral is the integrand.

The limiting process also can be applied to the covariance matrix difference equation (Eq. 4.2-51) to obtain the corresponding differential equation. Using Eq. 4.2-70, this is

$$\begin{aligned}\mathbf{P}_k &\simeq [\mathbf{I}_n + \mathbf{F}_{k-1}\Delta t]\mathbf{P}_{k-1}[\mathbf{I}_n + \mathbf{F}_{k-1}\Delta t]^T + \mathbf{Q}_{k-1} \\ &= \mathbf{P}_{k-1} + \mathbf{F}_{k-1}\mathbf{P}_{k-1}\Delta t + \mathbf{P}_{k-1}\mathbf{F}_{k-1}^T\Delta t + \mathbf{F}_{k-1}\mathbf{P}_{k-1}\mathbf{F}_{k-1}^T\Delta t^2 + \mathbf{Q}_{k-1}\end{aligned} \tag{4.2-74}$$

or

$$\frac{\mathbf{P}_k - \mathbf{P}_{k-1}}{\Delta t} \simeq \mathbf{F}_{k-1}\mathbf{P}_{k-1} + \mathbf{P}_{k-1}\mathbf{F}_{k-1}^T + \mathbf{F}_{k-1}\mathbf{P}_{k-1}\mathbf{F}_{k-1}^T\Delta t + \frac{\mathbf{Q}_{k-1}}{\Delta t} \quad (4.2\text{-}75)$$

As $\Delta t \to 0$,

$$\lim_{\Delta t \to 0} \frac{\mathbf{P}_k - \mathbf{P}_{k-1}}{\Delta t} = \dot{\mathbf{P}}(t) \quad (4.2\text{-}76)$$

$$\lim_{\Delta t \to 0} \mathbf{F}_{k-1}\mathbf{P}_{k-1}\mathbf{F}_{k-1}^T \, \Delta t = \mathbf{0} \quad (4.2\text{-}77)$$

$$\lim_{\Delta t \to 0} \frac{\mathbf{Q}_{k-1}}{\Delta t} = \mathbf{L}(t)\mathbf{Q}'_C(t)\mathbf{L}^T(t) \quad (4.2\text{-}78)$$

so Eq. 4.2-75 can be written as

$$\dot{\mathbf{P}}(t) = \mathbf{F}(t)\mathbf{P}(t) + \mathbf{P}(t)\mathbf{F}^T(t) + \mathbf{L}(t)\mathbf{Q}'_C(t)\mathbf{L}^T(t) \quad (4.2\text{-}79)$$

with $\mathbf{P}(0) = \mathbf{P}_o$. This is the desired result, and like the equivalent discrete-time equation, it is a symmetric, linear differential matrix equation. Note that the spectral density matrix of the disturbance input $\mathbf{Q}'_C$ appears as the forcing function in Eq. 4.2-79, a consequence of the limiting process.

Simulating Cross-Correlated White Noise

Having specified a covariance matrix of the symmetric form,

$$\begin{aligned}
\mathbf{Q}_k &= E[\mathbf{w}_k\mathbf{w}_k^T] \\
&= \begin{bmatrix} q_{11} & q_{12} & \cdots & q_{1s} \\ q_{12} & q_{22} & \cdots & q_{2s} \\ \vdots & & & \\ q_{1s} & q_{2s} & \cdots & q_{ss} \end{bmatrix}_k \\
&= \begin{bmatrix} \sigma_1^2 & \rho_{12}\sigma_1\sigma_2 & \cdots & \rho_{1s}\sigma_1\sigma_s \\ \rho_{12}\sigma_1\sigma_2 & \sigma_2^2 & \cdots & \\ \vdots & & & \\ \rho_{1s}\sigma_1\sigma_s & \rho_{2s}\sigma_2\sigma_s & \cdots & \sigma_s^2 \end{bmatrix}_k
\end{aligned} \quad (4.2\text{-}80)$$

it often is necessary to generate the random sequence $\mathbf{w}_k$, whose covariance is specified by $\mathbf{Q}_k$. Here, σ_i represents the standard deviation of the ith variable, and ρ_{ij} is the cross-correlation coefficient for the ith and jth variables. Given the covariance propagation equation,

$$\mathbf{P}_{k+1} = \mathbf{\Phi}_k \mathbf{P}_k \mathbf{\Phi}_k^T + \mathbf{Q}_k \tag{4.2-81}$$

the corresponding state propagation is to be simulated:

$$\mathbf{x}_{k+1} = \mathbf{\Phi}_k \mathbf{x}_k + \mathbf{w}_k \tag{4.2-82}$$

Computer routines that generate Gaussian random numbers are commonplace, but they do so for scalar variables. Hence it is easy to specify $\mathbf{w}_k$ if $\mathbf{Q}_k$ is diagonal ($\rho_{ij} = 0$); $\mathbf{w}_k$ is merely an ordered set of the outputs of the computer's random number generator. When $\rho_{ij} \neq 0$, as is frequently the case in sampled-data representation of continuous systems, $\mathbf{w}_k$ must reflect a combination of independent random sequences.

Let $\mathbf{v}_k$ represent an s-dimensional uncorrelated white-noise sequence, with

$$E[\mathbf{v}_k] = \mathbf{0} \tag{4.2-83}$$

$$\begin{aligned} E[\mathbf{v}_k \mathbf{v}_k^T] &= \mathbf{N}_k \\ &= \begin{bmatrix} \sigma_{n_1}^2 & 0 & \dots & 0 \\ 0 & \sigma_{n_2}^2 & \dots & 0 \\ \vdots & \vdots & & \vdots \\ 0 & 0 & \dots & \sigma_{n_s}^2 \end{bmatrix}_k \end{aligned} \tag{4.2-84}$$

The correlated sequence $\mathbf{w}_k$, whose covariance matrix is $\mathbf{Q}_k$, is to be derived from $\mathbf{v}_k$. The eigenvalues of $\mathbf{Q}_k$ define the $\sigma_{n_i}^2$, and the eigenvectors of $\mathbf{Q}_k$ define the modal matrix $\mathbf{D}_k$ that transforms $\mathbf{v}_k$ to $\mathbf{w}_k$. From Eqs. 2.2-72 and 2.2-73,

$$|\lambda \mathbf{I} - \mathbf{Q}_k| = (\lambda - \sigma_{n_1}^2)(\lambda - \sigma_{n_2}^2) \quad \dots \quad (\lambda - \sigma_{n_s}^2) = 0 \tag{4.2-85}$$

Because $\mathbf{Q}_k$ is real and symmetric, its eigenvalues are real variables, and they should be positive if $\mathbf{Q}_k$ has been specified properly. Furthermore, the corresponding eigenvectors, Eqs. 2.2-74 or 2.2-80, are orthogonal, so the transpose of the modal matrix (Eq. 2.2-75) equals the inverse. Then Eq. 2.2-78 indicates that

$$\begin{aligned} \mathbf{Q}_k &= \mathbf{D}_k \mathbf{N}_k \mathbf{D}_k^{-1} \\ &= \mathbf{D}_k \mathbf{N}_k \mathbf{D}_k^T \\ &= E[\mathbf{D}_k \mathbf{v}_k \mathbf{v}_k^T \mathbf{D}_k^T] \end{aligned} \tag{4.2-86}$$

and the state propagation equation is

$$\mathbf{x}_{k+1} = \mathbf{\Phi}_k \mathbf{x}_k + \mathbf{D}_k \mathbf{v}_k \tag{4.2-87}$$

The symmetric and diagonal spectral density matrices for continuous simulation of continuous-time systems could be related in the same way. Given a cross-correlated spectral density matrix $\mathbf{Q}_C$, the elements of a diagonal spectral density matrix $\mathbf{N}_C$ would be defined by the eigenvalues of $\mathbf{Q}_C$. The simulating differential equation driven by a cross-correlated, white vector random process $\mathbf{w}(t)$,

$$\dot{\mathbf{x}}(t) = \mathbf{F}(t)\mathbf{x}(t) + \mathbf{w}(t) \tag{4.2-88}$$

with

$$E[\mathbf{w}(t)] = \mathbf{0} \tag{4.2-89}$$

$$E[\mathbf{w}(t)\mathbf{w}(\tau)^T] = \mathbf{Q}_C(t)\delta(t-\tau) \tag{4.2-90}$$

would be implemented as

$$\dot{\mathbf{x}}(t) = \mathbf{F}(t)\mathbf{x}(t) + \mathbf{D}(t)\mathbf{v}(t) \tag{4.2-91}$$

where the independent random process $\mathbf{v}(t)$ is characterized by

$$E[\mathbf{v}(t)] = \mathbf{0} \tag{4.2-92}$$

$$E[\mathbf{v}(t)\mathbf{v}(\tau)^T] = \mathbf{N}_C(t)\delta(t-\tau) \tag{4.2-93}$$

and $\mathbf{D}(t)$ is the modal matrix of $\mathbf{Q}_C(t)$.

An alternate method of generating $\mathbf{w}_k$ that involves much less computation than the generation of eigenvalues and eigenvectors is based on *Cholesky decomposition* of $\mathbf{Q}_k$. This factorization technique generates the *matrix square root* of a positive definite matrix, $\mathbf{Q}_k$, which is defined implicitly by

$$\mathbf{Q}_k = \mathbf{S}_k\mathbf{S}_k^T \tag{4.2-94}$$

The matrix square root, $\mathbf{S}_k$, is not uniquely determined by Eq. 4.2-94, and many definitions of $\mathbf{S}_k$ could satisfy the equation. We can, however, specify a unique $\mathbf{S}_k$ of a given form, such as the lower (or upper) triangular matrix that results from the Cholesky decomposition:

$$\mathbf{S} = \begin{bmatrix} s_{11} & 0 & \dots & 0 \\ s_{21} & s_{22} & \dots & 0 \\ \vdots & \vdots & & \vdots \\ s_{n1} & s_{n2} & \dots & s_{nn} \end{bmatrix} \tag{4.2-95}$$

The elements of the lower triangular $\mathbf{S}$ are computed recursively, beginning in the upper left corner:

$$\mathbf{s}_{ij} = \begin{cases} \left(q_{ii} - \sum_{k=1}^{i-1} s_{ik}^2\right)^{1/2}, & j = i \\ 0, & j > i \\ \dfrac{\left[q_{ij} - \sum_{k=1}^{j-1} s_{ik}s_{jk}\right]}{s_{jj}}, & j = 1, 2, \ldots, i-1 \end{cases} \tag{4.2-96}$$

The sequence of nonzero computations is s_{11}, s_{21}, s_{22}, $s_{31}, \ldots$, and the summations are not evaluated when the upper limit is less than one. Because Eq. 4.2-94 could be written

$$\mathbf{Q}_k = \mathbf{S}_k \mathbf{I}_s \mathbf{S}_k^T \tag{4.2-97}$$

the application to cross-correlated white noise is apparent: define $\mathbf{D} = \mathbf{S}$ and $\mathbf{N} = \mathbf{I}_s$ in the white-noise simulation equations.

4.3 DISCRETE-TIME OPTIMAL FILTERS AND PREDICTORS

The objective addressed in this section is to calculate "best" estimates of a state vector sequence $\mathbf{x}_k$ in the discrete interval $0 \leq k \leq k_{\max}$. The estimates $\hat{\mathbf{x}}_k$ are to be conditioned on all available information about a linear dynamic process and a set of measurements taken during the interval. It is assumed that the discrete interval of indices k corresponds to a sampled interval of time $0 \leq t_k \leq t_{k_{\max}}$, where the time increment between successive samples, $t_k - t_{k-1}$, normally is constant.

Estimation problems fall into three categories (Fig. 4.3-1). If the time at which the calculations are made t_{calc} is greater than $t_{k\max}$, then all the measurements are available to estimate each sample $\mathbf{x}_k$ at t_k, and an algorithm called a *smoother* can be applied to the data. If the time of calculation is the same as the time of estimate t_k then the computation of $\hat{\mathbf{x}}_k$ must rely on past and present data only (i.e., on measurements in the "physically realizable" interval $[0, t_k]$. Such an estimation algorithm is called a *filter*, as it is meant to "filter out" the noise in the available signals. If an estimate of $\mathbf{x}_k$ at some future time—say $t_{k_{\max}}$—is to be made at time t_{calc}, the estimation algorithm is called a *predictor* for obvious reasons.

Because the smoother can use more data than the filter or predictor, it usually can produce better estimates (except at the endpoint, $t_k = t_{k\max}$, where the estimators necessarily have the same information); however, the smoother can only be applied "after the fact." The filter and predictor can be implemented for real-time operation because they do not require future

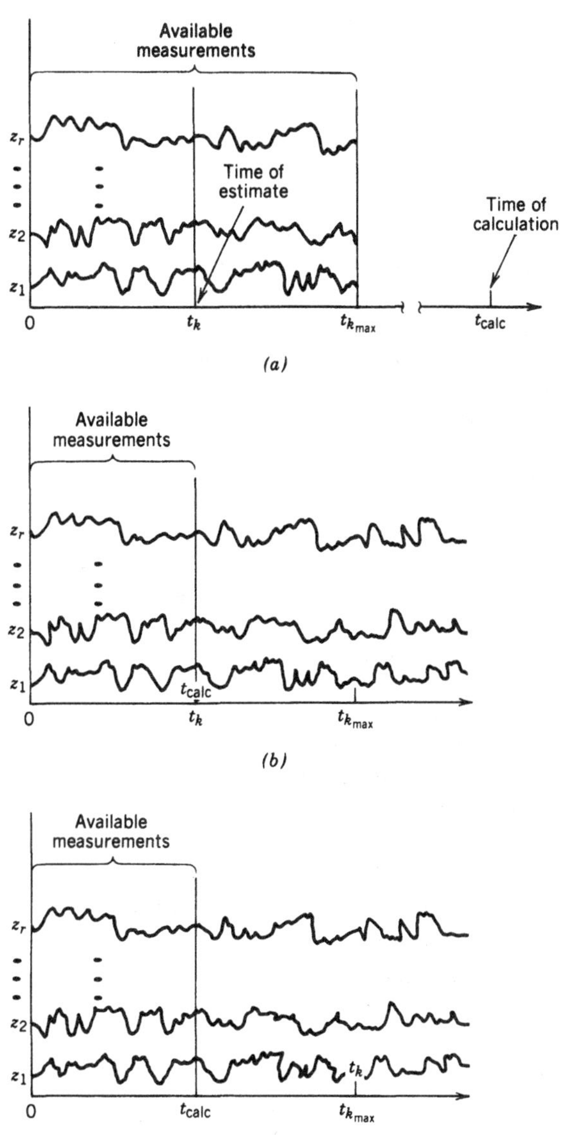

FIGURE 4.3-1 Classification of dynamic estimation problems. (*a*) Smoothing problem; (*b*) filtering problem; (*c*) prediction problem.

measurements. As our principal concern here is control and control systems rather than data processing, optimal filters and predictors are developed, but smoothers are not.

Kalman Filter

In keeping with the philosophy of the previous section, the estimate of $\mathbf{x}_k$ is specified by its conditional probability density function. The purpose of a filter is to compute the state estimate $\hat{\mathbf{x}}_k$, while an optimal filter minimizes the spread of the estimate-error probability density in the process. Minimizing the spread of a unimodal density function raises the peak of the function (or its likelihood), as the volume under the function is constant. (Recall that the cumulative probability in the range $(-\infty, +\infty)$ of all variables is one.) Hence, for a Gaussian distribution, a minimum-variance estimator produces the same result as a maximum-likelihood estimator.

A recursive optimal filter propagates the conditional probability density function from one sampling instant to the next, taking into account system dynamics and inputs, and it incorporates measurements and measurement-error statistics in the estimate. Computing the weighting factors (or filter gains) that optimally combine measurements and extrapolations is a crucial intermediate step in the computation.

It is consistent with the past discussion to take the mean (or expected) value of the conditional density function as the estimate of $\mathbf{x}_k$ and the covariance matrix as a measure of the spread (or uncertainty) in the estimate. The recursive generation of the mean and covariance in $0 \leq k \leq k_{\max}$ can be expressed in five equations:

1. State Estimate Extrapolation (Propagation)
2. Covariance Estimate Extrapolation (Propagation)
3. Filter Gain Computation
4. State Estimate Update
5. Covariance Estimate "Update"

The first two of these were introduced in the previous section, while the last three follow directly from recursive weighted-least-squares estimation (Section 4.1). Given the state estimate from a previous iteration, (1) uses the dynamic process model to propagate the estimate of the state mean value to the next sampling instant without regard to new measurements. (2) does the same thing for the state covariance matrix, assuming that "process noise" of known covariance is forcing the system. The result of (2) enters the computation of the optimal filter gains. The filter gain computation (3) weights prior knowledge of measurement error covariance with state estimate covariance on a purely statistical basis. The actual measurements have no effect on the gain computation. These measurements correct the state estimate in (4), adding the product of the gain matrix and the measurement residual to the state estimate propagated by (1). A similar correction is made to the covariance estimate (5), accounting for the known covariance of measurement errors.

Let's review what we know about the system before formulating the five equations of the estimator. The system's dynamic equation is

$$\mathbf{x}_k = \mathbf{\Phi}_{k-1}\mathbf{x}_{k-1} + \mathbf{\Gamma}_{k-1}\mathbf{u}_{k-1} + \mathbf{\Lambda}_{k-1}\mathbf{w}_{k-1} \tag{4.3-1}$$

with variables defined and dimensioned as before. $\mathbf{\Phi}_{k-1}$, $\mathbf{\Gamma}_{k-1}$, $\mathbf{\Lambda}_{k-1}$, and $\mathbf{u}_{k-1}$ are known without error in the interval $0 \le k \le k_{max}$, the expected values of the initial state and its covariance are known,

$$E(\mathbf{x}_0) = \hat{\mathbf{x}}_0 \tag{4.3-2}$$

$$E[(\mathbf{x}_0 - \hat{\mathbf{x}}_0)(\mathbf{x}_0 - \hat{\mathbf{x}}_0)^T] = \mathbf{P}_0 \tag{4.3-3}$$

and the disturbance input is a white, zero-mean Gaussian random sequence:

$$E(\mathbf{w}_k) = \mathbf{0} \tag{4.3-4}$$

$$E(\mathbf{w}_k\mathbf{w}_k^T) = \mathbf{Q}'_k \tag{4.3-5}$$

$$E(\mathbf{w}_k\mathbf{w}_j^T) = \mathbf{0}, \qquad (j \neq k) \tag{4.3-6}$$

The observation vector is

$$\mathbf{z}_k = \mathbf{H}_k\mathbf{x}_k + \mathbf{n}_k \tag{4.3-7}$$

where $\mathbf{H}_k$ is known and the measurement error is a white, zero-mean Gaussian random sequence that is uncorrelated with the disturbance input:

$$E(\mathbf{n}_k) = \mathbf{0} \tag{4.3-8}$$

$$E(\mathbf{n}_k\mathbf{n}_k^T) = \mathbf{R}_k \tag{4.3-9}$$

$$E(\mathbf{n}_k\mathbf{n}_j^T) = \mathbf{0} \qquad (j \neq k) \tag{4.3-10}$$

$$E(\mathbf{n}_k\mathbf{w}_j^T) = \mathbf{0} \qquad (\text{all } j \text{ and } k) \tag{4.3-11}$$

Rather than rederiving the Kalman filter equations from the start, we can make use of results from Sections 4.1 and 4.2. The dynamic system is linear, and the recursive least-squares estimator is linear; hence the principle of superposition applies. In the following equations, we must distinguish between estimates made before and after the updates occur; $\hat{\mathbf{x}}_k(-)$ is the state estimate that results from the propagation equation alone (i.e., *before* the measurements are considered), and $\hat{\mathbf{x}}_k(+)$ is the corrected state estimate that accounts for the measurements. $\mathbf{P}_k(-)$ and $\mathbf{P}_k(+)$ are defined similarly; for completeness, the initial conditions (Eqs. 4.3-2 and 4.3-3) should be treated as $\hat{\mathbf{x}}_0(+)$ and $\mathbf{P}_0(+)$.

From Eq. 4.2-14, the *state estimate extrapolation* is

$$\hat{\mathbf{x}}_k(-) = \mathbf{\Phi}_{k-1}\hat{\mathbf{x}}_{k-1}(+) + \mathbf{\Gamma}_{k-1}\mathbf{u}_{k-1} \tag{4.3-12}$$

while the *covariance estimate extrapolation* (Eq. 4.2-15) is

$$\begin{aligned}\mathbf{P}_k(-) &= \mathbf{\Phi}_{k-1}\mathbf{P}_{k-1}(+)\mathbf{\Phi}_{k-1}^T + \mathbf{Q}_{k-1} \\ &= \mathbf{\Phi}_{k-1}\mathbf{P}_{k-1}(+)\mathbf{\Phi}_{k-1}^T + \mathbf{\Lambda}_{k-1}\mathbf{Q}'_{k-1}\mathbf{\Lambda}_{k-1}^T\end{aligned} \tag{4.3-13}$$

The uncorrected mean value and covariance estimates are propagated from the previous corrected estimates, with no modifications to the equations (other than the nomenclature). The recursive mean-value estimator provides the *filter gain computation* (Eq. 4.1-50),

$$\mathbf{K}_k = \mathbf{P}_k(-)\mathbf{H}_k^T[\mathbf{H}_k\mathbf{P}_k(-)\mathbf{H}_k^T + \mathbf{R}_k]^{-1} \tag{4.3-14}$$

where $\mathbf{P}_k(-)$ replaces $\mathbf{P}_{k-1}$ as the pre-update covariance estimate. Similarly, the *state estimate update* equation derives from Eq. 4.1-49 [with $\mathbf{x}_k(-)$ replacing $\mathbf{x}_{k-1}$],

$$\hat{\mathbf{x}}_k(+) = \hat{\mathbf{x}}_k(-) + \mathbf{K}_k[\mathbf{z}_k - \mathbf{H}_k\hat{\mathbf{x}}_k(-)] \tag{4.3-15}$$

and the *covariance estimate update* comes from Eq. 4.1-51:

$$\mathbf{P}_k(+) = [\mathbf{P}_k(-)^{-1} + \mathbf{H}_k^T\mathbf{R}_k^{-1}\mathbf{H}_k]^{-1} \tag{4.3-16}$$

The functional relationships of these equations are illustrated by the block diagrams of Fig. 4.3-2, where the recursive nature of the algorithm is clearly evident (Notice the "delay" blocks).* The physical and observation processes also are shown; both the known inputs to the system and the observations "drive" the state estimator. The optimal gain $\mathbf{K}_k$ depends only on the relative strengths of the *stochastic* input and the measurement, and its computation is entirely separate from the two state equations. Thus the deterministic input $\mathbf{u}_{k-1}$ and the measurement $\mathbf{z}_k$ have no effect on $\mathbf{K}_k$. In fact, the filter gain could be precomputed and stored as a function of k if the time variations in the system and its statistics were known; real-time implementation of the Kalman filter then would consist only of the computations shown in Fig. 4.3-2*a*, with the equations of Fig. 4.3-2*b* computed before state estimation begins.

*The terms in the blocks of Fig. 4.3-2*a* multiply their inputs in conventional fashion; the expressions in each block of Fig. 4.3-2*b* are the block outputs.

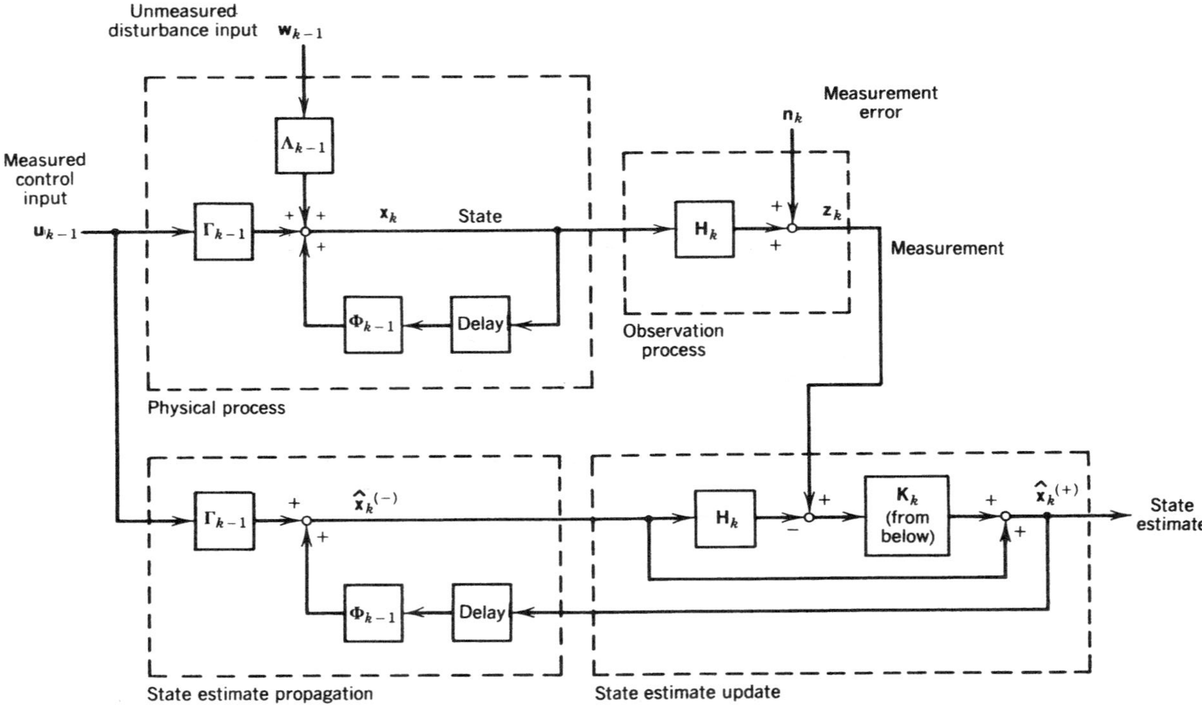

(a)

FIGURE 4.3-2 Discrete-time system and linear-optimal filter. (*a*) Dynamic system and state estimator.

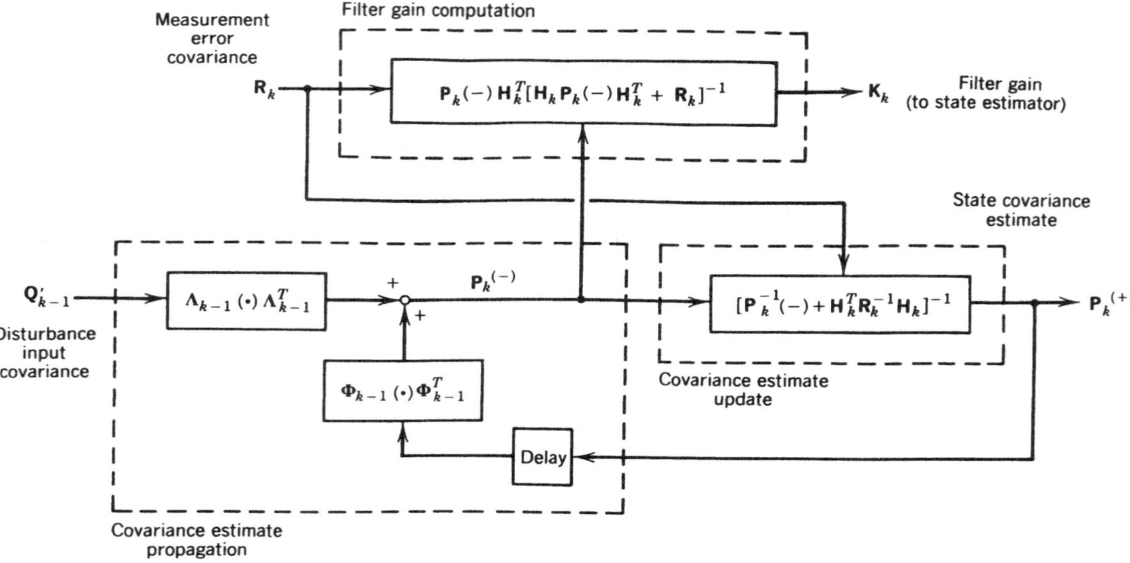

FIGURE 4.3-2 Discrete time system and linear-optimal filter. (*b*) Covariance estimator and gain computation.

The five Kalman filter equations (Eqs. 4.3-12 to 4.3-16) can be expressed as three equations, although this may not decrease the amount of computation needed. Substituting Eq. 4.3-13 in Eq. 4.3-16 provides a single *covariance estimation equation*:

$$\mathbf{P}_k(+) = \{[\mathbf{\Phi}_{k-1}\mathbf{P}_{k-1}(+)\mathbf{\Phi}_{k-1}^T + \mathbf{Q}_{k-1}]^{-1} + \mathbf{H}_k^T\mathbf{R}_k^{-1}\mathbf{H}_k\}^{-1} \quad (4.3\text{-}17)$$

The previous equation for $\mathbf{K}_k$ (Eq. 4.3-14) required the intermediate variable $\mathbf{P}_k(-)$, which does not appear in this solution; however, the gain computation can be reformulated in terms of $\mathbf{P}_k(+)$. Equation 4.3-14 can be manipulated as follows:

$$\begin{aligned}
\mathbf{K}_k &= \mathbf{P}_k(-)\mathbf{H}_k^T[\mathbf{H}_k\mathbf{P}_k(-)\mathbf{H}_k^T + \mathbf{R}_k]^{-1} \\
&= \mathbf{P}_k(-)\mathbf{H}_k^T\mathbf{R}_k^{-1}\mathbf{R}_k[\mathbf{H}_k\mathbf{P}_k(-)\mathbf{H}_k^T + \mathbf{R}_k]^{-1} \\
&= \mathbf{P}_k(-)\mathbf{H}_k^T\mathbf{R}_k^{-1}[\mathbf{I}_r + \mathbf{H}_k\mathbf{P}_k(-)\mathbf{H}_k^T\mathbf{R}_k^{-1}]^{-1} \qquad (4.3\text{-}18)
\end{aligned}$$

$$\mathbf{K}_k[\mathbf{I}_r + \mathbf{H}_k\mathbf{P}_k(-)\mathbf{H}_k^T\mathbf{R}_k^{-1}] = \mathbf{P}_k(-)\mathbf{H}_k^T\mathbf{R}_k^{-1} \qquad (4.3\text{-}19)$$

$$\begin{aligned}
\mathbf{K}_k &= \mathbf{P}_k(-)\mathbf{H}_k^T\mathbf{R}_k^{-1} - \mathbf{K}_k\mathbf{H}_k\mathbf{P}_k(-)\mathbf{H}_k^T\mathbf{R}_k^{-1} \\
&= (\mathbf{I}_n - \mathbf{K}_k\mathbf{H}_k)\mathbf{P}_k(-)\mathbf{H}_k^T\mathbf{R}_k^{-1} \qquad (4.3\text{-}20)
\end{aligned}$$

From the matrix inversion lemma (Eq. 2.2-67b), Eq. 4.3-16 can be written as

$$\begin{aligned}
\mathbf{P}_k(+) &= \mathbf{P}_k(-) - \mathbf{P}_k(-)\mathbf{H}_k^T[\mathbf{H}_k\mathbf{P}_k(-)\mathbf{H}_k^T + \mathbf{R}_k]^{-1}\mathbf{H}_k\mathbf{P}_k(-) \qquad (4.3\text{-}21a) \\
&= (\mathbf{I}_n - \mathbf{K}_k\mathbf{H}_k)\mathbf{P}_k(-) \qquad (4.3\text{-}21b)
\end{aligned}$$

with $\mathbf{K}_k$ expressed by Eq. 4.3-14. [Equation 4.3-21b implies that the update step reduces the magnitude of the covariance estimate, while the propagation step (Eq. 4.3-13) increases the magnitude.] Consequently, the *gain matrix equation* could be written in the alternate form

$$\mathbf{K}_k = \mathbf{P}_k(+)\mathbf{H}_k^T\mathbf{R}_k^{-1} \qquad (4.3\text{-}22)$$

Substituting Eq. 4.3-12 in Eq. 4.3-15 provides the third equation, the *state estimator*:

$$\hat{\mathbf{x}}_k(+) = [\mathbf{\Phi}_{k-1}\hat{\mathbf{x}}_{k-1}(+) + \mathbf{\Gamma}_{k-1}\mathbf{u}_{k-1}] + \mathbf{K}_k\{\mathbf{z}_k - \mathbf{H}_k[\mathbf{\Phi}_{k-1}\hat{\mathbf{x}}_{k-1}(+) + \mathbf{\Gamma}_{k-1}\mathbf{u}_{k-1}]\} \qquad (4.3\text{-}23)$$

The term in square brackets appears twice, but it is more efficient to compute it just once (i.e., to compute $\hat{\mathbf{x}}_k(+)$ in two steps as before).

EXAMPLE 4.3-1 WEATHERVANE ANGLE AND RATE ESTIMATION

The angle and rate of the weathervane considered in Example 4.2-3 can be estimated from noisy measurements of both the angle and the rate or from either measurement alone. As before, the system is examined at tenth-second intervals. Its state vector and state transition matrix are

$$\mathbf{x} = \begin{bmatrix} x_1 \\ x_2 \end{bmatrix} = \begin{bmatrix} \theta \\ \dot{\theta} \end{bmatrix} = \begin{bmatrix} \text{angle, deg} \\ \text{angular rate, deg/s} \end{bmatrix}$$

$$\mathbf{\Phi} = \begin{bmatrix} 0.8309 & 0.078 \\ -3.075 & 0.5372 \end{bmatrix}$$

and its input covariance matrix is

$$\mathbf{Q} = \omega_n^4 \begin{bmatrix} 0.0002 & 0.003 \\ 0.003 & 0.063 \end{bmatrix} = \begin{bmatrix} 0.31 & 4.67 \\ 4.67 & 97.99 \end{bmatrix}$$

$\mathbf{Q}$ is diagonalized to find the variances and modal matrix needed for simulation (Eqs. 4.2-85 and 2.2-75). Using single-precision arithmetic,

$$\mathbf{N} = \begin{bmatrix} 0.0873 & 0. \\ 0. & 98.21 \end{bmatrix}$$

$$\mathbf{D} = \begin{bmatrix} 0.999 & 0.476 \\ 0.476 & 0.999 \end{bmatrix}$$

and the weathervane motion is simulated by

$$\mathbf{x}_k = \mathbf{\Phi}\mathbf{x}_{k-1} + \mathbf{D}\mathbf{v}_{k-1}, \qquad \mathbf{x}_0 = \mathbf{0}$$

with

$$E[\mathbf{v}_k] = \mathbf{0}$$

$$E[\mathbf{v}_k \mathbf{v}_k^T] = \mathbf{N}$$

The state measurement is subject to error,

$$\mathbf{z}_k = \mathbf{H}\mathbf{x}_k + \mathbf{n}_k$$

with

$$E[\mathbf{n}_k] = \mathbf{0}$$

$$E[\mathbf{n}_k \mathbf{n}_k^T] = \mathbf{R}$$

$$\mathbf{R} = \begin{bmatrix} 0.1\ \text{deg}^2 & 0 \\ 0 & 0.1\ \text{deg}^2/\text{s}^2 \end{bmatrix}$$

Three measurement cases are considered:

1. Both measurements are used: $\mathbf{H}_1 = \mathbf{I}$, $\mathbf{R}_1$ as above
2. Angle measurement only: $\mathbf{H}_2 = [1 \quad 0]$, $\mathbf{R}_2 = 0.1$
3. Angle rate measurement only: $\mathbf{H}_3 = [0 \quad 1]$, $\mathbf{R}_3 = 0.1$

For all three cases, the covariance is calculated using Eqs. 4.3-13 and 4.3-21 with $\mathbf{P}_0 = \mathbf{0}$, and the gain matrix is calculated by Eq. 4.3-14. Because $\mathbf{\Phi}$, $\mathbf{D}$, $\mathbf{H}$, $\mathbf{N}$, and $\mathbf{R}$ are constant, $\mathbf{P}_k$ and $\mathbf{K}_k$ can be expected to reach steady-state values; evaluated to four digits, this takes about half a second in the first two cases and three seconds in the third case, with values shown in Table 4.3-1.

Two-second time histories are shown in the accompanying figure. Points at the sampling intervals are connected by straight lines to aid visualization of trends. The angle history (θ), angle measurement (θ_m), and Case 1 angle estimate (θ_1) are presented in Diagram 1. The estimate generally lies between the actual value and the measurement, indicating the improvement of the estimate over the measurement. Diagram 2 illustrates that the angle measurement alone provides an estimate (θ_2) that is nearly as good as the two-measurement estimate. The estimate based only on rate information (θ_3) captures the proper trends, but it drifts away from the actual value during this short simulation.

The rate measurement is virtually indistinguishable from the actual rate on the scale of Diagram 3 as are the two-measurement and rate-only estimates. Diagram 3 illustrates the rate history and the estimate based on angle information alone. As implied by p_{22}, the estimation error is very small for Cases 1 and 3; the large rate estimation error associated with Case 2 is corroborated by Diagram 3.

These are high-signal-to-noise-ratio examples, in that the levels of $\mathbf{x}$ are much larger than those of $\mathbf{n}$. Dividing $\mathbf{Q}$ by ω_n^4 provides three low-signal-

TABLE 4.3-1 High-Signal-to-Noise-Ratio Examples

	Case 1	Case 2	Case 3
p_{11}	0.0597	0.0893	0.8968
p_{12}	0.0018	0.752	0.0017
p_{22}	0.0999	60.68	0.0999
k_{11}	0.5972	0.8934	—
k_{12}	0.0183	—	0.0173
k_{21}	0.0183	7.52	—
k_{22}	0.9982	—	0.991

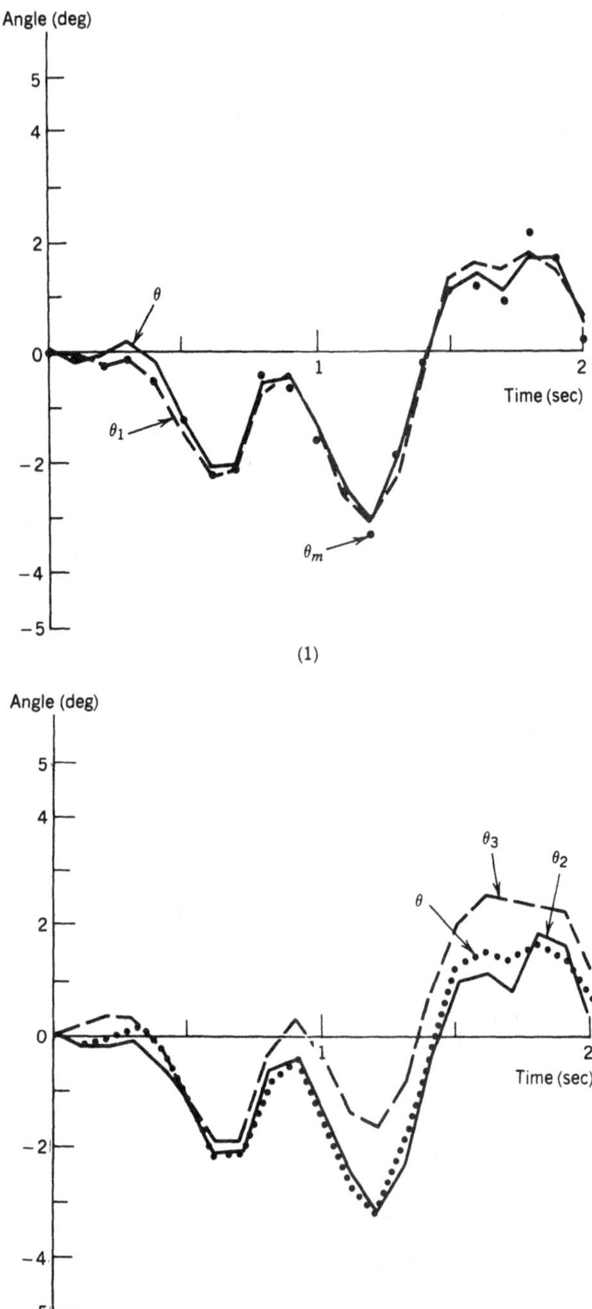

(1) Angle estimate with two measurements; (2) angle estimates with one measurement;

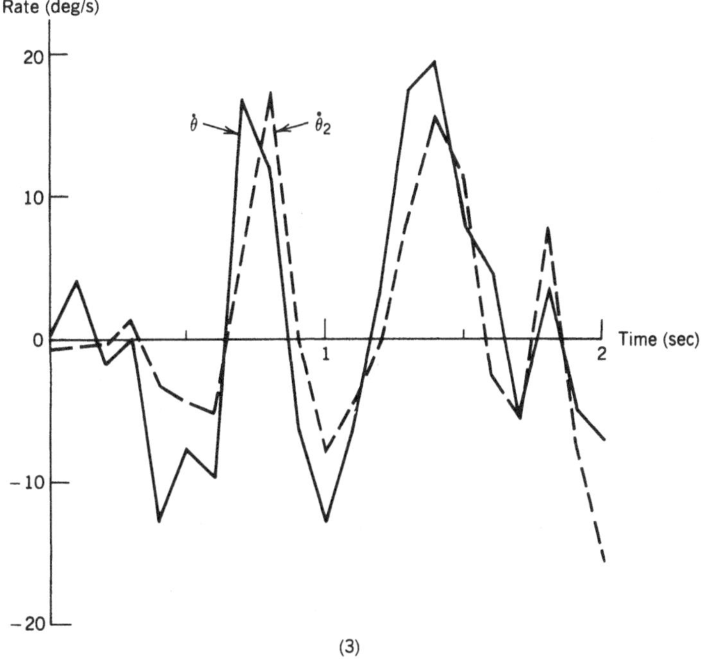

(3) angular rate estimate with angle measurement.

to-noise-ratio examples that can be compared with the previous results and with the "open-loop" uncertainty propagation of Example 4.2-3. Steady-state covariances and gains for the three measurement alternatives are given in Table 4.3-2. What implications can be drawn about the effect of the filter and the accuracy of the computation from this comparison?

TABLE 4.3-2 Low-Signal-to-Noise-Ratio Examples

	Case 1′	Case 2′	Case 3′
p_{11}	0.00158	0.003134	0.00164
p_{12}	0.00011	−0.00036	0.00003
p_{22}	0.04966	0.1511	0.04995
k_{11}	0.01584	0.03134	—
k_{12}	0.00109	—	0.00026
k_{21}	0.00109	−0.00365	—
k_{22}	0.4966	—	0.4995

Linear-Optimal Predictor

The propagation equations of Section 4.2 predict the state and its covariance on the basis of all the information that is available without measurement updating. A *linear-optimal predictor* extends the state and covariance estimates beyond the time of the last available measurement. It uses the results of linear-optimal filtering to provide the initial conditions for the propagation ahead in time; hence the prediction is "optimal" by the choice of optimal estimates for these initial conditions and by the use of the proper system model.

The predictor algorithm consists of the filter equations presented previously plus two additional equations for state and covariance propagation. Denoting the present sampling time by t_k and the time for which the prediction is made as t_K, the equation for *state prediction* is

$$\hat{\mathbf{x}}_K = \mathbf{\Phi}_k(t_K - t_k)\hat{\mathbf{x}}_k(+) + \mathbf{\Gamma}_k(\cdot)\mathbf{u}_k \tag{4.3-24}$$

$\mathbf{\Phi}(\cdot)$ is the state transition matrix for the prediction interval $(t_K - t_k)$ and $\mathbf{\Gamma}(\cdot)$ is the corresponding control effect matrix, $\hat{\mathbf{x}}(+)$ is the current linear-optimal filter estimate, and $\mathbf{u}_k$ is the current control, which is assumed to be held constant during the prediction interval.

The prediction interval need not be the same as the sampling interval of the optimal filter, and the deterministic control effect $\mathbf{\Gamma}_k(\cdot)\mathbf{u}_k$ can be specified differently. For example, it could be the integrated effect of a planned continuous control profile, as in Eq. 4.2-36.

The equation for discrete-time *covariance prediction* is

$$\mathbf{P}_K = \mathbf{\Phi}_k(t_K - t_k)\mathbf{P}_k(+)\mathbf{\Phi}_{k-1}^T(\cdot) + \mathbf{Q}_k(t_K - t_k) \tag{4.3-25}$$

where $\mathbf{Q}_k(\cdot)$ represents the covariance of uncertain inputs during the prediction interval, and other matrices are defined as before. It can be surmised from previous results that the uncertainty in the predicted state must be at least as great as the uncertainty in the present state, and it is likely to grow larger. Even noisy measurements tend to reduce the state estimate covariance below nonupdated values if the estimate is truly optimal, and the "known" $\mathbf{u}$ may not be certain in the future, contributing to higher values of $\mathbf{Q}_k$. Nevertheless, prediction has potential application in control when its merits outweigh the increased error. Examples include estimation of the future state for comparison with a target state and compensation for computation delay in a digital feedback control system.

Alternative Forms of the Linear-Optimal Filter

The two forms of the linear-optimal filter described above are conceptually straightforward, and they provide satisfactory results in many applications.

However, other forms may prove desirable for reducing computational requirements and/or improving numerical accuracy. These methods provide alternate means for propagating and updating the covariance matrix and for calculating the gain matrix; propagation and updating of the state vector proceed as before.

Matrix inversion is one of the most time-consuming operations demanded by the filter, and computational savings can be realized if the dimensions of variable matrices requiring inversion can be kept to a minimum. (Inverting constant matrices poses the same problem, but each inversion needs to be done only once, with the results stored for later use.) The original covariance update equation (Eq. 4.3-16) requires the inversion of two $(n \times n)$ matrices, while the second version (Eq. 4.3-21) requires a single inversion of an $(r \times r)$ matrix, where n and r are the dimensions of the state and observation vectors, respectively.

If the number of measurements is large, as in a massively redundant system, the original equation could prove more efficient than the second version. If the number of observations processed on any given update can be kept small, the second equation will be substantially faster than the first (i.e., it will require fewer arithmetic operations). Given r measurements with a diagonal covariance matrix $\mathbf{R}$, processing only one measurement at each sampling interval would allow Eqs. 4.3-14 and 4.3-21b to be written as

$$\mathbf{K}_{i_k} = \frac{\mathbf{P}_k(-)\mathbf{H}_{i_k}^T}{[\mathbf{H}_{i_k}\mathbf{P}_k(-)\mathbf{H}_{i_k}^T + r_{ik}]} \tag{4.3-26}$$

$$\mathbf{P}_k(+) = (\mathbf{I}_n - \mathbf{K}_{i_k}\mathbf{H}_{i_k})\mathbf{P}_k(-) \tag{4.3-27}$$

Note that the term in square brackets (Eq. 4.3-26) is a scalar. From Eq. 4.3-15, the corresponding state update equation would be

$$\hat{\mathbf{x}}_k(+) = \hat{\mathbf{x}}_k(-) + \mathbf{K}_{i_k}[z_{i_k} - \mathbf{H}_{i_k}\hat{\mathbf{x}}_k(-)] \tag{4.3-28}$$

Only the ith element of the observation vector $\mathbf{z}$ is considered, appearing as z_{i_k} in Eq. 4.3-28, and the gain matrix has dimension $(n \times 1)$. The observation matrix $\mathbf{H}_{i_k}$ is the appropriate row of $\mathbf{H}_k$, and its error variance is r_{i_k}. This example of *sequential processing* could be used without modification to minimize computation time, with state and covariance extrapolations performed by Eqs. 4.3-12 and 4.3-13. If, however, more data are available at each sampling instant, the method has the disadvantage of ignoring potentially useful information.

Sequential processing can be applied to *all* the available data at a sampling instant, overcoming the latter objection. In effect, the three preceding equations are repeated in sequence at each sampling instant (as if one measurement incrementally followed another) before proceeding to the next instant. Envision the procedure as a computer "DO loop" for $i = 1$ to r

with the following equations:

$$\mathbf{K}_{i_k} = \frac{\mathbf{P}_{i-1_k}(+)\mathbf{H}_{i_k}^T}{\mathbf{H}_i\mathbf{P}_{i-1_k}(+)\mathbf{H}_{i_k}^T + r_{i_k}} \tag{4.3-29}$$

$$\mathbf{P}_{i_k}(+) = (\mathbf{I}_n - \mathbf{K}_{i_k}\mathbf{H}_{i_k})\mathbf{P}_{i-1_k}(+), \qquad \mathbf{P}_{0_k}(+) = \mathbf{P}_k(-) \tag{4.3-30}$$

$$\hat{\mathbf{x}}_{i_k}(+) = \hat{\mathbf{x}}_{i-1_k}(+) + \mathbf{K}_{i_k}[z_{i_k} - \mathbf{H}_{i_k}\hat{\mathbf{x}}_{i-1_k}(+)], \qquad \hat{\mathbf{x}}_{0_k}(+) = \hat{\mathbf{x}}_k(-) \tag{4.3-31}$$

and $\mathbf{P}_k(+) = \mathbf{P}_{r_k}(+)$, $\hat{\mathbf{x}}_k(+) = \hat{\mathbf{x}}_{r_k}(+)$. Further simplification can occur by using Eq. 4.3-22 to bring the state update outside the "DO loop,"

$$\hat{\mathbf{x}}_k(+) = \hat{\mathbf{x}}_k(-) + \mathbf{P}_k(+)\mathbf{H}_k^T\mathbf{R}_k^{-1}[\mathbf{z}_k - \mathbf{H}_k\hat{\mathbf{x}}_k(-)] \tag{4.3-32}$$

reserving sequential processing for the calculation of $\mathbf{P}_k(+)$.

Sequential processing requires diagonal $\mathbf{R}$; if $\mathbf{R}$ is symmetric but not diagonal, the diagonalization and factorization methods described in Section 4.2 could be applied to find a linear transformation of the measurements $\boldsymbol{\xi}$, whose error covariance is diagonal. With the transformation matrix $\mathbf{C}$

$$\begin{aligned} \boldsymbol{\xi} &= \mathbf{Cz} = \mathbf{CHx} + \mathbf{Cn} \\ &= \mathbf{H}'\mathbf{x} + \boldsymbol{\mu} \end{aligned} \tag{4.3-33}$$

where

$$E(\boldsymbol{\mu}) = E(\mathbf{Cn}) = \mathbf{0} \tag{4.3-34}$$

$$E(\boldsymbol{\mu}\boldsymbol{\mu}^T) = \mathbf{N} = E(\mathbf{Cnn}^T\mathbf{C}^T) = \mathbf{CRC}^T \tag{4.3-35}$$

As before, there are two alternatives. Defining $\mathbf{N}$ as the diagonal matrix of $\mathbf{R}$'s eigenvalues, $\mathbf{C}$ is the modal matrix inverse. (Since $\mathbf{R}$ is real and symmetric, $\mathbf{C}^{-1} = \mathbf{C}^T$.) Defining $\mathbf{N}$ as an identity matrix, $\mathbf{C}$ is $\mathbf{S}^{-1}$, where $\mathbf{S}$ is the lower-triangular square root of $\mathbf{R}$. ($\mathbf{S}^{-1} \neq \mathbf{S}^T$, as $\mathbf{S}$ is not orthonormal.) $\boldsymbol{\xi}$, $\mathbf{H}'$, and $\mathbf{N}$ then replace $\mathbf{z}$, $\mathbf{H}$, and $\mathbf{R}$ in the filter equations.

Retaining the positive-definiteness and symmetry of $\mathbf{P}_k(+)$ can be a numerical problem when Eq. 4.3-21b is used for the covariance update. [The symmetry problem can be avoided by using Eq. 4.3-21a (i.e., by not substituting $\mathbf{K}_k$ in the equation). Then it is appropriate to calculate $\mathbf{K}_k$ using Eq. 4.3-22.] The *Joseph form* of the covariance update (also called the "stabilized Kalman form") assures positive-definiteness and symmetry. Equation 4.3-15 could be written as

$$\hat{\mathbf{x}}_k(+) = (\mathbf{I}_n - \mathbf{K}_k\mathbf{H}_k)\hat{\mathbf{x}}_k(-) + \mathbf{K}_k\mathbf{z}_k \tag{4.3-36}$$

Defining pre- and postupdate estimation errors as

$$\boldsymbol{\epsilon}_k(-) = \mathbf{x}_k - \hat{\mathbf{x}}_k(-) \tag{4.3-37}$$

$$\boldsymbol{\epsilon}_k(+) = \mathbf{x}_k - \hat{\mathbf{x}}_k(+) \tag{4.3-38}$$

and recalling Eq. 4.3-7, Eq. 4.3-36 indicates that the estimation error propagates as

$$\boldsymbol{\epsilon}_k(+) = (\mathbf{I}_n - \mathbf{K}_k\mathbf{H}_k)\boldsymbol{\epsilon}_k(-) - \mathbf{K}_k\mathbf{n}_k \tag{4.3-39}$$

Because

$$E[\boldsymbol{\epsilon}_k\boldsymbol{\epsilon}_k^T] = \mathbf{P}_k \tag{4.3-40}$$

$$E[\mathbf{n}_k\mathbf{n}_k^T] = \mathbf{R}_k \tag{4.3-41}$$

$$E[\boldsymbol{\epsilon}_k\mathbf{n}_k^T] = \mathbf{0} \tag{4.3-42}$$

the covariance update equation can be derived by multiplying the right- and left-hand sides of Eq. 4.3-39 by their respective transposes and taking expected values. Then

$$\mathbf{P}_k(+) = (\mathbf{I}_n - \mathbf{K}_k\mathbf{H}_k)\mathbf{P}_k(-)(\mathbf{I}_n - \mathbf{K}_k\mathbf{H}_k)^T + \mathbf{K}_k\mathbf{R}_k\mathbf{K}_k^T \tag{4.3-43}$$

which is seen to be a symmetric equation that will remain positive-definite as long as $\mathbf{P}_k(-)$ is positive-definite and $\mathbf{R}_k$ is positive-semidefinite. Equation 4.3-43 clearly requires more computation than Eq. 4.3-21b, and sequential processing could be used to reduce the burden. Equations 4.3-39 and 4.3-43 do not depend on $\mathbf{K}_k$ being an optimal gain; hence the Joseph form can be used in covariance error analyses of suboptimal filters. This is of particular value in evaluating the effects of optimal gain calculation errors on system performance and in designing reduced-order filters.

It may be recalled that the original covariance update equation (Eq. 4.3-16) requires the inversion of two $(n \times n)$ matrices, but the *inverse* of the covariance could be updated without further inversion. The inverse covariance matrix (or *information matrix*) is defined as

$$\mathcal{I} = \mathbf{P}^{-1} \tag{4.3-44}$$

and Eq. 4.3-16 gives its *update equation* as

$$\mathcal{I}_k(+) = \mathcal{I}_k(-) + \mathbf{H}_k^T\mathbf{R}_k^{-1}\mathbf{H}_k \tag{4.3-45}$$

Using the matrix inversion lemma, the *propagation equation* can be written as (A-1),

$$\mathscr{I}_k(-) = [\mathbf{I}_n - \mathbf{B}_{k-1}\mathbf{\Lambda}_{k-1}^T]\mathbf{A}_{k-1} \tag{4.3-46}$$

where

$$\mathbf{A}_{k-1} = [\mathbf{\Phi}_{k-1}^{-1}]^T \mathscr{I}_{k-1}(+)\mathbf{\Phi}_{k-1}^{-1} \tag{4.3-47}$$

$$\mathbf{B}_{k-1} = \mathbf{A}_{k-1}\mathbf{\Lambda}_{k-1}[\mathbf{\Lambda}_{k-1}^T\mathbf{A}_{k-1}\mathbf{\Lambda}_{k-1} + \mathbf{Q}_{k-1}'^{-1}]^{-1} \tag{4.3-48}$$

and

$$\mathscr{I}_0 = \mathbf{P}_0^{-1} \tag{4.3-49}$$

It is necessary to invert $\mathbf{\Phi}$, $\mathbf{Q}'$, and the $(s \times s)$ matrix of Eq. 4.3-48 to estimate the information matrix. Thus the update step is simplified but the propagation step is complicated by using the information matrix instead of the covariance matrix. From Eq. 4.3-22, the state estimate *gain matrix* could be expressed as

$$\mathbf{K}_k = \mathscr{I}_k^{-1}(+)\mathbf{H}_k^T\mathbf{R}_k^{-1} \tag{4.3-50}$$

requiring yet another inversion. The *information filter* (or inverse covariance filter) described by these equations may be more efficient than the conventional Kalman filter in problems that involve large measurement vectors. If there is no initial uncertainty about the state, $\mathbf{P}_0 = \mathbf{0}$, and the filter cannot be initialized. In such case, it would be necessary to begin estimation with a conventional Kalman filter.

Computational difficulty can be expected when the filter equations are *ill-conditioned*, that is, when there is a large disparity in the sensitivity of solutions to perturbations in initial conditions, inputs, and measurement errors. Ill-conditioning can occur when the system dynamics contain both slow and fast modes, when both noisy and near-perfect measurements must be processed at the same time, or when disturbance inputs force some modes more strongly than others.

One measure of the potential for difficulty is the *condition number*, $\kappa(\mathbf{P})$, which is defined as

$$\kappa(\mathbf{P}) = \left(\frac{\lambda_{\max}}{\lambda_{\min}}\right)^{1/2} \tag{4.3-51}$$

where $\lambda_{\max}$ and $\lambda_{\min}$ are the maximum and minimum eigenvalues of $\mathbf{PP}^T$, and their square roots are known as *singular values*. If $\kappa(\mathbf{P})$ is large [on the order of 10^x, denoted by $\mathcal{O}(10^x)$, where x is the number of significant digits in a computer word], computational difficulty can be expected, as very large and very small numbers must be represented in simultaneous solutions. Conditioning is improved by deriving estimation equations that solve for and use the *square roots* of **P** or $\mathcal{I}$ rather than **P** or $\mathcal{I}$ directly. Defining the square root of **P** as in Section 4.2,

$$\kappa(\mathbf{P}) = \kappa(\mathbf{S}\mathbf{S}^T) = [\kappa(\mathbf{S})]^2 \tag{4.3-52}$$

or

$$\kappa(\mathbf{S}) = [\kappa(\mathbf{P})]^{1/2} \tag{4.3-53}$$

If $\kappa(\mathbf{P})$ is $\mathcal{O}(10^x)$, then $\kappa(\mathbf{S})$ is $\mathcal{O}(10^{x/2})$, implying that the square-root formulation would require half the significant digits of the covariance formulation. If $\kappa(\mathbf{P})$ is $\mathcal{O}(1)$, little is gained by using the square-root filter.

A second useful property of square-root formulations is that **P** equals $\mathbf{SS}^T$ and is always positive-semidefinite, even when **S** is not. Finite word-length computation errors can cause the estimate of **P** generated by some implementations of the conventional or stabilized Kalman equations to become nonpositive-definite. This is avoided in square-root formulations at the expense of increased computer burden: the algorithms typically require 50–150% more time than that required by conventional Kalman filters.

Rather than present any of the square-root algorithms, we will skip to the *U-D filter*, which is based on the "U-D" factorization of the covariance matrix. This method has its origins in and many properties of square-root filters, but it is computationally more efficient, approaching the conventional Kalman filter in timing and storage requirements. The principal distinction is that the covariance matrix is factored as

$$\mathbf{P} = \mathbf{U}\mathbf{D}\mathbf{U}^T \tag{4.3-54}$$

U is an $(n \times n)$ unitary upper triangular matrix, and **D** is an $(n \times n)$ diagonal matrix, so the factorization can be likened to

$$\begin{aligned}\mathbf{P} &= (\mathbf{U}\mathbf{D}^{1/2})(\mathbf{U}\mathbf{D}^{1/2})^T \\ &= \mathbf{S}\mathbf{S}^T\end{aligned} \tag{4.3-55}$$

where **S** is an upper triangular square root of **P**. Nevertheless, the factorization is accomplished without taking square roots, which is one reason for the algorithm's efficiency.

U-D factorization is applied to both the covariance propagation and update equations, and the state estimation gain matrix is expressed in terms of these results. As the original motivation was to improve the covariance update, this aspect is considered first. The derivation is based on sequential processing of scalar measurements, which is more efficient than vector processing when $r < n$ for reasons cited earlier. For a scalar measurement,

z_i, the covariance update equation (Eq. 4.3-21a) is

$$\mathbf{P}_{i_k}(+) = \mathbf{P}_{i-1_k}(+) - \frac{\mathbf{P}_{i-1_k}(+)\mathbf{H}_{i_k}^T\mathbf{H}_{i_k}\mathbf{P}_{i-1_k}(+)}{\alpha_{i_k}} \tag{4.3-56}$$

where $\mathbf{H}_{i_k}$ is the ith row of $\mathbf{H}_k$, and

$$\alpha_{i_k} = \mathbf{H}_{i_k}\mathbf{P}_{i-1_k}(+)\mathbf{H}_{i_k}^T + r_{i_k} \tag{4.3-57}$$

is a scalar variable. Substituting Eq. 4.3-54 for $\mathbf{P}$ and temporarily dropping the subscripts for clarity,

$$\begin{aligned}
&\mathbf{U}(+)\mathbf{D}(+)\mathbf{U}^T(+) \\
&= \mathbf{U}(-)\mathbf{D}(-)\mathbf{U}^T(-) - \frac{[\mathbf{U}(-)\mathbf{D}(-)\mathbf{U}^T(-)\mathbf{H}^T][\mathbf{H}\mathbf{U}(-)\mathbf{D}(-)\mathbf{U}^T(-)]}{\alpha} \\
&= \mathbf{U}(-)\left\{\mathbf{D}(-) - \frac{[\mathbf{D}(-)\mathbf{U}^T(-)\mathbf{H}^T][\mathbf{D}(-)\mathbf{U}^T(-)\mathbf{H}^T]^T}{\alpha}\right\}\mathbf{U}^T(-)
\end{aligned} \tag{4.3-58}$$

Defining

$$\mathbf{f} = \mathbf{U}^T(-)\mathbf{H}^T \tag{4.3-59}$$

$$\mathbf{v} = \mathbf{D}(-)\mathbf{f} \tag{4.3-60}$$

this can be written as

$$\begin{aligned}
\mathbf{U}(+)\mathbf{D}(+)\mathbf{U}^T(+) &= \mathbf{U}(-)\left\{\mathbf{D}(-) - \frac{[\mathbf{D}(-)\mathbf{f}][\mathbf{D}(-)\mathbf{f}]^T}{\alpha}\right\}\mathbf{U}^T(-) \\
&= \mathbf{U}(-)\left[\mathbf{D}(-) - \frac{\mathbf{v}\mathbf{v}^T}{\alpha}\right]\mathbf{U}^T(-)
\end{aligned} \tag{4.3-61}$$

The matrix $[\mathbf{D}(-) - \mathbf{v}\mathbf{v}^T/\alpha]$ is symmetric, so it, too, can be factored as a product of unitary upper triangular and diagonal matrices $\mathbf{W}$ and $\mathbf{X}$:

$$\mathbf{W}\mathbf{X}\mathbf{W}^T = \left[\mathbf{D}(-) - \frac{\mathbf{v}\mathbf{v}^T}{\alpha}\right] \tag{4.3-62}$$

Therefore,

$$\mathbf{U}(+)\mathbf{D}(+)\mathbf{U}^T(+) = \mathbf{U}(-)\mathbf{W}\mathbf{X}\mathbf{W}^T\mathbf{U}^T(-) \tag{4.3-63}$$

and equating like terms,

$$\mathbf{U}_{i_k}(+) = \mathbf{U}_{i-1_k}(+)\mathbf{W}_{i_k} \tag{4.3-64}$$

$$\mathbf{D}_{i_k}(+) = \mathbf{X}_{i_k} \tag{4.3-65}$$

with $\mathbf{U}_{0_k}(+) = \mathbf{U}_k(-)$ and $\mathbf{D}_{0_k}(+) = \mathbf{D}_k(-)$, formally completing the *factor update*, which is equivalent to the covariance update. For r-vector measurements, these equations must be repeated r times, as in the earlier sequential processing application.

From Eq. 4.3-14 and the preceding equations, the *gain computation* is

$$\begin{aligned}\mathbf{K}_{i_k} &= \frac{\mathbf{U}_{i_k}(-)\mathbf{D}_{i_k}(-)\mathbf{U}_{i_k}^T(-)\mathbf{H}_{i_k}^T}{\alpha_{i_k}} \\ &= \frac{\mathbf{U}_{i_k}(-)\mathbf{D}_{i_k}(-)\mathbf{f}_{i_k}}{\alpha_{i_k}} \\ &= \frac{\mathbf{U}_{i_k}(-)\mathbf{v}_{i_k}}{\alpha_{i_k}}\end{aligned} \tag{4.3-66}$$

The gain is calculated in r steps, and as before, the gain computation does not depend explicitly on the factor update.

The covariance propagation (Eq. 4.3-13) can be written as

$$\begin{aligned}\mathbf{U}_k(-)\mathbf{D}_k(-)\mathbf{U}_k^T(-) = {} & \mathbf{\Phi}_{k-1}\mathbf{U}_{k-1}(+)\mathbf{D}_{k-1}(+)\mathbf{U}_{k-1}^T(+)\mathbf{\Phi}_{k-1}^T \\ & + \mathbf{\Lambda}_{k-1}\mathbf{Q}_{k-1}'^{-1}\mathbf{\Lambda}_{k-1}^T\end{aligned} \tag{4.3-67}$$

with $\mathbf{\Lambda}$ defined so that $\mathbf{Q}'$ is diagonal. The right side can be factored as $\mathbf{YZY}^T$, where

$$\mathbf{Y} = [\mathbf{\Phi}_{k-1}\mathbf{U}_{k-1}(+) \quad \mathbf{\Lambda}_{k-1}] \tag{4.3-68}$$

$$\mathbf{Z} = \begin{bmatrix}\mathbf{D}_{k-1}(+) & \mathbf{0} \\ \mathbf{0} & \mathbf{Q}_{k-1}'^{-1}\end{bmatrix} \tag{4.3-69}$$

Although the factorization has the same general form as Eq. 4.3-54, it does not have the same dimensions: $\mathbf{Y}$ is an $n \times (n+s)$ matrix, and $\mathbf{Z}$ is an $(n+s) \times (n+s)$ matrix.

The problem is to map $\mathbf{Y}$ and $\mathbf{Z}$ to the desired $(n \times n)$ unitary upper triangular and diagonal matrices, and a *weighted Gram–Schmidt orthogonalization* procedure serves this purpose. Define the $(n+s)$-vector $\mathbf{y}_i$ as the ith column of $\mathbf{Y}^T$, for $i = 1$–n. A corresponding "Z-orthogonal" set of vectors, $\mathbf{b}_i$, $i = 1 - n$, is defined as follows:

$$\begin{aligned}
\mathbf{b}_n &= \mathbf{y}_n \\
\mathbf{b}_{n-1} &= \mathbf{y}_{n-1} - \frac{\mathbf{y}_{n-1}^T \mathbf{Z} \mathbf{b}_n}{\mathbf{b}_n^T \mathbf{Z} \mathbf{b}_n} \mathbf{b}_n \\
&\vdots \\
\mathbf{b}_1 &= \mathbf{y}_1 - \sum_{k=2}^{n} \frac{\mathbf{y}_1^T \mathbf{Z} \mathbf{b}_k}{\mathbf{b}_k^T \mathbf{Z} \mathbf{b}_k} \mathbf{b}_k
\end{aligned} \tag{4.3-70}$$

Dropping the sampling index once again, the elements of $\mathbf{D}(-)$ and $\mathbf{U}(-)$ are

$$d_{jj} = \mathbf{b}_j^T \mathbf{Z} \mathbf{b}_j, \qquad j = 1, \ldots, n \tag{4.3-71}$$

$$u_{ij} = \begin{cases} (\mathbf{y}_i^T \mathbf{Z} \mathbf{b}_j)/d_{jj}, & i = 1, \ldots, j-1 \\ 0, & i > j \end{cases} \tag{4.3-72}$$

completing the $\mathbf{U}(-)$ and $\mathbf{D}(-)$ *propagation*.

The $\mathbf{U}$ and $\mathbf{D}$ factors corresponding to $\mathbf{P}$ are needed to specify initial conditions. The algorithm is similar to the Cholesky decomposition (Eq. 4.2-96). Beginning with the nth column,

$$\begin{aligned}
d_{nn} &= p_{nn} \\
u_{in} &= \begin{cases} 1, & i = n \\ p_{in}/d_{nn}, & i = 1, \ldots, n-1 \end{cases}
\end{aligned} \tag{4.3-73}$$

while for $j = (n-1)$ to 1,

$$\begin{aligned}
d_{jj} &= p_{jj} - \sum_{k=j+1}^{n} d_{kk} u_{jk}^2 \\
u_{ij} &= \begin{cases} 0, & i > j \\ 1, & i = j \\ \dfrac{1}{d_{jj}} \left[p_{ij} - \displaystyle\sum_{k=j+1}^{n} d_{kk} u_{ik} u_{jk} \right], & i = j-1, \ldots, 1 \end{cases}
\end{aligned} \tag{4.3-74}$$

This decomposition also must be applied repeatedly to obtain $\mathbf{W}$ and $\mathbf{X}$ (Eq. 4.3-62). Efficient algorithms for the U-D filter equations, including Fortran mechanizations, are contained in (B-4).

4.4 CORRELATED DISTURBANCE INPUTS AND MEASUREMENT NOISE

The filter equations presented in Section 4.3 can account for time-correlation in the disturbance inputs through the use of shaping filters (Section 4.2), but they do not account for time correlation in the measurement noise or for cross-correlation in the disturbance inputs and measurement noise. The former case occurs when the measurement noise is "colored," (i.e., when its spectrum is band-limited or the noise itself is the output of a dynamic process). The latter case occurs when the disturbances that force the system are sensed directly in the observations. Whereas the filter state dimension must be increased to account for disturbance time-correlation, the discrete-time minimum-variance filters for disturbance-measurement correlation or measurement time-correlation retain the dimension of the original state.

Cross-Correlation of Disturbance Input and Measurement Noise

We begin with the discrete-time system

$$\mathbf{x}_k = \mathbf{\Phi}_{k-1}\mathbf{x}_{k-1} + \mathbf{\Gamma}_{k-1}\mathbf{u}_{k-1} + \mathbf{w}_{k-1}, \qquad \mathbf{x}_0 \text{ given} \tag{4.4-1}$$

$$\mathbf{z}_k = \mathbf{H}_{\mathbf{x}_k}\mathbf{x}_k + \mathbf{H}_{\mathbf{u}_k}\mathbf{u}_k + \mathbf{n}_k \tag{4.4-2}$$

which provides deterministic and random forcing of the dynamics through $\mathbf{u}_{k-1}$ and $\mathbf{w}_{k-1}$, and which allows the observation to be a linear function of the control as well as the state. From previous results, we can expect a *linear minimum-variance filter* for this system to take the form

$$\hat{\mathbf{x}}_k(-) = \mathbf{\Phi}_{k-1}\hat{\mathbf{x}}_{k-1}(+) + \mathbf{\Gamma}_{k-1}\mathbf{u}_{k-1}, \qquad \hat{\mathbf{x}}_0(+) \text{ given} \tag{4.4-3}$$

$$\hat{\mathbf{x}}_k(+) = \hat{\mathbf{x}}_k(-) + \mathbf{K}_k[\mathbf{z}_k - \mathbf{H}_{\mathbf{x}_k}\hat{\mathbf{x}}_k(-) - \mathbf{H}_{\mathbf{u}_k}\mathbf{u}_k] \tag{4.4-4}$$

where the optimal filter gain $\mathbf{K}_k$ is to be determined, $\hat{\mathbf{x}}_k(-)$ is the state estimate before the measurement update, and $\hat{\mathbf{x}}_k(+)$ is the estimate after the update. Defining equivalent state residuals as

$$\boldsymbol{\epsilon}_k(-) = \mathbf{x}_k - \hat{\mathbf{x}}_k(-) \tag{4.4-5}$$

$$\boldsymbol{\epsilon}_k(+) = \mathbf{x}_k - \hat{\mathbf{x}}_k(+) \tag{4.4-6}$$

the dynamics of the estimation error can be described by subtracting Eq.

4.4-3 from Eq. 4.4-1 and substituting Eq. 4.4-2 in Eq. 4.4-4:

$$\boldsymbol{\epsilon}_k(-) = \boldsymbol{\Phi}_{k-1}\boldsymbol{\epsilon}_{k-1}(+) + \mathbf{w}_{k-1} \tag{4.4-7}$$

$$\boldsymbol{\epsilon}_k(+) = \boldsymbol{\epsilon}_k(-) - \mathbf{K}_k[\mathbf{H}_{\mathbf{x}_k}\boldsymbol{\epsilon}_k(-) + \mathbf{n}_k] \tag{4.4-8}$$

Thus the known deterministic control input has no effect on the estimation error, and $\mathbf{H}_{\mathbf{x}_k}$ is rewritten as $\mathbf{H}_k$ for simplicity.

The disturbance input and measurement noise are modeled as a time-skewed, white joint stochastic process such that

$$E\left\{\begin{bmatrix}\mathbf{w}_{k-1}\\ \mathbf{n}_k\end{bmatrix}[\mathbf{w}_{k-1}^T \quad \mathbf{n}_k^T]\right\} = \begin{bmatrix}\mathbf{Q}_{k-1} & \mathbf{M}_k\\ \mathbf{M}_k^T & \mathbf{R}_k\end{bmatrix} \tag{4.4-9}$$

and

$$E\begin{bmatrix}\mathbf{w}_{k-1}\\ \mathbf{n}_k\end{bmatrix} = \begin{bmatrix}\mathbf{0}\\ \mathbf{0}\end{bmatrix} \tag{4.4-10}$$

Note also that $E[\mathbf{x}\mathbf{w}^T] = \mathbf{0}$ and $E[\mathbf{x}\mathbf{n}^T] = \mathbf{0}$ for all sampling indices.

$\mathbf{Q}_{k-1}$ and $\mathbf{R}_k$ are defined as before, and $\mathbf{M}_k$ expresses the cross-correlation between the disturbance at $(k-1)$ and the measurement at k. The rationale for using this model is that the measurement residual of Eq. 4.4-8 includes $\mathbf{w}_{k-1}$ [substituting Eq. 4.4-7 for $\boldsymbol{\epsilon}_k(-)$] and $\mathbf{n}_k$. Hence the correlation between these two terms (rather than between $\mathbf{w}_k$ and $\mathbf{n}_k$) is of principal concern in the estimation.

The covariances of the state estimate error before and after the measurement update are expressed by forming the outer product of Eqs. 4.4-7 and 4.4-8 and taking expected values:

$$\begin{aligned}E[\boldsymbol{\epsilon}_k(-)\boldsymbol{\epsilon}_k(-)^T] &= \mathbf{P}_k(-)\\ &= E[\boldsymbol{\Phi}_{k-1}\boldsymbol{\epsilon}_{k-1}(+)\boldsymbol{\epsilon}_{k-1}^T(+)\boldsymbol{\Phi}_{k-1}^T + \mathbf{w}_{k-1}\mathbf{w}_{k-1}^T]\\ &= \boldsymbol{\Phi}_{k-1}\mathbf{P}_{k-1}(+)\boldsymbol{\Phi}_{k-1}^T + \mathbf{Q}_{k-1}, \qquad \mathbf{P}_0(+) \text{ given}\end{aligned} \tag{4.4-11}$$

$$\begin{aligned}E[\boldsymbol{\epsilon}_k(+)\boldsymbol{\epsilon}_k(+)^T] &= \mathbf{P}_k(+)\\ &= E\{(\boldsymbol{\epsilon}_k(-) - \mathbf{K}_k[\mathbf{H}_k\boldsymbol{\epsilon}_k(-) + \mathbf{n}_k])(\cdot)^T\}\\ &= \mathbf{P}_k - E[\boldsymbol{\epsilon}_k(-)\boldsymbol{\epsilon}_k(-)^T\mathbf{H}_k^T\mathbf{K}_k^T + \boldsymbol{\epsilon}_k(-)\mathbf{n}_k^T\mathbf{K}_k^T]\\ &\quad - E[\mathbf{K}_k\mathbf{H}_k\boldsymbol{\epsilon}_k(-)\boldsymbol{\epsilon}_k(-)^T + \mathbf{K}_k\mathbf{n}_k\boldsymbol{\epsilon}_k(-)^T]\\ &\quad + E[\mathbf{K}_k\mathbf{H}_k\boldsymbol{\epsilon}_k(-)\boldsymbol{\epsilon}_k(-)^T\mathbf{H}_k^T\mathbf{K}_k^T\\ &\quad + \mathbf{K}_k\mathbf{H}_k\boldsymbol{\epsilon}_k(-)\mathbf{n}_k^T\mathbf{K}_k^T + \mathbf{K}_k\mathbf{n}_k\boldsymbol{\epsilon}_k^T(-)\mathbf{H}_k^T\mathbf{K}_k^T\\ &\quad + \mathbf{K}_k\mathbf{n}_k\mathbf{n}_k^T\mathbf{K}_k^T]\end{aligned} \tag{4.4-12}$$

Substituting Eq. 4.4-7 in Eq. 4.4-12 and moving the expectation operation inside the deterministic system matrices, the postupdate covariance can be written as

$$\begin{aligned}\mathbf{P}_k(+) = (\mathbf{I}_n - \mathbf{K}_k\mathbf{H}_k)\mathbf{P}_k(-)(\mathbf{I}_n - \mathbf{K}_k\mathbf{H}_k)^T + \mathbf{K}_k\mathbf{R}_k\mathbf{K}_k^T \\ + \mathbf{K}_k(\mathbf{H}_k\mathbf{M}_k + \mathbf{M}_k^T\mathbf{H}_k^T)\mathbf{K}_k^T - \mathbf{M}_k\mathbf{K}_k^T - \mathbf{K}_k\mathbf{M}_k^T \qquad (4.4\text{-}13)\end{aligned}$$

This is written in the Joseph form (Eq. 4.3-43), and it does not rely on $\mathbf{K}_k$ being an optimal gain.

The optimal gain matrix minimizes the expected value of the state-residual-squared at each step; hence the following cost function is to be minimized by the choice of $\mathbf{K}_k$:

$$J_k = E[\boldsymbol{\epsilon}_k(+)^T\boldsymbol{\epsilon}_k(+)] = \mathrm{Tr}[\mathbf{P}_k(+)] \qquad (4.4\text{-}14)$$

The minimizing value occurs when

$$\frac{\partial J_k}{\partial \mathbf{K}_k} = \mathbf{0} \qquad (4.4\text{-}15)$$

Consequently, from Eqs. 2.2-53 and 2.2-54,

$$\frac{\partial J_k}{\partial \mathbf{K}_k} = 2\{\mathbf{K}_k[\mathbf{H}_k\mathbf{P}_k(-)\mathbf{H}_k^T + \mathbf{H}_k\mathbf{M}_k + \mathbf{M}_k^T\mathbf{H}_k^T + \mathbf{R}_k] - \mathbf{P}_k(-)\mathbf{H}_k^T - \mathbf{M}_k\} = \mathbf{0} \qquad (4.4\text{-}16)$$

and the optimal gain for Eq. 4.4-4 is

$$\mathbf{K}_k = [\mathbf{P}_k(-)\mathbf{H}_k^T + \mathbf{M}_k][\mathbf{H}_k\mathbf{P}_k(-)\mathbf{H}_k^T + \mathbf{H}_k\mathbf{M}_k + \mathbf{M}_k^T\mathbf{H}_k^T + \mathbf{R}_k]^{-1} \qquad (4.4\text{-}17)$$

Using the optimal gain in Eq. 4.4-13, the updated state error covariance matrix is

$$\begin{aligned}\mathbf{P}_k(+) &= \mathbf{P}_k(-) - [\mathbf{P}_k(-)\mathbf{H}_k^T + \mathbf{M}_k][\mathbf{H}_k\mathbf{P}_k(-)\mathbf{H}_k^T + \mathbf{H}_k\mathbf{M}_k + \mathbf{M}_k^T\mathbf{H}_k^T + \mathbf{R}_k]^{-1} \\ &\qquad \cdot [\mathbf{H}_k\mathbf{P}_k(-) + \mathbf{M}_k^T] \\ &= \mathbf{P}_k(-) - \mathbf{K}_k[\mathbf{M}_k^T + \mathbf{H}_k\mathbf{P}_k(-)] \qquad (4.4\text{-}18)\end{aligned}$$

As an example of a problem in which disturbance-measurement correlation could be significant, consider the sampled-data estimate of an aircraft's path through a turbulent wind field (Fig. 4.4-1). The effects of the random wind on the aircraft's acceleration are rather complex; suffice to say that there is an integrated effect on the lift, drag, and side force, as well as on the pitching, rolling, and yawing moments. This disturbance effect could

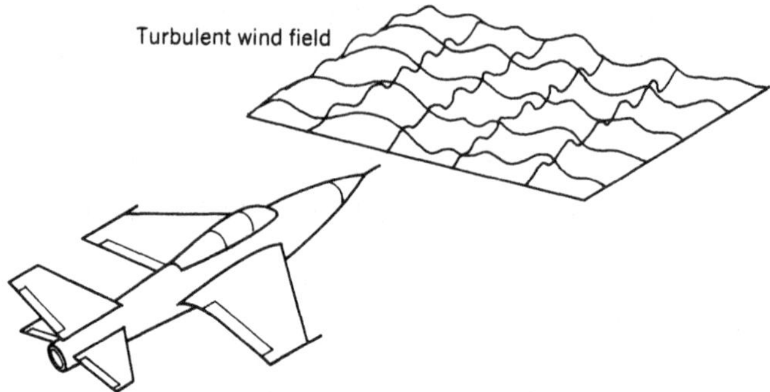

FIGURE 4.4-1 Aircraft flying through a field of random turbulence.

be modeled as $\mathbf{w}_{k-1}$. Sensors mounted on the aircraft can measure its angle and velocity relative to the air, and these measurements are corrupted by the same turbulent wind field that forces the aircraft. Thus the measurement error $\mathbf{n}_k$ may be correlated with $\mathbf{w}_{k-1}$. The value of $\mathbf{R}_k$ would be determined by the turbulent wind field's statistics and the forward velocity of the aircraft. The values of $\mathbf{Q}_{k-1}$ and $\mathbf{M}_k$ would be proportional to $\mathbf{R}_k$ and would depend upon the aircraft's aerodynamic characteristics.

Time-Correlated Measurement Noise

Consider the discrete-time system

$$\mathbf{x}_k = \mathbf{\Phi}_{k-1}\mathbf{x}_{k-1} + \mathbf{w}_{k-1}, \qquad \mathbf{x}_0 \text{ given} \tag{4.4-19}$$

$$\mathbf{z}_k = \mathbf{H}_k\mathbf{x}_k + \mathbf{n}_k \tag{4.4-20}$$

The deterministic controls are neglected, as the previous section illustrated that they have no effect on the calculation of the state error covariance or the optimal gain. It is assumed that the disturbance input is a white Gaussian process described by

$$E[\mathbf{w}_k] = \mathbf{0} \tag{4.4-21}$$

$$E[\mathbf{w}_k\mathbf{w}_k^T] = \mathbf{Q}_k \tag{4.4-22}$$

while the measurement noise is the result of a Gauss–Markov process,

$$\mathbf{n}_k = \mathbf{\Psi}_{k-1}\mathbf{n}_{k-1} + \boldsymbol{\nu}_{k-1} \tag{4.4-23}$$

for which

$$E[\boldsymbol{\nu}_k] = \mathbf{0} \tag{4.4-24}$$

$$E[\boldsymbol{\nu}_k\boldsymbol{\nu}_k^T] = \mathbf{Q}_{\nu_k} \tag{4.4-25}$$

$$E[\mathbf{w}_k\boldsymbol{\nu}_k^T] = \mathbf{0} \tag{4.4-26}$$

We seek a minimum-variance estimator that accounts for the time correlation in $\mathbf{n}_k$ defined by Eq. 4.4-23. Note that only the measurement error is subjected to a dynamic process. If sensor dynamics influence the state as well as the error, then the estimator's dynamic model should be augmented by the sensor model, and estimation should proceed as in Section 4.3.

Because the noise process is represented by a dynamic model, the latter could be appended to the original system model. Given $\mathbf{x}$ as an n-dimensional vector and $\mathbf{n}$ as an r-dimensional vector, the augmented state vector $\mathbf{x}_A$ has dimension $(n+r)$, and an equivalent discrete-time system can be expressed as

$$\begin{bmatrix}\mathbf{x}\\ \mathbf{n}\end{bmatrix}_k = \begin{bmatrix}\mathbf{\Phi} & \mathbf{0}\\ \mathbf{0} & \mathbf{\Psi}\end{bmatrix}_{k-1}\begin{bmatrix}\mathbf{x}\\ \mathbf{n}\end{bmatrix}_{k-1} + \begin{bmatrix}\mathbf{w}\\ \boldsymbol{\nu}\end{bmatrix}_{k-1} \tag{4.4-27}$$

or

$$\mathbf{x}_{A_k} = \mathbf{\Phi}_{A_{k-1}}\mathbf{x}_{A_{k-1}} + \mathbf{w}_{A_{k-1}} \tag{4.4-28}$$

with

$$\mathbf{Q}_{A_k} = \begin{bmatrix}\mathbf{Q}_k & \mathbf{0}\\ \mathbf{0} & \mathbf{Q}_{\nu_k}\end{bmatrix} \tag{4.4-29}$$

$$\mathbf{z}_k = [\mathbf{H}_k \quad \mathbf{I}_r]\begin{bmatrix}\mathbf{x}_k\\ \mathbf{n}_k\end{bmatrix} \tag{4.4-30}$$

$$\mathbf{z}_k = \mathbf{H}_{A_k}\mathbf{x}_{A_k} \tag{4.4-31}$$

The trouble with this system is that the equivalent noise covariance matrix $\mathbf{R}_{A_k}$ is singular because the observation is a noise-free function of the augmented state. Although the state error covariance update does not formally require a nonsingular noise covariance matrix (see Eq. 4.3-21a), numerical problems can be encountered. To see this, postmultiply Eq. 4.3-21a by $\mathbf{H}_k^T$ and evaluate the result with $\mathbf{R}_k = \mathbf{0}$.

The problem is overcome by using a measurement differencing approach due to Bryson and Henrikson (B-10), which restores the original dimension of the filter and introduces an equivalent measurement $\boldsymbol{\zeta}$, whose error is white. From Eqs. 4.4-19, 4.4-20, and 4.4-23.

$$\begin{aligned}\boldsymbol{\zeta}_{k-1} &\triangleq \mathbf{z}_k - \mathbf{\Psi}_{k-1}\mathbf{z}_{k-1}\\ &= \mathbf{H}_k(\mathbf{\Phi}_{n-1}\mathbf{x}_{k-1} + \mathbf{w}_{k-1}) + (\mathbf{\Psi}_{k-1}\mathbf{n}_{k-1} + \boldsymbol{\nu}_{k-1})\\ &\quad -\mathbf{\Psi}_{k-1}[\mathbf{H}_{k-1}\mathbf{x}_{k-1} + \mathbf{n}_{k-1}]\\ &= (\mathbf{H}_k\mathbf{\Phi}_{k-1} - \mathbf{\Psi}_{k-1}\mathbf{H}_{k-1})\mathbf{x}_{k-1} + \mathbf{H}_k\mathbf{w}_{k-1} + \boldsymbol{\nu}_{k-1}\end{aligned} \tag{4.4-32a}$$

This can be written in the standard form as

$$\boldsymbol{\zeta}_{k-1} = \mathbf{D}_{k-1}\mathbf{x}_{k-1} + \mathbf{n}_{D_{k-1}} \tag{4.4-32b}$$

where

$$\mathbf{D}_{k-1} = (\mathbf{H}_k\mathbf{\Phi}_{k-1} - \mathbf{\Psi}_{k-1}\mathbf{H}_{k-1}) \tag{4.4-33}$$

$$\mathbf{n}_{D_{k-1}} = \mathbf{H}_k\mathbf{w}_{k-1} + \boldsymbol{\nu}_{k-1} \tag{4.4-34}$$

Although the subscript $(k-1)$ is used, note that $\boldsymbol{\zeta}_{k-1}$ formally depends on $\mathbf{z}_k$ as well as $\mathbf{z}_{k-1}$, while $\mathbf{n}_{D_{k-1}}$ depends on $\mathbf{H}_k$. By its definition, the derived measurement noise is instantaneously correlated with the disturbance input,

$$E\left\{\begin{bmatrix}\mathbf{w}_{k-1}\\ \mathbf{n}_{D_{k-1}}\end{bmatrix}[\mathbf{w}_{k-1}^T\mathbf{n}_{D_{k-1}}^T]\right\} = \begin{bmatrix}\mathbf{Q}_{k-1} & \mathbf{Q}_{k-1}\mathbf{H}_k^T\\ \mathbf{H}_k\mathbf{Q}_{k-1} & (\mathbf{H}_k\mathbf{Q}_{k-1}\mathbf{H}_k^T + \mathbf{Q}_{\nu_{k-1}})\end{bmatrix}$$

$$\triangleq \begin{bmatrix}\mathbf{Q} & \mathbf{M}\\ \mathbf{M}^T & \mathbf{R}\end{bmatrix}_{k-1} \tag{4.4-35}$$

and $E(\mathbf{w}_k\mathbf{w}_j^T) = E(\mathbf{w}_k\mathbf{n}_{D_j}^T) = \mathbf{0}$ for $j \neq k$. The two-step state estimate of (B-10), presented without derivation, is

$$\hat{\mathbf{x}}_{k-1}(+) = \hat{\mathbf{x}}_{k-1}(-) + \mathbf{K}_{k-1}[\boldsymbol{\zeta}_{k-1} - \mathbf{D}_{k-1}\hat{\mathbf{x}}_{k-1}(-)] \tag{4.4-36}$$

$$\hat{\mathbf{x}}_k(-) = \mathbf{\Phi}_{k-1}\hat{\mathbf{x}}_{k-1}(+) + \mathbf{C}_{k-1}[\boldsymbol{\zeta}_{k-1} - \mathbf{D}_{k-1}\hat{\mathbf{x}}_{k-1}(+)] \tag{4.4-37}$$

where

$$\mathbf{K}_{k-1} = \mathbf{P}_{k-1}(-)\mathbf{D}_{k-1}^T[\mathbf{D}_{k-1}\mathbf{P}_{k-1}(-)\mathbf{D}_{k-1}^T + \mathbf{R}_{k-1}]^{-1} \tag{4.4-38}$$

$$\mathbf{C}_{k-1} = \mathbf{M}_{k-1}[\mathbf{D}_{k-1}\mathbf{P}_{k-1}(-)\mathbf{D}_{k-1}^T + \mathbf{R}_{k-1}]^{-1} \tag{4.4-39}$$

In contrast to the previous notation, the final estimate is denoted by $\hat{\mathbf{x}}_k(-)$ rather than $\hat{\mathbf{x}}_k(+)$, as the first update precedes the propagation step. The two-step covariance estimation is

$$\mathbf{P}_{k-1}(+) = (\mathbf{I}_n - \mathbf{K}_{k-1}\mathbf{D}_{k-1})\mathbf{P}_{k-1}(-)(\mathbf{I}_n - \mathbf{K}_{k-1}\mathbf{D}_{k-1})^T + \mathbf{K}_{k-1}\mathbf{R}_{k-1}\mathbf{K}_{k-1}^T \tag{4.4-40}$$

$$\mathbf{P}_k(-) = \mathbf{\Phi}_{k-1}\mathbf{P}_{k-1}(+)\mathbf{\Phi}_{k-1}^T + \mathbf{Q}_{k-1} - \mathbf{C}_{k-1}\mathbf{M}_{k-1}^T - \mathbf{\Phi}_{k-1}\mathbf{K}_{k-1}\mathbf{M}_{k-1} - \mathbf{M}_{k-1}^T\mathbf{K}_{k-1}^T\mathbf{\Phi}_{k-1}^T \tag{4.4-41}$$

As an example of time-correlated measurement noise, the natural atmospheric turbulence considered earlier is better modeled as a dynamic process with known spectral content than as white noise. If the disturbance effect of the turbulence is negligible, the spectral model could be transformed to a state-space model (Eq. 4.4-23) and could be used to compute an improved estimate of the original state. Depending on its coefficients, Eq. 4.4-23 can express band-limiting, resonant effects, coupling of noise sources, or generalized spectral shaping.

4.5 CONTINUOUS-TIME OPTIMAL FILTERS AND PREDICTORS

Our objectives in this section are similar to those of Section 4.3, except that dynamic processes, measurements, and estimates are assumed to be continuous functions of time rather than discrete sequences. We wish to calculate a "best" estimate of a state vector $\mathbf{x}(t)$ in the continuous interval, $[0, t_f]$. The estimated time history, $\hat{\mathbf{x}}(t_{\text{est}})$, is conditioned on all available information about a linear dynamic process and a history of measurements from zero to the "present" time, $t_{\text{calc}} \leq t_f$. If $t_{\text{est}} = t_{\text{calc}}$, the estimator is called a *filter*; if $t_{\text{calc}} < t_{\text{est}} \leq t_f$, the estimator is a *predictor*.

These linear estimators are optimal in the sense that the variance of the state estimate error is minimized "on the average." They are derived from the previous discrete-time results by letting the implied interval between sampling instants shrink to zero. Consequently, all of the Section 4.3 equations have their analogs in continuous-time filters and predictors. Certain assumptions are easier to justify or visualize in continuous time, while others are more difficult. For example, the notion of a random constant disturbance that jumps from one sampling instant to the next may be a poor approximation for a continuous disturbance. On the other hand, the mathematically rigorous discrete white-noise sequence must be replaced by the less-rigorous continuous white-noise process, invoking the caveats of Section 4.2.

In addition, there is the problem of implementing the estimator. Linear discrete-time estimators are naturally and precisely executed by digital computers (subject to quantization error and computation delay), while continuous-time estimators can only be approximated using digital computers. Nevertheless, this seeming advantage is irrelevant for analog computation, and it is of little consequence in digital (sampled-data) state estimation for nonlinear dynamic systems (Section 4.6). Furthermore, linear continuous-time estimators with constant coefficients can be analyzed using conventional frequency-domain concepts (Section 2.6 and Chapter 6), a distinct engineering advantage.

Kalman–Bucy Filter

The estimate of $\mathbf{x}(t)$ in $[0, t_f]$ is based on the conditional probability density function in $[0, t_f]$. The state estimate $\hat{\mathbf{x}}(t)$ is taken to be the mean value specified by the distribution while the covariance of the estimation error is the second central moment. The linear-optimal estimator minimizes the state error covariance for all t in $[0, t_f]$ by properly blending prior information and current measurements. Given knowledge of the system, its inputs, and its uncertainties, this amounts to minimizing the mean-square estimation error with respect to the choice of a time-varying filter gain matrix. It is assumed that minimizing this error makes $\hat{\mathbf{x}}(t)$ most like $\mathbf{x}(t)$.

The optimal values of $\hat{\mathbf{x}}(t)$ and the covariance matrix, $\mathbf{P}(t)$, can be computed using three equations:

1. State Estimate (Propagation and Update)
2. Covariance Estimate (Propagation and Update)
3. Filter Gain Computation

The *state estimate* uses a linear, ordinary differential equation based on the system model, with the actual residual measurement errors driving the state propagation through the optimal gain matrix. The *covariance estimate* is derived from nonlinear, ordinary differential equation driven by the statistics of the assumed measurement errors and disturbance inputs, under the assumption that measurements are incorporated in optimal fashion. (We will see later that this nonlinear equation can be replaced by a linear equation of increased dimension if the system matrices are constant.) The optimal *filter gain computation* is an algebraic equation that provides a continuing balance between state uncertainties due to disturbance inputs and those due to measurement errors.

The linear dynamic system is described by two equations—the dynamic process model,

$$\dot{\mathbf{x}}(t) = \mathbf{F}(t)\mathbf{x}(t) + \mathbf{G}(t)\mathbf{u}(t) + \mathbf{L}(t)\mathbf{w}(t), \qquad \mathbf{x}(0) = \mathbf{x}_0 \tag{4.5-1}$$

and the observation model,

$$\mathbf{z}(t) = \mathbf{H}(t)\mathbf{x}(t) + \mathbf{n}(t) \tag{4.5-2}$$

The vectors and matrices are defined and dimensioned as before: $\mathbf{x}$ is the n-dimensional state vector, $\mathbf{u}$ is the m-dimensional controlled input vector, $\mathbf{w}$ is the s-dimensional disturbance input vector, $\mathbf{z}$ is the r-dimensional observation (or measurement) vector, and $\mathbf{n}$ is the corresponding measurement error. The matrices $\mathbf{F}$, $\mathbf{G}$, $\mathbf{L}$, and $\mathbf{H}$ are conformable with these vectors. The observation vector could be a function of the control as well as the state (as in Eq. 4.4-2) with minimal modification to what follows.

The statistics of the initial conditions, disturbance inputs, and measurement errors are assumed to be known. The expected values of the initial state and covariance are

$$E[\mathbf{x}(0)] = \hat{\mathbf{x}}_0 \tag{4.5-3}$$

$$E\{[\mathbf{x}(0) - \hat{\mathbf{x}}_0][\mathbf{x}(0) - \hat{\mathbf{x}}_0]^T\} = \mathbf{P}_0 \tag{4.5-4}$$

The disturbance input is a white, zero-mean Gaussian random process

$$E[\mathbf{w}(t)] = \mathbf{0} \tag{4.5-5}$$

$$E[\mathbf{w}(t)\mathbf{w}^T(\tau)] = \mathbf{Q}_C(t)\delta(t-\tau) \tag{4.5-6}$$

which is specified by its spectral density matrix $\mathbf{Q}_C(t)$. The measurement error is a white, zero-mean Gaussian random process,

$$E[\mathbf{n}(t)] = \mathbf{0} \tag{4.5-7}$$

$$E[\mathbf{n}(t)\mathbf{n}^T(\tau)] = \mathbf{R}_C(t)\delta(t-\tau) \tag{4.5-8}$$

with the measurement uncertainty expressed by its spectral density matrix $\mathbf{R}_C(t)$. For the moment, it is assumed that the disturbance input and measurement are uncorrelated.

The discrete-time equivalent of this model is given by Eqs. 4.3-1 to 4.3-11, and the corresponding filter equations can be expressed by Eqs. 4.3-13, 4.3-21b, 4.3-22, and 4.3-23. The state covariance at $(k-1)$ is

$$\mathbf{P}_{k-1}(+) = [\mathbf{I}_n - \mathbf{K}_{k-1}\mathbf{H}_{k-1}]\mathbf{P}_{k-1}(-) \tag{4.5-9}$$

while for vanishingly small time interval, $\Delta t = t_k - t_{k-1}$, the pre-update covariance at k is approximately

$$\mathbf{P}_k(-) \simeq [\mathbf{I}_n + \mathbf{F}_{k-1}\Delta t]\mathbf{P}_{k-1}(+)[\mathbf{I}_n + \mathbf{F}_{k-1}\Delta t]^T + \mathbf{Q}_{k-1} \tag{4.5-10}$$

Substituting Eq. 4.5-9 in Eq. 4.5-10, expanding as in Eq. 4.2-74 and dividing by Δt,

$$\begin{aligned}\frac{\mathbf{P}_k(-) - \mathbf{P}_{k-1}(-)}{\Delta t} &= \mathbf{F}_{k-1}\mathbf{P}_{k-1}(-) + \mathbf{P}_{k-1}(-)\mathbf{F}_{k-1}^T + \frac{\mathbf{Q}_{k-1}}{\Delta t} \\ &\quad - \frac{\mathbf{K}_{k-1}}{\Delta t}\mathbf{H}_{k-1}\mathbf{P}_{k-1}(-) - \mathbf{F}_{k-1}\mathbf{K}_{k-1}\mathbf{H}_{k-1}\mathbf{P}_{k-1}(-) \\ &\quad - \mathbf{K}_{k-1}\mathbf{H}_{k-1}\mathbf{P}_{k-1}(-)\mathbf{F}_{k-1}^T\end{aligned} \tag{4.5-11}$$

[$\mathbf{F}_{k-1}$ denotes $\mathbf{F}(t_{k-1})$, and so on.] For small values of Δt, the integrated disturbance effect is (Eq. 4.2-53)

$$\mathbf{Q}_{k-1} \simeq \mathbf{L}_{k-1}\mathbf{Q}'_C(t_{k-1})\mathbf{L}_{k-1}^T\Delta t \tag{4.5-12}$$

For constant error spectral density, $\mathbf{R}_C$, in the interval, the error covariance is

$$\mathbf{R}_{k-1} \simeq \frac{\mathbf{R}_C(t_{k-1})}{\Delta t} \tag{4.5-13}$$

Here $\mathbf{R}_{k-1}\Delta t$ is equivalent to the integrated effect of the impulsive autocorrelation function, $\mathbf{R}_C(t_{k-1})\delta(t_{k-1}-\tau)$. The discrete-time optimal gain matrix (Eq. 4.3-22) is

$$\mathbf{K}_{k-1} = \mathbf{P}_{k-1}(+)\mathbf{H}_{k-1}^T\mathbf{R}_C^{-1}(t_{k-1})\Delta t \tag{4.5-14}$$

As Δt goes to zero, the measurement update becomes continuous, and there is no distinction between $\mathbf{P}_{k-1}(+)$ and $\mathbf{P}_{k-1}(-)$. In the limit, $t_{k-2} \to t_{k-1} \to t_k \to t$, and $[\mathbf{P}_k(-)-\mathbf{P}_{k-1}(-)]/\Delta t$ becomes $d\mathbf{P}(t)/dt$. The *covariance estimate* is found by integrating the following equation from 0 to t_{est}. Dropping the prime on $\mathbf{Q}_C$,

$$\begin{aligned}\dot{\mathbf{P}}(t) &= \mathbf{F}(t)\mathbf{P}(t) + \mathbf{P}(t)\mathbf{F}^T(t) + \mathbf{L}(t)\mathbf{Q}_C(t)\mathbf{L}^T(t) - \mathbf{K}_C(t)\mathbf{H}(t)\mathbf{P}(t)\\ &= \mathbf{F}(t)\mathbf{P}(t) + \mathbf{P}(t)\mathbf{F}^T(t) + \mathbf{L}(t)\mathbf{Q}_C(t)\mathbf{L}^T(t)\\ &\quad - \mathbf{P}(t)\mathbf{H}^T(t)\mathbf{R}_C^{-1}(t)\mathbf{H}(t)\mathbf{P}(t), \qquad \mathbf{P}(0) = \mathbf{P}_0\end{aligned} \tag{4.5-15}$$

Although this involves n^2 integrations, $\mathbf{P}(t)$ is symmetric, so only $n(n+1)/2$ integrations need be performed. From Eq. 4.5-14, the *optimal filter gain equation* is

$$\mathbf{K}_C(t) = \mathbf{P}(t)\mathbf{H}^T(t)\mathbf{R}_C^{-1}(t) \tag{4.5-16}$$

in the limit as $\Delta t \to 0$. It is clear that $\mathbf{K}_C(t)$ cannot exist if $\mathbf{R}_C(t)$ is singular, and for stable filter estimates, $\mathbf{R}_C(t)$ must be positive-definite.

The state estimation equation is derived similarly. For small Δt, Eq. 4.3-23 is approximately

$$\begin{aligned}\hat{\mathbf{x}}_k(+) &\simeq \hat{\mathbf{x}}_{k-1}(+) + [\mathbf{F}_{k-1}\hat{\mathbf{x}}_{k-1}(+) + \mathbf{G}_{k-1}\mathbf{u}_{k-1}]\,\Delta t + \mathbf{K}_k\{\mathbf{z}_k - \mathbf{H}_k[\hat{\mathbf{x}}_{k-1}(+)\\ &\quad + (\mathbf{F}_{k-1}\hat{\mathbf{x}}_{k-1}(+) + \mathbf{G}_{k-1}\mathbf{u}_{k-1})\,\Delta t]\}\end{aligned} \tag{4.5-17}$$

In the limit as Δt goes to zero, $[\hat{\mathbf{x}}_k(+) - \hat{\mathbf{x}}_{k-1}(+)]/\Delta t$ becomes $d\hat{\mathbf{x}}/dt$, and the *state estimate* is found by integrating

$$\dot{\hat{\mathbf{x}}}(t) = \mathbf{F}(t)\hat{\mathbf{x}}(\cdot) + \mathbf{G}(t)\mathbf{u}(t) + \mathbf{K}_C(t)[\mathbf{z}(t) - \mathbf{H}(t)\hat{\mathbf{x}}(t)], \qquad \hat{\mathbf{x}}(0) = \hat{\mathbf{x}}_0 \tag{4.5-18}$$

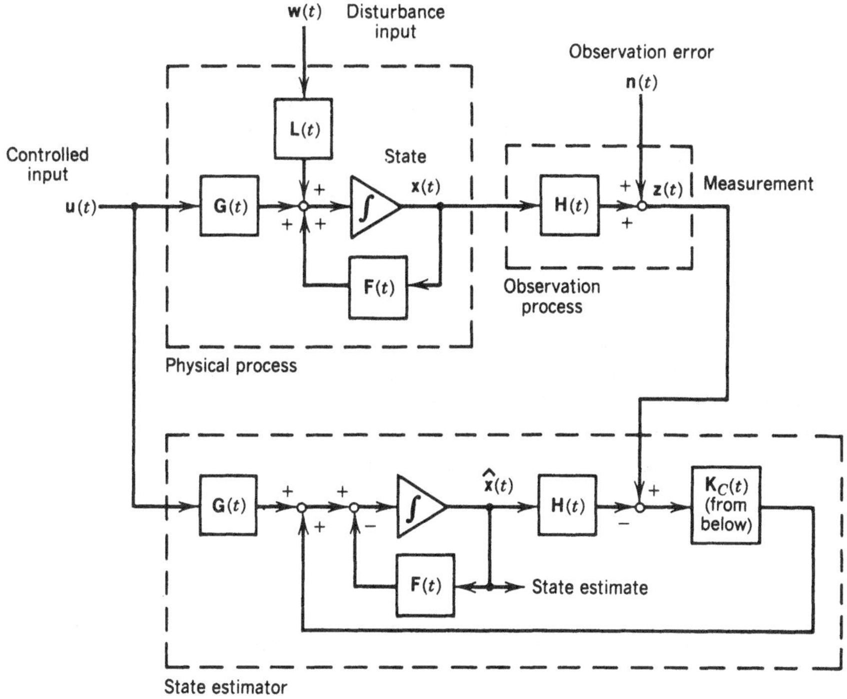

w(t) Disturbance input
Observation error
n(t)
L(t)
Controlled input
State
u(t)
G(t)
x(t)
H(t)
z(t)
Measurement
F(t)
Observation process
Physical process
G(t)
x̂(t)
H(t)
K_C(t) (from below)
F(t)
State estimate
State estimator

(a)

Filter gain computation
Measurement error spectral density
R_C(t)
P(t)H^T(t)R_C^-1(t)
K_C(t) filter gain (to state estimator)
P(t)H^T(t)R_C^-1(t)H(t)P(t)
Q'_C(t)
L(t)Q'_C L^T(t)
P(t)
State covariance estimate
Disturbance input spectral density
F(t)P(t) + P(t)F^T(t)

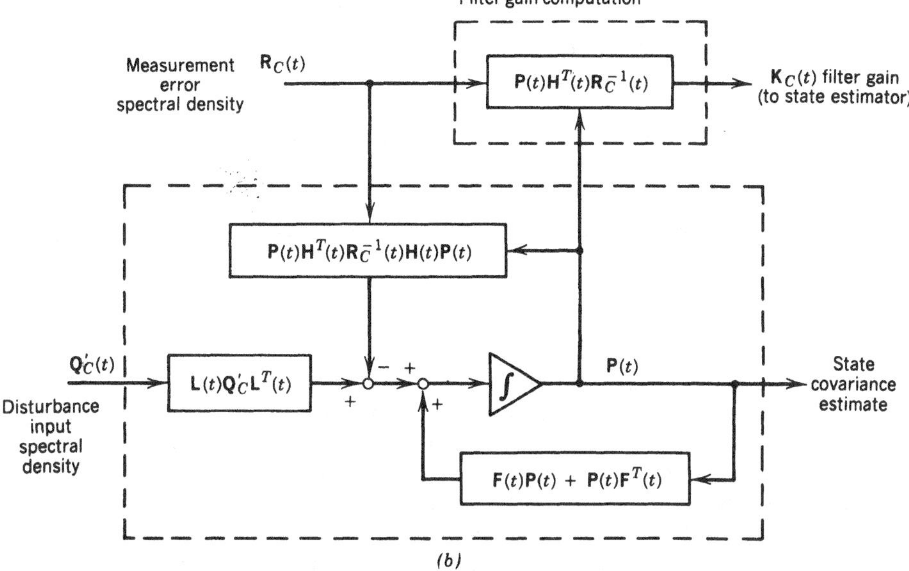

(b)

FIGURE 4.5-1 Continuous-time system and linear-optimal filter. (*a*) Dynamic system and state estimator; (*b*) covariance estimator and filter gain computation.

from 0 to t_{est}. The Kalman–Bucy algorithm consists of Eqs. 4.5-15, 4.5-16, and 4.5-18, all of which must be computed concurrently. Block diagrams of the Kalman–Bucy filter and its relationship to the physical system are shown in Fig. 4.5-1. As in the discrete-time case, there is a natural separation between the state and covariance estimates. The actual measurements and known inputs drive the state estimator (Fig. 4.5-1*a*), while the covariance estimator is driven by the assumed statistics (Fig. 4.5-1*b*). The state estimator is seen to feed back the estimate, $\hat{\mathbf{x}}(t)$; while the open-loop system dynamics are governed by $\mathbf{F}(t)$, the closed-loop filter dynamics are characterized by $[\mathbf{F}(t)-\mathbf{K}_C(t)\mathbf{H}(t)]$. The optimal gain depends only on the statistics and the system model, and it could be precomputed if the time variations of system coefficients and statistics were known ahead of time. Then only the state estimator (Fig. 4.5-1*a*) would be executed in "real time."

Duality

It has been shown that the linear-optimal filter covariance and gain equations are

$$\dot{\mathbf{P}}_{est} = \mathbf{F}(t)\mathbf{P}_{est}(t) + \mathbf{P}_{est}(t)\mathbf{F}^T(t) + \mathbf{L}(t)\mathbf{Q}_C(t)\mathbf{L}^T(t) - \mathbf{P}_{est}(t)\mathbf{H}^T(t)\mathbf{R}_C^{-1}(t)\mathbf{H}(t)\mathbf{P}_{est}(t), \qquad \mathbf{P}_{est}(0) \text{ given} \tag{4.5-19}$$

$$\mathbf{K}_C(t) = \mathbf{P}_{est}(t)\mathbf{H}^T(t)\mathbf{R}_C^{-1}(t) \tag{4.5-20}$$

In Section 3.7 it was found that the Riccati matrix and optimal gain calculations for the linear-optimal control law (neglecting cross-product weighting) are

$$\dot{\mathbf{P}}(t) = -\mathbf{F}^T(t)\mathbf{P}(t) - \mathbf{P}(t)\mathbf{F}(t) - \mathbf{Q}(t) + \mathbf{P}(t)\mathbf{G}(t)\mathbf{R}^{-1}(t)\mathbf{G}^T(t)\mathbf{P}(t), \qquad \mathbf{P}(t_f) = \mathbf{P}_f \tag{4.5-21}$$

$$\mathbf{C}(t) = -\mathbf{R}^{-1}(t)\mathbf{G}^T(t)\mathbf{P}(t) \tag{4.5-22}$$

The similarity between the two solutions is apparent; both Eq. 4.5-19 and Eq. 4.5-21 are Riccati equations, the principal differences being the definition of coefficient matrices and the point at which boundary conditions are specified. Perhaps this duality is not surprising, in that quadratic criteria are minimized, subject to linear dynamic constraints, in both cases.

Consequently, the *linear-optimal estimation problem is the mathematical dual of the linear-optimal regulator problem*. The structures of the two problems are the same; therefore, the structures of their solutions are the same. The significance of duality is that algorithms for solving one problem can be applied to the other using the definitions of Table 4.5-1.

TABLE 4.5-1 Matrix Dualities in Continuous-Time Linear Control and Estimation

Control	Estimation
$\mathbf{F}(t)$	$\mathbf{F}^T(t)$
$\mathbf{G}(t)$	$\mathbf{H}^T(t)$
$\mathbf{Q}(t)$	$\mathbf{L}(t)\mathbf{Q}_C(t)\mathbf{L}^T(t)$
$\mathbf{R}(t)$	$\mathbf{R}_C(t)$
$\mathbf{P}(\tau)^a$	$\mathbf{P}_{est}(t)$
$\mathbf{P}(t_f)$	$\mathbf{P}_{est}(0)$
$-\mathbf{C}(t)$	$\mathbf{K}^T(t)$

[a] $\tau = t_f - t$.

The estimator's Riccati equation is integrated forward in time, while the controller's Riccati equation is integrated with time running in reverse. (Note that $d\mathbf{P}/d\tau = -d\mathbf{P}/dt$.) The linear-optimal regulator presents a two-point boundary value problem, while all the boundary values of the linear-optimal filter are specified at the initial time. Estimation becomes a two-point boundary value problem for the smoothing case, in which a "best" estimate of the entire state history in $[t_0, t_f]$ is to be computed after the fact. Then, the starting values of a reverse-time recursion for Lagrange multipliers and the improved estimates of $\mathbf{x}(t)$ and $\mathbf{P}(t)$ are specified at t_f. For nonlinear smoothing problems, iterative solutions that are duals of the nonlinear optimal control solutions of Section 3.6 must be used to achieve a fully optimal estimate. Although not shown here, discrete-time estimation and control also are duals of each other, as can be deduced from the corresponding equations.

Linear-Optimal Predictor

As in the discrete-time case, propagation of the estimate beyond the time of the last available measurement is based on the system equation alone. The optimal filtering result, found by integrating Eq. 4.5-18 to t_{calc}, provides the initial condition for the state and covariance prediction at time t_{est}. The state prediction is the integral of

$$\dot{\hat{\mathbf{x}}}_p(t) = \mathbf{F}(t)\hat{\mathbf{x}}_p(t) + \mathbf{G}(t)\mathbf{u}(t), \qquad \hat{\mathbf{x}}_p(t_{calc}) \text{ given} \tag{4.5-23}$$

from t_{calc} to t_{est}. Clearly, Eq. 4.5-23 is Eq. 4.5-18 without observations; hence the predicted mean value necessarily neglects the measurable effects of future disturbances as well as the errors of future observations.

The predicted error covariance of the estimate is found by integrating

$$\dot{\mathbf{P}}_p(t) = \mathbf{F}(t)\mathbf{P}_p(t) + \mathbf{P}_p(t)\mathbf{F}^T(t) + \mathbf{L}(t)\mathbf{Q}_C\mathbf{L}^T(t), \qquad \mathbf{P}_p(t_{\text{calc}}) \text{ given} \tag{4.5-24}$$

from t_{calc} to t_{est}. It can be concluded that $\mathbf{P}_p(t_{\text{est}}) \geq \mathbf{P}(t_{\text{calc}})$, as $\mathbf{P}_p(t)$ is positive-semidefinite.

Alternative Forms of the Linear-Optimal Filter

The optimal gain matrix can be computed using equations other than Eqs. 4.5-15 and 4.5-16. The nonlinear Riccati equation that governs the state estimate error covariance dynamics can be replaced by a linear equation of higher dimension. In principle, this opens the way to analytic solutions for $\mathbf{P}(t)$ as matrix exponentials, although there may be practical restrictions on the utility of this result. Another possibility is to eliminate the Riccati equation entirely, replacing it and the algebraic gain equation with two differential equations of smaller dimension. Finally, a square-root formulation for the covariance equation reduces the problems of ill-conditioning discussed in Section 4.3.

We have seen that $\mathbf{P}(t)$ relates the state and its adjoint in the case of the linear-optimal controller (Section 3.7). Suppose a dual relationship holds for the estimator such that the following $(n \times n)$ matrix equation is true:

$$\mathbf{\Lambda}(t) = \mathbf{P}(t)\mathbf{X}(t) \tag{4.5-25a}$$

or

$$\mathbf{P}(t) = \mathbf{\Lambda}(t)\mathbf{X}^{-1}(t) \tag{4.5-25b}$$

Using Eq. 2.2-49, the derivative with respect to time is

$$\dot{\mathbf{P}}(t) = \dot{\mathbf{\Lambda}}(t)\mathbf{X}^{-1}(t) - \mathbf{\Lambda}(t)\mathbf{X}^{-1}(t)\dot{\mathbf{X}}(t)\mathbf{X}^{-1}(t) \tag{4.5-26}$$

and, from Eq. 4.5-15,

$$\begin{aligned}\dot{\mathbf{P}}(t) = {} & \mathbf{F}(t)\mathbf{\Lambda}(t)\mathbf{X}^{-1}(t) + \mathbf{\Lambda}(t)\mathbf{X}^{-1}(t)\mathbf{F}^T(t) + \mathbf{L}(t)\mathbf{Q}_C(t)\mathbf{L}^T(t) \\ & - \mathbf{\Lambda}(t)\mathbf{X}^{-1}(t)\mathbf{H}^T(t)\mathbf{R}_C^{-1}(t)\mathbf{H}(t)\mathbf{\Lambda}(t)\mathbf{X}^{-1}(t)\end{aligned} \tag{4.5-27}$$

Postmultiplying these equations by $\mathbf{X}(t)$, it can be seen that if

$$\dot{\mathbf{\Lambda}}(t) = \mathbf{F}(t)\mathbf{\Lambda}(t) + \mathbf{L}(t)\mathbf{Q}_C(t)\mathbf{L}^T(t)\mathbf{X}(t) \tag{4.5-28}$$

then

$$\dot{\mathbf{X}}(t) = \mathbf{H}^T(t)\mathbf{R}_C^{-1}(t)\mathbf{H}(t)\mathbf{\Lambda}(t) - \mathbf{F}^T(t)\mathbf{X}(t) \tag{4.5-29}$$

The last two equations are linear, and they can be combined as

$$\begin{bmatrix}\dot{\mathbf{\Lambda}}(t)\\ \dot{\mathbf{X}}(t)\end{bmatrix}=\begin{bmatrix}\mathbf{F}(t) & \mathbf{L}(t)\mathbf{Q}_C(t)\mathbf{L}^T(t)\\ \mathbf{H}^T(t)\mathbf{R}_C^{-1}(t)\mathbf{H}(t) & -\mathbf{F}^T(t)\end{bmatrix}\begin{bmatrix}\mathbf{\Lambda}(t)\\ \mathbf{X}(t)\end{bmatrix}$$

$$\triangleq \mathbf{A}(t)\begin{bmatrix}\mathbf{\Lambda}(t)\\ \mathbf{X}(t)\end{bmatrix} \tag{4.5-30}$$

The following initial conditions are suitable:

$$\begin{bmatrix}\mathbf{\Lambda}(0)\\ \mathbf{X}(0)\end{bmatrix}=\begin{bmatrix}\mathbf{P}_0\\ \mathbf{I}_n\end{bmatrix} \tag{4.5-31}$$

Then $\mathbf{\Lambda}(t)$ and $\mathbf{X}(t)$ are the integrals of Eq. 4.5-30, and $\mathbf{P}(t)$ is computed from Eq. 4.5-25b.

If it is acceptable to generate $\mathbf{P}(t)$ at discrete instants of time, then the state transition matrix of $\mathbf{A}(t)$ could be used in lieu of integration (Section 2.3). One such case occurs when $\mathbf{F}$, $\mathbf{G}$, $\mathbf{L}$, $\mathbf{Q}_c$, and $\mathbf{R}_C$ are constant, and steady-state values of $\mathbf{P}$ and $\mathbf{K}$ are sought. For an arbitrary time interval Δt, the state transition matrix is constant, and it can be partitioned in $(n \times n)$ blocks:

$$\begin{aligned}\mathbf{\Theta}(\Delta t) &= e^{\mathbf{A}\Delta t}\\ &=\begin{bmatrix}\mathbf{\Theta}_{11} & \mathbf{\Theta}_{12}\\ \mathbf{\Theta}_{21} & \mathbf{\Theta}_{22}\end{bmatrix}\end{aligned} \tag{4.5-32}$$

The state transition equation for $\mathbf{\Lambda}(t)$ and $\mathbf{X}(t)$ is

$$\begin{bmatrix}\mathbf{\Lambda}(t+\Delta t)\\ \mathbf{X}(t+\Delta t)\end{bmatrix}=\begin{bmatrix}\mathbf{\Theta}_{11} & \mathbf{\Theta}_{12}\\ \mathbf{\Theta}_{21} & \mathbf{\Theta}_{22}\end{bmatrix}\begin{bmatrix}\mathbf{\Lambda}(t)\\ \mathbf{X}(t)\end{bmatrix} \tag{4.5-33}$$

$\mathbf{\Lambda}(t+\Delta t)$ and $\mathbf{\Lambda}(t)$ can be eliminated using Eq. 4.5-25a, and Eq. 4.5-33 becomes equivalent to

$$\mathbf{P}(t+\Delta t)[\mathbf{\Theta}_{21}\mathbf{P}(t)\mathbf{X}(t)+\mathbf{\Theta}_{22}\mathbf{X}(t)]=\mathbf{\Theta}_{11}\mathbf{P}(t)\mathbf{X}(t)+\mathbf{\Theta}_{12}\mathbf{X}(t) \tag{4.5-34}$$

$\mathbf{X}(t)$ cancels, and the solution for $\mathbf{P}(t+\Delta t)$ is

$$\mathbf{P}(t+\Delta t)=[\mathbf{\Theta}_{11}\mathbf{P}(t)+\mathbf{\Theta}_{12}][\mathbf{\Theta}_{21}\mathbf{P}(t)+\mathbf{\Theta}_{22}]^{-1} \tag{4.5-35}$$

This provides a recursive relation for $\mathbf{P}(t)$, which is initialized with $\mathbf{P}(0)=\mathbf{P}_0$.

The *Chandrasekhar-type algorithm** provides an alternate means of finding the time-varying gain matrix for a system with constant $\mathbf{F}$, $\mathbf{H}$, $\mathbf{L}$, $\mathbf{Q}_C$, and $\mathbf{R}_C$. Neglecting control inputs, the state estimate equation (Eq. 4.5-18) can be expressed as

$$\dot{\hat{\mathbf{x}}}(t) = [\mathbf{F} - \mathbf{K}_C(t)\mathbf{H}]\hat{\mathbf{x}}(t) + \mathbf{K}_C(t)\mathbf{z}(t), \qquad \mathbf{x}(0) = \mathbf{x}_0 \tag{4.5-36}$$

From Eq. 2.3-66, the state transition matrix $\mathbf{\Phi}(t, 0)$ of the closed-loop filter matrix $[\mathbf{F} - \mathbf{K}_C(t)\mathbf{H}]$ satisfies its own differential equation

$$\dot{\mathbf{\Phi}}(t, 0) = [\mathbf{F} - \mathbf{K}_C(t)\mathbf{H}]\mathbf{\Phi}(t, 0), \qquad \mathbf{\Phi}(0, 0) = \mathbf{I}_n \tag{4.5-37}$$

and this relation can be used to compute $\mathbf{\Phi}(t, 0)$, given $\mathbf{K}_C(t)$. Differentiating Eq. 4.5-15,

$$\ddot{\mathbf{P}}(t) = \mathbf{F}\dot{\mathbf{P}}(t) + \dot{\mathbf{P}}(t)\mathbf{F}^T - \dot{\mathbf{P}}(t)\mathbf{H}^T\mathbf{R}_C^{-1}\mathbf{H}\mathbf{P}(t) - \mathbf{P}(t)\mathbf{H}^T\mathbf{R}_C^{-1}\mathbf{H}\dot{\mathbf{P}}(t) \tag{4.5-38a}$$

Using Eq. 4.5-16, this leads to a linear differential equation for $\dot{\mathbf{P}}(t)$,

$$\ddot{\mathbf{P}}(t) = [\mathbf{F} - \mathbf{K}_C(t)\mathbf{H}]\dot{\mathbf{P}}(t) + \dot{\mathbf{P}}(t)[\mathbf{F} - \mathbf{K}_C(t)\mathbf{H}]^T \tag{4.5-38b}$$

with the corresponding transition equation

$$\dot{\mathbf{P}}(t) = \mathbf{\Phi}(t, 0)\dot{\mathbf{P}}(0)\mathbf{\Phi}^T(t, 0) \tag{4.5-39}$$

At the initial time,

$$\begin{aligned}\dot{\mathbf{P}}(0) &= \mathbf{F}\mathbf{P}(0) + \mathbf{P}(0)\mathbf{F}^T + \mathbf{L}\mathbf{Q}_C\mathbf{L}^T - \mathbf{P}(0)\mathbf{H}^T\mathbf{R}_C^{-1}\mathbf{H}\mathbf{P}(0)\\ &\triangleq \mathbf{D}\end{aligned} \tag{4.5-40}$$

which is seen to be a symmetric $(n \times n)$ matrix. Symmetric matrices can be factored in the *LDU decomposition* (W-1),

$$\begin{aligned}\mathbf{D} &= [\mathbf{M}_1\mathbf{M}_2]\mathbf{S}[\mathbf{M}_1^T\mathbf{M}_2^T]\\ &= \mathbf{M}_1\mathbf{M}_1^T - \mathbf{M}_2\mathbf{M}_2^T\end{aligned} \tag{4.5-41}$$

where $\mathbf{S}$ is a diagonal $(\alpha \times \alpha)$ "signature" matrix defined as

$$\mathbf{S} = \text{Diag}[\underbrace{1, 1 \ldots 1}_{\beta}, \underbrace{-1, -1 \ldots -1}_{(\alpha-\beta)}] \tag{4.5-42}$$

*So-named because the differential equations are similar to equations for stellar radiative transfer that were presented by astrophysicist S. Chandrasekhar in 1948. The filter equations were developed by T. Kailath (K-2).

The rank of $\mathbf{D}$ is α, β is the number of positive eigenvalues of $\mathbf{D}$, and $\mathbf{M}_1$ and $\mathbf{M}_2$ are matrices of dimension $(n \times \alpha)$ and $[n \times (\alpha - \beta)]$, respectively. (Note the similarity to the matrix square roots and U-D factors of Sections 4.2 and 4.3.) Equation 4.5-39 then can be written

$$\begin{aligned}\dot{\mathbf{P}}(t) &= \mathbf{\Phi}(t, 0)\mathbf{M}_1\mathbf{M}_1^T\mathbf{\Phi}^T(t, 0) - \mathbf{\Phi}(t, 0)\mathbf{M}_2\mathbf{M}_2^T\mathbf{\Phi}^T(t, 0) \\ &\triangleq \mathbf{Y}_1(t)\mathbf{Y}_1^T(t) - \mathbf{Y}_2(t)\mathbf{Y}_2^T(t) \qquad (4.5\text{-}43)\end{aligned}$$

The definitions of $\mathbf{Y}_1(t)$ and $\mathbf{Y}_2(t)$ involve transition equations:

$$\mathbf{Y}_1(t) = \mathbf{\Phi}(t, 0)\mathbf{M}_1 = \mathbf{\Phi}(t, 0)\mathbf{Y}_1(0) \qquad (4.5\text{-}44)$$

$$\mathbf{Y}_2(t) = \mathbf{\Phi}(t, 0)\mathbf{M}_2 = \mathbf{\Phi}(t, 0)\mathbf{Y}_2(0) \qquad (4.5\text{-}45)$$

Consequently, the corresponding homogeneous differential equations are

$$\dot{\mathbf{Y}}_1(t) = [\mathbf{F} - \mathbf{K}_C(t)\mathbf{H}]\mathbf{Y}_1(t), \qquad \mathbf{Y}_1(0) = \mathbf{M}_1 \qquad (4.5\text{-}46)$$

$$\dot{\mathbf{Y}}_2(t) = [\mathbf{F} - \mathbf{K}_C(t)\mathbf{H}]\mathbf{Y}_2(t) \qquad \mathbf{Y}_2(0) = \mathbf{M}_2 \qquad (4.5\text{-}47)$$

Differentiating the gain equation (Eq. 4.5-16) and substituting for $\dot{\mathbf{P}}(t)$ (Eq. 4.5-43),

$$\begin{aligned}\dot{\mathbf{K}}_C(t) &= \dot{\mathbf{P}}(t)\mathbf{H}^T\mathbf{R}_C^{-1} \\ &= [\mathbf{Y}_1(t)\mathbf{Y}_1^T(t) - \mathbf{Y}_2(t)\mathbf{Y}_2^T(t)]\mathbf{H}^T\mathbf{R}_C^{-1} \qquad (4.5\text{-}48)\end{aligned}$$

with the $(n \times r)$ initial condition specified by Eq. 4.5-16:

$$\mathbf{K}_C(0) = \mathbf{P}(0)\mathbf{H}^T\mathbf{R}_C^{-1} \qquad (4.5\text{-}49)$$

Equations 4.5-46 to 4.5-49 comprise the Chandrasekhar-type algorithm. Whereas the solution for the symmetric $\mathbf{P}(t)$ requires the integration of $n(n+1)/2$ differential equations, the Chandrasekhar-type algorithm contains nr equations for the components of $\mathbf{K}_C(t)$ and $n\alpha$ equations for the components of $\mathbf{Y}_1(t)$ and $\mathbf{Y}_2(t)$. Thus if $n > [2(r + \alpha) - 1]$, the Chandrasekhar-type algorithm contains fewer integrations. This could be the case with few measurements (dimension r) and few disturbance inputs (dimension s). If $\mathbf{P}(0) = \mathbf{0}$ and $\mathbf{Q}_C$ is positive semidefinite, then

$$\mathbf{D} = \mathbf{L}\mathbf{Q}_C\mathbf{L}^T \qquad (4.5\text{-}50)$$

$\mathbf{D}$ is positive semidefinite, so $\beta = \alpha \le s$, $\mathbf{M}_2 = \mathbf{0}$, and

$$\mathbf{D} = \mathbf{M}_1\mathbf{M}_1^T \qquad (4.5\text{-}51)$$

$\mathbf{Y}_2(t)$ is zero, and the algorithm for computing the filter gain matrix is given by

$$\dot{\mathbf{Y}}_1(t) = [\mathbf{F} - \mathbf{K}_C(t)\mathbf{H}]\mathbf{Y}_1(t), \qquad \mathbf{Y}_1(0) = \mathbf{M}_1 \tag{4.5-52}$$

$$\dot{\mathbf{K}}_C(t) = \mathbf{Y}_1(t)\mathbf{Y}_1^T(t)\mathbf{H}^T\mathbf{R}_C^{-1}, \quad \mathbf{K}_C(0) = \mathbf{P}(0)\mathbf{H}^T\mathbf{R}_C^{-1} \tag{4.5-53}$$

A continuous-time *square-root formulation* for the time-varying Riccati equation can be derived for estimation with ill-conditioned matrices. Defining $\mathbf{S}(t)$ as the upper triangular square root of $\mathbf{P}(t)$,

$$\mathbf{P}(t) = \mathbf{S}(t)\mathbf{S}^T(t) \tag{4.5-54}$$

the elements of $\mathbf{S}(t)$ can be computed with Eq. 4.2-96 (interchanging the indices i and j). Then Eq. 4.5-15 can be written as

$$\begin{aligned}\dot{\mathbf{S}}(t)\mathbf{S}^T(t) &+ \mathbf{S}(t)\dot{\mathbf{S}}^T(t)\\ &= \mathbf{F}(t)\mathbf{S}(t)\mathbf{S}^T(t) + \mathbf{S}(t)\mathbf{S}^T(t)\mathbf{F}^T(t) + \mathbf{L}(t)\mathbf{Q}_C(t)\mathbf{L}^T(t)\\ &\quad - \mathbf{S}(t)\mathbf{S}^T(t)\mathbf{H}^T(t)\mathbf{R}_C^{-1}(t)\mathbf{H}(t)\mathbf{S}(t)\mathbf{S}^T(t), \qquad \mathbf{S}(0)\mathbf{S}^T(0) = \mathbf{P}(0)\end{aligned} \tag{4.5-55}$$

Premultiplying by $\mathbf{S}^{-1}(t)$ and postmultiplying by $\mathbf{S}^{-T}(t)$,

$$\begin{aligned}\mathbf{S}^{-1}(t)\dot{\mathbf{S}}(t) &+ \dot{\mathbf{S}}^T(t)\mathbf{S}^{-T}(t)\\ &= \mathbf{S}^{-1}(t)\mathbf{F}(t)\mathbf{S}(t) + \mathbf{S}^T(t)\mathbf{F}^T(t)\mathbf{S}^{-T}(t) + \mathbf{S}^{-1}(t)\mathbf{L}(t)\mathbf{Q}_C(t)\mathbf{L}^T(t)\mathbf{S}^{-T}(t)\\ &\quad - \mathbf{S}^T(t)\mathbf{H}^T(t)\mathbf{R}_C^{-1}(t)\mathbf{H}(t)\mathbf{S}(t)\\ &\triangleq \mathbf{M}(t) = \mathbf{M}_{\mathrm{LT}}(t) + \mathbf{M}_{\mathrm{UT}}(t)\end{aligned} \tag{4.5-56}$$

where $\mathbf{M}_{\mathrm{UT}}(t)$ is the upper triangular partition of $\mathbf{M}(t)$ defined by

$$(m_{ij})_{\mathrm{UT}} = \begin{cases} m_{ij}, & i<j \\ \frac{1}{2}m_{ij}, & i=j \\ 0, & i>j \end{cases} \tag{4.5-57}$$

$\mathbf{M}_{\mathrm{LT}}(t)$ is similarly defined as the lower triangular partition. Since $\mathbf{S}^{-1}(t)$ and $\mathbf{S}(t)$ are upper triangular, their product is upper triangular; hence,

$$\dot{\mathbf{S}}(t) = \mathbf{S}(t)\mathbf{M}_{\mathrm{UT}}(t), \qquad \mathbf{S}(0)\mathbf{S}^T(0) = \mathbf{P}(0) \tag{4.5-58}$$

which is the propagation equation for the covariance square root. $\mathbf{S}(0)$ can be computed by first defining the U-D factors of $\mathbf{P}$ (Eq. 4.3-73 and 4.3-74), then writing $\mathbf{S}(0)$ as $\mathbf{U}(0)\mathbf{D}^{1/2}(0)$ (Eq. 4.3-55). Because $\mathbf{S}^{-1}$ is involved in the definition of $\mathbf{M}$ (Eq. 4.5-56), $\mathbf{S}(0)$ [and, therefore, $\mathbf{P}(0)$] must be positive definite. The optimal filter gain matrix can be expressed in terms of $\mathbf{S}(t)$:

$$\mathbf{K}_C(t) = \mathbf{S}(t)\mathbf{S}^T(t)\mathbf{H}^T(t)\mathbf{R}_C^{-1}(t) \tag{4.5-59}$$

The complexity of the square-root approach is scarcely greater than that of the Riccati solution, and the result is likely to be somewhat more accurate when the system contains both fast and slow modes or when the components of $\mathbf{Q}_C(t)$ or $\mathbf{R}_C(t)$ differ by many orders of magnitude.

Correlation in Disturbance Inputs and Measurement Noise

Paralleling the discrete-time filters, continuous-time filters can account for time-correlated disturbance inputs, cross-correlated disturbance inputs and measurement noise, and time-correlated measurement noise. The first of these can be handled by augmenting the state vector and revising the system model to include the disturbance input dynamics. The remaining two are discussed below.

Assume that the continuous-time system is described by Eqs. 4.5-1 to 4.5-5 and 4.5-7, with cross-correlated disturbance input and measurement noise:

$$E\left\{\begin{bmatrix}\mathbf{w}(t)\\ \mathbf{n}(t)\end{bmatrix}[\mathbf{w}^T(\tau)\ \mathbf{n}^T(\tau)]\right\} = \begin{bmatrix}\mathbf{Q}_C(t) & \mathbf{M}_C(t)\\ \mathbf{M}_C^T(t) & \mathbf{R}_C(t)\end{bmatrix}\delta(t-\tau) \tag{4.5-60}$$

The corresponding discrete-time equations are given in Section 4.4, from which the following post- and pre-update covariance equations are obtained:

$$\mathbf{P}_{k-1}(+) = \mathbf{P}_{k-1}(-) - \mathbf{K}_{k-1}[\mathbf{M}_{k-1}^T + \mathbf{H}_{k-1}\mathbf{P}_{k-1}(-)] \tag{4.5-61}$$

$$\mathbf{P}_k(-) \simeq [\mathbf{I}_n + \mathbf{F}_{k-1}\,\Delta t]\mathbf{P}_{k-1}(+)[\mathbf{I}_n + \mathbf{F}_{k-1}\,\Delta t]^T + \mathbf{Q}_{k-1} \tag{4.5-62}$$

Substituting Eq. 4.5-61 and 4.5-62 and manipulating as in Eq. 4.5-11,

$$\begin{aligned}\frac{\mathbf{P}_k(-) - \mathbf{P}_{k-1}(-)}{\Delta t} &= \mathbf{FP} + \mathbf{PF}^T + \mathbf{FPF}^T\,\Delta t + \frac{\mathbf{Q}}{\Delta t} - \frac{\mathbf{K}}{\Delta t}\mathbf{M}^T\\ &\quad - \mathbf{KM}^T\mathbf{F}^T - \frac{\mathbf{K}}{\Delta t}\mathbf{HP} - \mathbf{KHPF}^T - \mathbf{FKM}^T - \mathbf{FKM}^T\mathbf{F}^T\,\Delta t\\ &\quad - \mathbf{FKHP} - \mathbf{FKHPF}^T\,\Delta t \end{aligned} \tag{4.5-63}$$

where the subscripts for all matrices on the right side are $(k-1)$ and where $\mathbf{P} = \mathbf{P}_{k-1}(-)$. From Eq. 4.4-17 and 4.5-13,

$$\mathbf{K}_{k-1} = [\mathbf{PH}^T + \mathbf{M}]\left[\mathbf{HPH}^T + \mathbf{HM} + \mathbf{M}^T\mathbf{H}^T + \frac{\mathbf{R}_C}{\Delta t}\right]^{-1} \tag{4.5-64}$$

with subscripts deleted as above. In the limit as $\Delta t \to 0$, the continuous-time optimal gain matrix is defined by

$$\mathbf{K}_C(t) = \lim_{\Delta t \to 0} \frac{\mathbf{K}_{k-1}}{\Delta t} = [\mathbf{P}(t)\mathbf{H}^T(t) + \mathbf{L}(t)\mathbf{M}(t)]\,\mathbf{R}_C^{-1}(t) \tag{4.5-65}$$

$\mathbf{M}_{k-1}$ goes to $\mathbf{L}(t)\mathbf{M}_C(t)$ as $\Delta t \to 0$. The differential equation for the state error covariance matrix is the limit of Eq. 4.5-63:

$$\begin{aligned}\dot{\mathbf{P}}(t) &= \mathbf{F}(t)\mathbf{P}(t) + \mathbf{P}(t)\mathbf{F}^T(t) + \mathbf{L}(t)\mathbf{Q}_C(t)\mathbf{L}^T(t) \\ &\quad - \mathbf{K}_C(t)[\mathbf{H}(t)\mathbf{P}(t) + \mathbf{M}_C^T(t)\mathbf{L}^T(t)] \\ &= \mathbf{F}(t)\mathbf{P}(t) + \mathbf{P}(t)\mathbf{F}^T(t) + \mathbf{L}(t)\mathbf{Q}_C(t)\mathbf{L}^T(t) \\ &\quad - [\mathbf{P}(t)\mathbf{H}^T(t) + \mathbf{L}(t)\mathbf{M}_C(t)]\mathbf{R}_C^{-1}(t)[\mathbf{P}(t)\mathbf{H}^T(t) + \mathbf{L}(t)\mathbf{M}_C(t)]^T\end{aligned} \tag{4.5-66}$$

These results are qualitatively similar to the discrete-time results, and the time-skewed correlation assumed in Eq. 4.4-9 vanishes in the limiting process.

The filtering solution for time-correlated measurement error makes use of these results. The continuous-time system is modeled by Eqs. 4.5-1 to 4.5-7, but the measurement error $\mathbf{n}(t)$ is the result of a Gauss–Markov process:

$$\dot{\mathbf{n}}(t) = \mathbf{N}(t)\mathbf{n}(t) + \boldsymbol{\nu}(t) \tag{4.5-67}$$

$$E[\boldsymbol{\nu}(t)] = \mathbf{0} \tag{4.5-68}$$

$$E[\boldsymbol{\nu}(t)\boldsymbol{\nu}^T(\tau)] = \mathbf{V}_C(t)\delta(t-\tau) \tag{4.5-69}$$

Thus the measurement error is not white; however, the error of a *derived* measurement is (B-9). In the limit as $\Delta t \to 0$, the differenced measurement of Eq. 4.4-32 is

$$\boldsymbol{\zeta}(t) = \dot{\mathbf{z}}(t) - \mathbf{N}(t)\mathbf{z}(t) \tag{4.5-70}$$

Since

$$\dot{\mathbf{z}}(t) = \dot{\mathbf{H}}(t)\mathbf{x}(t) + \mathbf{H}(t)\dot{\mathbf{x}}(t) + \dot{\mathbf{n}}(t) \tag{4.5-71}$$

substitution for $\dot{\mathbf{x}}(t)$ and $\mathbf{n}(t)$ leads to

$$\begin{aligned}\boldsymbol{\zeta}(t) &= [\dot{\mathbf{H}}(t) + \mathbf{H}(t)\mathbf{F}(t) - \mathbf{N}(t)\mathbf{H}(t)]\mathbf{x}(t) + [\mathbf{H}(t)\mathbf{L}(t)\mathbf{w}(t) + \boldsymbol{\nu}(t)] \\ &\triangleq \mathbf{D}(t)\mathbf{x}(t) + \boldsymbol{\eta}(t)\end{aligned} \tag{4.5-72}$$

Then the spectral density matrices for the stochastic input and derived measurement error are expressed by

$$E\left\{\begin{bmatrix}\mathbf{w}(t)\\ \boldsymbol{\eta}(t)\end{bmatrix}[\mathbf{w}^T(\tau)\ \boldsymbol{\eta}^T(\tau)]\right\}$$

$$=\begin{bmatrix}\mathbf{Q}_C(t) & \mathbf{Q}_C(t)\mathbf{L}^T(t)\mathbf{H}^T(t)\\ \mathbf{H}(t)\mathbf{L}(t)\mathbf{Q}_C(t) & [\mathbf{H}(t)\mathbf{L}(t)\mathbf{Q}_C(t)\mathbf{L}^T(t)\mathbf{H}^T(t)+\mathbf{V}_C(t)]\end{bmatrix}\delta(t-\tau)$$

$$\triangleq\begin{bmatrix}\mathbf{Q}_{CC}(t) & \mathbf{M}_{CC}(t)\\ \mathbf{M}^T{}_{CC}(t) & \mathbf{R}_{CC}(t)\end{bmatrix}\delta(t-\tau) \tag{4.5-73}$$

The gain and covariance matrices are computed using Eq. 4.5-73 in Eq. 4.5-65 and 4.5-66, and the state estimation equation is

$$\begin{aligned}\dot{\hat{\mathbf{x}}}(t) &= \mathbf{F}(t)\hat{\mathbf{x}}(t)+\mathbf{K}_C(t)[\boldsymbol{\zeta}(t)-\mathbf{D}(t)\hat{\mathbf{x}}(t)]\\ &= \mathbf{F}(t)\hat{\mathbf{x}}(t)+\mathbf{K}_C(t)[\dot{\mathbf{z}}(t)-\mathbf{N}(t)\mathbf{z}(t)-\mathbf{D}(t)\hat{\mathbf{x}}(t)]\end{aligned} \tag{4.5-74}$$

Because $\dot{\hat{\mathbf{x}}}(t)$ is a function of $\dot{\mathbf{z}}(t)$, $\hat{\mathbf{x}}(t)$ is a function of $\mathbf{z}(t)$ within an arbitrary constant. Thus the initial condition for Eq. 4.5-74 must account not only for $\hat{\mathbf{x}}_0$ but for the effect of $\mathbf{z}(0)$ as well. It is shown in (S-3) and (G-1) that this effect is analogous to the measurement update in the discrete-time filter:

$$\hat{\mathbf{x}}(0)=\hat{\mathbf{x}}_0+\mathbf{P}_0\mathbf{H}^T(0)[\mathbf{H}(0)\mathbf{P}_0\mathbf{H}^T(0)+\mathbf{V}_C(0)]^{-1}[\mathbf{z}(0)-\mathbf{H}(0)\hat{\mathbf{x}}_0] \tag{4.5-75}$$

The corresponding initial condition for covariance propagation is

$$\mathbf{P}(0)=\mathbf{P}_0-\mathbf{P}_0\mathbf{H}^T(0)[\mathbf{H}(0)\mathbf{P}_0\mathbf{H}^T(0)+\mathbf{V}_C(0)]^{-1}\mathbf{H}(0)\mathbf{P}_0 \tag{4.5-76}$$

Differentiating the measurement to obtain $\dot{\mathbf{z}}(t)$ for the state estimate is undesirable because $\mathbf{z}(t)$ contains noise; hence an alternate formulation that eliminates $\dot{\mathbf{z}}(t)$ could be useful. Because

$$\mathbf{K}_C(t)\dot{\mathbf{z}}(t)=\frac{d[\mathbf{K}_C(t)\mathbf{z}(t)]}{dt}-\dot{\mathbf{K}}_C(t)\mathbf{z}(t) \tag{4.5-77}$$

Eq. 4.5-74 could be written as

$$\begin{aligned}\frac{d\boldsymbol{\chi}(t)}{dt} &= \frac{d[\hat{\mathbf{x}}(t)-\mathbf{K}_C(t)\mathbf{z}(t)]}{dt}\\ &= [\mathbf{F}(t)-\mathbf{K}_C(t)\mathbf{D}(t)]\hat{\mathbf{x}}(t)-[\mathbf{K}_C(t)\mathbf{N}(t)+\dot{\mathbf{K}}_C(t)]\mathbf{z}(t)\end{aligned} \tag{4.5-78}$$

with the initial condition

$$\boldsymbol{\chi}(0)=\hat{\mathbf{x}}(0)-\mathbf{K}_C(0)\mathbf{z}(0) \tag{4.5-79}$$

Equation 4.5-78 is integrated to find $\hat{\chi}(t)$, and the state estimate is

$$\hat{\mathbf{x}}(t) = \hat{\chi}(t) + \mathbf{K}_C(t)\mathbf{z}(t) \tag{4.5-80}$$

$\dot{\mathbf{K}}_C(t)$ is found by differentiating Eq. 4.5-65, which presumably is smoother than $\mathbf{z}(t)$. For constant coefficients and input statistics,

$$\dot{\mathbf{K}}_C(t) = \dot{\mathbf{P}}(t)\mathbf{H}^T\mathbf{R}_C^{-1} \tag{4.5-81}$$

where $\dot{\mathbf{P}}(t)$ is the right side of Eq. 4.5-66.

4.6 OPTIMAL NONLINEAR ESTIMATION

Optimal state estimation is considerably more difficult when the system contains nonlinear elements because the probability density functions of signals and noise are altered as they are transmitted through these elements. Gaussian inputs cause non-Gaussian response, and the criteria for "best" state estimates are not quite so obvious as before. The mean and standard deviation are incomplete descriptors of the probability density function, and a state estimate based on the conditional mean may be different from one based on the mode or median. Consider the effects of linear and nonlinear transmission on "large" and "small" probability density functions (Fig. 4.6-1). Probability density functions are transmitted through linear elements with no change in the shapes of the distributions. The mean value of the function $\bar{\mathbf{y}}(\mathbf{x})$ is the function of the mean value $\mathbf{y}(\bar{\mathbf{x}})$ and the only possible variation is in the spread of the distribution. Although the probability density function of a small signal may undergo little change in nonlinear transmission, large-signal probability functions may be altered substantially, as suggested by Fig. 4.6-1*b*.

Fortunately, estimators for many nonlinear systems can be based on Kalman and Kalman-Bucy filters; though not precisely "optimum," they are "optimal" in the sense that they tend to the optimum. These modified linear-optimal estimators are useful when the stochastic effects are *additive* and *small*, either as a result of the original system's structure or of reasonable assumptions regarding magnitudes of these effects. Details of specific probability density functions may not be well portrayed, but the overall performance in state estimation can be satisfactory for two reasons. The first is that random signals are summed in the estimators, and the central limit theorem assures that the probability density functions of the sums tend to become Gaussian no matter what the individual distributions look like. The second is that the estimators contain integration or summation, which tend to average out the Gaussian-destroying effects of the nonlinearities in producing the state estimates.

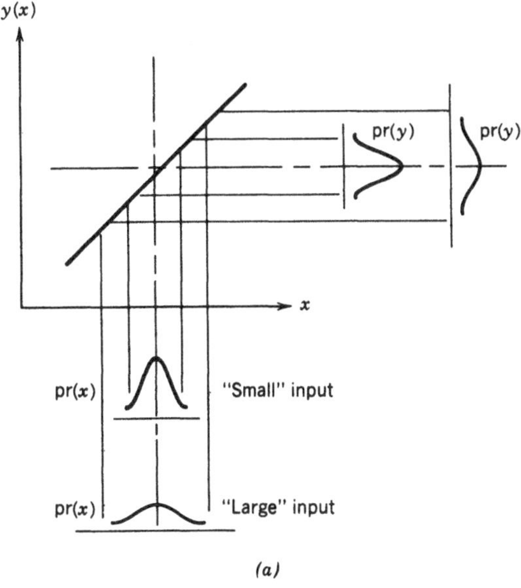

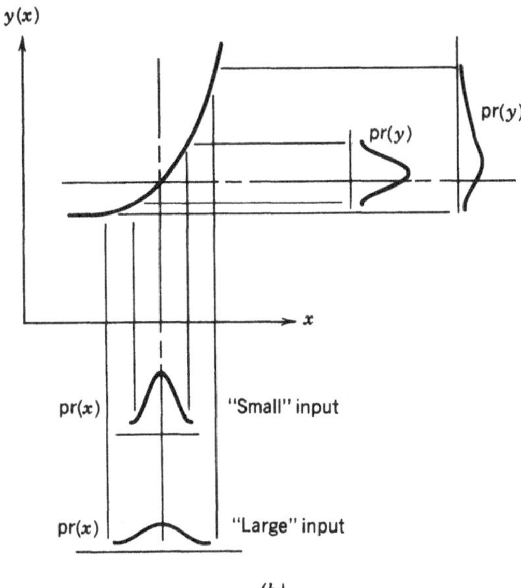

FIGURE 4.6-1 Transmission of probability density functions through linear and nonlinear elements (*a*) Linear transmission; (*b*) nonlinear transmission.

From Chapters 1 and 2, the general nonlinear system can be described by its dynamic, output, and observation equations,

$$\dot{\mathbf{x}}(t) = \mathbf{f}[\mathbf{x}(t), \mathbf{u}(t), \mathbf{w}(t), \mathbf{p}(t), t], \qquad \mathbf{x}(0) = \mathbf{x}_0 \tag{4.6-1}$$

$$\mathbf{y}(t) = \mathbf{h}[\mathbf{x}(t), \mathbf{u}(t), \mathbf{w}(t), \mathbf{p}(t), t] \tag{4.6-2}$$

$$\mathbf{z}(t) = \mathbf{j}[\mathbf{y}(t), \mathbf{n}(t), t] \tag{4.6-3}$$

where $\mathbf{w}(t)$ and $\mathbf{n}(t)$ are random processes, and $\mathbf{x}_0$ is a random variable. To avoid unnecessary detail, assume that explicit effects of the parameter vector $\mathbf{p}(t)$ can be neglected and that $\mathbf{u}(t)$ and $\mathbf{w}(t)$ do not appear in the output (Eq. 4.6-2). If the random effects are additive,

$$\dot{\mathbf{x}}(t) = \mathbf{f}[\mathbf{x}(t), \mathbf{u}(t), t] + \Delta\mathbf{f}[\mathbf{x}(t), \mathbf{w}(t), t], \qquad \mathbf{x}(0) = \mathbf{x}_0 \tag{4.6-4}$$

$$\mathbf{y}(t) = \mathbf{h}[\mathbf{x}(t), t] \tag{4.6-5}$$

$$\begin{aligned}\mathbf{z}(t) &= \mathbf{y}(t) + \mathbf{n}(t)\\ &= \mathbf{h}[\mathbf{x}(t), t] + \mathbf{n}(t)\end{aligned} \tag{4.6-6}$$

then optimal estimates can be formed with certain restrictions on the form of $\Delta\mathbf{f}[\cdot]$. It may be desirable to distinguish between nominal and perturbational solutions to these equations. Denoting perturbation-level values as $\Delta(\cdot)$ and nominal values as $(\cdot)_o$,

$$\begin{aligned}\dot{\mathbf{x}}_o(t) + \Delta\dot{\mathbf{x}}(t) = \mathbf{f}[\mathbf{x}_o(t), \mathbf{u}_o(t), t] &+ \Delta\mathbf{f}_1[\Delta\mathbf{x}(t), \Delta\mathbf{u}(t), t]\\ &+ \Delta\mathbf{f}_2[\Delta\mathbf{w}(t), t], \qquad \mathbf{x}_o(0) \text{ given}, \quad \Delta\mathbf{x}(0) = \Delta\mathbf{x}_0\end{aligned} \tag{4.6-7}$$

$$\mathbf{y}_o(t) + \Delta\mathbf{y}(t) = \mathbf{h}[\mathbf{x}_o(t), t] + \Delta\mathbf{h}[\Delta\mathbf{x}(t), t] \tag{4.6-8}$$

$$\mathbf{z}_o(t) + \Delta\mathbf{z}(t) = \mathbf{h}[\mathbf{x}_o(t), t] + \Delta\mathbf{h}[\Delta\mathbf{x}(t), t] + \Delta\mathbf{n}(t) \tag{4.6-9}$$

In this structure, perturbations are not necessarily "small," and they can be either deterministic or random in nature. The nominal values are deterministic, and their time variations may dictate the explicit time variations of $\Delta\mathbf{f}_1[\cdot]$, $\Delta\mathbf{f}_2[\cdot]$, and $\Delta\mathbf{h}[\cdot]$. If the perturbation variables remain "small," Taylor series expansions of the equations, truncated to first degree, can be used to advantage:

$$\begin{aligned}\dot{\mathbf{x}}_o(t) + \Delta\dot{\mathbf{x}}(t) = \mathbf{f}[\mathbf{x}_o(t), \mathbf{u}_o(t), t] &+ \mathbf{F}(t)\,\Delta\mathbf{x}(t) + \mathbf{G}(t)\,\Delta\mathbf{u}(t)\\ &+ \mathbf{L}(t)\Delta\mathbf{w}(t) + \text{higher-degree terms (H.D.T.)}\end{aligned} \tag{4.6-10}$$

$$\mathbf{y}_o(t) + \Delta\mathbf{y}(t) = \mathbf{h}[\mathbf{x}_o(t), t] + \mathbf{H}(t)\,\Delta\mathbf{x}(t) + \text{H.D.T.} \tag{4.6-11}$$

$$\mathbf{z}_o(t) + \Delta\mathbf{z}(t) = \mathbf{h}[\mathbf{x}_o(t), t] + \mathbf{H}(t)\,\Delta\mathbf{x}(t) + \Delta\mathbf{n}(t) + \text{H.D.T.} \tag{4.6-12}$$

Dropping the higher-degree terms and equating like terms, the equations can be partitioned in deterministic and stochastic parts. The time-varying coefficient matrices [$\mathbf{F}(t)$, $\mathbf{G}(t)$, $\mathbf{L}(t)$, and $\mathbf{H}(t)$] are defined as in Section 2.3. They may be functions of the nominal solution, and it is possible for the dynamic system to be only partially nonlinear (i.e., for either $\mathbf{f}[\cdot]$ or $\mathbf{h}[\cdot]$ to be zero).

In the remainder of this section, consideration is given to the neighboring-optimal linear estimator, the extended Kalman–Bucy filter, and a quasilinear variant of the latter. Primary attention is focused on continuous-time systems with continuous or sampled-data estimators.

Neighboring-Optimal Linear Estimator

If the system can be partitioned as in Eqs. 4.6-10 to 4.6-12, then the nominal (deterministic) solution in $[0, t_f]$ is found by integrating

$$\dot{\mathbf{x}}_o(t) = \mathbf{f}[\mathbf{x}_o(t), \mathbf{u}_o(t), t], \qquad \mathbf{x}_o(0) \text{ given} \tag{4.6-13}$$

and the nominal observation is

$$\mathbf{z}_o(t) = \mathbf{y}_o(t) = \mathbf{h}[\mathbf{x}_o(t), t] \tag{4.6-14}$$

The perturbation equations are

$$\Delta\dot{\mathbf{x}}(t) = \mathbf{F}(t)\,\Delta\mathbf{x}(t) + \mathbf{G}(t)\,\Delta\mathbf{u}(t) + \mathbf{L}(t)\,\Delta\mathbf{w}(t), \qquad \Delta\mathbf{x}(0) = \Delta\mathbf{x}_0 \tag{4.6-15}$$

$$\Delta\mathbf{z}(t) = \mathbf{H}(t)\,\Delta\mathbf{x}(t) + \Delta\mathbf{n}(t) \tag{4.6-16}$$

where

$$\mathbf{F}(t) = \mathbf{F}[\mathbf{x}_o(t), \mathbf{u}_o(t), t] = \frac{\partial\mathbf{f}[\cdot]}{\partial\mathbf{x}} \tag{4.6-17}$$

$$\mathbf{G}(t) = \mathbf{G}[\mathbf{x}_o(t), \mathbf{u}_o(t), t] = \frac{\partial\mathbf{f}[\cdot]}{\partial\mathbf{u}} \tag{4.6-18}$$

$$\mathbf{L}(t) = \mathbf{L}[\mathbf{x}_o(t), \mathbf{u}_o(t), t] = \frac{\partial\mathbf{f}[\cdot]}{\partial\mathbf{w}} \tag{4.6-19}$$

$$\mathbf{H}(t) = \mathbf{H}[\mathbf{x}_o(t), t] = \frac{\partial\mathbf{h}[\cdot]}{\partial\mathbf{x}} \tag{4.6-20}$$

as in Section 2.3, and these are used in the definitions of a neighboring-optimal linear estimator. Assuming

$$E[\Delta\mathbf{x}(0)] = \Delta\hat{\mathbf{x}}_0 \tag{4.6-21}$$

$$E\{[\Delta\mathbf{x}(0) - \Delta\hat{\mathbf{x}}_0][\Delta\mathbf{x}(0) - \Delta\hat{\mathbf{x}}_0]^T\} = \mathbf{P}_0 \tag{4.6-22}$$

$$E[\Delta\mathbf{w}^T(t)\,\Delta\mathbf{n}^T(t)] = \mathbf{0} \tag{4.6-23}$$

$$E\left\{\begin{bmatrix}\Delta\mathbf{w}(\tau)\\ \Delta\mathbf{n}(\tau)\end{bmatrix}[\Delta\mathbf{w}^T(t) \quad \Delta\mathbf{n}^T(t)]\right\} = \begin{bmatrix}\mathbf{Q}_C(t) & \mathbf{0}\\ \mathbf{0} & \mathbf{R}_C(t)\end{bmatrix}\delta(t-\tau) \tag{4.6-24}$$

the Kalman–Bucy filtering equations are given by Eqs. 4.5-15, 4.5-16, and 4.5-18:

$$\Delta\dot{\hat{\mathbf{x}}}(t) = \mathbf{F}(t)\,\Delta\hat{\mathbf{x}}(t) + \mathbf{G}(t)\,\Delta\mathbf{u}(t) + \mathbf{K}_C(t)[\Delta\mathbf{z}(t) - \mathbf{H}(t)\,\Delta\hat{\mathbf{x}}(t)], \quad \Delta\hat{\mathbf{x}}(0) = \Delta\hat{\mathbf{x}}_0 \tag{4.6-25}$$

$$\mathbf{K}_C(t) = \mathbf{P}(t)\mathbf{H}^T(t)\mathbf{R}_C^{-1}(t) \tag{4.6-26}$$

$$\dot{\mathbf{P}}(t) = \mathbf{F}(t)\mathbf{P}(t) + \mathbf{P}(t)\mathbf{F}^T(t) + \mathbf{L}(t)\mathbf{Q}_C(t)\mathbf{L}^T(t) - \mathbf{P}(t)\mathbf{H}^T(t)\mathbf{R}_C^{-1}(t)\mathbf{H}(t)\mathbf{P}(t), \qquad \mathbf{P}(0) = \mathbf{P}_0 \tag{4.6-27}$$

with

$$\Delta\mathbf{z}(t) = \mathbf{z}(t) - \mathbf{z}_o(t) \tag{4.6-28}$$

$$\Delta\mathbf{u}(t) = \mathbf{u}(t) - \mathbf{u}_0(t) \tag{4.6-29}$$

The total state estimate is simply the sum of the perturbation estimate and the nominal solution:

$$\hat{\mathbf{x}}(t) = \mathbf{x}_o(t) + \Delta\hat{\mathbf{x}}(t) \tag{4.6-30}$$

This state estimate is clearly approximate, and it assumes that there is a known nominal state trajectory; however, if the deterministic control perturbation, $\Delta\mathbf{u}(t)$, and filter residual, $[\Delta\mathbf{z}(t) - \mathbf{H}(t)\,\Delta\hat{\mathbf{x}}(t)]$, can be kept small, the state estimate will be close to the true value. The time-varying Jacobian matrices (Eqs. 4.6-17 to 4.6-20) and gain matrix (Eq. 4.6-26) can be precomputed if the nominal trajectory is known ahead of time, leaving just Eqs. 4.6-25 and 4.6-30 to be executed in real time. A corresponding sampled-data estimate of the state at periodic instants could be obtained using the discrete-time estimator of Section 4.3 in place of Eqs. 4.6-25 to 4.6-27.

Extended Kalman–Bucy Filter

An improved state estimate can be calculated with no prior knowledge of a nominal trajectory. The extended Kalman–Bucy filter retains the linear calculation of the covariance and gain matrices, and it updates the state estimate using a linear function of a filter residual; however, it uses the original nonlinear equations for state propagation and definition of the output vector. The estimate error statistics are approximated as Gaussian, and the filter equations are as follows:

$$\dot{\hat{\mathbf{x}}}(t) = \mathbf{f}[\hat{\mathbf{x}}(t), \mathbf{u}(t), t] + \mathbf{K}_C(t)\{\mathbf{z}(t) - \mathbf{h}[\hat{\mathbf{x}}(t), t]\}, \qquad \hat{\mathbf{x}}(0) = \hat{\mathbf{x}}_0 \tag{4.6-31}$$

$$\mathbf{K}_C(t) = \mathbf{P}(t)\mathbf{H}^T(t)\mathbf{R}_C^{-1}(t) \tag{4.6-32}$$

$$\begin{aligned}\dot{\mathbf{P}}(t) &= \mathbf{F}(t)\mathbf{P}(t) + \mathbf{P}(t)\mathbf{F}^T(t) + \mathbf{L}(t)\mathbf{Q}_C(t)\mathbf{L}^T(t)\\ &\quad - \mathbf{P}(t)\mathbf{H}^T(t)\mathbf{R}_C^{-1}(t)\mathbf{H}(t)\mathbf{P}(t), \qquad \mathbf{P}(0) = \mathbf{P}_0\end{aligned} \tag{4.6-33}$$

where

$$E[\mathbf{x}(0)] = \hat{\mathbf{x}}_0 \tag{4.6-34}$$

$$E\{[\mathbf{x}(0) - \hat{\mathbf{x}}_0][\mathbf{x}(0) - \hat{\mathbf{x}}_0]^T\} = \mathbf{P}_0 \tag{4.6-35}$$

$$\mathbf{F}(t) = \mathbf{F}[\mathbf{x}(t), \mathbf{u}(t), t] = \frac{\partial \mathbf{f}[\cdot]}{\partial \mathbf{x}} \tag{4.6-36}$$

$$\mathbf{G}(t) = \mathbf{G}[\mathbf{x}(t), \mathbf{u}(t), t] = \frac{\partial \mathbf{f}[\cdot]}{\partial \mathbf{u}} \tag{4.6-37}$$

$$\mathbf{L}(t) = \mathbf{L}[\mathbf{x}(t), \mathbf{u}(t), t] = \frac{\partial \mathbf{f}[\cdot]}{\partial \mathbf{w}} \tag{4.6-38}$$

$$\mathbf{H}(t) = \mathbf{H}[\mathbf{x}(t), t] = \frac{\partial \mathbf{h}[\cdot]}{\partial \mathbf{x}} \tag{4.6-39}$$

and the partial derivative matrices are evaluated at current values of the variables $\hat{\mathbf{x}}$ and $\mathbf{u}$. These time-varying Jacobian matrices cannot be precomputed because they are functions of the state estimate. Consequently, the gain matrix also depends on the state estimate, and all three filter equations (Eqs. 4.6-31 to 4.6-33) must be calculated in real time. Improved estimates could be obtained from a second- or higher-degree filter that retains more terms in the Taylor series expansions and accompanying derivations, but the quasilinear filter described below generally performs as well or better than such filters.

If sampled-data measurements of a continuous process are available, a hybrid extended Kalman filter is most appropriate for state estimation. The hybrid filter uses continuous-time models of dynamics, observation, and covariance propagation in combination with the discrete-time equations for measurement update and gain computation. It is convenient to express the filter with five equations as in Section 4.3. The differential equations are integrated incrementally from t_{k-1} to t_k, beginning at $k = 0$, using an established algorithm [e.g., the Euler or Runge–Kutta method (Section 2.3)]:

State Estimate Propagation

$$\hat{\mathbf{x}}[t_k(-)] = \hat{\mathbf{x}}[t_{k-1}(+)] + \int_{t_{k-1}}^{t_k} \mathbf{f}\{\hat{\mathbf{x}}[\tau(-)], \mathbf{u}(\tau), \tau\}\, d\tau \tag{4.6-40}$$

Covariance Estimate Propagation

$$\mathbf{P}[t_k(-)] = \mathbf{P}[t_{k-1}(+)] + \int_{t_{k-1}}^{t_k} [\mathbf{F}(\tau)\mathbf{P}(\tau) + \mathbf{P}(\tau)\mathbf{F}^T(\tau) + \mathbf{L}(\tau)\mathbf{Q}_C(\tau)\mathbf{L}^T(\tau)]\, d\tau \tag{4.6-41}$$

$\mathbf{F}(\tau)$ is evaluated in the interval for corresponding values of $\hat{\mathbf{x}}[\tau(-)]$, where $(-)$ indicates that the update has not occurred. The update and gain equations are analogous to the previous discrete-time filters, with the filter residual modified to account for nonlinear observations:

Filter Gain Computation

$$\mathbf{K}(t_k) = \mathbf{P}[t_k(-)]\mathbf{H}^T(t_k)\{\mathbf{H}(t_k)\mathbf{P}[t_k(-)]\mathbf{H}^T(t_k) + \mathbf{R}(t_k)\}^{-1} \tag{4.6-42}$$

State Estimate Update

$$\hat{\mathbf{x}}[t_k(+)] = \hat{\mathbf{x}}[t_k(-)] + \mathbf{K}(t_k)\{\mathbf{z}(t_k) - \mathbf{h}[\hat{\mathbf{x}}[t_k(-)], t_k]\} \tag{4.6-43}$$

Covariance Estimate "Update"

$$\mathbf{P}[t_k(+)] = [\mathbf{I}_n - \mathbf{K}(t_k)\mathbf{H}(t_k)]\mathbf{P}[t_k(-)] \tag{4.6-44}$$

$\mathbf{H}(t_k)$ is evaluated with $\mathbf{x} = \hat{\mathbf{x}}[t_k(-)]$. Note that the hybrid filter characterizes the stochastic disturbance input by its spectral density matrix and the measurement error by its covariance.

Although suboptimum, a constant-gain approximation (based on the steady-state solution for $\mathbf{P}(t)$, which is described in Chapter 6) can be used if the system is time-invariant (S-1). The extended Kalman–Bucy filter is robust, in part because system modeling is improved and the error dynamics are stable; hence the constant-gain version may be adequate for many applications.

Quasilinear Filter

Given a nonlinear scalar function of a scalar variable $f(x)$ the truncated Taylor series expansion about a nominal point x_0 provides a linear approximation to the nonlinearity

$$f(x) \simeq a_0 + a_1(x - x_0) \tag{4.6-45}$$

where $a_0 = f(x_0)$ and $a_1 = \partial f(x)/\partial x$ evaluated at $x = x_0$.* No consideration is given to the magnitude of $(x - x_0)$, and the approximation generally gets worse as $|x - x_0|$ increases. The saturation nonlinearity sketched in Fig. 4.6-2 makes the case. Expanding about $x_0 = 0$, the linear approximation is exact until $|x| > x_1$, at which point the error grows rapidly. If the magnitude of $(x - x_0)$ were known, the lines with lower slope could be closer to the output of the nonlinearity. These lines represent *describing functions*, which have the same form as Eq. 4.6-45 but whose coefficients are dependent on the magnitude of the independent variable.

More specifically, we would like to choose a_0 and a_1 so that the error between $f(x)$ and $[a_0 + a_1(x - x_0)]$ is minimized, where the possible magnitudes of $(x - x_0)$ are represented by its probability density function. Forming the mean-square-error cost function

$$J = E\{[f(x) - a_0 - a_1\tilde{x}]^2\} \tag{4.6-46}$$

where $\tilde{x} \triangleq x - x_0$, we wish to minimize J by our choice of a_0 and a_1. From Section 2.1, the minimum is characterized by

$$\frac{\partial J}{\partial a_0} = \frac{\partial J}{\partial a_1} = 0 \tag{4.6-47}$$

which leads to

$$a_0 = E[f(x)] + a_1 E[\tilde{x}] \tag{4.6-48}$$

$$a_1 = \frac{E[\tilde{x}f(x)] + a_0 E[\tilde{x}]}{E[\tilde{x}^2]} \tag{4.6-49}$$

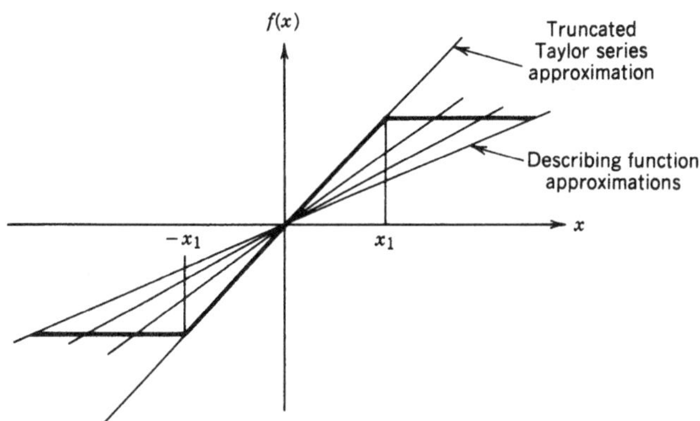

FIGURE 4.6-2 Saturation nonlinearity and various linear approximations.

*Strictly speaking, Eq. 4.6-45 is an *affine approximation* (or transformation) because it contains a bias as well as a scale factor, whereas a linear transformation contains only a scale factor; however, it is common usage to refer to affine transformations as "linear."

or

$$a_0 = \frac{E[\tilde{x}^2]E[f(x)] + E[\tilde{x}]E[\tilde{x}f(x)]}{E[\tilde{x}^2] - E^2[\tilde{x}]} \tag{4.6-50}$$

$$a_1 = \frac{E[\tilde{x}f(x)] + E[\tilde{x}]E[f(x)]}{E[\tilde{x}^2] - E^2[\tilde{x}]} \tag{4.6-51}$$

The expected value is a function of pr(x); if x is Gaussian, the describing function coefficients can be expressed as tabulated functions of the mean and standard deviation. If $f(x)$ and $\tilde{x}$ both have zero mean values, a_0 is zero, and the describing function is simply

$$a_1 = \frac{E[\tilde{x}f(x)]}{E[\tilde{x}^2]} \tag{4.6-52}$$

As an example, the describing function gain (a_1) for the symmetric saturation nonlinearity is shown in Fig. 4.6-3. The effective gain decreases as the

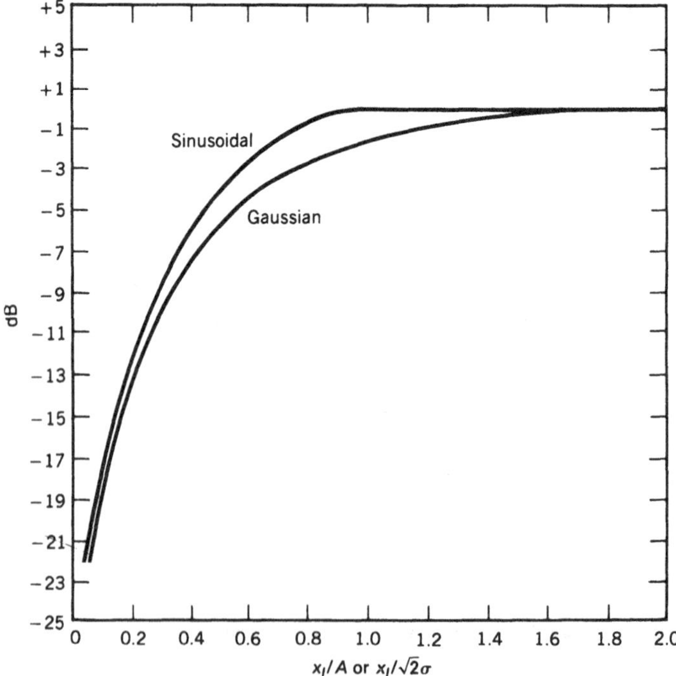

FIGURE 4.6-3 Sinusoidal and Gaussian-random-input describing functions for saturation [from (G-3)].

signal's standard deviation (σ) or sinusoidal amplitude (A) increases, in keeping with our qualitative observations of Fig. 4.6-2. Random input describing functions for a number of common nonlinearities are presented in (G-2).

These scalar concepts can be generalized to quasilinear approximations for nonlinear vector functions of vector variables $\mathbf{f}(\mathbf{x})$. We wish to approximate the r-vector $\mathbf{f}(\mathbf{x})$ as

$$\mathbf{f}(\mathbf{x}) \simeq \mathbf{b} + \mathbf{D}(\mathbf{x} - \hat{\mathbf{x}}) \tag{4.6-53}$$

where $\mathbf{b}$ is an $(r \times 1)$ quasilinear bias vector, $\mathbf{D}$ is an $(r \times n)$ quasilinear sensitivity matrix, $\mathbf{x}$ has dimension n, and $\hat{\mathbf{x}}$ is the expected value of $\mathbf{x}$. The objective is to minimize the trace of the error covariance matrix

$$\begin{aligned} J &= E\{\mathrm{Tr}[\mathbf{f}(\mathbf{x}) - \mathbf{b} - \mathbf{D}\tilde{\mathbf{x}}][\mathbf{f}(\mathbf{x}) - \mathbf{b} - \mathbf{D}\tilde{\mathbf{x}}]^T\} \\ &= E\{[\mathbf{f}(\mathbf{x}) - \mathbf{b} - \mathbf{D}\tilde{\mathbf{x}}]^T[\mathbf{f}(\mathbf{x}) - \mathbf{b} - \mathbf{D}\tilde{\mathbf{x}}]\} \end{aligned} \tag{4.6-54}$$

by proper choice of $\mathbf{b}$ and $\mathbf{D}$, with $\tilde{\mathbf{x}} \triangleq (\mathbf{x} - \hat{\mathbf{x}})$. The first component of the criterion for optimality is

$$\begin{aligned} \frac{\partial J}{\partial \mathbf{b}} &= \mathbf{0} = 2E[-\mathbf{f}(\mathbf{x}) + \mathbf{b} + \mathbf{D}\tilde{\mathbf{x}}] \\ &= 2\{-E[\mathbf{f}(\mathbf{x})] + \mathbf{b}\} \end{aligned} \tag{4.6-55}$$

which indicates that $\mathbf{b}$ is the expected value of the nonlinear function:

$$\mathbf{b} = E[\mathbf{f}(\mathbf{x})] \tag{4.6-56}$$

The minimizing value of $\mathbf{D}$ is defined by the second component of the optimality criterion:

$$\begin{aligned} \frac{\partial J}{\partial \mathbf{D}} &= \mathbf{0} = 2E\{-\mathbf{f}(\mathbf{x})\tilde{\mathbf{x}}^T + \mathbf{b}\tilde{\mathbf{x}}^T + \mathbf{D}\tilde{\mathbf{x}}\tilde{\mathbf{x}}^T\} \\ &= 2E\{-\mathbf{f}(\mathbf{x})\tilde{\mathbf{x}}^T + E[\mathbf{f}(\mathbf{x})]\tilde{\mathbf{x}}^T\} + 2\mathbf{D}\mathbf{P} \end{aligned} \tag{4.6-57}$$

where

$$\mathbf{P} = E[(\mathbf{x} - \hat{\mathbf{x}})(\mathbf{x} - \hat{\mathbf{x}})^T\} \tag{4.6-58}$$

Solving for the quasilinear sensitivity matrix,

$$\mathbf{D} = E\{\mathbf{f}(\mathbf{x})\tilde{\mathbf{x}}^T - E[\mathbf{f}(\mathbf{x})]\tilde{\mathbf{x}}^T\}\mathbf{P}^{-1} \tag{4.6-59}$$

These results can be applied to the dynamic system and observation vectors used in estimation:

$$\mathbf{f}(\mathbf{x}) \simeq E[\mathbf{f}(\mathbf{x}) + E\{\mathbf{f}(\mathbf{x})\tilde{\mathbf{x}}^T - E[\mathbf{f}(\mathbf{x})]\tilde{\mathbf{x}}^T\}\mathbf{P}^{-1}(\mathbf{x}-\hat{\mathbf{x}})$$
$$= \hat{\mathbf{f}}(\mathbf{x}) + \mathbf{D}(\mathbf{x}-\hat{\mathbf{x}}) \tag{4.6-60}$$
$$\mathbf{h}(\mathbf{x}) \simeq E[\mathbf{h}(\mathbf{x})] + E\{\mathbf{h}(\mathbf{x})\tilde{\mathbf{x}}^T - E[\mathbf{f}(\mathbf{x})]\tilde{\mathbf{x}}^T\}\mathbf{P}^{-1}(\mathbf{x}-\tilde{\mathbf{x}})$$
$$= \hat{\mathbf{h}}(\mathbf{x}) + \mathbf{H}_D(\mathbf{x}-\hat{\mathbf{x}}) \tag{4.6-61}$$

Consequently, the quasilinear extended Kalman filter uses $\hat{\mathbf{f}}(\mathbf{x})$ and $\hat{\mathbf{h}}(\mathbf{x})$ to propagate the state estimate and form the residual, and it uses $\mathbf{D}$ and $\mathbf{H}_D$ in the filter gain computation and covariance propagation. Including time variation in the system equations, deterministic control inputs, and linear-additive disturbance inputs, the continuous-time formulation is

$$\dot{\hat{\mathbf{x}}}(t) = \hat{\mathbf{f}}[\hat{\mathbf{x}}(t), \mathbf{u}(t), t] + \mathbf{K}_C(t)\{\mathbf{z}(t) - \hat{\mathbf{h}}[\hat{\mathbf{x}}(t), t]\}, \qquad \hat{\mathbf{x}}(0) = \hat{\mathbf{x}}_0 \tag{4.6-62}$$
$$\mathbf{K}_C(t) = \mathbf{P}(t)\mathbf{H}_D^T(t)\mathbf{R}_C^{-1}(t) \tag{4.6-63}$$
$$\dot{\mathbf{P}}(t) = \mathbf{D}(t)\mathbf{P}(t) + \mathbf{P}(t)\mathbf{D}^T(t) + \mathbf{L}(t)\mathbf{Q}_C(t)\mathbf{L}^T(t)$$
$$- \mathbf{P}(t)\mathbf{H}_D^T(t)\mathbf{R}_C^{-1}(t)\mathbf{H}_D(t)\mathbf{P}(t), \qquad \mathbf{P}(0) = \mathbf{P}_0 \tag{4.6-64}$$

Although these equations look like those of the extended Kalman–Bucy filter, they are somewhat more complex. Much auxiliary computation is needed to obtain the various vectors and matrices of expected values, which depend on the continually changing parameters of pr($\mathbf{x}$) [i.e., on $\hat{\mathbf{x}}(t)$ and $\mathbf{P}(t)$]. Tabulated values of the quasilinear functions (analogous to the describing function of Fig. 4.6-3) can be stored for "lookup" during estimation, greatly reducing computations.

An equivalent sampled-data filter is formed by using quasilinear approximations in Eqs. 4.6-40 to 4.6-44.

4.7 ADAPTIVE FILTERING

State estimators have been formulated under the assumption that dynamic system parameters and input/measurement-error statistics are known, but there usually is some degree of uncertainty in this knowledge. If the actual values of system coefficients and covariances are different from those used in estimation, then the filter is suboptimal; state estimates may contain more error than is necessary and, in some instances, diverge from the neighborhood of true values. State estimates could be improved by simultaneously estimating the uncertain parameters and statistics and using this additional information to *adapt* the filter gains and model coefficients to the measurements. Adaptive filters may perform as well as optimal filters in the limit; however, the less that is known about a system prior to estimation, the greater the error in the resulting estimates.

Our attention is restricted to adaptive state estimators for linear systems, although the general approaches could be applied to nonlinear systems with minor modifications. Furthermore, it is assumed that the unknown parameters and statistics are constant, although their estimates will vary in time. In any case, adaptive filters prove to be nonlinear, as can be inferred from inspecting a linear, continuous-time model:

$$\dot{\mathbf{x}}(t) = \mathbf{F}(\mathbf{p}, t)\mathbf{x}(t) + \mathbf{G}(\mathbf{p}, t)\mathbf{u}(t) + \mathbf{L}(\mathbf{p}, t)\mathbf{w}(t), \qquad \mathbf{x}(0) = \mathbf{x}_0 \tag{4.7-1}$$

$$\mathbf{z}(t) = \mathbf{H}(t)\mathbf{x}(t) + \mathbf{n}(t) \tag{4.7-2}$$

$$E[\mathbf{w}^T(t)\ \mathbf{n}^T(t)] = \mathbf{0} \tag{4.7-3}$$

$$E\left\{\begin{bmatrix}\mathbf{w}(t)\\ \mathbf{n}(t)\end{bmatrix}[\mathbf{w}^T(\tau)\ \mathbf{n}^T(\tau)]\right\} = \begin{bmatrix}\mathbf{Q}_C & \mathbf{0}\\ \mathbf{0} & \mathbf{R}_C\end{bmatrix}\delta(t-\tau) \tag{4.7-4}$$

A *parameter-adaptive filter* estimates $\mathbf{p}$ as well as $\mathbf{x}(t)$ by augmenting the state; even if $\mathbf{F}(\cdot)$, $\mathbf{G}(\cdot)$, and $\mathbf{L}(\cdot)$ are linear functions of $\mathbf{p}$, nonlinearity is introduced by products of the elements of $\mathbf{p}(t)$ and $\mathbf{x}(t)$ or $\mathbf{u}(t)$. A *noise-adaptive filter* bases its estimates of $\mathbf{Q}_C$ (sometimes called "process noise") and $\mathbf{R}_C$ on the measurement residual, $[\mathbf{z}(t) - \mathbf{H}(t)\hat{\mathbf{x}}(t)]$. The optimal filter gain $\mathbf{K}_C(t)$ is a function of $\mathbf{Q}_C$ and $\mathbf{R}_C$ (Eqs. 4.5-15 and 4.5-16),

$$\mathbf{K}_C(t) = \mathbf{P}(t)\mathbf{H}^T(t)\mathbf{R}_C^{-1} \tag{4.7-5}$$

$$\dot{\mathbf{P}}(t) = \mathbf{F}(t)\mathbf{P}(t) + \mathbf{P}(t)\mathbf{F}^T(t) + \mathbf{L}(t)\mathbf{Q}_C\mathbf{L}^T(t) - \mathbf{P}\mathbf{H}^T(t)\mathbf{R}_C^{-1}\mathbf{H}(t)\mathbf{P}(t), \qquad \mathbf{P}(0) = \mathbf{P}_0 \tag{4.7-6}$$

and $\mathbf{K}_C(t)$ multiplies the residual in the state estimator,

$$\dot{\hat{\mathbf{x}}}(t) = \mathbf{F}(t)\hat{\mathbf{x}}(t) + \mathbf{K}_C(t)[\mathbf{z}(t) - \mathbf{H}(t)\hat{\mathbf{x}}(t)], \qquad \hat{\mathbf{x}}(0) = \hat{\mathbf{x}}_0 \tag{4.7-7}$$

therefore, the state estimate is a nonlinear function of the residual.

Parameter-Adaptive Filtering

The extended Kalman–Bucy filter provides a basis for adaptive filtering that is straightforward and optimal in the sense of Section 4.6. This is not to say, however, that satisfactory results are guaranteed for every realization of the filter. The success of parameter estimation depends on many characteristics of the system: the number of uncertain parameters, the magnitude of uncertainty, the functional dependence of outputs on the uncertain parameters, the quality of output measurements, and the knowledge of system inputs. An *augmented state vector* $\mathbf{x}_A(t)$ is formed from the original

n-dimensional state vector $\mathbf{x}(t)$ and the l-vector of parameters to be estimated, $\mathbf{p}(t)$:

$$\mathbf{x}_A(t) = \begin{bmatrix} \mathbf{x}(t) \\ \mathbf{p}(t) \end{bmatrix} \tag{4.7-8}$$

The nonlinear dynamic equation has dimension $(l+n)$:

$$\dot{\mathbf{x}}_A(t) = \mathbf{f}_A[\mathbf{x}_A(t), \mathbf{u}(t), \mathbf{w}(t), t], \qquad \mathbf{x}_A(0) \text{ given} \tag{4.7-9a}$$

with m-dimensional $\mathbf{u}(t)$ and s-dimensional $\mathbf{w}(t)$. Alternatively,

$$\begin{bmatrix} \dot{\mathbf{x}}(t) \\ \dot{\mathbf{p}}(t) \end{bmatrix} = \begin{bmatrix} \mathbf{f}_1[\mathbf{x}(t), \mathbf{p}(t), \mathbf{u}(t), \mathbf{w}(t), t] \\ \mathbf{f}_2[\mathbf{p}(t), \mathbf{w}(t), t] \end{bmatrix}, \qquad \mathbf{x}(0) = \mathbf{x}_0, \mathbf{p}(0) = \mathbf{p}_0 \tag{4.7-9b}$$

where

$$\mathbf{f}_1[\cdot] = \mathbf{F}[\mathbf{p}(t), t]\mathbf{x}(t) + \mathbf{G}[\mathbf{p}(t), t]\mathbf{u}(t) + \mathbf{L}[\mathbf{p}(t), t]\mathbf{w}(t) \tag{4.7-10}$$

and $\mathbf{f}_2[\cdot]$ depends on the model for the parameter vector, described below. The observation equation is

$$\begin{aligned} \mathbf{z}(t) &= \mathbf{H}_A(t)\mathbf{x}_A(t) + \mathbf{n}(t) \\ &= [\mathbf{H}(t) \quad \mathbf{0}] \begin{bmatrix} \mathbf{x}(t) \\ \mathbf{p}(t) \end{bmatrix} + \mathbf{n}(t) \end{aligned} \tag{4.7-11}$$

assuming that no additional measurements are made. The noise statistics are defined as before, although $\mathbf{w}(t)$, $\mathbf{L}(t)$, and $\mathbf{Q}_C$ may be redefined to account for disturbances that drive the parameter dynamics.

Parametric variations in $\mathbf{H}(t)$ may not be distinguished from variations in $\mathbf{F}(t)$ without prior information because the measurement time history could result from any one of an infinite number of "similar" systems (see Section 2.2 for a discussion of similarity transformations). As long as the principal objective is an improved state estimate, neglecting parameter variations in $\mathbf{H}(t)$ may be of little significance, as the parameters of $\mathbf{F}(\mathbf{p}, t)$ can adapt to those of the best similar system.

To the extent that $\mathbf{p}$ represents an input-output sensitivity which in turn provides the basis for parameter estimation, specific knowledge of the input as well as the output is needed. Deterministic inputs can provide better parameter estimates than stochastic inputs, as long as the effects of parameter variations are observable in the output. If the input is only inferred by its statistics, and if the inference itself is uncertain, then the parameter estimates will suffer. For example, assuming that the unmeasured input is white noise when it actually is a sinusoid could produce spurious results. If, however, the sinusoid can be measured (as for a controlled input)

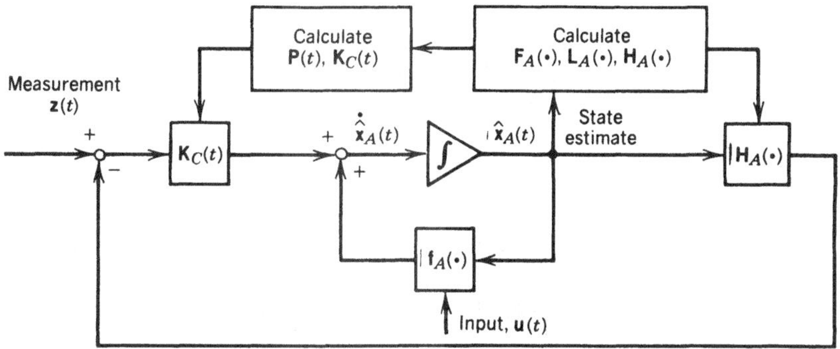

FIGURE 4.7-1 Parameter-adaptive extended Kalman–Bucy filter.

and its driving frequency can be varied to elicit frequency-dependent response, parameter estimation can be improved.

The extended Kalman–Bucy filter follows Section 4.6 directly (Fig. 4.7-1). The state estimate is found by integrating the nonlinear differential equation that models system and parameter dynamics with optimal "forcing" by the measurement residual:

$$\dot{\hat{\mathbf{x}}}_A(t) = \mathbf{f}_A[\hat{\mathbf{x}}_A(t), \mathbf{u}(t), \mathbf{0}, t] + \mathbf{K}_C(t)[\mathbf{z}(t) - \mathbf{H}_A(t)\hat{\mathbf{x}}_A(t)], \qquad \hat{\mathbf{x}}_A(0) \text{ given} \tag{4.7-12}$$

The zero in the third element of $\mathbf{f}_A[\cdot]$ represents the assumed zero mean value of disturbance inputs. The $[(l+n) \times r]$ filter gain matrix is

$$\mathbf{K}_C(t) = \mathbf{P}(t)\mathbf{H}_A^T(t)\mathbf{R}_C^{-1} \tag{4.7-13}$$

where $\mathbf{P}(t)$ is the integral of the $[(l+n) \times (l+n)]$ matrix Riccati equation:

$$\dot{\mathbf{P}}(t) = \mathbf{F}_A(t)\mathbf{P}(t) + \mathbf{P}(t)\mathbf{F}_A^T(t) + \mathbf{L}_A(t)\mathbf{Q}_A(t)\mathbf{L}_A^T(t) - \mathbf{P}(t)\mathbf{H}_A^T(t)\mathbf{R}_C^{-1}\mathbf{H}_A(t)\mathbf{P}(t), \qquad \mathbf{P}(0) = \mathbf{P}_0 \tag{4.7-14}$$

The sensitivity to process noise is described by

$$\mathbf{L}_A(t) = \left.\begin{bmatrix} \mathbf{L}[\mathbf{p}(t), t] \\ \partial \mathbf{f}_2/\partial \mathbf{w} \end{bmatrix}\right|_{\mathbf{x}=\hat{\mathbf{x}}(t),\ \mathbf{p}=\hat{\mathbf{p}}(t),\ \mathbf{u}=\mathbf{u}(t)} \tag{4.7-15}$$

and $\mathbf{Q}_A$ is a block-diagonal matrix containing $\mathbf{Q}_C$ and the assumed parameter process noise $\mathbf{Q}_P$. The system's Jacobian matrix $\mathbf{F}_A(t)$ is formed in the usual way:

$$\mathbf{F}_A(t) = \begin{bmatrix} \partial\mathbf{f}_1/\partial\mathbf{x} & \partial\mathbf{f}_1/\partial\mathbf{p} \\ \partial\mathbf{f}_2/\partial\mathbf{x} & \partial\mathbf{f}_2/\partial\mathbf{p} \end{bmatrix} \Bigg|_{\mathbf{x}=\hat{\mathbf{x}}(t),\ \mathbf{p}=\hat{\mathbf{p}}(t),\ \mathbf{u}=\mathbf{u}(t)} \tag{4.7-16}$$

From Eqs. 4.7-9 and 4.7-10,

$$\frac{\partial\mathbf{f}_1}{\partial\mathbf{x}} = \mathbf{F}[\mathbf{p}(t), t] \tag{4.7-17}$$

$$\frac{\partial\mathbf{f}_1}{\partial\mathbf{p}} = \left\{ \left[\frac{\partial\mathbf{F}(\cdot)}{\partial\mathbf{p}}\right]\mathbf{x}(t) + \left[\frac{\partial\mathbf{G}(\cdot)}{\partial\mathbf{p}}\right]\mathbf{u}(t) \right\} \tag{4.7-18}$$

$$\frac{\partial\mathbf{f}_2}{\partial\mathbf{x}} = \mathbf{0} \tag{4.7-19}$$

$\partial\mathbf{f}_2/\partial\mathbf{w}$ and $\partial\mathbf{f}_2/\partial\mathbf{p}$ depend upon the dynamic model for the parameter $\mathbf{p}$.

It should be noted that the dimension of $\mathbf{p}$ is problem-dependent and has no general relationship to the dimensions of $\mathbf{F}(\cdot)$, $\mathbf{G}(\cdot)$, and $\mathbf{L}(\cdot)$. The number of uncertain parameters must be small in some sense (e.g., on the order of n or less), because model uncertainty diminishes the filter's ability to distinguish between variations resulting from disturbance inputs and those due to measurement error. The magnitude of each parameter's uncertainty should be small for similar reasons.

Assume that $\mathbf{p}$ is an unknown *random constant*, specified only by prior estimates of its mean value $\mathbf{p}_0$ and covariance $\mathbf{P}_p(0)$. The appropriate dynamic model for estimation is

$$\begin{aligned} \dot{\mathbf{p}}(t) &= \mathbf{f}_2[\mathbf{p}(t), \mathbf{w}(t), t] \\ &= \mathbf{0}, \qquad \mathbf{p}(0) = \mathbf{p}_0 \end{aligned} \tag{4.7-20}$$

The augmented covariance matrix is calculated by integrating Eq. 4.7-14, with the corresponding elements of $\mathbf{P}(0)$ initialized as $\mathbf{P}_p(0)$. While Eq. 4.7-20 may appear to assure that $\hat{\mathbf{p}}(t)$ will always be $\mathbf{p}(0)$, recall that the estimate of $\mathbf{p}(t)$ is forced by the measurement residual in Eq. 4.7-12. With no process noise in Eq. 4.7-20, $\mathbf{Q}_p(t) = \mathbf{0}$, and the elements of $\mathbf{K}_C(t)$ that update $\mathbf{p}(t)$ should go to zero as time increases.

Example 4.7-1 Estimation of a Random Constant Parameter Using an Extended Kalman–Bucy Filter

This example is based on the weathervane dynamics of Example 4.3-1, and it demonstrates extended Kalman–Bucy filtering as well as adaptive filtering. It presents a difficult parameter estimation problem, in that the input is known only by its mean and covariance. The dynamic model is

$$\begin{bmatrix} \dot{x}_1 \\ \dot{x}_2 \end{bmatrix} = \begin{bmatrix} 0 & 1 \\ -\omega_n^2 & -2\zeta\omega_n \end{bmatrix} \begin{bmatrix} x_1 \\ x_2 \end{bmatrix} + \begin{bmatrix} 0 \\ \omega_n^2 \end{bmatrix} w$$

where x_1 represents the weathervane angle, x_2 represents angular rate, and w is the random wind input. The damping ratio ζ of the weathervane is 0.1, and the natural frequency ω_n is 2 rad/s. Both the angle and the rate are measured:

$$\begin{bmatrix} z_1 \\ z_2 \end{bmatrix} = \begin{bmatrix} 1 & 0 \\ 0 & 1 \end{bmatrix} \begin{bmatrix} x_1 \\ x_2 \end{bmatrix} + \begin{bmatrix} n_1 \\ n_2 \end{bmatrix}$$

The initial conditions for the state mean and covariance are zero, while the zero-mean white-noise wind input and measurement errors are represented by the following spectral density matrices:

$$\mathbf{Q}_C = \mathbf{Q}_A = q = 1000$$

$$\mathbf{R}_C = \begin{bmatrix} r_1 & 0 \\ 0 & r_2 \end{bmatrix} = \begin{bmatrix} 10 & 0 \\ 0 & 10 \end{bmatrix}$$

Assuming that the ω_n^2 is an uncertain constant, the system equation (Eq. 4.7-9) for the extended Kalman–Bucy filter is

$$\begin{bmatrix} \dot{x}_1 \\ \dot{x}_2 \\ \dot{a} \end{bmatrix} = \begin{bmatrix} 0 & 1 & 0 \\ a & b & 0 \\ 0 & 0 & 0 \end{bmatrix} \begin{bmatrix} x_1 \\ x_2 \\ a \end{bmatrix} + \begin{bmatrix} 0 \\ -a \\ 0 \end{bmatrix} w$$

and the sensitivity matrices for the Riccati equation (Eq. 4.7-14) are as follows:

$$\mathbf{F}_A = \begin{bmatrix} 0 & 1 & 0 \\ \hat{a} & b & \hat{x}_1 \\ 0 & 0 & 0 \end{bmatrix}$$

$$\mathbf{H}_A = \begin{bmatrix} 1 & 0 & 0 \\ 0 & 1 & 0 \end{bmatrix}$$

$$\mathbf{L}_A = \begin{bmatrix} 0 \\ -\hat{a} \\ 0 \end{bmatrix}$$

[$\hat{(\cdot)}$ denotes an estimated quantity.] The Riccati equation can be expressed as six scalar equations:

$$\dot{p}_{11} = p_{12}\left(2 - \frac{p_{12}}{r_2}\right) - \frac{p_{11}^2}{r_1}$$

$$\dot{p}_{12} = \hat{a}p_{11} + p_{12}\left(b - \frac{p_{11}}{r_1} - \frac{p_{22}}{r_2}\right) + \hat{x}_1 p_{13} + p_{22}$$

$$\dot{p}_{13} = p_{23}\left(1 - \frac{p_{12}}{r_1}\right) - \frac{p_{11}p_{13}}{r_1}$$

$$\dot{p}_{22} = p_{12}\left(2\hat{a} - \frac{p_{12}}{r_1}\right) + p_{22}\left(2b - \frac{p_{22}}{r_2}\right) + 2\hat{x}_1 p_{23} + \hat{a}^2/q$$

$$\dot{p}_{23} = p_{13}\left(\hat{a} - \frac{p_{12}}{r_1}\right) + p_{23}\left(b - \frac{p_{22}}{r_2}\right) + \hat{x}_1 p_{33}$$

$$\dot{p}_{33} = -\frac{p_{13}^2}{r_1} - \frac{p_{23}^2}{r_2}$$

The optimal gain matrix (Eq. 4.7-13) is

$$\begin{bmatrix} k_{11} & k_{12} \\ k_{21} & k_{22} \\ k_{31} & k_{32} \end{bmatrix} = \begin{bmatrix} p_{11}/r_1 & p_{12}/r_2 \\ p_{12}/r_1 & p_{22}/r_2 \\ p_{13}/r_1 & p_{23}/r_2 \end{bmatrix}$$

and the state estimator is

$$\begin{bmatrix} \dot{\hat{x}}_1 \\ \dot{\hat{x}}_2 \\ \dot{\hat{a}} \end{bmatrix} = \begin{bmatrix} \hat{x}_2 \\ \hat{a}\hat{x}_1 + b\hat{x}_2 \\ 0 \end{bmatrix} + \begin{bmatrix} k_{11} & k_{12} \\ k_{21} & k_{22} \\ k_{31} & k_{32} \end{bmatrix} \begin{bmatrix} z_1 - \hat{x}_1 \\ z_2 - \hat{x}_2 \end{bmatrix}$$

An example of this filter's performance in estimating the parameter a is given in the diagram; the initial expected value of $\hat{a}$ (or $-\omega_n^2$) is -4.4, and the assumed covariance of the uncertainty is 20. The average value of $\hat{a}$ during the 20-s simulation is closer to the actual value than is the initial estimate, and the estimated covariance of the uncertainty is reduced by 70%; however, convergence to the actual value is irregular and by no means guaranteed. Initially, the parameter estimate appears to be converging to the proper value. Nevertheless, it diverges because p_{33} is still large enough to allow continued variation and the short-term trend in forcing by the residual is degrading the estimate. Inspection of the Riccati equation reveals that the parameter estimate is strongly dependent on the angle estimate $\hat{x}_1$. The rate of change of p_{23}, which governs the angular rate gain, contains the product of $\hat{x}_1$ and p_{33}; hence there is substantial variation in not only the magnitude but the sign of this important gain due to the

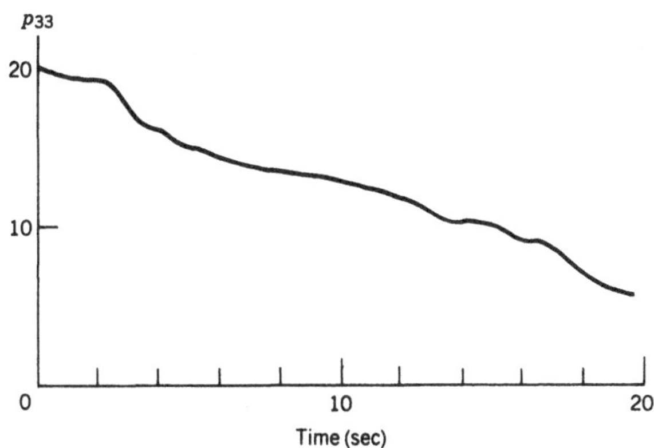
p_{33}
20
10
0
10
20
Time (sec)

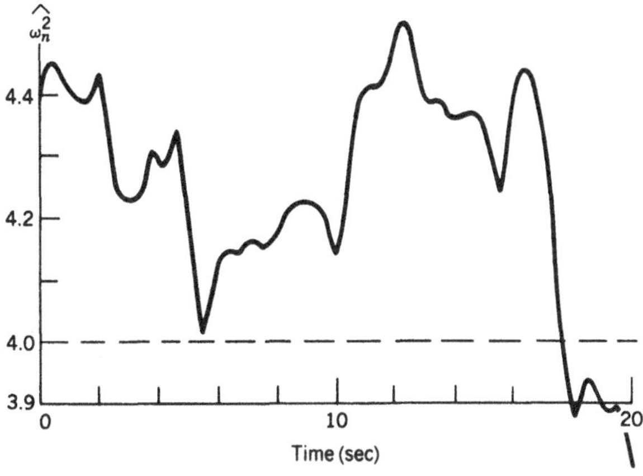
$\widehat{\omega_n^2}$
4.4
4.2
4.0
3.9
0
10
20
Time (sec)

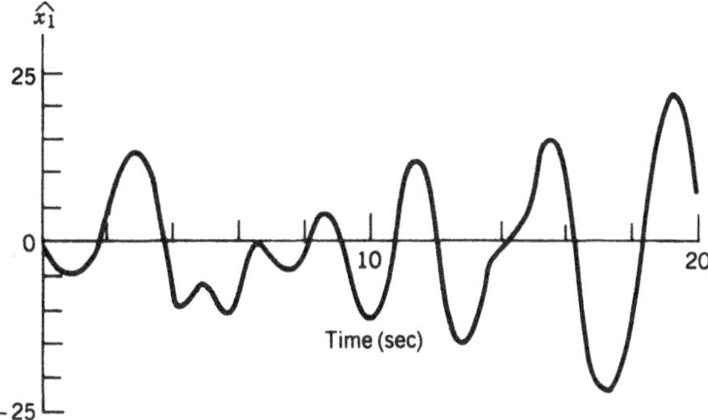
$\widehat{x_1}$
25
0
-25
10
20
Time (sec)

oscillations of the lightly damped system. k_{31} and k_{32} ultimately will go to zero as p_{33} goes to zero; however, if $\hat{a}$ is not in the neighborhood of its proper value as these gains vanish, a bias in $\hat{a}$ will remain.

Noise-Adaptive Filtering

The linear-optimal filter is said to have the *innovations property*; that is, the filter extracts all the available information from the measurement, leaving only zero-mean white noise in the measurement residual. Considering a discrete-time model with constant system and noise matrices, the residual is

$$\mathbf{r}_k = \mathbf{z}_k - \mathbf{H}\hat{\mathbf{x}}_k(-) \tag{4.7-21a}$$

$$= \mathbf{H}[\mathbf{x}_k - \hat{\mathbf{x}}_k(-)] + \mathbf{n}_k \tag{4.7-21b}$$

The optimal residual covariance is

$$E[\mathbf{r}_k\mathbf{r}_k^T] = \mathbf{S}_k = \mathbf{H}\mathbf{P}_k(-)\mathbf{H}^T + \mathbf{R} \tag{4.7-22}$$

because $E[\hat{\mathbf{x}}_k\mathbf{n}_k^T] = E[\mathbf{x}_k\mathbf{n}_k^T] = \mathbf{0}$. If the residual is not white or if its covariance does not equal Eq. 4.7-22, the filter is suboptimal. This could be the result of parameter uncertainty, or it may be the result of actual disturbance input or measurement-error covariances that are different from the $\mathbf{Q}$ and $\mathbf{R}$ used in filter implementation. Sample means of measurement and forcing residuals can indicate bias errors, and the results of these tests can be used to improve filter performance.

A *test for whiteness* can be based upon the autocorrelation function matrix $\mathbf{C}(k)$ of the measurement residual. If $\mathbf{r}_k$ is white, $\mathbf{C}(k)$ should be nonzero only when k is zero. Assuming that the random process represented by $\mathbf{r}_k$ is not only stationary but ergodic, an estimate of $\mathbf{C}(k)$ based on N samples (Eq. 4-7-21a) can be calculated as

$$\mathbf{C}(k) = \frac{1}{N}\sum_{n=k}^{N} \mathbf{r}_n\mathbf{r}_{n-k}^T \tag{4.7-23}$$

where N is large compared to k. Elements of the autocorrelation function $c_{ij}(k)$ can be normalized by their zero-lag values, producing elements, $\rho_{ij}(k)$, that theoretically range between plus and minus one. Assuming negligible measurement cross-correlation, the test for whiteness can be conducted on the diagonal elements alone:

$$\rho_{ii}(k) = \frac{c_{ii}(k)}{c_{ii}(0)} \tag{4.7-24}$$

The normalized diagonal elements should be one for $k = 0$ and zero otherwise; however, variations occur as a result of finite sample length. It can be shown that the 95% confidence limits on $\rho_{ii}(k)$ for $k \neq 0$ are

$$|\rho_{ii}(k)| \leq 1.96/N^{1/2} \tag{4.7-25}$$

Consequently, if less than 5% of the $\rho_{ii}(k)$ exceed this threshold, then the ith residual can be considered white, and the filter is processing the ith measurement in optimal fashion. The test is clearly more specific to errors in **R** than in **Q**; single-element errors in the latter presumably could cause nonwhiteness in all elements of $\mathbf{r}_k$.

Measurement bias and covariance can be estimated approximately with relatively simple batch-processing techniques (M-7). The sample mean of Eq. 4.7-21a is

$$\bar{\mathbf{r}} = \frac{1}{N} \sum_{k=1}^{N} \mathbf{r}_k \tag{4.7-26}$$

and the sample covariance matrix is

$$\hat{\mathbf{S}} = \frac{1}{N-1} \sum_{k=1}^{N} (\mathbf{r}_k - \bar{\mathbf{r}})(\mathbf{r}_k - \bar{\mathbf{r}})^T \tag{4.7-27}$$

From Eq. 4.7-22, which neglects state-noise correlation due to suboptimal filtering, the estimate of **R** is

$$\hat{\mathbf{R}} = \frac{1}{N-1} \sum_{k=1}^{N} \left\{ (\mathbf{r}_k - \bar{\mathbf{r}})(\mathbf{r}_k - \bar{\mathbf{r}})^T - \frac{N-1}{N} \mathbf{H}\mathbf{P}_k(-)\mathbf{H}^T \right\} \tag{4.7-28}$$

where $\mathbf{P}_k(-)$ is computed in the filter algorithm. If the Riccati equation has reached steady state,

$$\hat{\mathbf{R}} = \hat{\mathbf{S}} - \mathbf{H}\mathbf{P}_{ss}(-)\mathbf{H}^T \tag{4.7-29}$$

Similar approximate estimates can be formed for disturbance bias and covariance. Given the state equation

$$\mathbf{x}_{k+1} = \mathbf{\Phi}\mathbf{x}_k + \mathbf{w}_k \tag{4.7-30}$$

a forcing residual can be formed as

$$\mathbf{q}_k = \mathbf{x}_{k+1} - \mathbf{\Phi}\hat{\mathbf{x}}_k(+) \tag{4.7-31a}$$

$$= \mathbf{\Phi}[\mathbf{x}_k - \hat{\mathbf{x}}_k(+)] + \mathbf{w}_k \tag{4.7-31b}$$

with sample mean

$$\bar{\mathbf{q}} = \frac{1}{N} \sum_{k=1}^{N} \mathbf{q}_k \tag{4.7-32}$$

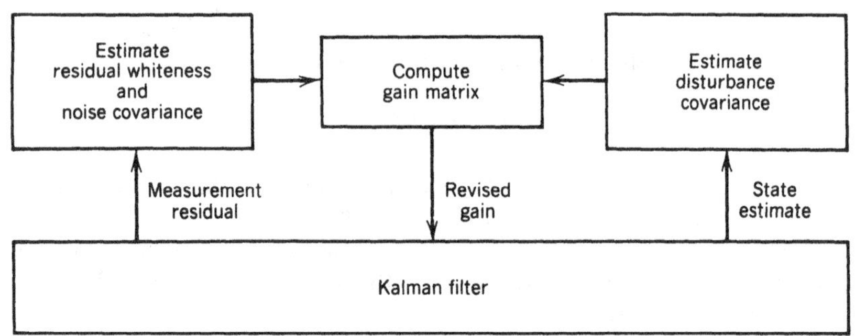

FIGURE 4.7-2 Organization of noise-adaptive filter.

Unlike the measurement residual, the forcing residual must be approximated, using $\hat{\mathbf{x}}_{k+1}(+)$ in Eq. 4.7-31a. Then the residual covariance matrix is estimated as

$$\mathbf{Q}_S = \frac{1}{N-1}\sum_{k=1}^{N}(\mathbf{q}_k - \bar{\mathbf{q}})(\mathbf{q}_k - \bar{\mathbf{q}})^T \tag{4.7-33}$$

From Eq. 4.7-31b, the theoretical forcing residual covariance matrix is approximately

$$\begin{aligned}\mathbf{Q}_{S_k} &= E[(\mathbf{q}_k - \bar{\mathbf{q}})(\mathbf{q}_k - \bar{\mathbf{q}})^T] \\ &= \mathbf{\Phi}\mathbf{P}_k(+)\mathbf{\Phi}^T + \mathbf{Q}\end{aligned} \tag{4.7-34}$$

leading to the following estimate for $\mathbf{Q}$:

$$\hat{\mathbf{Q}} = \frac{1}{N-1}\sum_{k=1}^{N}\{(\mathbf{q}_k - \bar{\mathbf{q}})(\mathbf{q}_k - \bar{\mathbf{q}})^T - [(N-1)/N]\mathbf{\Phi}\mathbf{P}_k(+)\mathbf{\Phi}^T\} \tag{4.7-35}$$

In steady state, this is equivalent to

$$\hat{\mathbf{Q}} = \mathbf{Q}_S - \mathbf{\Phi}\mathbf{P}_{ss}(+)\mathbf{\Phi}^T \tag{4.7-36}$$

It is possible to obtain the noise-adaptive gain directly, without explicit estimation of $\mathbf{Q}$ and $\mathbf{R}$. A full discussion is beyond the present scope, but further information can be found in (M-3), (M-4), and (A-1). The preceding results for $\hat{\mathbf{Q}}$, $\hat{\mathbf{R}}$, $\bar{\mathbf{r}}$, and $\bar{\mathbf{q}}$ can be recast in recursive form for continual estimates of slowly varying statistics (Fig. 4.7-2).

Multiple-Model Estimation

Suppose the system contains a single parameter that can take one of two values. Linear-optimal filters could be constructed using each hypothetical value of the parameter, and the filter that gave the "best" state estimates would identify the actual value of the parameter. Multiple-model estimation is a generalization of this principle which allows for many possible values of system parameters and noise levels. In this approach, a hypothesis is defined by specific values of the system matrices $(\mathbf{\Phi}, \mathbf{\Gamma}, \mathbf{H})$, disturbance-input matrix $(\mathbf{Q})$, measurement noise matrix $(\mathbf{R})$, or initial conditions $(\mathbf{x}_0, \mathbf{P}_0)$; and a separate nth-order filter is assigned to each hypothesis. The computational burden is great—given l independent parameters that each can have k possible values, $(k)^l$ parallel filters would be required—but the discrimination of multiple-model algorithms is somewhat better than that of the adaptive filtering techniques discussed earlier.

The success of multiple-model estimation is dependent not only on the initial choice of likely models but on the criterion chosen for hypothesis testing. Previous results suggest that the *statistics of the measurement residual* should be involved, but neither the whiteness test nor a comparison of residual covariances can discriminate between models with high confidence. A better approach is to use the measurement residual and residual covariance for each filter to estimate the *likelihood* that the associated hypothesis is the correct one. The optimal estimate is then computed as the sum of the individual estimates, with each estimate weighted by its hypothesis's probability. Recursive estimation of these probabilities is based on sequential, parallel application of *Bayes's rule* (Section 2.4). Bayes's rule was developed for scalar variables, and it can be extended to the vector variables of a parameter estimation problem. Given I hypothetical parameter vectors, $\mathbf{p}$, of dimension l and the r-component measurement vector, $\mathbf{z}_k$, (where the subscript denotes the kth sample of a sequence), the conditional probability mass function for the jth parameter set is

$$\Pr(\mathbf{p}_j \mid \mathbf{z}_k) = \frac{\mathrm{pr}(\mathbf{z}_k \mid \mathbf{p}_j) \Pr(\mathbf{p}_j)_{k-1}}{\sum_{i=1}^{I} [\mathrm{pr}(\mathbf{z}_k \mid \mathbf{p}_i) \Pr(\mathbf{p}_i)_{k-1}]} \tag{4.7-37}$$

The probability that the measurement has been obtained at $k-1$ presumably is one; hence,

$$\Pr(\mathbf{p}_j)_{k-1} = \Pr(\mathbf{p}_j \mid \mathbf{z}_{k-1}) \tag{4.7-38}$$

and Eq. 4.7-37 forms the basis of a recursion, beginning with a given value of $\Pr(\mathbf{p}_j \mid 0)$ and $1 \le j \le I$.

The probability density function of $\mathbf{z}_k$ conditioned by $\mathbf{p}_j$ remains to be found. Recall that the equations for a constant-coefficient system with true parameter set $\mathbf{p}$ are

$$\mathbf{z}_k = \mathbf{H}\mathbf{x}_k + \mathbf{n}_k \tag{4.7-39}$$

$$\mathbf{x}_k = \mathbf{\Phi}\mathbf{x}_{k-1} + \mathbf{\Gamma}\mathbf{u}_{k-1} + \mathbf{\Lambda}\mathbf{w}_{k-1}, \qquad \mathbf{x}_0 = \mathbf{x}(0) \tag{4.7-40}$$

If $\mathbf{x}_k$ were known, the conditional probability density function would be

$$\begin{aligned}\text{pr}(\mathbf{z}_k \mid \mathbf{p}) &= \text{pr}[\mathbf{z}_k \mid \mathbf{x}_k(\mathbf{p})]\\ &= \frac{1}{(2\pi)^{n/2}|\mathbf{R}_k|^{1/2}} e^{-(1/2)(\mathbf{z}_k-\mathbf{H}\mathbf{x}_k)^T\mathbf{R}^{-1}(\mathbf{z}_k-\mathbf{H}\mathbf{x}_k)}\\ &= \frac{1}{(2\pi)^{n/2}|\mathbf{R}_k|^{1/2}} e^{-(1/2)\mathbf{n}_k^T\mathbf{R}^{-1}\mathbf{n}_k}\end{aligned} \qquad (4.7\text{-}41)$$

where $\mathbf{R}_k$ is the covariance matrix of the measurement error $\mathbf{n}_k$. However, the true state is not known, so it is necessary to use the probability density function of $\mathbf{z}_k$ conditioned by an optimal estimate of the state $\hat{\mathbf{x}}_k$. The state estimate is generated by a Kalman filter (Section 4.3):

$$\hat{\mathbf{x}}_k(-) = \mathbf{\Phi}\hat{\mathbf{x}}_{k-1}(+) + \mathbf{\Gamma}\mathbf{u}_{k-1} \qquad (4.7\text{-}42)$$

$$\mathbf{P}_k(-) = \mathbf{\Phi}\mathbf{P}_{k-1}(-)\mathbf{\Phi}^T + \mathbf{\Lambda}\mathbf{Q}'_{k-1}\mathbf{\Lambda}^T \qquad (4.7\text{-}43)$$

$$\mathbf{K}_k = \mathbf{P}_k(-)\mathbf{H}^T[\mathbf{H}\mathbf{P}_k(-)\mathbf{H}^T + \mathbf{R}_k]^{-1} \qquad (4.7\text{-}44)$$

$$\hat{\mathbf{x}}_k(+) = \hat{\mathbf{x}}_k(-) + \mathbf{K}_k[\mathbf{z}_k - \mathbf{H}\hat{\mathbf{x}}_k(-)] \qquad (4.7\text{-}45)$$

$$\mathbf{P}_k(+) = [\mathbf{P}_k(-)^{-1} + \mathbf{H}^T\mathbf{R}_k^{-1}\mathbf{H}]^{-1} \qquad (4.7\text{-}46)$$

This allows $\text{pr}[\mathbf{z}_k \mid \hat{\mathbf{x}}_k(\mathbf{p})]$ to be estimated by

$$\text{pr}[\mathbf{z}_k \mid \hat{\mathbf{x}}_k(\mathbf{p})] = \frac{1}{(2\pi)^{n/2}|\mathbf{S}_k|^{1/2}} e^{-(1/2)\mathbf{r}_k^T\mathbf{S}_k^{-1}\mathbf{r}_k} \qquad (4.7\text{-}47)$$

with the measurement residual

$$\mathbf{r}_k = \mathbf{z}_k - \mathbf{H}\hat{\mathbf{x}}_k(-) \qquad (4.7\text{-}48)$$

and the corresponding residual covariance matrix

$$\mathbf{S}_k = \mathbf{H}\mathbf{P}_k(-)\mathbf{H}^T + \mathbf{R}_k \qquad (4.7\text{-}49)$$

$\mathbf{S}_k$ accounts for the estimation error in $\hat{\mathbf{x}}_k$, and $\text{pr}[\mathbf{z}_k \mid \hat{\mathbf{x}}(\mathbf{p})]$ has a greater spread than does $\text{pr}[\mathbf{z}_k \mid \mathbf{x}(\mathbf{p})]$.

Construction of the multiple-model algorithm is now clear. A Kalman filter (Eqs. 4.7-42 to 4.7-46) is formed for each hypothetical parameter set $\mathbf{p}_i$, providing state, residual, and covariance estimates denoted by $\hat{\mathbf{x}}_k[\mathbf{p}_i, (\pm)]$, $\mathbf{r}_k(\mathbf{p}_i)$, $\mathbf{P}_k(\mathbf{p}_i)$, and $\mathbf{S}_k(\mathbf{p}_i)$ (Fig. 4.7-3). The conditional probability densities of the observations based on each hypothesis are estimated as

$$\text{pr}[\mathbf{z}_k \mid \hat{\mathbf{x}}_k(\mathbf{p}_i)] = \frac{1}{(2\pi)^{n/2}|\mathbf{S}_k(\mathbf{p}_i)|^{1/2}} e^{-(1/2)\mathbf{r}_k^T(\mathbf{p}_i)\mathbf{S}_k^{-1}(\mathbf{p}_i)\mathbf{r}_k(\mathbf{p}_i)} \qquad (4.7\text{-}50)$$

$$\mathbf{r}_k(\mathbf{p}_i) = \mathbf{z}_k - \mathbf{H}\hat{\mathbf{x}}_k[\mathbf{p}_i, (-)] \qquad (4.7\text{-}51)$$

$$\mathbf{S}_k(\mathbf{p}_i) = \mathbf{H}\mathbf{P}_k(\mathbf{p}_i)\mathbf{H}^T + \mathbf{R}_k \qquad (4.7\text{-}52)$$

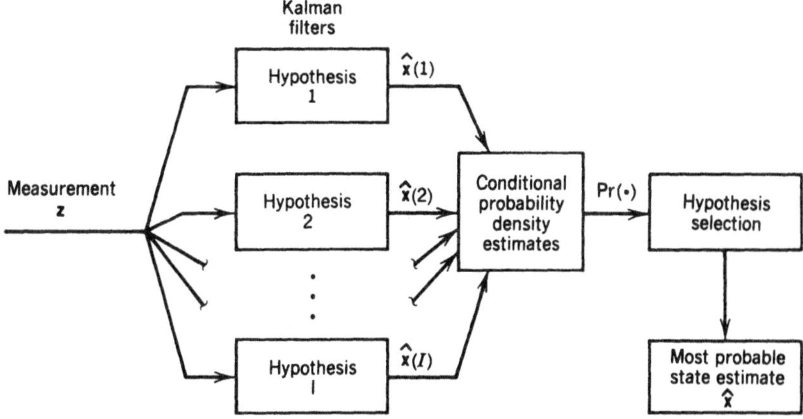

FIGURE 4.7-3 Organization for multiple-model estimation.

for $i = 1$ to I. The conditional probabilities of the $\mathbf{p}_j$, $1 \le j \le I$, are estimated as

$$\Pr(\mathbf{p}_j \mid \mathbf{z}_k) = \frac{\text{pr}[\mathbf{z}_k \mid \hat{\mathbf{x}}_k(\mathbf{p}_j)] \Pr(\mathbf{p}_j \mid \mathbf{z}_{k-1})}{\sum_{i=1}^{I} \{\text{pr}[\mathbf{z}_k \mid \hat{\mathbf{x}}_k(\mathbf{p}_i)] \Pr(\mathbf{p}_i \mid \mathbf{z}_{k-1})\}} \tag{4.7-53}$$

and the conditional mean estimate is the weighted sum of parallel filter estimates:

$$\hat{\mathbf{x}}_k(+) = \sum_{i=1}^{I} \{\Pr(\mathbf{p}_i \mid \mathbf{z}_k) \hat{\mathbf{x}}_k[\mathbf{p}_i, (+)]\} \tag{4.7-54}$$

If a specific estimate of $\mathbf{p}$ is desired, it is formed as the weighted sum of the hypothetical parameter sets:

$$\hat{\mathbf{p}}_k = \sum_{i=1}^{I} [\Pr(\mathbf{p}_i \mid \mathbf{z}_k) \hat{\mathbf{p}}_{i_k}] \tag{4.7-55}$$

The multiple-model algorithm can be simplified in a number of ways. It is always desirable to minimize the number of hypotheses; the choice of the number is a matter for engineering judgment. For example, if $\mathbf{Q}$ is written as $\alpha\mathbf{Q}'$, satisfactory adaptation may be afforded by varying the single parameter α rather than the individual elements of $\mathbf{Q}$. Once probability estimates have pinpointed a small number of most likely hypotheses, Kalman filter solutions for the less likely hypotheses can be dropped. The process of refining filter adaptation could be facilitated by introducing a new filter based on the most probable parameters, with commensurate "de-weighting" of the probabilities associated with the old filters. Furthermore, parallel-processing computers obviously would be well suited to multiple-model estimation.

EXAMPLE 4.7-2. ESTIMATION OF A RANDOM CONSTANT PARAMETER USING PARALLEL FILTERS

The weathervane natural frequency that was estimated using an extended Kalman–Bucy filter in Example 4.7-1 is estimated using three Kalman filters, each of which is tuned to a different assumed natural frequency. The initial assumption is that ω_n^2 is 4.4, when in reality it is 4. For the example, the three filters are implemented with the hypotheses that ω_n^2 is 4, 4.4, or 4.8; the predicted probabilities are 0.2, 0.6, and 0.2. Whereas a continuous-time algorithm was used in the previous example, the current example is based on discrete-time estimation of the continuous-time system. With a sampling interval of 0.1 s, the three hypothetical state transition matrices

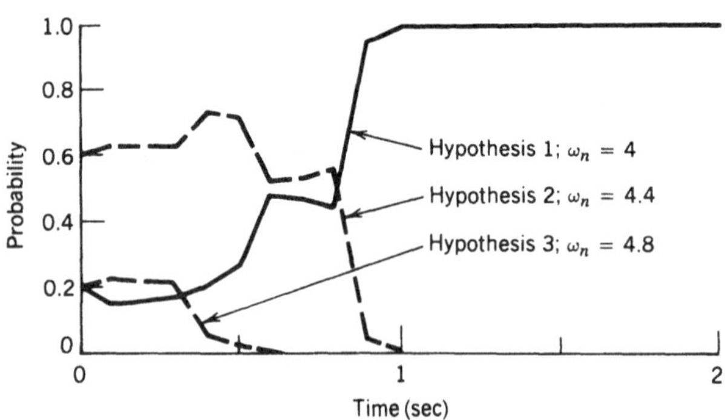

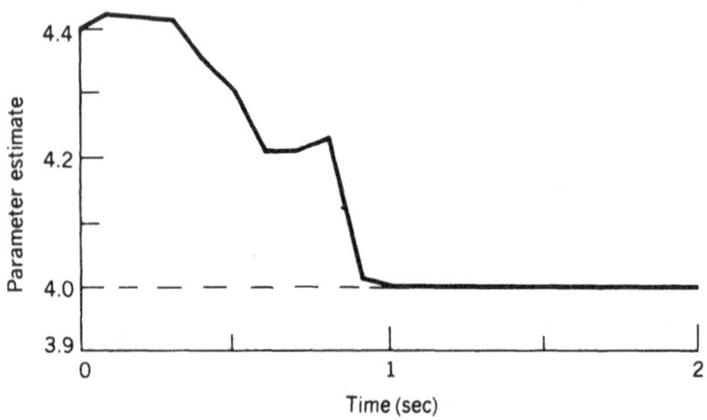

are

$$\mathbf{\Phi}_1 = \begin{bmatrix} 0.980 & 0.097 \\ -0.389 & 0.941 \end{bmatrix} \text{(correct)}$$

$$\mathbf{\Phi}_2 = \begin{bmatrix} 0.978 & 0.097 \\ -0.428 & 0.939 \end{bmatrix} \text{(assumed)}$$

$$\mathbf{\Phi}_3 = \begin{bmatrix} 0.976 & 0.097 \\ -0.467 & 0.938 \end{bmatrix}$$

The disturbance-input covariance matrices corresponding to the previous spectral density (Eq. 4.2-52) are

$$\mathbf{Q}_1 = \begin{bmatrix} 5.059 & 75.12 \\ 75.12 & 1518. \end{bmatrix}$$

$$\mathbf{Q}_2 = \begin{bmatrix} 40.66 & 246.5 \\ 246.5 & 1562. \end{bmatrix}$$

$$\mathbf{Q}_3 = \begin{bmatrix} 120.1 & 411.6 \\ 411.6 & 1440. \end{bmatrix}$$

while the measurement-error covariance,

$$\mathbf{R} = \begin{bmatrix} 10 & 0 \\ 0 & 10 \end{bmatrix}$$

is effectively 10 times higher than the continuous measurement error due to sampling effects (Eq. 4.5-13).

The problem as posed is almost too easy for the multiple-model algorithm because the parameter appears in the input matrix **L**, and therefore has a large effect on the sampled disturbance effect **Q**. This in turn has a large effect on $\mathbf{P}_k(-)$ and $\mathbf{S}_k$, and the accompanying simulation demonstrates convergence to the proper parameter in one sec. Had ω_n^2 not appeared in **L**, the convergence would have been much less certain.

PROBLEMS

Section 4.1

1. Using a least-squares algorithm, fit quadratic and cubic polynomials to the following time series:

   z: 1,27,33,45,12,16,83,67,54,39,23,6,14,15,19,31,37,44,56,60

   Compute the mean-square error in both cases, and plot the results.

2. Repeat Problem 1, assuming that the first data point is 10 times better than the last and that intermediate points are weighted in proportion to their positions in sequence.

3. One more piece of data is to be added to the sequence in Problem 1. Assuming equal weighting of all points, how would a new reading of 25 affect the quadratic curve fit? Use a recursive least-squares estimator to find the answer.

4. Apply recursive estimation to the entire time series of Problem 1, that is, compute a running estimate beginning with the first point and ending with the last.

5. The vector **x** is related to the vector **y** by the following equation:

$$\begin{bmatrix} y_1 \\ y_2 \end{bmatrix} = \begin{bmatrix} 0 & 1 \\ 3 & 4 \end{bmatrix} \begin{bmatrix} x_1 \\ x_2 \end{bmatrix}$$

Given the following noisy measurements ($\mathbf{z} = \mathbf{y} + \mathbf{n}$), what is the least-squares estimate of **x**?

$$z_1 = 0,1,7,8,5,7,9,10,6,4$$

$$z_2 = 10,7,4,5,5,3,0,2,2,4$$

6. The measurement sequence in (5) is taken to represent the nonlinear relationship

$$z_1 = x_1^2 + x_2 + 5 + n_1$$

$$z_2 = \frac{x_2^3}{4} + 4 + n_2$$

Use a Newton–Raphson algorithm to estimate x_1 and x_2. What are the statistics of n_1 and n_2?

Section 4.2

1. Under what conditions does the expected value of the state take a constant value? When can the covariance of the state take a constant value?

2. An oscillatory discrete-time system is described by

$$\begin{bmatrix} x_1 \\ x_2 \end{bmatrix}_k = \begin{bmatrix} 1 & 1 \\ -1 & 0.4 \end{bmatrix} \begin{bmatrix} x_1 \\ x_2 \end{bmatrix}_{k-1} + \begin{bmatrix} 0 \\ 1 \end{bmatrix} w_{k-1}$$

Assuming that

$$E(\mathbf{x}_0) = \mathbf{0} \quad E(\mathbf{x}_0\mathbf{x}_0^T) = \mathbf{0}$$
$$E(w_k) = 0 \quad E(w_k^2) = 1$$

generate 20 steps of a simulated state history, $\mathbf{x}_k$, and the corresponding covariance matrix, $\mathbf{P}_k$. What is the history of the mean value in this case?

3. The system in Problem 2 is a first-order discrete-time approximation to the continuous-time system

$$\begin{bmatrix} \dot{x}_1(t) \\ \dot{x}_2(t) \end{bmatrix} = \begin{bmatrix} 0 & 1 \\ -1 & -0.6 \end{bmatrix} \begin{bmatrix} x_1(t) \\ x_2(t) \end{bmatrix} + \begin{bmatrix} 0 \\ 1 \end{bmatrix} w(t)$$

for a sampling interval of 1 s. Suppose $E[w(t)] = 0$ and $E[w(t)w(\tau)] = \delta(t-\tau)$. What is the sampled-data disturbance covariance $\mathbf{Q}_k$?

4. Compute a 20-s simulated state history, $\mathbf{x}(t)$, and corresponding covariance matrix, $\mathbf{P}(t)$, for the continuous-time system defined in Problem 3.

5. Perform a Cholesky decomposition of the following matrix:

$$\mathbf{Q} = \begin{bmatrix} 1 & 0 & 1 \\ 0 & 9 & 9 \\ 1 & 9 & 35 \end{bmatrix}$$

Section 4.3

1. Write a computer program that solves the discrete-time Kalman filter equations for a system with a three-component state vector, one measurement, and no control inputs.

2. A third-order continuous-time system is described by the following

dynamic and observation equations:

$$\begin{bmatrix} \dot{x}_1 \\ \dot{x}_2 \\ \dot{x}_3 \end{bmatrix} = \begin{bmatrix} -0.8 & 0.1 & 0 \\ 0 & -0.5 & 0.1 \\ 0 & 0 & 0.1 \end{bmatrix} \begin{bmatrix} x_1 \\ x_2 \\ x_3 \end{bmatrix} + \begin{bmatrix} w_1 \\ w_2 \\ w_3 \end{bmatrix}$$

$$z = [1 \quad 0 \quad 0] \begin{bmatrix} x_1 \\ x_2 \\ x_3 \end{bmatrix} + n$$

Formulate the corresponding discrete-time model, including $\mathbf{\Phi}$ and $\mathbf{\Lambda}$, assuming that $T = 0.1$ s and

$$E(\mathbf{w}_k \mathbf{w}_k^T) = \begin{bmatrix} 1 & 0 & 0 \\ 0 & 1 & 0 \\ 0 & 0 & 1 \end{bmatrix} = \mathbf{Q}$$

$$R = 0.25$$

Generate an example state vector $\mathbf{x}$ and observation z for 2 s, with $\mathbf{x}(0) = \mathbf{0}$ and random inputs and measurement noise of the proper spectral densities.

3. Compute and plot the Kalman filter gain, and the estimate of $\mathbf{x}$, given the observation generated in Problem 2, with $\mathbf{P}(0) = \mathbf{0}$.

4. Repeat Problem 2, separately increasing the diagonal elements of $\mathbf{Q}$ and R by 10.

5. Find the U and D factors for the following:

$$\mathbf{P} = \begin{bmatrix} 20 & 10 & 3 \\ 10 & 6 & 2 \\ 3 & 2 & 1 \end{bmatrix}$$

6. Calculate $\sigma_{\max}$ and $\sigma_{\min}$ for $\mathbf{P}$. What is the condition number?

7. A third-order system has the following description:

$$\mathbf{\Phi} = \begin{bmatrix} 0.9 & 0.1 & 0 \\ 0 & 0.8 & 0.1 \\ 0 & 0 & 0.75 \end{bmatrix}$$

$$\mathbf{H} = \begin{bmatrix} 1 & 0 & 0 \\ 0 & 1 & 0 \end{bmatrix}$$

$$\mathbf{Q} = \begin{bmatrix} 1 & 0 & 0 \\ 0 & 1 & 0 \\ 0 & 0 & 1 \end{bmatrix}$$

$$\mathbf{R} = \begin{bmatrix} 1 & 0 \\ 0 & 1 \end{bmatrix}$$

With **P** from Problem 5 as the initial condition, use the sequential processing algorithm to calculate the filter gain.

8. Repeat Problem 7 using the Joseph form.

Section 4.4

1. Write the individual scalar equations of a minimum-variance filter for a second-order discrete-time system with correlated disturbance and measurement error. Assuming that all matrices of the system are constant, the filter gain matrix **K** should approach a constant value as the sampling index becomes large. How would you express the steady-state solution of **K**?

2. The first-order system

$$x_k = 0.8x_{k-1} + w_{k-1}$$

$$z_k = x_k + n_k$$

is forced by a white random disturbance w_{k-1}, and its measurement is corrupted by white random error n_k. The expected values of w_{k-1} and n_k are correlated and are given by the following for all j and k:

$$E\left\{\begin{bmatrix} w_{k-1} \\ n_k \end{bmatrix}\begin{bmatrix} w_{k-1} & n_k \end{bmatrix}\right\} = \begin{bmatrix} Q & M \\ M & R \end{bmatrix} = \begin{bmatrix} 1 & 0.25 \\ 0.25 & 0.1 \end{bmatrix}$$

$$E\begin{bmatrix} w_{k-1} \\ n_k \end{bmatrix} = 0, \quad E(x_k w_j) = E(x_k n_j) = 0$$

Generate 50 samples of z_k for typical values of w_{k-1} and n_k. Compute optimal filter gains and state estimates for the sequence. Repeat the calculations with $M = 0$ and $M = -0.25$.

3. The first-order system of Problem 2 is driven by a white random disturbance w_{k-1}, with $E(w_k) = 0$ and $E(w_k^2) = 1$. Its measurement is corrupted by the time-correlated error n_k, where

$$n_k = 0.2n_{k-1} + 0.8\nu_{k-1}$$

and ν_{k-1} is a white random sequence, with $E(\nu_k) = 0$ and $E(\nu_k^2) = 0.1$. Generate 50 samples of z_k for typical values of w_{k-1} and ν_k. Compute optimal filter gains and state estimates for the sequence.

4. A discrete-time model for the pitching motions of an aircraft (see Example 2.6-1 for an equivalent continuous-time model) is given by the following:

$$\begin{bmatrix} x_1 \\ x_2 \end{bmatrix}_k = \begin{bmatrix} 0.7 & -0.15 \\ 0.03 & 0.79 \end{bmatrix} \begin{bmatrix} x_1 \\ x_2 \end{bmatrix}_{k-1} + \begin{bmatrix} 0.15 \\ 0.21 \end{bmatrix} w_{k-1}$$

$$\begin{bmatrix} z_1 \\ z_2 \end{bmatrix}_k = \begin{bmatrix} 1 & 0 \\ 0 & 1 \end{bmatrix} \begin{bmatrix} x_1 \\ x_2 \end{bmatrix}_k + \begin{bmatrix} n_1 \\ n_2 \end{bmatrix}_k$$

Suppose that $E(w_k) = 0$, $E(w_k^2) = 1$, and that the measurement error is given by the following Gauss–Markov process:

$$\begin{bmatrix} n_1 \\ n_2 \end{bmatrix}_k = \begin{bmatrix} 0.5 & 0 \\ 0 & 0.5 \end{bmatrix} \begin{bmatrix} n_1 \\ n_2 \end{bmatrix}_{k-1} + \begin{bmatrix} \nu_1 \\ \nu_2 \end{bmatrix}_{k-1}$$

$$E(\boldsymbol{\nu}_k) = \mathbf{0}, \quad E(\boldsymbol{\nu}_k \boldsymbol{\nu}_k^T) = \begin{bmatrix} 0.05 & 0 \\ 0 & 0.05 \end{bmatrix}$$

Generate 50 samples of $\mathbf{z}_k$ for typical values of w_{k-1} and $\boldsymbol{\nu}_k$. Compute optimal filter gains and state estimates for the sequence.

5. For the discrete-time model of Problem 4, assume that w_k is a Gauss–Markov process described by

$$w_k = 0.5\, w_{k-1} + 0.5\, v_{k-1}$$

$$E(v_k) = 0, \quad E(v_k^2) = 1$$

and that the measurement error is given by

$$E(\mathbf{n}_k) = \mathbf{0}, \quad E(\mathbf{n}_k \mathbf{n}_k^T) = \begin{bmatrix} 0.1 & 0 \\ 0 & 0.1 \end{bmatrix}$$

Generate 50 samples of $\mathbf{z}_k$ for typical values of v_k and $\mathbf{n}_k$. Compute optimal filter gains and state estimates for the sequence.

Section 4.5

1. A robot arm has a natural frequency of 10 Hz and a damping ratio of 0.3 for motions in the vertical plane. Vertical position is denoted by x_1 and vertical velocity by x_2. The spectral density of velocity disturbances is 10. Calculate and plot $\mathbf{P}(t)$ and $\mathbf{K}(t)$ for a time interval of 0.2 sec, assuming $\mathbf{P}(0) = \mathbf{0}$ and using the conventional Kalman–Bucy filter equations for the following measurements:

   (a) Velocity only, $R = 1$.
   (b) Position only, $R = 1$.
   (c) Velocity and position, $r_{11} = r_{22} = 1$.

2. Predict the covariance matrix at $t = 0.4$ s, assuming that the measurements stop at 0.2 s in all three cases.

3. Repeat Problem 1 using the Chandrasekhar-type algorithm.

4. Repeat Problem 1 using the square-root algorithm.

5. In Problem 9, Section 3.5, the mass of drug in the bloodstream is monitored by measuring z continuously. What is the Kalman–Bucy filter gain for estimating x_1 and x_2, given $\mathbf{Q}$ and $\mathbf{R}$?

Section 4.6

1. As mentioned in a previous problem, the temperature of an oven can be described by the first-order differential equation,

$$\dot{x}(t) = -k_1\dot{x}(t) - k_2x^4(t) + k_3u + k_4$$

Assume that $x(0) = 0$, $P(0) = 0$, $k_1 = 1$, $k_2 = 0.1$, $k_3u = 10$, and that k_4 is a white random process with $E(k_4) = 1$, $E(k_4^2) = 1$. The temperature is measured by a probe whose error statistics are $E(n) = 0$, $R = 1$.
(a) Simulate 20 s of the noisy temperature measurement.
(b) Compute the filter gain and state estimate, neglecting the x^4 term.
(c) Formulate the extended Kalman–Bucy filter for this problem (including the x^4 term).
(d) Compute the extended Kalman–Bucy filter gain and state estimate.

2. A ship's motions are modeled by the fourth-order equation

$$\begin{bmatrix} \dot{x}_1 \\ \dot{x}_2 \\ \dot{x}_3 \\ \dot{x}_4 \end{bmatrix} = \begin{bmatrix} 0 & 0 & 1 & 0 \\ 0 & 0 & 0 & 1 \\ 0 & 0 & -1/\tau & 0 \\ 0 & 0 & 0 & -1/\tau \end{bmatrix} \begin{bmatrix} x_1 \\ x_2 \\ x_3 \\ x_4 \end{bmatrix} + \begin{bmatrix} w_1 \\ w_2 \\ u_3/\tau + w_4 \\ u_4/\tau + w_4 \end{bmatrix}$$

x_1 and x_2 are the northerly and easterly positions measured in meters, and x_3 and x_4 are the corresponding linear velocities measured in meters/sec; w_1 to w_4 represent disturbances due to currents, waves, and the wind, while u_3 and u_4 are the commanded velocity components. The ship is equipped to measure its range relative to two reference points, (r_{n_1}, r_{e_1}) and (r_{n_2}, r_{e_2}), its speed relative to the water, and its heading. The measurement equations are

$$\begin{bmatrix} z_1 \\ z_2 \\ z_3 \\ z_4 \end{bmatrix} = \begin{bmatrix} (x_1 - r_{n_1})^2 + (x_2 - r_{e_1})^2 \\ (x_1 - r_{n_2})^2 + (x_2 - r_{e_2})^2 \\ (x_3^2 + x_4^2)^{1/2} \\ \sin^{-1}[x_4/(x_3^2 + x_4^2)^{1/2}] \end{bmatrix} + \begin{bmatrix} n_1 \\ n_2 \\ n_3 \\ n_4 \end{bmatrix}$$

The dynamic equations are linear, but the measurement equations are nonlinear. The continuous-time problem will be transformed to discrete time, so an extended Kalman filter can be formulated. In the following, assume that the sampling interval is 2 s and that $\tau = 30$ s. The navigation reference points are (0,0) and $(10^5,0)$, respectively, while the covariances of the zero-mean disturbances and measurement errors are as follows:

$$\mathbf{Q}=\text{Diag}[4\ (\text{m/s})^2, 4\ (\text{m/s})^2, 0.25\ (\text{m/s}^2)^2, 0.25\ (\text{m/s}^2)^2]$$

$$\mathbf{R}=\text{Diag}[100\ \text{m}^2, 100\ \text{m}^2, 4\ (\text{m/s})^2, 0.001\ \text{rad}^2]$$

The initial conditions and commands are as follows:

$$\hat{\mathbf{x}}_0(+) = [0\ \text{m} \quad 25000\ \text{m} \quad 10\ \text{m/s} \quad 0\ \text{m/s}]^T$$

$$\mathbf{P}_0(+)=\text{Diag}[100\ \text{m}^2, 100\ \text{m}^2, 4\ (\text{m/s})^2, 4\ (\text{m/s})^2]$$

$$u_3=20\ \text{m/s},\ u_4=0\ \text{m/s}$$

(a) Generate the discrete-time model and formulate the extended Kalman filter for this problem.

(b) Compute a 5-min trajectory and measurement history for the given conditions.

(c) Compute the corresponding $\mathbf{P}_k(+)$ and estimation gain matrix, $\mathbf{K}_k$. Plot the square roots of the diagonal elements of $\mathbf{P}_k(+)$ as functions of time.

(d) Compute the state estimates, $\hat{\mathbf{x}}_k(+)$, as functions of time.

3. Repeat Problem 2c for the following initial conditions and commands, assuming that the ship follows the corresponding nominal path:

$$\hat{\mathbf{x}}_0(+) = [0\ \text{m} \quad 25000\ \text{m} \quad 10\ \text{m/s} \quad 0\ \text{m/s}]^T$$

$$\mathbf{P}_0(+)=\text{Diag}[0, 0, 0, 0]$$

$$u_3=0\ \text{m/s},\ u_4=20\ \text{m/s}$$

4. Determine the effects of disturbance and measurement-error statistics on the problem defined in Problem 3.

(a) Individually increase the diagonal elements of $\mathbf{Q}$ by 10 and note the effects on the estimates.

(b) Individually increase the diagonal elements of $\mathbf{R}$ by 10 and note the effects on the estimates.

5. Calculate and plot the random-input describing function for the following nonlinearity:

$$f(x)=\begin{cases}-1, & x<-a\\ 0, & -a\le x\le a\\ 1, & x>a\end{cases}$$

Assume that x is a zero-mean Gaussian process with standard deviation σ, and express the describing function as a function of a/σ.

Section 4.7

1. A second-order discrete-time system can be described as follows:

$$\begin{bmatrix}x_1\\ x_2\end{bmatrix}_{k+1}=\begin{bmatrix}0.9 & 1\\ 0.5 & 0.8\end{bmatrix}\begin{bmatrix}x_1\\ x_2\end{bmatrix}_k+\begin{bmatrix}w_1\\ w_2\end{bmatrix}_k$$

The initial conditions are $x_1(0)=1$ and $x_2(0)=-1$, and the disturbance is a zero-mean process with covariance matrix

$$\mathbf{Q}=\begin{bmatrix}1 & 0\\ 0 & 1\end{bmatrix}$$

The actual value of the state transition matrix's (2,1) element is unknown; however, it is thought to be a constant whose value is 0.6 with a standard deviation of 0.2. Both state components are measured; the measurement error is a zero-mean process with covariance matrix

$$\mathbf{R}=\begin{bmatrix}0.1 & 0\\ 0 & 0.1\end{bmatrix}$$

Simulate the actual system's response to a random disturbance input and the associated measurement. Design filters and document estimation performance for the following cases:

(a) Kalman filter with the actual system model.

(b) Kalman filter with the assumed system model.

(c) Parameter-adaptive filter with the assumed model.

2. Generate a 1000-number random sequence on a computer.

(a) Perform the test for whiteness on the sequence, and report your results.

(b) Filter the random-number sequence, u_k, with the recursive equation,

$$x_{k+1}=0.7x_k+u_k$$

Perform the whiteness test, and report your results.

3. The first-order system

$$x_{k+1}=0.7x_k+w_k$$

is observed through the equation

$$z_k=x_k+n_k$$

The actual values of zero-mean disturbance and measurement-error covariances are 10 and 1, respectively.

(a) Generate a 1000-number sequence of z_k.

(b) Filter the measurement sequence using correct values of Q and R to calculate filter gains, and estimate the noise covariances and biases from these results.

(c) Repeat Problem 3b with filter gains based on incorrect assumptions about Q and R, namely, (i) $Q=20$, $R=0.5$ and (ii) $Q=8$, $R=2$.

4. Simulate the response of the actual system described in Problem 1 to a random disturbance input, adding the appropriate measurement noise to the output.

(a) Implement a multiple-model estimation algorithm with three parallel filters, assuming that the (2, 1) element is 0.6, 0.5, or 0.4.

(b) Initialize the algorithm with the following assumptions about the (2, 1) element: 90% chance that it is 0.6, 5% chance that it is 0.5, and 5% chance that it is 0.4. Plot the simulation results, including the state estimate and the probabilities of each hypothesis.

REFERENCES

A-1 Alspach, D. L. A Parallel Filtering Algorithm for Linear Systems with Unknown Time-Varying Noise, *IEEE Transactions on Automatic Control*, **AC-19**(5), 552–556, October 1974.

A-2 Anderson, B.D.O. and Moore, J. B., *Optimal Filtering*, Prentice-Hall, Englewood Cliffs, 1979.

A-3 Athans, M., Wishner, R. P., and Bertolini, A., Suboptimal State Estimation for Continuous-Time Nonlinear Systems from Discrete Noisy Measurements, *IEEE Transactions on Automatic Control*, **AC-13**(5), 504–514, October 1968.

B-1 Bar-Itzhack, I. Y. and Medan, Y., Efficient Square Root Algorithm for Measurement Update in Kalman Filtering, *AIAA Journal of Guidance, Control, and Dynamics*, **6**(3), 129–134, May-June 1983.

B-2 Bendat, J. S. and Piersol, A. G., *Measurement and Analysis of Random Data*, Wiley, New York, 1966.

B-3 Bierman, G. J., A Comparison of Discrete Linear Filtering Algorithms, *IEEE Aerospace and Electronic Systems*, **AES-9**(1), 28–37, January 1973.

B-4 Bierman, G. J., Measurement Updating Using the U-D Factorization, *Automatica*, **12**(4), 375–382, July 1976.

B-5 Bierman, G. J., *Factorization Methods for Discrete Sequential Estimation*, Academic Press, New York, 1977.

B-6 Bierman, G. J. and Thornton, C. L., Numerical Comparison of Kalman Filter Algorithms: Orbital Determination Case Study, *Automatica*, **13**(1), 23–35, January 1977.

B-7 Brogan, W. L., *Modern Control Theory*, Prentice-Hall, Englewood Cliffs, NJ, 1985.

B-8 Bryson, A. E., Jr. and Frazier, M., Smoothing of Linear and Nonlinear Systems, in **ASD-TRD-63-119**, 354–364, February 1963.

B-9 Bryson, A. E., Jr. and Johansen, D. E., Linear Filtering for Time-Varying Systems Using Measurements Containing Colored Noise, *IEEE Transactions on Automatic Control*, **AC-10**(1), 4–10, January 1965.

B-10 Bryson, A. E., Jr. and Henrikson, L. J., Estimation Using Sampled Data Containing Sequentially Correlated Noise, *AIAA Journal of Spacecraft and Rockets*, **5**(6), 662–665, June 1968.

B-11 Bryson, A. E., Jr. and Ho, Y. C., *Applied Optimal Control*, Hemisphere Publishing, Washington, D.C., 1975.

B-12 Bucy, R. S., Global Theory of the Riccati Equation, *Journal of Computer and System Sciences*, **4**(7), 349–361, December 1967.

B-13 Bucy, R. S. and Joseph, P. D., *Filtering for Stochastic Processes with Applications to Guidance*, Wiley, New York, 1968.

C-1 Carlson, N. A., Fast Triangular Formulation of the Square Root Filter, *AIAA Journal*, **11**(9), 1259–1265, September 1973.

C-2 Cox, H. On the Estimation of State Variables and Parameters for Noisy Dynamic Systems, *IEEE Transactions on Automatic Control*, **AC-9**(1), 5–12, January 1964.

D-1 Davis, M.H.A., *Linear Estimation and Stochastic Control*, Halsted Press, New York, 1977.

E-1 Eykhoff, P., *System Identification*, Wiley, New York, 1974.

F-1 Forrester, J. W., Systems Analysis as a Tool for Urban Planning, *IEEE Transactions on Systems Science and Cybernetics*, **SSC-6**(4), October 1970.

F-2 Friedland, B. and Bernstein, I., Estimation of the State of a Nonlinear Process in the Presence of NonGaussian Noise and Disturbances, *Journal of the Franklin Institute*, **281**(6), 455–480, June 1966.

G-1 Gelb, A. (ed.), *Applied Optimal Estimation*, M.I.T. Press, Cambridge, MA, 1974.

G-2 Gelb, A. and VanderVelde, W. E., *Multi-Input Describing Functions and Nonlinear System Design*, McGraw-Hill, New York, 1968.

G-3 Graham, D. and McRuer, D., *Analysis of Nonlinear Control Systems*, Wiley, New York, 1961.

H-1 Henriksen, R., The Truncated Second-Order Nonlinear Filter Revisited, *IEEE Transactions on Automatic Control*, **AC-27**(1), 247–251, February 1982.

H-2 Householder, A., *The Theory of Matrices in Numerical Analysis*, Ginn-Blaisdell, Waltham, MA, 1964.

J-1 Jazwinski, A. H., *Stochastic Processes and Filtering Theory*, Academic Press, New York, 1970.

J-2 Jenkins, G. M. and Watts, D. G., *Spectral Analysis and Its Applications*, Holden-Day, San Francisco, 1968.

K-1 Kadanoff, L. P., From Simulation Model to Public Policy, *American Scientist*, **60**, January-February 1972, pp. 74–79.

K-2 Kailath, T., Some Algorithms for Recursive Estimation in Constant Linear Systems, *IEEE Transactions on Information Theory*, **IT-19**(6), 750–760, December 1971.

K-3 Kalman, R. E., A New Approach to Linear Filtering and Prediction Problems, *ASME Transactions, Journal of Basic Engineering*, **82D,** 35–50, March 1960.

K-4 Kalman, R. E. and Bucy, R. S., New Results in Linear Filtering and Prediction, *ASME Transactions, Journal of Basic Engineering*, **83D**, 95–108, March 1961.

K-5 Kaminski, P. G., Bryson, A. E., Jr., and Schmidt, S. F., Discrete Square Root Filtering: A Survey of Current Techniques, *IEEE Transactions on Automatic Control*, **AC-16**(6), 727–735, December 1971.

K-6 Krebs, V., Nonlinear Filtering Theory, in *Advances in the Techniques and Technology of the Application of Nonlinear Filters and Kalman Filters*, **AGARD-AG-256**, 1-1 to 1-26, March 1982.

K-7 Kwakernaak, H. and Sivan, R., *Linear Optimal Control Systems*, Wiley, New York, 1972.

L-1 Liang, D. F., Exact and Approximate Nonlinear Estimation Techniques, in **AGARD-AG-256**, 2-1 to 2-21, March 1982.

L-2 Liang, D. F., Comparisons of Nonlinear Filters for Systems with Nonnegligible Nonlinearities, in **AGARD-AG-256**, 16-1 to 16-34, March 1982.

L-3 Liebelt, P. B., *An Introduction to Optimal Estimation*, Addison-Wesley, Reading, MA, 1967.

L-4 Liu, C-H. and Marcus, S. I., Estimator Performance for a Class of Nonlinear Estimation Problems, *IEEE Transactions on Automatic Control*, **AC-25**(2), 299–302, April 1980.

M-1 Maybeck, P. S., *Stochastic Models, Estimation, and Control*, Academic Press, New York, Vol. 1, 1979, Vol. 2, 1982.

M-2 McGarty, T. P., *Stochastic Systems and State Estimation*, Wiley, New York, 1974.

M-3 Mehra, R. K., On the Identification of Variances and Adaptive Kalman Filtering, *IEEE Transactions on Automatic Control*, **AC-15**(2), 175–184, April 1970.

M-4 Mehra, R. K., Approaches to Adaptive Filtering, *IEEE Transactions on Automatic Control*, **AC-17**(5), 693–698, October 1972.

M-5 Mirsky, L., *An Introduction to Linear Algebra*, Oxford University Press, London, 1963.

M-6 Morf, M., Levy, B., and Kailath, T., Square-Root Algorithms for the Continuous-Time Linear Least-Square Estimation Problem, *IEEE Transactions on Automatic Control*, **AC-23**(5), 907–911, October 1978.

M-7 Myers, K. A. and Tapley, B. D., Adaptive Sequential Estimation with Unknown Noise Statistics, *IEEE Transactions on Automatic Control*, **AC-21**(4), 520–523, August 1976.

R-1 Rhodes, I. B., A Tutorial Introduction to Estimation and Filtering, *IEEE Transactions on Automatic Control*, **AC-16**(6), 688–706, December 1971.

S-1 Safonov, M. G. and Athans, M., Robustness and Computational Aspects of Nonlinear Stochastic Estimators and Regulators, *IEEE Transactions on Automatic Control*, **AC-23**(4), 717–725, August 1978.

S-2 Schweppe, F. C., *Uncertain Dynamic Systems*, Prentice-Hall, Englewood Cliffs, NJ, 1973.

S-3 Sorenson, H. W., *Parameter Estimation: Principles and Problems*, Marcel Dekker, New York, 1980.

S-4 Stear, E. B. and Stubberud, A. R., Optimal Filtering for Gauss-Markov Noise, *International Journal of Control*, **8**(2), 123–130, July 1968.

T-1 Tapley, B. D. and Peters, J. G., Sequential Estimation Algorithm Using a Continuous UDU^T Covariance Factorization, *AIAA Journal of Guidance and Control*, **3**(4), 326–331, July-August 1980.

T-2 Thornton, C. L. and Bierman, G. J., Gram-Schmidt Algorithms for Covariance Propagation, *International Journal of Control*, **25**(2), 243–260, February 1977.

U-1 Ursin, B., Asymptotic Convergence Properties of the Extended Kalman Filter Using Filtered State Estimates, *IEEE Transactions on Automatic Control*, **AC-25**(6), 1207–1216, December 1980.

W-1 Wilkinson, J. H. and Reinsch, C., *Linear Algebra*, Springer-Verlag, New York, 1971.

5 STOCHASTIC OPTIMAL CONTROL

When a dynamic system is subjected to disturbances or parameter variations that cannot be specified ahead of time, then a deterministic cost function of the type considered in Chapter 3 cannot be minimized by the choice of control. If, however, the statistics of these uncertain quantities are known, a cost function that is the *expected value* of the previous cost function can be minimized. Being unable to compute a time history of controls that accounts for known time histories of the disturbances and parameters, we must accept a control strategy that satisfies our objectives "on the average".

As time passes, the future becomes the present, and the uncertain disturbances and parameter variations become certain (i.e., they happen). The open-loop-optimal trajectory is based on expected values of disturbance and parameter-variation effects, so its control cannot be adjusted to these actualities; however, a closed-loop control strategy can account for these effects as they occur. This control strategy must include feedback information that describes the actual state of the system. Feedback entails measurements, and these, too, may be uncertain or indirect. With uncertain or indirect measurements, it is necessary to *estimate* the state history that is most likely to have caused the measurements.

Not surprisingly, the control principles of Chapter 3 and the estimation principles of Chapter 4 can be used together to solve the *stochastic optimal control problem*. In the most general case, the control and estimation strategies must be designed concurrently (i.e., one depends upon the other). However, in a wide range of problems, the control and estimation strategies can be derived independently, then concatenated to form the optimal solution. The ability to separate these designs depends upon the manner in which uncertainties enter the problem and the statistics of the uncertainties themselves.

Important elements of the theory of stochastic optimal control are presented in this chapter, beginning with the conditions for stochastic optimality in a class of nonlinear, continuous-time systems subject to additive random inputs. The theory for stochastic optimization of nonlinear systems with random variables of arbitrary probability distribution is extremely complex, and useful solutions to practical problems are not obtained readily. Consequently, emphasis is placed on systems whose random effects are assumed to be small compared to basic nonlinear effects. The need for measurements is introduced in the implementation of neighboring-optimal trajectories (Section 5.1). After considering the case of noise-free measurements, the problem of measurements containing uncertainty is addressed (Section 5.2). The conditions for separate formulation of optimal control and estimation laws are determined, including the subset for which the optimal control law can be derived without regard to uncertainty. The characteristics of both continuous- and discrete-time linear-quadratic-Gaussian (LQG) controllers, which possess the certainty-equivalence and separation properties, are discussed (Section 5.3). The stability and performance properties of linear, time-invariant systems with constant-coefficient controllers and estimators are presented in Section 5.4.

5.1 NONLINEAR SYSTEMS WITH RANDOM INPUTS AND PERFECT MEASUREMENTS

Following Section 3.4, we begin by defining a cost function that is analogous to the Bolza type, except that the expected values of the terminal and integral costs rather than their deterministic values are to be minimized:

$$J = E\left\{\phi[\mathbf{x}(t_f), t_f] + \int_{t_0}^{t_f} \mathscr{L}[\mathbf{x}(t), \mathbf{u}(t), t]\, dt\right\} \tag{5.1-1}$$

The nonlinear dynamic system that constrains the minimization is described by an n-dimensional ordinary differential equation. It is assumed that the effects of random inputs $\mathbf{w}(t)$ are small and additive, and that the initial conditions $\mathbf{x}(0)$ are known:

$$\dot{\mathbf{x}}(t) = \mathbf{f}[\mathbf{x}(t), \mathbf{u}(t), t] + \mathbf{L}(t)\mathbf{w}(t), \qquad \mathbf{x}(0) = \mathbf{x}_0 \tag{5.1-2}$$

The actual value of $\mathbf{w}(t)$ is unknown; however, it is assumed to be a white-noise process with mean value and spectral density matrix described by the following:

$$E[\mathbf{w}(t)] = \bar{\mathbf{w}} = \mathbf{0} \qquad (s \text{ vector}) \tag{5.1-3}$$

$$E[\mathbf{w}(t)\mathbf{w}(\tau)^T] = \mathbf{W}(t)\delta(t-\tau) \qquad (s \times s \text{ matrix}) \tag{5.1-4}$$

Because there is no single optimal trajectory, there is no stochastic equivalent to the Euler–Lagrange equations, and it is necessary to base stochastic optimization of nonlinear systems on the principle of optimality (Section 3.4).

Stochastic Principle of Optimality for Nonlinear Systems

Although the conditions for optimality of deterministic and stochastic problems are similar, the minimum values of the cost functions are different because uncertain inputs tend to increase the expected cost. This is most easily seen using a dynamic programming approach; hence we replace the cost function J by the value function V (with the distinction noted in Section 3.4). Furthermore, it is assumed that $\mathbf{L}(t)\mathbf{w}(t)$ represents the effects of zero-mean parameter variations as well as random disturbances. Given the stochastic cost function (Eq. 5.1-1), the optimal stochastic value function at time t_1 can be expressed as

$$V^*(t_1) = E\left\{ \phi[\mathbf{x}^*(t_f), t_f] - \int_{t_f}^{t_1} \mathscr{L}[\mathbf{x}^*(t), \mathbf{u}^*(t), t]\, dt \right\} \tag{5.1-5}$$

for t_1 in $[t_0, t_f]$. The total derivative of $V^*(t_1)$ with respect to time is

$$\frac{dV^*(t_1)}{dt} = -E\{\mathscr{L}[\mathbf{x}^*(t_1), \mathbf{u}^*(t_1), t_1]\} \tag{5.1-6a}$$

and, because $\mathbf{x}(t)$ and $\mathbf{u}(t)$ can be known without error at time t_1,

$$\frac{dV^*(t_1)}{dt} = -\mathscr{L}[\mathbf{x}^*(t_1), \mathbf{u}^*(t_1), t_1] \tag{5.1-6b}$$

As an alternative, this derivative can be expressed by a series expansion. Retaining second-degree terms, the incremental change in V^* can be written as follows (with partial derivatives evaluated at time t_1):

$$\begin{aligned} \frac{dV^*}{dt}\Delta t &= E\left\{ \frac{\partial V^*}{\partial t}\Delta t + \frac{\partial V^*}{\partial \mathbf{x}}\dot{\mathbf{x}}\,\Delta t + \tfrac{1}{2}\left[\dot{\mathbf{x}}^T \frac{\partial^2 V^*}{\partial \mathbf{x}^2}\dot{\mathbf{x}} \right]\Delta t^2 + \cdots \right\} \\ &= E[V_t^*\,\Delta t + V_{\mathbf{x}}^*(\mathbf{f}+\mathbf{Lw})\,\Delta t + \tfrac{1}{2}(\mathbf{f}+\mathbf{Lw})^T V_{\mathbf{xx}}^*(\mathbf{f}+\mathbf{Lw})\,\Delta t^2] \end{aligned} \tag{5.1-7}$$

Functions of $\mathbf{x}(t)$ equal their own expectations, and $E[\mathbf{w}(t)] = \mathbf{0}$. Dividing by Δt, and replacing the third term by its trace, the time derivative is

$$\begin{aligned} \frac{dV^*}{dt} &= V_t^* + V_{\mathbf{x}}^*\mathbf{f} + \tfrac{1}{2}\mathrm{Tr}\{E[(\mathbf{f}+\mathbf{Lw})^T V_{\mathbf{xx}}^*(\mathbf{f}+\mathbf{Lw})]\,\Delta t\} \\ &= V_t^* + V_{\mathbf{x}}^*\mathbf{f} + \tfrac{1}{2}\mathrm{Tr}\{E[V_{\mathbf{xx}}^*(\mathbf{f}+\mathbf{Lw})(\mathbf{f}+\mathbf{Lw})^T]\,\Delta t\} \end{aligned} \tag{5.1-8}$$

Noting that $\mathbf{f}$ and $\mathbf{w}$ are uncorrelated and taking the limit as Δt approaches zero,

$$\begin{aligned}\frac{dV^*}{dt} &= V_t^* + V_{\mathbf{x}}^*\mathbf{f} + \tfrac{1}{2}\lim_{\Delta t\to 0}\mathrm{Tr}\{V_{\mathbf{xx}}^*[E(\mathbf{ff}^T)\,\Delta t + \mathbf{L}E(\mathbf{ww}^T)\mathbf{L}^T\,\Delta t]\} \\ &= V_t^* + V_{\mathbf{x}}^*\mathbf{f} + \tfrac{1}{2}\mathrm{Tr}(V_{\mathbf{xx}}^*\mathbf{LWL}^T) \qquad (5.1\text{-}9)\end{aligned}$$

where $E(\mathbf{ww}^T)$ is defined by Eq. 5.1-4. The *stochastic principle of optimality* is obtained by combining Eqs. 5.1-6 and 5.1-9, then minimizing the time-rate-of-change of the value function by the choice of control. Letting $t = t_1$,

$$\begin{aligned}V_t^*(t) = -\min_{\mathbf{u}}\{&\mathscr{L}[(\mathbf{x}^*(t), \mathbf{u}(t), t] + V_{\mathbf{x}}^*\mathbf{f}[\mathbf{x}^*(t), \mathbf{u}(t), t] \\ &+ \tfrac{1}{2}\mathrm{Tr}[V_{\mathbf{xx}}^*\mathbf{L}(t)\mathbf{W}(t)\mathbf{L}^T(t)]\} \qquad (5.1\text{-}10)\end{aligned}$$

Thus minimization of the value function is achieved by pointwise minimization of the term in brackets for the time interval from t_f to t_0. From Eq. 5.1-5, the starting value for evaluation of $V^*(t)$ is $E\{\phi[\mathbf{x}(t_f), t_f]\}$, which is $\phi[\mathbf{x}(t_f), t_f]$ because $\mathbf{x}(t_f)$ can be measured without error. The minimization requires iteration to satisfy the initial condition on the state. Comparing the stochastic principle of optimality to its deterministic equivalent (Eq. 3.4-39) reveals an additional term containing the spectral density matrix of the random input. $V_{\mathbf{xx}}^*\mathbf{LWL}^T$ is positive semidefinite; integrating back from t_f to t_0, the stochastic $V_{\max}^*$ (i.e., the expected value of the cost) is greater than the deterministic $V_{\max}^*$ unless $\mathrm{Tr}(V_{\mathbf{xx}}^*\mathbf{LWL}^T)$ equals zero in the entire interval.

Stochastic Principle of Optimality for Linear-Quadratic Problems

For future reference, we note the continuous- and discrete-time *principles of optimality* for linear-quadratic problems with random disturbances. The continuous-time case is considered first, with the quadratic value function defined as

$$\begin{aligned}V(t_0) = \tfrac{1}{2}E\bigg\{&\mathbf{x}^T(t_f)\mathbf{S}_f\mathbf{x}(t_f) \\ &- \int_{t_f}^{t_0}[\mathbf{x}^T(t)\;\mathbf{u}^T(t)]\begin{bmatrix}\mathbf{Q} & \mathbf{M}\\ \mathbf{M}^T & \mathbf{R}\end{bmatrix}\begin{bmatrix}\mathbf{x}(t)\\ \mathbf{u}(t)\end{bmatrix}dt\bigg\} \qquad (5.1\text{-}11)\end{aligned}$$

The linear system equation is

$$\dot{\mathbf{x}}(t) = \mathbf{Fx}(t) + \mathbf{Gu}(t) + \mathbf{Lw}(t), \qquad \mathbf{x}(0) = \mathbf{x}_0 \qquad (5.1\text{-}12)$$

and **Q**, **M**, **R**, **F**, **G**, and **L** may be functions of time. As in Section 3.7, it can be assumed that the value function takes a quadratic form, but an additional scalar term $v(t)$ must be added to account for the random input:

$$V(t)=\tfrac{1}{2}\mathbf{x}^T(t)\mathbf{S}(t)\mathbf{x}(t)+v(t) \tag{5.1-13}$$

Here $\frac{1}{2}\mathbf{x}^T(t)\mathbf{S}(t)\mathbf{x}(t)$ is the *certainty-equivalent value function* and $v(t)$ is the *stochastic value function increment*:

$$v(t)=\tfrac{1}{2}\int_t^{t_f}\mathrm{Tr}[\mathbf{S}(\tau)\mathbf{L}(\tau)\mathbf{W}(\tau)\mathbf{L}^T(\tau)]\,d\tau \tag{5.1-14}$$

The first and second partial derivatives of the value function (with respect to the state) are

$$V_{\mathbf{x}}(t)=\mathbf{x}^T(t)\mathbf{S}(t) \tag{5.1-15}$$

$$V_{\mathbf{xx}}(t)=\mathbf{S}(t) \tag{5.1-16}$$

Hence a stochastic partial differential equation that is equivalent to the deterministic HJB equation is

$$\begin{aligned}V_t^*&=-\min_{\mathbf{u}}\tfrac{1}{2}\{E[\mathbf{x}^T\mathbf{Q}\mathbf{x}+2\mathbf{x}^T\mathbf{M}\mathbf{u}+\mathbf{u}^T\mathbf{R}\mathbf{u}\\&\quad+\mathbf{x}^T\mathbf{S}(\mathbf{F}\mathbf{x}+\mathbf{G}\mathbf{u})]+\mathrm{Tr}(\mathbf{S}\mathbf{L}\mathbf{W}\mathbf{L}^T)\}\\&=-\min_{\mathbf{u}}\tfrac{1}{2}\{\mathbf{x}^T\mathbf{Q}\mathbf{x}+2\mathbf{x}^T\mathbf{M}\mathbf{u}+\mathbf{u}^T\mathbf{R}\mathbf{u}+\mathbf{x}^T\mathbf{S}(\mathbf{F}\mathbf{x}+\mathbf{G}\mathbf{u})\\&\quad+\mathrm{Tr}(\mathbf{S}\mathbf{L}\mathbf{W}\mathbf{L}^T)\}\end{aligned} \tag{5.1-17}$$

This is subject to the terminal condition

$$V^*(t_f)=\tfrac{1}{2}\mathbf{x}^T(t_f)\mathbf{S}_f\mathbf{x}(t_f) \tag{5.1-18}$$

Following Section 3.7, the minimizing control law is found by differentiating the right side of Eq. 5.1-17 with respect to **u** and setting the result equal to zero:

$$\mathbf{u}(t)=-\mathbf{R}^{-1}[\mathbf{G}^T\mathbf{S}(t)+\mathbf{M}^T]\mathbf{x}(t) \tag{5.1-19}$$

Substituting in Eq. 5.1-17,

$$\begin{aligned}V_t^*(t)&=\tfrac{1}{2}\mathbf{x}^T(t)\dot{\mathbf{S}}(t)\mathbf{x}(t)+\dot{v}(t)\\&=-\tfrac{1}{2}\mathbf{x}^T(t)[(\mathbf{F}-\mathbf{G}\mathbf{R}^{-1}\mathbf{M}^T)^T\mathbf{S}(t)+\mathbf{S}(\mathbf{F}-\mathbf{G}\mathbf{R}^{-1}\mathbf{M}^T)+\mathbf{Q}\\&\quad-\mathbf{S}(t)\mathbf{G}\mathbf{R}^{-1}\mathbf{G}^T\mathbf{S}(t)-\mathbf{M}\mathbf{R}^{-1}\mathbf{M}^T]\mathbf{x}(t)+\tfrac{1}{2}\mathrm{Tr}[\mathbf{S}(t)\mathbf{L}\mathbf{W}(t)\mathbf{L}^T]\end{aligned} \tag{5.1-20}$$

Consequently, $V^*(t)$ takes the assumed form (Eq. 5.1-13), where $\mathbf{S}(t)$ is the solution to a matrix Riccati equation (Eq. 3.7-29), and $v(t)$ is defined by Eq. 5.1-14.

The corresponding discrete-time problem is to minimize the value function

$$V_0^* = \tfrac{1}{2}E\left\{\mathbf{x}_{k_f}^T\mathbf{S}_f\mathbf{x}_{k_f} + \sum_{k=0}^{k_f-1}[\mathbf{x}_k^T\ \mathbf{u}_k^T]\begin{bmatrix}\mathbf{Q} & \mathbf{M}\\ \mathbf{M}^T & \mathbf{R}\end{bmatrix}\begin{bmatrix}\mathbf{x}_k\\ \mathbf{u}_k\end{bmatrix}\right\} \qquad (5.1\text{-}21)$$

subject to the dynamic constraint

$$\mathbf{x}_{k+1} = \mathbf{\Phi}\mathbf{x}_k + \mathbf{\Gamma}\mathbf{u}_k + \mathbf{\Lambda}\mathbf{w}_k \qquad \mathbf{x}_0 \text{ given} \qquad (5.1\text{-}22)$$

$\mathbf{Q}$, $\mathbf{M}$, $\mathbf{R}$, $\mathbf{\Phi}$, $\mathbf{\Gamma}$, and $\mathbf{\Lambda}$ may be functions of the sampling index k, and $\mathbf{w}_k$ is a zero-mean Gaussian process with covariance matrix $\mathbf{W}_k$. The discrete-time stochastic principle of optimality for the linear-quadratic problem is expressed by

$$\begin{aligned} V_k^* &= \min_{\mathbf{u}_k} E[\mathscr{L}(\mathbf{x}_k^*, \mathbf{u}_k, k) + V_{k+1}^*] \\ &= \min_{\mathbf{u}_k} \tfrac{1}{2}E(\mathbf{x}_k^{*T}\mathbf{Q}_k\mathbf{x}_k^* + 2\mathbf{x}_k^{*T}\mathbf{M}_k\mathbf{u}_k + \mathbf{u}_k^T\mathbf{R}_k\mathbf{u}_k \\ &\quad + \mathbf{x}_{k+1}^{*T}\mathbf{S}_{k+1}\mathbf{x}_{k+1}^* + 2v_{k+1}) \end{aligned} \qquad (5.1\text{-}23)$$

The quadratic value function has two parts

$$V_k = \tfrac{1}{2}\mathbf{x}_k^T\mathbf{S}_k\mathbf{x}_k + v_k \qquad (5.1\text{-}24)$$

where $\frac{1}{2}\mathbf{x}_k^T\mathbf{S}_k\mathbf{x}_k$ is the *certainty-equivalent value function* and v_k is the *stochastic value function increment*,

$$v_k = \tfrac{1}{2}\operatorname{Tr}(\mathbf{S}_{k+1}\mathbf{\Lambda}\mathbf{W}_k\mathbf{\Lambda}^T) + v_{k+1}, \qquad v_{k_f} = 0 \qquad (5.1\text{-}25)$$

With no correlation between the disturbance and the state/control variables, substitution of Eq. 5.1-22 into Eq. 5.1-23 yields

$$\begin{aligned} V_k^* = \min_{\mathbf{u}_k} \tfrac{1}{2}\{&\mathbf{x}_k^T\mathbf{Q}\mathbf{x}_k + 2\mathbf{x}_k^T\mathbf{M}\mathbf{u}_k + \mathbf{u}_k^T\mathbf{R}\mathbf{u}_k \\ &+ (\mathbf{\Phi}\mathbf{x}_k + \mathbf{\Gamma}\mathbf{u}_k)^T\mathbf{S}_{k+1}(\mathbf{\Phi}\mathbf{x}_k + \mathbf{\Gamma}\mathbf{u}_k) + \operatorname{Tr}(\mathbf{S}_{k+1}\mathbf{\Lambda}_k\mathbf{W}_k\mathbf{\Lambda}_k^T) + 2v_{k+1}\} \end{aligned} \qquad (5.1\text{-}26)$$

Minimizing the right side of the equation with respect to the control leads to the control law

$$\mathbf{u}_k = -(\mathbf{R} + \mathbf{\Gamma}^T\mathbf{S}_{k+1}\mathbf{\Gamma})^{-1}(\mathbf{M}^T + \mathbf{\Gamma}^T\mathbf{S}_{k+1}\mathbf{\Phi})\mathbf{x}_k \qquad (5.1\text{-}27)$$

Substituting in Eq. 5.1-26,

$$
\begin{aligned}
V_k^* &= \tfrac{1}{2}\mathbf{x}_k^T\mathbf{S}_k\mathbf{x}_k + v_k \\
&= \mathbf{x}_k^T\{\boldsymbol{\Phi}^T\mathbf{S}_{k+1}\boldsymbol{\Phi} + \mathbf{Q}) \\
&\quad - (\mathbf{M}^T + \boldsymbol{\Gamma}^T\mathbf{S}_{k+1}\boldsymbol{\Phi})^T(\mathbf{R} + \boldsymbol{\Gamma}^T\mathbf{S}_{k+1}\boldsymbol{\Gamma})^{-1}(\mathbf{M}^T + \boldsymbol{\Gamma}^T\mathbf{S}_{k+1}\boldsymbol{\Phi})\}\mathbf{x}_k \\
&\quad + \tfrac{1}{2}\operatorname{Tr}(\mathbf{S}_{k+1}\boldsymbol{\Lambda}\mathbf{W}_k\boldsymbol{\Lambda}^T) + v_{k+1}
\end{aligned}
\tag{5.1-28}
$$

which confirms the form of V_k^* (Eq. 5.1-24).

Neighboring-Optimal Control

As noted in the chapter introduction, uncertain disturbances and parameter variations become certain as time passes. For real-time application, it is conceivable that the exact optimal solution could be updated continuously to provide the instantaneous control, letting t_0 slide along with the current time and adjusting $\mathbf{x}_0$ accordingly. This is a dynamic programming approach that could best be applied using a stored family of neighboring-optimal trajectories. If the storage requirement exceeds available resources, a practical alternative is to partition the solution into nominal and perturbational parts, where the former represents an off-line (prior) solution and the latter represents an on-line (real-time) solution.

The additive nature of the stochastic effects in Eq. 5.1-2 and the resulting optimal solution suggest that the optimal trajectory could be well approximated by the sum of a deterministic-optimal trajectory, computed with no stochastic disturbances or parameter variations, plus a stochastic-neighboring-optimal solution. The neighboring-optimal solution could be based on higher-degree or quasilinear expansions (Section 4.6), but conventional linearization is adequate to motivate the need for feedback control that accounts for disturbance and parameter-variation effects as they occur. The developments of Section 3.7 apply without change, so we need only summarize the procedure here. In the following, specific values of the variables are inserted *after* differentiations take place; for example,

$$\mathscr{H}_{\mathbf{u}}[\mathbf{x}^*(t), \mathbf{u}^*(t), \boldsymbol{\lambda}^*(t), t] = \mathscr{H}_{\mathbf{u}}[\mathbf{x}, \mathbf{u}, \boldsymbol{\lambda}, t]\big|_{\mathbf{x}=\mathbf{x}^*(t),\, \mathbf{u}=\mathbf{u}^*(t),\, \boldsymbol{\lambda}=\boldsymbol{\lambda}^*(t)} \tag{5.1-29}$$

Nominal (Deterministic) Solution

1. Specify the cost function:

$$J = \phi[\mathbf{x}(t_f), t_f] + \int_{t_0}^{t_f} \mathscr{L}[\mathbf{x}(t), \mathbf{u}(t), t]\, dt \tag{5.1-30}$$

2. Specify the dynamic constraint:

$$\dot{\mathbf{x}}(t) = \mathbf{f}[\mathbf{x}(t), \mathbf{u}(t), \mathbf{p}_o(t), \mathbf{w}_o(t), t], \qquad \mathbf{x}(0) = \mathbf{x}_0 \tag{5.1-31}$$

with $\mathbf{p}_o(t)$ and $\mathbf{w}_o(t)$ defined as nominal values.

3. Define the Hamiltonian:

$$\mathscr{H}[\mathbf{x}(t), \mathbf{u}(t), \boldsymbol{\lambda}(t), t] = \mathscr{L}[\mathbf{x}(t), \mathbf{u}(t), t] + \boldsymbol{\lambda}^T(t)\{\mathbf{f}[\mathbf{x}(t), \mathbf{u}(t), t]\} \quad (5.1\text{-}32)$$

4. Define the necessary and sufficient conditions for optimality:

a. $\dot{\boldsymbol{\lambda}}^*(t) = -\mathscr{H}_{\mathbf{x}}^T[\mathbf{x}^*(t), \mathbf{u}^*(t), \boldsymbol{\lambda}^*(t), t]$ (5.1-33)

b. $\boldsymbol{\lambda}^*(t_f) = \phi_{\mathbf{x}}^T[\mathbf{x}^*(t_f), t_f]$ (5.1-34)

c. $\mathscr{H}_{\mathbf{u}}[\mathbf{x}^*(t), \mathbf{u}^*(t), \boldsymbol{\lambda}^*(t), t] = \mathbf{0}$ (5.1-35)

d. $\mathscr{H}_{\mathbf{uu}}[\mathbf{x}^*(t), \mathbf{u}^*(t), \boldsymbol{\lambda}^*(t), t] \geq \mathbf{0}$ (5.1-36)

5. Compute the nominal solution $[\mathbf{x}_o^*(t), \mathbf{u}_o^*(t)]$ in $[t_0, t_f]$ using a numerical method from Section 3.6.

Continuous-Time, Linear-Quadratic, Neighboring-Optimal Solution

1. Specify the cost function:

$$\Delta^2 J = \tfrac{1}{2}\Delta\mathbf{x}^T(t_f)\phi_{\mathbf{xx}}(t_f)\,\Delta\mathbf{x}(t_f) + \frac{1}{2}\int_{t_0}^{t_f}\left\{[\Delta\mathbf{x}^T(t)\;\Delta\mathbf{u}^T(t)]\begin{bmatrix}\mathbf{Q}(t) & \mathbf{M}(t)\\ \mathbf{M}^T(t) & \mathbf{R}(t)\end{bmatrix}\begin{bmatrix}\Delta\mathbf{x}(t)\\ \Delta\mathbf{u}(t)\end{bmatrix}\right\}dt \quad (5.1\text{-}37)$$

where

$$\mathbf{Q}(t) = \mathscr{L}_{\mathbf{xx}}[\mathbf{x}_o^*(t), \mathbf{u}_o^*(t), t] \quad (5.1\text{-}38)$$

$$\mathbf{M}(t) = \mathscr{L}_{\mathbf{xu}}[\mathbf{x}_o^*(t), \mathbf{u}_o^*(t), t] \quad (5.1\text{-}39)$$

$$\mathbf{R}(t) = \mathscr{L}_{\mathbf{uu}}[\mathbf{x}_o^*(t), \mathbf{u}_o^*(t), t] \quad (5.1\text{-}40)$$

2. Specify the dynamic constraint:

$$\Delta\dot{\mathbf{x}}(t) = \mathbf{F}(t)\,\Delta\mathbf{x}(t) + \mathbf{G}(t)\,\Delta\mathbf{u}(t) + \mathbf{D}(t)\,\Delta\mathbf{p}(t) + \mathbf{L}(t)\,\Delta\mathbf{w}(t) \quad (5.1\text{-}41)$$

with

$$\mathbf{F}(t) = \mathbf{f}_{\mathbf{x}}[\mathbf{x}_o^*(t), \mathbf{u}_o^*(t), \mathbf{p}_o(t), \mathbf{w}_o(t), t] \quad (5.1\text{-}42)$$

$$\mathbf{G}(t) = \mathbf{f}_{\mathbf{u}}[\mathbf{x}_o^*(t), \mathbf{u}_o^*(t), \mathbf{p}_o(t), \mathbf{w}_o(t), t] \quad (5.1\text{-}43)$$

$$\mathbf{D}(t) = \mathbf{f}_{\mathbf{p}}[\mathbf{x}_o^*(t), \mathbf{u}_o^*(t), \mathbf{p}_o(t), \mathbf{w}_o(t), t] \quad (5.1\text{-}44)$$

$$\mathbf{L}(t) = \mathbf{f}_{\mathbf{w}}[\mathbf{x}_o^*(t), \mathbf{u}_o^*(t), \mathbf{p}_o(t), \mathbf{w}_o(t), t] \quad (5.1\text{-}45)$$

$$E[\Delta\mathbf{x}(0)] = \Delta\mathbf{x}_0 \quad (5.1\text{-}46)$$

$$E[\Delta\mathbf{p}(t)] = E[\mathbf{p}(t) - \mathbf{p}_o(t)] = \mathbf{0} \tag{5.1-47}$$

$$E[\Delta\mathbf{w}(t)] = E[\mathbf{w}(t) - \mathbf{w}_o(t)] = \mathbf{0} \tag{5.1-48}$$

$$E[\Delta\mathbf{p}(t)\,\Delta\mathbf{p}^T(\tau)] = \mathbf{U}(t)\delta(t-\tau) \tag{5.1-49}$$

$$E[\Delta\mathbf{w}(t)\,\Delta\mathbf{w}^T(\tau)] = \mathbf{W}(t)\delta(t-\tau) \tag{5.1-50}$$

$$E[\Delta\mathbf{p}(t)\,\Delta\mathbf{w}^T(\tau)] = \mathbf{0} \tag{5.1-51}$$

3. Define the deterministic Hamiltonian for the neighboring solution:

$$\mathscr{H}[\Delta\mathbf{x}(t), \Delta\mathbf{u}(t), \Delta\boldsymbol{\lambda}(t), t] = \tfrac{1}{2}[\Delta\mathbf{x}^T(t)\ \Delta\mathbf{u}^T(t)]\begin{bmatrix}\mathbf{Q}(t) & \mathbf{M}(t)\\ \mathbf{M}^T(t) & \mathbf{R}(t)\end{bmatrix}\begin{bmatrix}\Delta\mathbf{x}(t)\\ \Delta\mathbf{u}(t)\end{bmatrix} + \Delta\boldsymbol{\lambda}^T(t)[\mathbf{F}(t)\,\Delta\mathbf{x}(t) + \mathbf{G}(t)\,\Delta\mathbf{u}(t)] \tag{5.1-52}$$

4. Integrate the matrix Riccati equation to obtain $\mathbf{S}(t)$ in $[t_0, t_f]$. [In earlier chapters, $\mathbf{P}(t)$ was used to denote the solutions of Riccati equations for both control and estimation. In this and succeeding chapters, we will have cause to refer to both Riccati solutions concurrently, justifying the change in notation.]

$$\begin{aligned}\dot{\mathbf{S}}(t) = &-[\mathbf{F}(t) - \mathbf{G}(t)\mathbf{R}^{-1}(t)\mathbf{M}^T(t)]^T\mathbf{S}(t)\\ &- \mathbf{S}(t)[\mathbf{F}(t) - \mathbf{G}(t)\mathbf{R}^{-1}(t)\mathbf{M}^T(t)]\\ &+ \mathbf{S}(t)\mathbf{G}(t)\mathbf{R}^{-1}(t)\mathbf{G}^T(t)\mathbf{S}(t) - \mathbf{Q}(t) + \mathbf{M}(t)\mathbf{R}^{-1}(t)\mathbf{M}^T(t)\end{aligned} \tag{5.1-53}$$

with

$$\mathbf{S}(t_f) = \phi_{\mathbf{xx}}[\mathbf{x}_o^*(t_f), t_f] \tag{5.1-54}$$

5. Compute the optimal feedback gain matrix in $[t_0, t_f]$:

$$\mathbf{C}(t) = \mathbf{R}^{-1}(t)[\mathbf{G}^T(t)\mathbf{S}(t) + \mathbf{M}^T(t)] \tag{5.1-55}$$

6. Compute a member of the neighboring-optimal state and control history ensemble in $[t_0, t_f]$ by integrating

$$\Delta\dot{\mathbf{x}}^*(t) = \mathbf{F}(t)\,\Delta\mathbf{x}^*(t) + \mathbf{G}(t)\,\Delta\mathbf{u}^*(t) + \mathbf{D}(t)\,\Delta\mathbf{p}(t) + \mathbf{L}(t)\,\Delta\mathbf{w}(t), \qquad \Delta\mathbf{x}^*(0) = \Delta\bar{\mathbf{x}}_0 \tag{5.1-56}$$

with

$$\Delta\mathbf{u}^*(t) = -\mathbf{C}(t)\,\Delta\mathbf{x}^*(t) \tag{5.1-57}$$

7. Approximate a member of the optimal state and control history

ensemble in $[t_0, t_f]$ as

$$\mathbf{x}^*(t) = \mathbf{x}_o^*(t) + \Delta\mathbf{x}^*(t) \tag{5.1-58}$$

$$\mathbf{u}^*(t) = \mathbf{u}_o^*(t) + \Delta\mathbf{u}^*(t) \tag{5.1-59}$$

These equations might be applied in either of two ways: simulation or real-time control. In either case, the nominal solution is generated first. For *simulation*, Eq. 5.1-56 of the neighboring-optimal solution is integrated in the simulating computer; $\Delta\mathbf{p}(t)$ and $\Delta\mathbf{w}(t)$ are outputs of a suitable random process generator. For *real-time control*, the physical system itself provides the integration. The system must be driven by the total optimal control (Eq. 5.1-59), and the neighboring-optimal control is computed as a function of the *difference* between the measured state, $\mathbf{x}_M(t)$, and the nominal-optimal value, $\mathbf{x}_o^*(t)$. The total control is then

$$\mathbf{u}^*(t) = \mathbf{u}_o^*(t) - \mathbf{C}(t)[\mathbf{x}_M(t) - \mathbf{x}_o^*(t)] \tag{5.1-60}$$

This result is particularly significant for real-time control because the amount of computation that must be performed during control is minimal. $\mathbf{x}_o^*(t)$, $\mathbf{u}_o^*(t)$, and $\mathbf{C}(t)$ can be calculated before control begins, and only Eq. 5.1-60 need be computed as the process evolves.

Following Section 3.7, *sampled-data, linear-quadratic, neighboring-optimal control* could be implemented. The procedure is the same, except that $\mathbf{u}^*(t)$ is computed at discrete sampling instants t_k and

$$\mathbf{u}^*(t_k) = \mathbf{u}_o^*(t_k) - \mathbf{C}_k[\mathbf{x}_M(t_k) - \mathbf{x}_o^*(t_k)] \tag{5.1-61}$$

The discrete-time control gain matrix is computed as

$$\mathbf{C}_k = (\hat{\mathbf{R}}_k + \mathbf{\Gamma}_k^T\mathbf{S}_{k+1}\mathbf{\Gamma}_k)^{-1}(\hat{\mathbf{M}}_k^T + \mathbf{\Gamma}_k^T\mathbf{S}_{k+1}\mathbf{\Phi}_k) \tag{5.1-62}$$

where the Riccati matrix is found from

$$\begin{aligned}\mathbf{S}_k = {} & \hat{\mathbf{Q}}_k + \mathbf{\Phi}_k^T\mathbf{S}_{k+1}\mathbf{\Phi}_k \\ & - (\hat{\mathbf{M}}_k + \mathbf{\Phi}_k^T\mathbf{S}_{k+1}\mathbf{\Gamma}_k)(\hat{\mathbf{R}}_k + \mathbf{\Gamma}_k^T\mathbf{S}_{k+1}\mathbf{\Gamma}_k)^{-1}(\hat{\mathbf{M}}_k^T + \mathbf{\Gamma}_k^T\mathbf{S}_{k+1}\mathbf{\Phi}_k),\ \mathbf{S}_{k_f} = \phi_{\mathbf{xx}}(t_f)\end{aligned} \tag{5.1-63}$$

and $\mathbf{\Phi}_k$, $\mathbf{\Gamma}_k$, $\hat{\mathbf{Q}}_k$, $\hat{\mathbf{M}}_k$, and $\hat{\mathbf{R}}_k$, are derived from $\mathbf{F}$, $\mathbf{G}$, $\mathbf{Q}$, $\mathbf{M}$, and $\mathbf{R}$ as in Section 3.7.

Because the control will be held constant during each sampling interval $[t_k, t_{k+1}]$, the resulting trajectory will be closer to the continuous-time optimum if the nominal control approximates its average value in the interval. Using linear interpolation of the previously stored values, a discrete approximation to the nominal value would be

$$\mathbf{u}_A^*(t_k) = \frac{\mathbf{u}_o^*(t_k) + \mathbf{u}_o^*(t_{k+1})}{2} \tag{5.1-64}$$

and $\mathbf{u}_A^*$ would replace $\mathbf{u}_o^*$ in Eq. 5.1-61.

Evaluation of the Variational Cost Function

The continuous-time variational cost function (Eq. 5.1-37) has been defined as the second variation of the original cost function (Eq. 5.1-1), and it is instructive to evaluate the effect of linear-optimal control on this cost. For comparison, the cost function is first evaluated with no linear control at all; in both cases, it is assumed that cross-product weighting (**M**) is zero. $\Delta^2 J$ is a scalar, so the stochastic cost function can be expressed as the trace of expected values,

$$\begin{aligned}\Delta^2 J &= \tfrac{1}{2}\,\mathrm{Tr}\left\{E(\Delta\mathbf{x}_f^T \mathbf{S}_f\, \Delta\mathbf{x}_f) + E\int_{t_0}^{t_f} [\Delta\mathbf{x}^T\ \Delta\mathbf{u}^T]\begin{bmatrix}\mathbf{Q} & \mathbf{0}\\ \mathbf{0} & \mathbf{R}\end{bmatrix}\begin{bmatrix}\Delta\mathbf{x}\\ \Delta\mathbf{u}\end{bmatrix} dt\right\}\\ &= \tfrac{1}{2}\,\mathrm{Tr}\left\{\mathbf{S}_f E(\Delta\mathbf{x}_f\, \Delta\mathbf{x}_f^T) + \int_{t_0}^{t_f} [\mathbf{Q}E(\Delta\mathbf{x}\,\Delta\mathbf{x}^T) + \mathbf{R}E(\Delta\mathbf{u}\,\Delta\mathbf{u}^T)]\, dt\right\}\\ &= \tfrac{1}{2}\,\mathrm{Tr}\left[\mathbf{S}_f\mathbf{P}_f + \int_{t_0}^{t_f} (\mathbf{QP} + \mathbf{RU})\, dt\right]\end{aligned} \tag{5.1-65}$$

where

$$\mathbf{P} = E(\Delta\mathbf{x}\,\Delta\mathbf{x}^T) \tag{5.1-66}$$

$$\mathbf{U} = E(\Delta\mathbf{u}\,\Delta\mathbf{u}^T) \tag{5.1-67}$$

Integrating by parts,

$$\mathbf{SP}\Big|_{t_0}^{t_f} = \int_{t_0}^{t_f} \mathbf{S}\dot{\mathbf{P}}\, dt + \int_{t_0}^{t_f} \dot{\mathbf{S}}\mathbf{P}\, dt \tag{5.1-68a}$$

or

$$\mathbf{S}_f\mathbf{P}_f = \mathbf{S}_0\mathbf{P}_0 + \int_{t_0}^{t_f} \dot{\mathbf{S}}\mathbf{P}\, dt + \int_{t_0}^{t_f} \mathbf{S}\dot{\mathbf{P}}\, dt \tag{5.1-68b}$$

Equation 5.1-65 then can be written as

$$\Delta^2 J = \tfrac{1}{2}\,\mathrm{Tr}\left[\mathbf{S}_0\mathbf{P}_0 + \int_{t_0}^{t_f} (\mathbf{QP} + \mathbf{RU} + \dot{\mathbf{S}}\mathbf{P} + \mathbf{S}\dot{\mathbf{P}})\, dt\right] \tag{5.1-69}$$

Treating random parameter variations as random disturbances, the state equation is

$$\Delta\dot{\mathbf{x}} = \mathbf{F}\,\Delta\mathbf{x} + \mathbf{G}\,\Delta\mathbf{u} + \mathbf{L}\,\Delta\mathbf{w} \tag{5.1-70}$$

If the control is not exercised, $\mathbf{U} = \mathbf{0}$, and prior results indicate that the state covariance matrix is propagated by forward integration of

$$\dot{\mathbf{P}} = \mathbf{FP} + \mathbf{PF}^T + \mathbf{LWL}^T \qquad \mathbf{P}_0 \text{ given} \tag{5.1-71}$$

while the adjoint covariance matrix is found by backward integration of

$$\dot{\mathbf{S}} = -\mathbf{F}^T\mathbf{S} - \mathbf{SF} - \mathbf{Q} \qquad \mathbf{S}_f \text{ given} \tag{5.1-72}$$

Substituting in Eq. 5.1-69, and using $\mathrm{Tr}(\mathbf{ABC}) = \mathrm{Tr}(\mathbf{CAB})$, the open-loop variational cost function is found to be

$$\Delta^2 J = \tfrac{1}{2}\,\mathrm{Tr}\left[\mathbf{S}_0\mathbf{P}_0 + \int_{t_0}^{t_f} \mathbf{S}(t)\mathbf{L}(t)\mathbf{W}(t)\mathbf{L}^T(t)\,dt\right] \tag{5.1-73}$$

If the optimal control is implemented (Eqs. 5.1-53 to 5.1-55), the state equation can be written as

$$\Delta\dot{\mathbf{x}} = (\mathbf{F} - \mathbf{GC})\,\Delta\mathbf{x} + \mathbf{L}\,\Delta\mathbf{w} \tag{5.1-74}$$

and the expected value of control is

$$\mathbf{U} = \mathbf{CPC}^T = \mathbf{R}^{-1}\mathbf{G}^T\mathbf{SPSGR}^{-1} \tag{5.1-75}$$

Consequently, the differential equations for the state and adjoint covariance matrices are

$$\begin{aligned}\dot{\mathbf{P}} &= (\mathbf{F} - \mathbf{GC})\mathbf{P} + \mathbf{P}(\mathbf{F} - \mathbf{GC})^T + \mathbf{LWL}^T \\ &= \mathbf{FP} + \mathbf{PF}^T + \mathbf{LWL}^T - \mathbf{GR}^{-1}\mathbf{G}^T\mathbf{SP} - \mathbf{PSGR}^{-1}\mathbf{G}^T, \qquad \mathbf{P}_0 \text{ given}\end{aligned} \tag{5.1-76}$$

$$\dot{\mathbf{S}} = -\mathbf{F}^T\mathbf{S} - \mathbf{SF} + \mathbf{SGR}^{-1}\mathbf{G}^T\mathbf{S} - \mathbf{Q}, \qquad \mathbf{S}_f \text{ given} \tag{5.1-77}$$

Substituting in Eq. 5.1-69, the minimized cost function is

$$\Delta^2 J = \tfrac{1}{2}\,\mathrm{Tr}\left[\mathbf{S}_0\mathbf{P}_0 + \int_{t_0}^{t_f} \mathbf{S}(t)\mathbf{L}(t)\mathbf{W}(t)\mathbf{L}^T(t)\,dt\right] \tag{5.1-78}$$

which is identical in form to Eq. 5.1-73. It would appear that the neighboring-optimal control has no effect on the cost function; however, this appearance is deceiving, as the $\mathbf{S}_0$ and $\mathbf{S}(t)$ of Eqs. 5.1-73 and 5.1-78 are not the same, having been generated by Eqs. 5.1-72 and 5.1-77, respectively.

The variational cost function is seen to contain two parts. With no stochastic disturbance $[\mathbf{W}(t) = \mathbf{0}]$, the cost is determined by the initial condition covariance, $\mathbf{P}_0$, which may result from either a known constant or the expected value of a random constant variation in the initial state. With no initial condition, the cost is proportional to the weighted integral of the disturbance spectral density, $\mathbf{W}(t)$.

5.2 NONLINEAR SYSTEMS WITH RANDOM INPUTS AND IMPERFECT MEASUREMENTS

If a dynamic system is driven by uncertain disturbances, the stochastic optimal trajectory must be generated using measurements of the state. If the measurements contain random errors, then the feedback control responds not only to the effects of random inputs, but to the measurement errors as well. The control forces resulting from measurement errors are, of course, spurious, so it is desirable to minimize their effects, at the same time transmitting the maximum amount of information about the state. Therefore, the best neighboring control strategy involves optimal state estimation as well as optimal control. The estimator introduces *caution* or *hedging* in the feedback control by not responding to deviations that probably are due to measurement error.

In some systems, the converse also is true; that is, control actions affect the observability of the system. The control is then said to have a *dual effect*: it reveals information regarding the state and parameters of the system while achieving the control objective. Consequently, a net reduction in cost may result from using *adaptive* control strategies that test the system to determine which control will be most effective. Control strategies that perform a learning function are called *probing* strategies.

Not every system can benefit from probing controls. Probing controls normally have significant value only when the system is highly nonlinear and when uncertainties are large. *Neutral* systems, such as known linear plants with linear feedback controllers, do not exhibit the dual-control effect because no amount of probing changes our knowledge of the system's structure and parameters. On the other hand, systems with unknown parameters can evidence cost reductions as a consequence of probing controls.

Stochastic Principle of Optimality

A controller that uses feedback measurements must be *nonanticipative* to be physically realizable. It can use a *knowledge base* of past and present information to effect a control strategy, but future information is unavailable. Suppose that the "present" time is denoted by t_1; consider what can be known at t_1 about an optimal trajectory that begins at t_0 and ends at t_f, with $t_0 \leq t_1 \leq t_f$. For discussion, time profiles of a variable $a(t)$ in $[t_1, t_2]$ are denoted by $a[t_1, t_2]$.

Given the nominal disturbance input and parameter variation profiles, $\mathbf{w}_o[t_0, t_f]$ and $\mathbf{p}_0[t_0, t_f]$, the nominal-optimal trajectory can be calculated to yield $\mathbf{x}^*[t_0, t_f]$ and $\mathbf{u}^*[t_0, t_f]$. (If the disturbances and parameter variations have no known or expected components, their nominal values are zero.) Consequently, nominal values of all dynamic equations, measurement equations, and sensitivity matrices over the entire interval can be known at t_1. Profiles of the nominal probability density functions (or their significant descriptors, e.g., mean vectors and covariance matrices) for the entire trajectory also can be available at t_1.

Information regarding the actual trajectory is available only for the interval $[t_0, t_1]$, and it falls in two categories. The realized measurement vector profile, $\mathbf{z}[t_0, t_1]$ provides the only information regarding the actual state vector profile, $\mathbf{x}[t_0, t_1]$, and it is subject to error. The commanded control vector profile, $\mathbf{u}[t_0, t_1]$, indicates the manner in which the dynamic system was forced during the same interval. Consequently, the *information set* (frequently referred to as a "σ algebra") available at t_1 is

$$\mathfrak{I}[t_0, t_1] = \{\mathbf{z}[t_0, t_1], \mathbf{u}[t_0, t_1]\} \tag{5.2-1}$$

Future optimal control solutions must be *conditioned on* $\mathfrak{I}[t_0, t_1]$, and they can make use of all the aforementioned nominal data.

To be of any value in determining the future control, the information set must be transformed into quantities of direct relevance to the trajectory itself. A profile of the conditional probability density function estimate of $\mathbf{x}[t_0, t_1]$ would be more useful than the measurement and control histories. A precise estimate of the non-Gaussian statistics is needed in principle, although a state estimate (or conditional mean) profile, $\hat{\mathbf{x}}[t_0, t_1]$, and a state-estimate-error covariance profile, $\mathbf{P}[t_0, t_1]$, could provide a satisfactory alternative; this would allow the *derived information* set to be defined as follows:

$$\mathfrak{I}_D[t_0, t_1] = \{\hat{\mathbf{x}}[t_0, t_1], \mathbf{P}[t_0, t_1]\} \tag{5.2-2}$$

These estimates transform the original information set using, for example, one of the nonlinear estimators presented in Section 4.6: neighboring-optimal linear estimator, extended Kalman–Bucy filter, or quasilinear filter. Because the dynamic systems considered here exhibit the Markov property and measurement errors are "white" (uncorrelated in time), all the needed information about the past trajectory is assumed to be contained in the "present" post-update state and covariance estimates. Then, the *Markov derived information set,*

$$\mathfrak{I}_D(t_1) = [\hat{\mathbf{x}}(t_1), \mathbf{P}(t_1)] \tag{5.2-3}$$

is equivalent to Eq. 5.2-2 and is a suitable information set for the conditional stochastic optimization.

Of course, assumptions about the system and its noise statistics must be made to compute $\mathfrak{I}_D(t_1)$, and it is possible that these assumptions are themselves questionable. If alternate hypotheses for the system model are plausible, *multiple-model estimation* (Section 4.7) could be used to define an expanded, derived information set from $\mathfrak{I}[t_0, t_1]$. For I hypotheses, the multiple-model information set would be

$$\mathfrak{I}_{MM}(t_1) = [\hat{\mathbf{x}}_1(t_1) \ldots \hat{\mathbf{x}}_I(t_1); \quad \mathbf{P}_1(t_1) \ldots \mathbf{P}_I(t_1); \quad \Pr(H_1) \ldots \Pr(H_I)] \tag{5.2-4}$$

where $\hat{\mathbf{x}}_i(t_1)$ and $\mathbf{P}_i(t_i)$ are generated by the ith estimator, and $\Pr(H_i)$ is the probability of the ith hypothesis being true.

Having defined the knowledge base, an optimization procedure must be chosen. Dynamic programming provides a natural framework for stochastic optimization because the principle of optimality can be expressed readily in terms of expected values throughout the solution interval. Consider a stochastic cost function that is the expected value of a Bolza-type cost function J_D,

$$\begin{aligned} J &= E\{J_D\} \\ &= E\left\{\phi[\mathbf{x}(t_f), t_f] + \int_{t_0}^{t_f} \mathscr{L}[\mathbf{x}(t), \mathbf{u}(t), t]\, dt\right\} \end{aligned} \tag{5.2-5}$$

with dynamic constraint

$$\dot{\mathbf{x}}(t) = \mathbf{f}[\mathbf{x}(t), \mathbf{u}(t), t] + \mathbf{L}[\mathbf{x}(t), t]\mathbf{w}(t), \qquad \mathbf{x}(0) = \mathbf{x}_0 \tag{5.2-6}$$

$\mathbf{w}(t)$ is a zero-mean, white random process representing additive disturbance inputs and/or parameter variations with spectral density matrix $\mathbf{W}(t)$.

Given perfect knowledge of the state, the stochastic HJB equation for continuous-time systems (Eq. 5.1-10), which controls the evolution of the corresponding value function V, could be expressed as

$$\begin{aligned} 0 = \min_{\mathbf{u}} E &\left\{\frac{\partial V}{\partial t}(\mathbf{x}^*, t) + \mathscr{L}(\mathbf{x}^*, \mathbf{u}, t) + \frac{\partial V}{\partial \mathbf{x}}(\mathbf{x}^*, t)\mathbf{f}(\mathbf{x}^*, \mathbf{u}, t)\right. \\ &\left. + \tfrac{1}{2}\operatorname{Tr}\left[\frac{\partial^2 V}{\partial \mathbf{x}^2}(\mathbf{x}^*, t)\mathbf{L}(\mathbf{x}^*, \mathbf{u}, t)\mathbf{W}(t)\mathbf{L}^T(\mathbf{x}^*, \mathbf{u}, t)\right]\right\} \end{aligned} \tag{5.2-7}$$

In this equation, the expectation is evaluated over all possible values of $\mathbf{x}$, and the boundary conditions are specified at the terminal time:

$$E\{V[\mathbf{x}^*(t_f), t_f]\} = E\{\phi[\mathbf{x}^*(t_f), t_f]\} \tag{5.2-8}$$

Since

$$\begin{aligned}\min_{\mathbf{u}} E[\mathrm{fcn}(\mathbf{x}^*, \mathbf{u}, \mathbf{w}, t)] &= E\{\min_{\mathbf{u}}[\mathrm{fcn}(\mathbf{x}^*, \mathbf{u}, \mathbf{w}, t)]\}\\ &= E[\mathrm{fcn}(\mathbf{x}^*, \mathbf{u}^*, \mathbf{w}, t)]\end{aligned} \tag{5.2-9}$$

and $\mathbf{w}(t)$ is independent of $\mathbf{x}$, $\mathrm{fcn}(\mathbf{x}^*, \mathbf{u}^*, \mathbf{w}, t)$ can be calculated by ordinary minimization. Because $\mathbf{x}^*$ is assumed to be measurable without error, the only stochastic effect is due to $\mathbf{w}(t)$, yielding the form of the HJB equation given earlier. As in the deterministic case, solutions for nonlinear systems require iteration to satisfy boundary values at the beginning and end of each trajectory.

With imperfect knowledge of the state, the expectation in Eq. 5.2-7 must be conditioned on the available information set, $\mathfrak{I}[t_0, t]$. Therefore, the stochastic principle of optimality for the problem posed by Eqs. 5.2-5 and 5.2-6 involves conditional expected values:

$$\begin{aligned}0 = \min_{\mathbf{u}} E\{\langle V_t(\mathbf{x}^*, t) + \mathscr{L}(\mathbf{x}^*, \mathbf{u}, t) + V_{\mathbf{x}}(\mathbf{x}^*, t)\mathbf{f}(\mathbf{x}^*, \mathbf{u}, t)\\ + \tfrac{1}{2}\mathrm{Tr}[V_{\mathbf{xx}}(\mathbf{x}^*, t)\mathbf{L}(\mathbf{x}^*, \mathbf{u}, t)\mathbf{W}(t)\mathbf{L}^T(\mathbf{x}^*, \mathbf{u}, t)]\rangle \mid \mathfrak{I}[t_0, t]\}\end{aligned} \tag{5.2-10}$$

$$E\{V[\mathbf{x}^*(t_f), t_f] \mid \mathfrak{I}[t_0, t_f]\} = E\{\phi[\mathbf{x}^*(t_f), t_f] \mid \mathfrak{I}[t_0, t_f]\} \tag{5.2-11}$$

The principal difficulty in stochastic optimization of systems containing measurement uncertainty appears in Eq. 5.2-11. Propagation of the expected value function *back* from the final time t_f to some intermediate time t_1 is conditioned on $\mathfrak{I}[t_0, t_f]$; however, the available information set, $\mathfrak{I}[t_0, t_1]$, falls short.

By its definition (Eq. 5.2-1), $\mathfrak{I}[t_0, t_f]$ can be partitioned in two parts,

$$\mathfrak{I}[t_0, t_f] = \mathfrak{I}[t_0, t_1] + \mathfrak{I}[t_1+, t_f] \tag{5.2-12}$$

where t_1+ connotes a time infinitesimally greater than t_1. Then

$$E\{V[\mathbf{x}^*(t_f), t_f] \mid \mathfrak{I}[t_0, t_f]\} = E\{V(\mathbf{x}^*(t_f), t_f] \mid \langle \mathfrak{I}[t_0, t_1] + \mathfrak{I}[t_1+, t_f]\rangle\} \tag{5.2-13}$$

Success in stochastic optimization is commensurate with the ability to predict the conditioning effect of $\mathfrak{I}[t_1+, t_f]$ on the future value function. Qualitatively, $\mathfrak{I}[t_1+, t_f]$ contains known and unknown parts; for example, its statistical attributes may be known even though its actual time histories are unknown. If the conditioning effect is entirely predictable from the nominal knowledge base, as is the case for systems that possess the separation property described later, stochastic control can be optimum. If the future conditioning effect can only be approximated, then the stochastic control can only be suboptimum (but possibly very good in comparison to alternate strategies). Recalling that optimization of nonlinear systems *without* uncertainty requires iteration (Section 3.6) and that optimal nonlinear estimation *without* control is sufficiently difficult that only suboptimum solutions are normally practical (Section 4.6), it is not surprising that the stochastic control of nonlinear systems has similar limitations.

Dual Control

Uncertain system parameters degrade the nominal-optimal solution, so control performance can be improved by estimating these parameters in real time and adapting the control strategy accordingly. The simultaneous estimation of uncertain parameters as well as the state is a nonlinear problem, even if the dynamic system is linear in the parameters (Section 4.7); furthermore, parameter estimation depends on the input-output sensitivities established by the parameters. Consequently, parameter estimation can be aided by proper test inputs furnished by the control. The computation of control inputs that enhance parameter estimation, thereby improving state estimation and future cost reduction while providing current cost reduction is a dual-control problem.

Approximate dual-control solutions presented to date [e.g., (B-1) to (B-3), (T-5), and (T-6)] have been applied to discrete-time problems, with cost functions of the form

$$J_D(k_f) = \phi(\mathbf{x}_{k_f}, k_f) + \sum_{k=0}^{k_f-1} [\mathscr{L}(\mathbf{x}_k, k) + \mathscr{M}(\mathbf{u}_k, k)] \tag{5.2-14}$$

subject to the nonlinear dynamic constraint

$$\mathbf{x}_{k+1} = \mathbf{f}(\mathbf{x}_k, \mathbf{u}_k, k) + \mathbf{w}_k, \qquad \mathbf{x}_0 \text{ given} \tag{5.2-15}$$

and with measurement equation

$$\mathbf{z}_k = \mathbf{h}(\mathbf{x}_k, k) + \mathbf{n}_k \tag{5.2-16}$$

$\mathbf{w}_k$ and $\mathbf{n}_k$ are assumed to be zero-mean, white-noise sequences with covariance matrices, $\mathbf{W}_k$ and $\mathbf{N}_k$, respectively. While Eq. 5.2-15 can be, at best, only an approximation to a nonlinear continuous-time system (using Euler or rectangular integration, as in Section 2.3), it is an adequate model for eliciting the structural characteristics of probing control strategies in physical systems. It is the assumed model for many socioeconomic problems and for the estimation of parameters in linear discrete-time models. In the latter, $\mathbf{x}_k$ represents the original state augmented by the parameters to be estimated.

Probing feedback control logic cannot predict future measurements, but it can assess the probable value of future information to be obtained by probing, and this can be traded against the cost of probing.* In order to do this, the stochastic value function is expressed as

$$V(k) = V_D(k) + V_C(k) + V_P(k) \tag{5.2-17}$$

$V_D(k)$ represents the expected value function using the *deterministic nominal control* for the remainder of the trajectory; $V_C(k)$ is the value function increment due to *cautious feedback control* imposed by uncertain inputs and measurements; and $V_P(k)$ is the value function increment of *probing feedback control* that is intended to reduce uncertainty. $V_D(k)$ is the optimal deterministic value function only if the control at k is optimal and the state estimate $\hat{\mathbf{x}}_k$ has the nominal-optimal value $\mathbf{x}_k^*$.

The components of $V(k)$ are computed following (B-3). The deterministic value function estimate is expressed as

$$V_D(k) = \mathcal{M}(\mathbf{u}_k, k) + V_{D_0}(k+1) + \gamma_0(k+1) \tag{5.2-18}$$

Here $V_{D_0}(k+1)$ is the projected cost of applying $\mathbf{u}_k$ at k (to be determined) and the nominal-optimal control history for the remainder of the trajectory:

$$V_{D_0}(k+1) = \phi(\mathbf{x}_{k_f}, k_f) + \sum_{j=k+1}^{k_f-1} [\mathcal{L}(\mathbf{x}_j, j) + \mathcal{M}(\mathbf{u}_j^*, j)] \tag{5.2-19}$$

$\mathbf{x}_j$ is an open-loop prediction of the effect of control computed from the undisturbed state equation

$$\mathbf{x}_j = \mathbf{f}(\mathbf{x}_{j-1}, \mathbf{u}_{j-1}, j-1) \tag{5.2-20}$$

$$\mathbf{u}_{j-1} = \begin{cases} \mathbf{u}_k \text{ (to be determined)}, & j-1 = k \\ \mathbf{u}_{j-1}^* & j-1 > k \end{cases} \tag{5.2-21}$$

subject to an initial condition based on the current state estimate, $\hat{\mathbf{x}}_k(+)$. [$\hat{\mathbf{x}}_k(+)$ is obtained from a nonlinear estimator, such as an extended Kalman filter.]

*Probing feedback control strategies are sometimes referred to as "closed-loop" feedback control strategies (an obvious redundancy), leading to the contradictory appellation of "open-loop" feedback for the nonprobing feedback controllers. It also is called "active information storage" in (F-1) to distinguish it from the "passive information storage" that can occur without probing, as in an extended Kalman filter used for parameter estimation.

The remaining term in Eq. 5.2-18, $\gamma_o(k+1)$, accounts for the fact that $\mathbf{x}_j$ in $[k+1, k_f-1]$ generally is not an optimal trajectory, and its description is rather complex. [If $\mathbf{x}_j$ is optimal, $\gamma_o(k+1)$ is zero.] Define a discrete Hamiltonian for the remainder of the trajectory,

$$\begin{aligned}\mathscr{H}_j &= \mathscr{H}(\mathbf{x}_j, \mathbf{u}_j^*, \boldsymbol{\lambda}_{j+1}, j) \\ &= \mathscr{L}(\mathbf{x}_j, j) + \mathscr{M}(\mathbf{u}_j^*, j) + \boldsymbol{\lambda}_{j+1}^T \mathbf{f}(\mathbf{x}_j, \mathbf{u}_j^*, j)\end{aligned} \tag{5.2-22}$$

where the adjoint vector $\boldsymbol{\lambda}_j$ satisfies the following difference equation and boundary condition:

$$\boldsymbol{\lambda}_{j-1} = \left(\frac{\partial \mathscr{H}_j}{\partial \mathbf{x}} - \mathbf{A}_j \mathbf{B}_j^{-1} \frac{\partial \mathscr{H}_j}{\partial \mathbf{u}}\right)^T \tag{5.2-23}$$

$$\boldsymbol{\lambda}_{k_f} = \frac{(\partial \phi_{k_f})^T}{\partial \mathbf{x}} \tag{5.2-24}$$

The matrices $\mathbf{A}_j$ and $\mathbf{B}_j$ are Hessian matrices of $\mathscr{H}_j$ that account for second-degree perturbations from $\mathbf{x}_j$:

$$\mathbf{A}_j = \frac{\partial^2 \mathscr{H}_j}{\partial \mathbf{u}\, \partial \mathbf{x}} + \boldsymbol{\Gamma}_j^T \mathbf{S}_{j+1} \boldsymbol{\Phi}_j \tag{5.2-25}$$

$$\mathbf{B}_j = \frac{\partial^2 \mathscr{H}_j}{\partial \mathbf{x}^2} + \boldsymbol{\Gamma}_j^T \mathbf{S}_{j+1} \boldsymbol{\Gamma}_j \tag{5.2-26}$$

Here, state transition and control effect matrices appear,

$$\boldsymbol{\Phi}_j = \frac{\partial \mathbf{f}}{\partial \mathbf{x}} (\mathbf{x}_j, \mathbf{u}_j^*, j) \tag{5.2-27}$$

$$\boldsymbol{\Gamma}_j = \frac{\partial \mathbf{f}}{\partial \mathbf{u}} (\mathbf{x}_j, \mathbf{u}_j^*, j) \tag{5.2-28}$$

and $\mathbf{S}_{k+1}$ is the solution to the discrete-time matrix Riccati equation,

$$\mathbf{S}_j = \mathbf{Q}_j + \boldsymbol{\Phi}_j^T \mathbf{S}_{j+1} \boldsymbol{\Phi}_j - \boldsymbol{\Phi}_j^T \mathbf{S}_{j+1} \boldsymbol{\Gamma}_j (\mathbf{R}_j + \boldsymbol{\Gamma}_j^T \mathbf{S}_{j+1} \boldsymbol{\Gamma}_j)^{-1} \boldsymbol{\Gamma}_j^T \mathbf{S}_{j+1} \boldsymbol{\Phi}_j \tag{5.2-29}$$

with the following weighting function and terminal condition matrices:

$$\mathbf{Q}_j = \frac{\partial^2 \mathscr{L}}{\partial \mathbf{x}^2} (\mathbf{x}_j, \mathbf{u}_j^*, j) \tag{5.2-30}$$

$$\mathbf{R}_j = \frac{\partial^2 \mathscr{M}}{\partial \mathbf{u}^2} (\mathbf{x}_j, \mathbf{y}_j^*, j) \tag{5.2-31}$$

$$\mathbf{S}_{k_f} = \frac{\partial^2 \phi_{k_f}}{\partial \mathbf{x}^2} \tag{5.2-32}$$

The value function increment due to *cautious feedback control* is

$$V_C(k) = \tfrac{1}{2}\,\mathrm{Tr}[\mathbf{S}_{k+1}\mathbf{P}_{k+1}(-)] + \frac{1}{2}\sum_{j=k+1}^{k_f-1} \mathrm{Tr}(\mathbf{S}_{j+1}\mathbf{W}_j) \tag{5.2-33}$$

$\mathbf{P}_{k+1}(-)$ is a predicted pre-update state covariance estimate derived from the nonlinear state estimator mentioned above, and the remaining terms have been defined previously. $V_C(k)$ reflects the prior effects of measurement errors and random inputs as well as the predicted effects of random inputs.

The value function increment due to *probing feedback control* is

$$V_P(k) = \frac{1}{2}\sum_{j=k+1}^{k_f-1} \mathrm{Tr}[\mathbf{C}_k\mathbf{P}_k(+)] \tag{5.2-34}$$

where $\mathbf{P}_k(+)$ is the postupdate state covariance estimate derived from the nonlinear estimator. $\mathbf{C}_k$ weights the value of future information gained from probing inputs, and it is

$$\mathbf{C}_k = \mathbf{A}_k^T \mathbf{B}_k^{-1} \mathbf{A}_k \tag{5.2-35}$$

The probing control strategy is a direct application of dynamic programming; it seeks to minimize $V(k)$ by choice of the current control $\mathbf{u}_k$,

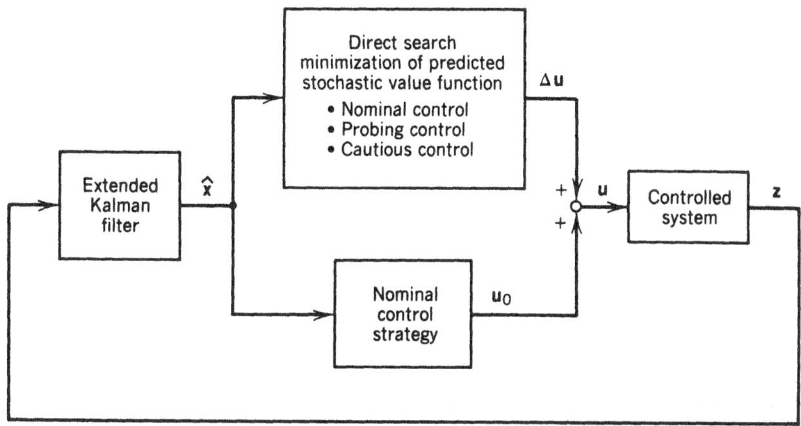

FIGURE 5.2-1 Organization of dual-control logic.

which can be written as the sum of nominal and variational controls (Fig. 5.2-1):

$$\mathbf{u}_k = \mathbf{u}_k^* + \Delta\mathbf{u}_k \tag{5.2-36}$$

A numerical search procedure must be used to find the optimizing value of $\mathbf{u}_k$ (or $\Delta\mathbf{u}_k$), and this is a computer-intensive task, particularly if the dimension of $\mathbf{u}_k$ (denoted by m) is much greater than one. At least three

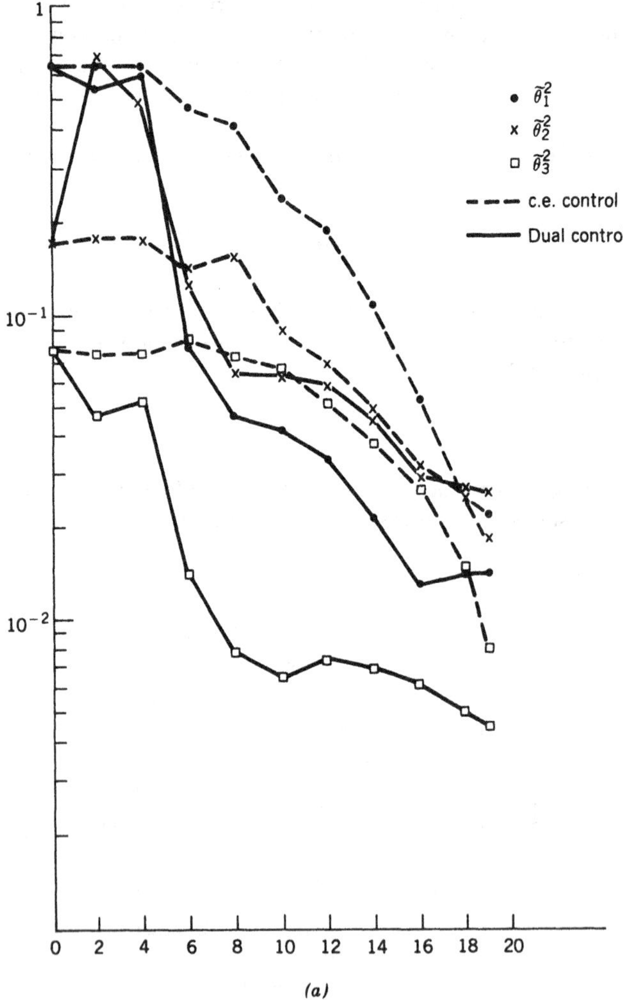

FIGURE 5.2-2 Example of missile tracking performance using optimal, certainty-equivalent (c.e.), and dual control [from (T-6)]. (*a*) Average estimation error in parameters 1 to 3 with c.e. and dual control.

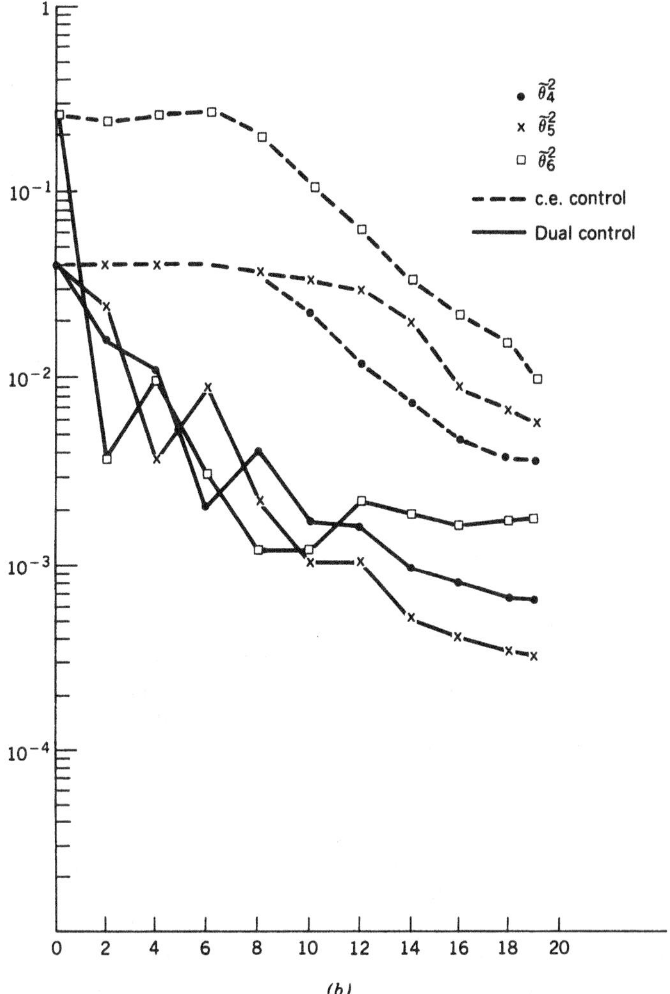

FIGURE 5.2-2 (*b*) Average estimation error in parameters 4 to 6 with c.e. and dual control.

Control Policy	Optimal Control with Known Parameters	CE Control with Unknown Parameters	Dual Control with Unknown Parameters
Average cost	6	114	14
Maximum cost in a sample of 20 runs	20	458	53
Standard deviation of the cost	6	140	16
Average miss distance squared	12	225	22
Weighted cumulative control energy prior to final stage	0.1	1.4	3.2

(*c*)

FIGURE 5.2-2 (*c*) Summary of performance results for the three controllers.

evaluations would be required to identify a quadratic minimum in scalar u_k, and the computations would be much more extensive for larger m. The search for a minimizing $\mathbf{u}_k$ could be implemented effectively in a parallel-processing computer, which would allow concurrent evaluation of future value functions. Derivations and further perspectives on the dual-control algorithm can be found in (B-2).

An example of dual control is reproduced from (T-6), in which the effects of probing are compared with the "exact" optimal solution (all parameters are known without error) and a "certainty-equivalent controller" (described below) that uses an extended Kalman filter for parameter estimation without explicit probing commands (Fig. 5.2-2). The example represents a third-order missile tracking model with six unknown parameters. The dual-control solution provides more accurate parameter estimates as a consequence of less cautious control activity, thereby allowing improved control. For a 20-run sample, the optimal cost was 6 units, the dual-control cost was 14 units, and the certainty-equivalent cost was 114 units.

Probing effects can be achieved with considerably less complexity using ad hoc means. If the ranges of possible parameter variations are known, *programmed identification maneuvers* could be scheduled at times when the summation penalty of Eq. 5.2-14 is relaxed or is of secondary importance. We can envision, for example, a stochastic controller consisting of a linear-optimal controller plus an extended Kalman filter that estimates parameters as well as the state. Ad hoc probing inputs (e.g., impulses, doublets, or other waveforms) would inject energy in the system to improve the identifiability of parameters, allowing better future state estimates and control actions.

Neighboring-Optimal Control

In the event that large parameter uncertainties are not present and random disturbances and measurement errors are small, neighboring-optimal trajectories will indeed stay close to the nominal-optimal trajectory, and the additional computations of dual control may be of marginal value. The differences between neighboring and nominal trajectories then can be characterized as small perturbations, allowing substantial simplification of control and estimation strategies. As in Section 5.1, the neighboring-optimal control history can be separated into two components: a nominal profile, $\mathbf{u}_o^*(t)$ in $[t_0, t_f]$, that represents the optimal deterministic solution, and a perturbational control, $\Delta\mathbf{u}^*(t)$ in $[t_0, t_f]$, that is the output of a feedback control law.

Generating $\mathbf{u}_o^*(t)$ is a control problem alone, involving no measurements or state estimation (Chapter 3). Measurements and estimation contribute only to the perturbational control; if the measurements contain uncertainty, $\Delta\mathbf{u}^*(t)$ cannot be a function of the true state perturbation, $\Delta\mathbf{x}(t)$, because $\Delta\mathbf{x}(t)$ is not known. The feedback control law must be based on the state estimate, $\Delta\hat{\mathbf{x}}(t)$, which is derived from the error-ridden measurements.

The approach to computing perturbational control and estimation strategies remains to be determined. In the previous section, we saw that a second-degree expansion about a nominal trajectory not only admits the possibility of dual control but couples the control and estimation solutions in a complex recursive process. Even if a linear control and estimation structure is forced on the problem as in (G-3), the second-degree expansion introduces *signal-dependent noise* that couples the neighboring-optimal solutions. Optimal estimation and control for the system defined by Eqs. 5.2-15 and 5.2-16 then takes the form

$$\hat{\mathbf{x}}_k = \mathbf{f}(\hat{\mathbf{x}}_{k-1}, \mathbf{u}_{k-1}, k-1) + \mathbf{K}_k[\mathbf{z}_k - \mathbf{h}(\hat{\mathbf{x}}_k, k)] \tag{5.2-37}$$

$$\mathbf{u}_k = \mathbf{u}_k^* - \mathbf{C}_k(\hat{\mathbf{x}}_k - \mathbf{x}_k^*) \tag{5.2-38}$$

where $\mathbf{K}_k$ and $\mathbf{C}_k$ are numerical solutions to a two-point boundary value problem involving both driving- and measurement-noise statistics.

If a first-degree expansion is adequate to describe perturbational dynamics, and if the corresponding variational cost function is adequately modeled by a quadratic function of the state and control, then the control and estimation logic can be derived separately, that is they obey a *separation property*. The separation property allows the stochastic optimal control profile to be expressed as

$$\Delta\mathbf{u}[t_0, t] = \boldsymbol{\psi}\{\Delta\hat{\mathbf{x}}[t_0, t]\} \tag{5.2-39}$$

for which the optimal estimation algorithm does not depend on the optimal control function, $\psi\{\cdot\}$, and vice versa. If $\psi\{\cdot\}$ is the *same* control function as the deterministic optimal control function, then the stochastic optimal control problem possesses the *certainty-equivalence property*.

Certainty equivalence is a stricter property than separation. Examples of stochastic optimal control problems that are separable but not certainty equivalent can be found. For example, the stochastic optimal control law for a linear system with a cost function of the form

$$J = E[\mu e^{(1/2)\mu J_D}] \tag{5.2-40}$$

where J_D is defined as in Eq. 5.2-5, consists of a Kalman filter that is designed without regard to the control function and a control function whose numerical values depends on the state covariance estimate of the filter (S-3). Certainty equivalence will be proven for the linear-quadratic problem in the next section, and the remainder of this section will be devoted to an outline of the use of the linear-quadratic structure in neighboring-optimal control.

In the following, it is assumed that the nonlinear, continuous-time dynamic equation is written as

$$\dot{\mathbf{x}}(t) = \mathbf{f}[\mathbf{x}(t), \mathbf{u}(t), t] + \mathbf{L}(t)\mathbf{w}(t) \tag{5.2-41}$$

Random parameter variations are contained in the disturbance input, $\mathbf{w}(t)$, which has a linear added effect on $\dot{\mathbf{x}}(t)$. Thus the corresponding local linearization of Eq. 5.2-41 is

$$\Delta\dot{\mathbf{x}}(t) = \mathbf{F}(t)\,\Delta\mathbf{x}(t) + \mathbf{G}(t)\,\Delta\mathbf{u}(t) + \mathbf{L}(t)\,\Delta\mathbf{w}(t) \tag{5.2-42}$$

Given the nonlinear measurement equation with additive noise,

$$\mathbf{z}(t) = \mathbf{h}[\mathbf{x}(t), \mathbf{u}(t), \mathbf{w}(t), t] + \mathbf{n}(t) \tag{5.2-43}$$

the locally linearized measurement is

$$\Delta\mathbf{z}(t) = \mathbf{H}_\mathbf{x}(t)\,\Delta\mathbf{x}(t) + \mathbf{H}_\mathbf{u}(t)\,\Delta\mathbf{u}(t) + \mathbf{H}_\mathbf{w}(t)\,\Delta\mathbf{w}(t) + \mathbf{n}(t) \tag{5.2-44}$$

Defining $[\mathbf{H}_\mathbf{w}(t)\,\Delta\mathbf{w}(t) + \Delta\mathbf{n}(t)]$ as $\Delta\mathbf{n}'(t)$ produces the correlated disturbance-input/measurement-noise case considered in Section 4.5. For simplicity, we shall assume that $\mathbf{H}_\mathbf{w}(t)$ is zero in this section.

The neighboring-optimal trajectory solution contains four parts: the nominal-optimal control, the neighboring-optimal control gain computation, the neighboring-optimal estimation gain computation, and the stochastic control implementation. All four components are summarized below. The first two are identical to the neighboring-optimal solution of

Section 5.1, and reference to that section should be made for the corresponding equations. Also, any of the alternate time-varying filter algorithms of Chapter 4 could be used in lieu of the following estimation algorithm. Although it may be assumed that random processes and sequences are Gaussian, it will be noted in the next section that less restrictive assumptions are admissible.

Nominal (Deterministic) Solution (Eqs. 5.1-30 to 5.1-36)

1. Specify the cost function.
2. Specify the dynamic constraint.
3. Define the Hamiltonian.
4. Define the necessary and sufficient conditions for optimality.
5. Compute the nominal solution using a numerical method from Section 3.6 to obtain $\mathbf{x}_o^*(t)$ and $\mathbf{u}_o^*(t)$ in $[t_0, t_f]$.

Continuous-Time, Linear-Quadratic, Neighboring-Optimal Control Gain Matrix (Eqs. 5.1-37 to 5.1-55)

1. Specify the cost function [$\Delta\mathbf{x}(t)$ should be replaced formally by the estimate $\Delta\hat{\mathbf{x}}(t)$. This has no effect on the control law structure or gain matrix computation.]
2. Specify the dynamic constraint.
3. Define the Hamiltonian.
4. Integrate the matrix Riccati equation that results from minimizing the Hamiltonian to obtain the adjoint covariance matrix, $\mathbf{S}(t)$, from t_f to t_0.
5. Compute the optimal control gain matrix in $[t_0, t_f]$.

Continuous-Time, Linear-Gaussian, Neighboring-Optimal Estimation Gain Matrix

1. Specify the initial condition, disturbance input, and measurement error statistics:

$$E[\mathbf{x}(t_0) - \mathbf{x}_o^*(t_0)] = E[\Delta\mathbf{x}(t_0)] = \Delta\hat{\mathbf{x}}_0 \tag{5.2-45}$$

$$E[\Delta\mathbf{x}(t_0)\,\Delta\mathbf{x}^T(t_0)] = \mathbf{P}_0 \tag{5.2-46}$$

$$E[\mathbf{w}(t) - \mathbf{w}_o(t)] = E[\Delta\mathbf{w}(t)] = \mathbf{0} \tag{5.2-47}$$

$$E[\mathbf{n}(t) - \mathbf{n}_o(t)] = E[\Delta\mathbf{n}(t)] = \mathbf{0} \tag{5.2-48}$$

$$E[\Delta\mathbf{w}(t)\,\Delta\mathbf{w}^T(\tau)] = \mathbf{W}(t)\,\delta(t-\tau) \tag{5.2-49}$$

$$E[\Delta\mathbf{n}(t)\,\Delta\mathbf{n}^T(\tau)] = \mathbf{N}(t)\,\delta(t-\tau) \tag{5.2-50}$$

2. Integrate the matrix Riccati equation to obtain the state-error

covariance matrix, $\mathbf{P}(t)$, from t_0 to t_f:

$$\dot{\mathbf{P}}(t) = \mathbf{F}(t)\mathbf{P}(t) + \mathbf{P}(t)\mathbf{F}^T(t) + \mathbf{L}(t)\mathbf{W}(t)\mathbf{L}^T(t) - \mathbf{P}(t)\mathbf{H}_\mathbf{x}^T(t)\mathbf{N}^{-1}(t)\mathbf{H}_\mathbf{x}(t)\mathbf{P}(t), \qquad \mathbf{P}(0) = \mathbf{P}_0 \quad (5.2\text{-}51)$$

3. Compute the optimal estimation gain matrix in $[t_0, t_f]$:

$$\mathbf{K}(t) = \mathbf{P}(t)\mathbf{H}_\mathbf{x}^T(t)\mathbf{N}^{-1}(t) \quad (5.2\text{-}52)$$

Continuous-Time, Neighboring-Optimal Stochastic Control Solution

1. Compute the neighboring-optimal state estimate by integrating

$$\Delta\dot{\hat{\mathbf{x}}}(t) = \mathbf{F}(t)\,\Delta\hat{\mathbf{x}}(t) + \mathbf{G}(t)\,\Delta\mathbf{u}^*(t) + \mathbf{K}(t)\{\Delta\mathbf{z}(t) - [\mathbf{H}_\mathbf{x}(t)\,\Delta\hat{\mathbf{x}}(t) + \mathbf{H}_\mathbf{u}(t)\,\Delta\mathbf{u}^*(t)]\} \quad (5.2\text{-}53a)$$

where

$$\Delta\mathbf{z}(t) = \mathbf{z}(t) - \mathbf{h}[\mathbf{x}_o^*(t), \mathbf{u}_o^*(t), t] \quad (5.2\text{-}54)$$

and $\mathbf{z}(t)$ is the measured output of the nonlinear system. Optionally, this differential equation could be driven by a nonlinear residual

$$\Delta\dot{\hat{\mathbf{x}}}(t) = \mathbf{F}(t)\,\Delta\hat{\mathbf{x}}(t) + \mathbf{G}(t)\,\Delta\mathbf{u}^*(t) + \mathbf{K}(t)\{\mathbf{z}(t) - \mathbf{h}[\hat{\mathbf{x}}(t), \mathbf{u}^*(t), t]\} \quad (5.2\text{-}53b)$$

with

$$\hat{\mathbf{x}}(t) \triangleq \mathbf{x}_o^*(t) + \Delta\hat{\mathbf{x}}(t) \quad (5.2\text{-}55)$$

$$\mathbf{u}^*(t) \triangleq \mathbf{u}_o^*(t) + \Delta\mathbf{u}^*(t) \quad (5.2\text{-}56)$$

The choice of linear or nonlinear residual can be made on the basis of the computational requirements of the specific implementation.

2. Concurrent with (1), compute the neighboring-optimal control vector:

$$\Delta\mathbf{u}^*(t) = -\mathbf{C}(t)\,\Delta\hat{\mathbf{x}}(t) \quad (5.2\text{-}57)$$

These equations can be used in conjunction with simulation or real-time control to generate a member of the ensemble of stochastic neighboring-optimal trajectories corresponding to the given cost function, dynamic constraint, and measurement, as illustrated in Fig. 5.2-3. Having computed the optimal gains, $\mathbf{K}(t)$ and $\mathbf{C}(t)$, the linear stochastic controller is seen to be driven by the nonlinear measurement, $\mathbf{z}(t)$. Its output is summed with the

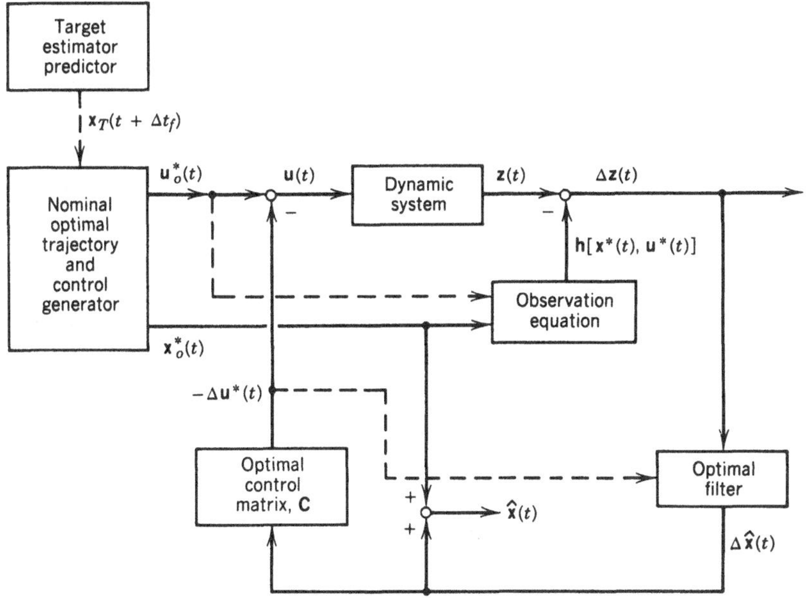

FIGURE 5.2-3 Stochastic neighboring-optimal control law.

nominal-optimal control, $\mathbf{u}_o^*(t)$; the nominal-optimal state, $\mathbf{x}_o^*(t)$, is used to derive either the perturbation measurement (Eq. 5.2-54) or the total state estimate (Eq. 5.2-55), depending on the choice of residual for state estimation.

As before, a corresponding discrete-time neighboring-optimal solution (summarized later) can be found. If the sampling interval is short, the discrete-time model can be approximated as piecewise-constant in each sampling interval, simplifying the computation of state transition and control effect matrices.

Nominal (Deterministic) Solution (Eqs. 5.1-30 to 5.1-36)

Same as above.

Discrete-Time, Linear-Quadratic, Neighboring-Optimal Control Gain Matrix

1. Specify the cost function

$$\Delta^2 J = \frac{1}{2}\left\{\Delta\mathbf{x}_f^T \mathbf{S}_f\,\Delta\mathbf{x}_f + \sum_{k=0}^{k_f-1} [\Delta\mathbf{x}_k^T\;\Delta\mathbf{u}_k^T]\begin{bmatrix}\mathbf{Q}_k & \mathbf{M}_k\\ \mathbf{M}_k^T & \mathbf{R}_k\end{bmatrix}\begin{bmatrix}\Delta\mathbf{x}_k\\ \Delta\mathbf{u}_k\end{bmatrix}\right\} \quad (5.2\text{-}58)$$

where $\mathbf{S}_f$ (denoted by $\mathbf{P}_f$ in Chapter 3), $\mathbf{Q}_k$, $\mathbf{M}_k$, and $\mathbf{R}_k$ are computed by Eqs. 3.7-30 and 3.7-37 to 3.7-39.

2. Specify the dynamic constraint and measurement equation:

$$\Delta\mathbf{x}_{k+1} = \mathbf{\Phi}_k \, \Delta\mathbf{x}_k + \mathbf{\Gamma}_k \, \Delta\mathbf{u}_k + \mathbf{\Lambda}_k \, \Delta\mathbf{w}_k, \qquad \Delta\mathbf{x}_0 \text{ given} \quad (5.2\text{-}59)$$

$$\Delta\mathbf{z}_k = \mathbf{H}_{\mathbf{x}_k} \, \Delta\mathbf{x}_k + \mathbf{H}_{\mathbf{u}_k} \, \Delta\mathbf{u}_k + \Delta\mathbf{n}_k \quad (5.2\text{-}60)$$

The matrices may be approximated as piecewise-constant during a suitably short sampling interval, $\Delta\mathbf{t}$:

$$\mathbf{\Phi}_k \simeq e^{\mathbf{F}(t_k') \Delta t} \quad (5.2\text{-}61)$$

$$\mathbf{\Gamma}_k \simeq [\mathbf{\Phi}_k - \mathbf{I}_n]\mathbf{F}^{-1}(t_k')\mathbf{G}(t_k') \quad (5.2\text{-}62)$$

$$\mathbf{\Lambda}_k \simeq [\mathbf{\Phi}_k - \mathbf{I}_n]\mathbf{F}^{-1}(t_k')\mathbf{L}(t_k') \quad (5.2\text{-}63)$$

Here

$$t_k' = \frac{t_k + t_{k+1}}{2} \quad (5.2\text{-}64)$$

This assumes that $\mathbf{F}$ and $\mathbf{\Phi}$ are nonsingular; otherwise, use the series expansion of Eq. 2.3-89 to compute $\mathbf{\Gamma}$ and $\mathbf{\Lambda}$. $\mathbf{F}(\cdot)$, $\mathbf{G}(\cdot)$, and $\mathbf{L}(\cdot)$ are evaluated by Eqs. 5.1-42, 5.1-43, and 5.1-45.

3. Define the Hamiltonian:

$$\mathscr{H}(\Delta\mathbf{x}_k, \Delta\mathbf{u}_k, \Delta\boldsymbol{\lambda}_k, k) = \tfrac{1}{2}(\Delta\mathbf{x}_k^T\hat{\mathbf{Q}}_k \, \Delta\mathbf{x}_k + 2\Delta\mathbf{x}_k^T\hat{\mathbf{M}}_k \, \Delta\mathbf{u}_k + \Delta\mathbf{u}_k^T \hat{\mathbf{R}}_k \, \Delta\mathbf{u}_k) + \Delta\boldsymbol{\lambda}_{k+1}^T(\mathbf{\Phi}_k \, \Delta\mathbf{x}_k + \mathbf{\Gamma}_k \, \Delta\mathbf{u}_k) \quad (5.2\text{-}65)$$

4. Solve the matrix Riccati difference equation that results from minimizing the Hamiltonian to obtain $\mathbf{S}_k$ in $[0, k_f]$,

$$\mathbf{S}_k = \hat{\mathbf{Q}}_k + \mathbf{\Phi}_k^T\mathbf{S}_{k+1}\mathbf{\Phi}_k - (\hat{\mathbf{M}}_k + \mathbf{\Phi}_k^T\mathbf{S}_{k+1}\mathbf{\Gamma}_k)(\hat{\mathbf{R}}_k + \mathbf{\Gamma}_k^T\mathbf{S}_{k+1}\mathbf{\Gamma}_k)^{-1}(\hat{\mathbf{M}}_k^T + \mathbf{\Gamma}_k^T\mathbf{S}_{k+1}\mathbf{\Phi}_k) \quad (5.2\text{-}66)$$

with

$$\mathbf{S}_{k_f} = \phi_{\mathbf{xx}}[\mathbf{x}_0^*(t_f), t_f] \quad (5.2\text{-}67)$$

5. Compute the optimal feedback gain matrix in $[0, k_f]$:

$$\mathbf{C}_k = (\hat{\mathbf{R}}_k + \mathbf{\Gamma}_k^T\mathbf{S}_{k+1}\mathbf{\Gamma}_k)^{-1}(\hat{\mathbf{M}}_k^T + \mathbf{\Gamma}_k^T\mathbf{S}_{k+1}\mathbf{\Phi}_k) \quad (5.2\text{-}68)$$

Discrete-Time, Linear-Gaussian, Neighboring-Optimal Estimation Gain Matrix

1. Specify the initial condition, disturbance input, and measurement error statistics:

$$E(\Delta\mathbf{x}_0) = \Delta\hat{\mathbf{x}}_0 \tag{5.2-69}$$

$$E(\Delta\mathbf{x}_0\,\Delta\mathbf{x}_0^T) = \mathbf{P}_0(+) \tag{5.2-70}$$

$$E(\Delta\mathbf{w}_k) = E(\Delta\mathbf{n}_k) = \mathbf{0} \tag{5.2-71}$$

$$E(\Delta\mathbf{w}_k\,\Delta\mathbf{w}_k^T) = \mathbf{W}_k \tag{5.2-72}$$

$$E(\Delta\mathbf{n}_k\,\Delta\mathbf{n}_k^T) = \mathbf{N}_k \tag{5.2-73}$$

$$E(\Delta\mathbf{w}_j\,\Delta\mathbf{w}_k^T) = E(\Delta\mathbf{n}_j\,\Delta\mathbf{n}_k^T) = \mathbf{0}, \qquad j \neq k \tag{5.2-74}$$

$$E(\Delta\mathbf{w}_j\,\Delta\mathbf{n}_k^T) = \mathbf{0}, \qquad \text{all } j \text{ and } k \tag{5.2-75}$$

2. Propagate the state-error covariance matrix $\mathbf{P}_k(-)$:

$$\mathbf{P}_k(-) = \mathbf{\Phi}_{k-1}\mathbf{P}_{k-1}(+)\mathbf{\Phi}_{k-1}^T + \mathbf{L}_{k-1}\mathbf{W}_{k-1}\mathbf{L}_{k-1}^T \tag{5.2-76}$$

3. Compute the estimation gain matrix:

$$\mathbf{K}_k = \mathbf{P}_k(-)\mathbf{H}_{\mathbf{x}_k}^T[\mathbf{H}_{\mathbf{x}_k}\mathbf{P}_k(-)\mathbf{H}_{\mathbf{x}_k}^T + \mathbf{N}_k]^{-1} \tag{5.2-77}$$

4. Update the state-error covariance matrix, $\mathbf{P}_k(+)$:

$$\mathbf{P}_k(+) = [\mathbf{P}_k(-)^{-1} + \mathbf{H}_{\mathbf{x}_k}^T\mathbf{N}_k^{-1}\mathbf{H}_{\mathbf{x}_k}]^{-1} \tag{5.2-78}$$

5. Cycle through steps 2–4 in $[0, k_f]$. (Note that steps 2–4 could be replaced by one of the several alternate forms of Section 4.3.)

Discrete-Time, Neighboring-Optimal Stochastic Control Solution

1. Propagate the state estimate, $\Delta\hat{\mathbf{x}}_k(-)$:

$$\Delta\hat{\mathbf{x}}_k(-) = \mathbf{\Phi}_{k-1}\,\Delta\hat{\mathbf{x}}_{k-1}(+) + \mathbf{\Gamma}_{k-1}\,\Delta\mathbf{u}_{k-1}^* \tag{5.2-79}$$

2. Update the state estimate, $\Delta\hat{\mathbf{x}}_k(+)$:

$$\Delta\hat{\mathbf{x}}_k(+) = \Delta\hat{\mathbf{x}}_k(-) + \mathbf{K}_k\{\Delta\mathbf{z}_k - [\mathbf{H}_{\mathbf{x}_k}\,\Delta\hat{\mathbf{x}}_k(-) + \mathbf{H}_{\mathbf{u}_k}\,\Delta\mathbf{u}_k^*]\} \tag{5.2-80a}$$

where

$$\Delta\mathbf{z}_k = \mathbf{z}(t_k) - \mathbf{h}[\mathbf{x}_o^*(t_k), \mathbf{u}_o^*(t_k), t_k] \tag{5.2-81}$$

Alternatively,

$$\Delta\hat{\mathbf{x}}_k(+) = \Delta\hat{\mathbf{x}}_{k-1}(-) + \mathbf{K}_k\{\mathbf{z}(t_k) - \mathbf{h}[\hat{\mathbf{x}}^*(t_k), \mathbf{u}^*(t_k), t_k]\} \quad (5.2\text{-}80b)$$

where

$$\hat{\mathbf{x}}^*(t_k) \triangleq \mathbf{x}_o^*(t_k) + \Delta\hat{\mathbf{x}}_k(-) \quad (5.2\text{-}82)$$

$$\mathbf{u}^*(t_k) \triangleq \mathbf{u}_o^*(t_k) + \Delta\mathbf{u}_k^* \quad (5.2\text{-}83)$$

3. Compute the neighboring-optimal control vector, $\Delta\mathbf{u}_k^*$:

$$\Delta\mathbf{u}_k^* = -\mathbf{C}_k\,\Delta\hat{\mathbf{x}}_k(+) \quad (5.2\text{-}84)$$

Before leaving stochastic neighboring-optimal control, several observations can be made about the linear portions of the solutions, relating principally to stability, robustness, commanded deviations from the nominal-optimal trajectory, and the state dimension. Once the optimal solution has been partitioned into nonlinear and linear solutions, the substantial literature of linear control system design and analysis can be applied to the latter solution; hence further developments in this book focus on linear systems. Criteria for stability, controllability, and observability are relevant, and the nature of closed-loop sensitivity to disturbances, parameter variations, and measurement errors can be deduced from linear models. This is important for assessing the effects of alternate weighting matrices, alternate measurement vectors, and resulting stability margins.

By the definition of $\Delta\mathbf{x}(t)$ and $\Delta\mathbf{x}_k$, the linear, neighboring-optimal stochastic control law acts as a *zero-set point controller*, in that it always tries to force $\Delta\mathbf{x}$ to zero. Purposeful deviations from a zero set point may be required, even though the results may be suboptimal. For example, it may not always be possible to provide a precise nominal-optimal trajectory; in such instance, a *non-zero-set-point controller* is of considerable interest, for it provides a feasible (if suboptimal) means of achieving the control objective.

Although the original state and control dimensions are n_o and m_o, respectively, it may prove useful to define perturbational vectors of different dimension. The dimension of the controller state and control, n_c and m_c, might be smaller if certain feedback variables have negligible effect or if open-loop command of some control variables provides acceptable performance. Their dimensions could be greater if added states would improve tracking error or robustness, or if perturbational controls are important but have negligible effect on the primary cost function.

Similarly, the estimator state dimension, n_e, may be larger or smaller than n_o, either because some measurements are essentially perfect representations of state variables or because disturbance and noise models are not white. The dimension of the observation vector r is only indirectly associated with n_o in that it must be chosen to provide adequate observability but otherwise is susceptible to many design constraints and objectives, as described in the next chapter. The concept of neighboring-optimal control provides an important foundation for making complex problems tractable by partitioning solutions along natural guidelines.

5.3 THE CERTAINTY-EQUIVALENCE PROPERTY OF LINEAR-QUADRATIC-GAUSSIAN CONTROLLERS

The control strategy that minimizes the expected value of a quadratic cost function for a linear dynamic system with random inputs and measurement errors function is described by a feedback control law that operates on optimal estimates of the state. The optimal control law can be formulated as if there is no uncertainty in the feedback measurements, and the optimal estimator can be derived as if there is no feedback control; hence the control and estimation design processes are separable, and the control law is said to be certainty equivalent. The certainty-equivalence property is demonstrated for Markov continuous-time and discrete-time systems, with the assumption that all random processes are white and Gaussian. Certainty equivalence can be retained with less restrictive assumptions, as noted later.

The Continuous-Time Case

In the preceding section, we saw that a Kalman–Bucy filter cascaded with a linear-quadratic control law forms a plausible stochastic controller for a linear dynamic system. In this section, we will see that this combination is indeed optimal and that its two major components can be designed independently (T-4). A stochastic quadratic cost function is formulated as

$$J = E\left\{\frac{1}{2}\mathbf{x}^T(t_f)\mathbf{S}(t_f)\mathbf{x}(t_f) + \frac{1}{2}\int_{t_0}^{t_f} [\mathbf{x}^T(t)\mathbf{Q}(t)\mathbf{x}(t) + \mathbf{u}^T(t)\mathbf{R}(t)\mathbf{u}(t)]\, dt\right\} \quad (5.3\text{-}1)$$

and the objective is to minimize J by proper choice of $\mathbf{u}[t_0, t_f]$, subject to a linear dynamic constraint

$$\dot{\mathbf{x}}(t) = \mathbf{F}(t)\mathbf{x}(t) + \mathbf{G}(t)\mathbf{u}(t) + \mathbf{L}(t)\mathbf{w}(t) \quad (5.3\text{-}2)$$

The s-dimensional disturbance input $\mathbf{w}(t)$ is a white random process with mean and spectral density given by

$$E[\mathbf{w}(t)] = \mathbf{0} \quad (5.3\text{-}3)$$

$$E[\mathbf{w}(t)\mathbf{w}^T(\tau)] = \mathbf{W}(t)\,\delta(t-\tau) \quad (5.3\text{-}4)$$

The state and control vectors, $\mathbf{x}(t)$ and $\mathbf{u}(t)$, have dimensions n and m, respectively. The initial condition is uncertain, but its mean and covariance are expressed by

$$E[\mathbf{x}(t_0)] = \bar{\mathbf{x}}_0 \tag{5.3-5}$$

$$E\{[\mathbf{x}(t_0) - \bar{\mathbf{x}}_0][\mathbf{x}(t_0) - \bar{\mathbf{x}}_0]^T\} = \mathbf{P}_0 \tag{5.3-6}$$

The state is observed through a noisy r-dimensional measurement

$$\mathbf{z}(t) = \mathbf{H}(t)\mathbf{x}(t) + \mathbf{n}(t) \tag{5.3-7}$$

where the white measurement noise statistics are given by

$$E[\mathbf{n}(t)] = \mathbf{0} \tag{5.3-8}$$

$$E[\mathbf{n}(t)\mathbf{n}^T(\tau)] = \mathbf{N}(t)\,\delta(t-\tau) \tag{5.3-9}$$

The stochastic cost function can only be evaluated in terms of expected values that are conditioned on the information set $\mathfrak{I}[t_0, t]$ (Eq. 5.2-1). The dynamic system has the Markov property, $\mathbf{n}(t)$ is assumed to be white, and it can be shown that the Markov derived information set, $\mathfrak{I}_D(t)$ (Eq. 5.2-3), provides all the information that is required to evaluate the cost function. Since

$$E\{E[\boldsymbol{\xi} \mid \mathfrak{I}_D(t_f)]\} = E(\boldsymbol{\xi}) \tag{5.3-10}$$

where $\boldsymbol{\xi}$ represents any random variable conditioned on $\mathfrak{I}_D(t_f)$, Eq. 5.3-1 can be expressed as

$$J = \tfrac{1}{2}E\left\{E[\mathbf{x}^T(t_f)\mathbf{S}(t_f)\mathbf{x}(t_f)|\mathfrak{I}_D(t_f)] + \int_{t_0}^{t_f} E\{[\mathbf{x}^T(t)\mathbf{Q}(t)\mathbf{x}(t) + \mathbf{u}^T(t)\mathbf{R}(t)\mathbf{u}(t)]|\mathfrak{I}_D(t)\}\,dt\right\} \tag{5.3-11a}$$

The expected value of $\mathbf{u}(t)$ is not conditioned on $\mathfrak{I}_D(t)$, but the first and second moments of $\mathbf{x}(t)$ are:

$$E[\mathbf{x}(t) \mid \mathfrak{I}_D(t)] = \hat{\mathbf{x}}(t) \tag{5.3-12}$$

$$E\{[\mathbf{x}(t) - \hat{\mathbf{x}}(t)][\mathbf{x}(t) - \hat{\mathbf{x}}(t)]^T \mid \mathfrak{I}_D(t)\} = \mathbf{P}(t) \tag{5.3-13}$$

These expected values are seen to be the conditional estimates of the state mean and covariance presented in Chapter 4. Because $\mathbf{x}^T\mathbf{Q}\mathbf{x} = \mathrm{Tr}(\mathbf{Q}\mathbf{x}\mathbf{x}^T)$ (Eq. 2.1-33), the cost function could be written (with more concise notation)

$$J=\tfrac{1}{2}E\left\{E[\mathrm{Tr}(\mathbf{S}_f\mathbf{x}_f\mathbf{x}_f^T)|\mathfrak{I}_{D_f}]+\int_{t_0}^{t_f}E[\mathrm{Tr}(\mathbf{Q}\mathbf{x}\mathbf{x}^T)|\mathfrak{I}_D]\,dt+\int_{t_0}^{t_f}\mathrm{Tr}(\mathbf{R}\mathbf{u}\mathbf{u}^T)\,dt\right\}$$

(5.3-11b)

The expectation and trace operations can be interchanged, and as $\mathbf{S}_f$ and $\mathbf{Q}$ are known matrices, this becomes

$$J=\tfrac{1}{2}E\left\{\mathrm{Tr}[\mathbf{S}_fE(\mathbf{x}_f\mathbf{x}_f^T\,|\,\mathfrak{I}_{D_f})]+\mathrm{Tr}\left[\int_{t_0}^{t_f}\mathbf{Q}E(\mathbf{x}\mathbf{x}^T\,|\,\mathfrak{I}_D)\,dt\right]+\mathrm{Tr}\left(\int_{t_0}^{t_f}\mathbf{R}\mathbf{u}\mathbf{u}^T\,dt\right)\right\}$$

(5.3-11c)

$E(\mathbf{x}\mathbf{x}^T\,|\,\mathfrak{I}_D)$ can be rewritten in terms of $\hat{\mathbf{x}}$ and $\mathbf{P}$. Expanding Eq. 5.3-13,

$$\begin{aligned}\mathbf{P}&=E[(\mathbf{x}\mathbf{x}^T-\hat{\mathbf{x}}\mathbf{x}^T-\mathbf{x}\hat{\mathbf{x}}^T+\hat{\mathbf{x}}\hat{\mathbf{x}}^T)\,|\,\mathfrak{I}_D]\\&=E(\mathbf{x}\mathbf{x}^T\,|\,\mathfrak{I}_D)-E[(\hat{\mathbf{x}}\mathbf{x}^T+\mathbf{x}\hat{\mathbf{x}}^T-\hat{\mathbf{x}}\hat{\mathbf{x}}^T)\,|\,\mathfrak{I}_D]\\&=E(\mathbf{x}\mathbf{x}^T\,|\,\mathfrak{I}_D)-\hat{\mathbf{x}}\hat{\mathbf{x}}^T\end{aligned}\qquad(5.3\text{-}14a)$$

or

$$E(\mathbf{x}\mathbf{x}^T\,|\,\mathfrak{I}_D)=\mathbf{P}+\hat{\mathbf{x}}\hat{\mathbf{x}}^T\qquad(5.3\text{-}14b)$$

because $E(\mathbf{x}\hat{\mathbf{x}}^T\,|\,\mathfrak{I}_D)=E(\hat{\mathbf{x}}\mathbf{x}^T\,|\,\mathfrak{I}_D)=\hat{\mathbf{x}}\hat{\mathbf{x}}^T$. Then the cost function can be written as

$$\begin{aligned}J&=\tfrac{1}{2}E\left\{\mathrm{Tr}[\mathbf{S}_f(\mathbf{P}_f+\hat{\mathbf{x}}_f\hat{\mathbf{x}}_f^T)]+\mathrm{Tr}\left[\int_{t_0}^{t_f}\mathbf{Q}(\mathbf{P}+\hat{\mathbf{x}}\hat{\mathbf{x}}^T)\,dt\right]+\mathrm{Tr}\left(\int_{t_0}^{t_f}\mathbf{R}\mathbf{u}\mathbf{u}^T\,dt\right)\right\}\\&=J_{\mathrm{CE}}+J_S\end{aligned}\qquad(5.3\text{-}11d)$$

where J_{CE} is the *certainty-equivalent stochastic cost function*,

$$J_{\mathrm{CE}}=\tfrac{1}{2}E\left[\mathrm{Tr}(\mathbf{S}_f\hat{\mathbf{x}}_f\hat{\mathbf{x}}_f^T)+\mathrm{Tr}\left(\int_{t_0}^{t_f}\mathbf{Q}\hat{\mathbf{x}}\hat{\mathbf{x}}^T\,dt\right)+\mathrm{Tr}\left(\int_{t_0}^{t_f}\mathbf{R}\mathbf{u}\mathbf{u}^T\,dt\right)\right]\qquad(5.3\text{-}15)$$

and J_S is the *cost function increment due to estimation error*:

$$J_S=\tfrac{1}{2}E\left[\mathrm{Tr}(\mathbf{S}_f\mathbf{P}_f)+\mathrm{Tr}\left(\int_{t_0}^{t_f}\mathbf{Q}\mathbf{P}\,dt\right)\right]\qquad(5.3\text{-}16)$$

In Chapter 4, we saw that the state-estimation-error covariance matrix, $\mathbf{P}(t)$, for a linear-optimal state estimator is independent of the actual control input, as

$$\dot{\mathbf{P}}(t) = \mathbf{F}(t)\mathbf{P}(t) + \mathbf{P}(t)\mathbf{F}^T(t) + \mathbf{L}(t)\mathbf{W}(t)\mathbf{L}^T(t) - \mathbf{P}(t)\mathbf{H}^T(t)\mathbf{N}^{-1}(t)\mathbf{H}(t)\mathbf{P}(t), \qquad \mathbf{P}(t_0) \text{ given} \tag{5.3-17}$$

Therefore, J_S is unaffected by $\mathbf{u}(t)$, and the control profile that minimizes J_{CE} also minimizes J:

$$\min_{\mathbf{u}[t_0, t_f]} J = \min_{\mathbf{u}[t_0, t_f]} J_{CE} + J_S \tag{5.3-18}$$

The certainty-equivalent cost function is identical in form to the quadratic cost function of Section 5.1, with $\hat{\mathbf{x}}(t)$ replacing $\mathbf{x}(t)$. J_{CE} must be minimized subject to a dynamic constraint on $\hat{\mathbf{x}}(t)$, which is seen to be the differential equation for the conditional mean estimate:

$$\dot{\hat{\mathbf{x}}}(t) = \mathbf{F}(t)\hat{\mathbf{x}}(t) + \mathbf{G}(t)\mathbf{u}(t) + \mathbf{K}(t)[\mathbf{z}(t) - \mathbf{H}(t)\hat{\mathbf{x}}(t)], \qquad \hat{\mathbf{x}}(0) \text{ given} \tag{5.3-19}$$

The optimal estimation gain matrix is

$$\mathbf{K}(t) = \mathbf{P}(t)\mathbf{H}^T(t)\mathbf{N}^{-1}(t) \tag{5.3-20}$$

and $\mathbf{P}(t)$ is computed by integrating Eq. 5.3-17.

Because $\hat{\mathbf{x}}(t)$ is an optimal estimate, the filter residual,

$$\boldsymbol{\epsilon}(t) = \mathbf{z}(t) - \mathbf{H}(t)\hat{\mathbf{x}}(t) \tag{5.3-21}$$

is a white-noise process with zero mean and spectral density matrix, $\mathbf{N}'(t) = \mathbf{N}(t) + \mathbf{H}\mathbf{P}(t)\mathbf{H}^T$. Consequently, the dynamic constraint on J_{CE},

$$\dot{\hat{\mathbf{x}}}(t) = \mathbf{F}(t)\hat{\mathbf{x}}(t) + \mathbf{G}(t)\mathbf{u}(t) + \mathbf{K}(t)\boldsymbol{\epsilon}(t) \tag{5.3-22}$$

is identical in form to the constraint of Section 5.1 (Eq. 5.1-41), with $\boldsymbol{\epsilon}(t)$ playing the role of process noise or disturbance input; therefore, the minimizing control law takes the same form as before. With no state-control weighting ($\mathbf{M} = \mathbf{0}$), the *optimal certainty-equivalent control law* is

$$\mathbf{u}(t) = -\mathbf{C}(t)\hat{\mathbf{x}}(t) \tag{5.3-23}$$

The optimal control gain matrix is

$$\mathbf{C}(t) = \mathbf{R}^{-1}(t)\mathbf{G}^T(t)\mathbf{S}(t) \tag{5.3-24}$$

and $\mathbf{S}(t)$ is found by integrating the following matrix Riccati equation:

$$\dot{\mathbf{S}}(t) = -\mathbf{F}^T(t)\mathbf{S}(t) - \mathbf{S}(t)\mathbf{F}(t) - \mathbf{Q}(t) + \mathbf{S}(t)\mathbf{G}(t)\mathbf{R}^{-1}\mathbf{G}^T(t)\mathbf{S}(t), \qquad \mathbf{S}(t_f) \text{ given} \tag{5.3-25}$$

In sum, the certainty-equivalence property of continuous-time linear-quadratic solutions has been shown, and it is apparent that the separation property pertains as well.

By analogy to the perfect-measurement result (Eq. 5.1-78), the certainty-equivalent cost function takes the following value with linear-optimal control and estimation:

$$\min_{\mathbf{u}[t_0, t_f]} J_{CE} = \tfrac{1}{2}\operatorname{Tr}\left[\mathbf{S}(t_0)E(\bar{\mathbf{x}}_0\bar{\mathbf{x}}_0^T) + \int_{t_0}^{t_f} \mathbf{S}(t)\mathbf{K}(t)\mathbf{N}'(t)\mathbf{K}^T(t)\,dt\right] \tag{5.3-26}$$

Then the minimum value of the total stochastic cost function (Eq. 5.3-11) is

$$\min_{\mathbf{u}[t_0, t_f]} J = \tfrac{1}{2}\operatorname{Tr}\left[\mathbf{S}(t_0)E(\bar{\mathbf{x}}_0\bar{\mathbf{x}}_0^T) + \mathbf{S}(t_f)\mathbf{P}(t_f) + \int_{t_0}^{t_f} \mathbf{Q}(t)\mathbf{P}(t)\,dt + \int_{t_0}^{t_f} \mathbf{S}(t)\mathbf{K}(t)\mathbf{N}'(t)\mathbf{K}^T(t)\,dt\right] \tag{5.3-27}$$

The Discrete-Time Case

Certainty equivalence of the discrete-time linear-quadratic controller could be demonstrated in much the same way that the continuous-time result has been developed, replacing the integral cost function term with a summation and the differential dynamic constraint by a difference equation. However, it is instructive to use the stochastic principle of optimality as an alternative (M-4). We consider a cost function that is the expected value of the quadratic form

$$J = \tfrac{1}{2}E\left[\mathbf{x}_{k_f}^T\mathbf{S}_{k_f}\mathbf{x}_{k_f} + \sum_{k=0}^{k_f-1}(\mathbf{x}_k^T\mathbf{Q}_k\mathbf{x}_k + \mathbf{u}_k^T\mathbf{R}_k\mathbf{u})\right] \tag{5.3-28}$$

subject to the linear dynamic constraint

$$\mathbf{x}_{k+1} = \mathbf{\Phi}_k\mathbf{x}_k + \mathbf{\Gamma}_k\mathbf{u}_k + \mathbf{\Lambda}_k\mathbf{w}_k \tag{5.3-29}$$

where $\mathbf{x}$, $\mathbf{u}$, and $\mathbf{w}$ have dimensions n, m, and s, respectively. The state is observed through the noisy r-dimensional vector

$$\mathbf{z}_k = \mathbf{H}_k\mathbf{x}_k + \mathbf{n}_k \tag{5.3-30}$$

The initial condition on $\mathbf{x}_k$ is a random variable with

$$E(\mathbf{x}_0) = \bar{\mathbf{x}}_0 \tag{5.3-31}$$

$$E[(\mathbf{x}_0 - \bar{\mathbf{x}}_0)(\mathbf{x}_0 - \bar{\mathbf{x}}_0)^T] = \mathbf{P}_0 \tag{5.3-32}$$

while $\mathbf{w}_k$ and $\mathbf{n}_k$ are uncorrelated, white, zero-mean random sequences with the following covariances:

$$E(\mathbf{w}_k \mathbf{w}_k^T) = \mathbf{W}_k \tag{5.3-33}$$

$$E(\mathbf{n}_k \mathbf{n}_k^T) = \mathbf{N}_k \tag{5.3-34}$$

The conditional estimates of the state's mean value and covariance before and after measurement updates are

$$E(\mathbf{x}_k \mid \mathfrak{I}_{D_{k-1}}) = \hat{\mathbf{x}}_k(-) \tag{5.3-35}$$

$$E(\mathbf{x}_k \mid \mathfrak{I}_{D_k}) = \hat{\mathbf{x}}_k(+) \tag{5.3-36}$$

$$E\{[\mathbf{x}_k - \hat{\mathbf{x}}_k(-)][\mathbf{x}_k - \hat{\mathbf{x}}_k(-)]^T \mid \mathfrak{I}_{D_k}\} = \mathbf{P}_k(-) \tag{5.3-37}$$

$$E\{[\mathbf{x}_k - \hat{\mathbf{x}}_k(+)][\mathbf{x}_k - \hat{\mathbf{x}}_k(+)]^T \mid \mathfrak{I}_{D_k}\} = \mathbf{P}_k(+) \tag{5.3-38}$$

From Section 4.3, we have seen that the conditional means and covariances are generated by the Kalman filter using the following equations:

$$\hat{\mathbf{x}}_k(-) = \mathbf{\Phi}_{k-1}\hat{\mathbf{x}}_{k-1}(+) + \mathbf{\Gamma}_{k-1}\mathbf{u}_{k-1} \tag{5.3-39}$$

$$\hat{\mathbf{x}}_k(+) = \hat{\mathbf{x}}_k(-) + \mathbf{K}_k[\mathbf{z}_k - \mathbf{H}_k\hat{\mathbf{x}}_k(-)] \tag{5.3-40}$$

$$\mathbf{P}_k(-) = \mathbf{\Phi}_{k-1}\mathbf{P}_{k-1}(+)\mathbf{\Phi}_{k-1}^T + \mathbf{\Lambda}_{k-1}\mathbf{W}_{k-1}\mathbf{\Lambda}_{k-1}^T \tag{5.3-41}$$

$$\mathbf{P}_k(+) = [\mathbf{P}_k^{-1}(-) + \mathbf{H}_k^T\mathbf{N}_k^{-1}\mathbf{H}_k]^{-1} \tag{5.3-42}$$

with

$$\mathbf{K}_k = \mathbf{P}_k(-)\mathbf{H}_k^T[\mathbf{H}_k\mathbf{P}_k(-)\mathbf{H}_k^T + \mathbf{N}_k]^{-1} \tag{5.3-43}$$

$$\hat{\mathbf{x}}_0(+) = \bar{\mathbf{x}}_0 \tag{5.3-44}$$

$$\mathbf{P}_0(+) = \mathbf{P}_0 \tag{5.3-45}$$

The covariance and filter gain computations are unaffected by deterministic control, although $\mathbf{u}_k$ forces the propagation of the mean (Eq. 5.3-39).

The stochastic principle of optimality corresponding to Eq. 5.3-28 is expressed as a recursive equation for the optimal quadratic value function, V_k^*, conditioned on the current information set; thus

$$V_k(-) = \tfrac{1}{2}\hat{\mathbf{x}}_k^T(-)\mathbf{S}_k\hat{\mathbf{x}}_k(-) + v_k \tag{5.3-46}$$

$$V_k(+) = \tfrac{1}{2}\hat{\mathbf{x}}_k^T(+)\mathbf{S}_k\hat{\mathbf{x}}_k(+) + v_k \tag{5.3-47}$$

where $\frac{1}{2}\hat{\mathbf{x}}_k^T\mathbf{S}_k\hat{\mathbf{x}}_k$ is the *certainty-equivalent value function* and v_k is the *stochastic value function increment* (to be determined). For the minimization problem defined by Eqs. 5.3-28 and 5.3-29, the stochastic principle of optimality is

$$V_k^*(+) = \min_{\mathbf{u}_k} E\{E\{[\tfrac{1}{2}(\mathbf{x}_K^{*T}\mathbf{Q}_k\mathbf{x}_k^* + \mathbf{u}_k^T\mathbf{R}_k\mathbf{u}_k) + V_{k+1}^*(-)] \mid \mathfrak{I}_{D_k}\}\},$$

$$V_{k_f}^*(-) = \tfrac{1}{2}\hat{\mathbf{x}}_{k_f}^{*T}(-)\mathbf{S}_{k_f}\hat{\mathbf{x}}_{k_f}^*(-) \tag{5.3-48}$$

or, suppressing the asterisks for clarity,

$$\tfrac{1}{2}\hat{\mathbf{x}}_k(+)\mathbf{S}_k\hat{\mathbf{x}}_k(+) + v_k = \min_{\mathbf{u}_k} E\{E\{[\tfrac{1}{2}(\mathbf{x}_k^T\mathbf{Q}_k\mathbf{x}_k + \mathbf{u}_k^T\mathbf{R}_k\mathbf{u}_k) + \tfrac{1}{2}\hat{\mathbf{x}}_{k+1}^T(-)\mathbf{S}_{k+1}\hat{\mathbf{x}}_{k+1}(-) + v_{k+1}] \mid \mathfrak{I}_{D_k}\}\} \tag{5.3-49}$$

Using Eq. 5.3-39 to express $\hat{\mathbf{x}}_{k+1}(-)$,

$$\tfrac{1}{2}\hat{\mathbf{x}}_k(+)\mathbf{S}_k\hat{\mathbf{x}}_k(+) + v_k = \min_{\mathbf{u}_k} E\{E\{[\tfrac{1}{2}(\mathbf{x}_k^T\mathbf{Q}_k\mathbf{x}_k + \mathbf{u}_k^T\mathbf{R}_k\mathbf{u}_k) + \tfrac{1}{2}(\mathbf{\Phi}_k\hat{\mathbf{x}}_k(+) + \mathbf{\Gamma}_k\mathbf{u}_k)^T\mathbf{S}_{k+1}(\mathbf{\Phi}_k\hat{\mathbf{x}}_k(+) + \mathbf{\Gamma}_k\mathbf{u}_k) + v_{k+1}] \mid \mathfrak{I}_{D_k}\}\} \tag{5.3-50}$$

From Eq. 5.2-9, the order of minimization and expectation can be reversed; representing the term in square brackets by $\boldsymbol{\xi}_k$,

$$\min_{\mathbf{u}_k} E\{E[\boldsymbol{\xi}_k \mid \mathfrak{I}_{D_k}]\} = E\{\min_{\mathbf{u}_k} E[\boldsymbol{\xi}_k \mid \mathfrak{I}_{D_k}]\} \tag{5.3-51}$$

Consequently, ordinary minimization of $E[\boldsymbol{\xi}_k \mid \mathfrak{I}_{D_k}]$ with respect to $\mathbf{u}_k$ can be used to find the optimizing control.

Following Section 2.1, the minimum is defined by

$$\begin{aligned}\frac{\partial}{\partial \mathbf{u}_k} E[\boldsymbol{\xi}_k \mid \mathfrak{I}_{D_k}] = \mathbf{0} \\ &= \frac{\partial}{\partial \mathbf{u}_k} E\{[\tfrac{1}{2}(\mathbf{x}_k^T\mathbf{Q}_k\mathbf{x}_k + \mathbf{u}_k^T\mathbf{R}_k\mathbf{u}_k) + v_{k+1} \\ &\quad + \tfrac{1}{2}(\mathbf{\Phi}_k\hat{\mathbf{x}}_k(+) + \mathbf{\Gamma}_k\mathbf{u}_k)^T\mathbf{S}_{k+1}(\mathbf{\Phi}_k\hat{\mathbf{x}}_k(+) + \mathbf{\Gamma}_k\mathbf{u}_k)] \mid \mathfrak{I}_{D_k}\}\} \\ &= \frac{1}{2}\frac{\partial}{\partial \mathbf{u}_k}\{\mathbf{u}_k^T\mathbf{R}_k\mathbf{u}_k \\ &\quad + E[(\mathbf{\Phi}_k\hat{\mathbf{x}}_k(+) + \mathbf{\Gamma}_k\mathbf{u}_k)^T\mathbf{S}_{k+1}(\mathbf{\Phi}_k\hat{\mathbf{x}}_k(+) + \mathbf{\Gamma}_k\mathbf{u}_k) \mid \mathfrak{I}_{D_k}]\}\end{aligned} \tag{5.3-52}$$

as $\mathbf{x}_k^T\mathbf{Q}_k\mathbf{x}_k$ and v_{k+1} are not functions of $\mathbf{u}_k$, and $\mathbf{u}_k$ is not a random variable. The remaining conditional expectation can be expanded as

$$
\begin{aligned}
&E\{[\boldsymbol{\Phi}_k\hat{\mathbf{x}}_k(+)+\boldsymbol{\Gamma}_k\mathbf{u}_k]^T\mathbf{S}_{k+1}[\boldsymbol{\Phi}_k\hat{\mathbf{x}}_k(+)+\boldsymbol{\Gamma}_k\mathbf{u}_k]|\mathfrak{I}_{D_k}\} \\
&\qquad = E[\hat{\mathbf{x}}_k^T(+)\boldsymbol{\Phi}_k^T\mathbf{S}_{k+1}\boldsymbol{\Phi}_k\hat{\mathbf{x}}_k(+)|\mathfrak{I}_{D_k}]+E[\hat{\mathbf{x}}_k^T(+)\boldsymbol{\Phi}_k^T\mathbf{S}_{k+1}\boldsymbol{\Gamma}_k\mathbf{u}_k|\mathfrak{I}_{D_k}] \\
&\qquad\quad + E[\mathbf{u}_k^T\boldsymbol{\Gamma}_k^T\mathbf{S}_{k+1}\boldsymbol{\Phi}_k\mathbf{x}_k(+)|\mathfrak{I}_{D_k}]+\mathbf{u}_k^T\boldsymbol{\Gamma}_k^T\mathbf{S}_{k+1}\boldsymbol{\Gamma}_k\mathbf{u}_k \qquad (5.3\text{-}53)
\end{aligned}
$$

The first term on the right side is not a function of $\mathbf{u}_k$, and

$$
E[\hat{\mathbf{x}}_k^T(+)\boldsymbol{\Phi}_k^T\mathbf{S}_{k+1}\boldsymbol{\Gamma}_k\mathbf{u}_k|\mathfrak{I}_{D_k}]=\hat{\mathbf{x}}_k^T(+)\boldsymbol{\Phi}_k^T\mathbf{S}_{k+1}\boldsymbol{\Gamma}_k\mathbf{u}_k \qquad (5.3\text{-}54)
$$

Therefore, Eq. 5.3-52 is equivalent to

$$
\begin{aligned}
\mathbf{0} &= \frac{1}{2}\frac{\partial}{\partial\mathbf{u}_k}[\mathbf{u}_k^T\mathbf{R}_k\mathbf{u}_k+\mathbf{u}_k^T\boldsymbol{\Gamma}_k^T\mathbf{S}_{k+1}\boldsymbol{\Gamma}_k\mathbf{u}_k \\
&\qquad +\hat{\mathbf{x}}_k^T(+)\boldsymbol{\Phi}_k^T\mathbf{S}_{k+1}\boldsymbol{\Gamma}_k\mathbf{u}_k+\mathbf{u}_k^T\boldsymbol{\Gamma}_k^T\mathbf{S}_{k+1}\boldsymbol{\Phi}_k\hat{\mathbf{x}}_k(+)] \\
&= \mathbf{u}_k^T(\mathbf{R}_k+\boldsymbol{\Gamma}_k^T\mathbf{S}_{k+1}\boldsymbol{\Gamma}_k)+\hat{\mathbf{x}}_k^T(+)\boldsymbol{\Phi}_k^T\mathbf{S}_{k+1}\boldsymbol{\Gamma}_k \qquad (5.3\text{-}55)
\end{aligned}
$$

leading to the optimal control law

$$
\begin{aligned}
\mathbf{u}_k &= -(\mathbf{R}_k+\boldsymbol{\Gamma}_k^T\mathbf{S}_{k+1}\boldsymbol{\Gamma}_k)^{-1}\boldsymbol{\Gamma}_k^T\mathbf{S}_{k+1}\boldsymbol{\Phi}_k\hat{\mathbf{x}}_k(+) \\
&= -\mathbf{C}_k\hat{\mathbf{x}}_k(+) \qquad (5.3\text{-}56)
\end{aligned}
$$

The stochastic optimal gain matrix takes the same form as the deterministic optimal gain matrix (Eq. 3.7-49b, with $\mathbf{M}=\mathbf{0}$ and $\mathbf{S}$ replacing $\mathbf{P}$):

$$
\mathbf{C}_k=(\mathbf{R}_k+\boldsymbol{\Gamma}_k^T\mathbf{S}_{k+1}\boldsymbol{\Gamma}_k)^{-1}\boldsymbol{\Gamma}_k^T\mathbf{S}_{k+1}\boldsymbol{\Phi}_k \qquad (5.3\text{-}57)
$$

The optimal feedback control law treats the state estimate as if it were the actual state, and the estimation algorithm is independent of the control law; hence the only remaining step in proving certainty equivalence is to show that $\mathbf{S}_{k+1}$ is independent of the problem's random variables. First, we note that

$$
E(\mathbf{x}_k^T\mathbf{Q}_k\mathbf{x}_k|\mathfrak{I}_{D_k})=\hat{\mathbf{x}}_k^T(+)\mathbf{Q}_k\hat{\mathbf{x}}_k(+)+\mathrm{Tr}[\mathbf{Q}_k\mathbf{P}_k(+)] \qquad (5.3\text{-}58)
$$

following Eq. 5.3-14. Substituting this and the optimal control law (Eq. 5.3-56) in Eq. 5.3-50, and with additional manipulations,

$$
\begin{aligned}
\tfrac{1}{2}\hat{\mathbf{x}}_k^T(+)\mathbf{S}_k\hat{\mathbf{x}}_k(+)+v_k &= \tfrac{1}{2}\{\hat{\mathbf{x}}_k^T(+)[\mathbf{Q}_k+\boldsymbol{\Phi}_k^T\mathbf{S}_{k+1}\boldsymbol{\Phi}_k]\hat{\mathbf{x}}_k(+) \\
&\qquad -\hat{\mathbf{x}}_k^T(+)[\boldsymbol{\Phi}_k^T\mathbf{S}_{k+1}\boldsymbol{\Gamma}_k(\mathbf{R}_k+\boldsymbol{\Gamma}_k\mathbf{S}_{k+1}\boldsymbol{\Gamma}_k)^{-1}\boldsymbol{\Gamma}_k^T\mathbf{S}_{k+1}\boldsymbol{\Phi}_k]\hat{\mathbf{x}}_k(+) \\
&\qquad +\mathrm{Tr}[\mathbf{Q}_k\mathbf{P}_k(+)]\}+v_{k+1} \qquad (5.3\text{-}59)
\end{aligned}
$$

For arbitrary $\hat{\mathbf{x}}_k(+)$, this equation can be partitioned into two separate solutions for $\mathbf{S}_k$ and v_k. The first of these is identical to the deterministic Riccati difference equation found previously (Eq. 3.7-51 with $\mathbf{S}$ replacing $\mathbf{P}$ and $\mathbf{M} = \mathbf{0}$):

$$\mathbf{S}_k = \mathbf{Q}_k + \mathbf{\Phi}_k^T \mathbf{S}_{k+1} \mathbf{\Phi}_k$$
$$- \mathbf{\Phi}_k^T \mathbf{S}_{k+1} \mathbf{\Gamma}_k (\mathbf{R}_k + \mathbf{\Gamma}_k^T \mathbf{S}_{k+1} \mathbf{\Gamma}_k)^{-1} \mathbf{\Gamma}_k^T \mathbf{S}_{k+1} \mathbf{\Phi}_k, \qquad \mathbf{S}_{k_f} \text{ given} \quad (5.3\text{-}60)$$

As in the continuous case, the certainty-equivalent cost function is minimized subject to a dynamic constraint expressed by the equation for the conditional mean estimate; combining Eqs. 5.3-39 and 5.3-40,

$$\hat{\mathbf{x}}_k(+) = \mathbf{\Phi}_{k-1}\hat{\mathbf{x}}_{k-1}(+) + \mathbf{\Gamma}_{k-1}\mathbf{u}_{k-1} + \mathbf{K}_k\{\mathbf{z}_k - \mathbf{H}_k[\mathbf{\Phi}_{k-1}\hat{\mathbf{x}}_{k-1}(+) + \mathbf{\Gamma}_{k-1}\mathbf{u}_{k-1}]\}$$
$$= \mathbf{\Phi}_{k-1}\hat{\mathbf{x}}_{k-1}(+) + \mathbf{\Gamma}_{k-1}\mathbf{u}_{k-1} + \mathbf{K}_k\boldsymbol{\epsilon}_k \quad (5.3\text{-}61)$$

where the optimal filter residual, $\boldsymbol{\epsilon}_k$, is a zero-mean white random sequence with

$$E(\boldsymbol{\epsilon}_k\boldsymbol{\epsilon}_k^T) = \mathbf{N}_k + \mathbf{H}_k\mathbf{P}_k(-)\mathbf{H}_k^T \quad (5.3\text{-}62)$$

The filter residual has the effect of a disturbance input on state estimate propagation; hence v_k is propagated as in Eq. 5.1-25, with the added effect of measurement error deduced from Eq. 5.3-59:

$$v_k = \tfrac{1}{2}\operatorname{Tr}\{\mathbf{S}_{k+1}\mathbf{K}_k[\mathbf{N}_k + \mathbf{H}_k\mathbf{P}_k(-)\mathbf{H}_k^T]\mathbf{K}_k^T + \tfrac{1}{2}\operatorname{Tr}[\mathbf{Q}_k\mathbf{P}_k(+)] + v_{k+1} \quad (5.3\text{-}63)$$

$$v_{k_f} = \tfrac{1}{2}\operatorname{Tr}[\mathbf{S}_{k_f}\mathbf{P}_{k_f}(-)] \quad (5.3\text{-}64)$$

The minimum cost then is found by evaluating Eq. 5.3-59 at $k = 0$:

$$J^* = V_0^* = \tfrac{1}{2}\bar{\mathbf{x}}_0^T\mathbf{S}_0\bar{\mathbf{x}}_0 + \tfrac{1}{2}\operatorname{Tr}[\mathbf{S}_{k_f}\mathbf{P}_{k_f}(-)$$
$$+ \sum_{k=0}^{k_f-1} \operatorname{Tr}\{\mathbf{S}_{k+1}\mathbf{K}_k[\mathbf{N}_k + \mathbf{H}_k\mathbf{P}_k(-)\mathbf{H}_k^T]\mathbf{K}_k^T + \mathbf{Q}_k\mathbf{P}_k(+)\} \quad (5.3\text{-}65)$$

Additional Cases Exhibiting Certainty Equivalence

Although it has been assumed that all random processes and sequences are white and Gaussian, certainty equivalence has been proven for linear dynamic systems and the indicated statistics in the cited references:

- Colored Gaussian measurement noise (T-7)
- Colored Gaussian disturbance input (T-7)
- White non-Gaussian measurement noise (R-1)

- White non-Gaussian disturbance input (R-1)
- Cross-correlated disturbance input and measurement noise (B-4)

Furthermore, it is concluded in (T-7) that neutral systems, which do not evidence a dual-control effect, have the certainty-equivalence property.

5.4 LINEAR, TIME-INVARIANT SYSTEMS WITH RANDOM INPUTS AND IMPERFECT MEASUREMENTS

Optimal control and estimation of linear, time-invariant systems—for which the system and cost function matrices are constant—are special cases of the problems considered up to now. The methods of optimizing trajectories of finite duration are identical to those for linear, time-varying systems, and, although the system and cost function matrices are constant, the optimal control and estimation gain matrices vary in time. Keep in mind that the actual optimal trajectory and control history depend on initial conditions, the disturbance profile, and the measurement error profile; however, the optimal gain profiles are fixed by the computations of Sections 5.2 and 5.3.

It will be seen that gain variations normally are confined to brief periods at the beginning and end of the trajectory. Estimation gains exhibit transient behavior at the beginning of the trajectory, when initial conditions may contain more (or less) information than the measurements, settling down to essentially constant values as the initial time recedes into the past. Conversely, control gains are very nearly constant until the final moments of a trajectory, when the terminal cost may become more significant than the integral cost.

Figure 5.4-1 provides a qualitative illustration of typical stochastic

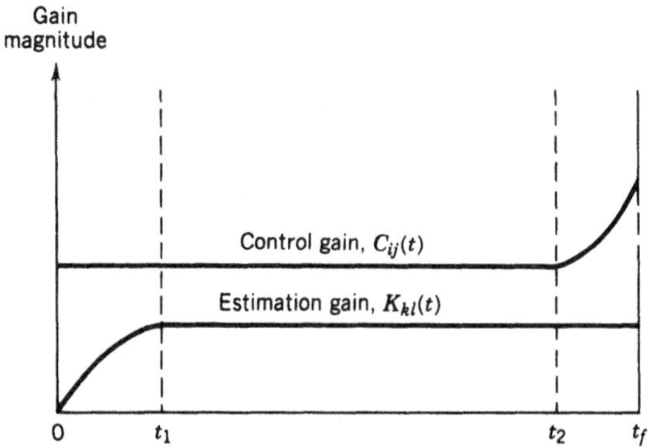

FIGURE 5.4-1 Typical trends in optimal control and estimation gains for a linear, time-invariant system.

optimal gain profiles. Let t_1 be the time at which estimation gains have essentially reached their final values and t_2 be the time at which control gain changes become noticeable. It can be inferred that the control and estimation Riccati equations are at equilibrium during the interval from t_1 to t_2, as $\mathbf{C}(t)$ and $\mathbf{K}(t)$ are unchanging. If $(t_2 - t_1)$ is large compared to the sum of t_1 and $(t_f - t_2)$, then the stochastic cost function is dominated by its integral terms, and constant control and estimation gains could be considered good approximations to the optimal gain profiles.

We consider the continuous- and discrete-time problems defined as before, with the notable exception that all coefficient matrices are assumed to be constant throughout the solution interval and that the final time approaches infinity in the limit. With continuing random inputs, even optimized stochastic cost functions of the form

$$J = \tfrac{1}{2}E\left\{\mathbf{x}^T(t_f)\mathbf{S}_f\mathbf{x}(t_f) + \int_0^{t_f} [\mathbf{x}^T(t)\mathbf{Q}x(t) + 2\mathbf{x}^T(t)\mathbf{M}\mathbf{u}(t) + \mathbf{u}^T(t)\mathbf{R}\mathbf{u}(t)]\,dt\right\} \tag{5.4-1}$$

become infinite in the limit, so it is useful to define an average cost or "cost rate"

$$J_A = \lim_{t_f \to \infty} \frac{J}{t_f} \tag{5.4-2}$$

whose value may be bounded as t_f increases. The optimality conditions are the same for J and J_A, as both pose fixed-end-time problems. The only difference is the cost attributed to the trajectory.

Some general properties of stochastic regulators for linear, time-invariant problems are presented in the remainder of the section. We begin by examining the stability and steady-state performance of continuous-time linear-optimal controllers and estimators separately, then jointly, introducing the limiting constant solutions that occur with infinite end time or "horizon." The same topics then are reviewed for discrete-time systems.

Asymptotic Stability of the Linear-Quadratic Regulator

The response stability of dynamic systems is of great practical significance, in that it determines whether or not the state tends to converge to a desired value with the passage of time. One of the great virtues of linear-optimal control is that it can guarantee closed-loop asymptotic stability for a wide variety of linear systems, including those that are naturally unstable. In Chapter 6, we shall see that there are limitations on the region of stable response, due principally to parameter uncertainty or nonlinearity (e.g., bounds on control amplitude); however, such modeling problems must be addressed in any control design procedure. The important point of this section is that the constant-gain, linear-optimal regulator provides a simple and elegant way to guarantee stability when a linear, time-invariant model with known coefficients is an accurate representation of the system and when perfect measurements of all states are available.

Consider the feedback controller that minimizes the stochastic quadratic cost function (with zero terminal cost),

$$J = \tfrac{1}{2}E\left\{\int_0^{t_f} [\mathbf{x}^T(t)\mathbf{Q}\mathbf{x}(t) + \mathbf{u}^T(t)\mathbf{R}\mathbf{u}(t)]\, dt\right\} \tag{5.4-3}$$

subject to the linear dynamic constraint

$$\dot{\mathbf{x}}(t) = \mathbf{F}\mathbf{x}(t) + \mathbf{G}\mathbf{u}(t) + \mathbf{L}\mathbf{w}(t), \qquad \mathbf{x}(0) = \mathbf{x}_0 \tag{5.4-4}$$

Q, **R**, **F**, **G**, and **L** are assumed to be constant matrices of appropriate dimension. As in previous problems, **Q** and **R** are positive definite matrices. The dynamic system is forced by a zero-mean disturbance input $\mathbf{w}(t)$ with constant spectral density matrix **W**, and its initial condition $\mathbf{x}_0$ is assumed to be known. With perfect measurements, the minimizing control law is

$$\mathbf{u}(t) = -\mathbf{C}(t)\mathbf{x}(t) \tag{5.4-5}$$

where

$$\mathbf{C}(t) = \mathbf{R}^{-1}\mathbf{G}^T\mathbf{S}(t) \tag{5.4-6}$$

and

$$\dot{\mathbf{S}}(t) = -\mathbf{F}^T\mathbf{S}(t) - \mathbf{S}(t)\mathbf{F} - \mathbf{Q} + \mathbf{S}(t)\mathbf{G}\mathbf{R}^{-1}\mathbf{G}^T\mathbf{S}(t), \qquad \mathbf{S}(t_f) = \mathbf{0} \tag{5.4-7}$$

When the optimal control law is applied to the system, it minimizes the cost function, which takes the following form (Eq. 5.1-78):

$$J^* = \frac{1}{2}\left[\mathbf{x}_0^T\mathbf{S}(0)\mathbf{x}_0 + \mathrm{Tr}\int_0^{t_f} \mathbf{S}(t)\mathbf{L}\mathbf{W}\mathbf{L}^T\, dt\right] \tag{5.4-8}$$

If the disturbance input is absent ($\mathbf{W} = \mathbf{0}$), the problem is deterministic, and the cost function takes the following value:

$$J^* = \tfrac{1}{2}\mathbf{x}_0^T\mathbf{S}(0)\mathbf{x}_0 \tag{5.4-9}$$

This result is unchanged as t_f approaches infinity, although the value of $\mathbf{S}(0)$ changes as t_f progresses from small to large values. Suppose that t_{f_2} is greater than t_{f_1} and that the corresponding cost functions are $J^*(t_{f_2})$ and

$J^*(t_{f_1})$, both of which are greater than or equal to zero. By the principle of optimality, the progression from $J^*(t_{f_1})$ to $J^*(t_{f_2})$ is monotonic and

$$J^*(t_{f_2}) \geq J^*(t_{f_1}) \geq 0 \tag{5.4-10}$$

Therefore, for arbitrary $\mathbf{x}_0$, Eq. 5.4-9 indicates that

$$\mathbf{S}(0, t_{f_2}) \geq \mathbf{S}(0, t_{f_1}) \geq \mathbf{0} \tag{5.4-11}$$

where the second argument signifies the corresponding end time of the optimal trajectory. Assuming that $\mathbf{x}_0$ is finite and $\mathbf{S}(0)$ is bounded, the integral term of Eq. 5.4-1 must be bounded. This is an important property, as shown below; therefore, it is of interest to assure that $\mathbf{S}(0)$ is bounded.

For finite t_f, the *existence theorem for ordinary differential equations* (also called the "calculus of limits" of Cauchy*) assures the existence of $\mathbf{S}(t)$ in $[0, t_f]$, as it is necessary only to show that the right side of Eq. 5.4-7 is an analytic function of $\mathbf{S}(t)$ and t.† This is the case for the matrix Riccati equation for completely controllable systems, and it implies that there is a continuum of solutions in the neighborhood of any given solution. Therefore, with bounded $\mathbf{Q}$ and zero starting conditions (at t_f), $\mathbf{S}(t)$ is bounded in $[0, t_f]$. As t_f approaches infinity, $\mathbf{S}(0)$ must converge to a limiting value, because it is monotonic as well as bounded (K-3). With $\mathbf{S}_f = \mathbf{0}$ and positive definite $\mathbf{Q}$, the limiting value of $\mathbf{S}(0)$ also must be positive definite, because $\mathbf{S}(t)$ is monotonic and $\dot{\mathbf{S}}(t_f) = -\mathbf{Q}$. The optimal gain is related algebraically to $\mathbf{S}(t)$ (Eq. 5.4-6), and it, too, takes a limiting value at $t = 0$.

Terminal cost presumably is of little concern as t_f approaches infinity, so the specification that $\mathbf{S}_f = \mathbf{0}$ is reasonable. Then the cost function (for zero disturbance input) can be expressed as

$$J^* = \frac{1}{2}\int_0^\infty [\mathbf{x}^{*T}(t)\mathbf{Q}\mathbf{x}^*(t) + \mathbf{u}^T(t)\mathbf{R}\mathbf{u}(t)]\, dt \tag{5.4-12}$$

Substituting for the optimal control (Eq. 5.4-6),

$$\begin{aligned} J^* &= \frac{1}{2}\int_0^\infty \mathbf{x}^{*T}(t)[\mathbf{Q} + \mathbf{C}^T(t)\mathbf{R}\mathbf{C}(t)]\mathbf{x}^*(t)\, dt \\ &= \frac{1}{2}\int_0^\infty \mathbf{x}^{*T}(t)\mathbf{Q}'(t)\mathbf{x}^*(t)\, dt \end{aligned} \tag{5.4-13}$$

*Augustin-Louis Cauchy (1789–1857) was a neighbor of Laplace and an acquaintance of Lagrange who began his career as an engineer. He relinquished his position as professor at L'Ecole Polytechnique in protest of the July 1830 revolution, exiling himself to various academic positions throughout Europe.

†That is, it is differentiable with respect to $\mathbf{S}(t)$ and t.

where $\mathbf{Q}'(t)$ must be a positive definite matrix defined as

$$\begin{aligned}\mathbf{Q}'(t) &= \mathbf{Q} + \mathbf{C}^T(t)\mathbf{R}\mathbf{C}(t) \\ &= \mathbf{Q} + \mathbf{S}(t)\mathbf{G}\mathbf{R}^{-1}\mathbf{G}^T\mathbf{S}(t) \end{aligned} \tag{5.4-14}$$

J^* is the integral of a weighted Euclidean (quadratic) norm-squared of $\mathbf{x}^*(t)$, and it is bounded as t_f approaches infinity (by Eq. 5.4-9). As long as $\mathbf{Q}'(t)$ remains positive definite and the system is completely controllable, the *global asymptotic stability* of $\mathbf{x}^*(t)$ is assured because if J^* is bounded, $\mathbf{x}^*(t)$ must approach zero as t_f approaches infinity (Section 2.5). Put another way, *if $\mathbf{Q}'(t) > \mathbf{0}$ throughout the interval, the linear-quadratic controller guarantees global asymptotic stability*. Thus the closed-loop system defined by

$$\begin{aligned}\dot{\mathbf{x}}(t) &= \mathbf{F}\mathbf{x}(t) - \mathbf{G}\mathbf{C}(t)\mathbf{x}(t) \\ &= [\mathbf{F} - \mathbf{G}\mathbf{C}(t)]\mathbf{x}(t) \end{aligned} \tag{5.4-15}$$

with $\mathbf{C}(t)$ defined by Eqs. 5.4-6 and 5.4-7 is stable. This result is true whether or not the open-loop system, whose stability is determined by the system matrix $\mathbf{F}$, is stable.

The limiting value of $\mathbf{S}(0)$ as t_f approaches infinity is obviously a constant; hence $\lim_{t\to\infty} \dot{\mathbf{S}}(0) = \mathbf{0}$, and $\mathbf{S}(0)$ is a solution to the so-called *algebraic Riccati equation*:

$$\mathbf{0} = -\mathbf{F}^T\mathbf{S}(0) - \mathbf{S}(0)\mathbf{F} - \mathbf{Q} + \mathbf{S}(0)\mathbf{G}\mathbf{R}^{-1}\mathbf{G}^T\mathbf{S}(0) \tag{5.4-16}$$

It can be inferred that the differential Riccati equation is stable with time running in reverse, and this result is proved in (K-2). The corresponding steady-state optimal control gain matrix is

$$\mathbf{C}(0) = \mathbf{R}^{-1}\mathbf{G}^T\mathbf{S}(0) \tag{5.4-17}$$

and feedback control using $\mathbf{C}(0)$ provides *global exponential stability* (as shown later). The constant-coefficient, feedback control law,

$$\mathbf{u}(t) = -\mathbf{C}(0)\mathbf{x}(t) \tag{5.4-18}$$

is referred to as the *linear-quadratic (LQ) regulator*.

Following Section 2.5, we define a positive definite Lyapunov function in which $\mathbf{S}\,[=\mathbf{S}(0)]$ is the defining matrix:

$$\mathcal{V}(\mathbf{x}, t) = \mathbf{x}^T(t)\mathbf{S}\mathbf{x}(t) \tag{5.4-19}$$

For asymptotic stability, $\dot{\mathcal{V}}(\mathbf{x}, t)$ must be negative definite at all times. Since $\mathbf{S}$ is a constant, the time-derivative of the Lyapunov function for the linear system with steady-state optimal control is

$$\dot{\mathcal{V}}(\mathbf{x}, t) = \frac{\partial \mathcal{V}}{\partial \mathbf{x}}(\mathbf{x}, t)\dot{\mathbf{x}}(t) = 2\mathbf{x}^T(t)\mathbf{S}[\mathbf{F}\mathbf{x}(t) + \mathbf{G}\mathbf{u}(t)]$$
$$= 2\mathbf{x}^T(t)\mathbf{S}[\mathbf{F} - \mathbf{G}\mathbf{R}^{-1}\mathbf{G}^T\mathbf{S}]\mathbf{x}(t)$$
$$= \mathbf{x}^T(t)\{\mathbf{S}[\mathbf{F} - \mathbf{G}\mathbf{R}^{-1}\mathbf{G}^T\mathbf{S}] + [\mathbf{F} - \mathbf{G}\mathbf{R}^{-1}\mathbf{G}^T\mathbf{S}]^T\mathbf{S}\}\mathbf{x}(t) \quad (5.4\text{-}20)$$

which, from Eq. 5.4-16, is equivalent to

$$\dot{\mathcal{V}}(\mathbf{x}, t) = -\mathbf{x}^T(t)[\mathbf{Q} + \mathbf{S}\mathbf{G}\mathbf{R}^{-1}\mathbf{G}^T\mathbf{S}]\mathbf{x}(t)$$
$$= -\mathbf{x}^T(t)\mathbf{Q}'\mathbf{x}(t) \quad (5.4\text{-}21)$$

As $\mathbf{Q}'$ is constant and positive definite, $\dot{\mathcal{V}}(x, t)$ is negative at all times, and the steady-state optimal control gain is stabilizing.

Exponential stability follows from the nature of linear, time-invariant systems. With no disturbance inputs, the state history is portrayed by the product of the closed-loop state transition matrix and the initial condition:

$$\mathbf{x}(t) = \mathbf{\Phi}_{\mathrm{CL}}(t - t_0)\mathbf{x}_0 = e^{[\mathbf{F}-\mathbf{G}\mathbf{C}](t-t_0)}\mathbf{x}_0 \quad (5.4\text{-}22)$$

Then the norm of $\mathbf{x}(t)$ is

$$\|\mathbf{x}(t)\| = \|e^{[\mathbf{F}-\mathbf{G}\mathbf{C}](t-t_0)}\mathbf{x}_0\| \leq \|\mathbf{x}_0\|e^{-\alpha(t-t_0)} \quad (5.4\text{-}23)$$

where α is a positive number smaller than $(1/\tau)$, and τ is the largest time constant of the stable closed-loop system. As a consequence, the eigenvalues of $(\mathbf{F} - \mathbf{G}\mathbf{C})$ must lie in the left half plane (i.e., they must have negative real parts).

Taking stock, we note that closed-loop stability is guaranteed if: (a) $[\mathbf{F}, \mathbf{G}]$ is a completely controllable pair; (b) $\mathbf{R}$ is positive definite; and (c) $\mathbf{Q}$ is positive definite. Complete controllability assures that the control can act on any unstable modes of the system. Positive definite $\mathbf{R}$ eliminates the possibility of singularity, which would render Eq. 5.4-6 unsolvable, and it assures that control activity has a positive effect on J. Positive definite $\mathbf{Q}$ assures that unstable modes are penalized in J and that all motions have positive effect on J.

Asymptotic stability can be guaranteed with less restrictive conditions that relate to system observability and controllability. It is possible, for example, to retain a positive definite cost function that penalizes potentially unstable modes with positive semidefinite $\mathbf{Q}$ if an additional observability criterion is satisfied. Specifically, if $\mathbf{F}$ and $\mathbf{D}$ form an observable pair, where $\mathbf{D}$ is a square root of $\mathbf{Q}$ such that

$$\mathbf{Q} = \mathbf{D}^T\mathbf{D} \quad (5.4\text{-}24)$$

then all states are observable in the cost function.* If the system contains naturally stable modes, then it is necessary only that **F** and **D** form a *detectable* pair and that **F** and **G** form a *stabilizable* pair, as the unstable modes can be adequately controlled while the stable modes take care of themselves.

EXAMPLE 5.4-1 LQ REGULATOR FOR A FIRST-ORDER SYSTEM

Consider a system described by the scalar, linear, time-invariant model

$$\dot{x}(t) = fx(t) + gu(t), \qquad x(0) = x_0$$

Forming the infinite-time quadratic cost function,

$$J = \frac{1}{2}\int_0^\infty (qx^2 + ru^2)\, dt$$

the corresponding algebraic Riccati equation,

$$0 = -2fs - q + \frac{g^2 s^2}{r}$$

has two solutions:

$$s = \frac{r}{g^2}\left[f \pm \sqrt{f^2 + g^2 \frac{q}{r}}\right]$$

Only one solution is positive, and the magnitude of the square root is always greater than $|f|$. As s must be positive, the positive sign is chosen. The optimal control gain is

$$c = \frac{gs}{r} = \frac{1}{g}\left[f + \sqrt{f^2 + g^2 \frac{q}{r}}\right]$$

Because c is positive in this example, and

$$u(t) = -cx(t)$$

the control always has a stabilizing effect on closed-loop response, unless c is identically zero. The closed-loop response is found by integrating

$$\dot{x}(t) = (f - gc)x(t), \qquad x(0) = x_0$$

*From Section 2.5, if $\mathbf{y} = \mathbf{Dx}$, and an observability matrix, $\mathbf{\Gamma}$, defined by $[\mathbf{D}^T\ \mathbf{F}^T\mathbf{D}^T \ldots \mathbf{F}^{n-1^T}\mathbf{D}^T]$, has rank n, then $\mathbf{x}$ is completely observable in $\mathbf{y}$. Since $\mathbf{y}^T\mathbf{y}$ corresponds to $\mathbf{x}^T\mathbf{D}^T\mathbf{Dx}$, all states are observable in the cost function.

Hence the order of the system is unchanged, and initial condition response is described by the stable exponential function

$$x(t) = x_0 e^{(f-gc)t}$$

If q approaches zero or r approaches infinity, the limiting values of the gain are

$$c = \frac{2f}{g} \quad \text{or} \quad 0$$

depending on the sign of f. In either case, the system response is the same as

$$(f - gc) = \begin{cases} -f, & f > 0, \quad \text{as } c = 2f/g \\ f, & f < 0, \quad \text{as } c = 0 \end{cases}$$

With positive f, the system is "closed-loop," while with negative f, it is "open-loop"! More will be said about this in Chapter 6. (What about the case when $f = 0$?)

EXAMPLE 5.4-2 INITIAL CONDITION RESPONSE FOR A LQ REGULATOR FOR A SECOND-ORDER SYSTEM

Doubling the dimension of the state raises the complexity of the regulator considerably; however, selected examples provide considerable insight into the nature of solutions. Limiting the problem to a single control variable and a diagonal state weighting matrix leads to the following:

$$\begin{bmatrix} \dot{x}_1(t) \\ \dot{x}_2(t) \end{bmatrix} = \begin{bmatrix} a & b \\ c & d \end{bmatrix} \begin{bmatrix} x_1(t) \\ x_2(t) \end{bmatrix} + \begin{bmatrix} e \\ f \end{bmatrix} u, \qquad \begin{matrix} x_1(0) = x_{1_0} \\ x_2(0) = x_{2_0} \end{matrix}$$

$$J = \frac{1}{2} \int_0^\infty [q_{11} x_1^2(t) + q_{22} x_2^2(t) + r u^2(t)]\, dt$$

$$\mathbf{0} = -\begin{bmatrix} a & c \\ b & d \end{bmatrix} \begin{bmatrix} s_{11} & s_{12} \\ s_{12} & s_{22} \end{bmatrix} - \begin{bmatrix} s_{11} & s_{12} \\ s_{12} & s_{22} \end{bmatrix} \begin{bmatrix} a & b \\ c & d \end{bmatrix} - \begin{bmatrix} q_{11} & 0 \\ 0 & q_{22} \end{bmatrix}$$

$$+ \frac{1}{r} \begin{bmatrix} s_{11} & s_{12} \\ s_{12} & s_{22} \end{bmatrix} \begin{bmatrix} e \\ f \end{bmatrix} [e \quad f] \begin{bmatrix} s_{11} & s_{12} \\ s_{12} & s_{22} \end{bmatrix}$$

$$\mathbf{C} = \frac{1}{r} [e \quad f] \begin{bmatrix} s_{11} & s_{12} \\ s_{12} & s_{22} \end{bmatrix}$$

$$= \frac{1}{r} [(e s_{11} + f s_{12}) \quad (e s_{12} + f s_{22})]$$

The algebraic Riccati equation preserves the symmetry of **S**; hence three scalar equations must be solved to find s_{11}, s_{12}, and s_{22}:

$$0 = -2(as_{11} + cs_{12}) - q_{11} + \frac{1}{r}(e^2 s_{11}^2 + 2efs_{11}s_{12} + f^2 s_{12}^2)$$

$$0 = -[bs_{11} + (a+d)s_{12} + cs_{22}] + \frac{1}{r}[e^2 s_{11}s_{12} + ef(s_{12}^2 + s_{11}s_{22}) + f^2 s_{12}s_{22}]$$

$$0 = -2(bs_{12} + ds_{22}) - q_{22} + \frac{1}{r}(e^2 s_{12}^2 + 2efs_{12}s_{22} + f^2 s_{22}^2)$$

Although only second degree, these equations are tightly coupled, and symbolic solution is formidable. There are multiple solutions, but only one yields a positive definite **S**, and that is the one sought.

If the problem is transformed to a canonical model with $a = e = 0$ and $b = 1$, the equations reduce to the following:

1. $0 = \frac{1}{r} f^2 s_{12}^2 - 2cs_{12} - q_{11}$

2. $0 = -s_{11} - ds_{12} - cs_{22} + \frac{1}{r} f^2 s_{12}s_{22}$

3. $0 = \frac{1}{r} f^2 s_{22}^2 - 2\,ds_{22} - 2s_{12} - q_{22}$

These equations can be solved in the sequence (1), (3), (2),

1. $s_{12} = \frac{r}{f^2}\left[c \pm \sqrt{c^2 + \frac{f^2 q_{11}}{r}}\right]$

3. $s_{22} = \frac{r}{f^2}\left[d \pm \sqrt{d^2 + (q_{22} + 2s_{12})\frac{f^2}{r}}\right]$

2. $s_{11} = \frac{f^2 s_{12}s_{22}}{r} - ds_{12} - cs_{22}$

such that **S** is positive definite $[s_{11} > 0, s_{22} > 0, (s_{11}s_{22} - s_{12}^2) > 0]$; then the optimal gain matrix is

$$\mathbf{C} = \left[\frac{fs_{12}}{r} \quad \frac{fs_{22}}{r}\right]$$

Let us apply the control law to the weathervane model of earlier problems, whose system equation is

$$\begin{bmatrix} \dot{x}_1(t) \\ \dot{x}_2(t) \end{bmatrix} = \begin{bmatrix} 0 & 1 \\ -39.4 & -3.8 \end{bmatrix} \begin{bmatrix} x_1(t) \\ x_2(t) \end{bmatrix} + \begin{bmatrix} 0 \\ f \end{bmatrix} u(t)$$

where x_1 represents angular position relative to the steady-state wind direction, x_2 represents angular rate, and f is the specific torque sensitivity of a control motor mounted on the weathervane's shaft. From the structure of the LQ regulator, it is apparent that the specific torque command is a linear function of the angular position and rate. In the following, f is set to

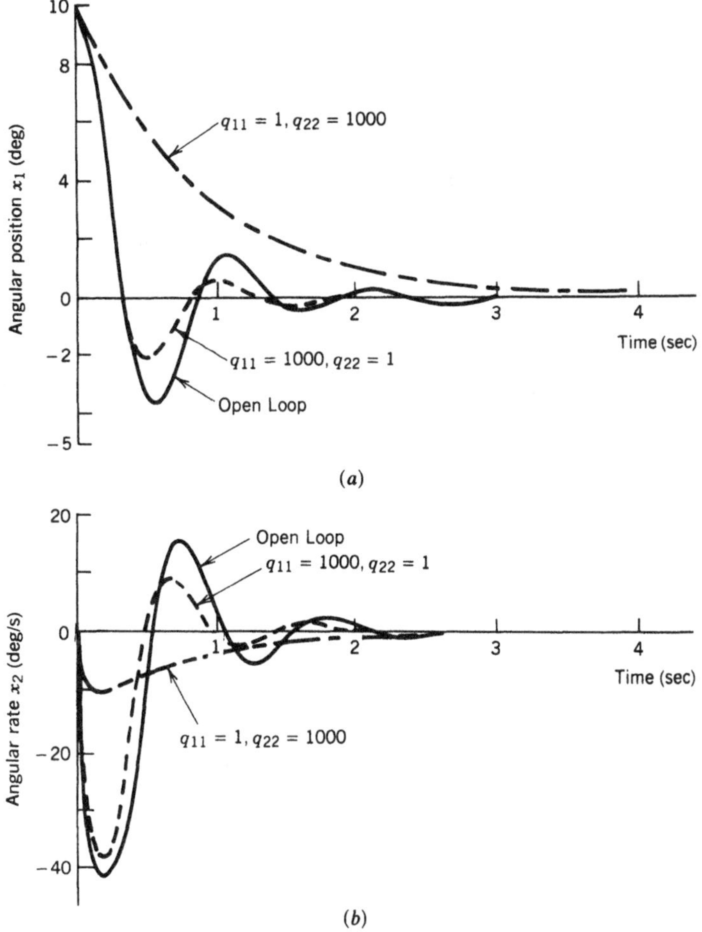

Diagram 1. (*a*) Angular position versus time; (*b*) angular rate versus time.

1; the effects of cost function weighting and system parameters on control gains and system response are examined.

If the weathervane is displaced 10° from the steady-state wind direction, it returns to its equilibrium value with the transient response marked Open Loop in Diagram 1. With perfect angle and rate sensors providing feedback to the control motor, the dynamic response could be modified, and closed-loop responses for two LQ regulators also are shown in the sketch. Taking $r = 1$, the separate effects of large displacement weighting ($q_{11} = 1000$) and large rate weighting ($q_{22} = 1000$) are plotted. A term such as

$$\int_0^\infty q_{11} x_1^2(t)\, dt$$

weights the square of the area between $x_1(t)$ and the horizontal axis in Diagram 1; it can be seen that increased weighting reduces the corresponding area. For $q_{11} = 1000$ and $q_{22} = 1$, transient response is quickened, and damping is increased slightly; for $q_{11} = 1$ and $q_{22} = 1000$, the response is slowed, but the peak angular rate and its square-integral value are greatly reduced.

Table 5.4-1 lists representative values of the control gains, c_1 and c_2, for various cost function weightings. Increased weighting on a particular state variable component tends to increase the corresponding gain. For this stable example, increased control weighting makes both gains become vanishingly small. Minimized cost function values J^* are computed from Eq. 5.4-9 as $100 s_{11}$. Direct comparisons from one case to the next are inappropriate, as the weightings vary; however, the tabulated J^* is lower than that which would be obtained for the associated weights and any other control law. Closed-loop damping ratio (ζ) and natural frequency (ω_n) also are listed; if the roots are real, the eigenvalues (λ_1, λ_2) are listed in parentheses. It is clear that all oscillatory cases are stable and at least as well-damped as the open-loop system. The single case with real roots has a very fast mode and a very slow mode.

TABLE 5.4-1 Cost Function Weighting Effects on Control Gains

q_{11}	q_{22}	r	c_1	c_2	J^*	$\zeta(\lambda_1)$	$\omega_n(\lambda_2)$
1	1	1	0.013	0.133	528	0.31	6.28
10	1	1	0.127	0.162	686	0.32	6.29
100	1	1	1.249	0.435	2,245	0.33	6.38
1,000	1	1	11.121	2.339	16,041	0.43	7.11
1	10	1	0.013	1.146	4,522	0.39	6.28
1	100	1	0.013	6.899	27,195	0.85	6.28
1	1,000	1	0.013	28.051	110,560	(−1.29)	(−30.56)
1	1	10	0.001	0.135	535	0.31	6.28
1	1	100	0.000	0.001	537	0.30	6.28

TABLE 5.4-2 System Parameter Effects on Control Gains

c	d	r	c_1	c_2	J^*	$\zeta(\lambda_1)$	$\omega_n(\lambda_2)$
−39.4	−3.8	1	0.013	0.133	528	0.31	6.28
39.4	−3.8	1	78.813	9.355	66,821	(−4.62)	(−8.54)
39.4	−3.8	1,000	78.800	9.316	66,650,800	(−4.66)	(−8.46)
−39.4	3.8	1	0.013	7.733	30,471	0.31	6.28
−39.4	3.8	1,000	0.000	7.600	29,944,400	0.30	6.28
39.4	3.8	1	78.813	16.955	36,877	(−4.62)	(−8.54)
39.4	3.8	1,000	78.800	16.916	36,706,600	(−4.66)	(−8.46)

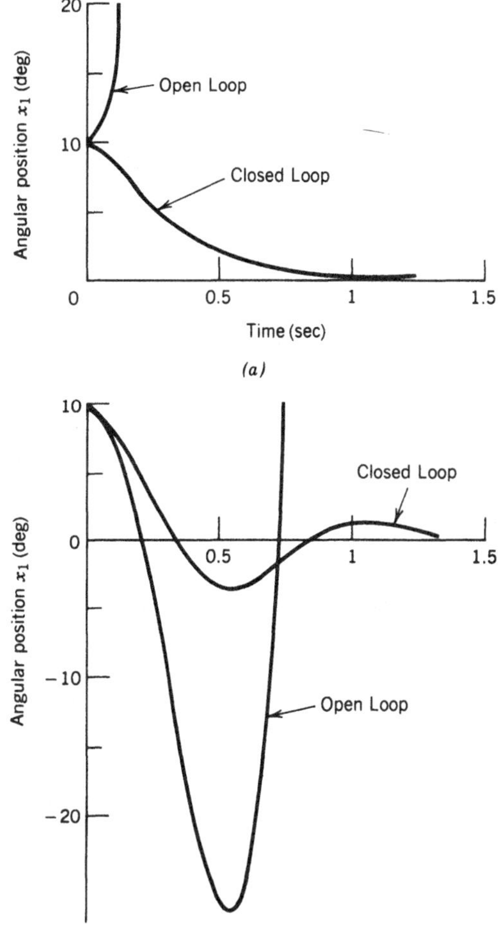

Diagram 2. (*a*) $c = 39.4$, $d = -3.8$, $r = 1$; (*b*) $c = -39.4$, $d = 3.8$, $r = 1$.

Increased control weighting does not necessarily lead to vanishingly small control gains if the system is unstable. The reason is that the LQ regulator guarantees stability; if the open-loop system is unstable, the regulator cannot do this with negligible gains. Consequently, the minimized cost function requires some control, even if the cost of control is high.

Table 5.4-2 provides a few examples of the control of unstable systems. Changing the sign of c is equivalent to orienting the weathervane with its tail headed into the wind. The necessary gains are very large in this case, whether the control weighting is low ($r = 1$) or high ($r = 1000$). Changing the sign of d (while restoring c to a negative value) is analogous to providing negative damping to the weathervane. In this case, rate feedback (c_2) is required, and its value is high with large or small r; however, c_1 becomes vanishingly small at large r. The combination of negative damping and static instability (c, $d > 0$) leads to high gains, although the cost of control is less than the cost for static instability alone. (In this case, cost functions have identical weightings and can be compared.)

Corresponding time histories appear in Diagram 2. The severe open-loop exponential divergence ($c = 39.4$) is restored to overdamped response by the LQ regulator, while open-loop oscillatory divergence is modified to yield a stable response that is very nearly the same as that of the original (stable) open-loop system. These seemingly different results fit a well-developed response pattern to be discussed further in Chapter 6. In any case, the stabilizing ability of the LQ regulator is demonstrated clearly in these examples.

The LQ design techniques might be considered unnecessarily elaborate for a second-order system, although the control gain equations are not that difficult; however, the real value of the optimal approach is that it applies equally well to high-order systems that may not be susceptible to easy stabilization using other methods.

Linear-quadratic problems that contain cross-product weighting in their cost functions are subject to similar criteria for guaranteed asymptotic stability. The easiest way to demonstrate this is through *transformations of variables* in the cost function and state equation. Given the coupled cost function

$$J_1 = \frac{1}{2}\int_0^\infty [\mathbf{x}_1^T(t)\mathbf{Q}_1\mathbf{x}_1(t) + 2\mathbf{x}_1^T(t)\mathbf{M}_1\mathbf{u}_1(t) + \mathbf{u}_1^T(t)\mathbf{R}_1\mathbf{u}_1(t)]\, dt \qquad (5.4\text{-}25)$$

an equivalent uncoupled cost function

$$J_2 = \frac{1}{2}\int_0^\infty [\mathbf{x}_2^T(t)\mathbf{Q}_2\mathbf{x}_2(t) + \mathbf{u}_2^T(t)\mathbf{R}_2\mathbf{u}_2(t)]\, dt \qquad (5.4\text{-}26)$$

is to be found. The coupled integrand can be rearranged as follows:

$$\begin{aligned}\mathbf{x}_1^T\mathbf{Q}_1\mathbf{x}_1 + 2\mathbf{x}_1^T\mathbf{M}_1\mathbf{u}_1 + \mathbf{u}_1^T\mathbf{R}_1\mathbf{u}_1 &= \mathbf{x}_1^T(\mathbf{Q}_1 - \mathbf{M}_1\mathbf{R}_1^{-1}\mathbf{M}_1^T)\mathbf{x}_1 \\ &\quad + (\mathbf{u}_1 + \mathbf{R}_1^{-1}\mathbf{M}_1^T\mathbf{x}_1)^T\mathbf{R}_1(\mathbf{u}_1 + \mathbf{R}_1^{-1}\mathbf{M}_1^T\mathbf{x}_1) \\ &= \mathbf{x}_1^T\mathbf{Q}_2\mathbf{x}_1 + \mathbf{u}_2^T\mathbf{R}_1\mathbf{u}_2 \qquad (5.4\text{-}27)\end{aligned}$$

Therefore, it can be deduced that

$$\mathbf{x}_2(t) = \mathbf{x}_1(t) \qquad (5.4\text{-}28)$$

$$\mathbf{u}_2(t) = \mathbf{u}_1(t) + \mathbf{R}_1^{-1}\mathbf{M}_1^T\mathbf{x}_1 \qquad (5.4\text{-}29)$$

$$\mathbf{Q}_2 = \mathbf{Q}_1 - \mathbf{M}_1\mathbf{R}_1^{-1}\mathbf{M}_1^T \qquad (5.4\text{-}30)$$

$$\mathbf{R}_2 = \mathbf{R}_1 \qquad (5.4\text{-}31)$$

The original dynamic system is

$$\dot{\mathbf{x}}_1(t) = \mathbf{F}_1\mathbf{x}_1(t) + \mathbf{G}_1\mathbf{u}_1(t) \qquad (5.4\text{-}32)$$

and the equivalent dynamic system is

$$\dot{\mathbf{x}}_2(t) = \mathbf{F}_2\mathbf{x}_2(t) + \mathbf{G}_2\mathbf{u}_2(t) \qquad (5.4\text{-}33)$$

As $\mathbf{x}_2(t) = \mathbf{x}_1(t)$, the equivalent system matrices must be

$$\mathbf{F}_2 = \mathbf{F}_1 - \mathbf{G}_1\mathbf{R}_1^{-1}\mathbf{M}_1^T \qquad (5.4\text{-}34)$$

$$\mathbf{G}_2 = \mathbf{G}_1 \qquad (5.4\text{-}35)$$

Consequently, we can summarize the four criteria that guarantee asymptotic stability for *all* the preceding cases as follows:

1. $[(\mathbf{F} - \mathbf{G}\mathbf{R}^{-1}\mathbf{M}^T), \mathbf{G}]$ is a *stabilizable* pair.
2. $[(\mathbf{F} - \mathbf{G}\mathbf{R}^{-1}\mathbf{M}^T), \mathbf{D}]$ is a *detectable* pair, where

$$\mathbf{Q} - \mathbf{M}\mathbf{R}^{-1}\mathbf{M}^T = \mathbf{D}^T\mathbf{D}$$

3. $(\mathbf{Q} - \mathbf{M}\mathbf{R}^{-1}\mathbf{M}^T)$ is a positive semidefinite matrix.
4. $\mathbf{R}$ is a positive definite matrix.

These critieria apply when $\mathbf{F}$ and $\mathbf{G}$ are known without error; a new problem occurs when $\mathbf{C}$ is computed for $\mathbf{F}$ and $\mathbf{G}$ but the actual system matrices are different, say $\mathbf{F}_A$ and $\mathbf{G}_A$. Stability margins, which express allowable deviations in system modeling while still guaranteeing stability, are discussed in Chapter 6.

The stability of the linear-quadratic regulator gives rise to an interesting observation: if a linear, time-invariant system with full state feedback is stable, then the control is optimal with respect to some quadratic cost function (K-3). It may not be apparent *which* quadratic cost function is minimized, but it is certain that the control gain matrix **C** corresponds to some choice of **Q**, **M**, and **R**. Constructive methods for determining **Q**, **M**, and **R**, given **C**, are considered in (J-3) and (M-5).

Asymptotic Stability of the Kalman–Bucy Filter

Because control and estimation are dual problems, stability of the Kalman–Bucy filter parallels that of the LQ regulator. Existence, boundedness, and positive-definiteness of the matrix Riccati equation for estimation follow directly, so the present discussion is restricted to an examination of estimation error stability. The estimation error is defined as

$$\boldsymbol{\epsilon}(t) = \mathbf{x}(t) - \hat{\mathbf{x}}(t) \tag{5.4-36}$$

where the estimate $\hat{\mathbf{x}}(t)$ is generated by a constant-gain Kalman–Bucy filter:

$$\dot{\hat{\mathbf{x}}}(t) = \mathbf{F}\hat{\mathbf{x}}(t) + \mathbf{G}\mathbf{u}(t) + \mathbf{K}[\mathbf{z}(t) - \mathbf{H}\hat{\mathbf{x}}(t)], \qquad \hat{\mathbf{x}}(0) = \hat{\mathbf{x}}_0 \tag{5.4-37}$$

The steady-state optimal gain matrix is defined by

$$\mathbf{K} = \mathbf{P}\mathbf{H}^T\mathbf{N}^{-1} \tag{5.4-38}$$

where $\mathbf{P} = \mathbf{P}(\infty)$ is the solution of the algebraic Riccati equation for estimation:

$$\mathbf{0} = \mathbf{F}\mathbf{P}(\infty) + \mathbf{P}(\infty)\mathbf{F}^T + \mathbf{L}\mathbf{W}\mathbf{L}^T - \mathbf{P}(\infty)\mathbf{H}^T\mathbf{N}^{-1}\mathbf{H}\mathbf{P}(\infty) \tag{5.4-39}$$

The state and measurement equations take the forms

$$\dot{\mathbf{x}}(t) = \mathbf{F}\mathbf{x}(t) + \mathbf{G}\mathbf{u}(t) + \mathbf{L}\mathbf{w}(t) \tag{5.4-40}$$

$$\mathbf{z}(t) = \mathbf{H}\mathbf{x}(t) + \mathbf{n}(t) \tag{5.4-41}$$

where $[\mathbf{F}, \mathbf{H}]$ is a completely observable pair, and $\mathbf{w}(t)$ and $\mathbf{n}(t)$ are zero-mean white processes with

$$E[\mathbf{w}(t)\mathbf{w}(\tau)^T] = \mathbf{W}\delta(t-\tau) \tag{5.4-42}$$

$$E[\mathbf{n}(t)\mathbf{n}(\tau)^T] = \mathbf{N}\delta(t-\tau) \tag{5.4-43}$$

$\mathbf{N}$ and $\mathbf{L}\mathbf{W}\mathbf{L}^T$ are assumed to be positive definite matrices. From Eqs.

5.4-36, 5.4-37, and 5.4-40, the estimation-error dynamics are expressed by

$$\dot{\boldsymbol{\epsilon}}(t) = (\mathbf{F} - \mathbf{KH})\boldsymbol{\epsilon}(t) + \mathbf{Lw}(t) - \mathbf{Kn}(t) \tag{5.4-44}$$

As $\mathbf{w}(t)$ and $\mathbf{n}(t)$ are random inputs, the stability of Eq. 5.4-44 is governed by the homogeneous equation

$$\dot{\boldsymbol{\epsilon}}(t) = (\mathbf{F} - \mathbf{KH})\boldsymbol{\epsilon}(t) \tag{5.4-45}$$

Stability is proved by examining the characteristics of a Lyapunov function

$$\mathscr{V}[\boldsymbol{\epsilon}(t)] = \boldsymbol{\epsilon}^T(t)\mathscr{I}\boldsymbol{\epsilon}(t) \tag{5.4-46}$$

in which $\mathscr{I}$ is the positive definite *information matrix* (i.e., the inverse of $\mathbf{P}$). From Eq. 2.2-49,

$$\dot{\mathscr{I}}(t) = \dot{\mathbf{P}}^{-1}(t) = -\mathscr{I}(t)\mathbf{P}(t)\mathscr{I}(t) \tag{5.4-47}$$

Substitution of $\mathscr{I}(t)$ and $\dot{\mathscr{I}}(t)$ in the matrix Riccati equation for estimation (Eq. 5.2-51) yields a differential equation for the information matrix

$$-\dot{\mathscr{I}}(t) = \mathscr{I}(t)\mathbf{F} + \mathbf{F}^T\mathscr{I}(t) + \mathscr{I}(t)\mathbf{LWL}^T\mathscr{I}(t) - \mathbf{H}^T\mathbf{N}^{-1}\mathbf{H} \tag{5.4-48}$$

that has the steady-state solution

$$\mathbf{0} = \mathscr{I}\mathbf{F} + \mathbf{F}^T\mathscr{I} + \mathscr{I}\mathbf{LWL}^T\mathscr{I} - \mathbf{H}^T\mathbf{N}^{-1}\mathbf{H} \tag{5.4-49}$$

The time derivative of the Lyapunov function is

$$\begin{aligned}\dot{\mathscr{V}}[\boldsymbol{\epsilon}(t)] &= 2\boldsymbol{\epsilon}^T\mathscr{I}\dot{\boldsymbol{\epsilon}} = 2\boldsymbol{\epsilon}^T\mathscr{I}[\mathbf{F} - \mathbf{KH}]\boldsymbol{\epsilon} \\ &= 2\boldsymbol{\epsilon}^T\mathscr{I}[\mathbf{F} - \mathbf{PH}^T\mathbf{N}^{-1}\mathbf{H}]\boldsymbol{\epsilon} \\ &= \boldsymbol{\epsilon}^T[(\mathscr{I}\mathbf{F} - \mathbf{H}^T\mathbf{N}^{-1}\mathbf{H}) + (\mathscr{I}\mathbf{F} - \mathbf{H}^T\mathbf{N}^{-1}\mathbf{H})^T]\boldsymbol{\epsilon}\end{aligned} \tag{5.4-50}$$

From Eq. 5.4-49, this is equivalent to

$$\dot{\mathscr{V}}[\boldsymbol{\epsilon}(t)] = -\boldsymbol{\epsilon}^T[\mathscr{I}\mathbf{LWL}^T\mathscr{I} + \mathbf{H}^T\mathbf{N}^{-1}\mathbf{H}]\boldsymbol{\epsilon} \tag{5.4-51}$$

As the term in square brackets is positive definite, $\dot{\mathscr{V}}[\boldsymbol{\epsilon}(t)]$ is negative definite at all times, verifying the asymptotic stability of the Kalman–Bucy filter. Hence the estimation error $\boldsymbol{\epsilon}(t)$ approaches zero as time approaches infinity. Because estimation-error dynamic characteristics are determined by $(\mathbf{F} - \mathbf{KH})$, this means that the eigenvalues of $(\mathbf{F} - \mathbf{KH})$ must lie in the left half plane.

The four criteria for asymptotic stability of the LQ regulator have duals for Kalman–Bucy filter stability:

1. $[\mathbf{F}, \mathbf{H}]$ is a *detectable* pair.
2. $[\mathbf{F}, \mathbf{D}]$ is a *stabilizable* pair, where

$$\mathbf{LWL}^T = \mathbf{D}^T\mathbf{D}$$

3. $\mathbf{LWL}^T$ is a positive semidefinite matrix.
4. $\mathbf{N}$ is a positive definite matrix.

In parallel with the earlier results, the estimate errors associated with unobserved, stable modes should approach zero asymptotically, so it is necessary for stability only that unstable modes be completely observable (1). Recasting the estimation problem as an equivalent control problem (A-1), $\mathbf{LWL}^T$ plays the same role as $\mathbf{Q}$ in the cost function, leading to (2) and (3). The need for (4) arises from Eqs. 5.4-38 and 5.4-39, which require finite $\mathbf{N}^{-1}$, and because noise variances are defined to be positive. From the analogy, it is clear that these criteria could be extended to the case of cross-correlated disturbance inputs and measurement noise (Section 4.4).

Example 5.4-3 Initial Condition Response of a Kalman–Bucy Filter for a Second-Order System

Paralleling the weathervane problem of the previous example, we take a look at the initial condition response of an optimal filter for the system described by

$$\begin{bmatrix} \dot{x}_1(t) \\ \dot{x}_2(t) \end{bmatrix} = \begin{bmatrix} 0 & 1 \\ c & d \end{bmatrix} \begin{bmatrix} x_1(t) \\ x_2(t) \end{bmatrix} + \begin{bmatrix} 0 \\ 1 \end{bmatrix} w(t)$$

$$z(t) = \begin{bmatrix} 0 & 1 \end{bmatrix} \begin{bmatrix} x_1(t) \\ x_2(t) \end{bmatrix} + n(t)$$

Only the angular rate is measured, and it drives a Kalman–Bucy filter of the form

$$\begin{bmatrix} \dot{\hat{x}}_1(t) \\ \dot{\hat{x}}_2(t) \end{bmatrix} = \begin{bmatrix} 0 & 1 \\ c & d \end{bmatrix} \begin{bmatrix} \hat{x}_1(t) \\ \hat{x}_2(t) \end{bmatrix} + \begin{bmatrix} k_1 \\ k_2 \end{bmatrix} [z(t) - \hat{x}_2(t)]$$

The filter gains are calculated as

$$\begin{bmatrix} k_1 \\ k_2 \end{bmatrix} = \begin{bmatrix} p_{11} & p_{12} \\ p_{12} & p_{22} \end{bmatrix} \begin{bmatrix} 0 \\ 1 \end{bmatrix} \frac{1}{N} = \begin{bmatrix} p_{12}/N \\ p_{22}/N \end{bmatrix}$$

where the p_{ij} are solutions to the following algebraic Riccati equation:

$$\mathbf{0} = \begin{bmatrix} 0 & 1 \\ c & d \end{bmatrix}\begin{bmatrix} p_{11} & p_{12} \\ p_{12} & p_{22} \end{bmatrix} + \begin{bmatrix} p_{11} & p_{12} \\ p_{12} & p_{22} \end{bmatrix}\begin{bmatrix} 0 & c \\ 1 & d \end{bmatrix} + \begin{bmatrix} 0 & 0 \\ 0 & W \end{bmatrix}$$
$$-\begin{bmatrix} p_{11} & p_{12} \\ p_{12} & p_{22} \end{bmatrix}\begin{bmatrix} 0 \\ 1 \end{bmatrix}\frac{1}{N}[0 \quad 1]\begin{bmatrix} p_{11} & p_{12} \\ p_{12} & p_{22} \end{bmatrix}$$

W and N are scalar spectral densities of the disturbance input specific torque and angular rate measurement error. The criteria for a stable estimator are satisfied, and three equations must be solved for the steady-state estimate covariance elements:

1. $0 = 2p_{12} - p_{12}^2/N$
2. $0 = cp_{11} + dp_{12} + p_{22} - p_{12}p_{22}/N$
3. $0 = 2cp_{12} + 2dp_{22} + W - p_{22}^2/N$

These equations can be solved in the sequence (1), (3), (2):

1. $p_{12} = 2N$ or 0
3. $p_{22} = N[d \pm \sqrt{d^2 + (2cp_{12} + W)/N}]$
2. $p_{11} = [p_{12}p_{22}/N - dp_{12} - p_{22}]/c$

such that $\mathbf{P}$ is positive definite.

With $N = 1$, the effects of varying W on the covariance elements for the nominal weathervane of Example 5.4-2 ($c = -39.4$, $d = -3.8$) are summarized in Table 5.4-3. It is apparent that the measurement acts directly on the angular rate estimate only, and the angular position estimate derives entirely from the modeled integration of angular rate.

Diagram 1 compares the estimates for the highest three disturbance spectral densities with the actual state components. The illustration confirms the stability of the estimation error, $\boldsymbol{\epsilon}(t)$, as all three estimates converge to the actual state. For $W = 100$, the filter is well tuned to the initial condition response. Noting the differences between estimates and

TABLE 5.4-3 Disturbance Spectral Density Effect on Estimate Covariance

W	p_{11}	p_{12}	p_{22}
1	0.003	0.000	0.129
10	0.029	0.000	1.144
100	0.175	0.000	6.898
1000	0.712	0.000	28.050

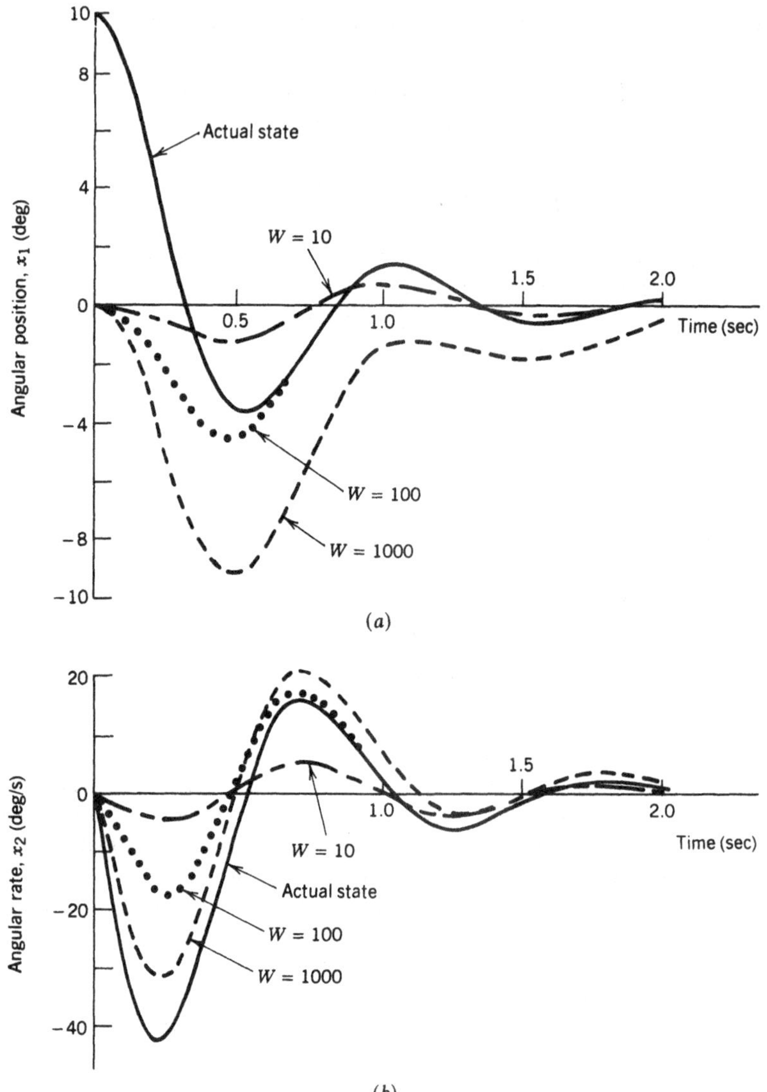

Diagram 1. (*a*) Angular position and estimates versus time; (*b*) angular rate and estimates versus time.

actual values, it is clear that the estimation error is oscillatory for $W = 10$ and 100, and it is overdamped for $W = 1000$. Although the rate estimate improves as W increases, the position estimate gets worse; the roots (or eigenvalues) of the "closed-loop" filter provide an explanation (Table 5.4-4). The damping ratio and natural frequency of the oscillatory modes are denoted by ζ and ω_n, while the real roots of overdamped modes are

TABLE 5.4-4 Roots of the Kalman–Bucy Filter

W	ζ, –	ω_n, rad/s	λ_1, rad/s	λ_2, rad/s
1	0.312	6.28	—	—
10	0.394	6.28	—	—
100	0.852	6.28	—	—
1000	—	—	−1.289	−30.56

denoted by λ_1 and λ_2. With lower values of W, the oscillatory dynamics cause both x_1 and x_2 to converge at the same rate. For $W = 1000$, there are two real roots; the convergence of x_1 is dominated by λ_1, while that of x_2 is dominated by λ_2.

If the dynamic system is unstable, the filter can do nothing to stabilize the actual state, but the estimate error is stable (i.e., the estimate approaches

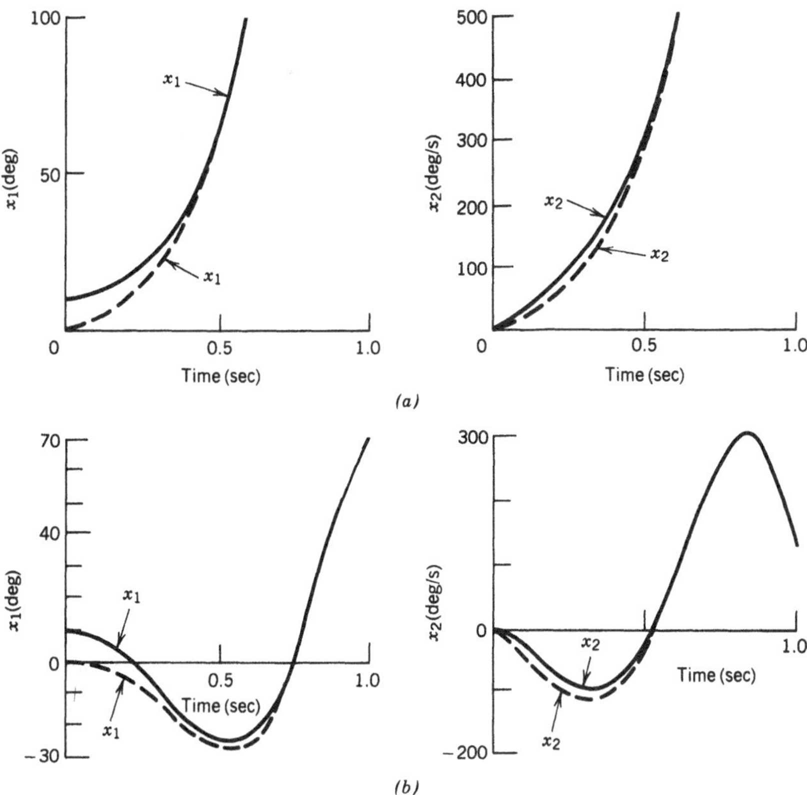

Diagram 2. (*a*) State and estimate with $c = 39.4$, $d = -3.8$, $W = 100$, $N = 1$; (*b*) state and estimate with $c = -39.4$, $d = 3.8$, $W = 100$, $N = 1$.

TABLE 5.4-5 Covariance and Roots of Filters for Unstable Systems $W = 100$, $N = 1$

c	d	p_{11}	p_{12}	p_{22}	$\lambda_1(\zeta)$	$\lambda_2(\omega_n)$
39.4	−3.8	0.515	2.	12.964	−2.898	−13.596
−39.4	3.8	0.368	0.	14.498	(0.852)	(6.28)
39.4	3.8	0.322	2.	20.294	−2.898	−13.596

the divergence of the actual state). This is illustrated in Diagram 2 for static ($c = 39.4$) and dynamic ($d = 3.8$) instabilities. In both cases, the estimates quickly follow the state. The corresponding covariances and roots are shown in Table 5.4-5. Several interesting results are revealed. The filter roots are the same for the first and third cases, and they are overdamped. In the second case, p_{12} and k_1 are zero, indicating that direct update of x_1 occurs only when there is static instability.

Asymptotic Stability of the Stochastic Regulator

The continuous-time stochastic regulator is a closed-loop system consisting of the constant-gain Kalman–Bucy filter and the constant-gain linear-

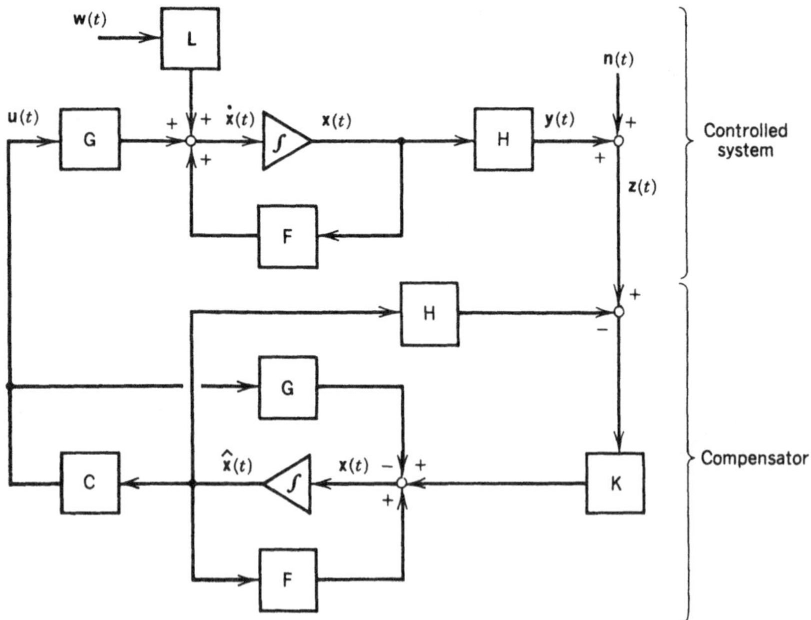

FIGURE 5.4-2 Block diagram of the stochastic regulator.

quadratic regulator, arranged as in Fig. 5.4-2. The operable equations are

$$\dot{\mathbf{x}}(t) = \mathbf{F}\mathbf{x}(t) + \mathbf{G}\mathbf{u}(t) + \mathbf{L}\mathbf{w}(t) \tag{5.4-52}$$

$$\mathbf{z}(t) = \mathbf{H}\mathbf{x}(t) + \mathbf{n}(t) \tag{5.4-53}$$

$$\dot{\hat{\mathbf{x}}}(t) = \mathbf{F}\hat{\mathbf{x}}(t) + \mathbf{G}\mathbf{u}(t) + \mathbf{K}[\mathbf{z}(t) - \mathbf{H}\hat{\mathbf{x}}(t)] \tag{5.4-54}$$

$$\mathbf{u}(t) = -\mathbf{C}\hat{\mathbf{x}}(t) \tag{5.4-55}$$

with

$$\mathbf{K} = \mathbf{P}\mathbf{H}^T\mathbf{N}^{-1} \tag{5.4-56}$$

$$\mathbf{C} = \mathbf{R}^{-1}\mathbf{G}^T\mathbf{S} \tag{5.4-57}$$

and **S** and **P** defined by Eqs. 5.4-16 and 5.4-39. The stability of the closed-loop system is most readily described in terms of the combined state and estimation error dynamics. Using Eqs. 5.4-36 and 5.4-55 the closed-loop state equation is

$$\begin{aligned}\dot{\mathbf{x}}(t) &= \mathbf{F}\mathbf{x}(t) - \mathbf{G}\mathbf{C}\hat{\mathbf{x}}(t) + \mathbf{L}\mathbf{w}(t) \\ &= (\mathbf{F} - \mathbf{G}\mathbf{C})\mathbf{x}(t) + \mathbf{G}\mathbf{C}\boldsymbol{\epsilon}(t) + \mathbf{L}\mathbf{w}(t)\end{aligned} \tag{5.4-58}$$

so the total system equation can be expressed as

$$\begin{bmatrix}\dot{\mathbf{x}}(t) \\ \dot{\boldsymbol{\epsilon}}(t)\end{bmatrix} = \begin{bmatrix}(\mathbf{F} - \mathbf{G}\mathbf{C}) & \mathbf{G}\mathbf{C} \\ \mathbf{0} & (\mathbf{F} - \mathbf{K}\mathbf{H})\end{bmatrix}\begin{bmatrix}\mathbf{x}(t) \\ \boldsymbol{\epsilon}(t)\end{bmatrix} + \begin{bmatrix}\mathbf{L} & \mathbf{0} \\ \mathbf{L} & \mathbf{K}\end{bmatrix}\begin{bmatrix}\mathbf{w}(t) \\ \mathbf{n}(t)\end{bmatrix} \tag{5.4-59a}$$

or

$$\dot{\mathbf{x}}'(t) = \mathbf{F}'\mathbf{x}'(t) + \mathbf{L}'\mathbf{w}'(t) \tag{5.4-59b}$$

Stability is determined by the characteristics of $\mathbf{F}'$, which is seen to be an upper-block-triangular matrix. There is one-way coupling from $\boldsymbol{\epsilon}(t)$ to $\mathbf{x}(t)$ but not from $\mathbf{x}(t)$ to $\boldsymbol{\epsilon}(t)$. Consequently, stability of the total system is described by the separately evaluated stability of the LQ regulator and the Kalman–Bucy filter. Because $\mathbf{F}'$ is an upper-block-triangular matrix, its eigenvalues are those of the diagonal blocks, that is, of $(\mathbf{F} - \mathbf{G}\mathbf{C})$ and $(\mathbf{F} - \mathbf{K}\mathbf{H})$. Since both have been proven stable, all their eigenvalues have negative real parts; hence the eigenvalues of $\mathbf{F}'$ have negative real parts, and the stochastic regulator is stable.

Example 5.4-4 Initial Condition Response of a Stochastic Regulator for a Second-Order System

A stochastic regulator for the weathervane problem is formed by feeding back the estimates of Example 5.4-3 to the regulator of Example 5.4-2

(Eqs. 5.4-52 to -57). Response to a 10° initial condition on angular position, x_1, is shown in the diagram for three cases: (a) the nominal case ($c = -39.4$, $d = -3.8$), (b) the statically unstable case ($c = 39.4$), and (c) the dynamically unstable case ($d = 3.8$). In all cases, $q_{11} = q_{22} = T = N = 1$, and $W = 100$; and the closed-loop roots can be deduced from earlier examples. The stochastic regulator is a coupled fourth-order system, with two roots deriving from the physical system and two from the Kalman–Bucy filter.

The closed-loop system is stable throughout, although there are differing degrees of oscillation and overshoot as the position and rate are regulated to zero. Case (a) is very nearly open-loop, as feedback gains are small (Example 5.4-2). The response of Case (b) initially is divergent; however,

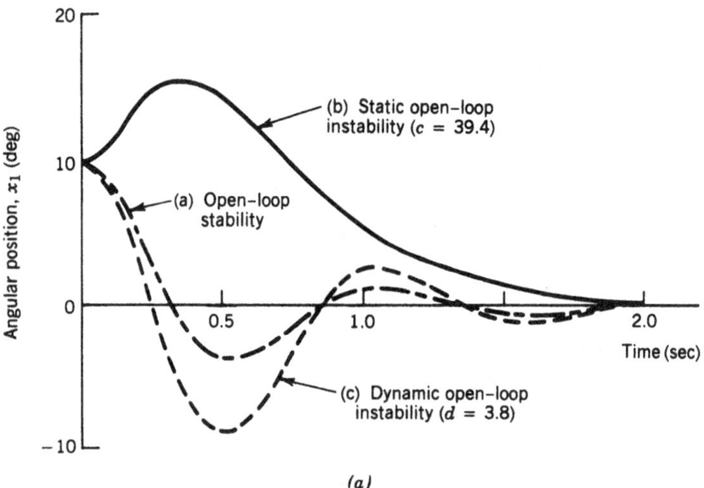

(a)

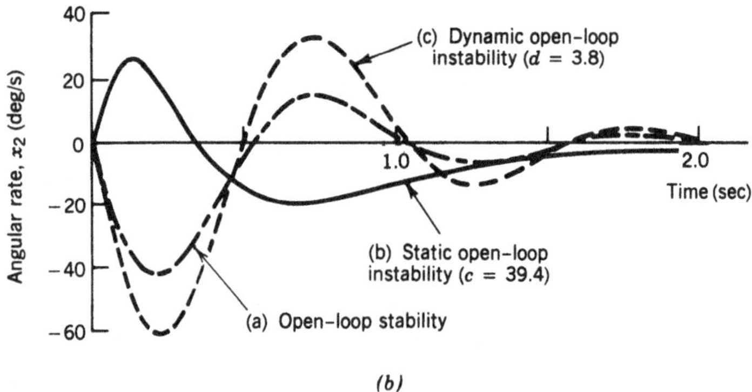

(b)

(*a*) Angular position versus time; (*b*) angular rate versus time.

the control produces a heavily damped response for times beyond 0.4 s, roughly corresponding to the time at which the state-estimate error has become negligible and the control has become effective. The ratio of successive peak amplitudes for the decaying oscillation of Case (c) is quite similar to the "logarithmic decrement" of Case (a), reflecting the similarity in closed-loop roots.

Steady-State Performance of the Stochastic Regulator

As the final time for control approaches infinity, the effect of initial conditions becomes negligible, and the *average cost* becomes the appropriate measure of performance. With perfect measurements, the cost function for the linear, time-invariant system is given by Eq. 5.4-8, and the minimized average cost is

$$J_A^* = \lim_{t_f \to \infty} \left(\frac{1}{2t_f}\right) \mathrm{Tr}\left[\mathbf{x}^T(0)\mathbf{S}\mathbf{x}(0) + \int_0^{t_f} \mathbf{SLWL}^T\, dt\right]$$

$$= \frac{\mathrm{Tr}(\mathbf{SLWL}^T)t_f}{2t_f} = \tfrac{1}{2}\,\mathrm{Tr}(\mathbf{SLWL}^T) \tag{5.4-60}$$

The average cost is proportional to the spectral density matrix of the disturbance input **W** and to the steady-state solution of the matrix Riccati equation for control **S**.

With imperfect measurements, the cost function is expressed by Eq. 5.3-27. With zero weight on the terminal error, the average cost is

$$J_A^* = \lim_{t_f \to \infty} \left(\frac{1}{2t_f}\right) \mathrm{Tr}\left\{\mathbf{S}[\mathbf{x}(0)\mathbf{x}(0)^T] + \int_0^{t_f} \mathbf{QP}\, dt + \int_0^{t_f} \mathbf{SKN'K}^T\, dt\right\}$$

$$= \tfrac{1}{2}\,\mathrm{Tr}(\mathbf{QP} + \mathbf{SKN'K}^T) \tag{5.4-61}$$

The effect of the random disturbance input is contained in the steady-state solution for state-estimate-error covariance **P**. The noise spectral density matrix **N** affects **P**, and it has direct effect on the average cost as well.

EXAMPLE 5.4-5 RESPONSE OF THE STOCHASTIC REGULATOR TO RANDOM INPUTS AND MEASUREMENT NOISE FOR A SECOND-ORDER SYSTEM

The weathervane problem of the previous examples is considered again, this time with random processes driving the system and corrupting the measurements of the stochastic regulator. With perfect measurements, the average cost J_A would be given by Eq. 5.4–60, while with imperfect measurements, it would be given by Eq. 5.4-61. J_A is tabulated for Cases

TABLE 5.4-6 Average Cost for the Stochastic Regulator of a Second-Order System

Case	Perfect Measurements J_A	Imperfect Measurements J_A
a	6.63	6.69
b	467.77	3403.24
c	386.63	820.06

(a), (b), and (c) of Example 5.4-4 in Table 5.4-6, assuming $q_{11} = q_{22} = 1$ and $W = 100$; with imperfect measurements, $N = 1$.

The open-loop-stable system [Case (a)] requires little regulation, and its cost is largely due to the disturbance input alone. Regulation stabilizes Cases (b) and (c), but it is achieved at considerably higher cost, both with and without measurement error.

Typical time histories for stochastic regulators with perfect and imperfect measurements are presented in Diagrams 1 and 2. The same profiles of $w(t)$ and $n(t)$ are used in all three cases, and while none of the systems have reached the steady state (oscillatory amplitudes are still growing in the interval), the relative trends explain the results of Table 5.4-6. Because q_{11},

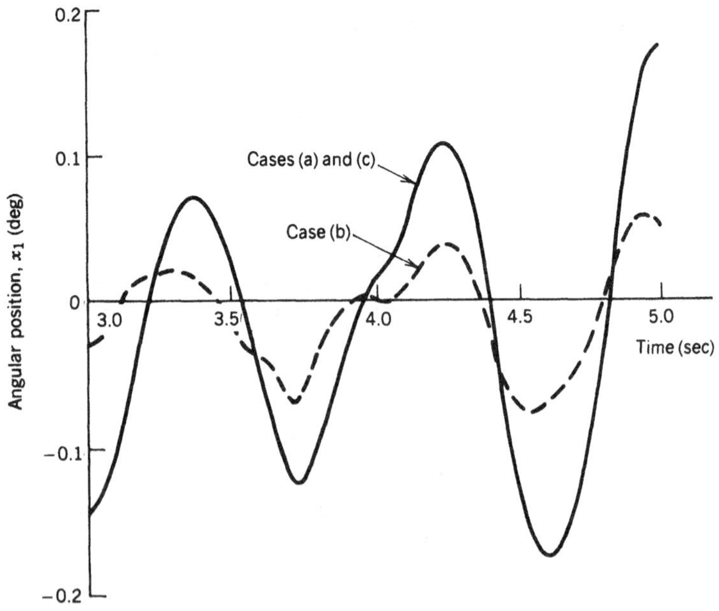

(a)

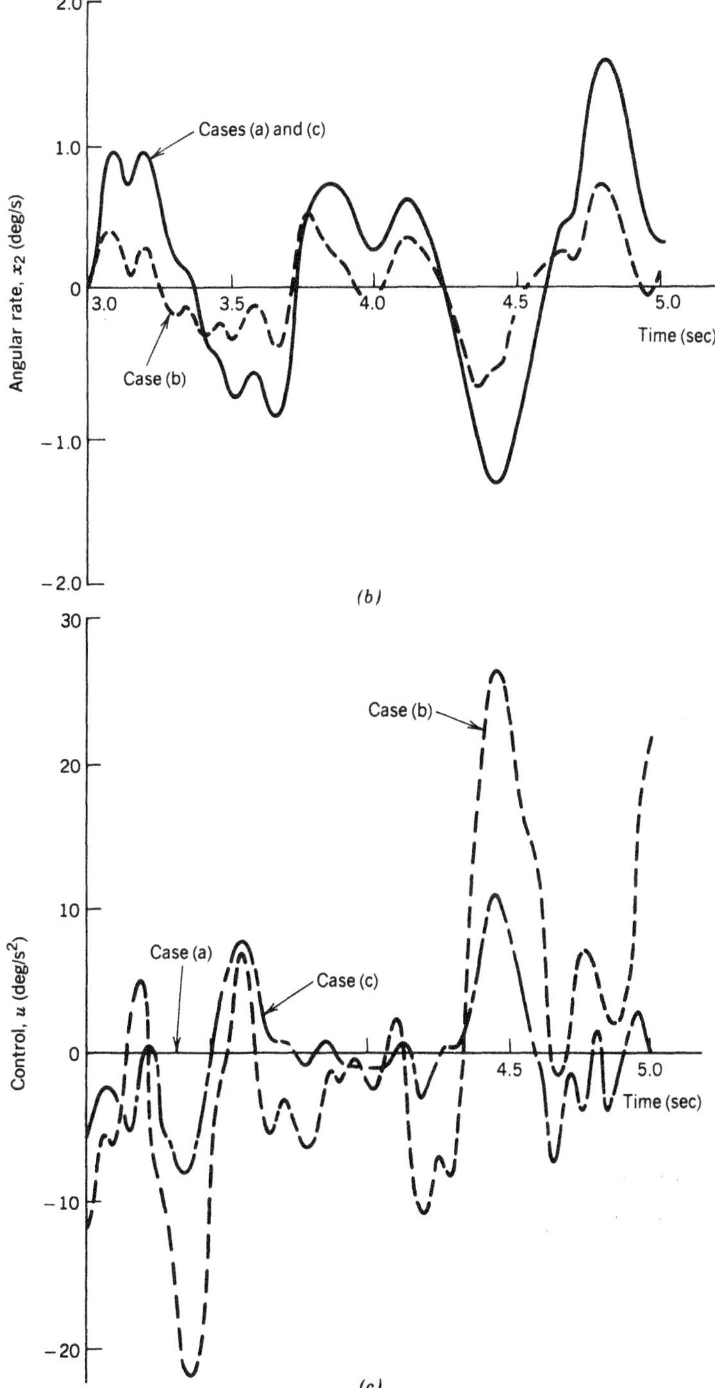

Diagram 1. (a) Angular position versus time; (b) angular rate versus time; (c) control versus time.

q_{22}, and r are identical, the individual cost functions are dominated by perturbations of the numerically largest variable. Case (a) uses little control; hence its cost is defined primarily by angular rate perturbations. With perfect measurements, Cases (a) and (c) have virtually identical responses (Diagram 1); however, the latter uses a great deal of control, so its cost is considerably larger. The state perturbations of Case (b) are lower than those of the other two cases, but its greater use of control also is evident in both the time history and J_A.

Imperfect measurements have minimal effects on the Case (a) response, while the Case (c) response is no longer identical to it (Diagram 2). Control

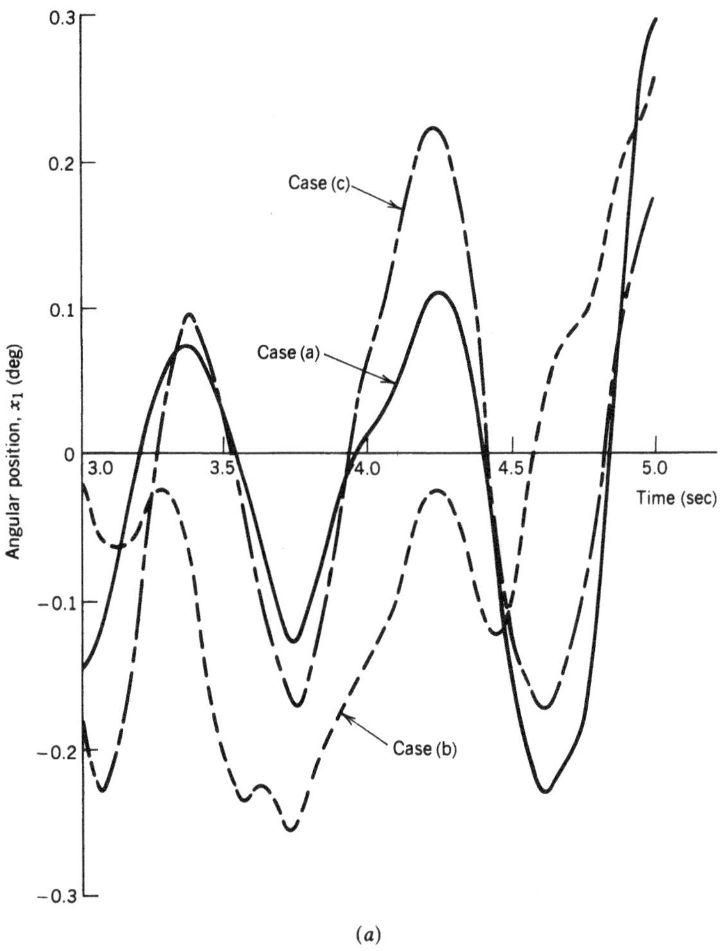

(a)

Diagram 2. (a) Angular position versus time.

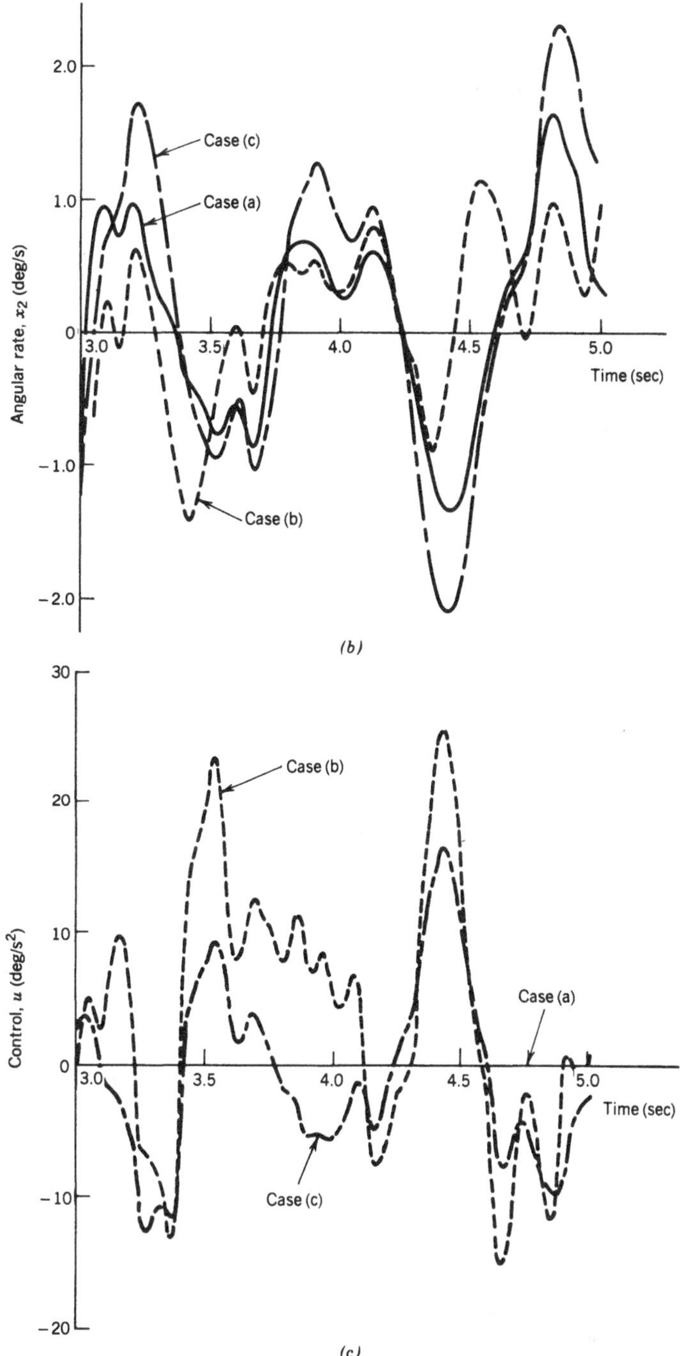

Diagram 2. (*b*) Angular rate versus time; (*c*) control versus time.

activity is increased materially for Cases (b) and (c), and their state perturbations are larger as well.

The Discrete-Time Case

The principal stability and steady-state performance results for continuous-time optimal regulators and filters apply to the discrete-time case with little modification. Recalling the notation used previously, the linear-quadratic regulator is guaranteed to be stable if:

1. $[(\mathbf{\Phi}-\mathbf{\Gamma R}^{-1}\mathbf{M}^T), \mathbf{\Gamma}]$ is a stabilizable pair.
2. $[(\mathbf{\Phi}-\mathbf{MR}^{-1}\mathbf{M}^T), \mathbf{D}]$ is a detectable pair, where

$$\mathbf{Q}-\mathbf{MR}^{-1}\mathbf{M}^T=\mathbf{D}^T\mathbf{D}$$

3. $(\mathbf{Q}-\mathbf{MR}^{-1}\mathbf{M}^T)$ is a positive semidefinite matrix.
4. $(\mathbf{R}+\mathbf{\Gamma}^T\mathbf{S\Gamma})$ is a positive definite matrix.

The Kalman filter for uncorrelated disturbance input and measurement noise produces a stable estimate error if:

1. $(\mathbf{\Phi}, \mathbf{H})$ is a detectable pair.
2. $(\mathbf{\Phi}, \mathbf{D})$ is a stabilizable pair, where

$$\mathbf{\Lambda W \Lambda}^T=\mathbf{D}^T\mathbf{D}$$

3. $\mathbf{\Lambda W \Lambda}^T$ is a positive semidefinite matrix.
4. $[\mathbf{N}+\mathbf{HP}(-)\mathbf{H}^T]$ is a positive definite matrix.

From Eq. 5.1-28 and with $\mathbf{S}_f = \mathbf{0}$, the average cost for perfect measurements is

$$\begin{aligned} J_A &= \lim_{k_f\to\infty}\left(\frac{1}{2k_f}\right)\left[\mathbf{x}_0^T\mathbf{S}_0\mathbf{x}_0+\sum_{k=0}^{k_f-1}\mathrm{Tr}(\mathbf{S\Lambda W\Lambda}^T)\right] \\ &= \tfrac{1}{2}\mathrm{Tr}(\mathbf{S\Lambda W\Lambda}^T) \end{aligned}$$

while from Eq. 5.3-65, the average cost with imperfect measurements is

$$\begin{aligned} J_A &= \lim_{k_f\to\infty}\frac{1}{2k_f}\left\{\mathbf{x}_0^T\mathbf{S}_0\mathbf{x}_0+\sum_{k=0}^{k_f-1}\mathrm{Tr}\{\mathbf{SK}[\mathbf{N}+\mathbf{HP}(-)\mathbf{H}^T]\mathbf{K}^T+\mathbf{QP}(+)\}\right\} \\ &= \tfrac{1}{2}\mathrm{Tr}\{\mathbf{SK}[\mathbf{N}+\mathbf{HP}(-)\mathbf{H}^T]\mathbf{K}^T+\mathbf{QP}(+)\} \end{aligned} \qquad (5.4\text{-}63)$$

PROBLEMS

Section 5.1

1. A first-order nonlinear system is described as

$$x_{k+1} = -10x_k^3 + u_k + w_k, \qquad x_0 = 1$$

where w_k is a Gaussian, white-noise process with zero mean. The following cost function is to be minimized using the principle of optimality:

$$J = E\left\{10x_4^2 + \sum_{k=0}^{3} [x_k^2 + u_k^2]\right\}$$

(a) What is the minimizing control history, state history, and cost, assuming that $E(w_k^2) = 0$?
(b) Repeat with $E(w_k^2) = 0.1$.

2. Express the stochastic principle of optimality for the high-temperature oven problem, in which the temperature can be described by

$$\dot{x}(t) = -k_1 x(t) - k_2 x^4(t) + k_3 u(t) + w(t)$$

with the quadratic stochastic value function,

$$V^*(t_1) = E\left\{q_1[x^*(t_f) - x_D]^2 - \int_0^{t_1} [q_2 x^{*2}(t) + r u^{*2}(t)]\, dt\right\}$$

3. Define a neighboring-optimal control solution for the rocket problem (1) of Section 3.6 (initially described in Problem 2 of Section 3.3).
(a) Formulate the equations that define the rocket's control gain matrix.
(b) Compute the control gain matrix as a function of time in (t_0, t_f).
(c) Wind gusts perturb the rocket's trajectory by causing random lift forces, so the flight path angle equation takes the form

$$\dot{\gamma}(t) = [1/mv(t)][T \sin\alpha - g\cos\gamma] + w(t)$$

where $w(t)$ represents the varying lift effect. Assume that $E(w) = 0$ and $E[w(t)w(\tau)] = 0.0001\,\delta(t-\tau)$ $(\text{rad/s})^2$, and compute an example of a neighboring-optimal trajectory. How does this compare with the optimal deterministic trajectory?

4. Evaluate the optimized cost function for the following second-order,

linear-quadratic problem:

$$\begin{bmatrix}\dot{x}_1(t)\\ \dot{x}_2(t)\end{bmatrix}=\begin{bmatrix}0 & 1\\ -9 & -3\end{bmatrix}\begin{bmatrix}x_1(t)\\ x_2(t)\end{bmatrix}+\begin{bmatrix}0\\ 1\end{bmatrix}u(t)+\begin{bmatrix}0\\ 1\end{bmatrix}w(t)$$

$$J=E\left\{10x_1^2(10)+\int_0^{10}[x_1^2(t)+x_2^3(t)+u^2(t)]\,dt\right\}$$

$$E[\mathbf{x}(0)\mathbf{x}^T(0)]=\begin{bmatrix}1 & 0\\ 0 & 1\end{bmatrix}$$

$$E[w(t)w(\tau)]=\delta(t-\tau)$$

Section 5.2

1. Outline the control requirements for a robot that has five degrees of freedom—for example, three linear positions and two angular orientations of the end effector. Postulate alternative information sets that could be used in stochastic optimal control. Express a dual-control algorithm that would allow the robot to operate with uncertain loads, disturbances, and measurement errors. How might the control algorithm be simplified by ad hoc probing inputs?

2. Formulate a linear-quadratic-Gaussian neighboring-optimal controller for the following problem:

$$J=E\left\{10x_1^4(t_f)+\int_{t_0}^{t_f}[5x_1^2(t)u^2(t)+x_2^2(t)]\,dt\right\}$$

$$\begin{bmatrix}\dot{x}_1(t)\\ \dot{x}_2(t)\end{bmatrix}=\begin{bmatrix}x_1(t)x_2(t)+w_1(t)\\ -x_2(t)+u(t)+w_2(t)\end{bmatrix}\qquad \mathbf{x}(0),\ \mathbf{P}(0)\text{ given}$$

$$z(t)=x_1(t)x_2(t)+x_1(t)+n(t)$$

$$\mathbf{W}(t)=\begin{bmatrix}w_{11} & 0\\ 0 & w_{22}\end{bmatrix}\delta(t-\tau)$$

$$N(t)=n(t)\,\delta(t-\tau)$$

3. The measurements that are necessary to implement the neighboring-optimal control for a rocket in Problem 3 of Section 5.1 include velocity, flight path angle, and altitude. Assuming that the spectral density matrix of measurement errors is given by

$$\mathbf{N}=\text{Diag}[10,\quad 0.001,\quad 100]$$

design a suitable neighboring-optimal controller, and compute an example of a neighboring-optimal trajectory.

Section 5.3

1. Evaluate the cost function for the following continuous-time linear-quadratic problem with zero-mean uncertain disturbances and

measurement errors:

$$\begin{bmatrix}\dot{x}_1(t)\\ \dot{x}_2(t)\end{bmatrix}=\begin{bmatrix}0 & 1\\ -9 & -3\end{bmatrix}\begin{bmatrix}x_1(t)\\ x_2(t)\end{bmatrix}+\begin{bmatrix}0\\ 1\end{bmatrix}u(t)+\begin{bmatrix}0\\ 1\end{bmatrix}w(t)$$

$$J=E\left\{10x_1^2(10)+\int_0^{10}[x_1^2(t)+x_2^3(t)+u^2(t)]\,dt\right\}$$

$$E[\mathbf{x}(0)\mathbf{x}^T(0)]=\begin{bmatrix}1 & 0\\ 0 & 1\end{bmatrix}$$

$$E[w(t)w(\tau)]=\delta(t-\tau)$$

(a)
$$z(t)=[1 \quad 0]\begin{bmatrix}x_1(t)\\ x_2(t)\end{bmatrix}+n(t)$$

$$E[n(t)n(\tau)]=0.1\,\delta(t-\tau)$$

(b) $E[n(t)n(\tau)]=\delta(t-\tau)$

(c)
$$z(t)=[1 \quad 1]\begin{bmatrix}x_1(t)\\ x_2(t)\end{bmatrix}+n(t)$$

$$E[n(t)n(\tau)]=\delta(t-\tau)$$

2. Evaluate the cost function for the following discrete-time linear-quadratic problem with zero-mean uncertain disturbances and measurement errors:

$$\begin{bmatrix}x_1\\ x_2\\ x_3\end{bmatrix}_k=\begin{bmatrix}1 & 1 & 0\\ -1 & 0.4 & 1\\ 0 & 0 & 0.2\end{bmatrix}\begin{bmatrix}x_1\\ x_2\\ x_3\end{bmatrix}_{k-1}+\begin{bmatrix}0\\ 0\\ 0.8\end{bmatrix}u_{k-1}+\begin{bmatrix}0\\ 0.6\\ 0\end{bmatrix}w_{k-1}$$

$$J=E\left\{10(x_1^2+x_2^2)_{100}+\sum_{k=1}^{99}(x_1^2+x_2^2+u^2)_k\right\}$$

$$E[\mathbf{x}(0)\mathbf{x}^T(0)]=\begin{bmatrix}1 & 0 & 0\\ 0 & 1 & 0\\ 0 & 0 & 1\end{bmatrix},\qquad E(w_k^2)=1$$

$$z_k=[1 \quad 0 \quad 0]\begin{bmatrix}x_1\\ x_2\\ x_3\end{bmatrix}+n_k,\qquad E(n_k^2)=0.1$$

3. Show that the certainty-equivalence property applies to linear, time-invariant systems with white, Gaussian disturbances and colored Gaussian measurement error.

4. Show that the certainty-equivalence property applies to linear, time-

invariant systems with white, non-Gaussian disturbances and white, Gaussian measurement error.

Section 5.4

1. Given the following system and cost function:

$$\begin{bmatrix} \dot{x}_1(t) \\ \dot{x}_2(t) \\ x_3(t) \end{bmatrix} = \begin{bmatrix} 1 & 0 & 0 \\ 0 & 2 & 1 \\ 0 & 1 & 3 \end{bmatrix} \begin{bmatrix} x_1(t) \\ x_2(t) \\ x_3(t) \end{bmatrix} + \begin{bmatrix} 1 \\ 0 \\ 1 \end{bmatrix} u$$

$$J = \frac{1}{2} \int_0^{\infty} [x_1^2(t) + x_3^2(t) + u^2(t)]\, dt$$

   (a) Can a linear-quadratic regulator stabilize this system?
   (b) What are the optimal gains and closed-loop eigenvalues?
   (c) How could the cost function be changed to reverse your answer to (a)?

2. A first-order system

$$\dot{x}(t) = -\,x(t) + u(t)$$

is to be controlled so that the following cost function is minimized:

$$J = \frac{1}{2} \int_0^{\infty} [x^2(t) + 2mx(t)u(t) + u^2(t)]\, dt$$

For what range of m is the closed-loop system stable?

3. Given the third-order system with two-component measurement:

$$\begin{bmatrix} \dot{x}_1(t) \\ \dot{x}_2(t) \\ \dot{x}_3(t) \end{bmatrix} + \begin{bmatrix} 1 & 0 & 0 \\ 0 & 2 & 1 \\ 0 & 1 & 3 \end{bmatrix} \begin{bmatrix} x_1(t) \\ x_2(t) \\ x_3(t) \end{bmatrix} + \begin{bmatrix} w_1(t) \\ w_2(t) \\ w_3(t) \end{bmatrix}$$

$$\begin{bmatrix} z_1(t) \\ z_2(t) \end{bmatrix} = \begin{bmatrix} 1 & 0 & 0 \\ 0 & 0 & 1 \end{bmatrix} \begin{bmatrix} x_1(t) \\ x_2(t) \\ x_3(t) \end{bmatrix} + \begin{bmatrix} n_1(t) \\ n_2(t) \end{bmatrix}$$

$$\mathbf{W} = \begin{bmatrix} 1 & 0 & 0 \\ 0 & 1 & 0 \\ 0 & 0 & 1 \end{bmatrix}, \qquad \mathbf{N} = \begin{bmatrix} 1 & 0 \\ 0 & 1 \end{bmatrix}$$

   (a) Can the Kalman–Bucy filter produce a stable estimate of $\mathbf{x}(t)$?
   (b) What are the optimal gains and closed-loop eigenvalues?
   (c) What modification to the measurement vector would reverse your answer to (a)?

4. A robot manipulator with a single degree of freedom is to be controlled

by feeding back a position measurement only. The manipulator is subject to zero-mean disturbance forces, and the position sensor contains zero-mean encoder error, leading to the following system model:

$$\begin{bmatrix} \dot{x}_1(t) \\ \dot{x}_2(t) \end{bmatrix} = \begin{bmatrix} 0 & 1 \\ -100 & -7 \end{bmatrix} \begin{bmatrix} x_1(t) \\ x_2(t) \end{bmatrix} + \begin{bmatrix} 0 \\ 100 \end{bmatrix} u(t) + \begin{bmatrix} 0 \\ 100 \end{bmatrix} w(t)$$

$$z(t) = x_1(t) + n(t), \qquad W = 1, N = 0.1$$

The control cost function to be minimized is

$$J = E\left\{\frac{1}{2}\int_0^\infty [x_1^2(t) + x_2^2(t) + u^2(t)]\, dt\right\}$$

(a) What is the steady state cost for the stochastic regulator?
(b) Repeat (a) with $W = 10$.
(c) Repeat (a) with $N = 1$.

5. The system in Problem 1 is to be controlled to minimize a quadratic cost function with the following weighting matrices:

$$\mathbf{Q} = \begin{bmatrix} 1 & 0 & 0 \\ 0 & 1 & 0 \\ 0 & 0 & 1 \end{bmatrix} \quad \mathbf{M} = \begin{bmatrix} 1 \\ 0 \\ 0 \end{bmatrix} \quad \mathbf{R} = 1$$

Find the transformations that define a similar minimization problem without the cross-weighting term **M**.

REFERENCES

A-1 Anderson, B.D.O. and Moore, J. B., *Linear Optimal Control*, Prentice-Hall, Englewood Cliffs, NJ, 1971.

A-2 Aoki, M., *Optimization of Stochastic Systems*, Academic Press, New York, 1967.

A-3 Åström, K. J., *Introduction to Stochastic Control Theory*, Academic Press, New York, 1970.

A-4 Athans, M., The Role and Use of the Stochastic Linear-Quadratic-Gaussian Problem in Control System Design, *IEEE Transactions on Automatic Control*, **AC-16**(6) 529–552, December 1971.

B-1 Bar-Shalom, Y., Stochastic Dynamic Programming: Caution and Probing, *IEEE Transactions on Automatic Control*, **AC-26**(5) 1184–1195, October 1981.

B-2 Bar-Shalom, Y. and Tse, E., Dual Effect, Certainty Equivalence, and Separation in Stochastic Control, *IEEE Transactions on Automatic Control*, **AC-19**(5), 494–500, October 1974.

B-3 Bar-Shalom, Y. and Tse, E., Caution, Probing, and the Value of Information in the Control of Uncertain Systems, *Annals of Economic and Social Measurement*, **5**(2), 323–337, Summer 1976.

B-4 Boland, J. S., Douglass, W. B., Dwivedi, N. P., and Hopkins, W. G., Filtering and Optimization of Linear Stochastic Systems with Cross-Correlated Noise, Auburn University Department of Electrical Engineering Report 20, 1969.

B-5 Boudarel, R., Delmas, J., and Guichet, P., *Dynamic Programming and Its Application to Optimal Control*, Academic Press, New York, 1971.

B-6 Bryson, A. E., Jr. and Ho, Y. C., *Applied Optimal Control*, Hemisphere Publishing Co., Washington, D.C., 1975.

C-1 Casler, R. J., Jr., Dual-Control Guidance Strategy for Homing Interceptors Taking Angle-Only Measurements, *AIAA Journal of Guidance and Control*, **1**(1), 63–70, January–February 1978.

D-1 Davis, M. H. A., *Linear Estimation and Stochastic Control*, Chapman and Hall, London, 1977.

D-2 Davis, H. T., *Introduction to Nonlinear Differential and Integral Equations*, Dover, New York, 1962.

D-3 Dreyfus, S. E., *Dynamic Programming and the Calculus of Variations*, Academic Press, New York, 1965.

F-1 Fel'dbaum, A. A., *Optimal Control Systems*, Academic Press, New York, 1965.

F-2 Fleming, W. H. and Rischel, R. W., *Deterministic and Stochastic Optimal Control*, Springer-Verlag, New York, 1975.

F-3 Florentin, J. J., Optimal Control of Continuous Time, Markov, Stochastic Systems, *Journal of Electronics and Control*, **10**, 473–488, 1961.

F-4 Friedland, B., On Solutions of the Riccati Equation in Optimization Problems, *IEEE Transactions on Automatic Control*, **AC-12**(3), 303–304, June 1967.

G-1 Goodwin, G. C. and Sin, K. S., *Adaptive Prediction and Control*, Prentice-Hall, Englewood Cliffs, NJ, 1984.

G-2 Gunckel, T. F. and Franklin, G. F., A General Solution for Linear Sampled-Data Control Systems, *Transactions of the ASME, Journal of Basic Engineering*, **85D**, 197, 1963.

G-3 Gustafson, D. E. and Speyer, J. L., Design of Linear Regulators for Nonlinear Stochastic Systems, *AIAA Journal of Spacecraft and Rockets*, **12**(6), 351–358, June 1975.

H-1 Hijab, O. B., The Adaptive LQG Problem-Part I, *IEEE Transactions on Automatic Control*, **AC-28**(2), 171–178, February 1983.

J-1 Jacobson, D. H., Martin, D. H., Pachter, M., and Geveci, T., *Extensions of Linear-Quadratic Control Theory*, Springer-Verlag, New York, 1980.

J-2 Johnson, G. W., A Deterministic Theory of Estimation and Control, *IEEE Transactions on Automatic Control*, **AC-14**(4), 380–384, August 1969.

J-3 Jameson, A. and Kreindler, E., Inverse Problem of Linear Optimal Control, *SIAM Journal on Control*, **11**(1), 1–19, February 1973.

J-4 Joseph, P. D. and Tou, J. T., On Linear Control Theory, *Transactions of the AIEE*, Part III, **80**(18), 193–196, September 1961.

K-1 Kailath, T. and Ljung, L., The Asymptotic Behavior of Constant-Coefficient Riccati Differential Equations, *IEEE Transactions on Automatic Control*, **AC-21**(3), 385–388, June 1976.

K-2 Kalman, R. E., Contributions to the Theory of Optimal Control, *Boletin de la Sociedad Matematica Mexicana*, **5**(1), 102–119, 1960.

K-3 Kalman, R. E., When is a Linear System Optimal? *Transactions of the ASME, Journal of Basic Engineering*, **86**, 51–60, March 1964.

K-4 Kaplan, W., *Advanced Mathematics for Engineers*, Addison-Wesley, Reading, MA, 1981.

K-5 Kramer, L. C. and Athans, M., On the Application of Deterministic Optimization

Methods to Stochastic Control Problems, *IEEE Transactions on Automatic Control*, **AC-19**(1), 22–30, February 1974.

K-6 Kwakernaak, H. and Sivan, R., *Linear Optimal Control Systems*, Wiley, New York, 1972.

L-1 Larson, R. E. and Casti, J. L., *Principles of Dynamic Programming*, Part 2, Marcel Dekker, New York, 1982.

M-1 Maybeck, P. S., *Stochastic Models, Estimation, and Control*, Vol. 3, Academic Press, New York, 1982.

M-2 McLane, P. J., Optimal Stochastic Control of Linear Systems with State- and Control-Dependent Disturbances, *IEEE Transactions on Automatic Control*, **AC-16**(6), 793–798, December 1971.

M-3 Meditch, J. S., *Stochastic Optimal Linear Estimation and Control*, McGraw-Hill, New York, 1969.

M-4 Meier, L., III, Larson, R. E., and Tether, A. J., Dynamic Programming for Stochastic Control of Discrete Systems, *IEEE Transactions on Automatic Control*, **AC-16**(6), 767–775, December 1971.

M-5 Molinari, B., The Stable Regulator Problem and Its Inverse, *IEEE Transactions on Automatic Control*, **AC-18**(5), 454–459, October 1973.

R-1 Root, J. G., Optimum Control of Non-Gaussian Linear Stochastic Systems with Inaccessible State Variables, *SIAM Journal on Control*, **7**, 317–323, May 1969.

S-1 Simon, H. A., Dynamic Programming Under Uncertainty with a Quadratic Criterion Function, *Econometrica*, **24**, 74–81, 1956.

S-2 Sorenson, H. W., An Overview of Filtering and Stochastic Control in Dynamic Systems, *Control and Dynamic Systems*, Vol. 12, C. T. Leondes, ed., Academic Press, New York, 1976.

S-3 Speyer, J. L., Deyst, J., and Jacobson, D. H., Optimization of Stochastic Linear Systems with Additive Measurement and Process Noise Using Exponential Performance Criteria, *IEEE Transactions on Automatic Control*, **AC-19**(4), 358–366, August 1974.

T-1 Thornton, C. L. and Jacobson, R. A., Linear Stochastic Control Using the UDU^T Matrix Factorization, *AIAA Journal of Guidance and Control*, **1**(4), 232–236, July–August 1978.

T-2 Titchmarsh, E. C., *The Theory of Functions*, Oxford University Press, London, 1960.

T-3 Tou, J. T., *Optimum Design of Control Systems*, Academic Press, New York, 1963.

T-4 Tse, E., On the Optimal Control of Linear Systems, *IEEE Transactions on Automatic Control*, **AC-16**(6), 776–785, December 1971.

T-5 Tse, E., Bar-Shalom, Y., and Meier, L., III, Wide-Sense Adaptive Dual Control for Nonlinear Stochastic Systems, *IEEE Transactions on Automatic Control*, **AC-18**(2), 98–108, April 1973.

T-6 Tse, E. and Bar-Shalom, Y., an Actively Adaptive Control for Linear Systems with Random Parameters via the Dual Control Approach, *IEEE Transactions on Automatic Control*, **AC-18**(2), 109–117, April 1973.

T-7 Tse, E. and Bar-Shalom, Y., Generalized Certainty Equivalence and Dual Effect in Stochastic Control, *IEEE Transactions on Automatic Control*, **AC-20**(6), 817–819, December 1975.

V-1 Van de Water, H. and Willems, J. C., The Certainty Equivalence Property in Stochastic Control Theory, *IEEE Transactions on Automatic Control*, **AC-26**(5), 1080–1087, October 1981.

W-1 Wonham, W. M., On the Separation Theorem of Stochastic Control, *SIAM Journal on Control* **6**(2), 312–326, May 1968.

W-2 Wonham, W. M., On a Matrix Riccati Equation of Stochastic Control, *SIAM Journal on Control*, **6**(4), 681–699, November 1968.

6 LINEAR MULTIVARIABLE CONTROL

Virtually all dynamic systems are nonlinear, yet an overwhelming majority of operating control laws have been designed as if their dynamic systems were linear and time-invariant. Earlier chapters have shown how to formulate a local linearization that characterizes perturbations from a nominal trajectory. If the nominal trajectory is, in fact, a *set point* [a static or quasi-static equilibrium point (Section 2.5)] or a sequence of set points, the locally linear model may be "essentially time-invariant."* As long as qualitative differences in response are minimal (or at least acceptable in some practical sense), the linear, time-invariant model facilitates control system design because of the direct manner in which response attributes can be associated with model parameters.

Although not the only approach to linear multivariable control system design, stochastic optimal control theory provides a particularly comprehensive, consistent, and flexible design approach. If we distinguish between methods of *analysis* and methods of *synthesis*, stochastic linear-optimal control theory clearly falls in the latter category. To some extent it is analytical, in that cost criteria can be evaluated; however, its principal virtue for linear multivariable control is that it provides a method of computing control and estimation gains explicitly, with certain guarantees regarding the stability of the closed-loop system.

Because it is a *constructive approach*, optimal control theory produces *feasible control system designs* that can be analyzed by criteria other than the cost function. For example, classical response criteria such as step response and frequency response are not used in the control gain algorithms, but such criteria may be helpful in determining what values to use

*A dynamic model can be considered *essentially time-invariant* if its parameters change only at discrete instants of time or if the rate-of-change of parameters is slow in comparison with the transient response of the state.

in quadratic cost function weighting matrices. These weighting factors have a powerful and direct effect on achieving desired response (i.e., on closed-loop system *performance*).

This chapter presents important results of linear multivariable control system design using stochastic optimal control theory. As in Section 5.4, attention is directed at linear, time-invariant dynamic systems with constant-coefficient control and estimation logic. Thus control and estimation gains derive from steady-state solutions of the appropriate Riccati equations (or equivalent algorithms). In some respects, this is a simplification of earlier results; however, it provides an opportunity to examine control structures, asymptotic properties, robustness, and digital control in some detail.

Numerical solutions to the algebraic Riccati equation are reviewed in Section 6.1 for both continuous-time and discrete-time cases. Discussion is based on the control forms of the equations; by duality, results also apply to the estimation equations. The steady-state response of a multivariable control system is governed by the physical system and its command vector, as shown in Section 6.2. Much of the flexibility afforded by the linear-optimal control design approach is due to the wide range of controller structures that result from alternate definitions of the cost function and state variables; some of these are presented in Section 6.3. Root locations for the linear-quadratic regulator and the Kalman–Bucy filter provide insights regarding the transient behavior and stability of the closed-loop system, as shown in Section 6.4. The stability and performance of a linear-quadratic closed-loop system with significant parameter uncertainties are presented in Section 6.5. This is an area in which extensions of "classical" control theory to multivariable systems prove valuable. Robustness of linear-quadratic-Gaussian systems is considered in Section 6.6, and an algorithmic approach based on the statistics of parameter uncertainties is presented. Summary comments on adaptive optimal control are contained in Section 6.7.

6.1 SOLUTION OF THE ALGEBRAIC RICCATI EQUATION

Numerical methods of finding the steady-state solution to the Riccati equation for control are considered here. Using the definitions of Table 4.5-1, these solutions apply to the Riccati equation for estimation as well. Cross-product weighting ($\mathbf{M} \neq \mathbf{0}$) can be included using the transformations of Section 5.4.

The Continuous-Time Case

The linear-quadratic regulator for the system,

$$\dot{\mathbf{x}}(t) = \mathbf{F}\mathbf{x}(t) + \mathbf{G}\mathbf{u}(t), \qquad \mathbf{x}(0) = \mathbf{x}_0 \tag{6.1-1}$$

with cost function

$$J = \lim_{t_f \to \infty} \tfrac{1}{2}\left\{\mathbf{x}^T(t_f)\mathbf{S}(t_f)\mathbf{x}(t_f) + \int_0^{t_f} [\mathbf{x}^T(t)\mathbf{Q}\mathbf{x}(t) + \mathbf{u}^T(t)\mathbf{R}\mathbf{u}(t)]\, dt\right\} \quad (6.1\text{-}2)$$

is given by

$$\begin{aligned}\mathbf{u}(t) &= -\mathbf{R}^{-1}\mathbf{G}^T\mathbf{S}\mathbf{x}(t)\\ &= -\mathbf{C}\mathbf{x}(t)\end{aligned} \quad (6.1\text{-}3)$$

S is the steady-state solution to the matrix Riccati equation

$$\dot{\mathbf{S}}(t) = -\mathbf{F}^T\mathbf{S}(t) - \mathbf{S}(t)\mathbf{F} - \mathbf{Q} + \mathbf{S}(t)\mathbf{G}\mathbf{R}^{-1}\mathbf{G}^T\mathbf{S}(t), \qquad \mathbf{S}(\infty) = \mathbf{S}_\infty \quad (6.1\text{-}4)$$

with symmetric matrices $\mathbf{Q} \geq \mathbf{0}$ and $\mathbf{R} > \mathbf{0}$. As long as $\mathbf{S}_\infty$ is nonnegative definite, the steady-state solution for **S** is symmetric and positive definite; hence we let $\dot{\mathbf{S}} = \mathbf{0}$ and seek a solution to the *algebraic Riccati equation*:

$$\mathbf{0} = -\mathbf{F}^T\mathbf{S} - \mathbf{S}\mathbf{F} - \mathbf{Q} + \mathbf{S}\mathbf{G}\mathbf{R}^{-1}\mathbf{G}^T\mathbf{S} \quad (6.1\text{-}5)$$

Of course, one approach is to write explicit scalar equations for the $n(n+1)/2$ independent terms of **S**, then solve them simultaneously. This approach was used for the examples of Section 5.4, but it is too complex for hand calculations when $n > 3$. (We will see that no approach is appropriate for hand calculations when $n > 3$.) The explicit algebraic approach could be extended to higher-order systems using a symbolic manipulation computer program such as MACSYMA (B-5).

Another alternative is to integrate Eq. 6.1-4 to steady state with $\mathbf{S}_\infty = \mathbf{0}$. The equation is nonlinear in $\mathbf{S}(t)$, so a numerical method such as Runge–Kutta integration must be used. Equation 6.1-4 is symmetric, but numerical errors can create asymmetry; hence it is customary to "symmetrize" the solution at regular intervals by replacing the computed values, $\mathbf{S}_c(t)$, with

$$\mathbf{S}(t) = \tfrac{1}{2}[\mathbf{S}_c(t) + \mathbf{S}_c^T(t)] \quad (6.1\text{-}6)$$

Direct integration of Eq. 6.1-4 involves about twice as much computation as is necessary due to the symmetry of **S**; a better approach would be to propagate the upper- or lower-triangular square root of **S** following the estimation algorithm presented as Eq. 4.5-56 to 4.5-58. Letting

$$\mathbf{S}(t) = \mathbf{D}(t)\mathbf{D}^T(t) \quad (6.1\text{-}7)$$

the lower-triangular differential equation would take the form

$$\dot{\mathbf{D}}(t) = \mathbf{D}(t)\mathbf{M}_{\mathrm{LT}}(t), \qquad \mathbf{D}(\infty)\mathbf{D}^T(\infty) = \mathbf{S}(\infty) \quad (6.1\text{-}8)$$

As before, $\mathbf{M}_{\mathrm{LT}}(t)$ is the lower-triangular partition of a matrix, $\mathbf{M}(t)$,

$$\begin{aligned}\mathbf{M}(t) &= \mathbf{M}_{\mathrm{LT}}(t) + \mathbf{M}_{\mathrm{UT}}(t)\\ &= -\mathbf{D}^{-1}(t)\mathbf{F}^T\mathbf{D}(t) - \mathbf{D}^T(t)\mathbf{F}\mathbf{D}^{-T}(t) - \mathbf{D}^{-1}(t)\mathbf{Q}\mathbf{D}^{-T}(t)\\ &\quad + \mathbf{D}^T(t)\mathbf{G}\mathbf{R}^{-1}\mathbf{G}^T\mathbf{D}(t)\end{aligned} \tag{6.1-9}$$

and the elements of $\mathbf{M}_{\mathrm{LT}}(t)$ are

$$(m_{ij})_{\mathrm{LT}} = \begin{cases} 0, & i<j \\ \frac{1}{2}m_{ij} & i=j \\ m_{ij} & i>j \end{cases} \tag{6.1-10}$$

The square-root algorithm cannot be initialized with $\mathbf{D}(\infty) = \mathbf{0}$, but any invertible positive definite lower-triangular matrix is allowable, so we can choose $\mathbf{D}(\infty) = \rho\mathbf{I}$, where ρ is an arbitrary constant. Better yet, $\mathbf{D}(\infty)$ could be initialized with a prior estimate of $\mathbf{D}(0)$, if available. The square-root approach improves accuracy in cases of ill-conditioning, and it has the advantage of preserving the symmetry of $\mathbf{S}$, which is calculated from Eq. 6.1-7 once the steady state has been reached.

Following the estimation examples (Section 4.5), alternate *linear integration methods* could be pursued. A *Chandrasekhar-type algorithm* is feasible; its development is left to the reader. A state-transition-matrix approach called the *Kalman–Englar method* proceeds directly from concurrent solution of the state and adjoint equations of Chapter 3. For the linear system and quadratic cost function given earlier and the adjoint vector,

$$\boldsymbol{\lambda}(t) = \mathbf{S}(t)\mathbf{x}(t) \tag{6.1-11}$$

the corresponding adjoint equation (Eq. 3.4-16) is

$$\dot{\boldsymbol{\lambda}}(t) = -\mathbf{Q}\mathbf{x}(t) - \mathbf{F}^T\boldsymbol{\lambda}(t), \qquad \boldsymbol{\lambda}(\infty) = \mathbf{S}_\infty\mathbf{x}(\infty) \tag{6.1-12}$$

and the necessary condition for optimality (Eq. 3.4-18) is

$$\mathbf{0} = \mathbf{R}\mathbf{u}(t) + \mathbf{G}^T\boldsymbol{\lambda}(t) \tag{6.1-13}$$

This specifies the optimal control law as

$$\mathbf{u}(t) = -\mathbf{R}^{-1}\mathbf{G}^T\boldsymbol{\lambda}(t) \tag{6.1-14}$$

The state and adjoint equations are coupled by the optimal control law and can be written jointly as

$$\begin{bmatrix} \dot{\mathbf{x}}(t) \\ \dot{\boldsymbol{\lambda}}(t) \end{bmatrix} = \begin{bmatrix} \mathbf{F} & -\mathbf{G}\mathbf{R}^{-1}\mathbf{G}^T \\ -\mathbf{Q} & -\mathbf{F}^T \end{bmatrix} \begin{bmatrix} \mathbf{x}(t) \\ \boldsymbol{\lambda}(t) \end{bmatrix} \tag{6.1-15a}$$

or

$$\dot{\boldsymbol{\chi}}(t) = \mathbf{F}'\boldsymbol{\chi}(t) \tag{6.1-15b}$$

with $\boldsymbol{\chi}^T(t) \triangleq [\mathbf{x}^T(t) \quad \boldsymbol{\lambda}^T(t)]$. $\mathbf{F}'$ has the properties of a Hamiltonian matrix, which is defined in a footnote to follow.

Equation 6.1-15 is a $(2n \times 2n)$ linear, time-invariant, ordinary differential equation that can be propagated forward or backward in time (at discrete instants) using the appropriate state transition matrix. Defining $\boldsymbol{\Theta}(-T)$ for backward propagation over a time interval T,

$$\boldsymbol{\Theta}(-T) = e^{-\mathbf{F}'\mathbf{T}} \tag{6.1-16}$$

the state transition equation could be written as

$$\boldsymbol{\chi}_{k-1} = \boldsymbol{\Theta}(-T)\boldsymbol{\chi}_k \tag{6.1-17a}$$

or

$$\begin{bmatrix} \mathbf{x} \\ \boldsymbol{\lambda} \end{bmatrix}_{k-1} = \begin{bmatrix} \boldsymbol{\Theta}_{11} & \boldsymbol{\Theta}_{12} \\ \boldsymbol{\Theta}_{21} & \boldsymbol{\Theta}_{22} \end{bmatrix} \begin{bmatrix} \mathbf{x} \\ \boldsymbol{\lambda} \end{bmatrix}_k \tag{6.1-17b}$$

where the $\boldsymbol{\Theta}_{ij}$ are $(n \times n)$ partitions of $\boldsymbol{\Theta}(-T)$. Since

$$\begin{aligned} \mathbf{x}_{k-1} &= \boldsymbol{\Theta}_{11}\mathbf{x}_k + \boldsymbol{\Theta}_{12}\boldsymbol{\lambda}_k \\ &= (\boldsymbol{\Theta}_{11} + \boldsymbol{\Theta}_{12}\mathbf{S}_k)\mathbf{x}_k \end{aligned} \tag{6.1-18}$$

and

$$\begin{aligned} \boldsymbol{\lambda}_{k-1} &= \mathbf{S}_{k-1}\mathbf{x}_{k-1} \\ &= \boldsymbol{\Theta}_{21}\mathbf{x}_k + \boldsymbol{\Theta}_{22}\boldsymbol{\lambda}_k \\ &= (\boldsymbol{\Theta}_{21} + \boldsymbol{\Theta}_{22}\mathbf{S}_k)\mathbf{x}_k \end{aligned} \tag{6.1-19}$$

we can write

$$\mathbf{S}_{k-1}(\boldsymbol{\Theta}_{11} + \boldsymbol{\Theta}_{12}\mathbf{S}_k)\mathbf{x}_k = (\boldsymbol{\Theta}_{21} + \boldsymbol{\Theta}_{22}\mathbf{S}_k)\mathbf{x}_k \tag{6.1-20}$$

As this holds for arbitrary $\mathbf{x}_k$, $\mathbf{S}_{k-1}$ can be propagated backward as

$$\mathbf{S}_{k-1} = (\boldsymbol{\Theta}_{21} + \boldsymbol{\Theta}_{22}\mathbf{S}_k)(\boldsymbol{\Theta}_{11} + \boldsymbol{\Theta}_{12}\mathbf{S}_k)^{-1} \tag{6.1-21}$$

starting at $\mathbf{S}_\infty = \mathbf{0}$ and proceeding to the steady state. Although matrix inversion is required at each step, there are fewer computations than in numerical integration. In practice, T must be relatively small to avoid numerical difficulty.

There are numerous iterative methods for solving Eq. 6.1-5; our discussion will be limited to Newton–Raphson iteration, solution by eigenvectors, and solution by Schur vectors. The iterations do not imply the passage of time; hence they are appropriate only for constant-coefficient equations.

The optimal control gain $\mathbf{C}^*$ takes the form

$$\mathbf{C}^* = \mathbf{R}^{-1}\mathbf{G}^T\mathbf{S}^* \tag{6.1-22}$$

$\mathbf{S}^*$ is the steady-state solution of Eq. 6.1-5; using Eq. 6.1-22, it can be written as

$$\mathbf{0} = -(\mathbf{F} - \mathbf{G}\mathbf{C}^*)^T\mathbf{S}^* - \mathbf{S}^*(\mathbf{F} - \mathbf{G}\mathbf{C}^*) - \mathbf{Q} - \mathbf{C}^{*T}\mathbf{R}\mathbf{C}^* \tag{6.1-23a}$$

or

$$(\mathbf{F} - \mathbf{G}\mathbf{C}^*)^T\mathbf{S}^* + \mathbf{S}^*(\mathbf{F} - \mathbf{G}\mathbf{C}^*) = -(\mathbf{Q} + \mathbf{C}^{*T}\mathbf{R}\mathbf{C}^*) \tag{6.1-23b}$$

(Verify this by substitution.) For given $\mathbf{C}$, this is a Lyapunov equation (Eq. 2.5-60); by Lyapunov's second theorem (Section 2.5), we can obtain a positive definite solution $\mathbf{S}$ if $(\mathbf{F} - \mathbf{G}\mathbf{C})$ is a stable matrix and if $(\mathbf{Q} + \mathbf{C}^T\mathbf{R}\mathbf{C})$ is positive definite.

Kleinman's method is an iterative procedure in which these two criteria are satisfied at each step; $\mathbf{C}_k$ and $\mathbf{S}_k$ approach $\mathbf{C}^*$ and $\mathbf{S}^*$, respectively, as k increases. The iteration must be initialized with a choice of $\mathbf{C}_0$ such that $(\mathbf{F} - \mathbf{G}\mathbf{C}_0)$ is stable. Beginning with $k = 0$, $\mathbf{S}_k$ is the solution of

$$(\mathbf{F} - \mathbf{G}\mathbf{C}_k)^T\mathbf{S}_k + \mathbf{S}_k(\mathbf{F} - \mathbf{G}\mathbf{C}_k) = -(\mathbf{Q} + \mathbf{C}_k^T\mathbf{R}\mathbf{C}_k) \tag{6.1-24}$$

found, for example, using the Bartels–Stewart algorithm (B-1) or the Hessenburg–Schur algorithm (G-1). An improved gain matrix is computed as

$$\mathbf{C}_{k+1} = \mathbf{R}^{-1}\mathbf{G}^T\mathbf{S}_k \tag{6.1-25}$$

and the iteration proceeds by successive solutions of these two equations. At each step, $\mathbf{S}_k$ is positive definite and $(\mathbf{F} - \mathbf{G}\mathbf{C}_k)$ is stable; the process itself is stable, and it converges quadratically to $\mathbf{S}^*$ (K-5).

If $\mathbf{F}$ is stable without closed-loop control, the iteration can begin with $\mathbf{C}_0 = \mathbf{0}$. If $\mathbf{F}$ is not stable, there are a number of ways to find $\mathbf{C}_0$. One possibility is to use a time-based method that computes $\mathbf{S}(t)$ to the point that significant transients have decayed but not long enough to get a precise solution. $\mathbf{C}_0$ then is calculated from the approximate $\mathbf{S}$. The principal difficulty is in knowing when to stop the starting solution. As long as $(\mathbf{F}, \mathbf{G})$ is completely controllable, a stabilizing value of $\mathbf{C}_0$ can be specified as (K-6),

$$\mathbf{C}_0 = \mathbf{G}^T \mathbf{M}^{-1}(\Delta t) \tag{6.1-26}$$

where $\mathbf{M}(\Delta t)$ is the nonsingular controllability Grammian matrix (Eq. 2.5-94) for an arbitrary interval $[0, \Delta t]$:

$$\mathbf{M}(\Delta t) = \int_0^{\Delta t} e^{-\mathbf{F}\tau} \mathbf{G}\mathbf{G}^T e^{-\mathbf{F}^T\tau}\, d\tau \tag{6.1-27}$$

If the system is stabilizable but not completely controllable, $\mathbf{M}(\Delta t)$ is singular; $\mathbf{C}_0$ then can be specified using a generalized inverse of $\mathbf{M}(\Delta t)$:

$$\mathbf{C}_0 = \mathbf{G}^T \mathbf{M}^{\mathrm{GI}}(\Delta t) \tag{6.1-28}$$

Defining the controllable and uncontrollable (but stable) parts of $\mathbf{x}$ as $\mathbf{x}_1$ and $\mathbf{x}_2$, respectively, the generalized inverse is

$$\mathbf{M}^{\mathrm{GI}}(\Delta t) = \begin{bmatrix} \mathbf{M}_1^{-1}(\Delta t) & 0 \\ 0 & 0 \end{bmatrix} \tag{6.1-29}$$

with

$$\mathbf{M}_1(\Delta t) = \int_0^{\Delta t} e^{-\mathbf{F}_1\tau} \mathbf{G}_1\mathbf{G}_1^T e^{-\mathbf{F}_1^T\tau}\, d\tau \tag{6.1-30}$$

$\mathbf{F}_1$ and $\mathbf{G}_1$ are defined by the implied partitioning of $\mathbf{F}$ and $\mathbf{G}$ (S-5).

The *MacFarlane–Potter method* appears to be a direct solution for $\mathbf{S}$ in terms of the eigenvectors to be presented below; however, iteration is required to find the eigenvalues of the $(2n \times 2n)$ Hamiltonian matrix, $\mathbf{F}'$ (Eq. 6.1-15) (M-1, P-4). In Section 6.4, it will be shown that the $2n$ eigenvalues of $\mathbf{F}'$ are symmetrically disposed about the imaginary axis of the s plane; therefore, n roots are stable and n roots are unstable. Assuming that the closed-loop roots are distinct, $\mathbf{F}'$ can be block diagonalized as in Section 2.2, and the eigenvalues can be arranged in $\mathbf{\Lambda}$ so that

$$\begin{aligned}\mathbf{F}' &= \begin{bmatrix} \mathbf{F} & -\mathbf{G}\mathbf{R}^{-1}\mathbf{G}^T \\ -\mathbf{Q} & -\mathbf{F}^T \end{bmatrix} \\ &= \begin{bmatrix} \mathbf{E}_{11} & \mathbf{E}_{12} \\ \mathbf{E}_{21} & \mathbf{E}_{22} \end{bmatrix} \begin{bmatrix} \mathbf{\Lambda} & \mathbf{0} \\ \mathbf{0} & -\mathbf{\Lambda} \end{bmatrix} \begin{bmatrix} \mathbf{E}_{11} & \mathbf{E}_{12} \\ \mathbf{E}_{21} & \mathbf{E}_{22} \end{bmatrix}^{-1} \\ &= \begin{bmatrix} \mathbf{E}_{11} & \mathbf{E}_{12} \\ \mathbf{E}_{21} & \mathbf{E}_{22} \end{bmatrix} \begin{bmatrix} \mathbf{\Lambda} & \mathbf{0} \\ \mathbf{0} & -\mathbf{\Lambda} \end{bmatrix} \begin{bmatrix} \mathbf{D}_{11}^T & \mathbf{D}_{21}^T \\ \mathbf{D}_{12}^T & \mathbf{D}_{22}^T \end{bmatrix}\end{aligned} \tag{6.1-31}$$

where $\mathbf{E}$ and $\mathbf{D}^T$ are the right and left modal matrices of $\mathbf{F}'$ (Section 2.2). All the unstable roots are in $\mathbf{\Lambda}$, all the stable roots are in $-\mathbf{\Lambda}$, and $\mathbf{E}$ and $\mathbf{D}^T$ are arranged accordingly.

The $2n$-dimensional normal state vector $\mathbf{q}(t)$ is a transformation of $\mathbf{x}(t)$ and $\boldsymbol{\lambda}(t)$,

$$\mathbf{q}(t)=\begin{bmatrix}\mathbf{q}_1(t)\\ \mathbf{q}_2(t)\end{bmatrix}=\begin{bmatrix}\mathbf{D}_{11}^T & \mathbf{D}_{21}^T\\ \mathbf{D}_{12}^T & \mathbf{D}_{22}^T\end{bmatrix}\begin{bmatrix}\mathbf{x}(t)\\ \boldsymbol{\lambda}(t)\end{bmatrix} \tag{6.1-32}$$

and the inverse also is true:

$$\boldsymbol{\chi}(t)=\begin{bmatrix}\mathbf{x}(t)\\ \boldsymbol{\lambda}(t)\end{bmatrix}=\begin{bmatrix}\mathbf{E}_{11} & \mathbf{E}_{12}\\ \mathbf{E}_{21} & \mathbf{E}_{22}\end{bmatrix}\begin{bmatrix}\mathbf{q}_1(t)\\ \mathbf{q}_2(t)\end{bmatrix} \tag{6.1-33}$$

The normal state vector satisfies the linear differential equation

$$\begin{bmatrix}\dot{\mathbf{q}}_1(t)\\ \dot{\mathbf{q}}_2(t)\end{bmatrix}=\begin{bmatrix}\mathbf{\Lambda} & \mathbf{0}\\ \mathbf{0} & -\mathbf{\Lambda}\end{bmatrix}\begin{bmatrix}\mathbf{q}_1(t)\\ \mathbf{q}_2(t)\end{bmatrix} \tag{6.1-34}$$

hence its value at t, given its value at zero time, is

$$\begin{bmatrix}\mathbf{q}_1(t)\\ \mathbf{q}_2(t)\end{bmatrix}=\begin{bmatrix}e^{\mathbf{\Lambda}t} & \mathbf{0}\\ \mathbf{0} & e^{-\mathbf{\Lambda}t}\end{bmatrix}\begin{bmatrix}\mathbf{q}_1(0)\\ \mathbf{q}_2(0)\end{bmatrix} \tag{6.1-35}$$

$\mathbf{q}_1(t)$ evidences a completely unstable response, while $\mathbf{q}_2(t)$ is entirely stable. If $\mathbf{x}(0)$ and $\boldsymbol{\lambda}(0)$ were known, this could be expressed as

$$\begin{bmatrix}\mathbf{q}_1(t)\\ \mathbf{q}_2(t)\end{bmatrix}=\begin{bmatrix}e^{\mathbf{\Lambda}t} & \mathbf{0}\\ \mathbf{0} & e^{-\mathbf{\Lambda}t}\end{bmatrix}\begin{bmatrix}\mathbf{D}_{11}^T & \mathbf{D}_{21}^T\\ \mathbf{D}_{12}^T & \mathbf{D}_{22}^T\end{bmatrix}\begin{bmatrix}\mathbf{x}(0)\\ \boldsymbol{\lambda}(0)\end{bmatrix} \tag{6.1-36}$$

Then

$$\begin{bmatrix}\mathbf{q}_1(t)\\ \mathbf{q}_2(t)\end{bmatrix}=\begin{bmatrix}e^{\mathbf{\Lambda}t} & \mathbf{0}\\ \mathbf{0} & e^{-\mathbf{\Lambda}t}\end{bmatrix}\begin{bmatrix}(\mathbf{D}_{11}^T+\mathbf{D}_{21}^T\mathbf{S})\mathbf{x}(0)\\ (\mathbf{D}_{12}^T+\mathbf{D}_{22}^T\mathbf{S})\mathbf{x}(0)\end{bmatrix} \tag{6.1-37}$$

since Eq. 6.1-11 relates $\boldsymbol{\lambda}$ and $\mathbf{x}$ by $\mathbf{S}$. The values of $\mathbf{x}(t)$ and $\boldsymbol{\lambda}(t)$ follow from Eq. 6.1-33:

$$\begin{aligned}\begin{bmatrix}\mathbf{x}(t)\\ \boldsymbol{\lambda}(t)\end{bmatrix}&=\begin{bmatrix}\mathbf{E}_{11} & \mathbf{E}_{12}\\ \mathbf{E}_{21} & \mathbf{E}_{22}\end{bmatrix}\begin{bmatrix}\mathbf{q}_1(t)\\ \mathbf{q}_2(t)\end{bmatrix}\\ &=\begin{bmatrix}[\mathbf{E}_{11}\mathbf{q}_1(t)+\mathbf{E}_{12}\mathbf{q}_2(t)]\\ [\mathbf{E}_{21}\mathbf{q}_1(t)+\mathbf{E}_{22}\mathbf{q}_2(t)]\end{bmatrix}\end{aligned} \tag{6.1-38}$$

Because $\mathbf{x}(t)$ is guaranteed to have stable response, it must be generated entirely by $\mathbf{q}_2(t)$. As $\mathbf{E}_{11}$ is unlikely to be a null matrix, this can occur only if $\mathbf{q}_1(t)$ is zero. This, in turn, requires that $\mathbf{q}_1(0)$ be zero. For arbitrary $\mathbf{x}(0)$, Eq. 6.1-37 indicates that the following must be true:

$$\mathbf{S} = -\mathbf{D}_{22}^{-T}\mathbf{D}_{12}^{T} \tag{6.1-39}$$

Then Eq. 6.1-38 becomes

$$\begin{bmatrix} \mathbf{x}(t) \\ \boldsymbol{\lambda}(t) \end{bmatrix} = \begin{bmatrix} \mathbf{E}_{12}\mathbf{q}_2(t) \\ \mathbf{E}_{22}\mathbf{q}_2(t) \end{bmatrix} \tag{6.1-40}$$

and, since Eq. 6.1-11 is still in force,

$$\boldsymbol{\lambda}(t) = \mathbf{E}_{22}\mathbf{q}_2(t) = \mathbf{S}\mathbf{E}_{12}\mathbf{q}_2(t) \tag{6.1-41}$$

For arbitrary $\mathbf{q}_2(t)$, this leads to

$$\mathbf{S} = \mathbf{E}_{22}\mathbf{E}_{12}^{-1} \tag{6.1-42}$$

Thus either Eq. 6.1-39 or 6.1-42 provides a solution for $\mathbf{S}$. Actually, "either" is misleading, as the two equations are equivalent; the structures of the left and right eigenvectors are such that (K-8)

$$\mathbf{D}^T = \begin{bmatrix} \mathbf{E}_{22}^T & -\mathbf{E}_{12}^T \\ -\mathbf{E}_{21}^T & \mathbf{E}_{11}^T \end{bmatrix} \tag{6.1-43}$$

Although details are omitted, it is remarked that computational advantages may accrue and repeated roots can be accommodated if $\mathbf{F}'$ is reduced to upper-block-triangular form* (L-1):

$$\mathbf{U}^T\mathbf{F}'\mathbf{U} = \boldsymbol{\Lambda}' \tag{6.1-44a}$$

or

$$\begin{bmatrix} \mathbf{U}_{11} & \mathbf{U}_{12} \\ \mathbf{U}_{21} & \mathbf{U}_{22} \end{bmatrix}^T \begin{bmatrix} \mathbf{F} & -\mathbf{G}\mathbf{R}^{-1}\mathbf{G}^T \\ -\mathbf{Q} & -\mathbf{F}^T \end{bmatrix} \begin{bmatrix} \mathbf{U}_{11} & \mathbf{U}_{12} \\ \mathbf{U}_{21} & \mathbf{U}_{22} \end{bmatrix} = \begin{bmatrix} \boldsymbol{\Lambda}_{11} & \boldsymbol{\Lambda}_{12} \\ \mathbf{0} & \boldsymbol{\Lambda}_{22} \end{bmatrix} \tag{6.1-44b}$$

where the columns of $\mathbf{U}$ are called *Schur vectors*. $\boldsymbol{\Lambda}'$ can be ordered so that all the eigenvalues with negative real parts (including multiple roots) belong

*This also is called a ***quasi-upper-triangular*** or ***real Schur form***, with (2×2) blocks and (1×1) elements on the (block) diagonal corresponding to complex and real eigenvalues. Denoting the elements of $\boldsymbol{\Lambda}'$ as λ_{ij}, it is apparent that $\lambda_{ij} = 0$ for $i > j+1$; hence this also is an ***upper Hessenburg form***.

to Λ_{11} (D-5); then *Laub's method* presents the desired solution as

$$\mathbf{S} = \mathbf{U}_{21}\mathbf{U}_{11}^{-1} \tag{6.1-45}$$

The Discrete-Time Case

The linear-quadratic regulator for the system

$$\mathbf{x}_{k+1} = \mathbf{\Phi}\mathbf{x}_k + \mathbf{\Gamma}\mathbf{u}_k, \qquad \mathbf{x}_0 \text{ given} \tag{6.1-46}$$

with cost function

$$J = \lim_{k_f \to \infty} \frac{1}{2k_f}\left[\mathbf{x}_{k_f}^T \mathbf{S}_{k_f} \mathbf{x}_{k_f} + \sum_{k=0}^{k_f} (\mathbf{x}_k^T \mathbf{Q}\mathbf{x}_k + \mathbf{u}_k^T \mathbf{R}\mathbf{u}_k)\right] \tag{6.1-47}$$

is given by

$$\begin{aligned}\mathbf{u}_k &= -(\mathbf{R} + \mathbf{\Gamma}^T\mathbf{S}\mathbf{\Gamma})^{-1}\mathbf{\Gamma}^T\mathbf{S}\mathbf{\Phi}\mathbf{x}_k \\ &= -\mathbf{C}\mathbf{x}_k \end{aligned} \tag{6.1-48}$$

S is the steady-state solution of the discrete-time matrix Riccati equation

$$\mathbf{S}_k = \mathbf{\Phi}^T\mathbf{S}_{k+1}\mathbf{\Phi} + \mathbf{Q} - \mathbf{\Phi}^T\mathbf{S}_{k+1}\mathbf{\Gamma}(\mathbf{R} + \mathbf{\Gamma}^T\mathbf{S}_{k+1}\mathbf{\Gamma})^{-1}\mathbf{\Gamma}^T\mathbf{S}_{k+1}\mathbf{\Phi}, \qquad \mathbf{S}_\infty \text{ given} \tag{6.1-49}$$

with the symmetric matrices $\mathbf{Q} \geq \mathbf{0}$ and $\mathbf{R} > \mathbf{0}$. $\mathbf{S}_\infty$ can be taken as any nonnegative definite matrix. In the steady state, $\mathbf{S}_{k+1} = \mathbf{S}_k = \mathbf{S}$, so the *discrete algebraic Riccati equation* can be expressed as

$$\mathbf{0} = \mathbf{\Phi}^T\mathbf{S}\mathbf{\Phi} - \mathbf{S} + \mathbf{Q} - \mathbf{\Phi}^T\mathbf{S}\mathbf{\Gamma}(\mathbf{R} + \mathbf{\Gamma}^T\mathbf{S}\mathbf{\Gamma})^{-1}\mathbf{\Gamma}^T\mathbf{S}\mathbf{\Phi} \tag{6.1-50a}$$

From the matrix inversion lemma (Eq. 2.2-67), this also could be written as

$$\mathbf{0} = \mathbf{\Phi}^T(\mathbf{S}^{-1} + \mathbf{\Gamma}\mathbf{R}^{-1}\mathbf{\Gamma})^{-1}\mathbf{\Phi} - \mathbf{S} + \mathbf{Q} \tag{6.1-50b}$$

Methods for computing the steady-state **S** for the discrete-time problem are analogous to the continuous-time alternatives. **S** can be obtained by recursion using Eq. 6.1-49 or any of the alternate forms presented in Section 4.3. The principal advantage of using one of these methods is commonality of algorithms for time-varying and time-invariant problems; the same subroutines or procedures can be applied to either. However, the algorithms to follow should provide improved precision for the time-invariant case, and they may be faster computationally.

A discrete-time equivalent to the Kalman–Englar method can be formulated, beginning with the system equation, the discrete Euler–Lagrange equations, and the adjoint relationship between $\mathbf{x}_k$ and $\boldsymbol{\lambda}_k$:

$$\mathbf{x}_{k+1} = \boldsymbol{\Phi}\mathbf{x}_k + \boldsymbol{\Gamma}\mathbf{u}_k \tag{6.1-51}$$

$$\boldsymbol{\lambda}_k = \mathbf{Q}\mathbf{x}_k + \boldsymbol{\Phi}^T\boldsymbol{\lambda}_{k+1} \tag{6.1-52}$$

$$\mathbf{u}_k = -\mathbf{R}^{-1}\boldsymbol{\Gamma}^T\boldsymbol{\lambda}_{k+1} \tag{6.1-53}$$

$$\boldsymbol{\lambda}_k = \mathbf{S}_k\mathbf{x}_k \tag{6.1-54}$$

We need a pair of equations relating $\mathbf{x}_k$ and $\boldsymbol{\lambda}_k$ to $\mathbf{x}_{k+1}$ and $\boldsymbol{\lambda}_{k+1}$. Rearranging Eq. 6.1-51 and incorporating Eq. 6.1-53:

$$\mathbf{x}_k = \boldsymbol{\Phi}^{-1}\mathbf{x}_{k+1} + \boldsymbol{\Phi}^{-1}\boldsymbol{\Gamma}\mathbf{R}^{-1}\boldsymbol{\Gamma}^T\boldsymbol{\lambda}_{k+1} \tag{6.1-55}$$

Substituting this result in Eq. 6.1-52:

$$\boldsymbol{\lambda}_k = \mathbf{Q}\boldsymbol{\Phi}^{-1}\mathbf{x}_{k+1} + (\boldsymbol{\Phi}^T + \mathbf{Q}\boldsymbol{\Phi}^{-1}\boldsymbol{\Gamma}\mathbf{R}^{-1}\boldsymbol{\Gamma}^T)\boldsymbol{\lambda}_{k+1} \tag{6.1-56}$$

Hence a symplectic matrix* $\boldsymbol{\Psi}$ can be formed so that

$$\boldsymbol{\chi}_k = \boldsymbol{\Psi}\boldsymbol{\chi}_{k+1} \tag{6.1-57a}$$

or

$$\begin{aligned}\begin{bmatrix}\mathbf{x}\\ \boldsymbol{\lambda}\end{bmatrix}_k &= \begin{bmatrix}\boldsymbol{\Phi}^{-1} & \boldsymbol{\Phi}^{-1}\boldsymbol{\Gamma}\mathbf{R}^{-1}\boldsymbol{\Gamma}^T\\ \mathbf{Q}\boldsymbol{\Phi}^{-1} & (\boldsymbol{\Phi}^T + \mathbf{Q}\boldsymbol{\Phi}^{-1}\boldsymbol{\Gamma}\mathbf{R}^{-1}\boldsymbol{\Gamma}^T)\end{bmatrix}\begin{bmatrix}\mathbf{x}\\ \boldsymbol{\lambda}\end{bmatrix}_{k+1}\\ &= \begin{bmatrix}\boldsymbol{\Psi}_{11} & \boldsymbol{\Psi}_{12}\\ \boldsymbol{\Psi}_{21} & \boldsymbol{\Psi}_{22}\end{bmatrix}\begin{bmatrix}\mathbf{x}\\ \boldsymbol{\lambda}\end{bmatrix}_{k+1}\end{aligned} \tag{6.1-57b}$$

The remainder of the development follows the continuous-time case. Consequently, from Eq. 6.1-21, we can write

$$\mathbf{S}_k = (\boldsymbol{\Psi}_{21} + \boldsymbol{\Psi}_{22}\mathbf{S}_{k+1})(\boldsymbol{\Psi}_{11} + \boldsymbol{\Psi}_{12}\mathbf{S}_{k+1})^{-1} \tag{6.1-58}$$

Iterative solutions to Eq. 6.1-50 are equivalent to the continuous-time solutions. *Hewer's method* is comparable to Kleinman's method. Using Eq.

*Define the $(2n \times 2n)$ matrix $\mathbf{J}$ as

$$\mathbf{J} = \begin{bmatrix}\mathbf{0} & \mathbf{I}_n\\ -\mathbf{I}_n & \mathbf{0}\end{bmatrix}$$

Then $\mathbf{J}^T = \mathbf{J}^{-1} = -\mathbf{J}$. $\mathbf{A}$ is a *symplectic matrix* if $\mathbf{J}^{-1}\mathbf{A}^T\mathbf{J} = \mathbf{A}^{-1}$. $\mathbf{A}$ is a *Hamiltonian matrix* if $\mathbf{J}^{-1}\mathbf{A}^T\mathbf{J} = -\mathbf{A}$.

6.1-48 for the optimal gain matrix, Eq. 6.1-50 can be expressed as a discrete-time Lyapunov equation (Eq. 2.5-134). Choosing a stabilizing starting gain [as in (K-7)], a convergent iteration to the optimal **S** and **C** can be accomplished (H-5).

The eigenvector and Schur vector approaches presented earlier can be applied to the discrete-time case as well. In *Vaughan's method* $\boldsymbol{\Psi}$ must have distinct eigenvalues, allowing it to be factored as

$$\boldsymbol{\Psi} = \begin{bmatrix} \mathbf{E}_{11} & \mathbf{E}_{12} \\ \mathbf{E}_{21} & \mathbf{E}_{22} \end{bmatrix} \begin{bmatrix} \boldsymbol{\Lambda} & \mathbf{0} \\ \mathbf{0} & \boldsymbol{\Lambda}^{-1} \end{bmatrix} \begin{bmatrix} \mathbf{E}_{11} & \mathbf{E}_{12} \\ \mathbf{E}_{21} & \mathbf{E}_{22} \end{bmatrix}^{-1} \tag{6.1-59}$$

where the eigenvalues of $\boldsymbol{\Lambda}$ are outside a unit circle centered at the origin of the z plane ($|\Lambda_{ii}| > 1$) (V-1). The eigenvalues of $\boldsymbol{\Lambda}^{-1}$ have magnitudes less than one and are, therefore, stable. The steady-state solution for **S** is obtained from the eigenvectors of $\boldsymbol{\Lambda}$ as

$$\mathbf{S} = \mathbf{E}_{21}\mathbf{E}_{11}^{-1} \tag{6.1-60}$$

With *Laub's method*, the symplectic matrix is put in real Schur form,

$$\begin{bmatrix} \mathbf{U}_{11} & \mathbf{U}_{12} \\ \mathbf{U}_{21} & \mathbf{U}_{22} \end{bmatrix}^T \begin{bmatrix} \boldsymbol{\Psi}_{11} & \boldsymbol{\Psi}_{12} \\ \boldsymbol{\Psi}_{21} & \boldsymbol{\Psi}_{22} \end{bmatrix} \begin{bmatrix} \mathbf{U}_{11} & \mathbf{U}_{12} \\ \mathbf{U}_{21} & \mathbf{U}_{22} \end{bmatrix} = \begin{bmatrix} \boldsymbol{\Lambda}_{11} & \boldsymbol{\Lambda}_{12} \\ \mathbf{0} & \boldsymbol{\Lambda}_{22} \end{bmatrix} \tag{6.1-61}$$

where $\boldsymbol{\Lambda}_{11}$ contains the stable closed-loop eigenvalues (L-1). The steady-state **S** then is found to be

$$\mathbf{S} = \mathbf{U}_{21}\mathbf{U}_{11}^{-1} \tag{6.1-62}$$

$\boldsymbol{\Phi}$ must be inverted to compute the symplectic matrix $\boldsymbol{\Psi}$, and this could pose a problem for the last three methods. $\boldsymbol{\Phi}$ normally is nonsingular if it can be specified as $e^{\mathbf{F}t}$; however, $\boldsymbol{\Phi}$ could be singular if the dynamic system contains pure time delays.* Low-order models of high-order systems often contain pure time delays, and digital computation of control logic introduces pure time delays.

When $\boldsymbol{\Phi}$ is singular, the direct recursive methods could be used to find **S**, but it also is possible to generate **S** as a function of *generalized eigenvectors* or *generalized Schur vectors*, as in (P-1). In both cases, the symplectic matrix $\boldsymbol{\Psi}$ is avoided by replacing Eq. 6.1-57 with

$$\begin{bmatrix} \mathbf{I}_n & \boldsymbol{\Gamma}\mathbf{R}^{-1}\boldsymbol{\Gamma}^T \\ \mathbf{0} & \boldsymbol{\Phi}^T \end{bmatrix} \begin{bmatrix} \mathbf{x} \\ \boldsymbol{\lambda} \end{bmatrix}_{k+1} = \begin{bmatrix} \boldsymbol{\Phi} & \mathbf{0} \\ -\mathbf{Q} & \mathbf{I}_n \end{bmatrix} \begin{bmatrix} \mathbf{x} \\ \boldsymbol{\lambda} \end{bmatrix}_k \tag{6.1-63a}$$

*A pure time delay is a dead time between the input and the output. In the time domain, it is symbolized, for example, as $y(t) = x(t-T)$, where T is the delay interval. In discrete-time systems, a delay of one sampling interval is modeled by the introduction of a new state element, such that $x_{i+1_{k+1}} = x_{i_k}$, where i is the component number and k is the sampling index. In the frequency domain, $y(s) = e^{-Ts}x(s)$, and $zy_{i+1} = x_i$.

or

$$\mathbf{L}\chi_{k+1} = \mathbf{M}\chi_k \tag{6.1-63b}$$

Paralleling the developments of Section 2.2, the *generalized eigenvalues*, λ, are defined as solutions to the equation

$$|\lambda\mathbf{L} - \mathbf{M}| = 0 \tag{6.1-64}$$

while distinct *generalized eigenvectors*, $\mathbf{e}_i$, satisfy the following condition:

$$\mathbf{M}\mathbf{e}_i = \lambda_i\mathbf{L}\mathbf{e}_i \tag{6.1-65}$$

[Added conditions for multiple roots are discussed in (P-1); for simplicity, they are not considered here.] Collecting the n stable eigenvalues in the diagonal matrix, $\mathbf{\Lambda}$, and defining the corresponding $(2n \times n)$ matrix of generalized eigenvectors as

$$\begin{bmatrix} \mathbf{E}_1 \\ \mathbf{E}_2 \end{bmatrix} = [\mathbf{e}_1, \ldots, \mathbf{e}_n] \tag{6.1-66}$$

Eq. 6.1-65 leads to

$$\mathbf{M}\begin{bmatrix} \mathbf{E}_1 \\ \mathbf{E}_2 \end{bmatrix} = \mathbf{L}\begin{bmatrix} \mathbf{E}_1\mathbf{\Lambda} \\ \mathbf{E}_2\mathbf{\Lambda} \end{bmatrix} \tag{6.1-67}$$

Following similar results presented earlier, the steady-state solution of $\mathbf{S}$ is

$$\mathbf{S} = \mathbf{E}_2\mathbf{E}_1^{-1} \tag{6.1-68}$$

The generalized Schur vector solution, which has computational advantages over the generalized eigenvector approach, has the same solution equations (Eqs. 6.1-67) and 6.1-68), except that $\mathbf{\Lambda}$ is defined in real Schur form (P-1).

6.2 STEADY-STATE RESPONSE TO COMMANDS

The nominal steady-state response to command inputs is decoupled from the calculation of feedback control gains. Given a desired set point, the corresponding equilibrium values of state and control depend only upon the coefficients of the dynamic system and the specification of the set point as a function of the state, control, and disturbance input. This is straightforward unless the definition of the command vector implies quasistatic equilibrium

(Section 2.5), in which constant commands cause some states to change continually.

The dynamic system is modeled by

$$\dot{\mathbf{x}}(t) = \mathbf{F}\mathbf{x}(t) + \mathbf{G}\mathbf{u}(t) + \mathbf{L}\mathbf{w}(t) \tag{6.2-1}$$

and the output is modeled as a linear combination of the state, control, and disturbance:

$$\mathbf{y}(t) = \mathbf{H}_{\mathbf{x}}\mathbf{x}(t) + \mathbf{H}_{\mathbf{u}}\mathbf{u}(t) + \mathbf{H}_{\mathbf{w}}\mathbf{w}(t) \tag{6.2-2}$$

The dimensions of $\mathbf{x}$, $\mathbf{u}$, $\mathbf{w}$, and $\mathbf{y}$ are n, m, s, and r, respectively, and the matrices are conformable with the vectors. To establish nominal relationships, it is assumed that $\mathbf{F}$, $\mathbf{G}$, $\mathbf{H}_{\mathbf{x}}$, $\mathbf{H}_{\mathbf{u}}$, and $\mathbf{H}_{\mathbf{w}}$ are known without error. Here $\mathbf{y}(t)$ is chosen as a *command variable*, which may or may not be synonymous with the measurement vector. As the equilibrium between command, state, and control is an open-loop characteristic, it is independent of any measurements.

Consider an example based on the control of an aircraft's longitudinal motions (S-16). The state vector could be chosen as

$$\mathbf{x} = \begin{bmatrix} x_1 \\ x_2 \\ x_3 \\ x_4 \end{bmatrix} \quad \begin{matrix} \text{forward velocity, m/s} \\ \text{angle of attack, deg} \\ \text{pitch rate, deg/s} \\ \text{pitch angle, deg} \end{matrix} \tag{6.2-3}$$

while the controls are

$$\mathbf{u} = \begin{bmatrix} u_1 \\ u_2 \end{bmatrix} \quad \begin{matrix} \text{elevator angle, deg} \\ \text{throttle setting, percent} \end{matrix} \tag{6.2-4}$$

With two controls, it is reasonable to command up to two variables independently; consequently, we might choose to command velocity and pitch angle, defining $\mathbf{y}(t)$ as

$$\begin{bmatrix} y_1(t) \\ y_2(t) \end{bmatrix} = \begin{bmatrix} 1 & 0 & 0 & 0 \\ 0 & 0 & 0 & 1 \end{bmatrix}\mathbf{x} + \begin{bmatrix} 0 & 0 \\ 0 & 0 \end{bmatrix}\mathbf{u} \tag{6.2-5}$$

or pitch rate and throttle setting, leading to

$$\begin{bmatrix} y_1(t) \\ y_2(t) \end{bmatrix} = \begin{bmatrix} 0 & 0 & 1 & 0 \\ 0 & 0 & 0 & 0 \end{bmatrix}\mathbf{x} + \begin{bmatrix} 0 & 0 \\ 0 & 1 \end{bmatrix}\mathbf{u} \tag{6.2-6}$$

Of course, the different specifications of $\mathbf{y}(t)$ would have fundamental impact on how the pilot might control the aircraft and on his opinion of the resulting response. Similar alternatives must be considered in designing regulators for all systems in which the set point varies.

Open-Loop Equilibrium

The "mixing logic" that converts desired values of the command variable $\mathbf{y}^*$ to the set point, defined by $\mathbf{x}^*$ and $\mathbf{u}^*$, must be determined (Fig. 6.2-1). At equilibrium, $\dot{\mathbf{x}}(t) = \mathbf{0}$, so the object is to satisfy Eqs. 6.2-1 and 6.2-2 simultaneously, with the left side of Eq. 6.2-1 set to zero. This can be written as the single equation

$$\begin{bmatrix} \mathbf{0} \\ \mathbf{y}^* \end{bmatrix} = \begin{bmatrix} \mathbf{F} & \mathbf{G} \\ \mathbf{H_x} & \mathbf{H_u} \end{bmatrix} \begin{bmatrix} \mathbf{x}^* \\ \mathbf{u}^* \end{bmatrix} + \begin{bmatrix} \mathbf{L} \\ \mathbf{H_w} \end{bmatrix} \mathbf{w}^* \tag{6.2-7a}$$

or

$$\begin{bmatrix} \mathbf{F} & \mathbf{G} \\ \mathbf{H_x} & \mathbf{H_u} \end{bmatrix} \begin{bmatrix} \mathbf{x}^* \\ \mathbf{u}^* \end{bmatrix} \triangleq \mathbf{A} \begin{bmatrix} \mathbf{x}^* \\ \mathbf{u}^* \end{bmatrix} = \begin{bmatrix} -\mathbf{L}\mathbf{w}^* \\ \mathbf{y}^* - \mathbf{H_w}\mathbf{w}^* \end{bmatrix} \tag{6.2-7b}$$

where $\mathbf{A}$ is an $(n+r) \times (n+m)$ matrix and $\mathbf{w}^*$ is a constant disturbance. If $r = m$, $\mathbf{A}$ is square; if $\mathbf{A}$ is nonsingular, the set point is readily defined by

$$\begin{bmatrix} \mathbf{x}^* \\ \mathbf{u}^* \end{bmatrix} = \mathbf{A}^{-1} \begin{bmatrix} -\mathbf{L}\mathbf{w}^* \\ \mathbf{y}^* - \mathbf{H_w}\mathbf{w}^* \end{bmatrix} \tag{6.2-8}$$

Evidently the set point depends on the constant disturbance as well as the command.

If $r < m$, $\mathbf{A}$ is not square, but the set point can be specified by the right pseudoinverse of $\mathbf{A}$ (Eq. 2.2-34),

$$\begin{bmatrix} \mathbf{x}^* \\ \mathbf{u}^* \end{bmatrix} = \mathbf{A}^R \begin{bmatrix} -\mathbf{L}\mathbf{w}^* \\ \mathbf{y}^* - \mathbf{H_w}\mathbf{w}^* \end{bmatrix} = \mathbf{A}^T(\mathbf{A}\mathbf{A}^T)^{-1} \begin{bmatrix} -\mathbf{L}\mathbf{w}^* \\ \mathbf{y}^* - \mathbf{H_w}\mathbf{w}^* \end{bmatrix} \tag{6.2-9}$$

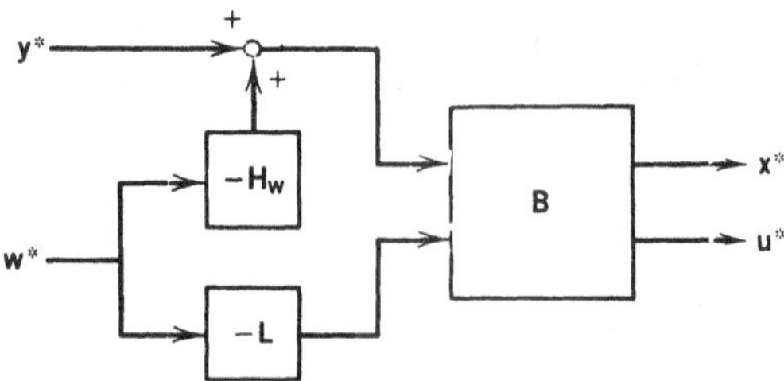

FIGURE 6.2-1 Determination of set point from command input and disturbance.

as long as $\mathbf{AA}^T$ has full rank $(n+r)$. Because this is the underdetermined case, there are an infinity of solutions; the solution of Eq. 6.2-9 is minimal in a least-squares sense. Thus there is an opportunity to specify some of the set-point control settings. One approach would be to fix $(m-r)$ controls at desired values and to calculate the remaining m controls using the "true" inverse of the resulting $\mathbf{A}$ (Eq. 6.2-8). Alternately, a weighted pseudo-inverse (Eq. 2.2-38) could be used to calculate all r control settings,

$$\begin{bmatrix}\mathbf{x}^*\\ \mathbf{u}^*\end{bmatrix} = \mathbf{A}^{WR}\begin{bmatrix}-\mathbf{Lw}^*\\ \mathbf{y}^*-\mathbf{H_w w}^*\end{bmatrix} = \mathbf{R}^{-1}\mathbf{A}^T(\mathbf{AR}^{-1}\mathbf{A}^T)^{-1}\begin{bmatrix}-\mathbf{Lw}^*\\ \mathbf{y}^*-\mathbf{H_w w}^*\end{bmatrix} \tag{6.2-10}$$

$\mathbf{R}$ is a symmetric, positive definite matrix that enhances or diminishes the significance of elements of $\mathbf{x}^*$ and $\mathbf{u}^*$ in achieving equilibrium.

Assuming that $r=m$, an expression for $\mathbf{A}^{-1}$ can be found from Eq. 6.2-7 by elimination. $\mathbf{B}$ is defined to be the inverse of $\mathbf{A}$,

$$\mathbf{B} = \begin{bmatrix}\mathbf{B}_{11} & \mathbf{B}_{12}\\ \mathbf{B}_{21} & \mathbf{B}_{22}\end{bmatrix} \triangleq \mathbf{A}^{-1} \tag{6.2-11}$$

and the partitions of $\mathbf{B}$ have the same dimensions as the partitions of $\mathbf{A}$. The set point is a function of the constant disturbance and the command input (Fig. 6.2-1),

$$\mathbf{x}^* = -\mathbf{B}_{11}\mathbf{Lw}^* + \mathbf{B}_{12}(\mathbf{y}^* - \mathbf{H_w w}^*) \tag{6.2-12}$$

$$\mathbf{u}^* = -\mathbf{B}_{21}\mathbf{Lw}^* + \mathbf{B}_{22}(\mathbf{y}^* - \mathbf{H_w w}^*) \tag{6.2-13}$$

with

$$\mathbf{B}_{11} = \mathbf{F}^{-1}(-\mathbf{GB}_{21} + \mathbf{I}_n) \tag{6.2-14}$$

$$\mathbf{B}_{12} = -\mathbf{F}^{-1}\mathbf{GB}_{22} \tag{6.2-15}$$

$$\mathbf{B}_{21} = -\mathbf{B}_{22}\mathbf{H_x F}^{-1} \tag{6.2-16}$$

$$\mathbf{B}_{22} = (-\mathbf{H_x F}^{-1}\mathbf{G} + \mathbf{H_u})^{-1} \tag{6.2-17}$$

From Section 2.5, it could have been anticipated that $\mathbf{A}$ would be singular if $\mathbf{F}$ was singular, but Eq. 6.2-17 indicates that there also is a problem if $(-\mathbf{H_x F}^{-1}\mathbf{G} + \mathbf{H_u})$ is singular. For the aircraft example presented earlier, $\mathbf{F}$ can be inverted and $\mathbf{B}_{22}$ exists if the command variable is specified by Eq. 6.2-5; however, if $\mathbf{y}(t)$ is given by Eq. 6.2-6, $\mathbf{B}_{22}$ cannot be computed. The distinction is readily described in physical terms. If pitch *angle* is a command variable, then constant y_1 signifies constant x_4 and zero x_3. If pitch *rate* is a command variable, then constant y_1 implies constant x_3 and steadily increasing x_4; therefore, $\mathbf{x}$ cannot be entirely constant, and $\mathbf{B}_{22}$

does not exist. Put another way, a row of $(\mathbf{F}, \mathbf{G})$ and a row of $(\mathbf{H_x}, \mathbf{H_u})$ are the same; therefore, $\mathbf{A}$ is singular.

The latter command variable implies that steady-state command response involves *quasistatic equilibrium*: the command vector, $\mathbf{y}^*$, is indeed constant but elements of $\mathbf{x}^*$ and $\mathbf{u}^*$ vary in time because one or more elements of $\mathbf{x}^*$ are integrals of $\mathbf{y}^*$. The following provides an example in which quasistatic equilibrium occurs:

$$\begin{bmatrix} \dot{\mathbf{x}}_1(t) \\ \dot{\mathbf{x}}_2(t) \end{bmatrix} = \begin{bmatrix} \mathbf{F}_{11} & \mathbf{F}_{12} \\ \mathbf{F}_{21} & \mathbf{0} \end{bmatrix} \begin{bmatrix} \mathbf{x}_1(t) \\ \mathbf{x}_2(t) \end{bmatrix} + \begin{bmatrix} \mathbf{G}_1 \\ \mathbf{G}_2 \end{bmatrix} \mathbf{u}(t) \tag{6.2-18}$$

$$\mathbf{y}^*(t) = \mathbf{H}_{\mathbf{x}_1}\mathbf{x}_1^*(t) + \mathbf{H_u}\mathbf{u}^*(t) \tag{6.2-19}$$

$\mathbf{x}_2$ is an n_2-vector of states that are integrals of the n_1-vector $\mathbf{x}_1$ (with $n_1 + n_2 = n$) and the m-dimensional control $\mathbf{u}$. $\mathbf{y}$ is an m-dimensional linear combination of $\mathbf{x}_1$ and $\mathbf{u}$. If $\mathbf{F}_{12} = \mathbf{0}$, the problem could be solved by computing the reduced-order equilibrium of $\mathbf{x}_1$. Then

$$\mathbf{x}_1^* = -\mathbf{F}_{11}^{-1}\mathbf{G}_1(-\mathbf{H}_{\mathbf{x}_1}\mathbf{F}_{11}^{-1}\mathbf{G}_1 + \mathbf{H_u})^{-1}\mathbf{y}^* \tag{6.2-20}$$

$$\mathbf{u}^* = (-\mathbf{H}_{\mathbf{x}_1}\mathbf{F}_{11}^{-1}\mathbf{G}_1 + \mathbf{H_u})^{-1}\mathbf{y}^* \tag{5.2-21}$$

and

$$\begin{aligned} \mathbf{x}_2^*(t) &= \mathbf{x}_2^*(0) + \int_0^t (\mathbf{F}_{21}\mathbf{x}_1^* + \mathbf{G}_2\mathbf{u}^*)\, dt \\ &= \mathbf{x}_2^*(0) + (\mathbf{F}_{21}\mathbf{x}_1^* + \mathbf{G}_2\mathbf{u}^*)t \end{aligned} \tag{6.2-22}$$

The quasistatic solution is more complicated when $\mathbf{F}_{12} \neq \mathbf{0}$. In order to balance Eq. 6.2-18, the following must hold:

$$\begin{bmatrix} \mathbf{x}_1^*(t) \\ \mathbf{u}^*(t) \end{bmatrix} = \begin{bmatrix} \mathbf{F}_{11} & \mathbf{G}_1 \\ \mathbf{H}_{\mathbf{x}_1} & \mathbf{H_u} \end{bmatrix}^{-1} \begin{bmatrix} \dot{\mathbf{x}}_1^*(t) - \mathbf{F}_{12}\mathbf{x}_2^*(t) \\ \mathbf{y}^* \end{bmatrix} \tag{6.2-23}$$

$$\mathbf{x}_2^*(t) = \mathbf{x}_2^*(0) + \int_0^t [\mathbf{F}_{21}\mathbf{x}_1^*(t) + \mathbf{G}_2\mathbf{u}^*(t)]\, dt \tag{6.2-24}$$

The coupling of $\mathbf{x}_2^*$ into the equation for $\mathbf{x}_1^*$ could cause some elements of $\mathbf{x}_1^*$ and $\mathbf{u}^*$ to vary in time, although $\mathbf{y}^*$ maintains a constant value. This means that $\dot{\mathbf{x}}_1^*(t)$ may be nonzero. Note also that $\mathbf{x}_2^*(0)$ affects the solution for $\mathbf{x}_1^*$ and $\mathbf{u}^*$, so it could be necessary to specify $\mathbf{x}_2^*(0)$ to satisfy constraints on $\mathbf{u}^*(0)$.

The problem is to solve the $(n_1 + m)$ scalar equations represented by Eq. 6.2-23 subject to the m constraints provided by Eq. 6.2-19. If $\mathbf{y}^*$ is constant, then $\dot{\mathbf{y}}^*$ is zero; differentiating Eq. 6.2-19 provides another m

constraints:

$$\begin{aligned}\dot{\mathbf{y}}^* = \mathbf{0} &= \mathbf{H}_{\mathbf{x}_1}\dot{\mathbf{x}}_1^* + \mathbf{H}_{\mathbf{u}}\dot{\mathbf{u}}^* \\ &= \mathbf{H}_{\mathbf{x}_1}(\mathbf{F}_{11}\mathbf{x}_1^* + \mathbf{F}_{12}\mathbf{x}_2^* + \mathbf{G}_1\mathbf{u}^*) + \mathbf{H}_{\mathbf{u}}\dot{\mathbf{u}}^* \end{aligned} \tag{6.2-25}$$

The number of additional constraints is increased to n_1 if $\dot{\mathbf{x}}_1^*$ is zero; this occurs only if

$$\mathbf{0} = \mathbf{F}_{11}\mathbf{x}_1^* + \mathbf{F}_{12}\mathbf{x}_2^* + \mathbf{G}_1\mathbf{u}^* \tag{6.2-26}$$

which requires, in turn, that

$$\mathbf{H}_{\mathbf{u}}\dot{\mathbf{u}}^* = \mathbf{0} \tag{6.2-27}$$

Equation 6.2-27 can be zero because either $\dot{\mathbf{u}}^*$ or $\mathbf{H}_{\mathbf{u}}$ is zero; the latter allows nonzero $\dot{\mathbf{u}}^*$ such that

$$\begin{aligned}\mathbf{G}_1\dot{\mathbf{u}}^* &= \mathbf{F}_{12}\dot{\mathbf{x}}_2^* \\ &= \mathbf{F}_{12}(\mathbf{F}_{21}\mathbf{x}_1^* + \mathbf{G}_2\mathbf{u}^*)\end{aligned} \tag{6.2-28}$$

The quasistatic solution then can be expressed (from Eq. 6.2-23) as

$$\begin{aligned}\begin{bmatrix}\mathbf{x}_1^* \\ \mathbf{u}^*(t)\end{bmatrix} &= \begin{bmatrix}\mathbf{F}_{11} & \mathbf{G}_1 \\ \mathbf{H}_{\mathbf{x}_1} & \mathbf{H}_{\mathbf{u}}\end{bmatrix}^{-1}\begin{bmatrix}-\mathbf{F}_{12}\mathbf{x}_2^*(t) \\ \mathbf{y}^*\end{bmatrix} \\ &\triangleq \begin{bmatrix}\mathbf{\Sigma}_{11} & \mathbf{\Sigma}_{12} \\ \mathbf{\Sigma}_{21} & \mathbf{\Sigma}_{22}\end{bmatrix}\begin{bmatrix}-\mathbf{F}_{12}\mathbf{x}_2^*(t) \\ \mathbf{y}^*\end{bmatrix}\end{aligned} \tag{6.2-29}$$

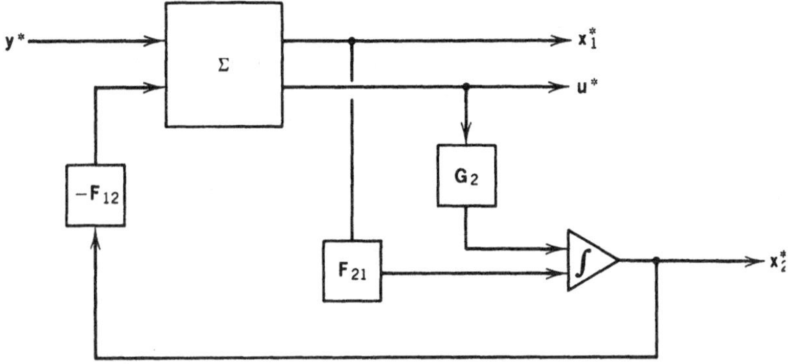

FIGURE 6.2-2 Determination of quasisteady set point from command input.

plus Eq. 6.2-24 (Fig. 6.2-2). If $\mathbf{x}_1^*$ cannot change but $\mathbf{x}_2^*$ is varying, $\mathbf{\Sigma}_{11}$ must be zero. Otherwise, at least some elements of $\mathbf{x}_1^*$ must vary in order to maintain constant $\mathbf{y}^*$. $\mathbf{\Sigma}_{11}$ is then nonzero, and higher derivatives of $\mathbf{y}^*$ establish the additional $(n_1 - m)$ constraints required to form the quasistatic solution.

Non-Zero-Set-Point Regulation

Having determined the set point $(\mathbf{x}^*, \mathbf{u}^*)$ as a function of the command $(\mathbf{y}^*)$, a controller that accepts the set point as the equilibrium point follows directly. Defining perturbations from equilibrium as

$$\tilde{\mathbf{x}}(t) = \mathbf{x}(t) - \mathbf{x}^* \tag{6.2-30}$$

$$\tilde{\mathbf{u}}(t) = \mathbf{u}(t) - \mathbf{u}^* \tag{6.2-31}$$

a linear regulator for the perturbation variables can be expressed as

$$\tilde{\mathbf{u}}(t) = -\mathbf{C}\tilde{\mathbf{x}}(t) \tag{6.2-32}$$

The gain matrix $\mathbf{C}$ need not be optimal for the following to hold, although our objective is to combine the requirement for holding a set point with minimization of a cost function that is quadratic in $\tilde{\mathbf{x}}$ and $\tilde{\mathbf{u}}$. Substituting for the perturbation variables,

$$[\mathbf{u}(t) - \mathbf{u}^*] = -\mathbf{C}[\mathbf{x}(t) - \mathbf{x}^*] \tag{6.2-33a}$$

or

$$\mathbf{u}(t) = \mathbf{u}^*(t) - \mathbf{C}[\mathbf{x}(t) - \mathbf{x}^*] \tag{6.2-33b}$$

When $\mathbf{x}(t) = \mathbf{x}^*$, the control is commanded to its steady-state value $[\mathbf{u}(t) = \mathbf{u}^*]$.

The *non-zero-set-point regulator* can be written directly in terms of $\mathbf{y}^*$, providing further insights into the partitioning of the feedback control and steady-state solutions. For nonsingular $\mathbf{A}$, static equilibrium can be achieved; substituting Eqs. 6.2-12 and 6.2-13 in Eq. 6.2-33b yields

$$\begin{aligned}\mathbf{u}(t) &= [-\mathbf{B}_{21}\mathbf{L}\mathbf{w}^* + \mathbf{B}_{22}(\mathbf{y}^* - \mathbf{H}_w\mathbf{w}^*)] - \mathbf{C}[\mathbf{x}(t) + \mathbf{B}_{11}\mathbf{L}\mathbf{w}^* - \mathbf{B}_{12}(\mathbf{y}^* - \mathbf{H}_w\mathbf{w}^*)] \\ &= (\mathbf{B}_{22} + \mathbf{C}\mathbf{B}_{12})(\mathbf{y}^* - \mathbf{H}_w\mathbf{w}^*) - \mathbf{C}\mathbf{x}(t) - (\mathbf{B}_{21} + \mathbf{C}\mathbf{B}_{11})\mathbf{L}\mathbf{w}^* \\ &= \mathbf{C}_F\mathbf{y}^* - \mathbf{C}_B\mathbf{x}(t) + \mathbf{C}_D\mathbf{w}^* \end{aligned} \tag{6.2-34}$$

where the forward, feedback, and disturbance gain matrices are defined as

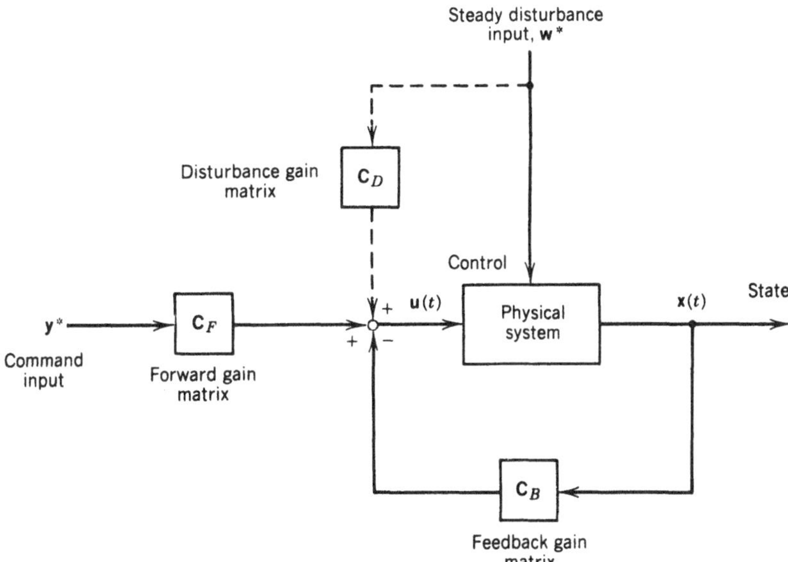

FIGURE 6.2-3 Block diagram of the non-zero-set-point regulator for nonsingular equilibrium.

follows:

$$\mathbf{C}_F = \mathbf{B}_{22} + \mathbf{C}\mathbf{B}_{12} \tag{6.2-35}$$

$$\mathbf{C}_B = \mathbf{C} \tag{6.2-36}$$

$$\mathbf{C}_D = -(\mathbf{B}_{21} + \mathbf{C}\mathbf{B}_{11})\mathbf{L} - (\mathbf{B}_{22} + \mathbf{C}\mathbf{B}_{12})\mathbf{H}_{\mathbf{w}} \tag{6.2-37}$$

As illustrated in Fig. 6.2-3, the result takes the form of a classical "servo regulator," with the principal distinction that all signal paths transmit vector rather than scalar quantities. $\mathbf{C}_F$, $\mathbf{C}_B$, and $\mathbf{C}_D$ are simply algebraic functions of the set-point sensitivities and the gain matrix $\mathbf{C}$.

When $\mathbf{A}$ is singular but Eqs. 6.2-26 and 6.2-27 are satisfied, the quasi-static solution given by Eq. 6.2-29 can be used in the non-zero-set-point regulator. Assuming that $\mathbf{w}^*$ is zero or cannot be measured, Eq. 6.2-33b becomes

$$\mathbf{u}(t) = [-\boldsymbol{\Sigma}_{21}\mathbf{F}_{12}\mathbf{x}_2^*(t) + \boldsymbol{\Sigma}_{22}\mathbf{y}^*] - \mathbf{C}\begin{bmatrix}\mathbf{x}_1(t) - \boldsymbol{\Sigma}_{12}\mathbf{y}^* \\ \mathbf{x}_2(t) - \mathbf{x}_2^*(t)\end{bmatrix} \tag{6.2-38}$$

with $\mathbf{x}_2^*(t)$ given by Eq. 6.2-24. Substituting for $\mathbf{x}_1^*$ and $\mathbf{u}^*(t)$ and assuming

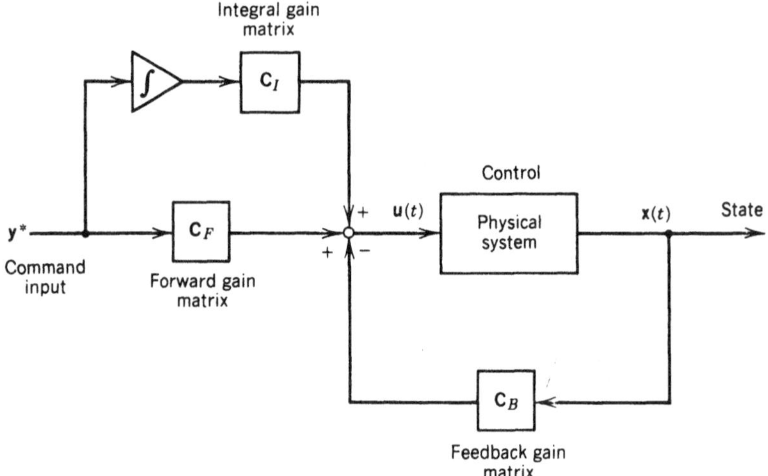

FIGURE 6.2-4 Block diagram of the non-zero-set-point regulator for singular equilibrium.

that $\mathbf{G}_2$ is zero for simplicity,

$$\mathbf{x}_2^*(t) = \mathbf{x}_2(0) + \mathbf{F}_{21}\mathbf{\Sigma}_{12}\int_0^t \mathbf{y}^*\,dt \tag{6.2-39}$$

This allows the control law to be written as

$$\begin{aligned}\mathbf{u}(t) &= (\mathbf{\Sigma}_{22} + \mathbf{C}_1\mathbf{\Sigma}_{12})\mathbf{y}^* - \mathbf{C}_1\mathbf{x}_1(t) - \mathbf{C}_2\mathbf{x}_2(t) + (-\mathbf{\Sigma}_{21}\mathbf{F}_{12} + \mathbf{C}_2)\mathbf{x}_2^*(t)\\ &= \mathbf{C}_F\mathbf{y}^* - \mathbf{C}_B\mathbf{x}(t) + \mathbf{C}_I\mathbf{x}_2^*(t)\end{aligned} \tag{6.2-40}$$

where $\mathbf{C}$ is partitioned into submatrices conformable with $\mathbf{x}_1$ and $\mathbf{x}_2$. By Eq. 6.2-39, $\mathbf{x}_2^*(t)$ is an integral of the command vector while $\mathbf{x}_2(t)$ is an integral of $\mathbf{x}_1(t)$; hence the non-zero-set-point regulator for quasistatic equilibrium contains *integral compensation* of the command input and the associated elements of $\mathbf{x}_1(t)$ (Fig. 6.2-4).

Of course, $\mathbf{y}^*$ cannot be expected to remain constant over the period of control, but the suggested control structures are applicable nonetheless. Just as a classical control design may seek to achieve a certain *scalar step function response*, the non-zero-set-point regulator is designed for *vector step function response*, with $\mathbf{y}^*$ taking a nonzero step at the start of the interval. In practice, $\mathbf{y}^*$ may vary continuously, and the regulator will adjust its set point accordingly.

6.3 COST FUNCTIONS AND CONTROLLER STRUCTURES

Given a linear, time-invariant system ($\mathbf{F}, \mathbf{G}, \mathbf{H}$), control cost function weighting ($\mathbf{Q}, \mathbf{M}, \mathbf{R}$), and input/measurement error statistics ($\mathbf{W}, \mathbf{N}$), the corresponding linear-quadratic-Gaussian (LQG) regulator processes the observation ($\mathbf{z}$) to provide an optimal estimate of the state ($\mathbf{x}$) which, when feed back to the control ($\mathbf{u}$), minimizes a stochastic quadratic cost function (J). Although there is a single optimal controller corresponding to a set of system and cost function parameters, optimality alone does not assure a satisfactory control system design. In linear multivariable control design, the elements of ($\mathbf{Q}, \mathbf{M}, \mathbf{R}$) and ($\mathbf{W}, \mathbf{N}$) are used as *design parameters* that are chosen to provide desirable closed-loop properties over and above the minimization of J (which characterizes *every* design alternative).

A linear multivariable controller may perform many tasks, but which tasks are most important depends on the specific application. Some processes are inherently unstable and depend on the controller for stability. Others are stable, but their dynamic characteristics are nevertheless unacceptable. Without a nominal optimal trajectory to specify a sequence of set points, the control law may be required to compute set points in response to command inputs. It may have to maintain a set point in the presence of unmodeled system parameter variations, measurement biases, or steady disturbances. It may have to accomplish its design objectives with restrictions on measurements, control rates, or computational complexity.

In the best of situations, there would be one control design parameter for each control design objective. Each parameter would have direct effect on one objective, and no objective would be affected by more than one parameter; therefore, the design procedure would be decoupled. If mean-square perturbations of the state and control were the only criteria, diagonal LQG weighting matrices would tend to have this characteristic, although coupling in the dynamic system often would preclude a strict 1:1 relationship between design parameters and objectives. It is less likely that such direct relationships will be found between basic weighting matrix elements and such common criteria as rise time, overshoot, settling time, cross-axis coupling, natural frequency, damping ratio, and bandwidth. Nevertheless, all of these considerations can be accommodated in LQG design by careful attention to the selection of cost functions and controller structures and to the specification of numerical values for the weighting matrices.

This section addresses the definition of cost functions and controller structures for continuous-time systems. Its results can be applied to sampled-data or discrete-time systems with minimal modification, following the developments of earlier chapters.

Specification of Cost Function Weighting Matrices

The linear-quadratic regulator $\mathbf{u}(t) = -\mathbf{C}\mathbf{x}(t)$ provides a feedback law that minimizes an infinite-time-horizon cost function. As noted in Section 5.4, this integral cost has a fixed end time T that approaches ∞ in the limit. If initial condition response is of primary concern, the cost function is best described by Eq. 6.1-2 with $T = t_f$ and $\mathbf{S}(T) = \mathbf{0}$. If the system is subject to persistent excitation, that cost function should be divided by T. The same optimal control gains are obtained either way; the principal difference is in the meaning attributed to the cost function weighting matrices. For this purpose, it is most convenient to use the latter cost function description:

$$J = \lim_{T\to\infty} \frac{1}{2T} \int_0^T \left\{ [\mathbf{x}^T(t) \quad \mathbf{u}^T(t)] \begin{bmatrix} \mathbf{Q} & \mathbf{M} \\ \mathbf{M}^T & \mathbf{R} \end{bmatrix} \begin{bmatrix} \mathbf{x}(t) \\ \mathbf{u}(t) \end{bmatrix} \right\} dt$$

$$= \lim_{T\to\infty} \frac{1}{2T} \int_0^T [\mathbf{x}^T(t)\mathbf{Q}\mathbf{x}(t) + 2\mathbf{x}^T(t)\mathbf{M}\mathbf{u}(t) + \mathbf{u}^T(t)\mathbf{R}\mathbf{u}(t)]\, dt \qquad (6.3\text{-}1)$$

Choosing $\mathbf{Q}$ and $\mathbf{R}$ as positive definite diagonal matrices and $\mathbf{M}$ as zero produces a cost function of the form

$$J = \lim_{T\to\infty} \frac{1}{2T} \int_0^T [q_{11}x_1^2(t) + \cdots + q_{nn}x_n^2(t) + r_{11}u_1^2(t) + \cdots + r_{mm}u_m^2(t)]\, dt \qquad (6.3\text{-}2)$$

If desired values of the zero-lag auto-covariance functions of x_i and u_j can be specified, for example,

$$\phi_{x_ix_i}(0) = \lim_{T\to\infty} \frac{1}{T} \int_0^T x_i^2(t)\, dt, \qquad i = 1 \text{ to } n \qquad (6.3\text{-}3)$$

$$\phi_{u_ju_j}(0) = \lim_{T\to\infty} \frac{1}{T} \int_0^T u_j^2(t)\, dt, \qquad j = 1 \text{ to } m \qquad (6.3\text{-}4)$$

then choosing

$$q_{ii} = \frac{1}{\phi_{x_ix_i}}(0) \qquad (6.3\text{-}5)$$

$$r_{jj} = \frac{1}{\phi_{u_ju_j}}(0) \qquad (6.3\text{-}6)$$

normalizes the cost function. If each integral takes its design value, $J = (n + m)/2$, and a trade-off between state perturbation and control usage is established. All r_{jj} must be positive to avoid singularity in **R**, while some q_{ii} can be zero as long as the stability criteria of Section 5.4 are satisfied.

Off-diagonal elements of **Q** and **R**, as well as all the elements of **M**, can be specified in similar fashion. Optimal control gains then are computed using either the cross-coupled Riccati equation of Section 3.7 or the uncoupled Riccati equation (Eq. 6.1-4) with the diagonalizing transformation of Section 5.4. With design objectives for zero-lag cross-covariance functions,

$$\phi_{x_ix_j}(0) = \lim_{T\to\infty} \frac{1}{T}\int_0^T x_i(t)x_j(t)\,dt, \qquad \begin{cases} i = 1 \text{ to } n \\ j = 1 \text{ to } n \\ j \neq i \end{cases} \tag{6.3-7}$$

$$\phi_{u_iu_j}(0) = \lim_{T\to\infty} \frac{1}{T}\int_0^T u_i(t)u_j(t)\,dt, \qquad \begin{cases} i = 1 \text{ to } m \\ j = 1 \text{ to } m \\ j \neq i \end{cases} \tag{6.3-8}$$

$$\phi_{x_iu_j}(0) = \lim_{T\to\infty} \frac{1}{T}\int_0^T x_i(t)u_j(t)\,dt, \qquad \begin{cases} i = 1 \text{ to } n \\ j = 1 \text{ to } m \end{cases} \tag{6.3-9}$$

The elements of **Q**, **R**, and **M** can be defined as

$$q_{ij} = q_{ji} = \frac{1}{\phi_{x_ix_j}(0)} \tag{6.3-10}$$

$$r_{ij} = r_{ji} = \frac{1}{\phi_{u_iu_j}(0)} \tag{6.3-11}$$

$$m_{ij} = \frac{1}{\phi_{x_iu_j}(0)} \tag{6.3-12}$$

Because the cross-products need not be positive, both positive and negative correlations must be considered in defining **M** and the off-diagonal elements of **Q** and **R**. Furthermore, it is particularly important to assure that the stability criteria of Section 5.4 are satisfied.

Output Weighting in the Cost Function

Variables that are linear combinations of the state and control can be weighted in the cost function. Quadratic weight on such a variable produces equivalent definitions of **Q**, **M**, and **R**, and the resulting control law feeds the full state vector back to the control vector, as in Eq. 6.2-32. Defining a general output vector as

$$\mathbf{y}(t) = \mathbf{H_x x}(t) + \mathbf{H_u u}(t) \tag{6.3-13}$$

with cost weighting matrix, $\mathbf{Q_y}$, the *output-weighted cost function* has the following integrand (or Lagrangian):

$$\tfrac{1}{2}\mathbf{y}^T(t)\mathbf{Q_y}\mathbf{y}(t)=\tfrac{1}{2}[\mathbf{x}^T(t)\quad \mathbf{u}^T(t)]\begin{bmatrix}\mathbf{H_x}^T\\ \mathbf{H_u}^T\end{bmatrix}\mathbf{Q_y}[\mathbf{H_x}\quad \mathbf{H_u}]\begin{bmatrix}\mathbf{x}(t)\\ \mathbf{u}(t)\end{bmatrix}$$

$$=\tfrac{1}{2}[\mathbf{x}^T(t)\quad \mathbf{u}^T(t)]\begin{bmatrix}\mathbf{H_x}^T\mathbf{Q_y}\mathbf{H_x} & \mathbf{H_x}^T\mathbf{Q_y}\mathbf{H_u}\\ \mathbf{H_u}^T\mathbf{Q_y}\mathbf{H_x} & \mathbf{H_u}^T\mathbf{Q_y}\mathbf{H_u}\end{bmatrix}\begin{bmatrix}\mathbf{x}(t)\\ \mathbf{u}(t)\end{bmatrix} \qquad (6.3\text{-}14)$$

where

$$\mathbf{Q}=\mathbf{H_x}^T\mathbf{Q_y}\mathbf{H_x} \qquad (6.3\text{-}15)$$

$$\mathbf{M}=\mathbf{H_x}^T\mathbf{Q_y}\mathbf{H_u} \qquad (6.3\text{-}16)$$

$$\mathbf{R}=\mathbf{H_u}^T\mathbf{Q_y}\mathbf{H_u}+\mathbf{R}_o \qquad (6.3\text{-}17)$$

and $\mathbf{R}_o$ represents the separate cost of control.

It would be reasonable to choose the cost function's $\mathbf{y}$ as the command vector for a non-zero-set-point regulator, but there are other possibilities. One alternative is to provide *state-rate weighting*, which penalizes a quadratic function of the state's derivative. This has the general effect of reducing response overshoots and reducing control activity in the closed-loop controller. Letting $\mathbf{y}(t)=\dot{\mathbf{x}}(t)$ and neglecting disturbance inputs*, $\mathbf{Q}$, $\mathbf{M}$, and $\mathbf{R}$ are defined as

$$\mathbf{Q}=\mathbf{F}^T\mathbf{Q}_{\dot{\mathbf{x}}}\mathbf{F} \qquad (6.3\text{-}18)$$

$$\mathbf{M}=\mathbf{F}^T\mathbf{Q}_{\dot{\mathbf{x}}}\mathbf{G} \qquad (6.3\text{-}19)$$

$$\mathbf{R}=\mathbf{G}^T\mathbf{Q}_{\dot{\mathbf{x}}}\mathbf{G}+\mathbf{R}_o \qquad (6.3\text{-}20)$$

In practice, elements of state, state-rate, control, and output weighting can be included in the same cost function by suitable definition of $\mathbf{Q}$, $\mathbf{M}$, and $\mathbf{R}$.

Implicit Model Following

The LQ regulator can force the controlled system to behave like an ideal system. In *implicit model following*, we want the actual state rate to simulate the state rate of an nth-order homogeneous ideal model specified by

$$\dot{\mathbf{x}}_M(t)=\mathbf{F}_M\mathbf{x}_M(t) \qquad (6.3\text{-}21)$$

*Recall from Section 5.1 that random disturbances increase the cost function but have no effect on the structure of the control law.

Given the same state for actual and ideal systems, the state-rate Lagrangian would be

$$\tfrac{1}{2}[\dot{\mathbf{x}}(t)-\dot{\mathbf{x}}_M(t)]^T\mathbf{Q}_M[\dot{\mathbf{x}}(t)-\dot{\mathbf{x}}_M(t)]$$
$$=\tfrac{1}{2}[\mathbf{x}^T(t) \quad \mathbf{u}^T(t)]\begin{bmatrix}(\mathbf{F}-\mathbf{F}_M)^T\\ \mathbf{G}^T\end{bmatrix}\mathbf{Q}_M[(\mathbf{F}-\mathbf{F}_M) \quad \mathbf{G}]\begin{bmatrix}\mathbf{x}(t)\\ \mathbf{u}(t)\end{bmatrix} \tag{6.3-22}$$

Hence the weighting matrices can be written as follows:

$$\mathbf{Q}=(\mathbf{F}-\mathbf{F}_M)^T\mathbf{Q}_M(\mathbf{F}-\mathbf{F}_M) \tag{6.3-23}$$
$$\mathbf{M}=(\mathbf{F}-\mathbf{F}_M)^T\mathbf{Q}_M\mathbf{G} \tag{6.3-24}$$
$$\mathbf{R}=\mathbf{G}^T\mathbf{Q}_M\mathbf{G}+\mathbf{R}_o \tag{6.3-25}$$

Reference E-1 shows that perfect model following can be achieved with finite controls only if

$$(\mathbf{G}\mathbf{G}^L-\mathbf{I}_n)(\mathbf{F}-\mathbf{F}_M)=\mathbf{0}. \tag{6.3-26}$$

In such case, control gains can be calculated by a simple algebraic manipulation (i.e, without cost function minimization). Otherwise, the LQ regulator provides implicit model following with minimum weighted-square error.*

The LQ cost function alternatives have been developed for zero set point, but they apply as well for non-zero-set point, with $\mathbf{x}^*$ and $\mathbf{u}^*$ derived from $\mathbf{y}^*$ as before. The steady-state response of an ideal model also can specify the set point. Let the model have a control input $\mathbf{u}_M$ such that

$$\dot{\mathbf{x}}_M(t)=\mathbf{F}_M\mathbf{x}(t)+\mathbf{G}_M\mathbf{u}_M(t) \tag{6.3-27}$$

Assuming that $\mathbf{F}_M$ is nonsingular, the equilibrium response to control is

$$\mathbf{x}_M^*=-\mathbf{F}_M^{-1}\mathbf{G}_M\mathbf{u}_M^* \tag{6.3-28}$$

Similarly, the controlled system's equilibrium response is

$$\mathbf{x}^*=-\mathbf{F}^{-1}\mathbf{G}\mathbf{u}^* \tag{6.3-29}$$

Since $\mathbf{x}^*=\mathbf{x}_M^*$ by design, the steady-state control is related to the model's control input by

$$\mathbf{u}^*=(\mathbf{F}^{-1}\mathbf{G})^L(\mathbf{F}_M^{-1}\mathbf{G}_M)\mathbf{u}_M^* \tag{6.3-30}$$

*This control design approach is called *implicit* model following because control gains are chosen such that system response will be the desired response if **F** and **G** are known without error; however, the ideal model appears only in the design equations. In *explicit* model following, ideal model dynamics are implemented in the control system, as described later.

and the non-zero-set-point regulator can be written as

$$\mathbf{u}(t) = \{[\mathbf{F}^{-1}\mathbf{G})^{L} - \mathbf{C}]\mathbf{F}_{M}^{-1}\mathbf{G}_{M}\}\mathbf{u}_{M}^{*} - \mathbf{C}\mathbf{x}(t) \tag{6.3-31}$$

with $\mathbf{C}$ chosen to minimize the time integral of Eq. 6.3-22. (Note that $\mathbf{u}$ and $\mathbf{u}_M$ need not have the same dimension.)

Dynamic Compensation Through State Augmentation

The non-zero-set-point regulator is designed under the assumption that the system to be controlled is modeled without error and that any system disturbances are white random processes. However, as noted in Section 3.7, the basic (or proportional) LQ regulator may be inadequate when these conditions are violated. Control system performance then may be improved by adding new states to the closed-loop system (i.e., by providing dynamic compensation). Just as a Kalman–Bucy filter can introduce dynamic compensation that improves performance in the face of measurement uncertainty, controller state augmentation provides dynamic compensation that improves performance when there are uncertain parameters and constant or slowly varying disturbances of uncertain magnitude. State augmentation can increase control system *robustness* (i.e., the ability to satisfy stability and command response objectives in the presence of uncertainty).

Four types of state augmentation are considered here, and each addresses a specific design problem. In all four cases, the control design procedure is straightforward: the state vector is augmented by new elements that are functions of $\mathbf{x}$, $\mathbf{u}$, and $\mathbf{y}^*$; corresponding differential equations are added to the system model; and the control law that minimizes a quadratic cost function is computed for the augmented systems using the algorithms presented earlier. The added differential equations must be implemented in the control law, providing the dynamic compensation. Minimizing integrals of the command vector error reduces the long-term effect of uncertain parameters or constant disturbances on the set point. Weighting the time-rate-of-change of $\mathbf{u}(t)$ in the cost function introduces low-pass filters* in the control law, reducing feedback gain that could adversely affect uncertain high-frequency modes of motion. The third case is a combination of the first two. Implementing the state equation of an ideal dynamic model provides a reference response against which actual response can be compared; the response error vector then becomes a controlled variable.

Proportional-Integral Compensation

The *proportional-integral* (*PI*) *controller* introduces command-error integrals into the LQ control law; here, it is developed for a nonsingular

*A low-pass filter attenuates high-frequency signals and leaves low-frequency signals relatively unchanged. It consists of an integrator with negative feedback from the output to the input.

command vector, that is, for a system in which **A** can be inverted to produce **B** (Eq. 6.2-11). The system to be controlled is described by the linear, time-invariant model

$$\dot{\mathbf{x}}(t) = \mathbf{F}\mathbf{x}(t) + \mathbf{G}\mathbf{u}(t) + \mathbf{L}\mathbf{w}(t) \tag{6.3-32}$$

and the m-dimensional command input vector, $\mathbf{y}^*$, represents a desired value of an output vector

$$\mathbf{y}(t) = \mathbf{H_x}\mathbf{x}(t) + \mathbf{H_u}\mathbf{u}(t) \tag{6.3-33}$$

The system possesses a set point $(\mathbf{x}^*, \mathbf{u}^*)$, and if **F** and **G** are known without error,

$$\mathbf{x}^* = -\mathbf{B}_{11}\mathbf{L}\mathbf{w}^* + \mathbf{B}_{12}\mathbf{y}^* \tag{6.3-34}$$

$$\mathbf{u}^* = -\mathbf{B}_{21}\mathbf{L}\mathbf{w}^* + \mathbf{B}_{22}\mathbf{y}^* \tag{6.3-35}$$

where $\mathbf{w}^*$ is a constant disturbance. The $\mathbf{B}_{ij}$ are defined by Eqs. 6.2-14 to 6.2-17. Perturbation variables are defined as

$$\tilde{\mathbf{x}}(t) = \mathbf{x}(t) - \mathbf{x}^* \tag{6.3-36}$$

$$\tilde{\mathbf{u}}(t) = \mathbf{u}(t) - \mathbf{u}^* \tag{6.3-37}$$

$$\tilde{\mathbf{y}}(t) = [\mathbf{H_x}\mathbf{x}(t) + \mathbf{H_u}\mathbf{u}(t)] - \mathbf{y}^* \tag{6.3-38a}$$

and $\tilde{\mathbf{y}}(t)$ is

$$\tilde{\mathbf{y}}(t) = \mathbf{H_x}\tilde{\mathbf{x}}(t) + \mathbf{H_u}\tilde{\mathbf{u}}(t) \tag{6.3-38b}$$

when $\mathbf{x}^*$ and $\mathbf{u}^*$ are properly described by Eqs. 6.3-34 and 6.3-35.

If we write the time-integral of $\tilde{\mathbf{y}}(t)$ as the m-vector $\boldsymbol{\xi}(t)$,

$$\boldsymbol{\xi}(t) = \boldsymbol{\xi}(0) + \int_0^t \tilde{\mathbf{y}}(\tau)\, d\tau \tag{6.3-39}$$

then an augmented state equation that includes the integral-state vector $\boldsymbol{\xi}(t)$ can be formed. Neglecting disturbances, this is

$$\begin{bmatrix} \dot{\tilde{\mathbf{x}}}(t) \\ \dot{\boldsymbol{\xi}}(t) \end{bmatrix} = \begin{bmatrix} \mathbf{F} & \mathbf{0} \\ \mathbf{H_x} & \mathbf{0} \end{bmatrix} \begin{bmatrix} \tilde{\mathbf{x}}(t) \\ \boldsymbol{\xi}(t) \end{bmatrix} + \begin{bmatrix} \mathbf{G} \\ \mathbf{H_u} \end{bmatrix} \tilde{\mathbf{u}}(t) \tag{6.3-40a}$$

Expressing this as an $(n + m)$-order differential equation,

$$\dot{\boldsymbol{\chi}}(t) = \mathbf{F}'\boldsymbol{\chi}(t) + \mathbf{G}'\tilde{\mathbf{u}}(t) \tag{6.3-40b}$$

with

$$\boldsymbol{\chi}(t) \triangleq [\tilde{\mathbf{x}}^T(t) \quad \boldsymbol{\xi}^T(t)] \tag{6.3-41}$$

provides a dynamic constraint for minimizing the following cost function:

$$J = \lim_{T\to\infty} \frac{1}{2T} \int_0^T [\boldsymbol{\chi}^T(t) \quad \tilde{\mathbf{u}}^T(t)] \begin{bmatrix} \mathbf{Q} & \mathbf{M} \\ \mathbf{M}^T & \mathbf{R} \end{bmatrix} \begin{bmatrix} \boldsymbol{\chi}(t) \\ \tilde{\mathbf{u}}(t) \end{bmatrix} \Bigg\} dt \tag{6.3-42}$$

The minimizing control law appears as

$$\begin{aligned} \tilde{\mathbf{u}}(t) &= -\mathbf{C}\boldsymbol{\chi}(t) \\ &= -\mathbf{C}_1\tilde{\mathbf{x}}(t) - \mathbf{C}_2\boldsymbol{\xi}(t) \end{aligned} \tag{6.3-43a}$$

or

$$\mathbf{u}(t) = \mathbf{u}^* - \mathbf{C}_1[\mathbf{x}(t) - \mathbf{x}^*] - \mathbf{C}_2 \left\{ \boldsymbol{\xi}(0) + \int_0^t [\mathbf{y}(\tau) - \mathbf{y}^*]\, d\tau \right\} \tag{6.3-43b}$$

For unmeasured (or zero) disturbance inputs, this becomes

$$\mathbf{u}(t) = \mathbf{C}_F\mathbf{y}^* - \mathbf{C}_B\mathbf{x}(t) - \mathbf{C}_I \left\{ \boldsymbol{\xi}(0) + \int_0^t [\mathbf{y}(\tau) - \mathbf{y}^*]\, d\tau \right\} \tag{6.3-44}$$

with $\mathbf{C}_F = (\mathbf{B}_{22} + \mathbf{C}_1\mathbf{B}_{12})$, $\mathbf{C}_B = \mathbf{C}_1$, and $\mathbf{C}_I = \mathbf{C}_2$ (Fig. 6.3-1). $\mathbf{F}'$ is singular because the integral state is included in $\boldsymbol{\chi}(t)$. However, the reduced-order equilibrium solution derived from $\mathbf{F}$ and $\mathbf{G}$ alone is sufficient to define $\mathbf{x}^*$ and $\mathbf{u}^*$.

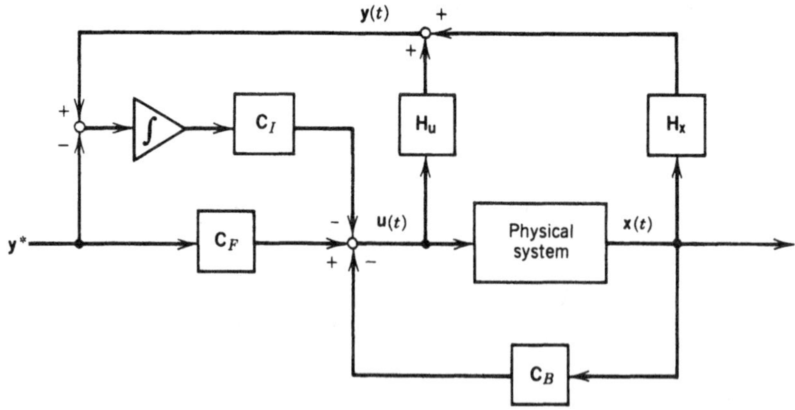

FIGURE 6.3-1 Proportional-integral (PI) regulator for nonsingular command vector.

The principle of the PI controller is that steady-state response error can be zeroed by integrating action. Persistent bias error causes an increasing control setting that ultimately reduces the error. This result is tied to the stability of the closed-loop system. With no disturbance, the closed-loop system perturbation equation (substituting Eq. 6.3-43a in Eq. 6.3-40a) is

$$\begin{bmatrix} \dot{\tilde{\mathbf{x}}}(t) \\ \dot{\boldsymbol{\xi}}(t) \end{bmatrix} = \begin{bmatrix} (\mathbf{F}-\mathbf{G}\mathbf{C}_1) & -\mathbf{G}\mathbf{C}_2 \\ (\mathbf{H}_\mathbf{x}-\mathbf{H}_\mathbf{u}\mathbf{C}_1) & -\mathbf{H}_\mathbf{u}\mathbf{C}_2 \end{bmatrix} \begin{bmatrix} \tilde{\mathbf{x}}(t) \\ \boldsymbol{\xi}(t) \end{bmatrix} \tag{6.3-45}$$

The system is guaranteed to be asymptotically stable if $\mathbf{Q}$, $\mathbf{M}$, and $\mathbf{R}$ are chosen to satisfy the Section 5.4 criteria; therefore, $\tilde{\mathbf{x}}$, $\boldsymbol{\xi}$, and $\tilde{\mathbf{u}}$ go to zero as time increases, indicating that $\mathbf{x}(t)\rightarrow\mathbf{x}^*$, $\mathbf{u}(t)\rightarrow\mathbf{u}^*$, $\tilde{\mathbf{y}}(t)\rightarrow\mathbf{0}$, and $\mathbf{y}(t)\rightarrow\mathbf{y}^*$.

Suppose that the controlled system is described by $(\mathbf{F}_A, \mathbf{G}_A)$ rather than the model assumed for control design $(\mathbf{F}, \mathbf{G})$. In steady-state response to a constant command, the proportional LQ regulator (Eq. 6.2-34) would drive the system to the wrong set point because $\mathbf{B}_{12}$ and $\mathbf{B}_{22}$ were computed incorrectly. The actual steady value of the command vector would be

$$\mathbf{y}_A^* = (-\mathbf{H}_\mathbf{x}\mathbf{F}_A^{-1}\mathbf{G}_A + \mathbf{H}_\mathbf{u})(-\mathbf{H}_\mathbf{x}\mathbf{F}^{-1}\mathbf{G} + \mathbf{H}_\mathbf{u})^{-1}\mathbf{y}^* \tag{6.3-46}$$

which is not equal to $\mathbf{y}^*$ in general. However, the PI controller will provide zero steady-state command error as long as the matrix

$$\mathbf{F}'_{\text{CL}} = \begin{bmatrix} (\mathbf{F}_A-\mathbf{G}_A\mathbf{C}_1) & -\mathbf{G}_A\mathbf{C}_2 \\ (\mathbf{H}_\mathbf{x}-\mathbf{H}_\mathbf{u}\mathbf{C}_1) & -\mathbf{H}_\mathbf{u}\mathbf{C}_2 \end{bmatrix} \tag{6.3-47}$$

is stable, because $\boldsymbol{\xi}(t)\rightarrow\mathbf{0}$ and $\dot{\boldsymbol{\xi}}(t)\rightarrow\mathbf{0}$. This can occur only if $\mathbf{y}(t)\rightarrow\mathbf{y}^*$.

When disturbance inputs are significant, there is a similar result. Let us assume that $\mathbf{F}$ and $\mathbf{G}$ represent both the actual system and the design model, that $\mathbf{L}\mathbf{w}^*$ must be considered, and that $\mathbf{H}_\mathbf{w}$ is zero. The proportional LQ regulator (Eq. 6.2-34) provides the correct steady state if $\mathbf{L}\mathbf{w}^*$ is included in the definition of $\mathbf{x}^*$ and $\mathbf{u}^*$; if it is neglected (i.e., if $\mathbf{C}_D$ is zero), the actual steady-state command response $\mathbf{y}_A^*$ is

$$\begin{aligned} \mathbf{y}_A^* &= (-\mathbf{H}_\mathbf{x}\mathbf{F}^{-1}\mathbf{G} + \mathbf{H}_\mathbf{u})\mathbf{B}_{22}\mathbf{y}^* - \mathbf{H}_\mathbf{x}\mathbf{F}^{-1}\mathbf{L}\mathbf{w}^* \\ &= \mathbf{y}^* - \mathbf{H}_\mathbf{x}\mathbf{F}^{-1}\mathbf{L}\mathbf{w}^* \end{aligned} \tag{6.3-48}$$

Adding the disturbance effect to the PI closed-loop system equation

$$\begin{bmatrix} \dot{\tilde{\mathbf{x}}}(t) \\ \dot{\boldsymbol{\xi}}(t) \end{bmatrix} = \mathbf{F}'_{\text{CL}} \begin{bmatrix} \tilde{\mathbf{x}}(t) \\ \boldsymbol{\xi}(t) \end{bmatrix} + \begin{bmatrix} \mathbf{L} \\ \mathbf{0} \end{bmatrix} \mathbf{w}^* \tag{6.3-49}$$

we see that the steady-state response is

$$\begin{bmatrix} \tilde{\mathbf{x}} \\ \boldsymbol{\xi} \end{bmatrix}_{\text{SS}} = (\mathbf{F}'_{\text{CL}})^{-1} \begin{bmatrix} \mathbf{L} \\ \mathbf{0} \end{bmatrix} \mathbf{w}^* \tag{6.3-50}$$

as long as $\mathbf{F}'_{\mathrm{CL}}$ is nonsingular. Hence the integral state reaches equilibrium, requiring that $\boldsymbol{\xi}(t) \to \mathbf{0}$ and $\mathbf{y}(t) \to \mathbf{y}^*$. The integrating action provides the proper offsetting control to balance the constant disturbance, and the desired command response is achieved.

Proportional-Filter Compensation

Integrators can be added in such a way that all control signals are processed by low-pass filters, restricting the control *rates* used by the LQ regulator. Control-rate weighting in the cost function

$$J = \lim_{T\to\infty} \frac{1}{2T} \int_0^T \left\{ [\tilde{\mathbf{x}}^T(t) \quad \tilde{\mathbf{u}}^T(t)] \begin{bmatrix} \mathbf{Q} & \mathbf{M} \\ \mathbf{M}^T & \mathbf{R}_1 \end{bmatrix} \begin{bmatrix} \tilde{\mathbf{x}}(t) \\ \tilde{\mathbf{u}}(t) \end{bmatrix} + \dot{\tilde{\mathbf{u}}}^T(t)\mathbf{R}_2\dot{\tilde{\mathbf{u}}}(t) \right\} dt \tag{6.3-51}$$

can be accommodated in the dynamic constraint by formally adding control rate equations to the state equation:

$$\dot{\tilde{\mathbf{x}}}(t) = \mathbf{F}\tilde{\mathbf{x}}(t) + \mathbf{G}\tilde{\mathbf{u}}(t) \tag{6.3-52}$$

$$\dot{\tilde{\mathbf{u}}}(t) = \dot{\tilde{\mathbf{u}}}_C(t) \triangleq \mathbf{v}(t) \tag{6.3-53}$$

or

$$\begin{bmatrix} \dot{\tilde{\mathbf{x}}}(t) \\ \dot{\tilde{\mathbf{u}}}(t) \end{bmatrix} = \begin{bmatrix} \mathbf{F} & \mathbf{G} \\ \mathbf{0} & \mathbf{0} \end{bmatrix} \begin{bmatrix} \tilde{\mathbf{x}}(t) \\ \tilde{\mathbf{u}}(t) \end{bmatrix} + \begin{bmatrix} \mathbf{0} \\ \mathbf{I}_m \end{bmatrix} \mathbf{v}(t) \tag{6.3-54}$$

with perturbation variables defined as

$$\tilde{\mathbf{x}}(t) = \mathbf{x}(t) - \mathbf{x}^* \tag{6.3-55}$$

$$\tilde{\mathbf{u}}(t) = \mathbf{u}(t) - \mathbf{u}^* \tag{6.3-56}$$

$$\dot{\tilde{\mathbf{u}}}(t) = \dot{\mathbf{u}}(t) - \dot{\mathbf{u}}^* = \dot{\mathbf{u}}(t) \tag{6.3-57}$$

The set point $(\mathbf{x}^*, \mathbf{u}^*)$ can be derived from a command vector $\mathbf{y}^*$, as before, and $\dot{\mathbf{u}}^* = \mathbf{0}$ by definition. The original control vector becomes a component of the $(n+m)$-dimensional augmented state vector

$$\boldsymbol{\chi}^T(t) \triangleq [\tilde{\mathbf{x}}^T(t) \quad \tilde{\mathbf{u}}^T(t)] \tag{6.3-58}$$

and the new control vector $\mathbf{v}$ is the original control rate. Expressing the cost function as

$$J = \lim_{T\to\infty} \frac{1}{2T} \int_0^T [\boldsymbol{\chi}^T(t)\mathbf{Q}'\boldsymbol{\chi}(t) + \mathbf{v}^T(t)\mathbf{R}'\mathbf{v}(t)]\, dt \tag{6.3-59}$$

the minimizing control law takes the form

$$\begin{aligned}\mathbf{v}(t) &= -\mathbf{C}\boldsymbol{\chi}(t)\\ &= -\mathbf{C}_1\tilde{\mathbf{x}}(t) - \mathbf{C}_2\tilde{\mathbf{u}}(t)\end{aligned} \tag{6.3-60}$$

This is equivalent to

$$\dot{\mathbf{u}}(t) = -\mathbf{C}_1[\mathbf{x}(t) - \mathbf{x}^*] - \mathbf{C}_2[\mathbf{u}(t) - \mathbf{u}^*] \tag{6.3-61}$$

which, for the nonsingular command vector and zero disturbance

$$\mathbf{y}^* = \mathbf{H}_x\mathbf{x}^* + \mathbf{H}_u\mathbf{u}^* \tag{6.3-62}$$

can be written as

$$\begin{aligned}\dot{\mathbf{u}}(t) &= (\mathbf{C}_1\mathbf{B}_{12} + \mathbf{C}_2\mathbf{B}_{22})\mathbf{y}^* - \mathbf{C}_1\mathbf{x}(t) - \mathbf{C}_2\mathbf{u}(t)\\ &= \mathbf{C}_F\mathbf{y}^* - \mathbf{C}_B\mathbf{x}(t) - \mathbf{C}_C\mathbf{u}(t)\end{aligned} \tag{6.3-63}$$

As indicated in the block diagram (Fig. 6.3–2), the control rate is integrated before it is sent to the system. Together with the $\mathbf{C}_C$ feedback to its input, the integrator provides the dynamic compensation in the *proportional-filter* (*PF*) *controller*.

That the PF controller introduces low-pass filtering is readily seen by taking the Laplace transform of Eq. 6.3-63 and rearranging terms. Neglec-

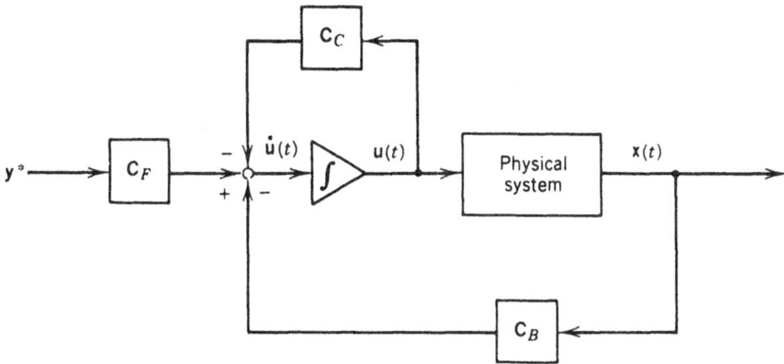

FIGURE 6.3-2 Proportional-filter (PF) regulator for nonsingular command vector.

ting initial conditions,

$$s\mathbf{u}(s) = \mathbf{C}_F\mathbf{y}^*(s) - \mathbf{C}_B\mathbf{x}(s) - \mathbf{C}_C\mathbf{u}(s) \qquad (6.3\text{-}64a)$$

$$(s\mathbf{I}_m + \mathbf{C}_C)\mathbf{u}(s) = \mathbf{C}_F\mathbf{y}^*(s) - \mathbf{C}_B\mathbf{x}(s) \qquad (6.3\text{-}64b)$$

$$\mathbf{u}(s) = (s\mathbf{I}_m + \mathbf{C}_C)^{-1}[\mathbf{C}_F\mathbf{y}^*(s) - \mathbf{C}_B\mathbf{x}(s)] \qquad (6.3\text{-}64c)$$

$(s\mathbf{I}_m + \mathbf{C}_C)^{-1}$ is an $(m \times m)$ strictly proper rational matrix that filters the command and feedback signals (i.e., it is a matrix low-pass filter). Consequently, high-frequency components of both $\mathbf{y}^*$ and $\mathbf{x}(t)$ are attenuated before being transmitted to the controlled system. This has a number of advantageous effects, including *de facto* filtering of measurement noise in the feedback signal, smooth commands to control actuators, and reduced forcing of uncertain high-frequency modes.

Proportional-Integral-Filter Compensation

In analyzing the behavior of electrical networks, Bodė identified desirable frequency response characteristics for single-input/single-output systems (B-3). For a unity feedback system, the open-loop transfer function described as $G(s)H(s)$ (Fig. 6.3-3*a*) should have high gain at low frequency and low gain at high frequency. The logarithm of the amplitude ratio should have a slope of about -20 dB/dec in the vicinity of the crossover frequency (where the amplitude ratio is one and its logarithm is zero)*, so that satisfactory phase margin is retained (Fig. 6.3-3*b*). Such a system evidences good low-frequency tracking, good high-frequency noise rejection, and reasonable tolerance of parameter variations.

Although proportional LQ regulators often provide comparable *multivariable* transfer characteristics for controlled systems, the desired high gain at low frequency and low gain at high frequency is enforced by using *proportional-integral-filter (PIF) compensation* (S-10). PIF compensation combines the integral and filter control structures presented earlier in a single controller, as shown in Fig. 6.3-4. Defining the system and cost variables as before, the augmented state equation is

$$\begin{bmatrix} \dot{\tilde{\mathbf{x}}}(t) \\ \dot{\tilde{\mathbf{u}}}(t) \\ \dot{\xi}(t) \end{bmatrix} = \begin{bmatrix} \mathbf{F} & \mathbf{G} & \mathbf{0} \\ \mathbf{0} & \mathbf{0} & \mathbf{0} \\ \mathbf{H}_\mathbf{x} & \mathbf{H}_\mathbf{u} & \mathbf{0} \end{bmatrix} \begin{bmatrix} \tilde{\mathbf{x}}(t) \\ \tilde{\mathbf{u}}(t) \\ \xi(t) \end{bmatrix} + \begin{bmatrix} \mathbf{0} \\ \mathbf{I}_m \\ \mathbf{0} \end{bmatrix} \mathbf{v}(t) \qquad (6.5\text{-}65)$$

*The crossover frequency establishes the *bandwidth* of the system.

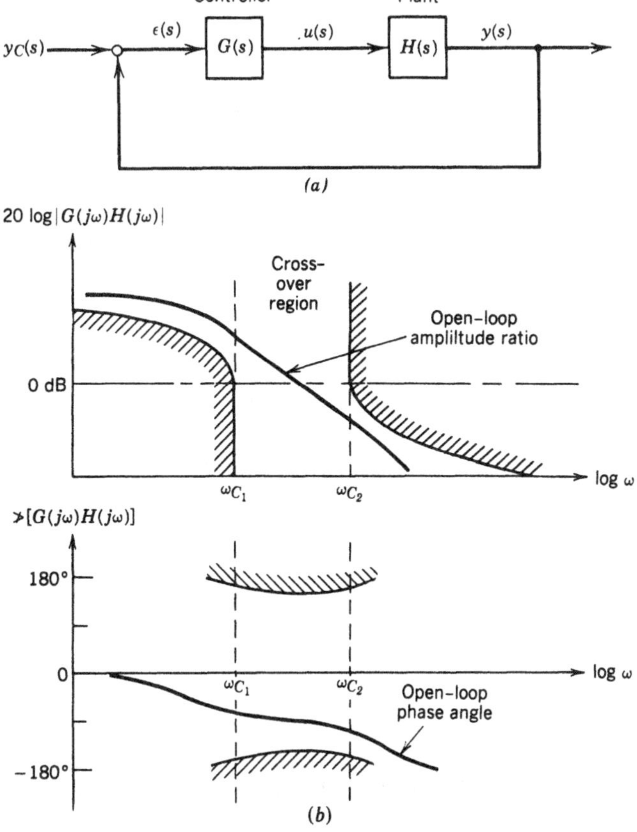

FIGURE 6.3-3 Review of single-input/single-output control design criteria. (*a*) Unity feedback, single-input/single-output closed-loop system; (*b*) suggested limitations on open-loop frequency response for satisfactory performance.

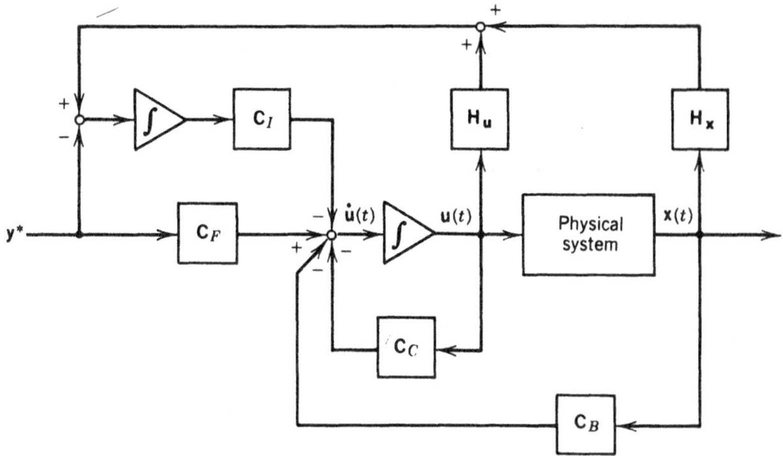

FIGURE 6.3-4 Proportional-integral-filter (PIF) regulator for nonsingular command vector.

and the quadratic cost function is

$$J = \lim_{T\to\infty} \frac{1}{2T} \int_0^T \left\{ [\tilde{\mathbf{x}}^T(t) \quad \tilde{\mathbf{u}}^T(t) \quad \boldsymbol{\xi}^T(t)] \begin{bmatrix} \mathbf{Q}_1 & \mathbf{M} & \mathbf{0} \\ \mathbf{M}^T & \mathbf{R}_1 & \mathbf{0} \\ \mathbf{0} & \mathbf{0} & \mathbf{Q}_2 \end{bmatrix} \begin{bmatrix} \tilde{\mathbf{x}}(t) \\ \tilde{\mathbf{u}}(t) \\ \boldsymbol{\xi}(t) \end{bmatrix} + \mathbf{v}^T(t)\mathbf{R}_2\mathbf{v}(t) \right\} dt \tag{6.3-66}$$

This leads to a control law of the form

$$\mathbf{v}(t) = -\mathbf{C}_1\tilde{\mathbf{x}}(t) - \mathbf{C}_2\tilde{\mathbf{u}}(t) - \mathbf{C}_3\boldsymbol{\xi}(t) \tag{6.3-67a}$$

or

$$\dot{\mathbf{u}}(t) = -\mathbf{C}_1[\mathbf{x}(t) - \mathbf{x}^*] - \mathbf{C}_2[\mathbf{u}(t) - \mathbf{u}^*] - \mathbf{C}_3 \left\{ \boldsymbol{\xi}(0) + \int_0^t [\mathbf{y}(\tau) - \mathbf{y}^*]\, d\tau \right\} \tag{6.3-67b}$$

For a nonsingular command vector, this can be rearranged as

$$\dot{\mathbf{u}}(t) = \mathbf{C}_F\mathbf{y}^* - \mathbf{C}_B\mathbf{x}(t) - \mathbf{C}_C\mathbf{u}(t) - \mathbf{C}_I \left\{ \boldsymbol{\xi}(0) + \int_0^t [\mathbf{y}(\tau) - \mathbf{y}^*]\, d\tau \right\} \tag{6.3-68}$$

and $\dot{\mathbf{u}}(t)$ is integrated before being transmitted to the system. With m controls and m commands, the PIF controller contains $2m$ integrators.

The open-loop relationship between $\mathbf{u}(s)$ and $\mathbf{y}^*(s)$ is analogous to the scalar $G(s)$ in Fig. 6.3-3, although the system of Fig. 6.3-4 has nonunity feedback. From Fig. 6.3-4 and Eq. 6.3-68, the Laplace transform of the command input provides the following control (with zero initial conditions):

$$s\mathbf{u}(s) = \mathbf{C}_F\mathbf{y}^*(s) - \mathbf{C}_C\mathbf{u}(s) + \frac{\mathbf{C}_I\mathbf{y}^*(s)}{s} \tag{6.3-69a}$$

or

$$\mathbf{u}(s) = \frac{1}{s}(s\mathbf{I}_m + \mathbf{C}_C)^{-1}(s\mathbf{C}_F + \mathbf{C}_I)\mathbf{y}^*(s) \tag{6.3-69b}$$

The open-loop controller transfer function is seen to contain pure integration $(1/s)$, low-pass filtering $(s\mathbf{I}_m + \mathbf{C}_C)^{-1}$, and lead compensation $(s\mathbf{C}_F + \mathbf{C}_I)$ in every channel between $\mathbf{y}^*$ and $\mathbf{u}$. This provides infinite gain at zero frequency, noise rejection, and transfer function shaping in the crossover region. Hence the PIF controller has a structure that is consistent with the desired open-loop transfer characteristics (Fig. 6.3-3).

The controlled system generally provides additional low-pass filtering. The Laplace transform of Eq. 6.2-1 is, with zero initial conditions and no disturbance,

$$s\mathbf{x}(s) = \mathbf{F}\mathbf{x}(s) + \mathbf{G}\mathbf{u}(s) \tag{6.3-70}$$

and the open-loop state-control transfer function matrix,

$$\mathbf{H}(s) = (s\mathbf{I}_n - \mathbf{F})^{-1}\mathbf{G} \tag{6.3-71}$$

is a strictly proper rational matrix. If the command vector is related to the state alone (i.e., if $\mathbf{H_u}$ is zero), then the open-loop relationship between $\mathbf{y}^*(s)$ and $\mathbf{y}(s)$ is

$$\mathbf{y}(s) = \frac{1}{s}\mathbf{H_x}(s\mathbf{I}_n - \mathbf{F})^{-1}\mathbf{G}(s\mathbf{I}_m + \mathbf{C}_C)^{-1}(s\mathbf{C}_F + \mathbf{C}_I)\mathbf{y}^*(s) \tag{6.3-72}$$

Of course, if $\mathbf{H_u} \neq \mathbf{0}$, there is feedthrough of $\mathbf{u}(s)$ around the system's $(s\mathbf{I}_n - \mathbf{F})^{-1}$ "filter." Furthermore, the gain-phase relationship can be non-minimum-phase if $\mathbf{H_x}(s\mathbf{I}_n - \mathbf{F})^{-1}\mathbf{G}$ contains positive poles or zeros. Nevertheless, the PIF controller is stabilizing as long as the criteria of Section 5.4 are satisfied.

Explicit Model Following

If a model with desirable dynamic properties can be specified, then an LQ regulator that follows the model can be constructed. The implicit-model-following controller introduced above is a proportional LQ regulator that accomplishes this task when there is negligible uncertainty in **F** and **G** and when disturbance levels are low. It does not require state augmentation, and the feedback gains often can be kept relatively small. With significant disturbances and system uncertainty, better model following can be obtained by *explicit model following*, in which the actual system is controlled as a function of the error between the ideal and actual responses (T-3). The equations of the ideal model can be viewed as providing dynamic compensation. Feedback gains are necessarily high, to allow good tracking of the model without "contamination" by the actual system's closed-loop response.

As illustrated in Fig. 6.3-5, both implicit and explicit model followers place poles (or eigenvalues) in desired locations; however, the latter introduces new poles in the desired locations and pushes the old poles to frequencies beyond the region of interest. Uncertainties in **F** and **G** may shift the actual locations of the closed-loop roots, but that has no effect on

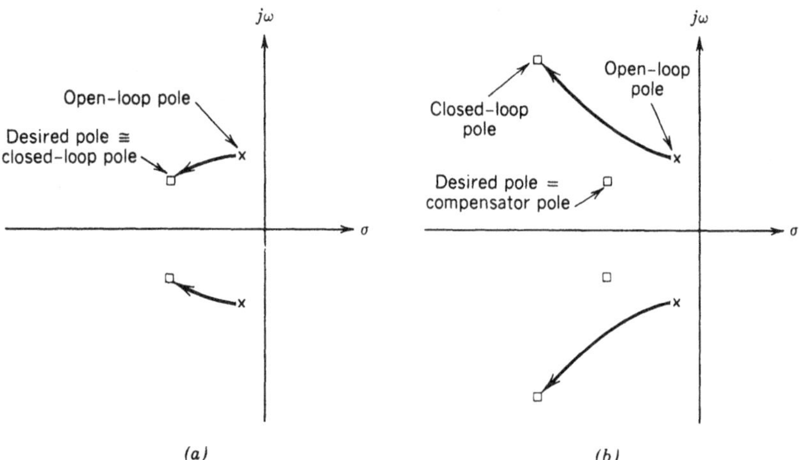

FIGURE 6.3-5 Comparison of closed-loop root loci for implicit and explicit model following.

the dominant compensator roots of the explicit model follower. Disturbances drive the actual system but not the ideal model; the high bandwidth of the actual closed-loop system lets controls act quickly to null resulting errors.

To derive the explicit-model-following control law, a $2n$-dimensional perturbation differential equation representing the nth-order model and system is written:

$$\begin{bmatrix} \dot{\tilde{\mathbf{x}}}(t) \\ \dot{\tilde{\mathbf{x}}}_M(t) \end{bmatrix} = \begin{bmatrix} \mathbf{F} & \mathbf{0} \\ \mathbf{0} & \mathbf{F}_M \end{bmatrix} \begin{bmatrix} \tilde{\mathbf{x}}(t) \\ \tilde{\mathbf{x}}_M(t) \end{bmatrix} + \begin{bmatrix} \mathbf{G} & \mathbf{0} \\ \mathbf{0} & \mathbf{G}_M \end{bmatrix} \begin{bmatrix} \tilde{\mathbf{u}}(t) \\ \tilde{\mathbf{u}}_M(t) \end{bmatrix} \tag{6.3-73}$$

An n-dimensional output vector representing the error between $\tilde{\mathbf{x}}(t)$ and $\tilde{\mathbf{x}}_M(t)$ is formed,

$$\begin{aligned} \tilde{\mathbf{y}}(t) &= \tilde{\mathbf{x}}(t) - \tilde{\mathbf{x}}_M(t) \\ &= [\mathbf{I}_n \quad -\mathbf{I}_n] \begin{bmatrix} \tilde{\mathbf{x}}(t) \\ \tilde{\mathbf{x}}_M(t) \end{bmatrix} \end{aligned} \tag{6.3-74}$$

and a quadratic cost function of the output and control is defined:

$$\begin{aligned} J &= \lim_{T\to\infty} \frac{1}{2T} \int_0^T [\tilde{\mathbf{y}}^T(t)\mathbf{Q}\tilde{\mathbf{y}}(t) + \tilde{\mathbf{u}}^T(t)\mathbf{R}\tilde{\mathbf{u}}(t)]\, dt \\ &= \lim_{T\to\infty} \frac{1}{2T} \int_0^T \left\{ [\tilde{\mathbf{x}}^T(t) \quad \tilde{\mathbf{x}}_M^T(t)] \begin{bmatrix} \mathbf{Q} & -\mathbf{Q} \\ -\mathbf{Q} & \mathbf{Q} \end{bmatrix} \begin{bmatrix} \tilde{\mathbf{x}}(t) \\ \tilde{\mathbf{x}}_M(t) \end{bmatrix} + \tilde{\mathbf{u}}^T(t)\mathbf{R}\tilde{\mathbf{u}}(t) \right\} dt \end{aligned} \tag{6.3-75}$$

The linear-quadratic regulator is

$$\tilde{\mathbf{u}}(t) = -\mathbf{C}_1\tilde{\mathbf{x}}(t) - \mathbf{C}_2\tilde{\mathbf{x}}_M(t) \tag{6.3-76a}$$

or

$$\mathbf{u}(t) = [\mathbf{u}^* + \mathbf{C}_1\mathbf{x}^* + \mathbf{C}_2\mathbf{x}_M^*] - \mathbf{C}_1\mathbf{x}(t) - \mathbf{C}_2\mathbf{x}_M(t) \tag{6.3-76b}$$

Presumably, we wish $\mathbf{x}^*$ to equal $\mathbf{x}_M^*$, which is, in turn, a function of $\mathbf{u}_M^*$:

$$\mathbf{x}^* = \mathbf{x}_M^* = -\mathbf{F}_M^{-1}\mathbf{G}_M\mathbf{u}_M^* \tag{6.3-77}$$

Because

$$\mathbf{x}^* = -\mathbf{F}^{-1}\mathbf{G}\mathbf{u}^* \tag{6.3-78}$$

the equilibrium value of the actual control is

$$\begin{aligned}\mathbf{u}(t) &= [(\mathbf{F}^{-1}\mathbf{G})^L - \mathbf{C}_1 - \mathbf{C}_2]\mathbf{F}_M^{-1}\mathbf{G}_M\mathbf{u}_M^* - \mathbf{C}_1\mathbf{x}(t) - \mathbf{C}_2\mathbf{x}_M(t)\\ &= \mathbf{C}_F\mathbf{u}_M^* - \mathbf{C}_B\mathbf{x}(t) - \mathbf{C}_M\mathbf{x}_M(t)\end{aligned} \tag{6.3-79}$$

as illustrated in Fig. 6.3-6. This controller is seen to consist of a conventional LQ feedback loop plus a prefilter containing the ideal model. Computation of the feedback gain, $\mathbf{C}_B$, is unaffected by the presence of the model (i.e., it is the gain matrix that would be found from $\mathbf{Q}$ and $\mathbf{R}$ in the simplest case). This occurs because the $(2n \times 2n)$ algebraic Riccati equation

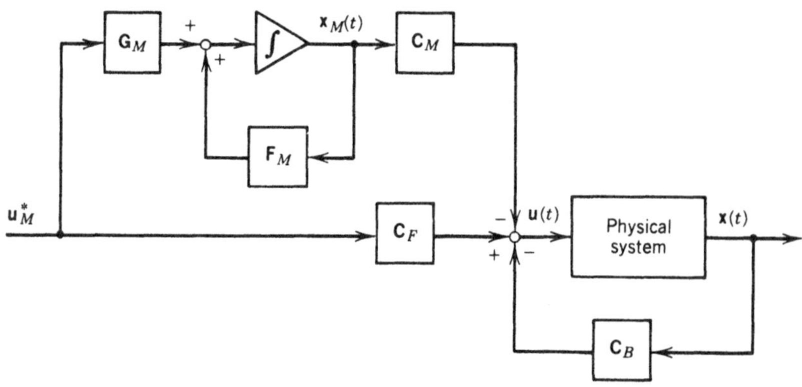

FIGURE 6.3-6 Explicit-model-following regulator.

specified by Eqs. 6.3-73 to 6.3-75 can be partitioned as follows:

$$\mathbf{0} = -\begin{bmatrix} \mathbf{F}^T & \mathbf{0} \\ \mathbf{0} & \mathbf{F}_M^T \end{bmatrix} \begin{bmatrix} \mathbf{S}_{11} & \mathbf{S}_{12} \\ \mathbf{S}_{12}^T & \mathbf{S}_{22} \end{bmatrix} - \begin{bmatrix} \mathbf{S}_{11} & \mathbf{S}_{12} \\ \mathbf{S}_{12}^T & \mathbf{S}_{22} \end{bmatrix} \begin{bmatrix} \mathbf{F} & \mathbf{0} \\ \mathbf{0} & \mathbf{F}_M \end{bmatrix}$$

$$- \begin{bmatrix} \mathbf{Q} & -\mathbf{Q} \\ -\mathbf{Q} & \mathbf{Q} \end{bmatrix} + \begin{bmatrix} \mathbf{S}_{11} & \mathbf{S}_{12} \\ \mathbf{S}_{12}^T & \mathbf{S}_{22} \end{bmatrix} \begin{bmatrix} \mathbf{G}\mathbf{R}^{-1}\mathbf{G}^T & \mathbf{0} \\ \mathbf{0} & \mathbf{0} \end{bmatrix} \begin{bmatrix} \mathbf{S}_{11} & \mathbf{S}_{12} \\ \mathbf{S}_{12}^T & \mathbf{S}_{22} \end{bmatrix} \tag{6.3-80}$$

This can be written in the following three equations:

$$\mathbf{0} = -\mathbf{F}^T\mathbf{S}_{11} - \mathbf{S}_{11}\mathbf{F} - \mathbf{Q} + \mathbf{S}_{11}\mathbf{G}\mathbf{R}^{-1}\mathbf{G}^T\mathbf{S}_{11} \tag{6.3-81}$$

$$\begin{aligned} \mathbf{0} &= -\mathbf{F}^T\mathbf{S}_{12} - \mathbf{S}_{12}\mathbf{F}_M + \mathbf{S}_{11}\mathbf{G}\mathbf{R}^{-1}\mathbf{G}^T\mathbf{S}_{12} + \mathbf{Q} \\ &= (-\mathbf{F}^T + \mathbf{S}_{11}\mathbf{G}\mathbf{R}^{-1}\mathbf{G}^T)\mathbf{S}_{12} - \mathbf{S}_{12}\mathbf{F}_M + \mathbf{Q} \end{aligned} \tag{6.3-82}$$

$$\mathbf{0} = -\mathbf{F}_M^T\mathbf{S}_{22} - \mathbf{S}_{22}\mathbf{F}_M - \mathbf{Q} + \mathbf{S}_{12}^T\mathbf{G}\mathbf{R}^{-1}\mathbf{G}^T\mathbf{S}_{12} \tag{6.3-83}$$

The $(m \times 2n)$ LQ control gain matrix is

$$\begin{aligned} \mathbf{C} &= \mathbf{R}^{-1}[\mathbf{G}^T \quad \mathbf{0}] \begin{bmatrix} \mathbf{S}_{11} & \mathbf{S}_{12} \\ \mathbf{S}_{12}^T & \mathbf{S}_{22} \end{bmatrix} \\ &= \mathbf{R}^{-1}\mathbf{G}^T[\mathbf{S}_{11} \quad \mathbf{S}_{12}] \\ &= [\mathbf{C}_B \quad \mathbf{C}_M] \end{aligned} \tag{6.3-84}$$

$\mathbf{C}_B$ depends only on $\mathbf{S}_{11}$, which is independent of $\mathbf{S}_{12}$ and $\mathbf{S}_{22}$; however, $\mathbf{C}_M$ is derived from $\mathbf{S}_{12}$, whose solution is affected by $\mathbf{S}_{11}$. $\mathbf{S}_{22}$ need not be calculated, as it does not affect the control gain.

It should be noted that zero set-point error depends on accurate knowledge of **F** and **G** (Eq. 6.3-79). To obtain zero set-point error with uncertain **F** and **G** or with disturbances, it would be necessary to add integral compensation to the explicit model follower.

The *tracking problem* can be posed as an extension of explicit model following. The objective is to minimize the quadratic error between an output vector of the controlled system and the output of a system whose motions are to be followed, such as a target or the output of a dynamic "command generator" that prescribes a planned maneuver. The definition of $\tilde{\mathbf{y}}(t)$ (Eq. 6.3-74) is replaced by $\tilde{\mathbf{y}}(t) = \mathbf{H}\tilde{\mathbf{x}}(t) - \mathbf{H}_M\tilde{\mathbf{x}}_M(t)$, where $\tilde{\mathbf{x}}(t)$ and $\tilde{\mathbf{x}}_M(t)$ are the states of the two systems. The systems need not have the same dimensions, and the dimension of $\tilde{\mathbf{y}}(t)$ may be less than n, as defined by the output matrices, **H** and $\mathbf{H}_M$. The tracking control law, which takes the form of Eq. 6.3-76, requires knowledge of $\tilde{\mathbf{x}}_M(t)$ as well as $\tilde{\mathbf{x}}(t)$. In practical application, it may be necessary to estimate $\tilde{\mathbf{x}}_M(t)$ from external measurements, and implementation clearly is simplified if the system to be tracked can be represented by a reduced-order model.

Transformed and Partitioned Solutions

The motions of a given physical system may be described most naturally by a particular state vector, but an infinity of mathematically equivalent state vectors could be defined using *similarity transformations*. For example, the angles and angular rates of a robot manipulator's arm are natural components for a state vector; however, the dynamic modes of motion may couple the angular responses, so a decomposition that identifies modal response variables could be applied. The control design procedure would begin by transforming variables and equations to modal coordinates; the control design algorithms would be applied to the transformed system; and the resulting control structure would be transformed back to the original coordinates. This approach would be advantageous if modal control was the principal concern or if a high-order system was to be controlled by a low-order regulator based on the dominant modes of motion.

Two equivalent procedures could be followed—transformation of the controlled system (as above) or transformation of the cost function alone—and both use the similarity transformations of Section 2.2. Given the system

$$\dot{\mathbf{x}}_1(t) = \mathbf{F}_1\mathbf{x}_1(t) + \mathbf{G}_1\mathbf{u}(t) \tag{6.3-85}$$

and the state vector transformation

$$\mathbf{x}_2(t) = \mathbf{T}\mathbf{x}_1(t) \tag{6.3-86}$$

the transformed system

$$\dot{\mathbf{x}}_2(t) = \mathbf{F}_2\mathbf{x}_2(t) + \mathbf{G}_2\mathbf{u}(t) \tag{6.3-87}$$

is equivalent to the original system with

$$\mathbf{F}_2 = \mathbf{T}\mathbf{F}_1\mathbf{T}^{-1} \tag{6.3-88}$$

$$\mathbf{G}_2 = \mathbf{T}\mathbf{G}_1 \tag{6.3-89}$$

Defining $\mathbf{T}$ as a modified modal matrix (Section 2.2), $\mathbf{F}_2$ is a block-diagonal matrix (for distinct roots), and $\mathbf{x}_2$ is a modal state vector. A quadratic cost function then is defined and minimized for the transformed system. Alternately, the system equations can be left unchanged, and the modal matrix can be used to transform the cost function

$$J = \lim_{T\to\infty} \frac{1}{2T} \int_0^T [\mathbf{x}^T(t)\mathbf{T}^T\mathbf{Q}\mathbf{T}\mathbf{x}(t) + \mathbf{u}^T(t)\mathbf{R}\mathbf{u}(t)]\, dt \tag{6.3-90}$$

where $\mathbf{Q}$ weights perturbations in $\mathbf{x}_2(t)$.

Whether or not the system has been diagonalized, *partitioned* or *reduced-order controllers* can be designed by applying *truncation* or *residualization* to the system equations. Suppose that the system

$$\dot{\mathbf{x}}(t) = \mathbf{F}\mathbf{x}(t) + \mathbf{G}\mathbf{u}(t) \tag{6.3-91}$$

is partitioned as

$$\begin{bmatrix} \dot{\mathbf{x}}_1(t) \\ \dot{\mathbf{x}}_2(t) \end{bmatrix} = \begin{bmatrix} \mathbf{F}_{11} & \mathbf{F}_{12} \\ \mathbf{F}_{21} & \mathbf{F}_{22} \end{bmatrix} \begin{bmatrix} \mathbf{x}_1(t) \\ \mathbf{x}_2(t) \end{bmatrix} + \begin{bmatrix} \mathbf{G}_{11} & \mathbf{G}_{12} \\ \mathbf{G}_{21} & \mathbf{G}_{22} \end{bmatrix} \begin{bmatrix} \mathbf{u}_1(t) \\ \mathbf{u}_2(t) \end{bmatrix} \tag{6.3-92}$$

with $\dim(\mathbf{x}_1) = n_1$, $\dim(\mathbf{u}_1) = m_1$, $\dim(\mathbf{x}_2) = n_2$, and $\dim(\mathbf{u}_2) = m_2$. If Subsystems 1 and 2 are weakly interacting ($\mathbf{F}_{12}$, $\mathbf{F}_{21}$, $\mathbf{G}_{12}$, and $\mathbf{G}_{21} \approx \mathbf{0}$), it may

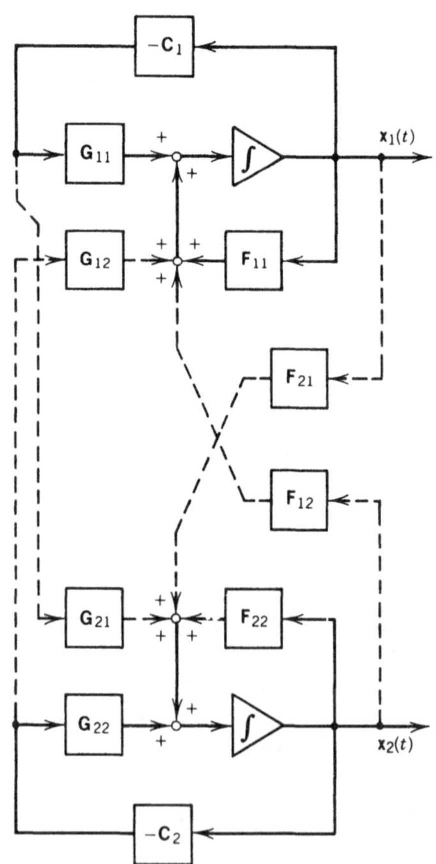

FIGURE 6.3-7 Reduced-order controller.

be possible to define a satisfactory (but suboptimal) controller as (Fig. 6.3-7),

$$\mathbf{u}_1(t) = -\mathbf{C}_1\mathbf{x}_1(t) \tag{6.3-93}$$

$$\mathbf{u}_2(t) = -\mathbf{C}_2\mathbf{x}_2(t) \tag{6.3-94}$$

where $\mathbf{C}_1$ and $\mathbf{C}_2$ are reduced-order control gain matrices calculated by minimizing separate cost functions for $(\mathbf{x}_1, \mathbf{u}_1)$ and $(\mathbf{x}_2, \mathbf{u}_2)$. There is a net saving in control-gain computation and in real-time implementation, as the number of scalar gains is $[(m_1 \times n_1) + (m_2 \times n_2)]$ rather than $[(m_1 + m_2) \times (n_1 + n_2)]$.

If a subsystem (say the second) is much faster than the other and is stable, it may be satisfactory to ignore that subsystem entirely (i.e., to *truncate* the system). The principal danger in this or any other model-order-reduction scheme is that neglected modes may be destabilized inadvertently. Assume that there is coupling between 1 and 2 as well as coupling in the output vector:

$$\mathbf{y}(t) = [\mathbf{H}_{x_1} \quad \mathbf{H}_{x_2}] \begin{bmatrix} \mathbf{x}_1(t) \\ \mathbf{x}_2(t) \end{bmatrix} \tag{6.3-95}$$

Replacing $\mathbf{x}_1(t)$ by an estimate $\hat{\mathbf{x}}_1(t)$ based on $\mathbf{y}(t)$ that neglects the dynamics of subsystem 2 "closes the loop" between $\mathbf{x}_2(t)$ and $\mathbf{u}_1(t)$. With nonzero $\mathbf{G}_{21}$ (or $\mathbf{F}_{12}$ and $\mathbf{F}_{21}$), this could result in unacceptable response.

Residualization provides a design alternative that may improve performance, subject to the previous warnings. If Subsystem 2 is sufficiently fast and stable, then it may always appear to have reached steady state on the time scale of Subsystem 1. Equation 6.3-92 then could be approximated as

$$\begin{bmatrix} \dot{\mathbf{x}}_1(t) \\ \mathbf{0} \end{bmatrix} = \begin{bmatrix} \mathbf{F}_{11} & \mathbf{F}_{12} \\ \mathbf{F}_{21} & \mathbf{F}_{22} \end{bmatrix} \begin{bmatrix} \mathbf{x}_1(t) \\ \mathbf{x}_2(t) \end{bmatrix} + \begin{bmatrix} \mathbf{G}_{11} & \mathbf{G}_{12} \\ \mathbf{G}_{21} & \mathbf{G}_{22} \end{bmatrix} \begin{bmatrix} \mathbf{u}_1(t) \\ \mathbf{u}_2(t) \end{bmatrix} \tag{6.3-96}$$

Letting $\mathbf{G}_{12}$ be zero for simplicity and assuming that Subsystem 2 is regulated by Eq. 6.3-94, the quasi-steady-state value of $\mathbf{x}_2(t)$ is

$$\mathbf{x}_2^* = -(\mathbf{F}_{22} - \mathbf{G}_{22}\mathbf{C}_2)^{-1}[\mathbf{F}_{21}\mathbf{x}_1(t) + \mathbf{G}_{21}\mathbf{u}_1(t)] \tag{6.3-97}$$

The response of Subsystem 1 can be approximated by the residualized model

$$\dot{\mathbf{x}}_1(t) = \mathbf{F}_1\mathbf{x}_1(t) + \mathbf{G}_1\mathbf{u}_1(t) \tag{6.3-98}$$

where $\mathbf{x}_2^*$ has been eliminated by defining

$$\mathbf{F}_1 = \mathbf{F}_{11} - \mathbf{F}_{12}(\mathbf{F}_{22} - \mathbf{G}_{22}\mathbf{C}_2)^{-1}\mathbf{F}_{21} \tag{6.3-99}$$

$$\mathbf{G}_1 = \mathbf{G}_{11} - \mathbf{F}_{12}(\mathbf{F}_{22} - \mathbf{G}_{22}\mathbf{C}_2)^{-1}\mathbf{G}_{21} \tag{6.3-100}$$

An LQ regulator for $(\mathbf{x}_1, \mathbf{u}_1)$ then is designed using $\mathbf{F}_1$ and $\mathbf{G}_1$ to represent the reduced-order dynamics.

If the system to be controlled is adequately modeled as a completely controllable and observable single-input/single-output system, *canonical transformations** provide additional design alternatives. Two examples are reviewed briefly. If the original n-dimensional system is given by

$$\dot{\mathbf{x}}(t) = \mathbf{F}\mathbf{x}(t) + \mathbf{G}u(t) \tag{6.3-101}$$

$$y(t) = \mathbf{H}\mathbf{x}(t) \tag{6.3-102}$$

where $u(t)$ and $y(t)$ are scalars, then the transfer function between $u(s)$ and $y(s)$ is

$$\frac{y(s)}{u(s)} = \mathbf{H}(s\mathbf{I}_n - \mathbf{F})^{-1}\mathbf{G}$$

$$= \frac{b_{n-1}s^{n-1} + \cdots + b_1 s + b_0}{s^n + c_{n-1}s^{n-1} + \cdots + c_1 s + c_0} \tag{6.3-103}$$

The *controller canonical form* for the system equations redefines the state vector such that

$$\mathbf{F} = \begin{bmatrix} -c_{n-1} & -c_{n-2} & \cdots & -c_1 & -c_0 \\ 1 & 0 & \cdots & 0 & 0 \\ \vdots & \vdots & & & \\ 0 & 0 & \cdots & 1 & 0 \end{bmatrix} \tag{6.3-104}$$

$$\mathbf{G} = \begin{bmatrix} 1 \\ 0 \\ \vdots \\ 0 \end{bmatrix} \tag{6.3-105}$$

$$\mathbf{H} = [\, b_{n-1} \quad b_1 \ldots \; b_0 \,] \tag{6.3-106}$$

while the *observer canonical form* defines $\mathbf{x}(t)$ such that

$$\mathbf{F} = \begin{bmatrix} -c_{n-1} & 1 & 0 & \cdots & 0 & 0 \\ -c_{n-2} & 0 & 1 & \cdots & 0 & 0 \\ \vdots & \vdots & \vdots & & & \\ -c_1 & 0 & 0 & \cdots & 0 & 1 \\ -c_0 & 0 & 0 & \cdots & 0 & 0 \end{bmatrix} \tag{6.3-107}$$

*A *canonical* transformation is any transformation to an accepted standard form. As used here, it refers to standard forms for input–output relationships of dynamic systems.

$$\mathbf{G} = \begin{bmatrix} b_{n-1} \\ b_1 \\ \cdot \\ \cdot \\ \cdot \\ b_0 \end{bmatrix} \tag{6.3-108}$$

$$\mathbf{H} = [1 \quad 0 \quad \dots \quad 0] \tag{6.3-109}$$

In addition to providing alternatives for cost function definition, a canonical form could be used to simplify real-time implementation of an LQG regulator. The canonical form matrices are "sparse," so the numbers of additions and multiplications are minimal; hence a canonical-form state estimator is an efficient expression of feedback information.

State Estimation and the LQG Regulator

All of the LQ control structures presented in this section are amenable to stochastic (LQG) regulation, with the feedback variable $\mathbf{x}(t)$ replaced by the estimate $\hat{\mathbf{x}}(t)$ obtained from a Kalman–Bucy filter. An estimator is required if measurements of the full state are noisy or if there are fewer measurements than state components. If $\mathbf{F}$ and $\mathbf{G}$ are known without error, the controller and estimator dynamics are uncoupled (Section 5.3 and 5.4). With uncertainty in $\mathbf{F}$ and $\mathbf{G}$, the robustness of the LQG regulator remains to be determined (Section 6.6).

As an example, a proportional, non-zero-set-point LQG regulator for the system defined by

$$\dot{\mathbf{x}}(t) = \mathbf{F}\mathbf{x}(t) + \mathbf{G}\mathbf{u}(t) + \mathbf{L}\mathbf{w}(t) \tag{6.3-110}$$

$$\mathbf{y}(t) = \mathbf{H}_x\mathbf{x}(t) + \mathbf{H}_u\mathbf{u}(t) \tag{6.3-111}$$

$$\mathbf{z}(t) = \mathbf{H}\mathbf{x}(t) + \mathbf{n}(t) \tag{6.3-112}$$

is shown in Fig. 6.3-8. It is assumed that disturbances are unmeasured and that the Kalman–Bucy filter estimates the state's perturbation from the set point $[\mathbf{x}(t) - \mathbf{x}^*]$ by integrating

$$\dot{\hat{\mathbf{x}}}(t) = \mathbf{F}\hat{\mathbf{x}}(t) + \mathbf{G}\tilde{\mathbf{u}}(t) + \mathbf{K}[\tilde{\mathbf{z}}(t) - \mathbf{H}\hat{\mathbf{x}}(t)] \tag{6.3-113}$$

with

$$\tilde{\mathbf{u}}(t) = \mathbf{u}(t) - \mathbf{u}^* \tag{6.3-114}$$

$$\tilde{\mathbf{z}}(t) = \mathbf{z}(t) - \mathbf{H}\mathbf{x}^* \tag{6.3-115}$$

The control law is defined by

$$\mathbf{u}(t) = \mathbf{u}^* - \mathbf{C}\hat{\mathbf{x}}(t) \tag{6.3-116}$$

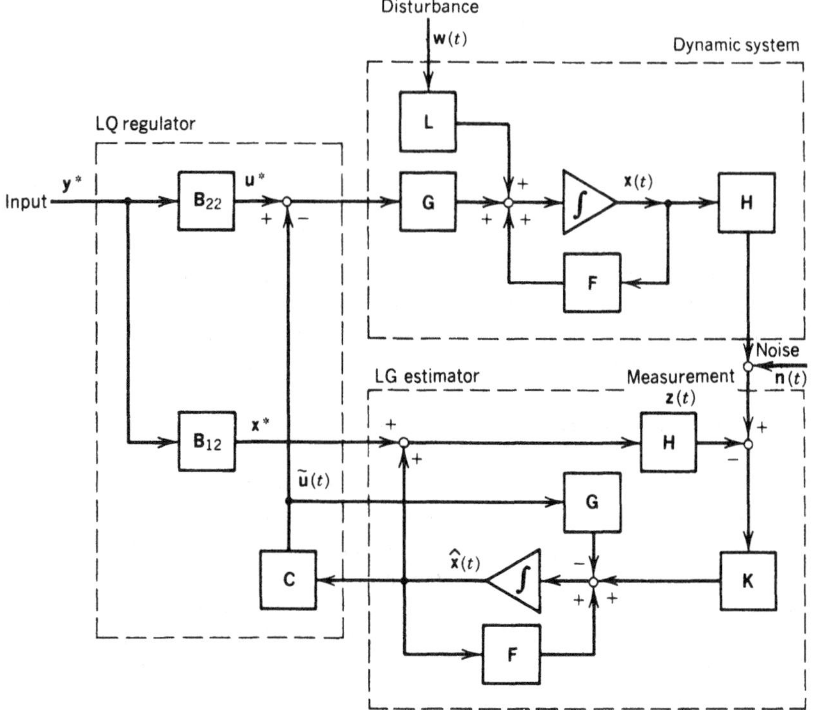

FIGURE 6.3-8 Non-zero-set-point linear-quadratic-Gaussian regulator.

and

$$\begin{bmatrix} \mathbf{x}^* \\ \mathbf{u}^* \end{bmatrix} = \begin{bmatrix} \mathbf{F} & \mathbf{G} \\ \mathbf{H}_\mathbf{x} & \mathbf{H}_\mathbf{u} \end{bmatrix}^{-1} \begin{bmatrix} \mathbf{0} \\ \mathbf{y}^* \end{bmatrix}$$

$$= \begin{bmatrix} \mathbf{B}_{12} \\ \mathbf{B}_{22} \end{bmatrix} \mathbf{y}^* \qquad (6.3\text{-}117)$$

Example 6.3-1 A "P-I-D" Controller Derived from the LQG Regulator

PI compensation and state estimation provide the elements of a classical proportional-integral-derivative (P-I-D) controller, as the following example shows. Consider a second-order system with a single command input, a single measurement, and a single control:

$$\begin{bmatrix} \dot{x}_1(t) \\ \dot{x}_2(t) \end{bmatrix} = \begin{bmatrix} 0 & 1 \\ -\omega_n^2 & -2\zeta\omega_n \end{bmatrix} \begin{bmatrix} x_1(t) \\ x_2(t) \end{bmatrix} + \begin{bmatrix} 0 \\ \omega_n^2 \end{bmatrix} u(t)$$

$$y^*(t) = [1 \quad 0]\begin{bmatrix} x_1^*(t) \\ x_2^*(t) \end{bmatrix} = x_1(t)$$

$$z(t) = [1 \quad 0]\begin{bmatrix} x_1(t) \\ x_2(t) \end{bmatrix} + n(t) = x_1(t) + n(t)$$

The optimal state estimator takes the form

$$\begin{bmatrix} \dot{\hat{x}}_1(t) \\ \dot{\hat{x}}_2(t) \end{bmatrix} = \begin{bmatrix} 0 & 1 \\ -\omega_n^2 & -2\zeta\omega_n \end{bmatrix}\begin{bmatrix} \hat{x}_1(t) \\ \hat{x}_2(t) \end{bmatrix} + \begin{bmatrix} 0 \\ \omega_n^2 \end{bmatrix} u(t) + \begin{bmatrix} k_1 \\ k_2 \end{bmatrix}[z(t) - \hat{x}_1(t)]$$

The filter generates an estimate of the displacement ($\hat{x}_1$) as well as its derivative ($\hat{x}_2$), and it provides feedback information for the PI control law. Neglecting the integrator's initial condition, the control signal is

$$u(t) = c_F y^*(t) + c_I \int_0^t [y^*(t) - \hat{x}_1(t)]\, dt - c_1\hat{x}_1(t) - c_2\hat{x}_2(t)$$

and its Laplace transform is

$$\begin{aligned} u(s) &= c_F y^*(s) + \frac{c_I[y^*(s) - \hat{x}_1(s)]}{s} - (c_1 + c_2 s)\hat{x}_1(s) \\ &= \left(c_F + \frac{c_I}{s}\right) y^*(s) - \left(c_1 + \frac{c_I}{s} + c_2 s\right)\hat{x}_1(s) \end{aligned}$$

The state estimator introduces dynamics between the measurement z and the estimate of displacement $\hat{x}_1$, but the P-I-D characteristic contained in $(c_1 + c_I/s + c_2 s)$ is clear. This characteristic would be retained if there were no estimator (i.e., if x_1 and x_2 were measured and fed back directly).

6.4 MODAL PROPERTIES OF OPTIMAL CONTROL SYSTEMS

Constant-coefficient optimal regulators and filters provide stable closed-loop response and state estimation when they are designed according to the rules of Section 5.4 and when knowledge of the linear, time-invariant plant parameters is perfect. Consequently, eigenvalues of the linear-quadratic controller and the linear-Gaussian filter are guaranteed to lie in stable regions of the s plane (for continuous-time systems) or z plane (for discrete-time systems). Although stability is guaranteed, choices of the weighting matrices and the parameters of the plant have significant effects on the locations of eigenvalues and on the state-component distributions of associated eigenvectors.

The eigenvalues and eigenvectors of regulated systems or state estimators define their respective *modal properties*. These properties may be primary design objectives, although taken alone they are insufficient to assure satisfactory response to command or measurement inputs. (This involves steady-state as well as transient response.) They may be secondary goals, as when eigenvalues and eigenvectors need only be placed in the neighborhoods of nominal values for satisfactory performance. It is shown below that modal properties assume well-defined asymptotic patterns as weighting factors take limiting values, establishing a direct link between linear-optimal design techniques and this classical system evaluation concept.

Eigenvalues of Optimally Regulated Systems

Following Section 6.1, we consider the continuous case, in which an infinite-time cost function

$$J = \frac{1}{2}\int_0^\infty (\mathbf{x}^T\mathbf{Q}\mathbf{x} + \mathbf{u}^T\mathbf{R}\mathbf{u})\,dt \tag{6.4-1}$$

is to be minimized subject to the n-dimensional linear, time-invariant dynamic constraint

$$\dot{\mathbf{x}}(t) = \mathbf{F}\mathbf{x}(t) + \mathbf{G}\mathbf{u}(t) \tag{6.4-2}$$

where $\mathbf{u}(t)$ has dimension m. The corresponding n-dimensional adjoint equation is

$$\dot{\boldsymbol{\lambda}}(t) = -\mathbf{Q}\mathbf{x}(t) - \mathbf{F}^T\boldsymbol{\lambda}(t) \tag{6.4-3}$$

Since the optimal zero-set-point control law can be written as

$$\mathbf{u}(t) = -\mathbf{R}^{-1}\mathbf{G}^T\boldsymbol{\lambda}(t) \tag{6.4-4}$$

the state and adjoint equations can be combined as follows:

$$\begin{bmatrix}\dot{\mathbf{x}}(t)\\ \dot{\boldsymbol{\lambda}}(t)\end{bmatrix} = \begin{bmatrix}\mathbf{F} & -\mathbf{G}\mathbf{R}^{-1}\mathbf{G}^T\\ -\mathbf{Q} & -\mathbf{F}^T\end{bmatrix}\begin{bmatrix}\mathbf{x}(t)\\ \boldsymbol{\lambda}(t)\end{bmatrix} \tag{6.4-5a}$$

or

$$\dot{\boldsymbol{\chi}}(t) = \mathbf{F}'\boldsymbol{\chi}(t) \tag{6.4-5b}$$

with $\boldsymbol{\chi}^T(t) \triangleq [\mathbf{x}^T(t) \quad \boldsymbol{\lambda}^T(t)]$. The $2n$ eigenvalues of $\mathbf{F}'$ represent the n

eigenvalues of the optimal closed-loop system and the n eigenvalues of its adjoint system; they are the roots of the polynomial equation defined by

$$|s\mathbf{I}_{2n} - \mathbf{F}'| = 0 \tag{6.4-6a}$$

or

$$\begin{vmatrix} (s\mathbf{I}_n - \mathbf{F}) & \mathbf{G}\mathbf{R}^{-1}\mathbf{G}^T \\ \mathbf{Q} & (s\mathbf{I}_n + \mathbf{F}^T) \end{vmatrix} = 0 \tag{6.4-6b}$$

Using Schur's formula (Eq. 2.2-101), this can be expressed as

$$|s\mathbf{I}_{2n} - \mathbf{F}'| = |s\mathbf{I}_n - \mathbf{F}|\,|(s\mathbf{I}_n + \mathbf{F}^T) - \mathbf{Q}(s\mathbf{I}_n - \mathbf{F})^{-1}\mathbf{G}\mathbf{R}^{-1}\mathbf{G}^T| \tag{6.4-7a}$$

and with Eq. 2.2-98, this is equivalent to

$$|s\mathbf{I}_{2n} - \mathbf{F}'| = |s\mathbf{I}_n - \mathbf{F}|\,|s\mathbf{I}_n + \mathbf{F}^T|\,|\mathbf{I}_n - \mathbf{Q}(s\mathbf{I}_n - \mathbf{F})^{-1}\mathbf{G}\mathbf{R}^{-1}\mathbf{G}^T(s\mathbf{I}_n + \mathbf{F}^T)^{-1}| \tag{6.4-7b}$$

An alternate form for the second determinant on the right is (Eq. 2.2-97)

$$|s\mathbf{I}_n + \mathbf{F}^T| = (-1)^n|-s\mathbf{I}_n - \mathbf{F}| \tag{6.4-8}$$

Using Eq. 2.2-103b, $|s\mathbf{I}_{2n} - \mathbf{F}'|$ can be expressed as a product of polynomials in s and $-s$:

$$\begin{aligned} |s\mathbf{I}_{2n} - \mathbf{F}'| &= \Delta_{\mathrm{CL}}(s)\Delta_{\mathrm{CL}}(-s) \\ &= (-1)^n|s\mathbf{I}_n - \mathbf{F}|\,|-s\mathbf{I}_n - \mathbf{F}|\,|\mathbf{I}_m + \mathbf{R}^{-1}\mathbf{G}^T(-s\mathbf{I}_n - \mathbf{F}^T)^{-1}\mathbf{Q}(s\mathbf{I}_n - \mathbf{F})^{-1}\mathbf{G}| \\ &= (-1)^n\Delta_{\mathrm{OL}}(s)\Delta_{\mathrm{OL}}(-s)|\mathbf{I}_m + \mathbf{R}^{-1}\mathbf{Y}_1^T(-s)\mathbf{Y}_1(s)| \end{aligned} \tag{6.4-9}$$

Here the *open-loop characteristic polynomial* is defined as

$$\Delta_{\mathrm{OL}}(s) = |s\mathbf{I}_n - \mathbf{F}| \tag{6.4-10}$$

the *closed-loop characteristic polynomial*, $\Delta_{\mathrm{CL}}(s)$, is defined below, and the *cost transfer function matrix*

$$\mathbf{Y}_1(s) = \mathbf{H}(s\mathbf{I}_n - \mathbf{F})^{-1}\mathbf{G} = \mathbf{V}_1(s)/\Delta_{\mathrm{OL}}(s) \tag{6.4-11}$$

is defined with

$$\mathbf{Q} = \mathbf{H}^T\mathbf{H} \tag{6.4-12}$$

where $\mathbf{H}$ is an $(r \times n)$ matrix. $\mathbf{Y}_1(s)$ implicitly defines an r-dimensional output vector $\mathbf{y}$ such that

$$\begin{aligned}\mathbf{y}(s) &= \mathbf{H}\mathbf{x}(s) \\ &= \mathbf{Y}_1(s)\mathbf{u}(s)\end{aligned} \tag{6.4-13}$$

so the original cost function can be described as

$$\begin{aligned}J &= \frac{1}{2}\int_0^\infty [\mathbf{x}^T(t)\mathbf{H}^T\mathbf{H}\mathbf{x}(t) + \mathbf{u}^T\mathbf{R}\mathbf{u}(t)]\,dt \\ &= \frac{1}{2}\int_0^\infty [\mathbf{y}^T(t)\mathbf{y}(t) + \mathbf{u}^T(t)\mathbf{R}\mathbf{u}(t)]\,dt\end{aligned} \tag{6.4-14}$$

Note that $\mathbf{H}$ need not be square, although it must assure observability of $\mathbf{x}$ in the cost function, as detailed in Section 5.4. The rank of $\mathbf{H}$ must equal the rank of $\mathbf{Q}$; if $\mathbf{Q}$ is rank p, then $\mathbf{H}$ must have dimension $(r \times n)$, $r \geq p$. The commonly used positive definite diagonal $\mathbf{Q}$ has rank n; therefore, $\mathbf{H}$ would have to have at least n rows in that case. If an $(m \times n)$ $\mathbf{H}$ can be formed, then $\mathbf{Y}_1(s)$ is square, and it possesses a determinant:

$$|\mathbf{Y}_1(s)| = |\mathbf{H}(s\mathbf{I}_n - \mathbf{F})^{-1}\mathbf{G}| = \frac{\psi_1(s)}{\Delta_{OL}(s)} \tag{6.4-15}$$

The numerator, $\psi_1(s)$, is a polynomial in s whose roots, z_i, are known as the *transmission zeros* of the system defined by Eq. 6.4-2 and 6.4-13:*

$$\begin{aligned}\psi_1(s) &= a_q s^q + a_{q-1}s^{q-1} + \cdots + a_1 s + a_0 \\ &= a_q(s - z_1)(s - z_2)\ldots(s - z_q) = 0\end{aligned} \tag{6.4-16}$$

The degree of the polynomial (and, therefore, the number of transmission zeros) can be shown to be less than or equal to $(n - m - d)$, where d is the rank deficiency of the matrix product $\mathbf{HG}$ (H-2).† The transmission zeros should not be confused with the zeros of the transfer functions between individual inputs and outputs, i.e., the zeros of the individual elements of the $(m \times m)$ matrix $\mathbf{V}_1(s)$. They are the same only when $m = 1$.

*A more general definition of transmission zeros for the system given by $\dot{\mathbf{x}} = \mathbf{F}\mathbf{x} + \mathbf{G}\mathbf{u}$, $\mathbf{y} = \mathbf{H}_x\mathbf{x} + \mathbf{H}_u\mathbf{u}$ is contained in (D-3). They are the set of complex numbers, z_i, for which

$$\text{Rank}\begin{bmatrix}\mathbf{F} - z_i\mathbf{I}_n & \mathbf{G} \\ \mathbf{H}_x & \mathbf{H}_u\end{bmatrix} < n + \min(r, m)$$

r need not equal m, although the condition leads to Eq. 6.4-15 when $r = m$.

†The *rank deficiency* of a matrix is the difference between full rank (in this case m) and actual rank.

Example 6.4-1 Transmission Zeros for Some Second-Order Systems

a. Consider the system

$$\begin{bmatrix} \dot{x}_1 \\ \dot{x}_2 \end{bmatrix} = \begin{bmatrix} a & 0 \\ 0 & d \end{bmatrix} \begin{bmatrix} x_1 \\ x_2 \end{bmatrix} + \begin{bmatrix} 1 & 0 \\ 0 & 1 \end{bmatrix} \begin{bmatrix} u_1 \\ u_2 \end{bmatrix}$$

$$\begin{bmatrix} y_1 \\ y_2 \end{bmatrix} = \begin{bmatrix} 1 & 0 \\ 0 & 1 \end{bmatrix} \begin{bmatrix} x_1 \\ x_2 \end{bmatrix}$$

with the following transfer matrix:

$$\begin{aligned} \mathbf{Y}(s) &= \mathbf{H}(s\mathbf{I}_n - \mathbf{F})^{-1}\mathbf{G} \\ &= \begin{bmatrix} 1 & 0 \\ 0 & 1 \end{bmatrix} \begin{bmatrix} (s-a) & 0 \\ 0 & (s-d) \end{bmatrix}^{-1} \begin{bmatrix} 1 & 0 \\ 0 & 1 \end{bmatrix} \\ &= \frac{\begin{bmatrix} (s-d) & 0 \\ 0 & (s-a) \end{bmatrix}}{(s-a)(s-d)} = \begin{bmatrix} 1/(s-a) & 0 \\ 0 & 1/(s-d) \end{bmatrix} \end{aligned}$$

The determinant of $\mathbf{Y}(s)$ is

$$|\mathbf{Y}(s)| = \frac{1}{(s-a)(s-d)} = \frac{\psi(s)}{\Delta_{\mathrm{OL}}(s)}$$

and, since $\psi(s) = \Delta_{\mathrm{OL}}|\mathbf{Y}(s)|$, there are no transmission zeros in the system.

b. Now let the system's dynamic equation be

$$\begin{bmatrix} \dot{x}_1 \\ \dot{x}_2 \end{bmatrix} = \begin{bmatrix} a & b \\ c & d \end{bmatrix} \begin{bmatrix} x_1 \\ x_2 \end{bmatrix} + \begin{bmatrix} 1 & 0 \\ 0 & 1 \end{bmatrix} \begin{bmatrix} u_1 \\ u_2 \end{bmatrix}$$

The transfer matrix is

$$\mathbf{Y}(s) = \frac{\begin{bmatrix} (s-d) & b \\ c & (s-a) \end{bmatrix}}{[s^2-(a+d)s+(ad-bc)]} = \frac{\begin{bmatrix} (s-d) & b \\ c & (s-a) \end{bmatrix}}{\Delta_{\mathrm{OL}}(s)}$$

its determinant is

$$\begin{aligned} |\mathbf{Y}(s)| &= [s^2-(a+d)s+(ad-bc)]/\Delta^2_{\mathrm{OL}}(s) \\ &= \Delta_{\mathrm{OL}}(s)/\Delta^2_{\mathrm{OL}}(s) = 1/\Delta_{\mathrm{OL}}(s) \end{aligned}$$

and again there are no transmission zeros in the system.

c. A system is described by

$$\begin{bmatrix} \dot{x}_1 \\ \dot{x}_2 \end{bmatrix} = \begin{bmatrix} a & 0 \\ c & d \end{bmatrix} \begin{bmatrix} x_1 \\ x_2 \end{bmatrix} + \begin{bmatrix} 1 & 0 \\ 0 & 1 \end{bmatrix} \begin{bmatrix} u_1 \\ u_2 \end{bmatrix}$$

$$\begin{bmatrix} y_1 \\ y_2 \end{bmatrix} = \begin{bmatrix} 1 & 1 \\ 0 & 1 \end{bmatrix} \begin{bmatrix} x_1 \\ x_2 \end{bmatrix}$$

The transfer matrix is

$$\mathbf{Y}(s) = \frac{\begin{bmatrix} [(s-d)+c] & (s-a) \\ c & (s-a) \end{bmatrix}}{(s-a)\,(s-d)}$$

and once again there are no transmission zeros

$$|\mathbf{Y}(s)| = \frac{s^2-(a+d)s+ad}{[(s-a)\,(s-d)]^2} = \frac{1}{\Delta_{\mathrm{OL}}(s)}$$

d. Finally, consider a system with scalar input and output $(m = r = 1)$:

$$\begin{bmatrix} \dot{x}_1 \\ \dot{x}_2 \end{bmatrix} = \begin{bmatrix} a & b \\ c & d \end{bmatrix} \begin{bmatrix} x_1 \\ x_2 \end{bmatrix} + \begin{bmatrix} e \\ f \end{bmatrix} u$$

$$\mathbf{y} = [g \quad h] \begin{bmatrix} x_1 \\ x_2 \end{bmatrix}$$

The transfer function is scalar,

$$\mathbf{Y}(s) = \frac{[g \quad h] \begin{bmatrix} (s-d) & b \\ c & (s-a) \end{bmatrix} \begin{bmatrix} e \\ f \end{bmatrix}}{s^2-(a+d)s+(ad-bc)}$$

$$= \frac{(eg+fh)s+[e(hc-gd)+f(gb-ha)]}{\Delta_{\mathrm{OL}}(s)}$$

and the system is seen to contain a transmission zero at

$$s = \frac{-[e(hc-gd)+f(gb-ha)]}{(eg+fh)}$$

The $2n$ eigenvalues of the optimally regulated system and its adjoint, $\mathbf{F}'$, are formed by setting Eq. 6.4-9 equal to zero and finding the roots. It is of interest to see what their asymptotic locations are as $\mathbf{R}$ becomes very large or very small. For this purpose, $\mathbf{R}$ can be expressed as

$$\mathbf{R} = \rho^2 \mathbf{R}_o \tag{6.4-17}$$

where ρ is a scalar constant that may approach zero or infinity in the limit.

It is convenient to redefine the system's transmission zeros to account for $\mathbf{R}_o$. Letting $\mathbf{U}$ be a positive definite $(m \times m)$ matrix square root of $\mathbf{R}_o^{-1}$ such that

$$\mathbf{U}\mathbf{U}^T = \mathbf{U}^T\mathbf{U} = \mathbf{R}_o^{-1} \tag{6.4-18}$$

Then Eq. 6.4-9 can be rewritten as :

$$|s\mathbf{I}_{2n} - \mathbf{F}'| = (-1)^n \Delta_{\text{OL}}(s)\Delta_{\text{OL}}(-s)|\mathbf{I}_m + \frac{1}{\rho^2}\mathbf{U}^T\mathbf{Y}_1^T(-s)\mathbf{Y}_1(s)\mathbf{U}|$$

$$= (-1)^n \Delta_{\text{OL}}(s)\Delta_{\text{OL}}(-s)|\mathbf{I}_m + \frac{1}{\rho^2}\mathbf{Y}_2^T(-s)\mathbf{Y}_2(s)| \tag{6.4-19}$$

$\mathbf{Y}_2(s)$ is

$$\mathbf{Y}_2(s) = \mathbf{H}(s\mathbf{I}_n - \mathbf{F})^{-1}\mathbf{G}\mathbf{U} = \frac{\mathbf{V}_2(s)}{\Delta_{\text{OL}}(s)} \tag{6.4-20}$$

and it represents the open-loop transfer function matrix between the weighted control input, $\mathbf{u}'(s) = \mathbf{U}^{-1}\mathbf{u}(s)$, and the output vector, $\mathbf{y}(s)$, defined by the quadratic cost function. The corresponding determinant takes the form,

$$|\mathbf{Y}_2(s)| = \frac{\psi_2(s)}{\Delta_{\text{OL}}(s)} \tag{6.4-21a}$$

so the transmission zeros can be defined as the roots of the polynomial

$$\psi_2(s) = \Delta_{\text{OL}}(s)|\mathbf{Y}_2(s)| \tag{6.4-21b}$$

Assuming that $\mathbf{Q}$ and $\mathbf{R}_o$ are constant, the transmission zeros of $\mathbf{Y}_2(s)$ are fixed; alternately, changing the values of $\mathbf{Q}$ and/or $\mathbf{R}_o$ shifts the transmission zero locations. An increase in ρ^2 is seen to be equivalent to a decrease in the magnitude of $\mathbf{Q}$. It can be anticipated that the transmission zeros of $\mathbf{Y}_2(s)$ have important effects on the eigenvalues of $\mathbf{F}'$ when ρ is small and that they have negligible effects when ρ is large.

As ρ approaches infinity, Eq. 6.4-19 becomes

$$\begin{aligned}|s\mathbf{I}_{2n}-\mathbf{F}'| &= (-1)^n\Delta_{\text{OL}}(s)\Delta_{\text{OL}}(-s)|\mathbf{I}_m| \\ &= (-1)^n\Delta_{\text{OL}}(s)\Delta_{\text{OL}}(-s)\end{aligned} \tag{6.4-22}$$

when $\Delta_{\text{OL}}(s)$ is defined as before (Eq. 6.4-10), and it may contain unstable eigenvalues (i.e., eigenvalues with positive real parts). The factors of $\Delta_{\text{OL}}(-s)$ are "mirror images" of the factors of $\Delta_{\text{OL}}(s)$. If

$$\Delta_{\text{OL}}(s) = (s-\lambda_1)(s-\lambda_2)\ldots(s-\lambda_n) \tag{6.4-23}$$

then

$$\begin{aligned}\Delta_{\text{OL}}(-s) &= (-s-\lambda_1)(-s-\lambda_1)\ldots(-s-\lambda_n) \\ &= (-1)^n(s+\lambda_1)(s+\lambda_2)\ldots(s+\lambda_n)\end{aligned} \tag{6.4-24}$$

Consequently, for every left-half-plane root of $\Delta_{\text{OL}}(s)$, there is a right-half-plane root of $\Delta_{\text{OL}}(-s)$ and vice versa, as shown in Fig. 6.4-1. [The roots of $\Delta_{\text{OL}}(s)$ are denoted by $\times$, while those of $\Delta_{\text{OL}}(-s)$ are given by $\square$.]

The optimal feedback control law can be expressed as (Eq. 6.1-3),

$$\mathbf{u}(t) = -\mathbf{R}^{-1}\mathbf{G}^T\mathbf{S}\mathbf{x}(t) = -\frac{1}{\rho^2}\mathbf{R}_o^{-1}\mathbf{G}^T\mathbf{S}\mathbf{x}(t) \tag{6.4-25a}$$

where $\mathbf{S}$ is the solution to Eq. 6.1-5:

$$\mathbf{0} = -\mathbf{F}^T\mathbf{S}-\mathbf{S}\mathbf{F}-\mathbf{Q}+\frac{1}{\rho^2}\mathbf{S}\mathbf{G}\mathbf{R}_o^{-1}\mathbf{G}^T\mathbf{S} \tag{6.4-26}$$

It possesses the Laplace transform,

$$\begin{aligned}\mathbf{u}(s) &= -\mathbf{R}^{-1}\mathbf{G}^T\mathbf{S}\mathbf{x}(s) = -\frac{1}{\rho^2}\mathbf{R}_o^{-1}\mathbf{G}^T\mathbf{S}\mathbf{x}(s) \\ &= -\mathbf{C}\mathbf{x}(s)\end{aligned} \tag{6.4-25b}$$

Substituting Eq. 6.4-25b in Eq. 6.4-2, the optimal closed-loop system eigenvalues are seen to be the roots of the equation

$$|s\mathbf{I}_n-(\mathbf{F}-\mathbf{G}\mathbf{C})| = \Delta_{\text{CL}}(s) = 0 \tag{6.4-27}$$

Having satisfied the rules of Section 5.4, the n closed-loop eigenvalues must be stable (i.e., they must lie in the left half of the s plane). As ρ approaches infinity, these n roots must approach n roots of Eq. 6.4-22.

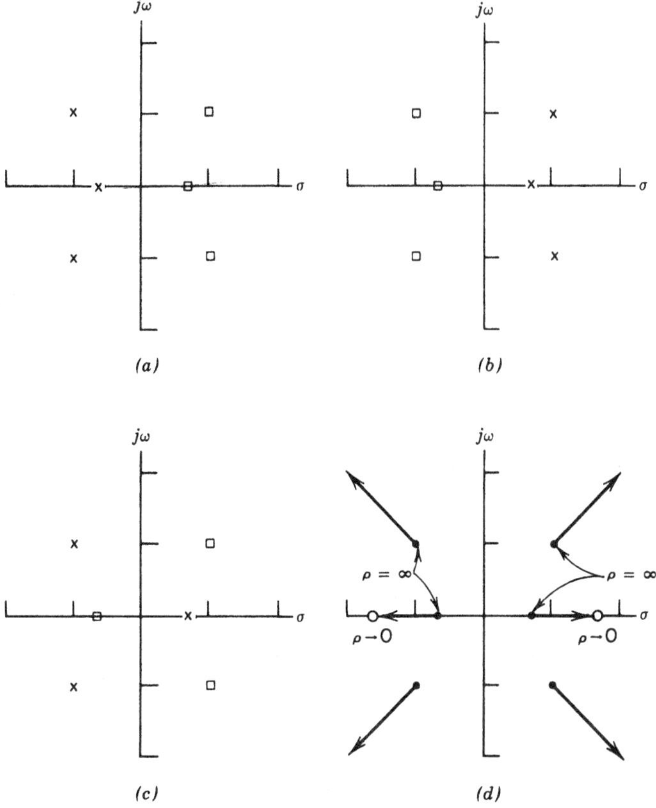

FIGURE 6.4-1 Examples of F' eigenvalues and corresponding root locus. (a) Stable system and its adjoint; (b) system with three unstable roots and its adjoint; (c) system with one unstable root and its adjoint; (d) root locus for all three cases (assuming one transmission zero).

Because stability is guaranteed, the closed-loop roots must be the left-half-plane roots of $(-1)^n\Delta_{OL}(s)\Delta_{OL}(-s)$, whether or not the stable roots appear in $\Delta_{OL}(s)$.

In this case, the "cost" of control is high because ρ is large. If the open-loop system is stable, the optimal control gains are vanishingly small. If the open-loop system is unstable, the optimal control gains achieve their smallest possible magnitudes but they do not vanish, as shown in Example 5.4-1.

As ρ approaches zero, the eigenvalues of F' separate into two groups: those whose magnitudes head for infinity and those which approach the transmission zeros. This behavior is most easily demonstrated for a scalar control variable ($m = 1$). Then Eq. 6.4-19 can be written as

$$|s\mathbf{I}_{2n} - \mathbf{F}'| = (-1)^n \Delta_{OL}(s)\Delta_{OL}(-s)\left[1 + \frac{1}{\rho^2}\frac{\mathbf{V}_2^T(-s)\mathbf{V}_2(s)}{(-1)^n\Delta_{OL}(s)\Delta_{OL}(-s)}\right]$$

$$= (-1)^n \Delta_{OL}(s)\Delta_{OL}(-s) + \frac{1}{\rho^2}\psi_2(-s)\psi_2(s) \qquad (6.4\text{-}28)$$

where $\mathbf{V}_2(s) = \psi_2(s)$, the scalar transmission polynomial:

$$\mathbf{V}_2(s) = \psi_2(s) = \mathbf{H}\,\mathrm{Adj}(s\mathbf{I}_n - \mathbf{F})\mathbf{G}\mathbf{U}, \qquad (m = 1) \qquad (6.4\text{-}29)$$

The characteristic polynomial of $\mathbf{F}'$ (Eq. 6.4-28) is seen to take the form

$$\begin{aligned}\Delta_{F'}(s) &= (-1)^n D(s) + KN(s)\\ &= (-1)^n[D(s) + (-1)^{-n}KN(s)]\end{aligned} \qquad (6.4\text{-}30)$$

$D(s)$ and $N(s)$ are polynomials whose roots are symmetric about the $j\omega$ axis of the s plane,

$$D(s) = \Delta_{OL}(s)\Delta_{OL}(-s) \qquad (6.4\text{-}31)$$

$$N(s) = \psi_2(-s)\psi_2(s) \qquad (6.4\text{-}32)$$

and K is $(1/\rho^2)$. Consequently, the root-locus analysis techniques of Section 2.6 can be used to chart the effects of ρ on the roots of the closed-loop system and its adjoint. If

$$\psi_2(s) = a_q(s - z_1)(s - z_2)\dots(s - z_q) \qquad (6.4\text{-}33)$$

then

$$\begin{aligned}\psi_2(-s) &= a_q(-s - z_1)(-s - z_2)\dots(-s - z_q)\\ &= (-1)^q a_q(s + z_1)(s + z_2)\dots(s + z_q)\end{aligned} \qquad (6.4\text{-}34)$$

and

$$\begin{aligned}(-1)^{-n}KN(s) &= (-1)^{(-n+q)}Ka_q^2\prod_{i=1}^{q}(s - z_i)(s + z_i)\\ &= (-1)^{(n-q)}Ka_q^2\prod_{i=1}^{q}(s - z_i)(s + z_i)\end{aligned} \qquad (6.4\text{-}35)$$

As the sign of the effective root-locus gain, $(-1)^{(n-q)}Ka_q^2$, depends on the system order and the number of transmission zeros, either the 180° or 0°

root locus criterion is used, as appropriate. In all cases, $2q$ roots go to the transmission zeros and their images. The "center of gravity" of the root locus is at the origin, and asymptotes for the remaining $2(n-q)$ eigenvalues of $\mathbf{F}'$ radiate from the origin along the following directions:

180° *Criterion* ($n-q$ *even*)

$$\frac{(2k+1)\pi}{2(n-q)}\text{ rad},\qquad k=0,1,\dots 2(n-q)-1$$

0° *Criterion* ($n-q$ *odd*)

$$\frac{k\pi}{(n-q)}\text{ rad},\qquad k=0,1,\dots 2(n-q)-1$$

There always are an even number of eigenvalues headed for infinity: $(n-q)$ in the left half plane, $(n-q)$ in the right half plane. This is referred to as a *Butterworth configuration of roots* in (G-3) and (C-1), after (B-9). The asymptotes are never aligned with the $j\omega$ axis (phase angle $=\pm\pi/2$ rad). The remaining $2q$ eigenvalues approach the zeros of $N(s)$ (i.e., the q transmission zeros of $\mathbf{Y}_2(s)$ and their q mirror images). By symmetry, roots are confined to a single half plane as K $(=1/\rho^2)$ varies from zero to infinity (Fig. 6.4-1d). Consequently, a root that originates in the left half plane $(K=0)$ terminates in the left half plane $(K\to\infty)$, either at a transmission zero (or its image) or at infinite radius, and it does not venture into the right half plane. *It is clear, therefore, that the left-half-plane root locus represents the optimally regulated system* and the right-half-plane root locus represents the adjoint system as ρ varies between infinity and zero.

Example 6.4-2 Root Locus for Optimally Regulated Single-Input Second-Order System

A stable second-order system

$$\begin{bmatrix}\dot{x}_1\\ \dot{x}_2\end{bmatrix}=\begin{bmatrix}0 & 1\\ -\omega_n^2 & -2\zeta\omega_n\end{bmatrix}\begin{bmatrix}x_1\\ x_2\end{bmatrix}+\begin{bmatrix}0\\ \omega_n^2\end{bmatrix}u$$

is to be regulated by the optimal control law

$$u=-\frac{1}{\rho^2}[0\quad \omega_n^2]\mathbf{S}\mathbf{x}$$

so as to minimize the quadratic cost function,

$$J=\frac{1}{2}\int_0^\infty (\mathbf{x}^T\mathbf{Q}\mathbf{x}+\rho^2u^2)\,dt$$

where **Q** has rank one. **S** is the solution to an algebraic Riccati equation (Eq. 6.4-26). Determine the locus of roots as ρ varies from infinity to zero.

There are three alternatives for a rank-1 state weighting matrix, and they are listed with their (1×2) matrix square roots:

a. $$\mathbf{Q} = \begin{bmatrix} q_{11} & 0 \\ 0 & 0 \end{bmatrix} = \mathbf{H}^T\mathbf{H}, \qquad \mathbf{H} = [\sqrt{q_{11}} \quad 0]$$

b. $$\mathbf{Q} = \begin{bmatrix} 0 & 0 \\ 0 & q_{22} \end{bmatrix} = \mathbf{H}^T\mathbf{H}, \qquad \mathbf{H} = [0 \quad \sqrt{q_{22}}]$$

c. $$\mathbf{Q} = \begin{bmatrix} q_{11} & \sqrt{q_{11}q_{22}} \\ \sqrt{q_{11}q_{22}} & q_{22} \end{bmatrix} = \mathbf{H}^T\mathbf{H}, \qquad \mathbf{H} = [\sqrt{q_{11}} \quad \sqrt{q_{22}}]$$

In all cases, the control weighting is a scalar

$$\mathbf{R} = \rho^2$$

and the cost transfer function "matrix" is a scalar:

$$\begin{aligned} Y_2(s) &= \mathbf{H}(s\mathbf{I}_n - \mathbf{F})^{-1}\mathbf{G}\mathbf{U} \\ &= \frac{[y_1 \quad y_2]}{(s^2 + 2\zeta\omega_n s + \omega_n^2)} \begin{bmatrix} (s + 2\zeta\omega_n) & 1 \\ -\omega_n^2 & s \end{bmatrix} \begin{bmatrix} 0 \\ \omega_n^2 \end{bmatrix} \\ &= (y_1 + y_2 s)\omega_n^2/\Delta_{\text{OL}}(s) \end{aligned}$$

These transfer functions correspond to the three choices of **Q**:

a. $$Y_2(s) = \frac{\psi_2(s)}{\Delta_{\text{OL}}(s)} = \frac{\sqrt{q_{11}}\,\omega_n^2}{\Delta_{\text{OL}}(s)}$$

b. $$Y_2(s) = \frac{\psi_2(s)}{\Delta_{\text{OL}}(s)} = \frac{\sqrt{q_{22}}\,\omega_n^2 s}{\Delta_{\text{OL}}(s)}$$

c. $$Y_2(s) = \frac{\psi_2(s)}{\Delta_{\text{OL}}(s)} = \sqrt{q_{22}}\,\omega_n^2 \frac{(s + \sqrt{q_{11}/q_{22}})}{\Delta_{\text{OL}}(s)}$$

The roots of the closed-loop system and its adjoint are the solutions to the equation

$$(-1)^n\Delta_{\text{OL}}(s)\Delta_{\text{OL}}(-s) + \frac{1}{\rho^2}\psi_2(s)\psi_2(-s) = 0$$

There are no transmission zeros in Case a, so this equation takes the form

$$(s^2 + 2\zeta\omega_n s + \omega_n^2)(s^2 - 2\zeta\omega_n s + \omega_n^2) + \frac{1}{\rho^2} q_{11}\omega_n^4 = 0$$

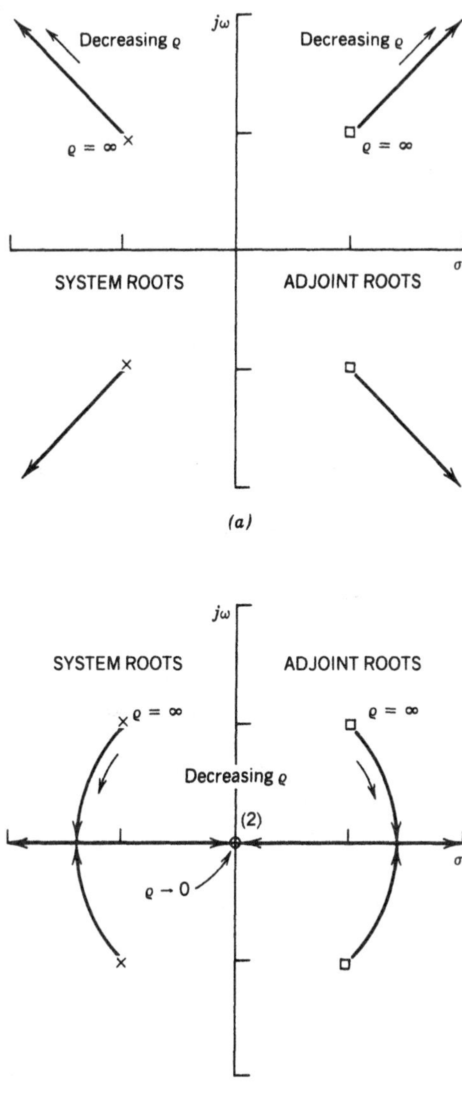

Effects of varying control weighting on system and adjoint eigenvalues. (*a*) Root locus for $\mathbf{Q} = \text{diag}\,[q_{11}\ 0]$; (*b*) root locus for $\mathbf{Q} = \text{diag}[0\ q_{22}]$.

and the root locus for $\infty > \rho > 0$ is sketched below, assuming a damping ratio (ζ) of 0.707. The roots follow the asymptotes, and decreasing ρ leads to increasing natural frequency (ω_n) with no change in damping ratio.

There is a transmission zero at the origin in Case b, and the characteristic equation for the state/adjoint system is

$$(s^2 + 2\zeta\omega_n s + \omega_n^2)(s^2 - 2\zeta\omega_n s + \omega_n^2) - \frac{1}{\rho^2} q_{22}\omega_n^4 s^2 = 0$$

The root locus satisfies the 0° phase criterion, as shown below. Decreasing ρ leads to increased ζ with no change in ω_n. At the critical damping point ($\zeta = 1$), the complex roots coalesce and split into two real roots; further decrease in ρ quickens one real mode of the closed-loop system and slows the other.

In the third case, the transmission zero is located at $z = -\sqrt{q_{11}/q_{22}}$. The root locus sketch is left to the reader. If the system had been defined with an unstable damping ratio of -0.707, the root loci would have been identical, although the optimal control gains would have been different. A comparison of these results with those of Example 5.4-1 provides an explanation of the effects of $\mathbf{Q}$, $\mathbf{R}$, and $\mathbf{F}$ variation shown there.

When there is more than one control variable, the details of the root locus are more complicated, but the general properties are the same. As ρ varies from infinity to zero, the n closed-loop system roots depart from the left-half-plane eigenvalues of $\mathbf{F}'$; q roots approach finite locations governed by $\mathbf{Y}_2(s)$ (whether or not it is square), and $(n-q)$ roots go to infinite radius. The finite root asymptotes are

$$\Delta_{F'}(s) \xrightarrow[\rho \to 0]{} (-1)^n \Delta_{OL}(s)\Delta_{OL}(-s)\frac{1}{\rho^2}|\mathbf{Y}_2^T(-s)\mathbf{Y}_2(s)| = (-1)^n \frac{1}{\rho^2}\psi_2(-s)\psi_2(s) \tag{6.4-36}$$

Hence the finite roots approach the transmission zeros of the system (or their images). The remaining $(n-q)$ closed-loop roots approach infinity in *multiple Butterworth configurations*. These configurations can have different numbers of roots and, therefore, different asymptotes (K-10). Their radii from the origin are proportional to $\rho^{-(1/k)}$, where k is the order of the Butterworth configuration (i.e., the number of left-half-plane roots headed for infinity).

Eigenvectors of Linearly Regulated Systems

The degree to which individual state components respond in each dynamic mode is described by the relative sizes of elements in the corresponding eigenvector. Regulation of the linear system

$$\dot{\mathbf{x}}(t) = \mathbf{F}\mathbf{x}(t) + \mathbf{G}\mathbf{u}(t) \tag{6.4-37}$$

by the constant-coefficient linear control law

$$\mathbf{u}(t) = -\mathbf{C}\mathbf{x}(t) \tag{6.4-38}$$

modifies both the eigenvalues and the eigenvectors. [As before, dim($\mathbf{x}$) = n, dim($\mathbf{u}$) = m, and all matrices are conformable.]

From Section 2.2, the closed-loop eigenvalues are solutions to the equation

$$|s\mathbf{I}_n - (\mathbf{F} - \mathbf{G}\mathbf{C})| = 0 \tag{6.4-39}$$

Denoting these solutions as λ_i, $i = 1\text{–}n$, and assuming that the roots are distinct, the corresponding eigenvectors, $\mathbf{x}_i$, satisfy the equation,

$$[\lambda_i\mathbf{I}_n - (\mathbf{F} - \mathbf{G}\mathbf{C})]\mathbf{x}_i = \mathbf{0} \tag{6.4-40}$$

and they can be computed using Eq. 2.2-74. If ($\mathbf{F}$, $\mathbf{G}$) is a completely controllable pair, the eigenvalues can be placed throughout the s plane, with complex roots always being accompanied by complex conjugates. Eigenvectors are real or complex, according to their associated eigenvalues.

With multiple controls, more than one gain matrix, $\mathbf{C}$, may provide the same set of eigenvalues, λ_i, $i = 1\text{–}n$. As an example, define the system

$$\begin{aligned}\dot{\mathbf{x}}(t) &= \mathbf{F}\mathbf{x}(t) + \mathbf{G}\mathbf{u}(t)\\ &= \mathbf{F}\mathbf{x}(t) + \mathbf{g}_1 u_1(t) + \mathbf{g}_2 u_2(t)\end{aligned} \tag{6.4-41}$$

such that either u_1 or u_2 provides complete controllability and $\mathbf{G}$ has a rank of two. Two single-input feedback controllers could be formed,

$$\dot{\mathbf{x}}(t) = \mathbf{F}\mathbf{x}(t) - \mathbf{g}_1\mathbf{C}_1\mathbf{x}(t) \tag{6.4-42}$$

$$\dot{\mathbf{x}}(t) = \mathbf{F}\mathbf{x}(t) - \mathbf{g}_2\mathbf{C}_2\mathbf{x}(t) \tag{6.4-43}$$

either of which could place the n closed-loop roots at arbitrary locations by suitable choice of its gain matrix (W-3). If both controls act together,

$$\dot{\mathbf{x}}(t) = \mathbf{F}\mathbf{x}(t) - \mathbf{G}\mathbf{C}\mathbf{x}(t) \tag{6.4-44}$$

$\mathbf{C}$ contains $2n$ elements—twice as many as needed to specify the eigenvalues. It can be concluded that more than one $\mathbf{C}$ could provide a given set of eigenvalues in this two-control example, and it is equally reasonable to expect the mode shapes to be different for different values of **C**.

The dimension of the control vector determines the freedom with which eigenvectors can be altered for a given set of eigenvalues (M-6). In the following, it is assumed that **G** has rank m and that there exists a *modal control vector*, $\mathbf{u}_i$, corresponding to any gain matrix, **C**, and any eigenvector, $\mathbf{x}_i$, through Eq. 6.4-38:

$$\mathbf{u}_i = -\mathbf{C}\mathbf{x}_i \tag{6.4-45}$$

Although we normally think of **C** as "mapping" the n-dimensional space of $\mathbf{x}$ into the m-dimensional space of $\mathbf{u}$, we can consider the reverse case: arbitrary variations in $\mathbf{u}_i$ can affect, at most, an m-dimensional subspace of $\mathbf{x}_i$. Following Eq. 6.4-40, if **C** provides a desired set of closed-loop eigenvalues, λ_i, $i = 1$–n, then,

$$(\lambda_i \mathbf{I}_n - \mathbf{F})\mathbf{x}_i = \mathbf{G}\mathbf{u}_i = -\mathbf{G}\mathbf{C}\mathbf{x}_i, \qquad i = 1 \text{ to } n \tag{6.4-46a}$$

or

$$\mathbf{x}_i = -(\lambda_i \mathbf{I}_n - \mathbf{F})^{-1}\mathbf{G}\mathbf{C}\mathbf{x}_i, \qquad i = 1 \text{ to } n \tag{6.4-46b}$$

This could be considered a specification that $\mathbf{x}_i$ must satisfy; however, it is a specification limited to an m-dimensional subspace by the matrix product **GC**. The point is made clearer by noting that the same equation defines the modal control vector

$$\mathbf{u}_i = -\mathbf{C}(\lambda_i \mathbf{I}_n - \mathbf{F})^{-1}\mathbf{G}\mathbf{u}_i, \qquad i = 1 \text{ to } n \tag{6.4-47a}$$

or

$$[\mathbf{I}_m + \mathbf{C}(\lambda_i \mathbf{I}_n - \mathbf{F})^{-1}\mathbf{G}]\mathbf{u}_i = \mathbf{0} \tag{6.4-47b}$$

using Eq. 6.4-45. Only m elements of $\mathbf{u}_i$ are specified, and although they map into an n-dimensional vector, $\mathbf{x}_i$, different values of $\mathbf{u}_i$ would affect, at most, an m-dimensional subspace of the n-dimensional state space. Different values of **C** that provide the same values of λ_i therefore have a restricted effect on eigenvector distribution for $m < n$.

An example of mode shape flexibility is sketched in Fig. 6.4-2. A single eigenvector of a three-state model is shown for varying numbers of controls, and it is assumed that the corresponding real eigenvalue is the same in all cases. With a single control, the eigenvector can change only in length. This is effectively no change at all, since eigenvectors are defined within an arbitrary scalar constant. Two controls allow the eigenvector to change orientation in a plane that passes through the origin. The orientation of the

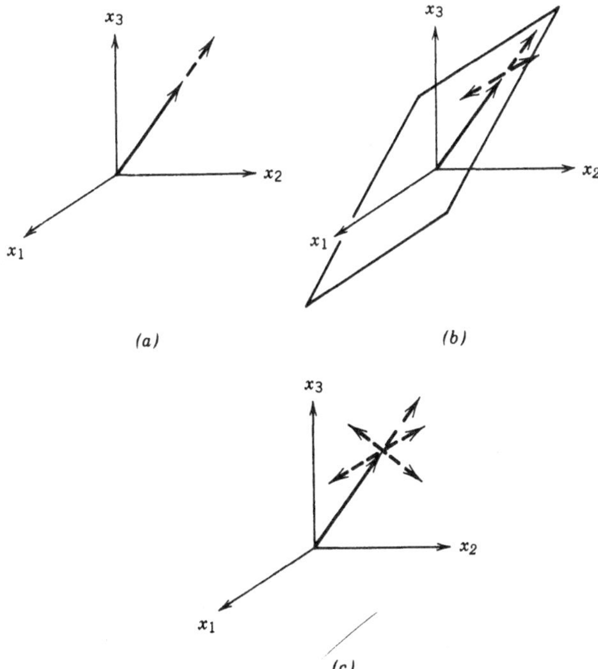

FIGURE 6.4-2 Freedom for eigenvector alteration in a three-state system. (*a*) One control; (*b*) two controls; (*c*) three controls.

plane is fixed by **F**, **G**, and λ_i. Three controls allow total freedom to alter the eigenvector, as every element in $(\mathbf{F}-\mathbf{GC})$ can be changed. (How does this relate to the perfect model-following criterion of Section 6.3?)

Eigenvectors of Optimally Regulated Systems

For any $(m \times n)$ gain matrix, **C**, the n eigenvalues of the closed-loop system can be found by solving

$$|s\mathbf{I}_n - (\mathbf{F}-\mathbf{GC})| = 0 \tag{6.4-48}$$

and the corresponding eigenvectors $\mathbf{x}_i$ satisfy the equation

$$[\lambda_i\mathbf{I}_n - (\mathbf{F}-\mathbf{GC})]\mathbf{x}_i = \mathbf{0}, \qquad i = 1\text{–}n \tag{6.4-49}$$

It has been shown that the closed-loop eigenvalues corresponding to limiting values of the control weighting matrix **R** can be found without solving for **C**, and the same is true for the closed-loop eigenvectors. The following discussion is restricted to the case of distinct eigenvalues.

As the ratio of the magnitudes of **Q** and **R** affects the gain matrix and eigenvalues, minimizing the cost function,

$$J = \frac{1}{2}\int_0^\infty (\mathbf{x}^T\mathbf{Q}\mathbf{x} + \rho^2\mathbf{u}^T\mathbf{R}_o\mathbf{u})\,dt \tag{6.4-50}$$

produces the same result as minimizing

$$J = \frac{1}{2}\int_0^\infty \left[\frac{1}{\rho^2}\mathbf{x}^T\mathbf{Q}\mathbf{x} + \mathbf{u}^T R_o\mathbf{u}\right] dt \tag{6.4-51}$$

for any value of ρ. The $2n$ eigenvalues of the closed-loop system and its adjoint are solutions to the following:

$$|s\mathbf{I}_{2n} - \mathbf{F}'| = \begin{bmatrix} (s\mathbf{I}_n - \mathbf{F}) & \mathbf{G}\mathbf{R}_0^{-1}\mathbf{G}^T \\ \dfrac{\mathbf{Q}}{\rho^2} & (s\mathbf{I}_n + \mathbf{F}^T) \end{bmatrix} = 0 \tag{6.4-52}$$

As ρ goes to infinity (the "minimum-control-energy" case), this becomes

$$\begin{aligned} |s\mathbf{I}_{2n} - \mathbf{F}'| &= \begin{vmatrix} (s\mathbf{I}_n - \mathbf{F}) & \mathbf{G}\mathbf{R}_o^{-1}\mathbf{G}^T \\ \mathbf{0} & (s\mathbf{I}_n + \mathbf{F}^T) \end{vmatrix} = (-1)^n \Delta_{\mathrm{OL}}(s)\Delta_{\mathrm{OL}}(-s) \\ &= 0 \end{aligned} \tag{6.4-53}$$

The eigenvectors for the state and adjoint variables, $\mathbf{x}_i$ and $\boldsymbol{\lambda}_i$, are then solutions to the equation

$$\begin{bmatrix} (s_i\mathbf{I}_n - \mathbf{F}) & \mathbf{G}\mathbf{R}_0^{-1}\mathbf{G}^T \\ \mathbf{0} & (s_i\mathbf{I}_n + \mathbf{F}^T) \end{bmatrix} \begin{bmatrix} \mathbf{x}_i \\ \boldsymbol{\lambda}_i \end{bmatrix} = \mathbf{0}, \qquad i = 1\text{-}2n \tag{6.4-54}$$

where $s_i < 0$ is a stable root of the closed-loop system and $s_i > 0$ is a root of the adjoint system.

Equation 6.4-54 and previous results can be used to compute $\mathbf{x}_i$ as ρ goes to infinity, considering only those roots of $\Delta_{\mathrm{OL}}(s)\Delta_{\mathrm{OL}}(-s)$ that are in the left half of the s plane. If s_i is a stable root of the open-loop system matrix **F**, then

$$|s_i\mathbf{I}_n - \mathbf{F}| = 0 \tag{6.4-55}$$

and the corresponding eigenvector is defined by

$$[s_i\mathbf{I}_n - \mathbf{F}]\mathbf{x}_i = \mathbf{0} \tag{6.4-56}$$

This indicates that $\mathbf{x}_i$ is an eigenvector of the open-loop system and that the mode is unaffected by control. The top row of Eq. 6.4-54 must be satisfied, subject to Eq. 6.4-56:

$$[\mathscr{s}_i\mathbf{I}_n-\mathbf{F}]\mathbf{x}_i+\mathbf{G}\mathbf{R}_o^{-1}\mathbf{G}^T\boldsymbol{\lambda}_i=\mathbf{0} \tag{6.4-57}$$

This is true only if

$$\mathbf{G}\mathbf{R}_o^{-1}\mathbf{G}^T\boldsymbol{\lambda}_i=\mathbf{0} \tag{6.4-58}$$

leading to

$$\boldsymbol{\lambda}_i=\mathbf{0} \tag{6.4-59}$$

Since $\boldsymbol{\lambda}_i$ is the adjoint of $\mathbf{x}_i$, it is related to $\mathbf{x}_i$ by

$$\boldsymbol{\lambda}_i=\mathbf{S}\mathbf{x}_i \tag{6.4-60}$$

where $\mathbf{S}$ is the solution to the algebraic Riccati equation,

$$\mathbf{0}=-\mathbf{F}^T\mathbf{S}-\mathbf{S}\mathbf{F}-\frac{\mathbf{Q}}{\rho^2}+\mathbf{S}\mathbf{G}\mathbf{R}_o^{-1}\mathbf{G}^T\mathbf{S} \tag{6.4-61}$$

as $\rho\rightarrow\infty$. $\mathbf{F}$ may contain unstable roots, so $\mathbf{S}$ may not be a null matrix; however, the product $\mathbf{S}\mathbf{x}_i$ is zero for stable $\mathscr{s}_i$. The modal control vector, $\mathbf{u}_i$, associated with $\mathbf{x}_i$ must be zero, as

$$\begin{aligned}\mathbf{u}_i&=-\mathbf{R}_o^{-1}\mathbf{G}^T\boldsymbol{\lambda}_i\\&=-\mathbf{R}_o^{-1}\mathbf{G}^T\mathbf{S}\mathbf{x}_i=-\mathbf{C}\mathbf{x}_i=\mathbf{0}\end{aligned} \tag{6.4-62}$$

If $-\mathscr{s}_i$ is an unstable root of $\mathbf{F}$, then $\mathscr{s}_i$ does not satisfy Eq. 6.4-55, but it is a root of the adjoint system:

$$|\mathscr{s}_i\mathbf{I}_n+\mathbf{F}^T|=0 \tag{6.4-63}$$

The adjoint vector then is specified by the system's open-loop characteristics,

$$[\mathscr{s}_i\mathbf{I}_n+\mathbf{F}^T]\boldsymbol{\lambda}_i=\mathbf{0} \tag{6.4-64}$$

and, from Eq. 6.4-57, the closed-loop state eigenvector is

$$\mathbf{x}_i=-(\mathscr{s}_i\mathbf{I}_n-\mathbf{F})^{-1}\mathbf{G}\mathbf{R}_o^{-1}\mathbf{G}^T\boldsymbol{\lambda}_i \tag{6.4-65}$$

This corresponds to the result for arbitrary linear controllers (Eq. 6.4-46b), with the optimal gain matrix specifying $\mathbf{C}$. $\mathbf{R}_o$ provides the flexibility to adjust $\mathbf{x}_i$ within an m-dimensional subspace, in keeping with the more general result.

We note in passing that Eq. 6.4-62 can be applied to the modal matrices $\mathbf{X}$ and $\mathbf{\Lambda}$ defined by $\mathbf{x}_i$ and $\boldsymbol{\lambda}_i$, $i = 1\text{–}n$:

$$\mathbf{C}[\mathbf{x}_1, \ldots, \mathbf{x}_n] = \mathbf{R}^{-1}\mathbf{G}^T[\boldsymbol{\lambda}_1, \ldots, \boldsymbol{\lambda}_n] \tag{6.4-66a}$$

or

$$\mathbf{CX} = \mathbf{R}^{-1}\mathbf{G}^T\mathbf{\Lambda} \tag{6.4-66b}$$

From previous discussion, we know that $\mathbf{X}$ is nonsingular; hence this equation can be used to compute the optimal gain matrix:

$$\begin{aligned}\mathbf{C} &= \mathbf{R}^{-1}\mathbf{G}^T\mathbf{\Lambda}\mathbf{X}^{-1} \\ &= \mathbf{R}^{-1}\mathbf{G}^T\mathbf{S}\end{aligned} \tag{6.4-67}$$

The equivalence between $\mathbf{S}$ and $\mathbf{\Lambda}\mathbf{X}^{-1}$ was shown in Section 6.1. In the limiting case, $\mathbf{\Lambda}$ contains columns of zeros corresponding to the stable roots of $\mathbf{F}$, confirming the lack of control effect on the stable modes in the minimum-control-energy case.

As control weighting goes to zero, the closed-loop eigenvalues go to the transmission zeros (or their images) or to infinity, and the eigenvectors go to asymptotic distributions as well. For the cost function defined by Eq. 6.4-50, the state and adjoint eigenvectors are defined by

$$\begin{bmatrix} (\lambda_i\mathbf{I}_n - \mathbf{F}) & (1/\rho^2)\mathbf{G}\mathbf{R}_o^{-1}\mathbf{G}^T \\ \mathbf{Q} & (\lambda_i\mathbf{I}_n + \mathbf{F}^T) \end{bmatrix} \begin{bmatrix} \mathbf{x}_i \\ \boldsymbol{\lambda}_i \end{bmatrix} = \mathbf{0} \tag{6.4-68}$$

which has the $2n$ nonsingular solutions

$$\boldsymbol{\lambda}_i = \frac{1}{\rho^2}(\lambda_i\mathbf{I}_n + \mathbf{F}^T)^{-1}\mathbf{Q}(\lambda_i\mathbf{I}_n - \mathbf{F})^{-1}\mathbf{G}\mathbf{R}_o^{-1}\mathbf{G}^T\boldsymbol{\lambda}_i \tag{6.4-69a}$$

or

$$\left[\mathbf{I}_n - \frac{1}{\rho^2}(\lambda_i\mathbf{I}_n + \mathbf{F}^T)^{-1}\mathbf{Q}(\lambda_i\mathbf{I}_n - \mathbf{F})^{-1}\mathbf{G}\mathbf{R}_o^{-1}\mathbf{G}^T\right]\boldsymbol{\lambda}_i = \mathbf{0}, \qquad i = 1\text{–}2n \tag{6.4-69b}$$

assuming λ_i is not a root of $\mathbf{F}$ or $-\mathbf{F}^T$. This is a general solution for $\boldsymbol{\lambda}_i$ that is not limited to extreme values of ρ, and since

$$\mathbf{u}_i = -\frac{1}{\rho^2}\mathbf{R}_o^{-1}\mathbf{G}^T\boldsymbol{\lambda}_i \tag{6.4-70}$$

we can write the modal control vector solution as well. Considering only the stable closed-loop roots, λ_i:

$$\left[\mathbf{I}_m + \frac{1}{\rho^2}\mathbf{R}_o^{-1}\mathbf{G}^T(\lambda_i\mathbf{I}_n + \mathbf{F}^T)^{-1}\mathbf{Q}(\lambda_i\mathbf{I}_n - \mathbf{F})^{-1}\mathbf{G}\right]\mathbf{u}_i = \mathbf{0}, \qquad i = 1\text{–}n \quad (6.4\text{-}71)$$

Defining $\mathbf{Y}_2(s)$ as in Eq. 6.4-20, this can be expressed as

$$\left[\mathbf{I}_m + \frac{1}{\rho^2}\mathbf{Y}_2^T(-\lambda_i)\mathbf{Y}_2(\lambda_i)\right]\mathbf{u}_i = \mathbf{0}, \qquad i = 1\text{–}n \quad (6.4\text{-}72)$$

Once the $\mathbf{u}_i$ are known, the state eigenvectors for the stable closed-loop modes can be computed using Eq. 6.4-46:

$$\mathbf{x}_i = (\lambda_i\mathbf{I}_n - \mathbf{F})^{-1}\mathbf{G}\mathbf{u}_i, \qquad i = 1\text{–}n \quad (6.4\text{-}73)$$

Example 6.4-3 Eigenvectors of Optimally Regulated Single-Input Second-Order Systems

The second-order system

$$\begin{bmatrix}\dot{x}_1\\ \dot{x}_2\end{bmatrix} = \begin{bmatrix}0 & 1\\ -a_1a_2 & -(a_1+a_2)\end{bmatrix}\begin{bmatrix}x_1\\ x_2\end{bmatrix} + \begin{bmatrix}0\\ a_1a_2\end{bmatrix}u$$

has the characteristic equation

$$\begin{aligned}0 &= s^2 + (a_1+a_2)s + a_1a_2\\ &= (s+a_1)(s+a_2)\end{aligned}$$

and its real roots $(-a_1, \ -a_2)$ are stable or unstable, depending on the signs of a_1 and a_2. In the following, assume that a_1 is positive, a_2 is negative, and $a_1 \neq a_2$. We wish to minimize the quadratic cost function

$$J = \frac{1}{2}\int_0^\infty \left\{[x_1 \ \ x_2]\begin{bmatrix}0 & 0\\ 0 & q_{22}\end{bmatrix}\begin{bmatrix}x_1\\ x_2\end{bmatrix} + \rho u^2\right\} dt$$

and to examine the asymptotic behavior of the closed-loop eigenvalues as ρ varies between ∞ and 0. From Eqs. 6.4-12 and 6.4-20,

$$\mathbf{H} = [0 \ \ \sqrt{q_{22}}]$$

and

$$Y_2(s) = \frac{\sqrt{q_{22}}\,a_1a_2s}{s^2 + (a_1+a_2)s + a_1a_2}$$

so the closed-loop roots satisfy the equation (Eq. 6.4-28)

$$(-1)^n(s+a_1)(s+a_2)(-s+a_1)(-s+a_2)-\frac{1}{\rho^2}q_{22}a_1^2a_2^2s^2=0$$

or

$$s^4-\left(a_1^2+a_2^2+\frac{q_{22}a_1^2a_2^2}{\rho^2}\right)s^2+a_1^2a_2^2=0$$

which takes the form

$$s^4+bs^2+c=0$$

The four real roots of this biquadratic equation are

$$\mathfrak{s}_{1,2,3,4}=\pm\left[-\frac{b}{2}\pm\sqrt{\left(\frac{b}{2}\right)^2-c}\right]^{1/2}$$

The two negative roots belong to the closed-loop system, and the two positive roots belong to the adjoint system.

As ρ approaches infinity, the controlled-system roots approach $-a_1$ and $+a_2$. From Eqs. 6.4-56 and 2.2-74, the first state eigenvector is defined by $\text{Adj}[-a_1\mathbf{I}_n-\mathbf{F}]$ (i.e., by its open-loop value), and it is

$$\mathbf{x}_1=\begin{bmatrix}1\\-a_1\end{bmatrix}$$

Because the open-loop $\mathfrak{s}_2$ is unstable, $-\mathfrak{s}_2$ is a stable root of $-\mathbf{F}^T$; hence we must first compute the adjoint variable eigenvector from Eq. 6.4-64:

$$[\mathfrak{s}_2\mathbf{I}_n+\mathbf{F}^T]\boldsymbol{\lambda}_2=\mathbf{0}$$

This gives

$$\boldsymbol{\lambda}_2=\begin{bmatrix}a_1\\1\end{bmatrix}$$

Then $\mathbf{x}_2$ is defined by Eq. 6.4-65 as

$$\mathbf{x}_2=\begin{bmatrix}1\\a_2\end{bmatrix}$$

As ρ approaches zero, the closed-loop system roots approach $-|\sqrt{q_{22}}a_1a_2|/\rho$ and 0. The control actively determines both of these values, and the scalar modal control vector can be computed from Eq. 6.4-72. Because u_i is scalar and is defined within an arbitrary constant, this computation can be bypassed in determining the relative components of x_1 and x_2. From Eq. 6.4-73, for diminishing values of ρ,

$$\mathbf{x}_1 = \left(\frac{-|\sqrt{q_{22}}\, a_1 a_2|}{\rho \mathbf{I}_n - \mathbf{F}}\right)^{-1} \mathbf{G}$$

$$= \alpha_1 \begin{bmatrix} 1 \\ -|\sqrt{q_{22}}\, a_1 a_2|/\rho \end{bmatrix}$$

and

$$\mathbf{x}_2 = -\mathbf{F}^{-1}\mathbf{G} = \alpha_2 \begin{bmatrix} 1 \\ 0 \end{bmatrix}$$

Eigenvalues and Eigenvectors of Optimal Estimators

By duality, it is clear that the continuous-time LQ regulator results apply to the continuous-time linear-Gaussian (LG) estimator, with the spectral density matrices of the disturbance input and measurement error, **W** and **N**, replacing the state and control weighting matrices, **Q** and **R** (Section 5.4). The system is forced by an s-dimensional disturbance, $\mathbf{w}(t)$,

$$\dot{\mathbf{x}}(t) = \mathbf{F}\mathbf{x}(t) + \mathbf{G}\mathbf{u}(t) + \mathbf{L}\mathbf{w}(t) \tag{6.4-74}$$

and the state is to be estimated from the r-dimensional measurement vector

$$\mathbf{z}(t) = \mathbf{H}\mathbf{x}(t) + \mathbf{n}(t) \tag{6.4-75}$$

The spectral density matrices are

$$E[\mathbf{w}(t)\mathbf{w}^T(\tau)] = \mathbf{W}\delta(t-\tau) \tag{6.4-76}$$

$$E[\mathbf{n}(t)\mathbf{n}^T(\tau)] = \mathbf{N}\delta(t-\tau) = \rho^2 \mathbf{N}_o \delta(t-\tau) \tag{6.4-77}$$

and there is zero correlation between the disturbance input and measurement error. The filter covariance and gain equations are given by Eqs. 4.5-19 and 4.5-20; from Table 5.4-1, we can define a cost transfer function matrix that is equivalent to Eq. 6.4-20,

$$\mathbf{Y}_2(s) = \mathbf{Z}(s\mathbf{I}_n - \mathbf{F}^T)^{-1}\mathbf{H}^T\mathbf{M} \tag{6.4-78}$$

with **Z** defined as a $(p \times n)$ "square root" of $\mathbf{LWL}^T$ $(p \geq s)$,

$$\mathbf{LWL}^T = \mathbf{Z}^T\mathbf{Z} \tag{6.4-79}$$

and $\mathbf{M}$ representing the $(r \times r)$ symmetric positive square root of $\mathbf{N}_o$:

$$\mathbf{N}_o = \mathbf{M}^T\mathbf{M} = \mathbf{M}\mathbf{M}^T \tag{6.4-80}$$

$\mathbf{Y}_2(s)$ is square only if $p = r$, which could be the case if the number of independent disturbances and measurements are equal.

Using $\mathbf{Y}_2(s)$ as defined earlier, the stable eigenvalues of Eq. 6.4-19 correspond to the LG estimator roots; that is, they determine the damping, natural frequencies, and time constants of the estimation errors. As ρ approaches ∞, signifying very noisy measurements, the state estimates corresponding to stable modes are made on the basis of initial conditions and known control inputs alone. Estimation gains remain nonzero only for those state components affected by unstable modes. As measurements get better, ρ approaches zero; the estimator roots approach finite locations determined by $\mathbf{Y}_2(s)$ or go to ∞ in one or more Butterworth configurations. These characteristics can be used to decompose a single high-order filter into a bank of "fast" (i.e., high-bandwidth) low-order filters plus a moderate-order filter that is equivalent to a *Luenberger observer* (S-8, L-5).

The eigenvectors of the estimation error are affected by the choice of $\mathbf{W}$ and $\mathbf{N}$. For specific, fixed eigenvalues, the mode shapes can be adjusted within r-dimensional subspaces of the dimensional eigenvectors by variations in the $(n \times r)$ estimation gain matrix $\mathbf{K}$. Consequently, optimal filters with similar bandwidths can have different multidimensional "bandpass" characteristics.

Modal Properties for the Stochastic Optimal Regulator

The linear-quadratic regulator and the linear-Gaussian estimator are combined within the stochastic optimal regulator; hence there is the potential for modal interaction between the two. If the controlled system's dynamic model $(\mathbf{F}, \mathbf{G}, \mathbf{H})$ is known without error and the LQG controller/estimator gains are computed accordingly, the eigenvalues of the controller and the estimator can be computed separately, and stability is retained unequivocally. If, however, there is a difference between the actual model and that which is used for design, controller and estimator dynamics are coupled, and there is the possiblity of instability.

We begin with the assumption that modeled and actual dynamic characteristics are the same. From Section 5.4, we can describe the $2n$-dimensional state and state-estimate dynamics of a zero-set-point LQG regulator as follows (Eq. 5.4-52 to 5.4-59):

$$\dot{\mathbf{x}}(t) = \mathbf{F}\mathbf{x}(t) - \mathbf{G}\mathbf{C}\hat{\mathbf{x}}(t) + \mathbf{L}\mathbf{w}(t) \tag{6.4-81}$$

$$\begin{aligned}\dot{\hat{\mathbf{x}}}(t) &= \mathbf{F}\hat{\mathbf{x}}(t) - \mathbf{G}\mathbf{C}\hat{\mathbf{x}}(t) + \mathbf{K}[\mathbf{z}(t) - \mathbf{H}\hat{\mathbf{x}}(t)] \\ &= (\mathbf{F} - \mathbf{G}\mathbf{C} - \mathbf{K}\mathbf{H})\hat{\mathbf{x}}(t) + \mathbf{K}\mathbf{z}(t)\end{aligned} \tag{6.4-82}$$

The measurement vector is

$$\mathbf{z}(t) = \mathbf{H}\mathbf{x}(t) + \mathbf{n}(t) \tag{6.4-83}$$

and all other terms are described as before. Further insight can be gained if the state-estimate equation is replaced by the state-estimate-error equation. Defining the estimate error as

$$\boldsymbol{\epsilon}(t) = \mathbf{x}(t) - \hat{\mathbf{x}}(t) \tag{6.4-84}$$

its differential equation is (neglecting the disturbance input)

$$\begin{aligned}\dot{\boldsymbol{\epsilon}}(t) &= \dot{\mathbf{x}}(t) - \dot{\hat{\mathbf{x}}}(t)\\ &= \mathbf{F}\mathbf{x}(t) - \mathbf{G}\mathbf{C}\hat{\mathbf{x}}(t) - (\mathbf{F} - \mathbf{G}\mathbf{C} - \mathbf{K}\mathbf{H})\hat{\mathbf{x}}(t) - \mathbf{K}[\mathbf{H}\mathbf{x}(t) + \mathbf{n}(t)]\\ &= (\mathbf{F} - \mathbf{K}\mathbf{H})\boldsymbol{\epsilon}(t) - \mathbf{K}\mathbf{n}(t)\end{aligned} \tag{6.4-85}$$

The combined homogeneous state and estimate-error equation is

$$\begin{bmatrix}\dot{\mathbf{x}}(t)\\ \dot{\boldsymbol{\epsilon}}(t)\end{bmatrix} = \begin{bmatrix}(\mathbf{F} - \mathbf{G}\mathbf{C}) & \mathbf{G}\mathbf{C}\\ \mathbf{0} & (\mathbf{F} - \mathbf{K}\mathbf{H})\end{bmatrix}\begin{bmatrix}\mathbf{x}(t)\\ \boldsymbol{\epsilon}(t)\end{bmatrix} \tag{6.4-86}$$

and the $2n$ eigenvalues, λ_i, for this equation can be found by solving the following:

$$\begin{vmatrix}[s\mathbf{I}_n - (\mathbf{F} - \mathbf{G}\mathbf{C})] & -\mathbf{G}\mathbf{C}\\ \mathbf{0} & [s\mathbf{I}_n - (\mathbf{F} - \mathbf{K}\mathbf{H})]\end{vmatrix} = 0 \tag{6.4-87}$$

Because this equation is upper-block-triangular, n eigenvalues are defined by

$$|s\mathbf{I}_n - (\mathbf{F} - \mathbf{G}\mathbf{C})| = 0 \tag{6.4-88}$$

and n eigenvalues are obtained from

$$|s\mathbf{I}_n - (\mathbf{F} - \mathbf{K}\mathbf{H})| = 0 \tag{6.4-89}$$

Although the eigenvalues are uncoupled, the eigenvectors are coupled, because $\mathbf{GC}$ induces estimator modal response in the state $\mathbf{x}(t)$. With distinct eigenvalues, this system of equations could be block-diagonalized by a transformation of the form

$$\mathbf{T} = \begin{bmatrix}\mathbf{I}_n & \mathbf{T}_2\\ \mathbf{0} & \mathbf{I}_n\end{bmatrix} \tag{6.4-90}$$

The $2n$ diagonalized eigenvectors, $\mathbf{e}_i$, each could be separated into two n-dimensional parts defined by

$$\begin{bmatrix} [\lambda_i \mathbf{I}_n - (\mathbf{F} - \mathbf{GC})] & \mathbf{0} \\ \mathbf{0} & [\lambda_i \mathbf{I}_n - (\mathbf{F} - \mathbf{KH})] \end{bmatrix} \begin{bmatrix} \mathbf{e}_{1_i} \\ \mathbf{e}_{2_i} \end{bmatrix} = \mathbf{0}, \quad i = 1 \text{ to } 2n \quad (6.4\text{-}91)$$

Assuming that the first n eigenvalues belong to $(\mathbf{F} - \mathbf{GC})$ and the remaining ones belong to $(\mathbf{F} - \mathbf{KH})$, the corresponding eigenvectors partition as $[\mathbf{e}_{1_i}^T \ \ \mathbf{0}]^T$, $i = 1$ to n, and $[\mathbf{0}, \ \ \mathbf{e}_{2_i}^T]^T$, $i = (n+1)$ to $2n$, respectively. Then the eigenvectors of the original system take the form

$$\mathbf{x}_i = \begin{bmatrix} \mathbf{e}_{1_i} \\ -\mathbf{T}_2 \mathbf{e}_{2_i} \end{bmatrix}, \quad \begin{matrix} i = 1 \text{ to } n \\ i = (n+1) \text{ to } 2n \end{matrix} \quad (6.4\text{-}92)$$

Suppose that the actual system $(\mathbf{F}_A, \mathbf{G}_A, \mathbf{H}_A)$ is different from the system $(\mathbf{F}, \mathbf{G}, \mathbf{H})$ used to design the controller and the estimator. Neglecting the disturbance input, the state and state-estimate equations become

$$\dot{\mathbf{x}}(t) = \mathbf{F}_A \mathbf{x}(t) - \mathbf{G}_A \mathbf{C} \hat{\mathbf{x}}(t) \quad (6.4\text{-}93)$$

$$\begin{aligned} \dot{\hat{\mathbf{x}}}(t) &= \mathbf{F}\hat{\mathbf{x}}(t) - \mathbf{GC}\hat{\mathbf{x}}(t) + \mathbf{K}[\mathbf{z}(t) - \mathbf{H}\hat{\mathbf{x}}(t)] \\ &= (\mathbf{F} - \mathbf{GC} - \mathbf{KH})\hat{\mathbf{x}}(t) + \mathbf{Kz}(t) \end{aligned} \quad (6.4\text{-}94)$$

where

$$\mathbf{z}(t) = \mathbf{H}_A \mathbf{x}(t) + \mathbf{n}(t) \quad (6.4\text{-}95)$$

The state-estimate-error equation is then

$$\begin{aligned} \dot{\boldsymbol{\epsilon}}(t) &= \mathbf{F}_A \mathbf{x}(t) - \mathbf{G}_A \mathbf{C}[\mathbf{x}(t) - \boldsymbol{\epsilon}(t)] \\ &\quad - (\mathbf{F} - \mathbf{GC} - \mathbf{KH})[\mathbf{x}(t) - \boldsymbol{\epsilon}(t)] - \mathbf{K}[\mathbf{H}_A \mathbf{x}(t) + \mathbf{n}(t)] \\ &= [(\mathbf{F}_A - \mathbf{F}) - (\mathbf{G}_A - \mathbf{G})\mathbf{C} - \mathbf{K}(\mathbf{H}_A - \mathbf{H})]\mathbf{x}(t) \\ &\quad + [\mathbf{F} + (\mathbf{G}_A - \mathbf{G})\mathbf{C} - \mathbf{KH}]\boldsymbol{\epsilon}(t) - \mathbf{Kn}(t) \end{aligned} \quad (6.4\text{-}96)$$

Combining Eqs. 6.4-93 and 6.4-96, the eigenvalues of the LQG regulator with system mismatch are defined by

$$\begin{vmatrix} [s\mathbf{I}_n - (\mathbf{F}_A - \mathbf{G}_A \mathbf{C})] & -\mathbf{G}_A \mathbf{C} \\ [(\mathbf{F} - \mathbf{F}_A) - (\mathbf{G} - \mathbf{G}_A)\mathbf{C} - \mathbf{K}(\mathbf{H} - \mathbf{H}_A)] & \{s\mathbf{I}_n - [\mathbf{F} + (\mathbf{G}_A - \mathbf{G})\mathbf{C} - \mathbf{KH}]\} \end{vmatrix} = 0 \quad (6.4\text{-}97)$$

The state and error-estimate dynamics are clearly coupled, and distinct effects of mismatch in **F**, **G**, and **H** are presented. The control gain matrix

$\mathbf{C}$ is computed from $(\mathbf{F}, \mathbf{G})$ rather than $(\mathbf{F}_A, \mathbf{G}_A)$; hence the control loop alone could be unstable unless $\mathbf{C}$ is robust enough to stabilize the actual system. $\mathbf{K}$ is computed for $(\mathbf{F}, \mathbf{H})$ rather than $(\mathbf{F}_A, \mathbf{H}_A)$; this alone does not allow primary instability, but mismatched control sensitivity does allow that possibility. Because no generic pattern has been specified for the mismatch, the effects on eigenvalues and eigenvectors can be determined only by direct solution.

Eigenvalues for the Discrete-Time Linear-Quadratic Regulator

The eigenvalues and eigenvectors of discrete-time regulators and estimators vary systematically with changes in the optimal design parameters. Modal properties are determined in much the same way that they are determined for continuous-time systems; however, the frequency-domain regions of stability and asymptotic locations of closed-loop roots are different, following Section 2.5. The principal differences can be understood by examining the closed-loop eigenvalues for the linear-quadratic regulator.

From Section 6.1, the infinite-time quadratic cost function

$$J = \frac{1}{2} \sum_{k=0}^{\infty} (\mathbf{x}_k^T \mathbf{Q} \mathbf{x}_k + \mathbf{u}_k^T \mathbf{R} \mathbf{u}_k) \tag{6.4-98}$$

for the n-dimensional, linear, discrete-time dynamic system

$$\mathbf{x}_{k+1} = \mathbf{\Phi} \mathbf{x}_k + \mathbf{\Gamma} \mathbf{u}_k \tag{6.4-99}$$

is minimized by the control law

$$\mathbf{u}_k = -\mathbf{R}^{-1} \mathbf{\Gamma}^T \boldsymbol{\lambda}_{k+1} \tag{6.4-100}$$

where $\boldsymbol{\lambda}_k$ is an m-dimensional adjoint vector determined by

$$\boldsymbol{\lambda}_k = \mathbf{Q} \mathbf{x}_k + \mathbf{\Phi}^T \boldsymbol{\lambda}_{k+1} \tag{6.4-101}$$

Substituting Eq. 6.4-100 in Eq. 6.4-99 and taking the z transforms of the state and adjoint equations leads to the following (for zero initial conditions):

$$z\mathbf{x}(z) = \mathbf{\Phi} \mathbf{x}(z) - \mathbf{\Gamma} \mathbf{R}^{-1} \mathbf{\Gamma}^T z \boldsymbol{\lambda}(z) \tag{6.4-102}$$

$$\boldsymbol{\lambda}(z) = \mathbf{Q} \mathbf{x}(z) + \mathbf{\Phi}^T z \boldsymbol{\lambda}(z) \tag{6.4-103}$$

or

$$\begin{bmatrix} (z\mathbf{I}_n - \mathbf{\Phi}) & \mathbf{\Gamma} \mathbf{R}^{-1} \mathbf{\Gamma}^T \\ -\mathbf{Q} & (z^{-1} \mathbf{I}_n - \mathbf{\Phi}^T) \end{bmatrix} \begin{bmatrix} \mathbf{x}(z) \\ z \boldsymbol{\lambda}(z) \end{bmatrix} = \mathbf{0} \tag{6.4-104}$$

The $2n$ eigenvalues can be found by solving the determinant equation

$$\begin{vmatrix} (z\mathbf{I}_n - \mathbf{\Phi}) & \mathbf{\Gamma}\mathbf{R}^{-1}\mathbf{G}^T \\ -\mathbf{Q} & (z^{-1}\mathbf{I}_n - \mathbf{\Phi}^T) \end{vmatrix} \triangleq \Delta_{\Phi}(z) = 0 \tag{6.4-105}$$

The similarity of the last equations to the s-transform equation for linear-optimal continuous systems (Eq. 6.4-6b) is apparent, and the determinant identities of Section 2.2 can be applied as before:

$$\begin{aligned} \Delta_{\Phi}(z) &= |z\mathbf{I}_n - \mathbf{\Phi}|\,|(z^{-1}\mathbf{I}_n - \mathbf{\Phi}^T) + \mathbf{Q}(z\mathbf{I}_n - \mathbf{\Phi})^{-1}\mathbf{\Gamma}\mathbf{R}^{-1}\mathbf{\Gamma}^T| \\ &= |z\mathbf{I}_n - \mathbf{\Phi}|\,|z^{-1}\mathbf{I}_n - \mathbf{\Phi}^T|\,|\mathbf{I}_n + \mathbf{Q}(z\mathbf{I}_n - \mathbf{\Phi})^{-1}\mathbf{\Phi}\mathbf{R}^{-1}\mathbf{\Gamma}^T(z^{-1}\mathbf{I}_n - \mathbf{\Phi}^T)^{-1}| \\ &= |z\mathbf{I}_n - \mathbf{\Phi}|\,|z^{-1}\mathbf{I}_n - \mathbf{\Phi}^T|\,\mathbf{I}_m + \mathbf{R}^{-1}\mathbf{\Gamma}^T(z^{-1}\mathbf{I}_n - \mathbf{\Phi}^T)^{-1}\mathbf{Q}(z\mathbf{I}_n - \mathbf{\Phi})^{-1}\mathbf{\Gamma}| \\ &= \Delta_{\mathrm{OL}}(z)\Delta_{\mathrm{OL}}(z^{-1})|\mathbf{I}_m + \mathbf{R}^{-1}\mathbf{Y}_1^T(z^{-1})\mathbf{Y}_1(z)| \end{aligned} \tag{6.4-106}$$

The *open-loop characteristic polynomial* is

$$\Delta_{\mathrm{OL}}(z) = |z\mathbf{I}_n - \mathbf{\Phi}| \tag{6.4-107}$$

and $\Delta_{\mathrm{OL}}(z^{-1})$ is the same function with z^{-1} replacing z. If the factors of $\Delta_{\mathrm{OL}}(z)$ are β_i, $i = 1$–n,

$$\Delta_{\mathrm{OL}}(z) = (z - \beta_1)\,(z - \beta_2)\ldots(z - \beta_n) = 0 \tag{6.4-108}$$

then the factors of $\Delta_{\mathrm{OL}}(z^{-1})$ are similar:

$$\Delta_{\mathrm{OL}}(z^{-1}) = (z^{-1} - \beta_1)\,(z^{-1} - \beta_2)\ldots(z^{-1} - \beta_n) = 0 \tag{6.4-109a}$$

Examining a single factor of Eq. 6.4-109a, $(1/z - \beta_1)$ is the same as $-\beta_1(z - 1/\beta_1)/z$, and an equivalent expression for $\Delta_{\mathrm{OL}}(z^{-1})$ is as follows:

$$\Delta_{\mathrm{OL}}(z^{-1}) = (-z)^{-n}\prod_{i=1}^{n}\left[\beta_i\left(z - \frac{1}{\beta_i}\right)\right] = 0 \tag{6.4-109b}$$

Note that z and β_i are complex variables; if β_i is an eigenvalue whose magnitude is less than one (i.e., if it lies within the unit circle), than $1/\beta_i$ is outside the unit circle. Recalling the stability criteria for discrete-time systems (Section 2.5), each stable (unstable) root of $\Delta_{\mathrm{OL}}(z)$ is accompanied by an unstable (stable) root of $\Delta_{\mathrm{OL}}(z^{-1})$.

The *cost transfer function matrix* from the control to the state components that are weighted in the cost function is

$$\mathbf{Y}_1(z) = \mathbf{H}(z\mathbf{I}_n - \mathbf{\Phi})^{-1}\mathbf{\Gamma} = \mathbf{V}_1(z)/\Delta_{\mathrm{OL}}(z) \tag{6.4-110}$$

where $\mathbf{H}$ is an $(r \times n)$ "square root" of $\mathbf{Q}$,

$$\mathbf{Q} = \mathbf{H}^T\mathbf{H} \tag{6.4-111}$$

for $r \geq p$, $p = \text{rank}(\mathbf{Q})$. The dimensions of $\mathbf{Y}_1(z)$ and $\mathbf{V}_1(z)$ are $(r \times m)$, where m is the dimension of the control; $\mathbf{Y}_1(z^{-1})$ and $\mathbf{V}_1(z^{-1})$ are defined accordingly. The determinant of $\mathbf{Y}_1(z)$ takes the form

$$|\mathbf{Y}_1(z)| = \psi_1(z)/\Delta_{\text{OL}}(z) \tag{6.4-112a}$$

leading to the following definition of the scalar numerator:

$$\psi_1(z) = \Delta_{\text{OL}}(z)|\mathbf{Y}_1(z)| \tag{6.4-112b}$$

The zeros of the nr scalar numerators of $\mathbf{Y}_1(z)$ can be plotted in the z plane. The transmission zeros of $\psi_1(z)$ that plot within the unit circle are accompanied by zeros of $\psi_1(z^{-1})$ that plot outside the unit circle and vice versa. As before, the cost transfer function matrix can be defined to include control weighting effects

$$\mathbf{Y}_2(z) = \mathbf{H}(z\mathbf{I}_n - \mathbf{\Phi})^{-1}\mathbf{\Gamma}\mathbf{U} = \frac{\mathbf{V}_2(z)}{\Delta_{\text{OL}}(z)} \tag{6.4-113}$$

with

$$\mathbf{R} = \rho^2\mathbf{R}_o = \rho^2\mathbf{U}^T\mathbf{U} = \rho^2\mathbf{U}\mathbf{U}^T \tag{6.4-114}$$

and

$$|\mathbf{Y}_2(z)| = \frac{\psi_2(z)}{\Delta_{\text{OL}}(z)} \tag{6.4-115}$$

Then, Eq. 6.4-106 is

$$\Delta_{\Phi}(z) = \Delta_{\text{OL}}(z)\Delta_{\text{OL}}(z^{-1})\left|\mathbf{I}_m + \frac{1}{\rho^2}\mathbf{Y}_2^T(z^{-1})\mathbf{Y}_2(z)\right| \tag{6.4-116}$$

As control weighting approaches infinity, Eq. 6.4-116 becomes

$$\Delta_{\Phi}(z) = \Delta_{\text{CL}}(z)\Delta_{\text{CL}}(z^{-1}) = \Delta_{\text{OL}}(z)\Delta_{\text{OL}}(z^{-1}) = 0 \tag{6.4-117}$$

This does not say that $\Delta_{\text{CL}}(z) = \Delta_{\text{OL}}(z)$. Because stability of the closed-loop system is guaranteed, the eigenvalues within the unit circle belong to the closed-loop system while those outside belong to the adjoint system. The feedback control matrix needed to assure this condition may or may not go to zero, depending on the stability of $\boldsymbol{\Phi}$.

Conventional root-locus techniques can be used to determine the locations of the eigenvalues. We know that the closed-loop roots must be confined within the unit circle to remain stable, which indicates, in turn, that there must be as many zeros as poles with magnitudes less than one.

In the scalar case, the transmission zeros of $\psi_2(z)$ and $\psi_2(z^{-1})$ are defined by

$$\psi_2(z) = a_q(z-\alpha_1)(z-\alpha_2)\dots(z-\alpha_q) \tag{6.4-118}$$

$$\begin{aligned}\psi_2(z^{-1}) &= a_q(z^{-1}-\alpha_1)(z^{-1}-\alpha_2)\dots(z^{-1}-\alpha_q)\\ &= a_q(-z)^{-q}\prod_{i=1}^{q}\left[\alpha_i\left(z-\frac{1}{\alpha_i}\right)\right]\end{aligned} \tag{6.4-119}$$

so $\Delta_{\Phi}(z)$ is

$$\begin{aligned}\Delta_{\Phi}(z) = \Delta_{CL}(z)\Delta_{CL}(z^{-1}) &= (-z)^{-n}\prod_{i=1}^{n}\left[\beta_i(z-\beta_i)\left(z-\frac{1}{\beta_i}\right)\right.\\ &\quad + (-z)^{-q}Ka_q^2\prod_{i=1}^{q}\left[\alpha_i(z-\alpha_i)\left(z-\frac{1}{\alpha_i}\right)\right] = 0\end{aligned} \tag{6.4-120}$$

with $K = (1/\rho^2)$. Then

$$\prod_{i=1}^{n}\left[(z-\beta_i)\left(z-\frac{1}{\beta_i}\right)\right] + \frac{(-z)^{n-q}Ka_q^2}{\prod_{i=1}^{n}\beta_i}\prod_{i=1}^{q}\left[\alpha_i(z-\alpha_i)\left(z-\frac{1}{\alpha_i}\right)\right] = 0 \tag{6.4-121}$$

which takes a form suitable for root-locus analysis:

$$D(z) + KN(z) = 0 \tag{6.4-122}$$

Note that $(n-q)$ zeros are at the origin and another q zeros are within the unit circle (either at α_i or $1/\alpha_i$, $i = 1$ to q); hence there are precisely n zeros within the unit circle as required. The adjoint system roots go to the q zeros outside the unit circle or to infinity.

For multiple controls, similar asymptotic behavior can be expected, with q closed-loop roots approaching the transmission zeros (or their inverses) and $(n-q)$ roots approaching the origin in groups that are analogous to the multiple Butterworth configurations of the continuous-time case. Multiple controls provide the same ability to alter eigenvectors (for fixed closed-loop eigenvalues) as in the continuous-time case.

6.5 ROBUSTNESS OF LINEAR-QUADRATIC REGULATORS

The design techniques and analytical results presented in this chapter have been predicated on perfect knowledge of the system to be controlled, yet we know that virtually all physical systems contain parametric uncertainty. Even if this uncertainty is negligible, it may be desirable to design a constant-coefficient controller for a plant whose parameters vary deterministically as a consequence of environmental factors or component failures. In either case, the controller should perform its essential functions satisfactorily; that is, it should be *robust* in the presence of parameter variations.

As noted earlier, the principal functions of a fixed-gain controller are maintenance of stability, proper response to commands, and reduction of response perturbations caused by disturbance inputs. Ability to perform the first function under parameter variation is called *stability robustness*, while ability to conduct the remaining two functions is termed *performance robustness*. The former normally is required for the latter, and it is important in its own right. Stability is a fundamental requirement of any continuing process, as small disturbances should not grow to large deviations from the desired set point or trajectory.*

Because the linear-quadratic-Gaussian regulator guarantees stability under nominal operating conditions, it would appear to be a superb choice for practical, realizable control systems. To some extent, performance robustness is addressed in the development of optimal controller structures (Section 6.3). However, stability robustness is not considered explicitly in establishing the optimality criteria or controller structures, and one might rightfully expect it to pose a problem under some circumstances. For example, a naturally stable system might seem to afford a greater closed-loop stability robustness than an open-loop unstable system; a system whose open- and closed-loop modal properties are more alike than another's might be more robust; and a closed-loop system with many good measurements should be more robust than the same system with few noisy measurements. These intuitive notions may or may not be correct, depending upon the parameter variations being considered, inherent system coupling, the magnitudes of controller gains, and, in the case of discrete-time systems, the sampling rate(s) for control.

Classical design procedures for single-input/single-output systems typic-

*Modest instability in an automatically regulated system may be tolerated if some external controller (e.g., a human operator) can augment stability. For example, commonly accepted specifications for aircraft flying qualities permit slight instability under some conditions (such as system failure), recognizing that the pilot can act as a stabilizing closed-loop controller. Nevertheless, aircraft stability is a prerequisite for satisfactory flying qualities in normal operation.

ally have included stability robustness in the design process, and it would be desirable to identify similar procedures for multi-input/multi-output systems. *Stability margins*—in the form of gain and phase margins (Section 2.6)—describe robustness with just two parameters, so it is not surprising that a number of investigations have sought to quantify the robustness of multivariable systems in similar terms. Usually it is helpful, if not imperative, to impose some implied structure on the types of parameter variations to be considered, and it is, of course, necessary to work in the frequency domain, that is, with transfer functions rather than differential (or difference) equations.

Spectral Characteristics of Optimally Regulated Systems

Frequency-domain descriptions of the controlled system and its cost function are central to the study of stability robustness. If the feedback control gain matrix is derived from an algebraic Riccati equation solution (thereby minimizing an associated cost function), a direct relationship between control-loop transfer properties and the optimized cost function can be found. This relationship reveals the importance of a matrix equivalent to the scalar return difference function, first introduced in Section 2.6.

As we have seen many times before, the optimal control law for a linear system described as

$$\dot{\mathbf{x}}(t) = \mathbf{F}\mathbf{x}(t) + \mathbf{G}\mathbf{u}(t) + \mathbf{L}\mathbf{w}(t) \tag{6.5-1}$$

is simply

$$\mathbf{u}(t) = -\mathbf{C}\mathbf{x}(t) \tag{6.5-2}$$

The linear-quadratic optimal gain matrix is

$$\mathbf{C} = \mathbf{R}^{-1}\mathbf{G}^T\mathbf{S} \tag{6.5-3}$$

and $\mathbf{S}$ is the steady-state solution of the $(n \times n)$ algebraic Riccati equation,

$$\mathbf{0} = -\mathbf{S}\mathbf{F} - \mathbf{F}^T\mathbf{S} - \mathbf{Q} + \mathbf{S}\mathbf{G}\mathbf{R}^{-1}\mathbf{G}^T\mathbf{S} \tag{6.5-4a}$$

where $\mathbf{Q}$ and $\mathbf{R}$ are the symmetric weighting matrices of a quadratic cost function (Eq. 6.4-1). Substituting Eq. 6.5-3 in Eq. 6.5-4a, and rearranging terms leads to the following:

$$-\mathbf{S}\mathbf{F} - \mathbf{F}^T\mathbf{S} + \mathbf{C}^T\mathbf{R}\mathbf{C} = \mathbf{Q} \tag{6.5-4b}$$

The quantity $s\mathbf{S}$, where s is the Laplace operator, can be added and subtracted on the left side of this equation such that

$$\mathbf{S}(s\mathbf{I}_n - \mathbf{F}) + (-s\mathbf{I}_n - \mathbf{F}^T)\mathbf{S} + \mathbf{C}^T\mathbf{R}\mathbf{C} = \mathbf{Q} \tag{6.5-5}$$

Each term of the resulting equation is then premultiplied by $\mathbf{G}^T(-s\mathbf{I}_n - \mathbf{F}^T)^{-1}$ and postmultiplied by $(s\mathbf{I}_n - \mathbf{F})^{-1}\mathbf{G}$ to yield the $(m \times m)$ equation

$$\begin{aligned}&\mathbf{G}^T(-s\mathbf{I}_n - \mathbf{F}^T)^{-1}\mathbf{S}\mathbf{G} + \mathbf{G}^T\mathbf{S}(s\mathbf{I}_n - \mathbf{F})^{-1}\mathbf{G}\\ &\quad + \mathbf{G}^T(-s\mathbf{I}_n - \mathbf{F}^T)^{-1}\mathbf{C}^T\mathbf{R}\mathbf{C}(s\mathbf{I}_n - \mathbf{F})^{-1}\mathbf{G}\\ &= \mathbf{G}^T(-s\mathbf{I}_n - \mathbf{F}^T)^{-1}\mathbf{Q}(s\mathbf{I}_n - \mathbf{F})^{-1}\mathbf{G}\end{aligned} \tag{6.5-6a}$$

Because $\mathbf{C}$ is specified by Eq. 6.5-3, Eq. 6.5-6a becomes

$$\begin{aligned}&\mathbf{G}^T(-s\mathbf{I}_n - \mathbf{F}^T)^{-1}\mathbf{C}^T\mathbf{R} + \mathbf{R}\mathbf{C}(s\mathbf{I}_n - \mathbf{F})^{-1}\mathbf{G}\\ &\quad + \mathbf{G}^T(-s\mathbf{I}_n - \mathbf{F}^T)^{-1}\mathbf{C}^T\mathbf{R}\mathbf{C}(s\mathbf{I}_n - \mathbf{F})^{-1}\mathbf{G}\\ &= \mathbf{G}^T(-s\mathbf{I}_n - \mathbf{F}^T)^{-1}\mathbf{Q}(s\mathbf{I}_n - \mathbf{F})^{-1}\mathbf{G}\end{aligned} \tag{6.5-6b}$$

Adding $\mathbf{R}$ to both sides, the left side can be factored so that

$$\begin{aligned}&[\mathbf{I}_m + \mathbf{G}^T(-s\mathbf{I}_n - \mathbf{F}^T)^{-1}\mathbf{C}^T]\mathbf{R}[\mathbf{I}_m + \mathbf{C}(s\mathbf{I}_n - \mathbf{F})^{-1}\mathbf{G}]\\ &\quad = \mathbf{R} + \mathbf{G}^T(-s\mathbf{I}_n - \mathbf{F}^T)^{-1}\mathbf{Q}(s\mathbf{I}_n - \mathbf{F})^{-1}\mathbf{G}\end{aligned} \tag{6.5-6c}$$

Defining the *loop transfer function matrix* as

$$\mathbf{A}(s) = \mathbf{C}(s\mathbf{I}_n - \mathbf{F})^{-1}\mathbf{G} \tag{6.5-7}$$

and the *cost transfer function matrix* as

$$\mathbf{Y}(s) = \mathbf{H}(s\mathbf{I}_n - \mathbf{F})^{-1}\mathbf{G} \tag{6.5-8}$$

where $\mathbf{Q} = \mathbf{H}^T\mathbf{H}$, the principal result is seen to take the para-Hermitian form:

$$[\mathbf{I}_m + \mathbf{A}(-s)]^T\mathbf{R}[\mathbf{I}_m + \mathbf{A}(s)] = \mathbf{R} + \mathbf{Y}^T(-s)\mathbf{Y}(s) \tag{6.5-6d}$$

The dimension of $\mathbf{A}(s)$ is $(m \times m)$, and $\mathbf{Y}(s)$ is a $(p \times m)$ matrix, $p \geq r$, where r is the rank of $\mathbf{Q}$.

The significance of $\mathbf{Y}(s)$ was brought out in the previous section: it reflects the effect of control-induced state variations on the cost function, and it governs the closed-loop modal properties as $\mathbf{R}$ becomes small. $\mathbf{Y}(s)$

can be computed without knowing **C** or **S**, and, for fixed values of **Q**, **F**, and **G**, it expresses an *invariant system characteristic*.

$\mathbf{A}(s)$ was shown to define the modal control vector (Eq. 6.4-47) associated with each eigenvalue, and it has a direct physical interpretation as well. The Laplace transforms of Eq. 6.5-1 and 6.5-2 (with zero initial conditions),

$$s\mathbf{x}(s) = \mathbf{F}\mathbf{x}(s) + \mathbf{G}[\mathbf{u}(s) + \mathbf{u}_c(s)] + \mathbf{L}\mathbf{w}(s)$$

$$= \mathbf{F}\mathbf{x}(s) + \mathbf{G}\boldsymbol{\delta}(s) + \mathbf{L}\mathbf{w}(s) \tag{6.5-9}$$

$$\mathbf{u}(s) = -\mathbf{C}\mathbf{x}(s) \tag{6.5-10}$$

can be combined to produce,

$$\mathbf{u}(s) = -\mathbf{C}(s\mathbf{I}_n - \mathbf{F})^{-1}[\mathbf{G}\boldsymbol{\delta}(s) + \mathbf{L}\mathbf{w}(s)] \tag{6.5-11}$$

as shown in Fig. 6.5-1. In this diagram, the LQ regulator is characterized as a negative-unity-feedback system. The commanded input, $\mathbf{u}_c(s)$, and disturbance input, $\mathbf{w}(s)$, can be ignored for the moment, and the n eigenvalues specified by

$$\Delta_{\mathrm{OL}}(s)|\mathbf{I}_m + \mathbf{C}(s\mathbf{I}_n - \mathbf{F})^{-1}\mathbf{G}| = \Delta_{\mathrm{OL}}(s)|\mathbf{I}_m + \mathbf{A}(s)| = \Delta_{\mathrm{CL}}(s) = 0 \tag{6.5-12}$$

define the closed-loop roots of the controlled system. $[\mathbf{I}_m + \mathbf{A}(s)]$ is an $(m \times m)$ matrix, but its elements are complex fractions whose denominators (prior to any pole-zero cancellations) are the nth-degree characteristic polynomial of $\mathbf{F}$, $\Delta_{\mathrm{OL}}(s)$; hence the closed-loop roots derive from the nth-degree polynomial specified by the determinant

$$|\Delta_{\mathrm{OL}}(s)\mathbf{I}_m + \mathbf{C}\,\mathrm{Adj}(s\mathbf{I}_n - \mathbf{F})\mathbf{G}| = 0 \tag{6.5-13}$$

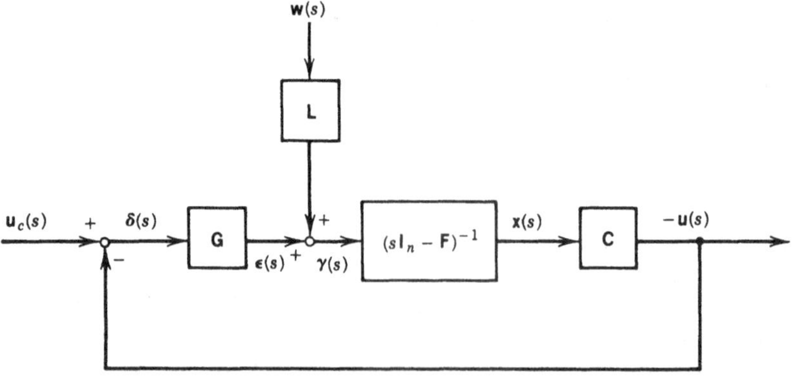

FIGURE 6.5-1 Linear-quadratic regulator portrayed as a negative-unity-feedback system.

The matrix $[\mathbf{I}_m + \mathbf{A}(s)]$ is called the *return difference function matrix*, and by performing simple block diagram algebra on Fig. 6.5-1, we can find the following transfer function relationships between the commanded input and the variable at various points in the loop:

$$\boldsymbol{\delta}(s) = [\mathbf{I}_m + \mathbf{A}(s)]^{-1}\mathbf{u}_c(s) \tag{6.5-14}$$

$$\boldsymbol{\epsilon}(s) = \boldsymbol{\gamma}(s) = \mathbf{G}[\mathbf{I}_m + \mathbf{A}(s)]^{-1}\mathbf{u}_c(s) \tag{6.5-15}$$

$$\mathbf{x}(s) = (s\mathbf{I}_n - \mathbf{F})^{-1}\mathbf{G}[\mathbf{I}_m + \mathbf{A}(s)]^{-1}\mathbf{u}_c(s) \tag{6.5-16}$$

$$\begin{aligned}\mathbf{u}(s) &= -\mathbf{C}(s\mathbf{I}_n - \mathbf{F})^{-1}\mathbf{G}[\mathbf{I}_m + \mathbf{A}(s)]^{-1}\mathbf{u}_c(s)\\ &= -\mathbf{A}(s)\,[\mathbf{I}_m + \mathbf{A}(s)]^{-1}\mathbf{u}_c(s)\end{aligned} \tag{6.5-17}$$

Defining

$$\mathbf{B}(s) = \mathbf{G}\mathbf{C}(s\mathbf{I}_n - \mathbf{F})^{-1} \tag{6.5-18}$$

the effects of disturbance inputs can be expressed as follows:

$$\boldsymbol{\gamma}(s) = [\mathbf{I}_n + \mathbf{B}(s)]^{-1}\mathbf{L}\mathbf{w}(s) \tag{6.5-19}$$

$$\mathbf{x}(s) = (s\mathbf{I}_n - \mathbf{F})^{-1}[\mathbf{I}_n + \mathbf{B}(s)]^{-1}\mathbf{L}\mathbf{w}(s) \tag{6.5-20}$$

$$\mathbf{u}(s) = \boldsymbol{\delta}(s) = -\mathbf{C}(s\mathbf{I}_n - \mathbf{F})^{-1}[\mathbf{I}_n + \mathbf{B}(s)]^{-1}\mathbf{L}\mathbf{w}(s) \tag{6.5-21}$$

$$\begin{aligned}\boldsymbol{\epsilon}(s) &= -\mathbf{G}\mathbf{C}(s\mathbf{I}_n - \mathbf{F})^{-1}[\mathbf{I}_n + \mathbf{B}(s)]^{-1}\mathbf{L}\mathbf{w}(s)\\ &= -\mathbf{B}(s)\,[\mathbf{I}_n + \mathbf{B}(s)]^{-1}\mathbf{L}\mathbf{w}(s)\end{aligned} \tag{6.5-22}$$

The dynamic modes of disturbance response may appear to be different from those of command response; however, they are identical, as indicated in the following. The modes are defined by the determinant of $[\mathbf{I}_n + \mathbf{B}(s)]$ as

$$[\mathbf{I}_n + \mathbf{B}(s)]^{-1} = \frac{\operatorname{Adj}[\mathbf{I}_n + \mathbf{B}(s)]}{|\mathbf{I}_n + \mathbf{B}(s)|} \tag{6.5-23}$$

but a determinant identity (Eq. 2.2-103b) shows that the characteristic roots must be the same as those of $|\mathbf{I}_m + \mathbf{A}(s)|$:

$$\begin{aligned}\Delta_{\mathrm{OL}}(s)|\mathbf{I}_n + \mathbf{B}(s)| &= \Delta_{\mathrm{OL}}(s)|\mathbf{I}_n + \mathbf{G}\mathbf{C}(s\mathbf{I}_n - \mathbf{F})^{-1}|\\ &= \Delta_{\mathrm{OL}}(s)|\mathbf{I}_m + \mathbf{C}(s\mathbf{I}_n - \mathbf{F})^{-1}\mathbf{G}|\\ &= \Delta_{\mathrm{OL}}(s)|\mathbf{I}_m + \mathbf{A}(s)|\end{aligned} \tag{6.5-24}$$

Consequently, the inverse of $[\mathbf{I}_n + \mathbf{B}(s)]$, which plays the role of a return difference function matrix for disturbance inputs, could be written

$$[\mathbf{I}_n + \mathbf{B}(s)]^{-1} = \frac{\mathrm{Adj}[\mathbf{I}_n + \mathbf{B}(s)]}{|\mathbf{I}_m + \mathbf{A}(s)|} \tag{6.5-25}$$

It is well known from Bode's work (B-3) that a scalar return difference function with magnitude greater than one at low frequency (Fig. 6.3-3) has the desirable properties of reducing sensitivity to disturbance inputs and improving response to commands, and these equations provide confirmation for both scalar and multivariate systems. With all other coefficients fixed and assuming that stability is maintained, values of $\mathbf{C}$ that increase $\mathbf{B}(s)$ cause disturbance response (Eq. 6.5-22) to decrease because

$$\boldsymbol{\epsilon}(s) = -\mathbf{B}(s)\,[\mathbf{I}_n + \mathbf{B}(s)]^{-1}\mathbf{L}\mathbf{w}(s) \rightarrow -\mathbf{L}\mathbf{w}(s) \tag{6.5-26}$$

Therefore, the input to $(s\mathbf{I}_n - \mathbf{F})^{-1}$ in Fig. 6.5-1 goes to zero, and the disturbance is effectively canceled. Similarly, values of $\mathbf{C}$ that increase $\mathbf{A}(s)$ lead to (Eqs. 6.5-2 and 6.5-17)

$$\mathbf{u}(s) = -\mathbf{A}(s)\,[\mathbf{I}_m + \mathbf{A}(s)]^{-1}\mathbf{u}_c(s) \rightarrow -\mathbf{u}_c(s) = \mathbf{C}\mathbf{x}_c(s) \tag{6.5-27}$$

where $\mathbf{x}_c(s)$ is the value of the state corresponding to $\mathbf{u}_c(s)$, as defined, for example, in Section 6.2. What is meant by "large" $\mathbf{A}(s)$ or $\mathbf{B}(s)$ in the multivariable case remains to be determined, but the qualitative notion is clear without embellishment.

Stability Margins of Scalar Linear-Quadratic Regulators

For scalar control, there is nothing arbitrary about the definition of large $\mathbf{A}(s)$, as it is scalar. In a seminal 1964 paper (K-2), Kalman established the basis for "classical" evaluation of optimal control laws, proving that the return difference function for a scalar LQ regulator must equal or exceed one if the Section 5.4 stability criteria are satisfied. Anderson and Moore (A-1) then interpreted this result in terms of gain and phase margins.

The relationship between the optimal return difference function and the cost transfer function is as follows, with unemphasized uppercase letters signifying the complex scalar equivalents of the matrices in Eq. 6.5-6d:

$$[1 + A(-s)]r[1 + A(s)] = r + \mathbf{Y}^T(-s)\mathbf{Y}(s) \tag{6.5-28a}$$

or

$$[1 + A(-s)][1 + A(s)] = 1 + \frac{\mathbf{Y}^T(-s)\mathbf{Y}(s)}{r} \tag{6.5-28b}$$

$A(s)$ and $\mathbf{Y}(s)$ are distinguished by the optimal gain matrix, $\mathbf{C}$, and the cost function output matrix, $\mathbf{H}$,

$$A(s) = \mathbf{C}(s\mathbf{I}_n - \mathbf{F})^{-1}\mathbf{G} \tag{6.5-29}$$

$$\mathbf{Y}(s) = \mathbf{H}(s\mathbf{I}_n - \mathbf{F})^{-1}\mathbf{G} \tag{6.5-30}$$

with $\dim(\mathbf{C}) = (1 \times n)$, $\dim(\mathbf{H}) = (p \times n)$, $\dim(\mathbf{F}) = (n \times n)$, and $\dim(\mathbf{G}) = (n \times 1)$. $\mathbf{Y}(s)$ is a complex $(p \times 1)$ vector, so $\mathbf{Y}^T(-s)\mathbf{Y}(s)$ is a real scalar. As we are interested in frequency domain characteristics, let $s = j\omega$;

$$[1 + A(-j\omega)][1 + A(j\omega)] = 1 + \frac{\mathbf{Y}^T(-j\omega)\mathbf{Y}(j\omega)}{r} \tag{6.5-31}$$

Writing $A(j\omega)$ as $c(\omega) + jd(\omega)$, the left side of this equation is seen to be equivalent to

$$\{[1 + c(\omega)] - jd(\omega)\}\{[1 + c(\omega)] + jd(\omega)\} = [1 + c(\omega)]^2 + d^2(\omega)$$
$$= |1 + A(j\omega)|^2 \tag{6.5-32}$$

where $|\cdot|$ denotes the absolute value. For $Y_i(j\omega) = l_i(\omega) + jm_i(\omega)$, the right side becomes

$$1 + \frac{\mathbf{Y}^T(-j\omega)\mathbf{Y}(j\omega)}{r} = 1 + \sum_{i=1}^{p} \frac{[l_i^2(\omega) + m_i^2(\omega)]}{r} \tag{6.5-33}$$

Equation 6.5-33 is always greater than or equal to one; hence the absolute value of the return difference function of the scalar LQ regulator is always greater than or equal to one:

$$|1 + A(j\omega)| \geq 1 \tag{6.5-34}$$

This is called the *Kalman inequality*.

If the complex-valued $A(j\omega)$ is graphed for all ω in $(-\infty, \infty)$, a Nyquist plot of the loop transfer function is obtained (Section 2.6). Equation 6.5-34 indicates that the $A(j\omega)$ plot is *excluded from a unit circle centered on the point* $(-1, 0)$, as shown in Fig. 6.5-2. Furthermore, we know that the closed-loop system is stable, so it must satisfy the Nyquist stability criterion: the number of counterclockwise encirclements of the point $(-1, 0)$ by the $A(j\omega)$ locus as ω varies from $-\infty$ to $+\infty$ equals the number of unstable poles in the system's open-loop characteristic equation. The degree to which the $(1 \times n)$ gain matrix can deviate from its optimal value and still retain stability is prescribed by the condition that the number of encirclements not change with variations in $A(j\omega)$. More precisely, it specifies allowable variations in the *scalar loop gain*, as described later.

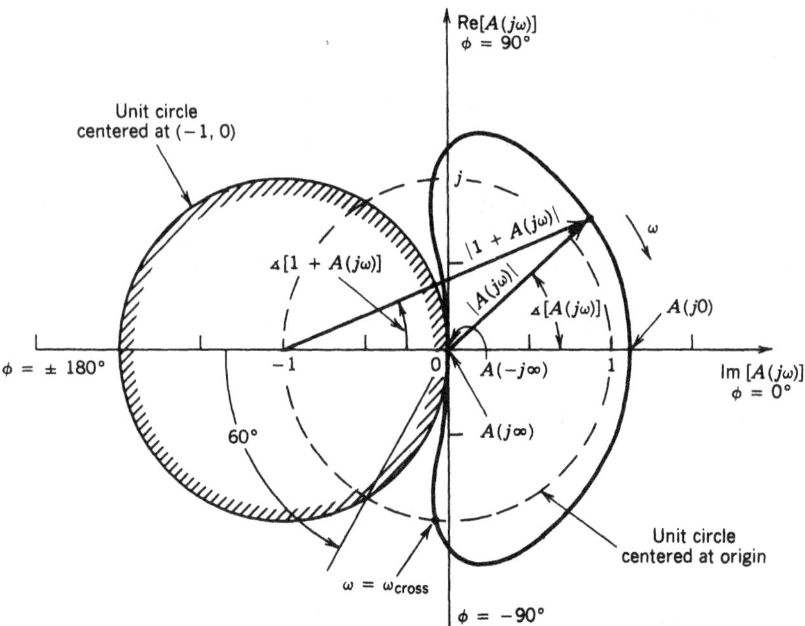

FIGURE 6.5-2 Relationships between scalar loop transfer function $[A(j\omega)]$, return difference function $[1 + A(j\omega)]$, and unit circles centered at (0, 0) and (−1, 0).

EXAMPLE 6.5-1 LOOP TRANSFER FUNCTION FOR AN OPTIMALLY REGULATED SECOND-ORDER SYSTEM

The optimal scalar regulator gain for a second-order system,

$$\begin{bmatrix} \dot{x}_1 \\ \dot{x}_2 \end{bmatrix} = \begin{bmatrix} 0 & 1 \\ -\omega_n^2 & -2\zeta\omega_n \end{bmatrix} \begin{bmatrix} x_1 \\ x_2 \end{bmatrix} + \begin{bmatrix} 0 \\ \omega_n^2 \end{bmatrix} u$$

is given by

$$\mathbf{c} = \frac{1}{r}\mathbf{G}^T\mathbf{S} = \frac{1}{r}[0 \quad \omega_n^2] \begin{bmatrix} s_{11} & s_{12} \\ s_{12} & s_{22} \end{bmatrix}$$

$$= \frac{\omega_n^2[s_{12} \quad s_{22}]}{r} = [c_1 \quad c_2]$$

The corresponding loop transfer function is

$$A(s) = \mathbf{C}(s\mathbf{I}_n - \mathbf{F})^{-1}\mathbf{G} = [c_1 \quad c_2] \begin{bmatrix} s & -1 \\ \omega_n^2 & (s + 2\zeta\omega_n) \end{bmatrix}^{-1} \begin{bmatrix} 0 \\ \omega_n^2 \end{bmatrix}$$

$$= \frac{c_2\omega_n^2(s + c_1/c_2)}{s^2 + 2\zeta\omega_n s + \omega_n^2}$$

and the Nyquist plot is described by plotting the real and imaginary parts of $A(s)$ for $s = j\omega$, $-\infty < \omega < +\infty$.

EXAMPLE 6.5-2 OPTIMAL GAIN COMPUTATION USING SPECTRAL FACTORIZATION

The results of Sections 6.4 and 6.5 can be used to compute the optimal control gains without solving a matrix Riccati equation. The eigenvalues of $(s\mathbf{I}_{2n} - \mathbf{F}')$ (Eq. 6.4-6 and following) are the eigenvalues of $[\mathbf{R} + \mathbf{Y}^T(-s)\mathbf{Y}(s)]$ (Eq. 6.5-6d). The n left-half-plane eigenvalues also are the closed-loop roots of $\Delta_{OL}(s)|\mathbf{I}_m + \mathbf{A}(s)|$, whose eigenvectors, $\mathbf{u}_i$ (the modal control vectors of Section 6.4), are the eigenvectors of $[\mathbf{R} + \mathbf{Y}^T(-s)\mathbf{Y}(s)]$ corresponding to the stable roots. Denoting the stable roots by λ_i, $i = 1$–n, the following equality is satisfied:

$$[\mathbf{I}_m + \mathbf{C}(\lambda_i\mathbf{I}_n - \mathbf{F})^{-1}\mathbf{G}]\mathbf{u}_i = [\mathbf{R} + \mathbf{Y}^T(-\lambda_i)\mathbf{Y}(\lambda_i)]\mathbf{u}_i$$

This provides nm equations for the elements of $\mathbf{C}$ (R-2, W-1, B-6).

For the previous example of a second-order system with scalar control, the gains can be computed by equating the coefficients of similar powers of s in $[1 + A(s)]$ and the stable part of $[1 + Y(-s)Y(s)/r]$. Multiplying by $\Delta(-s)\Delta(s)$, the latter takes the form

$$\begin{aligned}&(s^2 - 2\zeta\omega_n s + \omega_n^2)(s^2 + 2\zeta\omega_n s + \omega_n^2) + \omega_n^4(-h_2 s + h_1)(h_2 s + h_1)/r\\&\quad = s^4 + \omega_n^2(2 - 4\zeta^2 - \omega_n^2 h_2^2/r)s^2 + \omega_n^4(1 + h_1^2/r) = 0\end{aligned}$$

where h_1 and h_2 are defined for a rank-one $\mathbf{Q}$ as in Example 6.4-1. This is a biquadratic equation whose roots are

$$\lambda_{1,2,3,4} = \pm\,\omega_n\left[\left(\frac{\omega_n^2 h_2^2}{2r} + 2\zeta^2 - 1\right) \pm \sqrt{\left(1 - 2\zeta^2 - \frac{\omega_n^2 h_2^2}{2r}\right)^2 - \left(1 + \frac{h_1^2}{r}\right)}\,\right]^{1/2}$$

Calling the left-half-plane poles λ_1 and λ_2, the closed-loop characteristic equation is

$$s^2 - (\lambda_1 + \lambda_2)s + (\lambda_1\lambda_2) = 0$$

From Example 6.5-1, the closed-loop characteristic equation also can be expressed as

$$\Delta(s) + \omega_n^2(c_2 s + c_1) = s^2 + (2\zeta\omega_n + \omega_n^2 c_2)s + \omega_n^2(1 + c_1) = 0$$

Equating the two, the optimal gains are found to be

$$c_1 = \frac{\not{s}_1 \not{s}_2}{\omega_n^2} - 1$$

$$c_2 = -\frac{(\not{s}_1 + \not{s}_2 + 2\zeta\omega_n)}{\omega_n^2}$$

For illustration of the LQ regulator's robustness, consider the second-order loop transfer function (with $s = j\omega$),

$$A(j\omega) = \frac{c(j\omega + z)}{(j\omega)^2 + 2\zeta\omega_n(j\omega) + \omega_n^2} \tag{6.5-35}$$

with $z > \omega_n$ and $\zeta < 1$. The Nyquist plot is shown in Fig. 6.5-2, and it is representative of the systems discussed in the previous two examples. The phase angle of $A(j\omega)$ is $0°$ for $\omega = 0$ and $-90°$ for $\omega = \infty$, reaching a minimum value of $-96°$ at intermediate ω. Its phase angle is $-93°$ at the crossover frequency, ω_{cross} [i.e., at the frequency for which $|A(j\omega)| = 1$], so the loop transfer function's phase margin is $-87°$. The phase angle for the corresponding return difference function, $[1 + A(j\omega)]$, begins and ends at $0°$. More generally, it can be seen that the phase angle of $A(j\omega)$ must lie within $\pm 120°$ for $|\omega| > \omega_{\text{cross}}$ and within $\pm 90°$ as ω approaches $\pm\infty$. The return difference phase angle rotates through $-360q°$ as ω varies from $-\infty$ to ∞, where q is the number of right-half-plane poles in $A(s)$. The scalar LQ regulator's return difference phase angle always approaches $-90°$ at infinite frequency.

Equation 6.5-34 and the guaranteed stability property assure that *any scalar LQ regulator's phase margin is at least* $60°$ ($\pi/3$ rad). The point at which the two unit circles meet in the lower half plane (Fig. 6.5-2) sets the minimum phase margin, because $A(j\omega)$ is excluded from the circle centered at $(-1, 0)$, and its phase margin is measured at unit amplitude ratio. This means that a *pure phase lag* [i.e., a negative value of ϕ (rad)], appearing anywhere in the scalar control loop (as in the feedback of Fig. 6.5-3*a*) can be at least as great as $60°$ without causing instability. In such case, the loop transfer function of the nonoptimal controller would be

$$A_{no}(j\omega) = e^{j\phi} A_o(j\omega) = e^{j\phi} \mathbf{C}(s\mathbf{I}_n - \mathbf{F})^{-1}\mathbf{G} \tag{6.5-36}$$

By appearing in the scalar loop, the phase lag modifies the effect of each of the n elements of $\mathbf{C}$ to the same degree. The shape of the corresponding Nyquist plot would be unchanged, but it would be rotated by ϕ rad (Fig. 6.5-4*a*). Because the transfer function of a *pure time delay*, T sec, is e^{-sT}, a time delay of at least $\pi/3\omega_{\text{cross}}$ can be tolerated without instability (ω_{cross} expressed in rad/s).

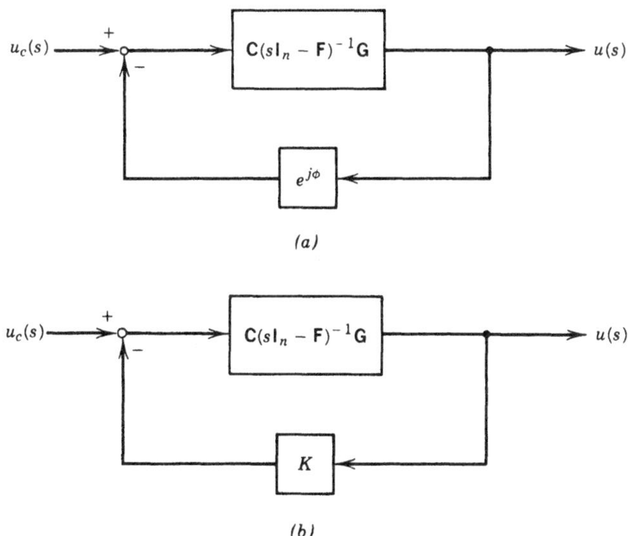

FIGURE 6.5-3 Multiplicative variations in the control loop of a scalar linear-quadratic regulator. (*a*) Phase lag in the control loop; (*b*) constant multiplier in the control loop.

If the scalar control loop contains a positive real multiplier, K (Fig. 6.5-3*b*), such that

$$A_{no}(j\omega) = KA_o(j\omega) = K\mathbf{C}(s\mathbf{I}_n - \mathbf{F})^{-1}\mathbf{G} \tag{6.5-37}$$

the loop transfer function locus expands or shrinks in comparison to its optimal *size* without changing its *shape* (Fig. 6.5-4*b*). If $A_o(j\omega)$ does not encircle the $(-1, 0)$ point (i.e., if the system is *open-loop stable* as in Fig. 6.5-2), K can take any value in $(0, \infty)$ without violating the Nyquist criterion; hence there is a *reduced-gain margin of* 100% and an *increased-gain margin of infinity*. If the system is *open-loop unstable*, with one or more encirclements of the $(-1, 0)$ point, then the *increased-gain margin is infinity*. The locus stays outside the corresponding unit circle, leaving the number of encirclements unchanged and the return difference function magnitude greater than one. The locus can shrink by at least one-half before the number of $(-1, 0)$ encirclements changes, providing a *reduced-gain margin of* 50%. In this nonoptimal case, the return difference function magnitude would become less than one over a range of frequencies, indicating degraded disturbance rejection and command response.

Ad hoc variations in the control gains can be made, although once the gains assume nonoptimal values, the stability margins are altered. The phase margin *or* the gain margin guarantees obtain separately; if part of one margin is "used up," the other margin is affected. Nevertheless, the results indicate that gains can be uniformly adjusted to achieve given robustness

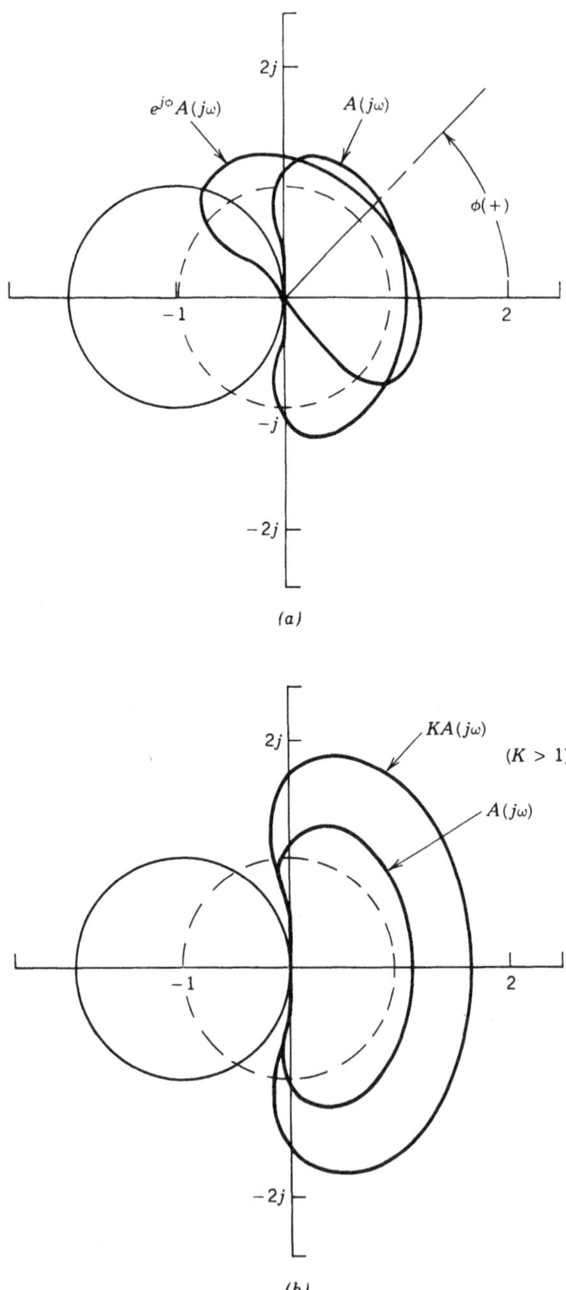

FIGURE 6.5-4 Effects of multiplicative variations on the scalar linear-quadratic regulator Nyquist plot.

objectives. For example, 60° phase margin could be considered "over-design" in some instances—40°–45° phase margin is a classical rule of thumb—and gains might be reduced accordingly. Conversely, if the gains were uniformly amplified, there would be, by definition, an increased guaranteed tolerance to gain reduction (relative to the amplified gains).

The scalar LQ regulator remains stable for a class of dynamic loop variations characterized by a linear transfer function $K(s)$. In this case, the loop transfer function is $K(s)A(s)$, and the system can be represented by Fig. 6.5-3b, with K defined as a function of s. Qualitatively, if $K(j\omega)$ has unit amplitude ratio and negligible phase lag at low frequency, is essentially "flat" to a bandwidth (ω_b) that is somewhat greater than ω_{cross}, and decreases as ω^{-k} $(k > 1)$ above ω_b, then the scalar LQ regulator designed without regard for $K(s)$ will not be destabilized by $K(s)$.* In other words, "low-pass filters" that do not change the number of encirclements of the $(-1, 0)$ point can be found.

The relationship between ω_b and ω_{cross} for which stability is guaranteed depends on the form of $K(s)$ and the definition of ω_b. Nevertheless, it is important to note that such relationships exist because real control actuators have low-pass characteristics. In particular, their amplitude ratios attenuate with increasing frequency due to inertial (or equivalent) effects and finite control power. Light damping would preclude flat actuator response, but stability would be retained if $|K(j\omega)A(j\omega)| < 1$ at the resonant peak.

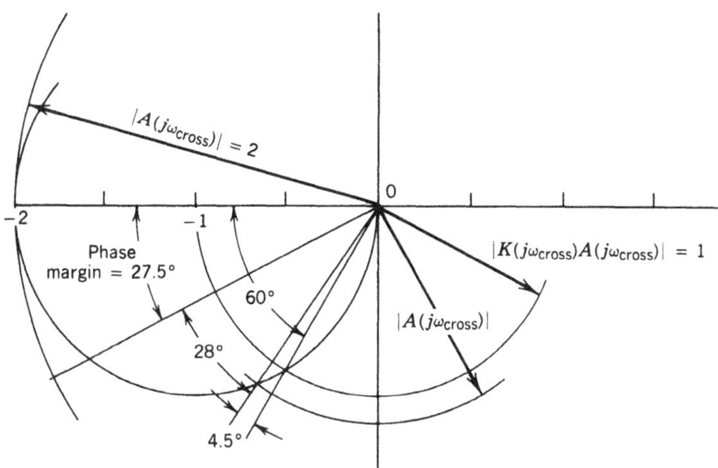

FIGURE 6.5-5 Construction to determine "worst case" phase margin of a second-order actuator in a scalar linear-quadratic regulator.

*Of course, full stability margins would be retained if the $K(s)$ dynamics were considered in the LQ design process.

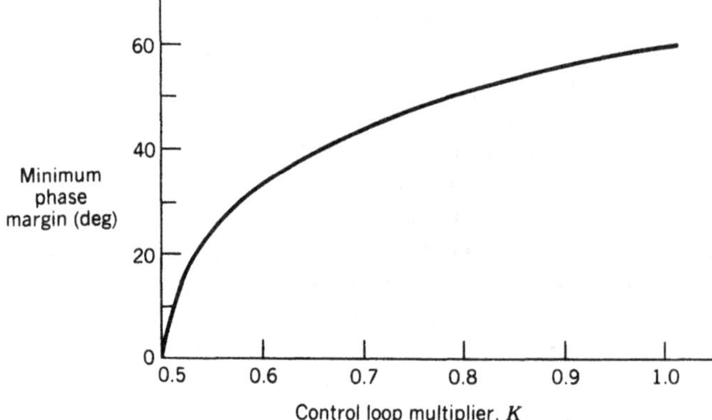

FIGURE 6.5-6 Trade-off between guaranteed phase and reduced-gain margins for a scalar linear-quadratic regulator.

A second-order actuator with $\zeta = 0.707$ and no zeros in its transfer function provides an example of relatively flat frequency response, and it does not have a resonant peak. Suppose that the actuator's natural frequency is three times greater than the nominal crossover frequency ω_{cross}. The phase lag and attenuation of $K(j\omega_{\text{cross}})$ are 28° and 11%, respectively. A "worst case" analysis of the phase margin is shown in Fig. 6.5-5. In this case, the worst $A(j\omega)$ locus would hug the unit circle centered at $(-1, 0)$ in the vicinity of the nominal crossover point (i.e., at the lower intersection of the two unit circles). Unit amplitude ratio for $K(j\omega)A(j\omega)$ with $|K| = 0.89$ requires $|A| = 1/0.89 = 1.12$; hence attenuation reduces the crossover frequency, producing a maximum phase lag of 4.5°. With an additional 28° phase lag contributed by the actuator (a conservative estimate due to the reduced crossover frequency), the minimum phase margin of the closed-loop system is 27.5°.

The construction of Fig. 6.5-5 can be extended to indicate the "worst case" trade-off between pure scalar gain reduction and pure phase lag. A 50% gain reduction causes $|A(j\omega_{\text{cross}})|$ to increase by 100%; this can be represented by a circle of radius 2 centered at the origin. The $|A|$ circle just touches the $|1 + A|$ circle at the point $(-2, 0)$, indicating zero phase margin, and there is a continuum of intersections of the two circles for K in $(1, 0.5)$ and phase margin in (60°, 0), as shown by Fig. 6.5-6.

Effects of System Variations on Stability

While the scalar LQ regulator possesses impressive stability margins against scalar control loop variations, these results imply but do not guarantee equivalent robustness against system parameter variations. Robustness can be shown to be small for parameter variations in some systems; however, it can be recovered by suitable definition of the weighting matrices and/or uniform (nonoptimal) adjustment of control gains.

The principal analytical difficulty is that the *shape* of the Nyquist plot for $\mathbf{C}(s\mathbf{I}_n - \mathbf{F})^{-1}\mathbf{G}$ changes when elements of $\mathbf{C}$, $\mathbf{F}$, or $\mathbf{G}$ vary independently; therefore, the generalities about gain and phase margins of the previous section may not readily apply. The inversion of $(s\mathbf{I}_n - \mathbf{F})$ and the multiplication by $\mathbf{G}$ introduce the possibility that small variations in the elements of $\mathbf{C}$, $\mathbf{F}$, or $\mathbf{G}$ can cause large changes in $A(s)$. This is not a failing of the LQ regulator but a fundamental property of linear system transfer functions. Nevertheless, it is reasonable to seek nominal values of $\mathbf{C}$ that minimize potentially destabilizing effects of such variations while achieving primary design objectives.

One might expect a satisfactory feedback controller for an *open-loop-stable* system to retain closed-loop system stability with realizable parameter variations; however, at least one example in which the LQ regulator purportedly fails this requirement has appeared in the literature (S-7). The following example illustrates that robustness can be enhanced by using a large-enough value of control weighting in the cost function. Optimal control theory provides a method for solving the problem by trading performance for robustness.

EXAMPLE 6.5-3 EFFECTS OF CONTROL MATRIX VARIATION ON A SECOND-ORDER LQ REGULATOR

The example of (S-7) can be stated as follows: minimize the quadratic cost function

$$J = \frac{1}{2T}\int_0^\infty (x_1^2 + 2x_1x_2 + x_2^2 + ru^2)\, dt$$

subject to the following (stable) dynamic constraint:

$$\begin{bmatrix} \dot{x}_1 \\ \dot{x}_2 \end{bmatrix} = \begin{bmatrix} -1 & 0 \\ 0 & -2 \end{bmatrix} \begin{bmatrix} x_1 \\ x_2 \end{bmatrix} + \begin{bmatrix} 1 \\ 1 \end{bmatrix} u$$

The minimizing scalar control law takes the form

$$u = -\mathbf{C}\mathbf{x} = -c_1x_1 - c_2x_2$$

where the gains can be found by spectral factorization (Example 6.5-2). The loop and cost transfer functions are

$$A(s) = \mathbf{C}(s\mathbf{I}_n - \mathbf{F})^{-1}\mathbf{G}$$

$$= \frac{(c_1 + c_2)s + (2c_1 + c_2)}{(s+1)(s+2)}$$

$$Y(s) = \frac{2s+3}{(s+1)(s+2)}$$

and Eq. 6.5-28b applies:

$$[1 + A(-s)][1 + A(s)] = 1 + \frac{Y(-s)\,Y(s)}{r}$$

Proceeding as in the previous example, the stable roots of $[1 + Y(-s)\,Y(s)/r]$ are

$$s_{1,2} = -\left[\frac{(5+4/r)}{2} \pm \sqrt{\frac{(5+4/r)^2}{4} - \left(4 + \frac{9}{r}\right)}\right]^{1/2}$$

The nominal closed-loop characteristic equation is

$$\begin{aligned}(s+1)(s+2)[1 + A(s)] &= s^2 + (3 + c_1 + c_2)s + (2 + 2c_1 + c_2)\\ &= \Delta_{\text{nom}}(s) = 0\end{aligned}$$

and because this must equal

$$s^2 - (s_1 + s_2)s + s_1 s_2 = 0$$

the LQ control gains are

$$c_1 = s_1 s_2 + (s_1 + s_2) + 1$$

$$c_2 = -s_1 s_2 - 2(s_1 + s_2) - 4$$

As r approaches ∞, s_1 and s_2 approach -2 and -1; therefore, the optimal gains approach zero. When r approaches zero, the magnitudes of the optimal control gains and of one root grow at $r^{-1/2}$, while the other root goes to the transmission zero at $-3/2$ (Table 6.5-1). As a consequence, closed-loop stability is increasingly sensitive to parameter variations as r becomes small.

TABLE 6.5-1 Effects of Control Weighting on Nominal Eigenvalues and Control Gains

r	s_1	s_2	c_1	c_2
10^5	−1	−2	−0.	−0.
100	−1.01	−2.01	0.01	0.01
1	−1.34	−2.68	0.58	0.45
0.01	−1.50	−20.07	9.54	9.04
0.001	−1.50	−63.27	31.14	30.64

Suppose that the actual control effect matrix, **G**, contains unknown variations, ϵ_1 and ϵ_2:

$$\mathbf{G} = \begin{bmatrix} g_1 \\ g_2 \end{bmatrix} = \begin{bmatrix} (1+\epsilon_1) \\ (1+\epsilon_2) \end{bmatrix}$$

Using the control gains designed for nominal control effects, the closed-loop characteristic equation is

$$[s^2 + (3 + c_1 + c_2)s + (2 + 2c_1 + c_2)] + [(\epsilon_1 c_1 + \epsilon_2 c_2)s + (2\epsilon_1 c_1 + \epsilon_2 c_2)] = 0$$

By observation, it is clear that values of ϵ_1 or ϵ_2 which destabilize this equation can be found and that the larger the magnitudes of c_1 and c_2, the smaller the destabilizing magnitude of ϵ_1 or ϵ_2.

This equation is amenable to root-locus analysis, separately choosing ϵ_1 ($\epsilon_2 = 0$), ϵ_2 ($\epsilon_1 = 0$), or $\epsilon_1 = \epsilon_2$ as the root-locus gain k. In these three cases, the closed-loop characteristic equation takes the following forms:

a. $\Delta_{\text{nom}}(s) + kc_1(s+2) = 0 \qquad (k = \epsilon_1)$

b. $\Delta_{\text{nom}}(s) + kc_2(s+2) = 0 \qquad (k = \epsilon_2)$

c. $\Delta_{\text{nom}}(s) + k(c_1 + c_2)\left[s + \dfrac{(2c_1 + c_2)}{c_1 + c_2}\right] = 0 \qquad (k = \epsilon_1 = \epsilon_2)$

The root loci for positive k are confined to the real axis (see diagram), and stability is retained for $0 < k < \infty$ in all three cases. Negative values of k leading to neutral stability are given in Table 6.5-2, and they indicate decreasing robustness for decreasing values of r.

Note that a subtractive gain reduction of k in either g_i or c_i corresponds to a multiplicative gain change of $K = (1 + k)$; hence the range $-1 < k < \infty$ corresponds to the range $0 < K < \infty$. Case (c) confirms the previous result

TABLE 6.5-2 Parameter Variations That Cause Neutral Stability

	k		
r	Case (a)	Case (b)	Case (c)
$\cdot 10^5$		—Tends to $-\infty$—	
100	−101	−203	−67.67
1	−3.11	−8.02	−2.24
0.01	−1.58	−3.33	−1.07
0.001	−1.52	−3.10	−1.02

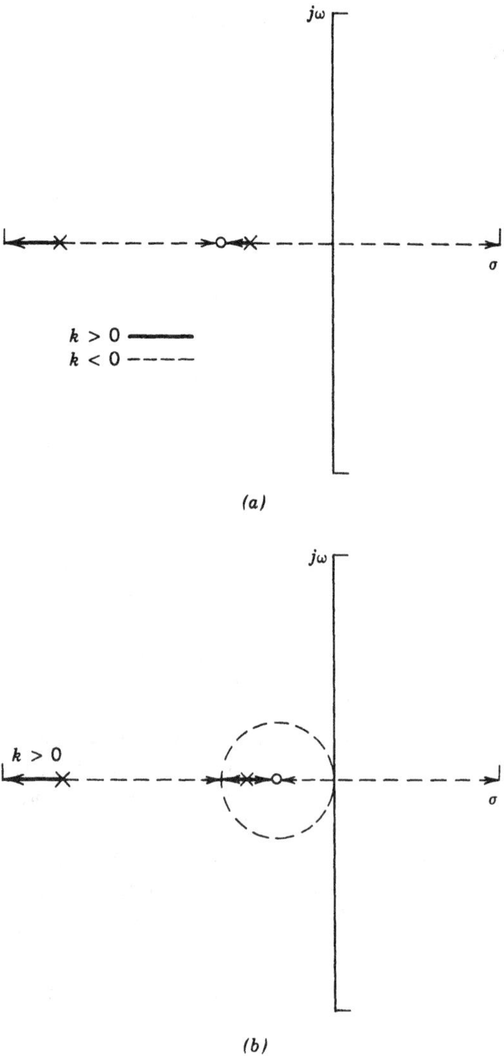

Typical root loci for parameter variations in a second-order LQ requlator (*a*) Cases a and c; (*b*) Case b.

for scalar loop gain variations in open-loop-stable systems; in this particular case, even greater gain reductions are allowable for individual gains. Nevertheless, for the worst case considered here, the control-gain/control-effect product(s) must change sign before instability occurs. The example confirms that it is unwise to implement a high-gain controller if there are large uncertainties in control effect. This is true whether or not the control is optimal.

Multivariable Nyquist Stability Criterion

Frequency-domain descriptions of parameter variations in multi-input/multi-output systems fall into two broad categories. For the $(m \times m)$ nominal loop transfer function matrix, $m > 1$,

$$\mathbf{A}_o(s) = \mathbf{C}_o(s\mathbf{I}_n - \mathbf{F}_o)^{-1}\mathbf{G}_o \tag{6.5-38}$$

additive variations can be expressed as follows (Fig. 6.5-7*a*):

$$\mathbf{A}(s) = \mathbf{A}_o(s) + \Delta\mathbf{A}(s) \tag{6.5-39}$$

Each element of $\mathbf{A}(s)$ can vary independently, providing maximum flexibility for specifying model variations. The connection between $\Delta\mathbf{A}(s)$ and

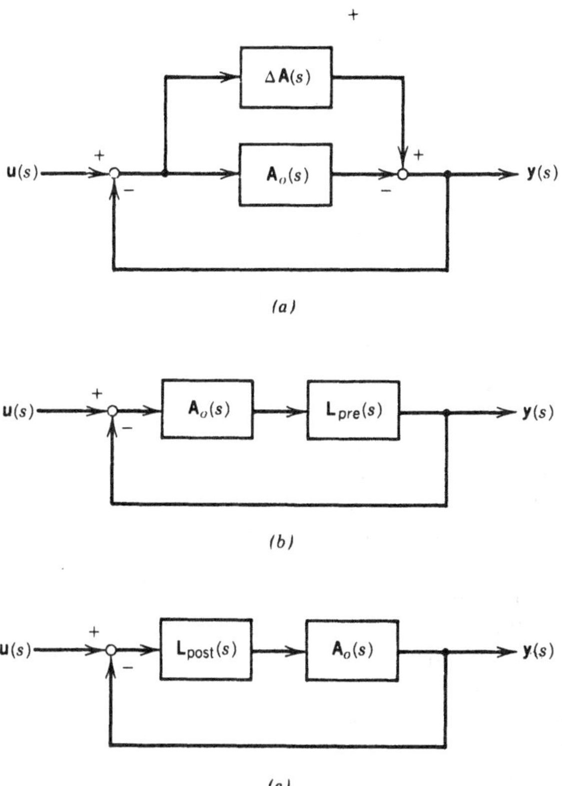

FIGURE 6.5-7 Frequency-domain descriptions of unity-feedback-system model variations. (*a*) Additive model variation; (*b*) premultiplicative model variation; (*c*) postmultiplicative model variation.

parameter variations in **C**, **F**, and **G** is relatively direct, as we could define

$$\Delta\mathbf{A}_C(s) = \Delta\mathbf{C}(s\mathbf{I}_n - \mathbf{F}_o)^{-1}\mathbf{G}_o \tag{6.5-40}$$

$$\Delta\mathbf{A}_F(s) = \mathbf{C}_o\{[s\mathbf{I}_n - (\mathbf{F}_o + \Delta\mathbf{F})]^{-1} - (s\mathbf{I}_n - \mathbf{F}_o)^{-1}\}\mathbf{G}_o \tag{6.5-41}$$

$$\Delta\mathbf{A}_G(s) = \mathbf{C}_o(s\mathbf{I}_n - \mathbf{F}_o)^{-1}\Delta\mathbf{G} \tag{6.5-42}$$

It is easier to specify $\Delta\mathbf{A}(s)$, given $\Delta\mathbf{C}$, $\Delta\mathbf{F}$, or $\Delta\mathbf{G}$, than to specify $\Delta\mathbf{C}$, $\Delta\mathbf{F}$, or $\Delta\mathbf{G}$ from $\Delta\mathbf{A}(s)$.

Multiplicative variations (Fig. 6.5-7*b*, *c*) are more structured, with single elements of the perturbing matrix, $\mathbf{L}(s)$, multiplying many elements of $\mathbf{A}_o(s)$. Pre- and postmultiplicative forms,

$$\mathbf{A}(s) = \mathbf{L}_{\text{pre}}(s)\mathbf{A}_o(s) \tag{6.5-43}$$

$$\mathbf{A}(s) = \mathbf{A}_o(s)\mathbf{L}_{\text{post}}(s) \tag{6.5-44}$$

have differing effects. For illustration, assume that

$$\begin{aligned}\mathbf{L}(s) &= \mathbf{I}_m + \Delta\mathbf{L}(s)\\ &= \mathbf{I}_m + \begin{bmatrix} l_{11}(s) & 0 & \dots & 0\\ 0 & 0 & & 0\\ \vdots & \vdots & & \vdots\\ 0 & 0 & \dots & 0\end{bmatrix}\end{aligned} \tag{6.5-45}$$

For the first case, l_{11} affects every element in the first row of $\mathbf{A}(s)$, while for the second, it affects every element in the first column of $\mathbf{A}(s)$. Nonzero elements in the remaining diagonal elements of $\Delta\mathbf{L}(s)$ would affect the remaining columns or rows accordingly, and off-diagonal elements of $\Delta\mathbf{L}(s)$ would have a coupled effect. It appears most natural to associate a premultiplicative $\mathbf{L}(s)$ with variations in **C** and a postmultiplicative $\mathbf{L}(s)$ with variations in **G**, although individual variations in the original system parameters are not readily portrayed by multiplicative variation. $\mathbf{L}(s)$ has the same effect on stability robustness whether it pre- or postmultiplies $\mathbf{A}(s)$; the principle difference between the two is in command response.

The open-loop stability of $\mathbf{A}(s)$ and the closed-loop stability of the corresponding unity-feedback system are specified by the locations of their respective n eigenvalues, but a criterion that represents the *degree* of stability is needed to define the multivariable system's stability *margins*. A most obvious choice is to extend the scalar Nyquist stability criterion for $A(s)/[1 + A(s)]$ to multi-input/multi-output systems, $\mathbf{A}(s)[\mathbf{I}_m + \mathbf{A}(s)]^{-1}$. This is easily done, because the determinant of the return difference matrix can be expressed by the same ratio of closed- and open-loop characteristic polynomials that appears in the scalar case. From Eq. 6.5-12,

$$|\mathbf{I}_m + \mathbf{A}(s)| = \Delta_{\text{CL}}(s)/\Delta_{\text{OL}}(s) \tag{6.5-46}$$

In other words, the determinant of the multivariable return difference matrix has one feature in common with the scalar return difference function. Evaluating $|\mathbf{I}_m + \mathbf{A}(s)|$ as s follows a closed $\mathbb{D}$ contour in the s plane,* the number of counterclockwise encirclements of the $|\mathbf{I}_m + \mathbf{A}(s)|$-plane origin equals the number of unstable roots of $\Delta_{\mathrm{OL}}(s)$ if the unity-feedback system is stable. This is the *multivariable Nyquist stability criterion*.

Unfortunately, $|\mathbf{I}_m + \mathbf{A}(s)|$ does not share a second important feature with its scalar counterpart: the determinant is not a good measure of the magnitude of a matrix, as the following example shows. Define $\mathbf{X}_1$, $\mathbf{X}_2$, and $\mathbf{X}_3$ as

$$\mathbf{X}_1 = \begin{bmatrix} 1 & 0 \\ 0 & 2 \end{bmatrix}; \qquad |\mathbf{X}_1| = 2 \tag{6.5-47}$$

$$\mathbf{X}_2 = \begin{bmatrix} 1 & 100 \\ 0 & 2 \end{bmatrix}; \qquad |\mathbf{X}_2| = 2 \tag{6.5-48}$$

$$\mathbf{X}_3 = \begin{bmatrix} 1 & 100 \\ 0.02 & 2 \end{bmatrix}; \qquad |\mathbf{X}_3| = 0 \tag{6.5-49}$$

The first two are seen to have the same determinant, although $\mathbf{X}_2$ might be considered to be bigger than $\mathbf{X}_1$ in some sense. Furthermore, $\mathbf{X}_2$ is "nearly the same" as $\mathbf{X}_3$, which is a singular matrix. Consequently, the return difference magnitude of the multivariable system is not indicated with any assurance by $|\mathbf{I}_m + \mathbf{A}(s)|$.

Matrix Norms and Singular Value Analysis

What we seek is a good *norm* of a matrix (i.e., a good measure of size in the sense of Section 2.1). Recall that the Euclidean norm was found to be a good measure of vector length (Eq. 2.1-29),

$$\|\mathbf{x}\| = (\mathbf{x}^T\mathbf{x})^{1/2} \tag{6.5-50a}$$

while the weighted Euclidean norm allowed certain components of $\mathbf{x}$ to be stressed (Eq. 2.1-32):

$$\|\mathbf{Dx}\| = (\mathbf{x}^T\mathbf{D}^T\mathbf{Dx})^{1/2} \tag{6.5-51a}$$

Conversely, if $\|\mathbf{x}\|$ takes a given value, Eq. 6.5-51 provides a relative measure of the size of $\mathbf{D}$.

In this regard, the *spectral norm* (or matrix norm) of a square matrix $\mathbf{D}$ is

*As in Section 2.6, the $\mathbb{D}$ contour extends along the $j\omega$ axis from $-jR$ to jR, indenting to avoid singularities; this segment of the contour is called Ω_R below. The contour closes in a right-half-plane semicircle of radius R, $R \to \infty$.

defined as

$$\|\mathbf{D}\| = \max_{\|\mathbf{x}\|=1} \|\mathbf{Dx}\| \tag{6.5-52}$$

where $\|\mathbf{Dx}\|$ is given by Eq. 6.5-51a if $\mathbf{D}$ and $\mathbf{x}$ are real. If $\mathbf{D}$ and $\mathbf{x}$ are complex, the transpose is replaced by the Hermitian (complex conjugate) transpose; then

$$\|\mathbf{x}\| = (\mathbf{x}^H\mathbf{x})^{1/2} \tag{6.5-50b}$$

$$\|\mathbf{Dx}\| = (\mathbf{x}^H\mathbf{D}^H\mathbf{Dx})^{1/2} \tag{6.5-51b}$$

$\|\mathbf{Dx}\|$ has more than one size; in fact, it can take an infinity of values. For $\dim(\mathbf{x}) = n$ and $\dim(\mathbf{D}) = (n \times n)$, $\mathbf{D}$'s magnitude is defined along n principal directions by the *eigenvalues* of $\mathbf{D}^T\mathbf{D}$ (or $\mathbf{D}^H\mathbf{D}$). These eigenvalues are real because $\mathbf{D}^T\mathbf{D}$ (or $\mathbf{D}^H\mathbf{D}$) is symmetric (or Hermitian), and the eigenvalues' positive square roots are the *singular values* of $\mathbf{D}$ (Section 2.2). The maximum and minimum singular values of $\mathbf{D}$, denoted by $\bar{\sigma}(\mathbf{D})$ and $\underline{\sigma}(\mathbf{D})$, equal the spectral norms of $\mathbf{D}$ and its inverse:

$$\bar{\sigma}(\mathbf{D}) = \|\mathbf{D}\| \tag{6.5-53}$$

$$\underline{\sigma}(\mathbf{D}) = \frac{1}{\|\mathbf{D}^{-1}\|} = \min_{\|\mathbf{x}\|=1} \|\mathbf{Dx}\| \tag{6.5-54}$$

Hence the size of $\mathbf{D}$ depends upon how we look at it, and it has positive-real maximum and minimum values.

The three matrices presented in Eqs. 6.5-47 to 6.5-49 can be used to compare determinants and singular values as measures of size (Table 6.5-3). The singular values provide more reasonable measures of size; the mini-

TABLE 6.5-3 Determinants and Singular Values of Three Matrices

Matrix, $\mathbf{X}$	Determinant, $\lvert\mathbf{X}\rvert$	$\bar{\sigma}(\mathbf{X})$	$\underline{\sigma}(\mathbf{X})$
$\begin{bmatrix} 1 & 0 \\ 0 & 2 \end{bmatrix}$	2	2	1
$\begin{bmatrix} 1 & 100 \\ 0 & 2 \end{bmatrix}$	2	100.025	0.02
$\begin{bmatrix} 1 & 100 \\ 0.02 & 2 \end{bmatrix}$	0	100.025	0.

mum singular value, $\underline{\sigma}$, indicates that the third matrix is singular and that the second is almost singular, while the maximum singular values are linked to the largest value in each matrix. Both $\bar{\sigma}$ and $\underline{\sigma}$ have utility in characterizing stability robustness, as will be shown.

Conservative bounds for acceptable additive and multiplicative variations in $\mathbf{A}(s)$ can be based on singular values. If the original closed-loop system, $\mathbf{A}_o(s)[\mathbf{I}_m + \mathbf{A}_o(s)]^{-1}$, is stable, then the *additive perturbation* $\Delta\mathbf{A}(s)$ does not destabilize the system as long as (S-6)

$$\bar{\sigma}[\Delta\mathbf{A}(j\omega)] < \underline{\sigma}[\mathbf{I}_m + \mathbf{A}_o(j\omega)], \qquad 0 \le \omega \le \infty \tag{6.5-55}$$

Note that the *maximum* singular value of the perturbation matrix is compared with the *minimum* singular value of the return difference matrix. The comparison must be made for all ω in $[0, \infty]$, so the singular values can be graphed in the form of Bode plots. If the test is failed at any ω, stability is not guaranteed; nevertheless, the test is conservative, and stability still is possible.

Similarly, it is shown in (S-6) that if the original closed-loop system, $\mathbf{A}_o(s)[\mathbf{I}_m + \mathbf{A}_o(s)]^{-1}$, is stable, then the *multiplicative perturbation*, $[\mathbf{I}_m + \Delta\mathbf{L}(s)]$, does not destabilize the system for

$$\bar{\sigma}[\Delta\mathbf{L}(j\omega)] < \underline{\sigma}[\mathbf{I}_m + \mathbf{A}_o^{-1}(j\omega)], \qquad 0 \le \omega \le \infty \tag{6.5-56}$$

This comparison also can be made in Bode plot form.

An alternate criterion for stability with multiplicative variation of $\mathbf{A}(s)$ is given in (L-3). As before, the nominal open-loop characteristic polynomial is signified by $\Delta_{OL}(s)$, while the characteristic polynomial for the perturbed system is $\tilde{\Delta}_{OL}(s)$. It is assumed that the nominal closed-loop system is stable, that $\Delta_{OL}(s)$ and $\tilde{\Delta}_{OL}(s)$ have the same number of right-half-plane roots, and that $\tilde{\Delta}_{OL}(j\omega_o) = 0$ implies $\Delta_{OL}(j\omega_o) = 0$ for any ω_o. Defining α as

$$\alpha = \underline{\sigma}[\mathbf{I}_m + \mathbf{A}_o(s)] \tag{6.5-57}$$

with s on Ω_R, the perturbed closed-loop system is stable if

$$\bar{\sigma}[\mathbf{L}^{-1}(s) - \mathbf{I}_m] < \alpha \tag{6.5-58}$$

and if *at least one* of the following is satisfied:

$$\alpha \le 1 \tag{6.5-59}$$

$$\mathbf{L}^H(s) + \mathbf{L}(s) \ge 0 \tag{6.5-60}$$

$$4(\alpha^2 - 1)\underline{\sigma}^2[\mathbf{L}(s) - \mathbf{I}_m] > \alpha^2\bar{\sigma}^2[\mathbf{L}(s) + \mathbf{L}^H(s) - 2\mathbf{I}_m] \tag{6.5-61}$$

From Eqs. 6.5-55 and 6.5-58, it can be deduced that the minimum singular value of the return difference matrix is itself an important measure of stability margin for both additive and multiplicative perturbations. The smaller $\underline{\sigma}[\mathbf{I}_m + \mathbf{A}_o(s)]$, the closer the return difference matrix is to singularity and system instability. Furthermore, if

$$\underline{\sigma}[\mathbf{I}_m + \mathbf{A}_o(s)] \geq \alpha_o, \qquad s \text{ on } \Omega_R \qquad \text{and } \alpha_o \leq 1 \tag{6.5-62}$$

then there is a *guaranteed gain margin*, K, of

$$K = \frac{1}{1 \pm \alpha_o} \tag{6.5-63}$$

or a *guaranteed phase margin*, ϕ, of

$$\phi = \pm \cos^{-1}\left(1 - \frac{\alpha_o^2}{2}\right) \tag{6.5-64}$$

in each of the control loops (i.e., for one to m controls) (L-3). If Eq. 6.5-60 is satisfied and, in addition,

$$\mathbf{A}_o(s) + \mathbf{A}_o^H(s) \geq 0, \qquad s \text{ on } \Omega_R \tag{6.5-65}$$

the guaranteed gain and phase margins are increased to $[0, \infty]$ and $\pm 90°$, respectively. Phase margins are reduced by gain variations, and vice versa.

Earlier in this section, the question of defining "large" $\mathbf{A}(s)$ and $\mathbf{B}(s)$ was raised; singular values provide a means of answering this question, and they allow a multivariable interpretation of Bode's feedback design guidelines (Section 6.3). At low frequencies, the minimum singular value of the return difference matrix should exceed some minimum value for good tracking performance:

$$\underline{\sigma}[\mathbf{I}_m + \mathbf{A}(j\omega)] > \sigma_{\min}(\omega) > 1, \qquad \omega < \omega_{c_1} \tag{6.5-66}$$

This implies that $\underline{\sigma}[\mathbf{A}(j\omega)] \gg 1$ in the same region (Fig. 6.5-8).

At high frequencies, the maximum singular values of the disturbance-response and control-response matrices (Eqs. 6.5-26 and 6.5-27) should be small enough for good noise and disturbance rejection. Using the matrix inversion identities of Section 2.2,

$$\begin{aligned} \mathbf{A}(s)\,[\mathbf{I}_m + \mathbf{A}(s)]^{-1} &= [\mathbf{I}_m + \mathbf{A}(s)]^{-1}\mathbf{A}(s) \\ &= [\mathbf{I}_m + \mathbf{A}^{-1}(s)]^{-1} \end{aligned} \tag{6.5-67}$$

$$\begin{aligned} \mathbf{B}(s)[\mathbf{I}_n + \mathbf{B}(s)]^{-1} &= [\mathbf{I}_n + \mathbf{B}(s)]^{-1}\mathbf{B}(s) \\ &= [\mathbf{I}_n + \mathbf{B}^{-1}(s)]^{-1} \end{aligned} \tag{6.5-68}$$

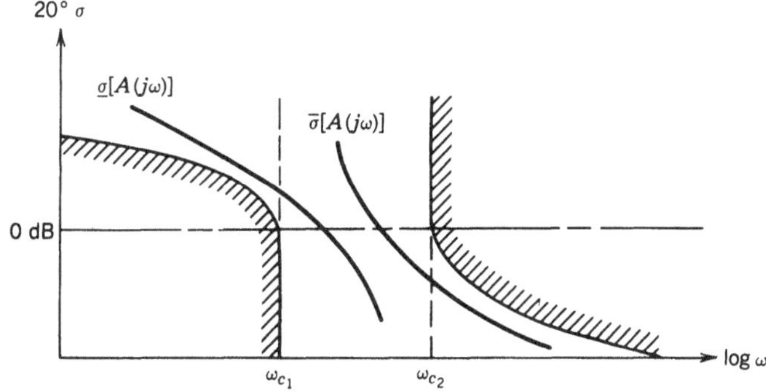

FIGURE 6.5-8 Desirable trends in maximum and minimum singular values of the loop transfer function matrix, $\bar{\sigma}[A(j\omega)]$ and $\underline{\sigma}[A(j\omega)]$.

For good noise and disturbance rejection, it is desired that

$$\bar{\sigma}\{[\mathbf{I}_m + \mathbf{A}^{-1}(j\omega)]^{-1}\} = 1/\underline{\sigma}[\mathbf{I}_m + \mathbf{A}^{-1}(j\omega)] < \sigma_{\mathrm{A}}(\omega), \quad \omega > \omega_{c_2} \tag{6.5-69}$$

$$\bar{\sigma}\{[\mathbf{I}_n + \mathbf{B}^{-1}(j\omega)]^{-1}\} = 1/\underline{\sigma}[\mathbf{I}_n + \mathbf{B}^{-1}(j\omega)] < \sigma_{B}(\omega), \quad \omega > \omega_{c_2} \tag{6.5-70}$$

This implies that the maximum singular values of $\mathbf{A}(j\omega)$ and $\mathbf{B}(j\omega)$ should be limited at high frequency (Fig. 6.5-8):

$$\bar{\sigma}[\mathbf{A}(j\omega)] < \sigma'_{\mathrm{A}}(\omega), \qquad \omega > \omega_{c_2} \tag{6.5-71}$$

$$\bar{\sigma}[\mathbf{B}(j\omega)] < \sigma'_{B}(\omega), \qquad \omega > \omega_{c_2} \tag{6.5-72}$$

The maximum and minimum singular values of LQ regulators generally fall within bounds of the type specified by Eqs. 6.5-66, 6.5-71, and 6.5-72.

Stability Margins of Multivariable Linear-Quadratic Regulators

The development of stability margins for LQ regulators with postmultiplicative variations, $\mathbf{L}(s)$, parallels the earlier development of LQ regulator spectral characteristics (L-3). Beginning at Eq. 6.5-4b, the quantity $s\mathbf{S}$ and its complex conjugate, $s^*\mathbf{S}$, are added to both sides of the algebraic Riccati equation

$$\mathbf{S}(s\mathbf{I}_n - \mathbf{F}) + (s^*\mathbf{I}_n - \mathbf{F}^T)\mathbf{S} + \mathbf{C}^T\mathbf{R}\mathbf{C} = \mathbf{Q} + 2\mu\mathbf{S} \tag{6.5-73}$$

where $s = \mu + j\omega$. Premultiplying by $\mathbf{G}^T(s^*\mathbf{I}_n - \mathbf{F}^T)^{-1}$, post multiplying by $(s\mathbf{I}_n - \mathbf{F})^{-1}\mathbf{G}$, and adding $\mathbf{R}$ to both sides leads to the equation

$$[\mathbf{I}_m + \mathbf{A}(s)]^H\mathbf{R}[\mathbf{I}_m + \mathbf{A}(s)] = \mathbf{R} + \mathbf{Y}(s) \tag{6.5-74}$$

where

$$\mathbf{A}(s) = \mathbf{C}(s\mathbf{I}_n - \mathbf{F})^{-1}\mathbf{G} \tag{6.5-75}$$

$$\mathbf{Y}(s) = [(s\mathbf{I}_n - \mathbf{F})^{-1}\mathbf{G}]^H(\mathbf{Q} + 2\mu\mathbf{S})[(s\mathbf{I}_n - \mathbf{F})^{-1}\mathbf{G}] \tag{6.5-76}$$

$\mathbf{S}$ is the solution to the algebraic Riccati equation, $\mathbf{C}$ is the corresponding optimal gain matrix, and s is confined to the Nyquist $\mathbb{D}$ contour. If the conditions of Section 5.4 are met and $\mathbf{G}$ has full rank, then

$$[\mathbf{I}_m + \mathbf{A}(s)]^H\mathbf{R}[\mathbf{I}_m + \mathbf{A}(s)] > \mathbf{R} \tag{6.5-77}$$

[$\mathbf{A} > \mathbf{B}$ means that $(\mathbf{A} - \mathbf{B})$ is a positive definite matrix, while $\mathbf{A} \geq \mathbf{B}$ means that $(\mathbf{A} - \mathbf{B})$ is a positive semidefinite matrix.] Furthermore, if there are no singularities on the $j\omega$ axis,

$$[\mathbf{I}_m + \mathbf{A}(j\omega)]^H\mathbf{R}[\mathbf{I}_m + \mathbf{A}(j\omega)] \geq \mathbf{R} \tag{6.5-78}$$

which is a *multivariable equivalent to the Kalman inequality*.

This is seen more clearly if the optimal loop transfer function matrix is transformed using the control weighting matrix. Taking $\mathbf{R}$ as a diagonal matrix with

$$\mathbf{U}\mathbf{U}^T = \mathbf{U}^T\mathbf{U} = \mathbf{R}^{-1} \tag{6.5-79}$$

the transformed matrix, $\tilde{\mathbf{A}}(s)$, is

$$\tilde{\mathbf{A}}(s) = \mathbf{U}^{-1}\mathbf{A}(s)\mathbf{U} \tag{6.5-80}$$

On the Nyquist $\mathbb{D}$ contour, $j\omega$ is replaced by s, and Eq. 6.5-78 becomes

$$[\mathbf{I}_m + \tilde{\mathbf{A}}(s)]^H[\mathbf{I}_m + \tilde{\mathbf{A}}(s)] \geq \mathbf{I}_m \tag{6.5-81}$$

Then Eq. 6.5-81 guarantees that the minimum singular value of the return difference matrix is

$$\underline{\sigma}[\mathbf{I}_m + \tilde{\mathbf{A}}(s)] \geq 1, \qquad \sigma \text{ on } \Omega_R \tag{6.5-82}$$

which is directly comparable to the scalar Kalman inequality (Eq. 6.5-34).

Multiplicative variations of the loop transfer function matrix

$$\mathbf{A}(s) = \mathbf{A}_0(s)\mathbf{L}(s) \tag{6.5-83}$$

also can be normalized by the control weighting, so that

$$\hat{\mathbf{L}}(s) = \mathbf{U}^{-1}\mathbf{L}(s)\mathbf{U} \tag{6.5-84}$$

$$\hat{\mathbf{A}}(s) = \mathbf{U}^{-1}\mathbf{A}_o(s)\mathbf{L}(s)\mathbf{U} = \hat{\mathbf{A}}_o(s)\hat{\mathbf{L}}(s) \tag{6.5-85}$$

Denoting the maximum singular value of the weighted perturbation by γ,

$$\gamma(s) = \bar{\sigma}[\hat{\mathbf{L}}(s)^{-1} - \mathbf{I}_m] \tag{6.5-86}$$

the perturbed closed-loop system remains stable if either

$$\mathbf{Q} > \mathbf{0} \quad \text{and} \quad \gamma(s) \leq 1 \text{ on } \Omega_R \tag{6.5-87}$$

or

$$\Delta_{\mathrm{OL}}(s) \neq 0 \quad \text{and} \quad \gamma(s) < 1 \text{ on } \Omega_R \tag{6.5-88}$$

$\gamma(s) \leq 1$ is equivalent to the condition (S-3)

$$\mathbf{RL}(s) + \mathbf{L}^H(s)\mathbf{R} - \mathbf{R} \geq \mathbf{0} \tag{6.5-89}$$

For diagonal $\mathbf{R}$ and $\mathbf{L}(s)$, satisfying these conditions guarantees the LQ regulator's minimum gain margin to be ($\frac{1}{2}$ to ∞) in each of the m feedback loops. The amplification or attenuation in each loop need not be the same (i.e., each gain may vary independently within the margin). Alternately, minimum simultaneous phase margins of $\pm 60°$ are guaranteed in the feedback loops.

The caveats of scalar LQ robustness apply as well to multivariable LQ control. The gain and phase margins are derived under assumptions that fix the qualitative shape of the multivariable Nyquist plot of $\mathbf{A}(s)$, and it is difficult to draw an equivalence between $\mathbf{L}(s)$ and variations in $\mathbf{C}$, $\mathbf{F}$, and $\mathbf{G}$. Furthermore, these stability margins are derived for a diagonal control weighting matrix; if $\mathbf{R}$ is nondiagonal, arbitrarily small margins may result (L-3).

It would be desirable to assure that feedback control of an open-loop-stable plant does not lead to instability when one or more control loops are broken (which would require an associated gain margin of zero) yet the minimum assured gain margin of the LQ regulator with diagonal $\mathbf{R}$ is 50%.* Ad hoc modifications to the LQ controller for an open-loop-stable plant can increase robustness, although there is necessarily a trade-off with the original performance objectives. First, it may be recalled that the stable plant's control gains become vanishingly small as $\mathbf{R}$ approaches infinity; although the gain margin does not vary (as a percentage of the nominal gain), there will be a value of $\mathbf{R}$ for which stability is retained with open control loops. Second, under the usual assumptions, scalar LQ regulators could be designed separately for each control, and the multivariable control could simply be a summation of the individual control loops. Then it would

*(R-1) presents an example in which a *neutrally stable* plant is destabilized by a two-control LQ regulator when one of the controls disappears and the gains for the remaining control are not changed.

be necessary to verify that all possible combinations of multiple controls provide satisfactory response, so the burden of analysis and evaluation would shift from the failed to the unfailed system. Neither of these approaches appears attractive in a general sense, although each could be considered for specific system designs.

A preferred (but nevertheless ad hoc) approach is to use the proven stability margins of the LQ regulator, first designing a high-control-weight, low-bandwidth controller, then restoring the bandwidth by increasing gains. (The second step takes advantage of the LQ regulator's infinite increased-gain margin.) This procedure is called *robustness recovery*, and it is described as follows. Let $\mathbf{Q}$ and $\mathbf{R}_o$ be the specified state and diagonal control weighting matrices, and let ρ^2 be an arbitrary scalar parameter between one and infinity that modifies the control weighting:

$$\mathbf{R} = \rho^2 \mathbf{R}_o \tag{6.5-90}$$

The control gain is calculated in the usual fashion:

$$\mathbf{C} = \mathbf{R}^{-1}\mathbf{G}^T\mathbf{S} = \frac{1}{\rho^2}\mathbf{R}_o^{-1}\mathbf{G}^T\mathbf{S} \tag{6.5-91}$$

where $\mathbf{S}$ is the solution to the algebraic Riccati equation:

$$\begin{aligned} &\mathbf{F}^T\mathbf{S} + \mathbf{S}\mathbf{F} + \mathbf{Q} - \mathbf{S}\mathbf{G}\mathbf{R}^{-1}\mathbf{G}^T\mathbf{S} = \\ &\quad \mathbf{F}^T\mathbf{S} + \mathbf{S}\mathbf{F} + \mathbf{Q} - \frac{1}{\rho^2}\mathbf{S}\mathbf{G}\mathbf{R}_o^{-1}\mathbf{G}^T\mathbf{S} = \mathbf{0} \end{aligned} \tag{6.5-92}$$

If $\mathbf{F}$ is stable, $\mathbf{C}$ approaches zero as ρ^2 approaches infinity, and the closed loop gain margin is $(\frac{1}{2}, \infty)$. If, however, the factor $(1/\rho^2)$ is omitted in the gain calculation,

$$\mathbf{C}_R = \mathbf{R}_o^{-1}\mathbf{G}^T\mathbf{S} \tag{6.5-93}$$

then $\mathbf{C}_R$ is $\rho^2\mathbf{C}$, and its gain margin is $(1/2\rho^2, \infty)$. The corresponding phase margin is $\pm\cos^{-1}(1/2\rho^2)$. In the limit as ρ goes to infinity, Eq. 6.5-92 becomes a *Lyapunov equation* for $\mathbf{S}$:

$$\mathbf{F}^T\mathbf{S} + \mathbf{S}\mathbf{F} + \mathbf{Q} = \mathbf{0} \tag{6.5-94}$$

It is shown in (W-2) that a control law using Eqs. 6.5-94 and 6.5-93 to compute $\mathbf{S}$ and $\mathbf{C}_R$ has gain margins of $(0, \infty)$ or phase margins of $\pm 90°$ in each of the m control loops, assuming stable $\mathbf{F}$ and diagonal $\mathbf{R}_o$. The Lyapunov design procedure retains many of the design flexibilities of the LQ regulator—particularly in the ability to affect closed-loop characteristics through choice of $\mathbf{Q}$ and $\mathbf{R}$—although a quadratic cost function is not minimized.

The Discrete-Time Case

Development of stability margins for discrete-time LQ regulators parallels the previous methodology, with z transforms replacing the Laplace transforms of the continuous case. An important departure from the earlier findings is that gain margins are narrower, reflecting the effective band-limiting and phase lag of the sampling process. This is demonstrated for a single-input regulator that minimizes

$$J = \lim_{k_f \to \infty} \frac{1}{2k_f} \sum_{k=0}^{k_f} (\mathbf{x}_k^T \mathbf{Q} \mathbf{x}_k + \mathbf{u}_k^T \mathbf{R} \mathbf{u}_k) \tag{6.5–95}$$

subject to the dynamic constraint

$$\mathbf{x}_{k+1} = \mathbf{\Phi} \mathbf{x}_k + \mathbf{\Gamma} \mathbf{u}_k, \qquad \mathbf{x}_o \text{ given} \tag{6.5-96}$$

The LQ regulator for this problem takes the form

$$\begin{aligned} \mathbf{u}_k &= -(\mathbf{R} + \mathbf{\Gamma}^T \mathbf{S} \mathbf{\Gamma})^{-1} \mathbf{\Gamma}^T \mathbf{S} \mathbf{\Phi} \mathbf{x}_k \\ &= -\mathbf{C} \mathbf{x}_k \end{aligned} \tag{6.5-97}$$

where $\mathbf{S}$ is the solution to the algebraic Riccati equation:

$$\begin{aligned} \mathbf{0} &= \mathbf{\Phi}^T \mathbf{S} \mathbf{\Phi} - \mathbf{S} + \mathbf{Q} - \mathbf{\Phi}^T \mathbf{S} \mathbf{\Gamma} (\mathbf{R} + \mathbf{\Gamma}^T \mathbf{S} \mathbf{\Gamma})^{-1} \mathbf{\Gamma}^T \mathbf{S} \mathbf{\Phi} \\ &= \mathbf{\Phi}^T \mathbf{S} \mathbf{\Phi} - \mathbf{S} + \mathbf{Q} - \mathbf{C}^T (\mathbf{R} + \mathbf{\Gamma}^T \mathbf{S} \mathbf{\Gamma}) \mathbf{C} \end{aligned} \tag{6.5-98}$$

Following (A-2) and previous results of this section, the z transform operator can be introduced, yielding

$$\begin{aligned} \mathbf{0} = (z^{-1} \mathbf{I}_n - \mathbf{\Phi})^T \mathbf{S} (z \mathbf{I}_n - \mathbf{\Phi}) + (z^{-1} \mathbf{I}_n - \mathbf{\Phi})^T \mathbf{S} \mathbf{\Phi} \\ + \mathbf{\Phi}^T \mathbf{S} (z \mathbf{I}_n - \mathbf{\Phi}) + \mathbf{Q} - \mathbf{C}^T (\mathbf{R} + \mathbf{\Gamma}^T \mathbf{S} \mathbf{\Gamma}) \mathbf{C} \end{aligned} \tag{6.5-99}$$

Discrete-time loop and cost transfer function matrices are defined in accordance with their continuous-time counterparts:

$$\mathbf{A}(z) = \mathbf{C}(z \mathbf{I}_n - \mathbf{\Phi})^{-1} \mathbf{\Gamma} \tag{6.5-100}$$

$$\mathbf{Y}(z) = \mathbf{H}(z \mathbf{I}_n - \mathbf{\Phi})^{-1} \mathbf{\Gamma} \tag{6.5-101}$$

$$\mathbf{H}^T \mathbf{H} = \mathbf{Q} \tag{6.5-102}$$

Equation 6.5-99 then can be expressed in spectral-factorized form:

$$[\mathbf{I}_m + \mathbf{A}(z^{-1})]^T (\mathbf{R} + \mathbf{\Gamma}^T \mathbf{S} \mathbf{\Gamma}) [\mathbf{I}_m + \mathbf{A}(z)] = \mathbf{R} + \mathbf{Y}^T(z^{-1})\mathbf{Y}(z) \quad (6.5\text{-}103)$$

This is equivalent to the continuous-time result (Eq. 6.5-6d), and it forms the basis for stability margin evaluation.

If the rank of $\mathbf{Q}$ is one, the dimension of $\mathbf{H}$ can be chosen as $(1 \times n)$, and, because the control is scalar, R, $A(z)$, and $Y(z)$ are scalar; hence Eq. 6.5-103 becomes

$$[1 + A(z^{-1})][1 + A(z)] = \frac{R + Y(z^{-1})\,Y(z)}{R + \mathbf{\Gamma}^T \mathbf{S} \mathbf{\Gamma}} \quad (6.5\text{-}104)$$

where $[1 + A(z)]$ is the optimal return difference function. Because $A(z)$ is stable, its roots lie within the unit circle, while those of $A(z^{-1})$ lie outside. Just as the $j\omega$ axis forms a significant part of the Nyquist contour for continuous-time systems, the unit circle is important in evaluating discrete-time system stability. For z on the unit circle, z^{-1} is the complex conjugate of z; hence $Y(z^{-1}) = Y^*(z)$, and $Y(z^{-1})\,Y(z)$ is a positive-real scalar. This leads to the conclusion that

$$[1 + A(z^{-1})][1 + A(z)] = |1 + A(z)|^2 \geq \frac{R}{R + \mathbf{\Gamma}^T \mathbf{S} \mathbf{\Gamma}}, \quad |z| = 1 \quad (6.5\text{-}105a)$$

or

$$|1 + A(z)| \geq \sqrt{\frac{R}{R + \mathbf{\Gamma}^T \mathbf{S} \mathbf{\Gamma}}}, \quad |z| = 1 \quad (6.5\text{-}105b)$$

Because $\mathbf{\Gamma}^T \mathbf{S} \mathbf{\Gamma}$ also is a positive-real scalar, a Nyquist plot of $A(z)$ is excluded from a circle centered at $(-1, 0)$ whose radius is less than or equal to one. Furthermore, since the LQ regulator is stabilizing, the Nyquist criterion is satisified.

If the control loop contains a positive-real multiplier K, the return difference function is $[1 + KA(z)]$; stability will be retained only as long as the zeros of $[1 + KA(z)]$ remain within the unit circle. Because

$$[1 + KA(z)] = \frac{\Delta_{\mathrm{CL}}(z)}{\Delta_{\mathrm{OL}}(z)} \quad (6.5\text{-}106)$$

the closed-loop polynomial is

$$\Delta_{\mathrm{CL}}(z) = \Delta_{\mathrm{OL}}(z) + KN(z) \quad (6.5\text{-}107)$$

where $N(z)$ contains the zeros of the optimal loop transfer function.

As K varies from its optimal value of one, the roots of $\Delta_{CL}(z)$ follow a locus that can be calculated in the usual fashion (Section 2.6). For some value of K less than one, the closed-loop roots will leave the interior of the unit circle if $\Delta_{OL}(z)$ contains unstable roots, so there is a lower guaranteed stability bound on K; if the open-loop system is stable, the reduced-gain margin is 100%.

The root locus also must leave the unit circle for some K greater than one because the degree of $N(z)$ must be less than the degree of $\Delta_{OL}(z)$, forcing one or more roots toward infinity with increasing K. This follows directly from Eq. 6.5-107. The degrees of $\Delta_{OL}(z)$ and $\Delta_{CL}(z)$ are n, and the coefficients of z^n in both are one.* But

$$\begin{aligned} N(z) &= [\Delta_{CL}(z)]_{opt} - \Delta_{OL}(z) \\ &= [z^n + a_{n-1}z^{n-1} + \cdots] - [z^n + b_{n-1}z^{n-1} + \cdots] \\ &= [(a_{n-1} - b_{n-1})z^{n-1} + \cdots] \end{aligned} \quad (6.5\text{-}108)$$

which is of $(n-1)$st degree or less. Unlike the continuous-time case, the increased-gain margin of the discrete-time regulator is, therefore, bounded.

The guaranteed stability margins can be derived by manipulating $|1 + KA(z)|$ and evaluating the results for $|z| = 1$. To retain stability as K departs from its optimal value, the Nyquist criterion requires that

$$|1 + KA(z)| \geq 0, \qquad |z| = 1 \quad (6.5\text{-}109)$$

It also follows that

$$\begin{aligned} |1 + KA(z)| &= K\left|\frac{1}{K} - 1 + 1 + A(z)\right| \\ &\geq K\left|\left|\frac{1}{K} - 1\right| - |1 + A(z)|\right| \end{aligned} \quad (6.5\text{-}110)$$

Substituting Eq. 6.5-105b in this inequality, stability is retained as long as

$$\left|\left|\left(\frac{1}{K}\right) - 1\right| - \sqrt{R/(R + \mathbf{\Gamma}^T\mathbf{S}\mathbf{\Gamma})}\right| \geq 0 \quad (6.5\text{-}111)$$

This leads to the sought-after bounds on K:

$$\frac{1}{1 + \sqrt{R/(R + \mathbf{\Gamma}^T\mathbf{S}\mathbf{\Gamma})}} \leq K \leq \frac{1}{1 - \sqrt{R/(R + \mathbf{\Gamma}^T\mathbf{S}\mathbf{\Gamma})}} \quad (6.5\text{-}112)$$

*Hence the polynomials are *monic*.

Because $[R/(R+\Gamma^T S\Gamma)] \leq 1$, the guaranteed gain margins are more restrictive than the continuous-time case i.e., they are within $(\frac{1}{2}, \infty)$. For decreasing sampling interval, $[R/(R+\Gamma^T S\Gamma)] \rightarrow 1$, and the bounds approach the continuous-time values.

6.6 ROBUSTNESS OF STOCHASTIC-OPTIMAL REGULATORS

Adding an estimator to the control logic complicates the robustness issue because the estimator itself is a dynamic system that is subject to instability. Because an isolated linear-Gaussian estimator is the mathematical dual of a linear-quadratic regulator, it possesses the same sorts of stability margins discussed in the last section. When the estimator is part of the feedback loop of a stochastic regulator, dynamic interactions between the estimator and the controlled system increase the potential for instability, as introduced in Section 6.4. It has been shown that estimate-error and state dynamic modes are uncoupled and stable when the actual system and the mathematical model used to design the optimal controller are the same (Section 6.4); instability is possible only if some aspect of the system description is wrong.

There are three ways in which the system description might be wrong. In the first case, system parameters are known without error and the stochastic regulator is designed properly, but signal paths are broken, as might occur when sensors or actuators fail. In the second case, signal paths are intact, but assumed system parameters are incorrect, as might result from inadequate prior data or from a change in operating point. In the third case, the structure of the mathematical model is deficient, as in the design of a low-order controller for a high-order system. This last possibility is qualitatively different from the first two and is not considered here.

It might be expected that the stability robustness of a stochastic regulator approaches that of a deterministic regulator if the estimator is "fast" (i.e., if the lag between the estimate and the actual state is small). However, a high-bandwidth estimator is no guarantee of robustness; in fact, it can make matters worse. A better supposition is ". . . if the estimator is *correct*," implying not only speed but accuracy in portraying the state vector's time-varying mean.

If the robustness of a stochastic regulator is inadequate, various methods can be used for robustness recovery. It is logical to assure first that the corresponding deterministic (full-state feedback) regulator is sufficiently robust using methods of the previous section. The estimator is then designed to provide a state estimate that is "good enough" from the robustness viewpoint while not yielding too large an increase in state-estimate covariance. A dual approach is to design first a sufficiently robust estimator, then close the loop with a "low-sensitivity" feedback controller. The two methods can provide similar stability margins, but the former is more closely aligned with achieving the control system's performance objectives.

Robustness itself may be a stochastic issue. To guard against destabilizing effects of parameter uncertainty, it is useful to minimize the probability of instability, given the probability distributions of uncertain parameters. This approach is amenable to state-space problem formulations, as uncertainties in individual elements of the system matrices can be considered directly without the intermediate step of defining and analyzing transfer function matrices.

Compensator Stability

The linear-quadratic Gaussian (LQG) regulator takes the form shown in Fig. 6.6-1; with the control vector specified as a linear function of the estimated state

$$\mathbf{u}(t) = -\mathbf{C}\hat{\mathbf{x}}(t) \tag{6.6-1}$$

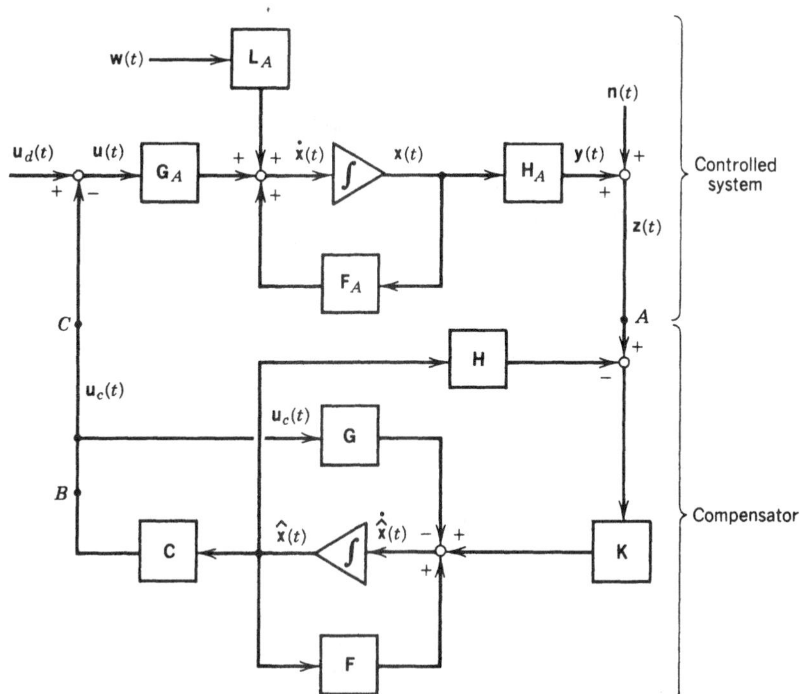

FIGURE 6.6-1 Block diagram of the stochastic regulator (in the time domain).

the dynamic equations of the controlled system and the state estimator are

$$\dot{\mathbf{x}}(t) = \mathbf{F}_A\mathbf{x}(t) - \mathbf{G}_A\mathbf{C}\hat{\mathbf{x}}(t) + \mathbf{L}_A\mathbf{w}(t) \tag{6.6-2}$$

$$\begin{aligned}\dot{\hat{\mathbf{x}}}(t) &= \mathbf{F}\hat{\mathbf{x}}(t) - \mathbf{G}\mathbf{C}\hat{\mathbf{x}}(t) + \mathbf{K}[\mathbf{z}(t) - \mathbf{H}\hat{\mathbf{x}}(t)] \\ &= (\mathbf{F} - \mathbf{G}\mathbf{C} - \mathbf{K}\mathbf{H})\hat{\mathbf{x}}(t) + \mathbf{K}\mathbf{z}(t)\end{aligned} \tag{6.6-3}$$

Vectors and matrices are defined as before; the subscript A connotes a matrix of the controlled (or actual) system, and the unsubscripted matrices are those used in or resulting from control system design. Consequently, the measurement is a function of the actual state through the matrix $\mathbf{H}_A$,

$$\mathbf{z}(t) = \mathbf{H}_A\mathbf{x}(t) + \mathbf{n}(t) \tag{6.6-4}$$

although the estimate is transformed to the measurement space by $\mathbf{H}$. The estimation and control logic represented by Eq. 6.6-1 and 6.6-3 define an $(m \times r)$ *feedback compensator* whose transfer function matrix from the measurement transform $\mathbf{z}(s)$ to the control transform $\mathbf{u}_c(s)$ is

$$\mathbf{B}(s) = -\mathbf{C}[s\mathbf{I}_n - (\mathbf{F} - \mathbf{G}\mathbf{C} - \mathbf{K}\mathbf{H})]^{-1}\mathbf{K} \tag{6.6-5}$$

The stability of the compensator taken alone (i.e., with the loop broken at A and C) is determined by

$$|s\mathbf{I}_n - (\mathbf{F} - \mathbf{G}\mathbf{C} - \mathbf{K}\mathbf{H})| = 0 \tag{6.6-6}$$

and it is affected by three internal feedback loops through $\mathbf{F}$, $\mathbf{GC}$, and $\mathbf{KH}$.

When the corresponding actual and modeled matrices are identical, the regulator and estimator modes are stable and uncoupled, with roots defined by

$$|s\mathbf{I}_n - (\mathbf{F} - \mathbf{G}\mathbf{C})| = 0 \tag{6.6-7}$$

$$|s\mathbf{I}_n - (\mathbf{F} - \mathbf{K}\mathbf{H})| = 0 \tag{6.6-8}$$

however, the fact that Eqs. 6.6-7 and 6.6-8 indicate stability does not guarantee that the isolated compensator is stable. Compensator instability is possible if the *net effect* of state estimate feedback in Eq. 6.6-6 yields right-half-plane roots. This might occur, for example, if $\mathbf{GC}$ and $\mathbf{KH}$ are individually stabilizing but the mode shapes (eigenvectors) corresponding to Eqs. 6.6-7 and 6.6-8 are very different. The regulator and estimator then are mismatched even though the LQ, LG, and LQG descriptions are stable.

Compensator stability is retained if the loop is broken at A and B because this also breaks the internal control gain feedback in the estimator. Similarly, if $\mathbf{C}\hat{\mathbf{x}}(t)$ were not fed back in the estimator, this result would be assured; however, the state-estimation error in the nominal closed-loop condition would be degraded. Not only is the covariance of the error increased, but there is the possibility that the nominal system will be destabilized. $\mathbf{G}$ would be set to zero in Eq. 6.4-97, leaving the positive feedback term $\mathbf{G}_A\mathbf{C}$ in the lower-right element.

As noted in (J-4), isolated compensator instability does not imply stochastic regulator instability (for which the feedback loop is *not* broken), but it can have practical consequences during system design or as a result of component failure. Conventional testing of the isolated compensator (e.g., determination of its frequency response) is difficult—if not impossible—with compensator instability. If the loop is broken at C, unstable compensator modes cannot affect the plant, although the system then is robbed of feedback control that may be needed for stability. Breaking the loop at A but not at C also eliminates the feedback; however, it has the added effect of driving the system with the output of the (possibly unstable) estimator.

Of course, total loss of feedback is, by definition, not a failure which any compensator can be expected to control. If the likelihood of such a failure is high and the corresponding results would be unacceptable, a backup control strategy is warranted. If partial failures such as losses of individual sensors and/or actuators must be accommodated, then the compensator does not become truly isolated. Consideration then must be given to issues of stabilizability, detectability, and robustness, as discussed in the remainder of this section.

Stability Margins of LQG Regulators

Many of the robustness concepts introduced earlier can be applied to the LQG regulator. Taking the Laplace transforms of Eqs. 6.6-1 to 6.6-3 while neglecting initial conditions, disturbances, and measurement errors leads to the relationships

$$\mathbf{y}(s) = \mathbf{D}(s)\mathbf{u}(s) = \mathbf{D}(s)[\mathbf{u}_d(s) + \mathbf{u}_c(s)] \tag{6.6-9}$$

$$\mathbf{u}_c(s) = \mathbf{B}(s)\mathbf{y}(s) \tag{6.6-10}$$

where

$$\mathbf{D}(s) = \mathbf{H}_A(s\mathbf{I}_n - \mathbf{F}_A)^{-1}\mathbf{G}_A \tag{6.6-11}$$

$$\mathbf{B}(s) = -\mathbf{C}[s\mathbf{I}_n - (\mathbf{F} - \mathbf{GC} - \mathbf{KH})]^{-1}\mathbf{K} \tag{6.6-12}$$

as illustrated by Fig. 6.6-2. The *return difference function matrix* of the LQG regulator is seen to be

$$[\mathbf{I}_m - \mathbf{B}(s)\mathbf{D}(s)] = [\mathbf{I}_m + \mathbf{A}(s)] \tag{6.6-13}$$

Identifying $-\mathbf{B}(s)\mathbf{D}(s)$ as $\mathbf{A}(s)$ allows the singular-value analysis of the

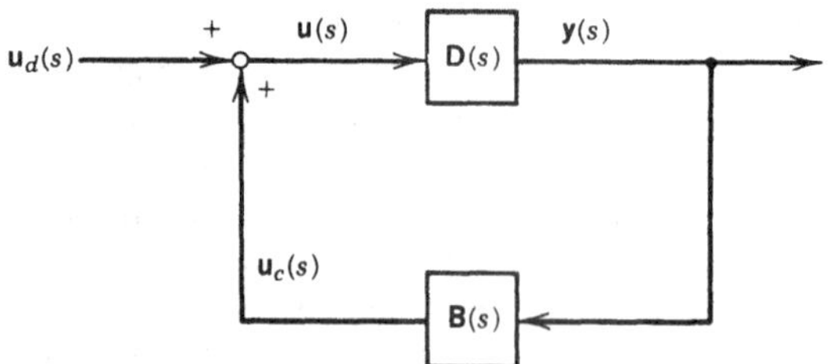

FIGURE 6.6-2 Simplified block diagram of the stochastic regulator (in the frequency domain).

preceding section to be applied here. Guaranteed gain and phase margins then are dependent on the relative magnitudes of the maximum singular value of the system perturbation and the minimum singular value of the return difference matrix.

While it is possible to attain good stability margins with the LQG regulator, it is not possible to do so with the global guarantees obtained for the LQ regulator. This was demonstrated by Doyle (D-7) with a counter-example in which the gain margin became vanishingly small as control and estimator gains became large. Consider the unstable single-input/single-output system

$$\begin{bmatrix} \dot{x}_1 \\ \dot{x}_2 \end{bmatrix} = \begin{bmatrix} 1 & 1 \\ 0 & 1 \end{bmatrix} \begin{bmatrix} x_1 \\ x_2 \end{bmatrix} + \begin{bmatrix} 0 \\ 1 \end{bmatrix} u + \begin{bmatrix} 1 \\ 1 \end{bmatrix} w \tag{6.6-14}$$

$$z = [1 \quad 0] \begin{bmatrix} x_1 \\ x_2 \end{bmatrix} + n \tag{6.6-15}$$

with controller cost function matrices

$$\mathbf{Q} = Q \begin{bmatrix} 1 & 1 \\ 1 & 1 \end{bmatrix} \tag{6.6-16}$$

$$\mathbf{R} = R = 1 \tag{6.6-17}$$

and disturbance and measurement-error spectral density matrices

$$\mathbf{W} = W \begin{bmatrix} 1 & 1 \\ 1 & 1 \end{bmatrix} \tag{6.6-18}$$

$$\mathbf{N} = 1 \tag{6.6-19}$$

The algebraic Riccati equation for control (Eq. 6.1-5) has the steady-state, positive-definite solution

$$s_{12} = s_{22} = 2 + \sqrt{4 + Q} \tag{6.6-20}$$

$$s_{11} = \frac{s_{12}^2 - Q}{2} \tag{6.6-21}$$

and the corresponding control gain matrix is

$$\begin{aligned}\mathbf{C} &= R^{-1}\mathbf{G}^T\mathbf{S} = [s_{12} \quad s_{22}] = [c \quad c] \\ &= (2 + \sqrt{4 + Q})[1 \quad 1]\end{aligned} \tag{6.6-22}$$

The estimator is posed as the mathematical dual of the controller; hence the estimator gain matrix is

$$K = (2 + \sqrt{4 + W})\begin{bmatrix} 1 \\ 1 \end{bmatrix} = \begin{bmatrix} k \\ k \end{bmatrix} \tag{6.6-23}$$

Stability of the closed-loop system is determined by the coupled roots of Eqs. 6.6-2 and 6.6-3. Our concern here is the effect of control gain variations on overall stability; therefore, $\mathbf{F}_A = \mathbf{F}$ and $\mathbf{H}_A = \mathbf{H}$. Assuming that the actual control effect matrix $\mathbf{G}_A$ is $[0 \quad \mu]^T$, the roots are the solutions to the following equation:

$$\begin{vmatrix} (s-1) & -1 & 0 & 0 \\ 0 & (s-1) & \mu c & \mu c \\ -k & 0 & (s-1+k) & -1 \\ -k & 0 & (c+k) & (s-1+c) \end{vmatrix} = s^4 + c_3 s^3 + c_2 s^2 + c_1 s + c_0 = 0 \tag{6.6-24}$$

By Routh's criterion (Section 2.5), the coefficients of this characteristic equation must be positive if it is stable; however, Doyle showed that values of μ other than one can force coefficients to change sign. The last two terms of the polynomial are

$$c_1 = k + c - 4 + 2(\mu - 1)ck \tag{6.6-25}$$

$$c_0 = 1 + (1 - \mu)ck \tag{6.6-26}$$

Both c and k are positive; if μ is greater than $(1 + 1/ck)$, then c_0 is negative, and if μ is less than $[1 - (k + c - 4)/2ck]$, c_1 is negative; hence the gain margin becomes smaller as ck becomes larger. Increasing values of c and k are related to increasing Q and W, so robustness is enhanced by low weighting on the state in the cost function and/or low disturbance spectral density. For fixed c, increasing the value k increases ck, confirming the earlier statement that higher bandwidth does not necessarily improve robustness. The result also reflects the point made in Example 6.5-3: when control effects are uncertain, keeping control loop gains low (consistent with the need for closed-loop stability) improves robustness.

Robustness Recovery

The LQG regulator's robustness would be the same as that of the LQ regulator if the control vector $\mathbf{u}(t)$ had identical effect on the state $\mathbf{x}(t)$, and the state estimate $\hat{\mathbf{x}}(t)$, for then $\mathbf{Cx}(t)$ and $\mathbf{C}\hat{\mathbf{x}}(t)$ would produce the same expected value of control, $E[\hat{\mathbf{u}}(t)] = \mathbf{u}(t)$. Of course, the covariance of the control, $E\{[\mathbf{u}(t) - \hat{\mathbf{u}}(t)][\mathbf{u}(t) - \hat{\mathbf{u}}(t)]^T\}$, would not be the same, as $\hat{\mathbf{x}}(t)$ contains measurement errors and $\mathbf{x}(t)$ does not. However, the random components have no effect on the stability of the closed-loop linear system. It may be concluded that an estimator which perfectly restores the state's time-varying mean from the measurements $\mathbf{z}(t)$ restores the system's robustness to that obtainable by the LQ regulator.

There are practical limitations on this process. Estimator bandwidth is limited by allowable levels of control covariance, and perfect restoration of the mean response to control is not possible if the plant contains non-minimum-phase (right-half-plane) transmission zeros. Nevertheless, the procedure developed by Doyle and Stein (D-8) allows the performance-robustness trade-off to be made in a systematic way.

Assuming that the actual and assumed model matrices of Fig. 6.6-1 are identical, the transform relationship between the control and the actual state is

$$\mathbf{x}(s) = (s\mathbf{I}_n - \mathbf{F})^{-1}\mathbf{Gu}(s) \tag{6.6-27}$$

We want to identify the conditions under which the effect of $\mathbf{u}(s)$ on $\hat{\mathbf{x}}(s)$ is the same. The estimator's frequency-domain block diagram can be redrawn as in Fig. 6.6-3, where uncertain inputs have been ignored. With minor

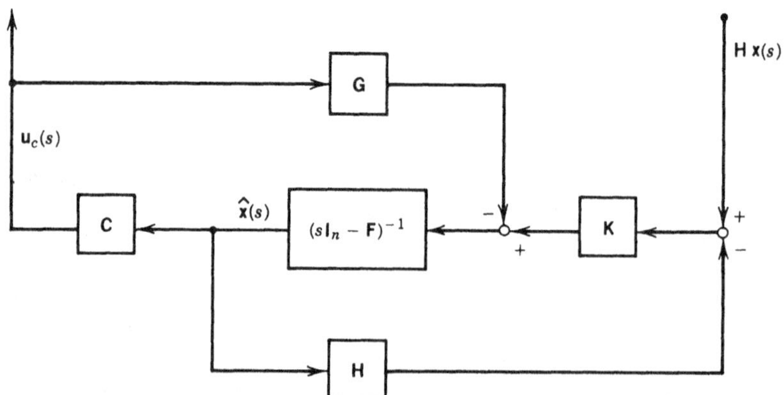

FIGURE 6.6-3 Block diagram of the state estimator.

manipulation, the state estimate is

$$\begin{aligned}\hat{\mathbf{x}}(s) &= [(s\mathbf{I}_n - \mathbf{F}) + \mathbf{KH}]^{-1}[\mathbf{KHx}(s) - \mathbf{Gu}_c(s)] \\ &= [(s\mathbf{I}_n - \mathbf{F}) + \mathbf{KH}]^{-1}[\mathbf{KH}(s\mathbf{I}_n - \mathbf{F})^{-1}\mathbf{Gu}(s) - \mathbf{Gu}_c(s)] \quad (6.6\text{-}28a)\end{aligned}$$

Denoting the characteristic matrix of **F** by **A**,

$$\mathbf{A}(s) = (s\mathbf{I}_n - \mathbf{F}) \quad (6.6\text{-}29)$$

the first term on the right of Eq. 6.6-28a can be manipulated as follows, assuming that the dimensions of the measurement and control are the same ($r = m$):

$$\begin{aligned}[\mathbf{A}(s) + \mathbf{KH}]^{-1} &= \{\mathbf{A}(s) - \mathbf{K}[\mathbf{HA}^{-1}(s)\mathbf{K} - \mathbf{I}_m + \mathbf{HA}^{-1}(s)\mathbf{K}]^{-1}\mathbf{H}\}^{-1} \\ &= \mathbf{A} - \mathbf{A}(\mathbf{A}^{-1}\mathbf{K})[(\mathbf{HA}^{-1})\mathbf{A}(\mathbf{A}^{-1}\mathbf{K}) - (\mathbf{I}_m + \mathbf{HA}^{-1}\mathbf{K})]^{-1}\mathbf{HA}^{-1}\mathbf{A} \quad (6.6\text{-}30a)\end{aligned}$$

Using the matrix inversion lemma (Eq. 2.2-67a), this becomes

$$[\mathbf{A} + \mathbf{KH}]^{-1} = \mathbf{A}^{-1} - \mathbf{A}^{-1}\mathbf{K}(\mathbf{I}_m + \mathbf{HA}^{-1}\mathbf{K})^{-1}\mathbf{HA}^{-1} \quad (6.6\text{-}30b)$$

and Eq. 6.6-28a evolves to the following:

$$\begin{aligned}\hat{\mathbf{x}}(s) &= [\mathbf{A}^{-1} - \mathbf{A}^{-1}\mathbf{K}(\mathbf{I}_m + \mathbf{HA}^{-1}\mathbf{K})^{-1}\mathbf{HA}^{-1}][\mathbf{KHA}^{-1}\mathbf{Gu}(s) - \mathbf{Gu}_c(s)] \\ &= [\mathbf{A}^{-1}\mathbf{K} - \mathbf{A}^{-1}\mathbf{K}(\mathbf{I}_m + \mathbf{HA}^{-1}\mathbf{K})^{-1}\mathbf{HA}^{-1}\mathbf{K}]\mathbf{HA}^{-1}\mathbf{Gu}(s) \\ &\quad - [\mathbf{A}^{-1} - \mathbf{A}^{-1}\mathbf{K}(\mathbf{I}_m + \mathbf{HA}^{-1}\mathbf{K})^{-1}\mathbf{HA}^{-1}]\mathbf{Gu}_c(s) \\ &= \mathbf{A}^{-1}\mathbf{K}(\mathbf{I}_m + \mathbf{HA}^{-1}\mathbf{K})^{-1}[\mathbf{I}_m + \mathbf{HA}^{-1}\mathbf{K} - \mathbf{HA}^{-1}\mathbf{K}]\mathbf{HA}^{-1}\mathbf{Gu}(s) \\ &\quad - [\mathbf{A}^{-1}\mathbf{G}(\mathbf{HA}^{-1}\mathbf{G})^{-1} - \mathbf{A}^{-1}\mathbf{K}(\mathbf{I}_m + \mathbf{HA}^{-1}\mathbf{K})^{-1}\mathbf{HA}^{-1}\mathbf{G}]\mathbf{u}_c(s) \\ &= \mathbf{A}^{-1}\mathbf{K}(\mathbf{I}_m + \mathbf{HA}^{-1}\mathbf{K})^{-1}\mathbf{HA}^{-1}\mathbf{Gu}(s) \\ &\quad - \mathbf{A}^{-1}[\mathbf{G}(\mathbf{HA}^{-1}\mathbf{G})^{-1} - \mathbf{K}(\mathbf{I}_m + \mathbf{HA}^{-1}\mathbf{K})^{-1}]\mathbf{HA}^{-1}\mathbf{Gu}_c(s) \quad (6.6\text{-}28b)\end{aligned}$$

If

$$\mathbf{G}(\mathbf{HA}^{-1}\mathbf{G})^{-1} = \mathbf{K}(\mathbf{I}_m + \mathbf{HA}^{-1}\mathbf{K})^{-1} \quad (6.6\text{-}31)$$

then $\mathbf{u}_c(s)$ has no effect on $\hat{\mathbf{x}}(s)$, and Eq. 6.6-28b becomes identical to Eq. 6.6-27:

$$\begin{aligned}\hat{\mathbf{x}}(s) &= \mathbf{A}^{-1}\mathbf{G}(\mathbf{HA}^{-1}\mathbf{G})^{-1}\mathbf{HA}^{-1}\mathbf{Gu}(s) \\ &= \mathbf{A}^{-1}\mathbf{Gu}(s) \\ &= (s\mathbf{I}_n - \mathbf{F})^{-1}\mathbf{Gu}(s) \quad (6.6\text{-}32)\end{aligned}$$

Equation 6.6-31 can be rewritten as

$$\mathbf{K}[\mathbf{I}_m + \mathbf{H}(s\mathbf{I}_n - \mathbf{F})^{-1}\mathbf{K}]^{-1} = \mathbf{G}[\mathbf{H}(s\mathbf{I}_n - \mathbf{F})^{-1}\mathbf{G}]^{-1} \tag{6.6-33}$$

Proper specification of the estimator's gain matrix provides the desired result. The poles of the estimator are the roots of the characteristic polynomial

$$\Delta_{\text{est}}(s) = |s\mathbf{I}_n - \mathbf{F}|\,|\mathbf{I}_m + \mathbf{H}(s\mathbf{I}_n - \mathbf{F})^{-1}\mathbf{K}| \tag{6.6-34}$$

If **K** were chosen as

$$\frac{\mathbf{K}}{\rho} = \mathbf{G} \tag{6.6-35}$$

then the estimator's characteristic polynomial would be given by

$$\Delta_{\text{est}}(s) = |s\mathbf{I}_n - \mathbf{F}||\mathbf{I}_m + \rho\mathbf{H}(s\mathbf{I}_n - \mathbf{F})^{-1}\mathbf{G}| \tag{6.6-36}$$

q of the estimator roots would approach the q transmission zeros of $\mathbf{H}(s\mathbf{I}_n - \mathbf{F})^{-1}\mathbf{G}$ as ρ approaches ∞, with the remaining $(n-q)$ roots going to ∞. Consequently, if $\mathbf{H}(s\mathbf{I}_n - \mathbf{F})^{-1}\mathbf{G}$ contains non-minimum-phase transmission zeros, the **K** satisfying Eq. 6.6-35 would produce instability.

It is not obvious that this **K** might be a Kalman–Bucy filter gain matrix; however, it is possible to identify a procedure by which the Kalman–Bucy filter gain approaches **K** in a limiting process. The disturbance spectral density matrix is increased by a positive-definite term, $\rho^2\mathbf{G}\mathbf{G}^T$, where ρ^2 approaches ∞ in the limit. Because the stability conditions are not violated by this addition, the estimator remains stable. It can be surmised that recovery of state-estimate response to control cannot be perfect in the non-minimum-phase case although it remains to be seen whether or not robustness is improved by the redesigned Kalman–Bucy filter. With minimum-phase transmission zeros, the limiting state-estimate response to control provides perfect recovery of the mean and deterministic LQ robustness.

The steady-state Kalman–Bucy filter gain is

$$\mathbf{K} = \mathbf{P}\mathbf{H}^T\mathbf{N}^{-1} \tag{6.6-37}$$

where **P** is the solution to an algebraic Riccati equation

$$\begin{aligned} \mathbf{0} &= \mathbf{F}\mathbf{P} + \mathbf{P}\mathbf{F}^T + \mathbf{W} - \mathbf{P}\mathbf{H}^T\mathbf{N}^{-1}\mathbf{H}\mathbf{P} \\ &= \mathbf{F}\mathbf{P} + \mathbf{P}\mathbf{F}^T + \mathbf{W} - \mathbf{K}\mathbf{N}\mathbf{K}^T \end{aligned} \tag{6.6-38}$$

and all matrices are defined as before. Suppose that the disturbance spectral density matrix is defined with additional "process noise"

$$\mathbf{W} = \mathbf{W}_o + \rho^2 \mathbf{G}\mathbf{G}^T \tag{6.6-39}$$

and that the measurement spectral density matrix $\mathbf{N}$ is taken as the identity matrix. (A more general $\mathbf{N}$ can be accommodated, but the assumption simplifies the discussion.) Substituting Eq. 6.6-39 in Eq. 6.6-38 and dividing by ρ^2

$$\mathbf{0} = \frac{\mathbf{FP}}{\rho^2} + \frac{\mathbf{PF}^T}{\rho^2} + \frac{\mathbf{W}_o}{\rho^2} + \mathbf{GG}^T - \frac{\mathbf{KK}^T}{\rho^2} \tag{6.6-40}$$

As ρ^2 grows very large,

$$\mathbf{KK}^T \rightarrow \rho^2 \mathbf{GG}^T \tag{6.6-41a}$$

or

$$\mathbf{K} \rightarrow \rho \mathbf{G} \tag{6.6-41b}$$

assuming that $\mathbf{K}$ and $\mathbf{G}$ are dimensionally equivalent. Thus the procedure approaches the desired result (Eq. 6.6-35) asymptotically, maintaining estimator stability but increasing measurement error transmission. It may be inferred that the LQG system is most robust when the assumed disturbance input and control input have similar effects on the system (i.e., when $\mathbf{G}$ and $\mathbf{L}$ are the same in Equation 6.6-2).

A dual approach to robustness recovery has been presented by Kwakernaak (K-9), in which the estimator is developed as a dynamic feedback compensator. The design is completed from $\hat{\mathbf{x}}$ to $\mathbf{u}$ using an LQ regulator whose control-weighting matrix is proportional to the disturbance spectral density matrix with a vanishingly small multiplier.

Probability of Instability

Given the probability distributions of uncertain parameters in $\mathbf{F}$, $\mathbf{G}$, and $\mathbf{H}$, it is possible to compute the probability that the open-loop system or any corresponding closed-loop system will become unstable. Efficient algorithms that relate the *probability of instability* to control system design have yet to be developed; however, the concept provides an analytical framework for defining the *stochastic robustness* of controlled systems, with a direct scalar measure of the effects of arbitrary parameter variations.

The plant, estimator, and regulator of Fig. 6.6-1 can be aggregated in a single $2n$-dimensional equation of the form

$$\dot{\mathbf{x}}(t) = \mathbf{A}(\mathbf{p})\mathbf{x}(t) + \mathbf{u}_d(t) \tag{6.6-42}$$

with stability determined by the eigenvalues of $\mathbf{A}(\mathbf{p})$:

$$|\lambda_i \mathbf{I}_{2n} - \mathbf{A}(\mathbf{p})| = 0, \qquad i = 1\text{–}2n \tag{6.6-43}$$

The constant-but-uncertain parameters of the system are contained in the l-vector $\mathbf{p}$, which is described by the probability density function pr($\mathbf{p}$) as in Section 2.4. Although the relationship is complex, there is a multivariate probability density function of the $2n$ eigenvalues pr($\boldsymbol{\lambda}$) resulting from pr($\mathbf{p}$) (Fig. 6.6-4). Of more particular concern is pr[Re($\boldsymbol{\lambda}$)] = pr($\boldsymbol{\sigma}$), where $\lambda_i = \sigma_i + j\omega_i$, as the probability of right-half-plane eigenvalues is to be evaluated,

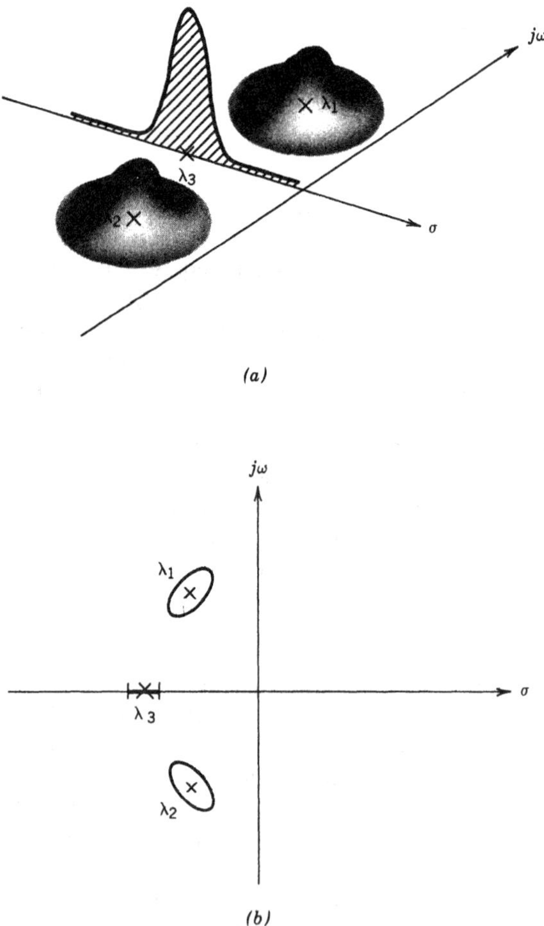

FIGURE 6.6-4 Effects of parameter uncertainty on eigenvalue location. (a) Qualitative illustration of joint probability distribution (λ_3 confined to the real axis); (b) "one-sigma" contours for eigenvalue distribution (λ_3 confined to the real axis).

subject to

$$\int_{-\infty}^{0} \operatorname{pr}(\boldsymbol{\sigma})\, d\boldsymbol{\sigma} + \int_{0}^{\infty} \operatorname{pr}(\boldsymbol{\sigma})\, d\boldsymbol{\sigma} = 1 \qquad (6.6\text{-}44\text{a})$$

or

$$\text{Probability(stability)} + \text{Probability(instability)} = 1 \qquad (6.6\text{-}44\text{b})$$

A numerical estimation of these probabilities can be made using "Monte Carlo" computation of the eigenvalues. Prior to evaluating Eq. 6.6-44, each element of $\mathbf{p}$ is computed by a random number generator, with appropriate shaping of the probability distribution. The eigenvalues are calculated, and if one or more have positive real parts, the case is declared unstable. After a sufficient number of evaluations, an estimate of the probability of instability can be made from these ensemble statistics. A trivial addition is to evaluate various degrees of stability or instability in the process, as in (S-13). For example, the "time to half amplitude" or "time to double amplitude" often is as important as the boundary between stability and instability.

If the parameter statistics are Gaussian, a less accurate but computationally simpler estimate can be based on the linear sensitivity of the eigenvalues to the parameters at the nominal operating point $\boldsymbol{\lambda}(\mathbf{p})$ specified by the mean value of the parameter vector $\mathbf{p}$. This approach neglects possible interactions between roots, such as the breakaway of coalescing real roots to form a complex pair. Defining the linear sensitivity of the eigenvalues' real parts as the $(2n \times l)$ matrix,

$$\mathbf{S} = \left.\frac{\partial \boldsymbol{\sigma}}{\partial \mathbf{p}}\right|_{\mathbf{p}=\bar{\mathbf{p}}} \qquad (6.6\text{-}45)$$

and the covariance of parameter variations as the symmetric $(l \times l)$ matrix

$$\mathbf{P} = E[(\mathbf{p} - \bar{\mathbf{p}})(\mathbf{p} - \bar{\mathbf{p}})^T] \qquad (6.6\text{-}46)$$

the covariance of $\operatorname{Re}(\boldsymbol{\lambda})$ can be approximated by

$$\boldsymbol{\Sigma} = E[(\boldsymbol{\sigma} - \bar{\boldsymbol{\sigma}})(\boldsymbol{\sigma} - \bar{\boldsymbol{\sigma}})^T] = \mathbf{SPS}^T \qquad (6.6\text{-}47)$$

The corresponding probability density function estimate for the $2n$ components of $\boldsymbol{\sigma}$ is

$$\operatorname{pr}(\boldsymbol{\sigma}) = \frac{1}{(2\pi)^n |\boldsymbol{\Sigma}|^{1/2}}\, e^{-(1/2)(\boldsymbol{\sigma}-\bar{\boldsymbol{\sigma}})^T \boldsymbol{\Sigma}^{-1} (\boldsymbol{\sigma}-\bar{\boldsymbol{\sigma}})} \qquad (6.6\text{-}48)$$

allowing Eq. 6.6-44 to be evaluated.

The probability of instability could be minimized with respect to the control system parameters by static minimization (Section 2.1). This, in turn, requires repeated evaluation of the probability of instability using either the Monte Carlo or linear-Gaussian approach. To integrate this procedure with LQG design algorithms, the control parameters used in minimization would be elements of $\mathbf{Q}$, $\mathbf{R}$, $\mathbf{W}$, and $\mathbf{N}$; hence each $\text{pr}(\boldsymbol{\sigma})$ would be associated with a fixed set of regulator and estimator gains, $\mathbf{C}$ and $\mathbf{K}$, that were optimal and stabilizing for $[\mathbf{F}(\bar{\mathbf{p}}), \mathbf{G}(\bar{\mathbf{p}}), \mathbf{H}(\bar{\mathbf{p}})]$.

6.7 FOOTNOTE ON ADAPTIVE CONTROL

It is appropriate to reiterate the many possibilities for designing adaptive controllers based on linear-quadratic-Gaussian control theory. The LQG regulator has evolved as a simplification of control logic for stochastic, nonlinear, time-varying systems and if parameter variations are likely to cause significant stability and performance variations, we need only "back up" to reconsider concepts presented in earlier chapters. For example, neighboring-optimal control is adaptive to programmed variations in parameters while dual control is both optimal and adaptive to evolving parameter variations. The present state of digital computation allows control logic that is somewhat more sophisticated than the constant-coefficient controllers of this chapter to be implemented in "real time." In an adaptive system, robustness has both a different meaning and a different function as the controller adjusts to keep the difference between actual and assumed system characteristics small.

While there are numerous ways of categorizing adaptive controllers, for present purposes we may identify two principal types: those whose gains and controller structures are scheduled according to exogenous (or external) variables and those which adapt as a consequence of changes in endogenous (or internal) variables. Exogenous variables include dynamic variables that are not formally contained in the state vector of the plant model as well as time or any other independent variable of the problem. For example, the gains of a satellite's attitude control system might vary with position in orbit, while the gains of a missile autopilot could vary with time- or range-to-go to target impact. Endogenous variables are synonymous with state and measurement vectors $\mathbf{x}(t)$ and $\mathbf{z}(t)$. We may estimate the parameters of the system recursively, as in Sections 4.6 and 4.7, and change the controller accordingly. For example, following sensed failures, alternate pre-computed gain sets could be brought "on line".

Many factors must be considered in real-time implementation, and the right answer for a given control problem depends upon availability and reliability of control resources. Perhaps the most important factors are the trade-offs between data storage and adaptation algorithm complexity, protection against sensor and actuator failures, and speed of adaptation. One must choose between storing a multitude of precomputed gains for a range of operating points or computing new gains on-line. For many systems, it is more practical to protect against component failures by storing alternate control strategies than by relying on constant-coefficient robustness, and it is virtually certain that proper adaptation can provide better performance following identified failures. The rate at which adaptation must occur is problem-dependent. It is clear that slower or less frequent adaptation provides additional time for executing more sophisticated logic, but rapidly varying parameters may require fast adaptation.

As long as the dynamic effects of parameter variations are slow by comparison to state variations, control design can be based on an ensemble of time-invariant dynamic models. Fast parameters may be indistinguishable from state components, in which case the parameters should be included in an augmented state vector for estimation and possibly for control.

PROBLEMS

Section 6.1

1. Use the Kalman–Englar method to solve the algebraic Riccati equation for the problem defined in Problem 5 of Section 5.4.

2. Use Kleinman's method to solve the algebraic Riccati equation for a continuous-time problem described by the following matrices:

$$\mathbf{F} = \begin{bmatrix} 1 & 1 \\ 0 & -1 \end{bmatrix} \quad \mathbf{G} = \begin{bmatrix} 0.5 \\ 1 \end{bmatrix} \quad \mathbf{Q} = \begin{bmatrix} 1 & 0 \\ 0 & 1 \end{bmatrix} \quad R = 1$$

3. Use the MacFarlane–Potter method to solve Problem 2. Compute the control gains and compare the eigenvalues for the open- and closed-loop systems.

4. Repeat Problem 3 with $R = 0.1$ and 10. Discuss the trends in your results.

5. Use the discrete-time Kalman–Englar method to find the algebraic Riccati equation solution for the following:

$$\mathbf{\Phi} = \begin{bmatrix} 0.8 & 0 \\ 0.1 & 0.5 \end{bmatrix} \quad \mathbf{\Gamma} = \begin{bmatrix} 1 \\ 1 \end{bmatrix} \quad \mathbf{Q} = \begin{bmatrix} 1 & 0 \\ 0 & 1 \end{bmatrix} \quad R = 10$$

Section 6.2

1. A second-order robot arm is controlled by a first-order actuator, resulting in the following mathematical model:

$$\begin{bmatrix} \dot{x}_1(t) \\ \dot{x}_2(t) \\ \dot{x}_3(t) \end{bmatrix} = \begin{bmatrix} 0 & 1 & 0 \\ -\omega_n^2 & -2\zeta\omega_n & \omega_n^2 \\ 0 & 0 & -1/\tau \end{bmatrix} \begin{bmatrix} x_1(t) \\ x_2(t) \\ x_3(t) \end{bmatrix} + \begin{bmatrix} 0 \\ 0 \\ 1/\tau \end{bmatrix} u(t)$$

$$y = x_1$$

   (a) Compute the matrices $\mathbf{S}_{11}$, $\mathbf{S}_{12}$, $\mathbf{S}_{21}$, and $\mathbf{S}_{22}$ that relate the steady-state vector to a constant control input.

   (b) Using the algebraic Riccati equation solution of your choice, compute the gain matrix, $\mathbf{C}$, given

$$\mathbf{Q} = \begin{bmatrix} 1 & 0 & 0 \\ 0 & 1 & 0 \\ 0 & 0 & 1 \end{bmatrix} \qquad R = 1 \qquad \zeta = 0.3 \qquad \omega_n = \pi \qquad \tau = 1$$

   (c) Compute $\mathbf{C}_F$ and $\mathbf{C}_B$ for a command-response controller.

   (d) Compute the response to a unit step command input.

2. Determine the feedback gains and closed-loop eigenvalues in Problem 1 for $R = 0.01$, 1, and 100.

3. A fourth-order model of an aircraft's longitudinal motions can use axial and normal velocities, pitch rate, and pitch angle as state vector components and elevator and throttle as control vector components. The corresponding system matrices for a high-performance aircraft flying at an altitude of 20,000 ft and airspeed of 700 ft/s are as follows:

$$\mathbf{F} = \begin{bmatrix} -0.04 & 0.1 & 54.3 & -32.1 \\ -0.02 & -0.87 & 698. & -2.62 \\ 0 & 0.003 & -0.74 & 0 \\ 0 & 0 & 1 & 0 \end{bmatrix}$$

$$\mathbf{G} = \begin{bmatrix} 0 & 14.5 \\ 0 & 0 \\ -7.22 & 0 \\ 0 & 0 \end{bmatrix}$$

The desired output vector, which is equivalent to the commanded input vector, takes the general form

$$\mathbf{y} = \mathbf{H}_x\mathbf{x} + \mathbf{H}_u\mathbf{u}$$

(a) Compute the steady-state state response to a unit elevator input and a unit throttle input (separately).

Compute the static equilibrium responses in $\mathbf{x}$ and $\mathbf{u}$ to separate constant command inputs assuming that

(b) $\mathbf{H}_x = \mathbf{0}, \quad \mathbf{H}_u = \begin{bmatrix} 1 & 0 \\ 0 & 1 \end{bmatrix}$

(c) $\mathbf{H}_x = \begin{bmatrix} 0 & 0 & 0 & 1 \\ 1 & 0 & 0 & 0 \end{bmatrix}, \quad \mathbf{H}_u = \mathbf{0}$

(d) $\mathbf{H}_x = \begin{bmatrix} 0 & 0 & 0 & 1 \\ 0 & 0 & 0 & 0 \end{bmatrix}, \quad \mathbf{H}_u = \begin{bmatrix} 0 & 0 \\ 0 & 1 \end{bmatrix}$

(e) Compute the quasi-static equilibrium responses at $t = 0$, 1 and 2 s for the following definition of the command input matrices:

$$\mathbf{H}_x = \begin{bmatrix} 0 & 0 & 1 & 0 \\ 1 & 0 & 0 & 0 \end{bmatrix}, \quad \mathbf{H}_u = \mathbf{0}$$

4. Design a command-response controller for Problem 3C using the following weighting matrices:

$$\mathbf{Q} = \text{Diag}[1, \quad 1, \quad 1, \quad 1]$$
$$\mathbf{R} = \text{Diag}[100, \quad 1]$$

Section 6.3

1. Examine the effects of cost function weighting on the initial-condition response and eigenvalues of the aircraft model given in Problem 3 of Section 6.2. The nominal values of $\mathbf{Q}$ and $\mathbf{R}$ are as follows:

$$\mathbf{Q} = \text{Diag}[1, 1, 1, 1]$$
$$\mathbf{R} = \text{Diag}[100, 1]$$

(a) Individually increase and decrease the elements of $\mathbf{Q}$ by factors of 10. Compute the effects of these modifications on the closed-loop initial-condition response. For variations in q_{ii}, choose the initial condition as $x_i(0) = 1$, $x_j = 0$ $(j \neq i)$.

(b) Individually and jointly increase and decrease the elements of $\mathbf{R}$ by factors of 10. Compute and plot the closed-loop eigenvalues.

2. Examine the effects of state-rate weighting on the model of Problem (1). Let the nominal weighting matrices be

$$\mathbf{Q}_{\dot{x}} = \text{Diag}[1, 1, 1, 1]$$

$$\mathbf{R}_o = \text{Diag}[0.1, 0.1]$$

Individually increase and decrease the elements of $\mathbf{Q}_{\dot{x}}$ as in (1a). Compute the effects of these changes on the closed-loop initial-condition responses and eigenvalues.

3. Examine the effects of implicit model-following on the robot system described in Problem 1 of Section 6.2. Assume that the model is described by $\zeta = 0.707$, $\omega_n = 2\pi$, and $\tau = 0.2$, and that $\mathbf{Q}$ and $\mathbf{R}_o$ are identity matrices.

4. Design a proportional-integral controller for the system used in Problem 3, assuming that the command input is the desired value of x_1 and the integral state is $\xi = \int x_1\, dt$. Assume that $\mathbf{Q}$ and $\mathbf{R}$ nominally are identity matrices.
   (a) Examine the effects of varying integral-state weighting on eigenvalues and initial-condition responses.
   (b) Compute the command response of the system with and without constant external disturbance on $\dot{x}_2(t)$.

5. Design a proportional-filter compensator for the system used in (3). Assume that $\mathbf{Q}$ and $\mathbf{R}$ nominally are identity matrices.
   (a) Examine the effects of varying control-rate weighting on eigenvalues and initial-condition responses.
   (b) Compute the command response of the system with and without constant external disturbance on $\dot{x}_2(t)$.

6. Design an explicit-model-following controller for the system used in (3), assuming that the desired model is as specified there.
   (a) Examine the effects of $\mathbf{Q}$ and $\mathbf{R}$ variations on eigenvalues and initial-condition responses.
   (b) Compute the command response of the system with and without constant external disturbance on $\dot{x}_2(t)$.

7. The actuator modeled in Problem 1 of Section 6.2 is stable and is faster than the rigid-body motion of the robot arm; therefore, the actuator is a candidate for elimination in producing a reduced-order controller for the system.
   (a) Repeat Problem 1 of Section 6.2, assuming that actuator dynamics have been eliminated by truncation.
   (b) Would the results be different if you had used residualization rather than truncation?

8. Derive the implicit-model-following control equations for the case in which the system and the model do not have the same dimensions. What requirements must be placed on relative dimension, controllability, and cost observability of the two systems?

9. The robot arm of Problem 3 is required to track a first-order Gauss–Markov process with a correlation time of 1 sec and a unit-spectral-density white random input. Design and simulate the tracking controller, noting the effects of varying **Q** and **R** on the solution.

Section 6.4

1. Angular rate perturbations of a spinning axisymmetric satellite can be modeled as

$$\begin{bmatrix} \dot{x}_1(t) \\ \dot{x}_2(t) \end{bmatrix} = \begin{bmatrix} 0 & an \\ -an & 0 \end{bmatrix} \begin{bmatrix} x_1(t) \\ x_2(t) \end{bmatrix} + \begin{bmatrix} 0 \\ b \end{bmatrix} u(t)$$

where $x_1(t)$ and $x_2(t)$ are pitch and yaw rates, respectively, a is a ratio of moments of inertia, and n is the nominal spin rate. It can be seen that the satellite has neutral stability without the effect of control, and it is assumed that a single control moment acts about the yaw axis only.

(a) Given the cost function,

$$J = \frac{1}{2} \int_0^\infty [Qx_1^2(T) + Ru^2(t)]\, dt$$

write the cost transfer function of the system, and determine whether or not there are any transmission zeros.

(b) Plot the system's root locus as R varies from 0 to ∞.

(c) Redefine the cost function as,

$$J = \frac{1}{2} \int_0^\infty [Qx_2^2(t) + Ru^2(t)]\, dt$$

and repeat (a) and (b).

2. Find the transmission zeros and root locus for the robot problem (1) of Section 6.2, assuming

(a) $\mathbf{Q} = \text{Diag}[1, 0, 0] \quad 0 < R < \infty$

(b) $\mathbf{Q} = \text{Diag}[0, 1, 0] \quad 0 < R < \infty$

3. Determine the eigenvectors for Problem 2, given $R = 0.01$, 1, and 100.

4. Feedback for the robot arm is to be generated by a Kalman–Bucy filter

with direct measurement of $x_1(t)$, and

$$\mathbf{W} = \text{Diag}[0, 100, 1] \qquad N = 1$$

As in Problem 1 of Section 6.2,

$$\mathbf{Q} = \text{Diag}[1, 1, 1] \qquad R = 1$$

(a) Compute the eigenvalues and eigenvectors of the closed-loop system.

(b) Holding the control gains, estimation gains, and estimation model fixed, vary ζ between -0.3 and 0.6. Evaluate the resulting changes in closed-loop eigenvalues and eigenvectors. How might $\mathbf{Q}$, R, $\mathbf{W}$, and/or N be modified to desensitize the closed-loop system to these mismatches?

5. A discrete-time linear-quadratic regulator is to be designed for the system

$$\begin{bmatrix} x_1 \\ x_2 \\ x_3 \end{bmatrix}_{k+1} = \begin{bmatrix} 1 & 0.02 & 0 \\ -0.2 & 0.98 & 0.2 \\ 0 & 0 & 0.9 \end{bmatrix} \begin{bmatrix} x_1 \\ x_2 \\ x_3 \end{bmatrix}_k + \begin{bmatrix} 0 \\ 0 \\ 0.1 \end{bmatrix} u_k$$

which is a first-order approximation to the robot problem with a sampling interval of 0.02 sec. Find the transmission zeros and root locus, assuming

(a) $\mathbf{Q} = \text{Diag}[1, 0, 0] \qquad 0 < R < \infty$

(b) $\mathbf{Q} = \text{Diag}[0, 1, 0] \qquad 0 < R < \infty$

Section 6.5

1. The single-degree-of-freedom robot manipulator with first-order actuator is modeled as

$$\begin{bmatrix} \dot{x}_1(t) \\ \dot{x}_2(t) \\ \dot{x}_3(t) \end{bmatrix} = \begin{bmatrix} 0 & 1 & 0 \\ -\omega_n^2 & -2\zeta\omega_n & \omega_n^2 \\ 0 & 0 & -1/\tau \end{bmatrix} \begin{bmatrix} x_1(t) \\ x_2(t) \\ x_3(t) \end{bmatrix} + \begin{bmatrix} 0 \\ 0 \\ 1/\tau \end{bmatrix} [u(t) + u_c(t)] + \begin{bmatrix} 0 \\ a \\ b \end{bmatrix} w(t)$$

and a full-state feedback controller provides

$$u(t) = -[c_1 \quad c_2 \quad c_3] \begin{bmatrix} x_1(t) \\ x_2(t) \\ x_3(t) \end{bmatrix}$$

(a) Express the return difference function matrix for this system.

(b) Referring to Figure 6.5-1, express $\boldsymbol{\delta}(s)$, $\boldsymbol{\epsilon}(s)$, $\mathbf{x}(s)$, and $\mathbf{u}(s)$ as functions of the command input, $\mathbf{u}_c(s)$. Sketch the corresponding Bode plots.

2. Use spectral factorization to determine the optimal control gains for (1), with $\zeta = 0.3$, $\omega_n = \pi$, $\tau = 0.2$, $R = 1$, and

(a) $\mathbf{Q} = \text{Diag}[1, 0, 0]$

(b) $\mathbf{Q} = \text{Diag}[0, 1, 0]$

(c) Determine the gain and phase margins in (a) and (b). What is the corresponding allowable pure time delay?

3. Compute the maximum and minimum singular values of the return difference function matrix as functions of frequency for Problems 2a and 2b. How large could additive and multiplicative perturbations in $\mathbf{A}(s)$ be without destabilizing the system?

4. Apply the multivariable Nyquist stability criterion to the aircraft dynamic model presented in Problem 3 of Section 6.2, assuming that a linear-quadratic regulator is to be designed with

$$\mathbf{Q} = \text{Diag}[1, 0, 1, 0]$$

$$\mathbf{R} = \text{Diag}[100, 1]$$

5. Compute the maximum and minimum singular values of the return difference function matrix as functions of frequency for (4). How large could additive and multiplicative perturbations in $\mathbf{A}(s)$ be without destabilizing the system?

6. Apply the robustness recovery procedure described in Section 6.5 to Problem 4. What effects does varying ρ have on the nominal eigenvalues and the stability margins of the system?

7. Determine the gain margins of the discrete-time problem described in Problem 5 of Section 6.4.

Section 6.6

1. Apply Doyle's and Stein's robustness recovery technique to the problem posed by Doyle (Eq. 6.6-14 to 6.6-19). Determine how much process noise must be added to W to stabilize the system when $\mu = 0.5$ and 1.5.

2. Compute maximum and minimum singular values for the return difference function matrices (Eq. 6.6-13) of the LQG regulators defined in Problem 1. How do the singular values indicate the effects of robustness recovery?

3. Apply the robustness recovery technique for an unstable solution of Problem 4, Section 6.4. Evaluate your design using singular value analysis.

4. Assume that the robot arm's damping ratio takes its nominal value. The purpose of this problem is to compare the nominal characteristics of the original controller design and that resulting from Problem 3.
   (a) Compute the eigenvalues of the two systems.
   (b) Compute the initial-condition response of the two systems, assuming that disturbance inputs and measurement errors have been "turned off" for the simulation.
   (c) Compute the response of the two systems with nominal disturbances and measurement errors "turned on."

5. Compute the probability of instability for the robot arm in Problem 4, Section 6.4. Assume that the mean values of ζ, ω_n, and τ are those given in the problem, while their standard deviations are 0.3, $\pi/2$, and 0.1, respectively. Use Monte Carlo computation of the eigenvalues (500 samples) and compare these results to those of the simplified approach suggested in Section 6.6.

REFERENCES

A-1 Anderson, B.D.O. and Moore, J. B., *Linear Optimal Control*, Prentice-Hall, Englewood Cliffs, NJ, 1971.

A-2 Åström, K. J. and Wittenmark, B., *Computer Controlled Systems*, Prentice-Hall, Englewood Cliffs, NJ, 1984.

A-3 Atzhorn, D. and Stengel, R. F., Design and Flight Test of a Lateral-Directional Command Augmentation System, *AIAA Journal of Guidance, Control, and Dynamics*, **7**(3), 361–368, May–June 1984.

B-1 Bartels, R. H. and Stewart, G. W., Solution of the Matrix Equation $AX + XB = C$, *Communications of the ACM*, **15**, 820–826, September 1972.

B-2 Berman, H. and Gran, R., Design Principles for Digital Autopilot Synthesis, *AIAA Journal of Aircraft*, **11**(7), 414–422, July 1974.

B-3 Bode, H. W., *Network Analysis and Feedback Amplifier Design*, Van Nostrand, Princeton, NJ, 1945.

B-4 Bodson, M. and Athans, M., Multivariable Control of VTOL Aircraft for Shipboard Landing, *Proceedings of the AIAA Guidance, Navigation and Control Conference*, Snowmass, Colorado, August 1985, pp. 473–481.

B-5 Bogen, R., et al. *MACSYMA Reference Manual*, Vols. I & II, MIT Laboratory for Computer Science, Cambridge, 1983.

B-6 Bryson, A. E., Jr., Control Theory for Random Systems, *Proceedings of the Thirteenth International Congress on Theoretical and Applied Mechanics*, Springer-Verlag, Berlin, 1973, pp. 1–19.

B-7 Bryson, A. E., Jr., Some Connections Between Modern and Classical Control Concepts, *ASME Journal of Dynamic Systems, Measurement, and Control*, **101**, 91–98, June 1979.

B-8 Bryson, A. E., Jr. and Ho, Y. C., *Applied Optimal Control*, Hemisphere Publishing, Washington, DC, 1975.

B-9 Butterworth, S., On the Theory of Filter Amplifiers, *Wireless Engineer*, **7**, 536–541, 1930.

C-1 Chang, S.S.L., *Synthesis of Optimal Control Systems*, McGraw-Hill, New York, 1961.

C-2 Cooper, D. J. and Graham, A., Errors in State Variable Reconstruction When an Observer Is Subject to Constant Input Disturbances and Measurement Noise, *IEEE Transactions on Automatic Control*, **AC-22**(1), 121–123, February 1977.

D-1 Davison, E. J., The Steady-State Invertibility and Feedforward Control of Linear Time-Invariant Systems, *IEEE Transactions on Automatic Control*, **AC-211**(4), 529–534, August 1976.

D-2 Davison, E. J. and Chow, S. G., Perfect Control in Linear Time-Invariant Multivariable Systems: The Control Inequality Principal, in *Control System Design by Pole-Zero Placement*, F. Fallside, ed., Academic Press, London, 1977, pp. 1–15.

D-3 Davison, E. J. and Wang, S. H., Properties and Calculation of Transmission Zeros of Linear Multivariable Systems, in D-2, pp. 16–42.

D-4 DiPietro, R. C. and Farrar, F. A., Comparative Evaluation of Numerical Methods for Solving the Algebraic Riccati Equation, in *Proceedings of the 1977 Joint Automatic Control Conference*, San Francisco, June 1977.

D-5 Dongarra, J. J., Moler, C. B., Bunch, J. R., and Stewart, G. W., *LINPACK User's Guide*, SIAM, Philadelphia, 1980.

D-6 Dorato, P. and Levis, A. H., Optimal Linear Regulators: The Discrete-Time Case, *IEEE Transactions on Automatic Control*, **AC-16**(6), 613–620, December 1971.

D-7 Doyle, J. C., Guaranteed Margins for LQG Regulators, *IEEE Transactions on Automatic Control*, **AC-23**(4), 756–757, August 1978.

D-8 Doyle, J. C. and Stein, G., Robustness with Observers, *IEEE Transactions on Automatic Control*, **AC-24**(4), 607–611, August 1979.

D-9 Doyle, J. C. and Stein, G., Multivariable Feedback Design: Concepts for a Classical/Modern Synthesis, *IEEE Transactions on Automatic Control*, **AC-26**(1), 4–16, February 1981.

E-1 Erzberger, H., Analysis and Design of Model Following Control Systems by State Space Techniques, *Proceedings of the 1968 Joint Automatic Control Conference*, June 1968, pp. 572–581.

F-1 Foxgrover, J. A., Design and Flight Test of a Digital Flight Control System for General Aviation Aircraft, Princeton University M.S.E. Thesis, MAE Report 1559-T, June 1982.

G-1 Golub, G. H., Nash, S., and Van Loan, C., A Hessenburg-Schur Method for the Problem $AX + XB = C$, *IEEE Transactions on Automatic Control*, **AC-24**(6), 909–913, December 1979.

G-2 Golub, G. H. and Van Loan, C. F., *Matrix Computations*, Johns Hopkins University Press, Baltimore, 1983.

G-3 Graham, D. and Lathrop, R. C., The Synthesis of "Optimum" Transient Response: Criteria and Standard Forms, *Transactions of the AIEE*, **72**, 273–288, 1953 [also included in (O-1)].

H-1 Hanson, G. D. and Stengel, R. F., Effects of Displacement and Rate Saturation on the Control of Statically Unstable Aircraft, *AIAA Journal of Guidance, Control, and Dynamics*, **7**(2), 197–205, March–April 1984.

H-2 Harvey, C. A., Safonov, M. G., Stein, G., and Doyle, J. C., Optimal Linear Control, Office of Naval Research Report ONR CR215-238-4F, August 1979.

H-3 Harvey, C. A. and Stein, G., Quadratic Weights for Asymptotic Regulator Properties, *IEEE Transactions on Automatic Control*, **AC-23**(3), 378–387, June 1978.

H-4 Harvey, C. A. and Pope, R. E., Design Techniques for Multivariable Flight Control Systems, *Theory and Applications of Optimal Control in Aerospace Systems*, **AGARD-AG-251**, 5-1–5-33, July 1981.

H-5 Hewer, G. A., An Iterative Technique for the Computation of the Steady State Gains for the Discrete Optimal Regulator, *IEEE Transactions on Automatic Control*, **AC-16**(4), 382–384, August 1971.

H-6 Holley, W. E., and Bryson, A. E., Jr., Multi-Input, Multi-Output Regulator Design for Constant Disturbances and Non-Zero Inputs with Application to Automatic Landing in a Crosswind, Stanford University SUDAAR No. 465, August 1973.

J-1 Jacobson, D. H., Martin, D. H., Pachter, M., and Geveci, T., *Extensions of Linear-Quadratic Control Theory*, Springer-Verlag, New York, 1980.

J-2 Jameson, A., Optimization of Linear Systems of Constrained Configurations, *International Journal of Control*, **11**(3), 409–421, 1970.

J-3 Jameson, A. and Rothschild, D., A Direct Approach to the Design of Asymptotically Optimal Controllers, *International Journal of Control*, **13**(6), 1041–1050, 1971.

J-4 Johnson, C. D., State-Variable Design Methods May Produce Unstable Feedback Controllers, *International Journal of Control*, **29**(4), 607–619, April 1979.

K-1 Kailath, T., *Linear Systems*, Prentice-Hall, Englewood Cliffs, NJ, 1980.

K-2 Kalman, R. E., When Is a Linear Control System Optimal?, *Transactions of the ASME*, *Journal of Basic Engineering*, **86**, 51–60, March 1964.

K-3 Kalman, R. E. and Englar, T. S., A User's Manual for the Automatic Synthesis Program, NASA CR-475, Washington, DC, 1966.

K-4 Kimura, H., Pole Assignment by Gain Output Feedback, *IEEE Transactions on Automatic Control*, **AC-20**(4), 509–516, August 1975.

K-5 Kleinman, D. L., On an Iterative Method for Riccati Equation Computations, *IEEE Transactions on Automatic Control*, **AC-13**(1), 114–115, February 1968.

K-6 Kleinman, D. L., An Easy Way to Stabilize a Linear Constant System, *IEEE Transactions on Automatic Control*, **AC-15**(6), 692, December 1970.

K-7 Kleinman, D. L., Stabilizing a Discrete, Constant Linear System with Application to Iterative Methods for Solving the Riccati Equation, *IEEE Transactions on Automatic Control*, **AC-19**(3), 252–254, June 1974.

K–8 Kwakernaak, H. and Sivan, R., *Linear Optimal Control Systems*, Wiley, New York, 1972.

K-9 Kwakernaak, H., Optimal Low-Sensitivity Linear Feedback Systems, *Automatica*, **5**(3), 279–285, May 1969.

K-10 Kwakernaak, H., Asymptotic Root Loci of Multivariable Linear Optimal Regulators, *IEEE Transactions on Automatic Control*, **AC-21**(3), 378-382, June 1976.

L-1 Laub, A. J., A Schur Method for Solving Algebraic Riccati Equations, *IEEE Transactions on Automatic Control*, **AC-24**(6), 913–921, December 1979.

L-2 Layton, J. M., *Multivariable Control Theory*, Peter Peregrinus, Stevenage, England, 1979.

L-3 Lehtomaki, N. A., Sandell, N. R., Jr., and Athans, M., Robustness Results in Linear-Quadratic-Gaussian Based Multivariable Control Designs, *IEEE Transactions on Automatic Control*, **AC-26**(1), 75–93, February 1981.

L-4 LeShack, A. R., Singular Value Decomposition: Theory and Applications, Interoffice Memorandum, The Analytic Sciences Corp., Reading, MA, December 1976.

L-5 Luenberger, D., An Introduction to Observers, *IEEE Transactions on Automatic Control*, **AC-16**(6), 596–602, December 1971.

M-1 MacFarlane, A.G.J., An Eigenvector Solution of the Optimal Linear Regulator, *Journal of Electronics and Control*, **14**(6), 643–654, June 1963.

M-2 MacFarlane, A.G.J. and Postlethwaite, I., The Generalized Nyquist Stability Criterion and Multivariable Root Loci, *International Journal of Control*, **25**(1), 81–127, January 1977.

M-3 Markland, C. A., Design of Optimal and Suboptimal Stability Augmentation Systems, *AIAA Journal*, **8**(4), 673–679, April 1970.

M-4 Maybeck, P. S., *Stochastic Models, Estimation, and Control*, Vol. 3, Academic Press, New York, 1982.

M-5 Michelsen, M. L., On the Eigenvalue–Eigenvector Method for Solution of the Stationary Discrete Matrix Riccati Equation, *IEEE Transactions on Automatic Control*, **AC-24**(3), 480–481, June 1979.

M-6 Moore, B. C., On the Flexibility Offered by State Feedback in Multivariable Systems Beyond Closed-Loop Eigenvalue Assignment, *IEEE Transactions on Automatic Control*, **AC-21**(5), 689–692, October 1976.

M-7 Munro, N., *Modern Approaches to Control System Design*, Peter Peregrinus, Stevenage, England, 1979.

O-1 Oldenburger, R., ed., *Optimal and Self-Optimizing Control*, MIT Press, Cambridge, MA, 1966.

O-2 Owens, D. H., *Feedback and Multivariable Systems*, Peter Peregrinus, Stevenage, England, 1978.

P-1 Pappas, T., Laub, A. J., and Sandell, N. R., Jr., On the Numerical Solution of the Discrete-Time Algebraic Riccati Equation, *IEEE Transactions on Automatic Control*, **AC-25**(4), 631–641, August 1980.

P-2 Pope, R. E., ed., Multi-Variable Analysis and Design Techniques, AGARD-LS-117, NATO, Advisory Group for Aerospace Research and Development, Neuilly-sur-Seine, France, September 1981.

P-3 Postlethwaite, I., Edmunds, J. M., and MacFarlane, A.G.J., Principal Gains and Principal Phases in the Analysis of Linear Multivariable Feedback Systems, *IEEE Transactions on Automatic Control*, **AC-26**(1), 32–46, February 1981.

P-4 Potter, J. E., Matrix Quadratic Solutions, *SIAM Journal of Applied Mathematics*, **14**(3), 496–501, May 1966.

R-1 Rosenbrock, H. H. and McMorran, P. D., Good, Bad, or Optimal?, *IEEE Transactions on Automatic Control*, **AC-16**(6), 552–554, December 1971.

R-2 Rynaski, E. G. and Whitbeck, R. F., Theory and Application of Linear Optimal Control, USAF AFFDL-TR-65-28, October 1965.

S-1 Safonov, M. G. and Athans, M., Gain and Phase Margin for Multiloop LQG Regulators, *IEEE Transactions on Automatic Control*, **AC-22**(2), 173–179, April 1977.

S-2 Safonov, M. G. and Athans, M., Robustness and Computational Aspects of Nonlinear Stochastic Estimators and Regulators, *IEEE Transactions on Automatic Control*, **AC-23**(4), 717–725, August 1978.

S-3 Safonov, M. G., *Stability and Robustness of Multivariable Feedback Systems*, MIT Press, Cambridge, MA, 1980.

S-4 Safonov, M. G., Laub, A. J., and Hartmann, G. L., Feedback Properties of Multivariable Systems: The Role and Use of the Return Difference Matrix, *IEEE Transactions on Automatic Control*, **AC-26**(1), 47–65, February 1981.

S-5 Sandell, N. R., Jr., On Newton's Method for Riccati Equation Solution, *IEEE Transactions on Automatic Control*, **AC-19**(3), 254–255, June 1974.

S-6 Sandell, N. R., Jr., ed., Recent Developments in the Robustness Theory of Multivariable Systems, Office of Naval Research Report ONR-CR215-271-1F, August 1979.

S-7 Soroka, E. and Shaked, U., On the Robustness of LQ Regulators, *IEEE Transactions on Automatic Control*, **AC-29**(7), 664–665, July 1984.

S-8 Stein, G., Asymptotic Eigenstructures of Filters, *Proceedings of the 1979 Conference on Decision and Control*, Ft. Lauderdale, pp. 297–301, December 1979.

S-9 Stein, G. and Athans, M., Robustness Properties of Discrete Time Regulators, LQG Regulators and Hybrid Systems, MIT LIDS-FR-960, October 1979.

S-10 Stengel, R. F., Broussard, J. R., and Berry, P. W., Digital Controllers for VTOL Aircraft, *IEEE Transactions on Aerospace and Electronic Systems*, **AES-14**(1), 54–63, January 1978.

S-11 Stengel, R. F., Broussard, J. R., and Berry, P. W., Digital Flight Control Design for a Tandem-Rotor Helicopter, *Automatica*, **14**(4), 301–311, July 1978.

S-12 Stengel, R. F. and Broussard, J. R., Prediction of Pilot-Aircraft Stability Boundaries and Performance Contours, *IEEE Transactions on Systems, Man, and Cybernetics*, **SMC-8**(5), 349–356, May 1978.

S-13 Stengel, R. F., Some Effects of Parameter Variations on the Lateral-Directional Stability of Aircraft, *AIAA Journal of Guidance and Control*, **3**(2), 124–131, March–April 1980.

S-14 Stengel, R. F., A Unifying Framework for Longitudinal Flying Qualities Criteria, *AIAA Journal of Guidance, Control, and Dynamics*, **6**(2), 84–90, March–April 1983.

S-15 Stengel, R. F. and Miller, G. E., Flight Tests of a Microprocessor Control System, *AIAA Journal of Guidance and Control*, **3**(6), 494–500, November–December 1980.

S-16 Stengel, R. F., Equilibrium Response of Flight Control Systems, *Automatica*, **18**(3), 343–348, May 1982.

T-1 Thornton, C. L. and Jacobson, R. A., Linear Stochastic Control Using the UDU^T Matrix Factorization, *AIAA Journal of Guidance and Control*, **1**(4), 232–236, July–August 1978.

T-2 Trankle, T. L., and Bryson, A. E., Jr., Control Logic to Track Outputs of a Command Generator, *AIAA Journal of Guidance and Control*, **1**(2), 130–135.

T-3 Tyler, J. S., Jr., The Characteristics of Model-Following Systems as Synthesized by Optimal Control, *IEEE Transactions on Automatic Control*, **AC-9**(5), 485–498, October 1964.

V-1 Vaughan, D. R., A Nonrecursive Algebraic Solution for the Discrete Riccati Equation, *IEEE Transactions on Automatic Control*, **AC-15**(5), 597–599, October 1970.

W-1 Whitbeck, R. F., A Frequency Domain Approach to Linear Optimal Control, *AIAA Journal of Aircraft*, **5**(4), 395–401, July–August 1968.

W-2 Wong, P. K. and Athans, M., Closed-Loop Structural Stability for Linear-Quadratic Optimal Systems, *IEEE Transactions on Automatic Control*, **AC-22**(1), 94–99, February 1977.

W-3 Wonham, W. M., On Pole Assignment in Multi-Input Controllable Linear Systems, *IEEE Transactions on Automatic Control*, **AC-12**(6), 660–665, December 1967.

EPILOGUE

Stochastic optimal control theory encompasses a wide range of mathematical principles, only a few of which could be addressed here. As must be the case in any treatment of a complex subject, selected facts, rules, and examples have been reviewed, but there are exceptions, alternative methods, and special cases that could not be included. The principal benefit to be derived from stochastic optimal control methods is that they provide a systematic way of describing *feasible solutions* that can be expanded or simplified to match the control design problem and its practical constraints. The theory provides the equations and algorithms that generate answers once the system model and performance indices are specified; however, it normally does not give guidance as to which indices are good and what numerical values are satisfactory. It must rely on the user's good judgment to specify models and objectives properly, and it is quite literal in its response: ask the wrong question, and it most surely will give the wrong answer. Pose the problem accurately, and it provides a practical solution, no matter how counterintuitive the solution may be. The challenge in applying stochastic optimal control methods is to match understanding of the methodology, knowledge of the system to be controlled, and reasonable expectations of optimal system performance.

INDEX

ABOUT THE AUTHOR

Robert Stengel is a Professor of Mechanical and Aerospace Engineering at Princeton University, where he directs the Topical Program on Robotics and Intelligent Systems and the Laboratory for Control and Automation. Dr. Stengel's current research focuses on system dynamics, control, and machine intelligence. He teaches courses on control and estimation, aircraft dynamics, space flight engineering, and aerospace guidance. Prior to his 1977 Princeton appointment, he was with The Analytic Sciences Corporation, Charles Stark Draper Laboratory, United States Air Force, and National Aeronautics and Space Administration. A principal designer of the Project Apollo Lunar Module manual-attitude control logic, he also contributed to the design of the Space Shuttle guidance and control system.

Dr. Stengel received degrees from the Massachusetts Institute of Technology (Aeronautics & Astronautics, S.B., 1960) and Princeton University (Aerospace and Mechanical Sciences, M.S.E., M.A., Ph.D., 1965, 1966, 1968). He is a Fellow of the Institute of Electrical and Electronics Engineers, an Associate Fellow of the American Institute of Aeronautics and Astronautics, and a Member of the Society of Automotive Engineers Aerospace Control and Guidance Systems Committee. Professional positions include Associate Editor at Large of the *IEEE Transactions on Automatic Control*, North American editor of the Cambridge University Press Aerospace Series, and Member of the Program Council for the New Jersey Space Grant Consortium. He was Vice Chairman of the Congressional Aeronautical Advisory Committee and Chairman of the American Automatic Control Council Awards Committee, and he has served on numerous governmental advisory committees.

A CATALOG OF SELECTED

DOVER BOOKS

IN SCIENCE AND MATHEMATICS

Astronomy

BURNHAM'S CELESTIAL HANDBOOK, Robert Burnham, Jr. Thorough guide to the stars beyond our solar system. Exhaustive treatment. Alphabetical by constellation: Andromeda to Cetus in Vol. 1; Chamaeleon to Orion in Vol. 2; and Pavo to Vulpecula in Vol. 3. Hundreds of illustrations. Index in Vol. 3. 2,000pp. 6⅛ x 9¼.

Vol. I: 0-486-23567-X
Vol. II: 0-486-23568-8
Vol. III: 0-486-23673-0

EXPLORING THE MOON THROUGH BINOCULARS AND SMALL TELESCOPES, Ernest H. Cherrington, Jr. Informative, profusely illustrated guide to locating and identifying craters, rills, seas, mountains, other lunar features. Newly revised and updated with special section of new photos. Over 100 photos and diagrams. 240pp. 8¼ x 11. 0-486-24491-1

THE EXTRATERRESTRIAL LIFE DEBATE, 1750–1900, Michael J. Crowe. First detailed, scholarly study in English of the many ideas that developed from 1750 to 1900 regarding the existence of intelligent extraterrestrial life. Examines ideas of Kant, Herschel, Voltaire, Percival Lowell, many other scientists and thinkers. 16 illustrations. 704pp. 5⅜ x 8½. 0-486-40675-X

THEORIES OF THE WORLD FROM ANTIQUITY TO THE COPERNICAN REVOLUTION, Michael J. Crowe. Newly revised edition of an accessible, enlightening book re-creates the change from an earth-centered to a sun-centered conception of the solar system. 242pp. 5⅜ x 8½. 0-486-41444-2

ARISTARCHUS OF SAMOS: The Ancient Copernicus, Sir Thomas Heath. Heath's history of astronomy ranges from Homer and Hesiod to Aristarchus and includes quotes from numerous thinkers, compilers, and scholasticists from Thales and Anaximander through Pythagoras, Plato, Aristotle, and Heraclides. 34 figures. 448pp. 5⅜ x 8½. 0-486-43886-4

A COMPLETE MANUAL OF AMATEUR ASTRONOMY: TOOLS AND TECHNIQUES FOR ASTRONOMICAL OBSERVATIONS, P. Clay Sherrod with Thomas L. Koed. Concise, highly readable book discusses: selecting, setting up and maintaining a telescope; amateur studies of the sun; lunar topography and occultations; observations of Mars, Jupiter, Saturn, the minor planets and the stars; an introduction to photoelectric photometry; more. 1981 ed. 124 figures. 25 halftones. 37 tables. 335pp. 6½ x 9¼. 0-486-42820-8

AMATEUR ASTRONOMER'S HANDBOOK, J. B. Sidgwick. Timeless, comprehensive coverage of telescopes, mirrors, lenses, mountings, telescope drives, micrometers, spectroscopes, more. 189 illustrations. 576pp. 5⅜ x 8¼. (Available in U.S. only.) 0-486-24034-7

STAR LORE: Myths, Legends, and Facts, William Tyler Olcott. Captivating retellings of the origins and histories of ancient star groups include Pegasus, Ursa Major, Pleiades, signs of the zodiac, and other constellations. "Classic."—Sky & Telescope. 58 illustrations. 544pp. 5⅜ x 8½. 0-486-43581-4

Chemistry

THE SCEPTICAL CHYMIST: THE CLASSIC 1661 TEXT, Robert Boyle. Boyle defines the term "element," asserting that all natural phenomena can be explained by the motion and organization of primary particles. 1911 ed. viii+232pp. 5$^3/_8$ x 8$^1/_2$. 0-486-42825-7

RADIOACTIVE SUBSTANCES, Marie Curie. Here is the celebrated scientist's doctoral thesis, the prelude to her receipt of the 1903 Nobel Prize. Curie discusses establishing atomic character of radioactivity found in compounds of uranium and thorium; extraction from pitchblende of polonium and radium; isolation of pure radium chloride; determination of atomic weight of radium; plus electric, photographic, luminous, heat, color effects of radioactivity. ii+94pp. 5$^3/_8$ x 8$^1/_2$. 0-486-42550-9

CHEMICAL MAGIC, Leonard A. Ford. Second Edition, Revised by E. Winston Grundmeier. Over 100 unusual stunts demonstrating cold fire, dust explosions, much more. Text explains scientific principles and stresses safety precautions. 128pp. 5$^3/_8$ x 8$^1/_2$. 0-486-67628-5

MOLECULAR THEORY OF CAPILLARITY, J. S. Rowlinson and B. Widom. History of surface phenomena offers critical and detailed examination and assessment of modern theories, focusing on statistical mechanics and application of results in mean-field approximation to model systems. 1989 edition. 352pp. 5$^3/_8$ x 8$^1/_2$. 0-486-42544-4

CHEMICAL AND CATALYTIC REACTION ENGINEERING, James J. Carberry. Designed to offer background for managing chemical reactions, this text examines behavior of chemical reactions and reactors; fluid-fluid and fluid-solid reaction systems; heterogeneous catalysis and catalytic kinetics; more. 1976 edition. 672pp. 6$^1/_8$ x 9$^1/_4$. 0-486-41736-0 $31.95

ELEMENTS OF CHEMISTRY, Antoine Lavoisier. Monumental classic by founder of modern chemistry in remarkable reprint of rare 1790 Kerr translation. A must for every student of chemistry or the history of science. 539pp. 5$^3/_8$ x 8$^1/_2$. 0-486-64624-6

MOLECULES AND RADIATION: An Introduction to Modern Molecular Spectroscopy. Second Edition, Jeffrey I. Steinfeld. This unified treatment introduces upper-level undergraduates and graduate students to the concepts and the methods of molecular spectroscopy and applications to quantum electronics, lasers, and related optical phenomena. 1985 edition. 512pp. 5$^3/_8$ x 8$^1/_2$. 0-486-44152-0

A SHORT HISTORY OF CHEMISTRY, J. R. Partington. Classic exposition explores origins of chemistry, alchemy, early medical chemistry, nature of atmosphere, theory of valency, laws and structure of atomic theory, much more. 428pp. 5$^3/_8$ x 8$^1/_2$. (Available in U.S. only.) 0-486-65977-1

GENERAL CHEMISTRY, Linus Pauling. Revised 3rd edition of classic first-year text by Nobel laureate. Atomic and molecular structure, quantum mechanics, statistical mechanics, thermodynamics correlated with descriptive chemistry. Problems. 992pp. 5$^3/_8$ x 8$^1/_2$. 0-486-65622-5

ELECTRON CORRELATION IN MOLECULES, S. Wilson. This text addresses one of theoretical chemistry's central problems. Topics include molecular electronic structure, independent electron models, electron correlation, the linked diagram theorem, and related topics. 1984 edition. 304pp. 5$^3/_8$ x 8$^1/_2$. 0-486-45879-2

Engineering

DE RE METALLICA, Georgius Agricola. The famous Hoover translation of greatest treatise on technological chemistry, engineering, geology, mining of early modern times (1556). All 289 original woodcuts. 638pp. $6^3/_4$ x 11. 0-486-60006-8

FUNDAMENTALS OF ASTRODYNAMICS, Roger Bate et al. Modern approach developed by U.S. Air Force Academy. Designed as a first course. Problems, exercises. Numerous illustrations. 455pp. $5^3/_8$ x $8^1/_2$. 0-486-60061-0

DYNAMICS OF FLUIDS IN POROUS MEDIA, Jacob Bear. For advanced students of ground water hydrology, soil mechanics and physics, drainage and irrigation engineering and more. 335 illustrations. Exercises, with answers. 784pp. $6^1/_8$ x $9^1/_4$. 0-486-65675-6

THEORY OF VISCOELASTICITY (SECOND EDITION), Richard M. Christensen. Complete consistent description of the linear theory of the viscoelastic behavior of materials. Problem-solving techniques discussed. 1982 edition. 29 figures. xiv+364pp. $6^1/_8$ x $9^1/_4$. 0-486-42880-X

MECHANICS, J. P. Den Hartog. A classic introductory text or refresher. Hundreds of applications and design problems illuminate fundamentals of trusses, loaded beams and cables, etc. 334 answered problems. 462pp. $5^3/_8$ x $8^1/_2$. 0-486-60754-2

MECHANICAL VIBRATIONS, J. P. Den Hartog. Classic textbook offers lucid explanations and illustrative models, applying theories of vibrations to a variety of practical industrial engineering problems. Numerous figures. 233 problems, solutions. Appendix. Index. Preface. 436pp. $5^3/_8$ x $8^1/_2$. 0-486-64785-4

STRENGTH OF MATERIALS, J. P. Den Hartog. Full, clear treatment of basic material (tension, torsion, bending, etc.) plus advanced material on engineering methods, applications. 350 answered problems. 323pp. $5^3/_8$ x $8^1/_2$. 0-486-60755-0

A HISTORY OF MECHANICS, René Dugas. Monumental study of mechanical principles from antiquity to quantum mechanics. Contributions of ancient Greeks, Galileo, Leonardo, Kepler, Lagrange, many others. 671pp. $5^3/_8$ x $8^1/_2$. 0-486-65632-2

STABILITY THEORY AND ITS APPLICATIONS TO STRUCTURAL MECHANICS, Clive L. Dym. Self-contained text focuses on Koiter postbuckling analyses, with mathematical notions of stability of motion. Basing minimum energy principles for static stability upon dynamic concepts of stability of motion, it develops asymptotic buckling and postbuckling analyses from potential energy considerations, with applications to columns, plates, and arches. 1974 ed. 208pp. $5^3/_8$ x $8^1/_2$. 0-486-42541-X

BASIC ELECTRICITY, U.S. Bureau of Naval Personnel. Originally a training course; best nontechnical coverage. Topics include batteries, circuits, conductors, AC and DC, inductance and capacitance, generators, motors, transformers, amplifiers, etc. Many questions with answers. 349 illustrations. 1969 edition. 448pp. $6^1/_2$ x $9^1/_4$. 0-486-20973-3

ROCKETS, Robert Goddard. Two of the most significant publications in the history of rocketry and jet propulsion: "A Method of Reaching Extreme Altitudes" (1919) and "Liquid Propellant Rocket Development" (1936). 128pp. 5³/₈ x 8¹/₂. 0-486-42537-1

STATISTICAL MECHANICS: PRINCIPLES AND APPLICATIONS, Terrell L. Hill. Standard text covers fundamentals of statistical mechanics, applications to fluctuation theory, imperfect gases, distribution functions, more. 448pp. 5³/₈ x 8¹/₂. 0-486-65390-0

ENGINEERING AND TECHNOLOGY 1650–1750: ILLUSTRATIONS AND TEXTS FROM ORIGINAL SOURCES, Martin Jensen. Highly readable text with more than 200 contemporary drawings and detailed engravings of engineering projects dealing with surveying, leveling, materials, hand tools, lifting equipment, transport and erection, piling, bailing, water supply, hydraulic engineering, and more. Among the specific projects outlined-transporting a 50-ton stone to the Louvre, erecting an obelisk, building timber locks, and dredging canals. 207pp. 8³/₈ x 11¹/₄. 0-486-42232-1

THE VARIATIONAL PRINCIPLES OF MECHANICS, Cornelius Lanczos. Graduate level coverage of calculus of variations, equations of motion, relativistic mechanics, more. First inexpensive paperbound edition of classic treatise. Index. Bibliography. 418pp. 5³/₈ x 8¹/₂. 0-486-65067-7

PROTECTION OF ELECTRONIC CIRCUITS FROM OVERVOLTAGES, Ronald B. Standler. Five-part treatment presents practical rules and strategies for circuits designed to protect electronic systems from damage by transient overvoltages. 1989 ed. xxiv+434pp. 6¹/₈ x 9¹/₄. 0-486-42552-5

ROTARY WING AERODYNAMICS, W. Z. Stepniewski. Clear, concise text covers aerodynamic phenomena of the rotor and offers guidelines for helicopter performance evaluation. Originally prepared for NASA. 537 figures. 640pp. 6¹/₈ x 9¹/₄. 0-486-64647-5

INTRODUCTION TO SPACE DYNAMICS, William Tyrrell Thomson. Comprehensive, classic introduction to space-flight engineering for advanced undergraduate and graduate students. Includes vector algebra, kinematics, transformation of coordinates. Bibliography. Index. 352pp. 5³/₈ x 8¹/₂. 0-486-65113-4

HISTORY OF STRENGTH OF MATERIALS, Stephen P. Timoshenko. Excellent historical survey of the strength of materials with many references to the theories of elasticity and structure. 245 figures. 452pp. 5³/₈ x 8¹/₂. 0-486-61187-6

ANALYTICAL FRACTURE MECHANICS, David J. Unger. Self-contained text supplements standard fracture mechanics texts by focusing on analytical methods for determining crack-tip stress and strain fields. 336pp. 6¹/₈ x 9¹/₄. 0-486-41737-9

STATISTICAL MECHANICS OF ELASTICITY, J. H. Weiner. Advanced, self-contained treatment illustrates general principles and elastic behavior of solids. Part 1, based on classical mechanics, studies thermoelastic behavior of crystalline and polymeric solids. Part 2, based on quantum mechanics, focuses on interatomic force laws, behavior of solids, and thermally activated processes. For students of physics and chemistry and for polymer physicists. 1983 ed. 96 figures. 496pp. 5³/₈ x 8¹/₂. 0-486-42260-7

Mathematics

FUNCTIONAL ANALYSIS (Second Corrected Edition), George Bachman and Lawrence Narici. Excellent treatment of subject geared toward students with background in linear algebra, advanced calculus, physics and engineering. Text covers introduction to inner-product spaces, normed, metric spaces, and topological spaces; complete orthonormal sets, the Hahn-Banach Theorem and its consequences, and many other related subjects. 1966 ed. 544pp. $6^{1}/_{8}$ x $9^{1}/_{4}$. 0-486-40251-7

DIFFERENTIAL MANIFOLDS, Antoni A. Kosinski. Introductory text for advanced undergraduates and graduate students presents systematic study of the topological structure of smooth manifolds, starting with elements of theory and concluding with method of surgery. 1993 edition. 288pp. $5^{3}/_{8}$ x $8^{1}/_{2}$. 0-486-46244-7

VECTOR AND TENSOR ANALYSIS WITH APPLICATIONS, A. I. Borisenko and I. E. Tarapov. Concise introduction. Worked-out problems, solutions, exercises. 257pp. $5^{5}/_{8}$ x $8^{1}/_{4}$. 0-486-63833-2

AN INTRODUCTION TO ORDINARY DIFFERENTIAL EQUATIONS, Earl A. Coddington. A thorough and systematic first course in elementary differential equations for undergraduates in mathematics and science, with many exercises and problems (with answers). Index. 304pp. $5^{3}/_{8}$ x $8^{1}/_{2}$. 0-486-65942-9

FOURIER SERIES AND ORTHOGONAL FUNCTIONS, Harry F. Davis. An incisive text combining theory and practical example to introduce Fourier series, orthogonal functions and applications of the Fourier method to boundary-value problems. 570 exercises. Answers and notes. 416pp. $5^{3}/_{8}$ x $8^{1}/_{2}$. 0-486-65973-9

COMPUTABILITY AND UNSOLVABILITY, Martin Davis. Classic graduate-level introduction to theory of computability, usually referred to as theory of recurrent functions. New preface and appendix. 288pp. $5^{3}/_{8}$ x $8^{1}/_{2}$. 0-486-61471-9

AN INTRODUCTION TO MATHEMATICAL ANALYSIS, Robert A. Rankin. Dealing chiefly with functions of a single real variable, this text by a distinguished educator introduces limits, continuity, differentiability, integration, convergence of infinite series, double series, and infinite products. 1963 edition. 624pp. $5^{3}/_{8}$ x $8^{1}/_{2}$. 0-486-46251-X

METHODS OF NUMERICAL INTEGRATION (SECOND EDITION), Philip J. Davis and Philip Rabinowitz. Requiring only a background in calculus, this text covers approximate integration over finite and infinite intervals, error analysis, approximate integration in two or more dimensions, and automatic integration. 1984 edition. 624pp. $5^{3}/_{8}$ x $8^{1}/_{2}$. 0-486-45339-1

INTRODUCTION TO LINEAR ALGEBRA AND DIFFERENTIAL EQUATIONS, John W. Dettman. Excellent text covers complex numbers, determinants, orthonormal bases, Laplace transforms, much more. Exercises with solutions. Undergraduate level. 416pp. $5^{3}/_{8}$ x $8^{1}/_{2}$. 0-486-65191-6

RIEMANN'S ZETA FUNCTION, H. M. Edwards. Superb, high-level study of landmark 1859 publication entitled "On the Number of Primes Less Than a Given Magnitude" traces developments in mathematical theory that it inspired. xiv+315pp. $5^{3}/_{8}$ x $8^{1}/_{2}$. 0-486-41740-9

CALCULUS OF VARIATIONS WITH APPLICATIONS, George M. Ewing. Applications-oriented introduction to variational theory develops insight and promotes understanding of specialized books, research papers. Suitable for advanced undergraduate/graduate students as primary, supplementary text. 352pp. 5³/₈ x 8¹/₂. 0-486-64856-7

MATHEMATICIAN'S DELIGHT, W. W. Sawyer. "Recommended with confidence" by *The Times Literary Supplement,* this lively survey was written by a renowned teacher. It starts with arithmetic and algebra, gradually proceeding to trigonometry and calculus. 1943 edition. 240pp. 5³/₈ x 8¹/₂. 0-486-46240-4

ADVANCED EUCLIDEAN GEOMETRY, Roger A. Johnson. This classic text explores the geometry of the triangle and the circle, concentrating on extensions of Euclidean theory, and examining in detail many relatively recent theorems. 1929 edition. 336pp. 5³/₈ x 8¹/₂. 0-486-46237-4

COUNTEREXAMPLES IN ANALYSIS, Bernard R. Gelbaum and John M. H. Olmsted. These counterexamples deal mostly with the part of analysis known as "real variables." The first half covers the real number system, and the second half encompasses higher dimensions. 1962 edition. xxiv+198pp. 5³/₈ x 8¹/₂. 0-486-42875-3

CATASTROPHE THEORY FOR SCIENTISTS AND ENGINEERS, Robert Gilmore. Advanced-level treatment describes mathematics of theory grounded in the work of Poincaré, R. Thom, other mathematicians. Also important applications to problems in mathematics, physics, chemistry and engineering. 1981 edition. References. 28 tables. 397 black-and-white illustrations. xvii + 666pp. 6¹/₈ x 9¹/₄. 0-486-67539-4

COMPLEX VARIABLES: Second Edition, Robert B. Ash and W. P. Novinger. Suitable for advanced undergraduates and graduate students, this newly revised treatment covers Cauchy theorem and its applications, analytic functions, and the prime number theorem. Numerous problems and solutions. 2004 edition. 224pp. 6¹/₂ x 9¹/₄. 0-486-46250-1

NUMERICAL METHODS FOR SCIENTISTS AND ENGINEERS, Richard Hamming. Classic text stresses frequency approach in coverage of algorithms, polynomial approximation, Fourier approximation, exponential approximation, other topics. Revised and enlarged 2nd edition. 721pp. 5³/₈ x 8¹/₂. 0-486-65241-6

INTRODUCTION TO NUMERICAL ANALYSIS (2nd Edition), F. B. Hildebrand. Classic, fundamental treatment covers computation, approximation, interpolation, numerical differentiation and integration, other topics. 150 new problems. 669pp. 5³/₈ x 8¹/₂. 0-486-65363-3

MARKOV PROCESSES AND POTENTIAL THEORY, Robert M. Blumental and Ronald K. Getoor. This graduate-level text explores the relationship between Markov processes and potential theory in terms of excessive functions, multiplicative functionals and subprocesses, additive functionals and their potentials, and dual processes. 1968 edition. 320pp. 5³/₈ x 8¹/₂. 0-486-46263-3

ABSTRACT SETS AND FINITE ORDINALS: An Introduction to the Study of Set Theory, G. B. Keene. This text unites logical and philosophical aspects of set theory in a manner intelligible to mathematicians without training in formal logic and to logicians without a mathematical background. 1961 edition. 112pp. 5³/₈ x 8¹/₂. 0-486-46249-8

INTRODUCTORY REAL ANALYSIS, A.N. Kolmogorov, S. V. Fomin. Translated by Richard A. Silverman. Self-contained, evenly paced introduction to real and functional analysis. Some 350 problems. 403pp. $5\frac{3}{8}$ x $8\frac{1}{2}$. 0-486-61226-0

APPLIED ANALYSIS, Cornelius Lanczos. Classic work on analysis and design of finite processes for approximating solution of analytical problems. Algebraic equations, matrices, harmonic analysis, quadrature methods, much more. 559pp. $5\frac{3}{8}$ x $8\frac{1}{2}$. 0-486-65656-X

AN INTRODUCTION TO ALGEBRAIC STRUCTURES, Joseph Landin. Superb self-contained text covers "abstract algebra": sets and numbers, theory of groups, theory of rings, much more. Numerous well-chosen examples, exercises. 247pp. $5\frac{3}{8}$ x $8\frac{1}{2}$. 0-486-65940-2

QUALITATIVE THEORY OF DIFFERENTIAL EQUATIONS, V. V. Nemytskii and V.V. Stepanov. Classic graduate-level text by two prominent Soviet mathematicians covers classical differential equations as well as topological dynamics and ergodic theory. Bibliographies. 523pp. $5\frac{3}{8}$ x $8\frac{1}{2}$. 0-486-65954-2

THEORY OF MATRICES, Sam Perlis. Outstanding text covering rank, nonsingularity and inverses in connection with the development of canonical matrices under the relation of equivalence, and without the intervention of determinants. Includes exercises. 237pp. $5\frac{3}{8}$ x $8\frac{1}{2}$. 0-486-66810-X

INTRODUCTION TO ANALYSIS, Maxwell Rosenlicht. Unusually clear, accessible coverage of set theory, real number system, metric spaces, continuous functions, Riemann integration, multiple integrals, more. Wide range of problems. Undergraduate level. Bibliography. 254pp. $5\frac{3}{8}$ x $8\frac{1}{2}$. 0-486-65038-3

MODERN NONLINEAR EQUATIONS, Thomas L. Saaty. Emphasizes practical solution of problems; covers seven types of equations. ". . . a welcome contribution to the existing literature. . . ."—*Math Reviews*. 490pp. $5\frac{3}{8}$ x $8\frac{1}{2}$. 0-486-64232-1

MATRICES AND LINEAR ALGEBRA, Hans Schneider and George Phillip Barker. Basic textbook covers theory of matrices and its applications to systems of linear equations and related topics such as determinants, eigenvalues and differential equations. Numerous exercises. 432pp. $5\frac{3}{8}$ x $8\frac{1}{2}$. 0-486-66014-1

LINEAR ALGEBRA, Georgi E. Shilov. Determinants, linear spaces, matrix algebras, similar topics. For advanced undergraduates, graduates. Silverman translation. 387pp. $5\frac{3}{8}$ x $8\frac{1}{2}$. 0-486-63518-X

MATHEMATICAL METHODS OF GAME AND ECONOMIC THEORY: Revised Edition, Jean-Pierre Aubin. This text begins with optimization theory and convex analysis, followed by topics in game theory and mathematical economics, and concluding with an introduction to nonlinear analysis and control theory. 1982 edition. 656pp. $6\frac{1}{8}$ x $9\frac{1}{4}$. 0-486-46265-X

SET THEORY AND LOGIC, Robert R. Stoll. Lucid introduction to unified theory of mathematical concepts. Set theory and logic seen as tools for conceptual understanding of real number system. 496pp. $5\frac{3}{8}$ x $8\frac{1}{4}$. 0-486-63829-4

TENSOR CALCULUS, J.L. Synge and A. Schild. Widely used introductory text covers spaces and tensors, basic operations in Riemannian space, non-Riemannian spaces, etc. 324pp. $5^3/_8$ x $8^1/_4$. 0-486-63612-7

ORDINARY DIFFERENTIAL EQUATIONS, Morris Tenenbaum and Harry Pollard. Exhaustive survey of ordinary differential equations for undergraduates in mathematics, engineering, science. Thorough analysis of theorems. Diagrams. Bibliography. Index. 818pp. $5^3/_8$ x $8^1/_2$. 0-486-64940-7

INTEGRAL EQUATIONS, F. G. Tricomi. Authoritative, well-written treatment of extremely useful mathematical tool with wide applications. Volterra Equations, Fredholm Equations, much more. Advanced undergraduate to graduate level. Exercises. Bibliography. 238pp. $5^3/_8$ x $8^1/_2$. 0-486-64828-1

FOURIER SERIES, Georgi P. Tolstov. Translated by Richard A. Silverman. A valuable addition to the literature on the subject, moving clearly from subject to subject and theorem to theorem. 107 problems, answers. 336pp. $5^3/_8$ x $8^1/_2$. 0-486-63317-9

INTRODUCTION TO MATHEMATICAL THINKING, Friedrich Waismann. Examinations of arithmetic, geometry, and theory of integers; rational and natural numbers; complete induction; limit and point of accumulation; remarkable curves; complex and hypercomplex numbers, more. 1959 ed. 27 figures. xii+260pp. $5^3/_8$ x $8^1/_2$. 0-486-42804-8

THE RADON TRANSFORM AND SOME OF ITS APPLICATIONS, Stanley R. Deans. Of value to mathematicians, physicists, and engineers, this excellent introduction covers both theory and applications, including a rich array of examples and literature. Revised and updated by the author. 1993 edition. 304pp. $6^1/_8$ x $9^1/_4$. 0-486-46241-2

CALCULUS OF VARIATIONS, Robert Weinstock. Basic introduction covering isoperimetric problems, theory of elasticity, quantum mechanics, electrostatics, etc. Exercises throughout. 326pp. $5^3/_8$ x $8^1/_2$. 0-486-63069-2

THE CONTINUUM: A CRITICAL EXAMINATION OF THE FOUNDATION OF ANALYSIS, Hermann Weyl. Classic of 20th-century foundational research deals with the conceptual problem posed by the continuum. 156pp. $5^3/_8$ x $8^1/_2$. 0-486-67982-9

CHALLENGING MATHEMATICAL PROBLEMS WITH ELEMENTARY SOLUTIONS, A. M. Yaglom and I. M. Yaglom. Over 170 challenging problems on probability theory, combinatorial analysis, points and lines, topology, convex polygons, many other topics. Solutions. Total of 445pp. $5^3/_8$ x $8^1/_2$. Two-vol. set.
Vol. I: 0-486-65536-9 Vol. II: 0-486-65537-7

INTRODUCTION TO PARTIAL DIFFERENTIAL EQUATIONS WITH APPLICATIONS, E. C. Zachmanoglou and Dale W. Thoe. Essentials of partial differential equations applied to common problems in engineering and the physical sciences. Problems and answers. 416pp. $5^3/_8$ x $8^1/_2$. 0-486-65251-3

STOCHASTIC PROCESSES AND FILTERING THEORY, Andrew H. Jazwinski. This unified treatment presents material previously available only in journals, and in terms accessible to engineering students. Although theory is emphasized, it discusses numerous practical applications as well. 1970 edition. 400pp. $5^3/_8$ x $8^1/_2$. 0-486-46274-9

Math—Decision Theory, Statistics, Probability

INTRODUCTION TO PROBABILITY, John E. Freund. Featured topics include permutations and factorials, probabilities and odds, frequency interpretation, mathematical expectation, decision-making, postulates of probability, rule of elimination, much more. Exercises with some solutions. Summary. 1973 edition. 247pp. $5^3/_8$ x $8^1/_2$. 0-486-67549-1

STATISTICAL AND INDUCTIVE PROBABILITIES, Hugues Leblanc. This treatment addresses a decades-old dispute among probability theorists, asserting that both statistical and inductive probabilities may be treated as sentence-theoretic measurements, and that the latter qualify as estimates of the former. 1962 edition. 160pp. $5^3/_8$ x $8^1/_2$. 0-486-44980-7

APPLIED MULTIVARIATE ANALYSIS: Using Bayesian and Frequentist Methods of Inference, Second Edition, S. James Press. This two-part treatment deals with foundations as well as models and applications. Topics include continuous multivariate distributions; regression and analysis of variance; factor analysis and latent structure analysis; and structuring multivariate populations. 1982 edition. 692pp. $5^3/_8$ x $8^1/_2$. 0-486-44236-5

LINEAR PROGRAMMING AND ECONOMIC ANALYSIS, Robert Dorfman, Paul A. Samuelson and Robert M. Solow. First comprehensive treatment of linear programming in standard economic analysis. Game theory, modern welfare economics, Leontief input-output, more. 525pp. $5^3/_8$ x $8^1/_2$. 0-486-65491-5

PROBABILITY: AN INTRODUCTION, Samuel Goldberg. Excellent basic text covers set theory, probability theory for finite sample spaces, binomial theorem, much more. 360 problems. Bibliographies. 322pp. $5^3/_8$ x $8^1/_2$. 0-486-65252-1

GAMES AND DECISIONS: INTRODUCTION AND CRITICAL SURVEY, R. Duncan Luce and Howard Raiffa. Superb nontechnical introduction to game theory, primarily applied to social sciences. Utility theory, zero-sum games, n-person games, decision-making, much more. Bibliography. 509pp. $5^3/_8$ x $8^1/_2$. 0-486-65943-7

INTRODUCTION TO THE THEORY OF GAMES, J. C. C. McKinsey. This comprehensive overview of the mathematical theory of games illustrates applications to situations involving conflicts of interest, including economic, social, political, and military contexts. Appropriate for advanced undergraduate and graduate courses; advanced calculus a prerequisite. 1952 ed. x+372pp. $5^3/_8$ x $8^1/_2$. 0-486-42811-7

FIFTY CHALLENGING PROBLEMS IN PROBABILITY WITH SOLUTIONS, Frederick Mosteller. Remarkable puzzlers, graded in difficulty, illustrate elementary and advanced aspects of probability. Detailed solutions. 88pp. $5^3/_8$ x $8^1/_2$. 0-486-65355-2

PROBABILITY THEORY: A CONCISE COURSE, Y. A. Rozanov. Highly readable, self-contained introduction covers combination of events, dependent events, Bernoulli trials, etc. 148pp. $5^3/_8$ x $8^1/_4$. 0-486-63544-9

THE STATISTICAL ANALYSIS OF EXPERIMENTAL DATA, John Mandel. First half of book presents fundamental mathematical definitions, concepts and facts while remaining half deals with statistics primarily as an interpretive tool. Well-written text, numerous worked examples with step-by-step presentation. Includes 116 tables. 448pp. $5^3/_8$ x $8^1/_2$. 0-486-64666-1

Math—Geometry and Topology

ELEMENTARY CONCEPTS OF TOPOLOGY, Paul Alexandroff. Elegant, intuitive approach to topology from set-theoretic topology to Betti groups; how concepts of topology are useful in math and physics. 25 figures. 57pp. 5³/₈ x 8¹/₂. 0-486-60747-X

A LONG WAY FROM EUCLID, Constance Reid. Lively guide by a prominent historian focuses on the role of Euclid's Elements in subsequent mathematical developments. Elementary algebra and plane geometry are sole prerequisites. 80 drawings. 1963 edition. 304pp. 5³/₈ x 8¹/₂. 0-486-43613-6

EXPERIMENTS IN TOPOLOGY, Stephen Barr. Classic, lively explanation of one of the byways of mathematics. Klein bottles, Moebius strips, projective planes, map coloring, problem of the Koenigsberg bridges, much more, described with clarity and wit. 43 figures. 210pp. 5³/₈ x 8¹/₂. 0-486-25933-1

THE GEOMETRY OF RENÉ DESCARTES, René Descartes. The great work founded analytical geometry. Original French text, Descartes's own diagrams, together with definitive Smith-Latham translation. 244pp. 5³/₈ x 8¹/₂. 0-486-60068-8

EUCLIDEAN GEOMETRY AND TRANSFORMATIONS, Clayton W. Dodge. This introduction to Euclidean geometry emphasizes transformations, particularly isometries and similarities. Suitable for undergraduate courses, it includes numerous examples, many with detailed answers. 1972 ed. viii+296pp. 6¹/₈ x 9¹/₄. 0-486-43476-1

EXCURSIONS IN GEOMETRY, C. Stanley Ogilvy. A straightedge, compass, and a little thought are all that's needed to discover the intellectual excitement of geometry. Harmonic division and Apollonian circles, inversive geometry, hexlet, Golden Section, more. 132 illustrations. 192pp. 5³/₈ x 8¹/₂. 0-486-26530-7

THE THIRTEEN BOOKS OF EUCLID'S ELEMENTS, translated with introduction and commentary by Sir Thomas L. Heath. Definitive edition. Textual and linguistic notes, mathematical analysis. 2,500 years of critical commentary. Unabridged. 1,414pp. 5³/₈ x 8¹/₂. Three-vol. set.
Vol. I: 0-486-60088-2 Vol. II: 0-486-60089-0 Vol. III: 0-486-60090-4

SPACE AND GEOMETRY: IN THE LIGHT OF PHYSIOLOGICAL, PSYCHOLOGICAL AND PHYSICAL INQUIRY, Ernst Mach. Three essays by an eminent philosopher and scientist explore the nature, origin, and development of our concepts of space, with a distinctness and precision suitable for undergraduate students and other readers. 1906 ed. vi+148pp. 5³/₈ x 8¹/₂. 0-486-43909-7

GEOMETRY OF COMPLEX NUMBERS, Hans Schwerdtfeger. Illuminating, widely praised book on analytic geometry of circles, the Moebius transformation, and two-dimensional non-Euclidean geometries. 200pp. 5⁵/₈ x 8¹/₄. 0-486-63830-8

DIFFERENTIAL GEOMETRY, Heinrich W. Guggenheimer. Local differential geometry as an application of advanced calculus and linear algebra. Curvature, transformation groups, surfaces, more. Exercises. 62 figures. 378pp. 5³/₈ x 8¹/₂. 0-486-63433-7

History of Math

THE WORKS OF ARCHIMEDES, Archimedes (T. L. Heath, ed.). Topics include the famous problems of the ratio of the areas of a cylinder and an inscribed sphere; the measurement of a circle; the properties of conoids, spheroids, and spirals; and the quadrature of the parabola. Informative introduction. clxxxvi+326pp. $5^3/_8 \times 8^1/_2$. 0-486-42084-1

A SHORT ACCOUNT OF THE HISTORY OF MATHEMATICS, W. W. Rouse Ball. One of clearest, most authoritative surveys from the Egyptians and Phoenicians through 19th-century figures such as Grassman, Galois, Riemann. Fourth edition. 522pp. $5^3/_8 \times 8^1/_2$. 0-486-20630-0

THE HISTORY OF THE CALCULUS AND ITS CONCEPTUAL DEVELOPMENT, Carl B. Boyer. Origins in antiquity, medieval contributions, work of Newton, Leibniz, rigorous formulation. Treatment is verbal. 346pp. $5^3/_8 \times 8^1/_2$. 0-486-60509-4

THE HISTORICAL ROOTS OF ELEMENTARY MATHEMATICS, Lucas N. H. Bunt, Phillip S. Jones, and Jack D. Bedient. Fundamental underpinnings of modern arithmetic, algebra, geometry and number systems derived from ancient civilizations. 320pp. $5^3/_8 \times 8^1/_2$. 0-486-25563-8

THE HISTORY OF THE CALCULUS AND ITS CONCEPTUAL DEVELOPMENT, Carl B. Boyer. Fluent description of the development of both the integral and differential calculus—its early beginnings in antiquity, medieval contributions, and a consideration of Newton and Leibniz. 368pp. $5^3/_8 \times 8^1/_2$. 0-486-60509-4

GAMES, GODS & GAMBLING: A HISTORY OF PROBABILITY AND STATISTICAL IDEAS, F. N. David. Episodes from the lives of Galileo, Fermat, Pascal, and others illustrate this fascinating account of the roots of mathematics. Features thought-provoking references to classics, archaeology, biography, poetry. 1962 edition. 304pp. $5^3/_8 \times 8^1/_2$. (Available in U.S. only.) 0-486-40023-9

OF MEN AND NUMBERS: THE STORY OF THE GREAT MATHEMATICIANS, Jane Muir. Fascinating accounts of the lives and accomplishments of history's greatest mathematical minds—Pythagoras, Descartes, Euler, Pascal, Cantor, many more. Anecdotal, illuminating. 30 diagrams. Bibliography. 256pp. $5^3/_8 \times 8^1/_2$. 0-486-28973-7

HISTORY OF MATHEMATICS, David E. Smith. Nontechnical survey from ancient Greece and Orient to late 19th century; evolution of arithmetic, geometry, trigonometry, calculating devices, algebra, the calculus. 362 illustrations. 1,355pp. $5^3/_8 \times 8^1/_2$. Two-vol. set. Vol. I: 0-486-20429-4 Vol. II: 0-486-20430-8

A CONCISE HISTORY OF MATHEMATICS, Dirk J. Struik. The best brief history of mathematics. Stresses origins and covers every major figure from ancient Near East to 19th century. 41 illustrations. 195pp. $5^3/_8 \times 8^1/_2$. 0-486-60255-9

Physics

OPTICAL RESONANCE AND TWO-LEVEL ATOMS, L. Allen and J. H. Eberly. Clear, comprehensive introduction to basic principles behind all quantum optical resonance phenomena. 53 illustrations. Preface. Index. 256pp. $5^3/_8$ x $8^1/_2$. 0-486-65533-4

QUANTUM THEORY, David Bohm. This advanced undergraduate-level text presents the quantum theory in terms of qualitative and imaginative concepts, followed by specific applications worked out in mathematical detail. Preface. Index. 655pp. $5^3/_8$ x $8^1/_2$. 0-486-65969-0

ATOMIC PHYSICS (8th EDITION), Max Born. Nobel laureate's lucid treatment of kinetic theory of gases, elementary particles, nuclear atom, wave-corpuscles, atomic structure and spectral lines, much more. Over 40 appendices, bibliography. 495pp. $5^3/_8$ x $8^1/_2$. 0-486-65984-4

A SOPHISTICATE'S PRIMER OF RELATIVITY, P. W. Bridgman. Geared toward readers already acquainted with special relativity, this book transcends the view of theory as a working tool to answer natural questions: What is a frame of reference? What is a "law of nature"? What is the role of the "observer"? Extensive treatment, written in terms accessible to those without a scientific background. 1983 ed. xlviii+172pp. $5^3/_8$ x $8^1/_2$. 0-486-42549-5

AN INTRODUCTION TO HAMILTONIAN OPTICS, H. A. Buchdahl. Detailed account of the Hamiltonian treatment of aberration theory in geometrical optics. Many classes of optical systems defined in terms of the symmetries they possess. Problems with detailed solutions. 1970 edition. xv + 360pp. $5^3/_8$ x $8^1/_2$. 0-486-67597-1

PRIMER OF QUANTUM MECHANICS, Marvin Chester. Introductory text examines the classical quantum bead on a track: its state and representations; operator eigenvalues; harmonic oscillator and bound bead in a symmetric force field; and bead in a spherical shell. Other topics include spin, matrices, and the structure of quantum mechanics; the simplest atom; indistinguishable particles; and stationary-state perturbation theory. 1992 ed. xiv+314pp. $6^1/_8$ x $9^1/_4$. 0-486-42878-8

LECTURES ON QUANTUM MECHANICS, Paul A. M. Dirac. Four concise, brilliant lectures on mathematical methods in quantum mechanics from Nobel Prize-winning quantum pioneer build on idea of visualizing quantum theory through the use of classical mechanics. 96pp. $5^3/_8$ x $8^1/_2$. 0-486-41713-1

THIRTY YEARS THAT SHOOK PHYSICS: THE STORY OF QUANTUM THEORY, George Gamow. Lucid, accessible introduction to influential theory of energy and matter. Careful explanations of Dirac's anti-particles, Bohr's model of the atom, much more. 12 plates. Numerous drawings. 240pp. $5^3/_8$ x $8^1/_2$. 0-486-24895-X

ELECTRONIC STRUCTURE AND THE PROPERTIES OF SOLIDS: THE PHYSICS OF THE CHEMICAL BOND, Walter A. Harrison. Innovative text offers basic understanding of the electronic structure of covalent and ionic solids, simple metals, transition metals and their compounds. Problems. 1980 edition. 582pp. $6^1/_8$ x $9^1/_4$. 0-486-66021-4

HYDRODYNAMIC AND HYDROMAGNETIC STABILITY, S. Chandrasekhar. Lucid examination of the Rayleigh-Benard problem; clear coverage of the theory of instabilities causing convection. 704pp. $5^3/_8$ x $8^1/_4$. 0-486-64071-X

INVESTIGATIONS ON THE THEORY OF THE BROWNIAN MOVEMENT, Albert Einstein. Five papers (1905–8) investigating dynamics of Brownian motion and evolving elementary theory. Notes by R. Fürth. 122pp. $5^3/_8$ x $8^1/_2$. 0-486-60304-0

THE PHYSICS OF WAVES, William C. Elmore and Mark A. Heald. Unique overview of classical wave theory. Acoustics, optics, electromagnetic radiation, more. Ideal as classroom text or for self-study. Problems. 477pp. $5^3/_8$ x $8^1/_2$. 0-486-64926-1

GRAVITY, George Gamow. Distinguished physicist and teacher takes reader-friendly look at three scientists whose work unlocked many of the mysteries behind the laws of physics: Galileo, Newton, and Einstein. Most of the book focuses on Newton's ideas, with a concluding chapter on post-Einsteinian speculations concerning the relationship between gravity and other physical phenomena. 160pp. $5^3/_8$ x $8^1/_2$. 0-486-42563-0

PHYSICAL PRINCIPLES OF THE QUANTUM THEORY, Werner Heisenberg. Nobel Laureate discusses quantum theory, uncertainty, wave mechanics, work of Dirac, Schroedinger, Compton, Wilson, Einstein, etc. 184pp. $5^3/_8$ x $8^1/_2$. 0-486-60113-7

ATOMIC SPECTRA AND ATOMIC STRUCTURE, Gerhard Herzberg. One of best introductions; especially for specialist in other fields. Treatment is physical rather than mathematical. 80 illustrations. 257pp. $5^3/_8$ x $8^1/_2$. 0-486-60115-3

AN INTRODUCTION TO STATISTICAL THERMODYNAMICS, Terrell L. Hill. Excellent basic text offers wide-ranging coverage of quantum statistical mechanics, systems of interacting molecules, quantum statistics, more. 523pp. $5^3/_8$ x $8^1/_2$. 0-486-65242-4

THEORETICAL PHYSICS, Georg Joos, with Ira M. Freeman. Classic overview covers essential math, mechanics, electromagnetic theory, thermodynamics, quantum mechanics, nuclear physics, other topics. First paperback edition. xxiii + 885pp. $5^3/_8$ x $8^1/_2$. 0-486-65227-0

PROBLEMS AND SOLUTIONS IN QUANTUM CHEMISTRY AND PHYSICS, Charles S. Johnson, Jr. and Lee G. Pedersen. Unusually varied problems, detailed solutions in coverage of quantum mechanics, wave mechanics, angular momentum, molecular spectroscopy, more. 280 problems plus 139 supplementary exercises. 430pp. $6^1/_2$ x $9^1/_4$. 0-486-65236-X

THEORETICAL SOLID STATE PHYSICS, Vol. 1: Perfect Lattices in Equilibrium; Vol. II: Non-Equilibrium and Disorder, William Jones and Norman H. March. Monumental reference work covers fundamental theory of equilibrium properties of perfect crystalline solids, non-equilibrium properties, defects and disordered systems. Appendices. Problems. Preface. Diagrams. Index. Bibliography. Total of 1,301pp. $5^3/_8$ x $8^1/_2$. Two volumes. Vol. I: 0-486-65015-4 Vol. II: 0-486-65016-2

WHAT IS RELATIVITY? L. D. Landau and G. B. Rumer. Written by a Nobel Prize physicist and his distinguished colleague, this compelling book explains the special theory of relativity to readers with no scientific background, using such familiar objects as trains, rulers, and clocks. 1960 ed. vi+72pp. $5^3/_8$ x $8^1/_2$. 0-486-42806-0

CATALOG OF DOVER BOOKS

A TREATISE ON ELECTRICITY AND MAGNETISM, James Clerk Maxwell. Important foundation work of modern physics. Brings to final form Maxwell's theory of electromagnetism and rigorously derives his general equations of field theory. 1,084pp. $5\frac{3}{8} \times 8\frac{1}{2}$. Two-vol. set. Vol. I: 0-486-60636-8 Vol. II: 0-486-60637-6

MATHEMATICS FOR PHYSICISTS, Philippe Dennery and Andre Krzywicki. Superb text provides math needed to understand today's more advanced topics in physics and engineering. Theory of functions of a complex variable, linear vector spaces, much more. Problems. 1967 edition. 400pp. $6\frac{1}{2} \times 9\frac{1}{4}$. 0-486-69193-4

INTRODUCTION TO QUANTUM MECHANICS WITH APPLICATIONS TO CHEMISTRY, Linus Pauling & E. Bright Wilson, Jr. Classic undergraduate text by Nobel Prize winner applies quantum mechanics to chemical and physical problems. Numerous tables and figures enhance the text. Chapter bibliographies. Appendices. Index. 468pp. $5\frac{3}{8} \times 8\frac{1}{2}$. 0-486-64871-0

METHODS OF THERMODYNAMICS, Howard Reiss. Outstanding text focuses on physical technique of thermodynamics, typical problem areas of understanding, and significance and use of thermodynamic potential. 1965 edition. 238pp. $5\frac{3}{8} \times 8\frac{1}{2}$. 0-486-69445-3

THE ELECTROMAGNETIC FIELD, Albert Shadowitz. Comprehensive under- graduate text covers basics of electric and magnetic fields, builds up to electromagnetic theory. Also related topics, including relativity. Over 900 problems. 768pp. $5\frac{5}{8} \times 8\frac{1}{4}$. 0-486-65660-8

GREAT EXPERIMENTS IN PHYSICS: FIRSTHAND ACCOUNTS FROM GALILEO TO EINSTEIN, Morris H. Shamos (ed.). 25 crucial discoveries: Newton's laws of motion, Chadwick's study of the neutron, Hertz on electromagnetic waves, more. Original accounts clearly annotated. 370pp. $5\frac{3}{8} \times 8\frac{1}{2}$. 0-486-25346-5

EINSTEIN'S LEGACY, Julian Schwinger. A Nobel Laureate relates fascinating story of Einstein and development of relativity theory in well-illustrated, nontechnical volume. Subjects include meaning of time, paradoxes of space travel, gravity and its effect on light, non-Euclidean geometry and curving of space-time, impact of radio astronomy and space-age discoveries, and more. 189 b/w illustrations. xiv+250pp. $8\frac{3}{8} \times 9\frac{1}{4}$. 0-486-41974-6

THE VARIATIONAL PRINCIPLES OF MECHANICS, Cornelius Lanczos. Philosophic, less formalistic approach to analytical mechanics offers model of clear, scholarly exposition at graduate level with coverage of basics, calculus of variations, principle of virtual work, equations of motion, more. 418pp. $5\frac{3}{8} \times 8\frac{1}{2}$. 0-486-65067-7